WHERE TO WATCH BIRDS IN BRITAIN

SECOND EDITION

SIMON HARRAP AND NIGEL REDMAN

CHRISTOPHER HELM
LONDON

For Eleanor and Emily

Second edition published 2010 by Christopher Helm,
an imprint of A & C Black Publishers Ltd,
36 Soho Square, London W1D 3QY

www.acblack.com

First edition 2003

ISBN 978-1-4081-10591

A CIP catalogue record for this book is available from the British Library

This book is produced using paper that is made from wood grown in managed sustainable forests. It is natural, renewable and recyclable. The logging and manufacturing processes conform to the environmental regulations of the country of origin.

Commissioning Editor: Nigel Redman
Project Editor: Julie Bailey

Layout by Jocelyn Lucas
Maps by Simon Harrap and Brian Southern

Puffins illustration on title page and Cranes illustration on page 15 by Dan Powell

Cover photo credits:
Front cover, top, and spine: Bearded Tit by Dean Eades
Front cover, bottom: Minsmere by David Tipling (rspb-images.com)
Back cover: Hawfinch, Black Grouse and Smew all by David Tipling

Printed in Spain by GraphyCems

10 9 8 7 6 5 4 3 2 1

CONTENTS

PERTH & KINROSS (NORTH)

ANGUS & DUNDEE

NORTH-EAST SCOTLAND

MORAY & NAIRN

HIGHLAND

ARGYLL

INTRODUCTION

A bird-finding guide is now a recognised part of any birder's kit – to be kept in the car alongside field guide, map (or satnav) and pub guide. Most birdwatchers are hungry to see as much as possible, and the majority will wish to visit new sites in search of new or unusual birds. *Where to Watch Birds in Britain* is intended to provide an efficient guide for today's mobile birder who wants to visit Britain's best birding areas. This is a book for every birdwatcher, but especially for those not 'in the know' about all those special places for birds.

Only 40 years ago, there was no comprehensive guide to the best birding sites in Britain. Since then, there have been several, each with a different emphasis, and a series of 14 guides covering every region of Britain in depth. In addition, other local guides cover specific sites or areas, and many nature reserves produce their own leaflets and have their own websites. Indeed, we have witnessed a positive explosion of information in the last four decades.

Our own guide to the best sites in Britain, *Birdwatching in Britain: a site by site guide*, was first published in 1987. It treated 115 sites or groups of sites of national interest, arranged on a regional basis, as well as a further 88 subsidiary sites. *Where to Watch Birds in Britain*, first published in 2003, was its successor and, at over 600 pages, was a considerably expanded volume. All the 'subsidiary' sites of the original book were expanded into full entries, and many new places were included. For this new edition, all the information has been fully revised, with several new sites added (and a few others that are now of less interest have been dropped). The total number of sites has now risen to 454 but this is a misleading figure, as in some cases a whole island or a long stretch of coastline is considered to be a single site, while a series of localities in close proximity (such as in Breckland) may be treated as separate sites. All the maps have been revised and updated as necessary, and many new maps have been drawn. There are now almost 300 maps in the book – but these cover many more sites than this as some of the maps feature more than one site.

It is just a few years since the first edition was published in 2003, yet there have been tremendous changes in the British avifauna. On the positive side, Little Egret is now a common bird in many parts of England and Wales, and Red Kite too is now thriving in several areas following highly successful reintroduction programmes. Marsh Harriers are commonplace in East Anglia and spreading elsewhere, Common Buzzard is ever more widespread and Peregrines, despite record levels of persecution, continue to gain ground, while Ravens have at last returned to south-east England. Both Chough and Osprey have trumped reintroduction programmes by recolonising England naturally, while Avocet has spread as a breeding bird and even moved to inland sites. Perhaps more surprisingly, Bitterns are now much more widespread in winter, with several sites regularly holding one or more birds (although its success as a breeding species has not noticeably increased). Amongst the passerines both Dartford and Cetti's Warblers have spread, presumably benefiting (like Little Egret), from a long run of mild winters, though it would not take much for their fortunes to change; Dartford Warbler declined by as much as 80 per cent in some areas following hard weather in February 2009, and there are grave concerns for its future following the hard winter of 2009-10. Conversely, global warming seems to be pushing some wintering species northwards, and Purple Sandpiper, Shore Lark and Twite have declined markedly in the south of Britain, with far fewer reliable sites than just a decade ago. Mild winters may also be responsible for a marked decline in the numbers of wintering White-fronted Geese (of both the nominate and 'Greenland' forms), although wintering Barnacle and Pink-footed Geese are doing very well indeed – perhaps as these come from the north and west, rather than northern Russia, they do not have the option of shifting their wintering grounds northwards, as they are limited by the North Atlantic and North Sea? Amongst resident species, however, all is not rosy. Lesser Spotted Woodpecker, Willow Tit, Lesser Redpoll, Tree Sparrow and Corn Bunting have all undergone marked declines and vanished from many sites (as too have Starling and House Sparrow, neither of which features strongly in this book). Summer visitors seem to be faring even worse.

Cuckoo, Tree Pipit, Yellow Wagtail, Wood and Willow Warblers and Spotted Flycatcher have all disappeared from many sites. In short, Britain's birds seem to be more dynamic than ever and, while we have made every effort to bring our site accounts up-to-date, we cannot guarantee to have kept up with the arrival (or disappearance) of every species!

The aims of *Where to Watch Birds in Britain* are very simple, and remain much as in the original guide. Our purpose is to provide a single-volume reference to the very best birding sites in Britain, covering as wide a range of habitats and species as possible. It will, we hope, be equally useful for resident birders visiting unfamiliar parts of the country as to visitors from overseas who wish to identify and visit the best areas that Britain has to offer. With this guide, we hope that birders will be able to see almost all of the British breeding species, many of the scarce migrants to the country, and even find a few rarities at key migration sites.

It is inevitable that a site guide such as this will become out of date as soon as it is published. In order to help us prepare future editions, we would welcome any corrections or further information. Please write to the authors at Christopher Helm Publishers, 36 Soho Square, London W1D 3QY.

ACKNOWLEDGEMENTS

The compilation of a birding site guide inevitably involves tapping into many sources of information. We have trawled the literature (regional and local site guides, county bird reports, magazine articles, reserves information leaflets, websites, etc.), as well as seeking the priceless knowledge and assistance of a great many people. The considerable number of people and organisations who helped provide and check information for our original guide need to be thanked again. In addition, many others kindly helped us by checking final drafts of the first edition of *Where to Watch Birds in Britain*. We are very grateful for all their comments and advice, and hope that no names have been omitted from the following list: Annette Adams (Abberton Reservoir), Peter Alker (Pennington Flash), Mike Bailey (Dyfi Estuary), Dawn Balmer (Thetford), Bill Boyd (Holme and Hunstanton), Bob Bullock (Northamptonshire), P. Burns (Pagham Harbour), Dave Bromwich (Gibraltar Point), Jo Calvert (Grafham Water), Phil Chantler (Kent), Nick Collinson (Walberswick), Simon Cox (Essex), Paul Culyer (Cors Caron), Mark Dennis (Trent Valley Pits, Bleasby and Hoveringham Pits), Paul Doyle (Cruden Bay), Bob Elliot (Threave Estate), Pete Ellis (Shetland), Peter Exley (Exe Estuary, Hayle and Sedgemoor), Richard Fairbank (Sussex), Brian Fellows (Thorney Island), Paul Fisher (Snettisham and Titchwell), Laurie Forsyth (Fingringhoe Wick), Nick Gardiner, Andrew Grieve (Blacktoft), Ren Hathway (Scilly and Cornwall), Ray Hawley (Ken–Dee Marshes), Norman Holton (Campfield Marsh, Hodbarrow and St Bees Head), Tony How (Paxton Pits), Michael Hughes (Oxwich, Gower Coast, Worms Head and Burry Inlet), Martin Humphreys (Cwm Clydach), Colin Jakes (Lackford Wildfowl Reserve), Pete Jennings (Elan Valley), Stephen John (Welsh Wildlife Centre), Russell Jones (Ynys-hir), Bill Kenmir (Haweswater), Charlie Kitchin (Nene Washes), Paul Laurie (St Margaret's Bay), Russell Leavett (Stour Estuary), Martin Lester (Wicken Fen), Ian Lewington (Oxfordshire), Eddie Maguire (Machrihanish Seabird Observatory), Pete Marsh (Heysham and Morecambe Bay), Eric Meek (Orkney), Richard Millington (Norfolk), Carl Mitchell (Welney), Steve Moon (Kenfig, Eglwys Nunydd Reservoir and Ogmore Estuary), Derek Moore (Suffolk), Pete Naylor (Staines Reservoirs), Malcolm Ogilvie (Islay), Scott Paterson (Lochwinnoch), Richard Porter (Grafham Water), Geoff Proffitt (Penclacwydd), Doug Radford (Fowlmere), Mick Rogers (Portland), Mike Rogers (county recorders), Roger Riddington (Fair Isle), Peter Robinson (Scilly), Bill Rutherford (Holme), David Saunders (Pembrokeshire), Charlie Self (Coll), Julia Sheehan (Lake Vyrnwy), Brian Small (Suffolk), Graham Smith (Bradwell), Andy Tasker (Brandon Marsh), Chris Tyas (Old Hall Marshes), Steve Whitehouse (Worcestershire), John Wilson (Leighton Moss and Morecambe Bay) and W. Wright (Caerlaverock).

For this new edition, we are indebted to the following for help with specific sites as follows: John Bowler (Tiree), Brian Gibbs (Avalon Marshes), Jon Green (Wales), Gavin Haig (Axe Estuary), William Legge (Fleet Pond), Marianne Taylor (Sevenoaks) and Howard Vaughan (Rainham). In addition, a number of others kindly checked our site accounts or provided additional information: John Ash, Chris Batty, Keith Betton, Sean Cole, Greg Conway, Dr Simon Cox, Rob Fray, Roy Frost, Steve Gantlett, Dave Jeynes, Andrew Lassey, Dr David Leech, Ian Lewington, Pete Marsh, Eric Meek, Derek Moore, Stephen Moss, Malcolm Ogilvie, Nigel Peace, Heather Rowe, Roger Riddington, Mark Roberts, Steve Rooke, Brian Small, Andy Stoddart, Darren Starkey, Peter Van der Veken, Marcus Ward and Mike Watson. We thank them all. We are also grateful to Bo Beolens for assistance with disabled access.

We would like to thank our editor, Julie Bailey, for her patience and understanding over the long gestation of this book and for her skill in putting it together, Marianne Taylor for her copy editing and for much other help and advice, Brian Southern for diligently producing more than one hundred maps and making corrections to many others (the rest were prepared by SH), and Joc Lucas for designing and laying out the book.

Last but not least we owe a huge debt of gratitude to our families, Anne and Eleanor Harrap, Cheryle Sifontes and Emily Redman, for their unswerving support and love while we spent considerable periods of time researching and writing.

HOW TO USE THIS BOOK

SITE SELECTION

The choice of sites in a guide such as this is obviously highly subjective. We have tried to select sites that cover a full range of habitats and species, and which provide something that you might not find at your local patch (unless your local patch is one of the sites listed). The principles for site selection remain unchanged from the original edition:

1. Quality. Does the area provide good birding, whether for a few hours or several days?
2. Species range. Can you see some of the more localised and uncommon species? A few sites have been included which have just one or two of these, but most will have a range of interesting birds, be they regional specialities, important concentrations at certain seasons, or passage migrants.
3. Accessibility. Is it possible to visit the area without making special arrangements? For some sites this can take time, which rules out a casual trip. We have made an exception in the case of offshore islands, which usually require advance planning, because their attractions often easily outweigh any inconvenience.

Our choices have been made irrespective of scenic considerations or the possibility of good walking, and there is inevitably a bias towards coastal areas. Most birdwatchers live inland, but the coast offers the most interesting variety of habitats and birds, as well as being the best area to see migrants. The original edition had a bias towards certain regions, for example East Anglia, but this book now has a full range of sites for each region. There is still an emphasis on wetland areas, and reservoirs in particular, as these are more likely to provide a greater variety of species than, say, a stand of mature woodland. We have not tried to give equal weight to each county, and if any county has few sites with the necessary qualifications, then there are few entries for it in this book. This does not mean that there are no good places for birds in these places, but that they may be inaccessible to the casual visitor or of a local, rather than national, interest. In the end, however, the final selection has been a personal one, based on our own experience and the advice of others.

For reasons of space and practicality, we could not include every site, and some readers may complain that their own particular favourites have been excluded (as a few did after the original edition was published). In some cases they may be justified. However, we have now included many sites that were previously omitted, having had some of these brought to our attention by readers and reviewers, and have tried to strike a reasonable balance overall. We hope that birders will write to us with their views and comments, and that the content will continue to evolve and improve in subsequent editions.

SITE ACCOUNTS

Sites have been grouped into seven regions. The book begins with the South-west and ends with Scotland. Each regional section commences with an overview map on which each site (or group of sites) is marked with a number that corresponds to the key beside it and to the main text entry. County (or regional) boundaries are marked on the overview maps to help determine the approximate locations of sites. A few sites that are close together and share similar birds are given a slightly different treatment: there is a general introduction to such 'site clusters' and a Habitat and Species section covering all the sites in the group. Under the Access heading, the different sites within a cluster are given their own sections, followed by the Further Information section that covers all the sites together. Examples of site clusters include the Isles of Scilly, New Forest, Breckland, Outer Hebrides, Orkney and Shetland.

After the site name, a county name is given in parentheses. This may not actually be the official local authority for the site (especially as there are now many unitary authorities for our more urban areas) but relates to the county recorder to whom unusual bird records should be sent. In the last 40 years some county boundaries have changed several times and, in a few cases, local bird groups have elected to use the old county boundaries for recording purposes. The 'counties' largely affected by these changes are Avon (no longer officially exists and now divided into North Somerset, South Gloucestershire and the unitary authorities of Bath and Bristol), Merseyside (the Wirral is included with Cheshire and North Merseyside with Lancashire), Cleveland (still retained for bird recording although officially it now belongs with Yorkshire and Durham), Tyne & Wear (now divided between Northumberland and Durham), and much of Wales and Scotland. The Welsh counties used in this book are those given in *Where to Watch Birds in Wales* (2008). In Scotland, we follow the recording areas of the Scottish Ornithologists' Club (www.the-soc.org.uk/recorders). A few areas in Britain are covered by more than one county recorder. Greater London is a good example of double recording; the London Bird Report covers a circle of 20 miles from St Paul's, but most areas (except the old county of Middlesex) are now also covered by their respective bordering counties. Yorkshire is another region affected by border changes; excluding part of Cleveland, it is divided into four recording regions: East (the former North Humberside), South, West and North. Names and addresses of county recorders can be found in *The Birdwatcher's Yearbook* (see Further Reading), or on the *British Birds* website (www.britishbirds.co.uk). An excellent paper 'Recording areas of Great Britain' by David Ballance and Judith Smith (*British Birds* 101: 364-375), summarises the current situation and gives much historical information; it is also freely downloadable from the BB website.

For each site, the relevant Ordnance Survey (OS) sheet number is given. These refer to the OS Landranger series, at a scale of 1:50,000. While any good road atlas will guide birders to every site when used in conjunction with the access information in this book, it is recommended that an OS map is used to locate footpaths and other more precise topographical features, particularly in the case of large areas that are not nature reserves (e.g. Sutherland). Larger scale OS maps (1:25,000) may be useful for greater detail. Ordnance Survey maps are regarded as the ultimate authority when it comes to the spelling of place names and road numbers. A number of grid references are provided within the text for specific locations. These will help locate places on an OS map, or can be used in conjunction with satellite navigation systems.

Each site account begins with a short introduction, highlighting a few key facts and the main points of ornithological interest. This may also include the existence of a reserve (and who owns it), and the most profitable period(s) for a visit. Sometimes, the approximate area of the reserve is given; we have used hectares rather than acres in such cases.

This introductory paragraph is followed by several headed sections as follows (sometimes combined for shorter site entries).

Habitat

The Habitat section briefly outlines the habitats represented and the general layout of the location. Where relevant, historical changes in the habitats may be mentioned and features of the surrounding area if they affect the sites concerned.

Species

In this section we have tried to highlight the more interesting species you may see, as well as mentioning some of the commoner ones; very common and widespread species are generally omitted. It should be stressed that you will not see all the birds listed on a single visit or even after several visits. At migration sites in particular, many days can be quite birdless if the weather conditions are wrong, and good 'falls' of migrants should be regarded as unusual. Scarce migrants are frequently specified where their occurrence is reasonably regular, although some may only be recorded once or twice a year. However, as a rule, rarities have only been mentioned if they have occurred in a more or less predictable pattern. In some instances, a few impressive rarities that have occurred in the past have been mentioned to whet the appetite.

The Species section has been roughly divided into seasons, but the treatment has been kept flexible. Rather than give a bald list of species that may be found, we have attempted to provide more detail to help locate particular birds, such as an indication of a species' abundance, as well as any specific locations within the site and, in some cases, tactics (such as timing) which may help in tracking down the bird. Where relevant, general weather considerations are also mentioned.

Access and Maps

The Access section is particularly important. For those sites where the use of the text in conjunction with any of the standard road atlases is sufficient, no map is provided. For the majority, however, a specially drawn map accompanies the text to illustrate the features described. These may help locate the reserve and/or show the position of trails and hides. The exceptions are very large sites, such as some areas in Scotland, where access away from roads and main tracks in such extensive and complicated areas is impossible for us to depict adequately. Reference to larger-scale Ordnance Survey maps is recommended in these cases. More than 290 maps are included in the book, with some maps covering more than one site, thus more than three-quarters of the sites described are mapped.

Directions from the nearest town or main road are given to the point(s) where you can park and exploration on foot begins. Distances are in imperial units to conform with current road signs and atlases. Thus we use miles not kilometres, and yards not metres. Take care to park in designated areas wherever possible, and if there are none use common sense and do not obstruct farm gates or private roads. It will be noted that directions invariably assume the use of a car. Most birders use cars to go birding, but it is still possible to use public transport to get to many sites and details can be obtained easily from rail and bus companies, tourist offices and websites. Many reserves have special access arrangements for the disabled and should be contacted for specific advice. For offshore islands, brief information on ferries and flights is provided.

Details of the best trails, vantage points and hides are given, where appropriate, together with notes on any specialities to be found. Any restrictions on access and permit arrangements, if needed, are also specified. Prices have generally been excluded as they rarely remain valid for long.

Visiting arrangements for reserves vary a great deal and are frequently subject to changes. We have specified the most vital information but a few points apply to most reserves and have not been constantly repeated. You should assume that all reserves are closed on Christmas and Boxing Days, and that dogs are not allowed onto any reserve. Larger groups are normally required to make special arrangements with the warden prior to a visit. Where members of various organisations are allowed free entry (or reductions on permit prices), they should expect to produce their membership card (if issued). Non-members will be expected to purchase a permit. Remember also that information/visitor centres often have more restricted times of opening than the reserve itself.

Birding may take you into unfamiliar situations where there is an element of risk for the inexperienced. We hardly need to point out the dangers presented by cliffs, but saltmarshes and tidal flats are also potentially lethal. In some places the tide can advance extremely quickly and unless you have expert local knowledge it is best to leave these well before high water or to avoid them altogether.

Further Information

The final section of the site accounts provides a variety of additional information. At least one, and sometimes several, grid references are given for each site, depending on the size or complexity of the site. These can be used in conjunction with OS maps or satellite navigation systems. Generally grid references pinpoint car parks or entrance points to a site rather than an arbitrary central point. Occasionally, a grid reference may refer to a visitor centre or a hide. If the grid reference quoted does not refer to an obvious starting point, then an explanation will be given as to what it refers to. We have not usually provided postcodes for sites unless they have a visitor centre or an on-site warden where there is a building that will have a full 6- or 7-digit postcode. It is of course possible to give a postcode minus the final two digits. This will enable those with sat navs to get to the general area, but not to a precise location. Grid references are, in our opinion, accurate enough to locate any site.

In this edition we have not included names and addresses of reserve wardens, partly for privacy reasons but also because they can quickly become out of date. In some cases telephone numbers and email addresses have been included, but often the reserves' own websites are the best means to obtain further information. Where a reserve has a visitor centre we have given the full address, postcode and telephone number, but note that these are only open during 'office hours'.

Accommodation is outside the scope of this book (though such information is provided for bird observatories), but its availability on islands is sometimes indicated. Full details can be obtained from the relevant tourist office or from the listed web links.

Other sources of further information (such as local wildlife trusts, local authorities, water companies and bird clubs) may also be given. Details on obtaining permits are mentioned where necessary but, with the exception of some reservoirs, most permits are available on site. Books or checklists devoted entirely to the site are occasionally listed but county avifaunas and local bird reports, both useful sources for anyone interested in a particular area, are not. We give relevant web links where we think they are useful, and encourage readers to use them.

FINDING THE SPECIALITIES

Although this book is firmly site-based, describing the very best sites in Britain and indicating what you can see at them, we recognise that many birders will want to target certain species or specialities to fill in gaps on their lists. Unlike the previous edition of the book, we have not provided an index of bird species in this volume as it wastes too much space listing innumerable sites for common or widespread species. Instead, we have included a new section at the end entitled 'Top 100 species to see in Britain'. The range and status of each of the species listed are briefly summarised, followed by a selection of the best sites to see the species concerned. We have not attempted to detail every site at which the species can be seen, or even every site for which the species is mentioned in the book.

USING THE INDEX

A comprehensive site index is the only index provided. It not only lists each of the 454 'sites' in the book, but also many of the 'sub-sites' listed within a main entry, as well as a number of important places or geographical features. Thus, any significant site name or birding location described in the book will be indexed, but not place names that are purely of a local nature. All entries are indexed by page number.

ENGLISH NAMES AND TAXONOMY

We have been conservative in our approach to the use of English names. While we generally approve of recent proposals to make all English names for species unique (but not necessarily for them to indicate precise taxonomic affinities), we have chosen to use simpler names in many cases. Thus, we have omitted the now familiar prefixes for many species listed in the site accounts. This is partly for simplicity (it takes up less space), but also because in practice these are the names in use in the field in Britain. So, we have Swallow, not Barn Swallow, and Kingfisher, not Common Kingfisher. One cannot deny that the majority of birders still use these names without the prefixes in the field, even if they do include them in their notebooks or in published work. We hope that visitors from overseas will not be confused by this, nor that they will accuse us of imperialism. The principle involved is that only one species of swallow or kingfisher is likely to be seen in Britain, with the other species on the British list being vagrants. An exception has been made for those birds that have a more regularly occurring sibling species. Thus, we have Common Buzzard (Rough-legged is a regular winter visitor, albeit a scarce one) and Common Redstart (because of the less common Black Redstart). We admit that this decision is arbitrary, and that birders in Britain still frequently refer to Common Buzzard as a Buzzard, and to Common Redstart as a Redstart, using the prefixes only for the scarcer siblings. Please note that we sometimes use established English names for well-marked subspecies, e.g. Dark-bellied and Pale-bellied Brent Geese, and Black Brant for the three forms of Brent Goose.

Taxonomy is not really an issue in a book such as this. We generally follow the British Ornithologists' Union in such matters, as they are keepers of the official British List. Fortuitously, the latest BOU deliberations are all incorporated into the new edition of *The Status of Birds in Britain & Ireland* (Parkin & Knox 2010), and therefore our taxonomy follows this book (but not English names – see above). Thus, we treat Yellow-legged Gull *Larus michahellis* and Caspian Gull *L. cachinnans* as separate species (distinct from Herring Gull), both of which occur with increasing regularity in Britain. Other recent splits include Green-winged Teal *Anas carolinensis*, now separated from Teal (officially Common or Eurasian Teal), and Redpoll, which has been split into Lesser Redpoll *Carduelis cabaret* and Common Redpoll *Carduelis flammea*. In the former case, Green-winged Teal is only a vagrant and cannot be guaranteed at any site; indeed, it is only mentioned in passing for a couple of sites. In the case of the redpolls, the form that commonly breeds in Britain is Lesser Redpoll, while Common is generally a migrant or winter visitor (but small numbers breed in the highlands of Scotland and the northern isles). In the text, we have specified Lesser Redpoll where the form involved is clearly the resident species, but if a flock may comprise either Lesser or Common (or both), for example in winter, then we simply refer to them as 'redpolls'.

ABBREVIATIONS

Abbreviations used throughout the book (without explanation) are: CCW (Countryside Council for Wales), CP (Country Park), FC (Forestry Commission), LNR (Local Nature Reserve), NE (Natural England), NNR (National Nature Reserve), NP (National Park), NR (Nature Reserve), NT (National Trust), NTS (National Trust for Scotland), OS (Ordnance Survey), RSNC (Royal Society for Nature Conservation), RSPB (Royal Society for the Protection of Birds), SNH (Scottish National Heritage), SPA (Special Protection Area), SSSI (Site of Special Scientific Interest), SWT (Scottish Wildlife Trust), and WWT (Wildfowl & Wetlands Trust). Other acronyms have only been employed on a site-by-site basis, and in each case the name is given in full at the first mention in the site account.

FURTHER READING

Probably the single most important additional source of information is the indispensable *Birdwatcher's Yearbook*. Published annually by Buckingham Press, it includes a full listing of county recorders, names and addresses of local bird clubs and national organisations, websites and tide tables.

For more detailed site information, and for sites of more local interest rather than national interest, we recommend the excellent series of regional *Where to Watch* guides, also published by Christopher Helm. Fourteen guides cover the whole of Britain (plus another for Ireland), and these are listed opposite the title page at the front of this book.

There are many other sources of information for birders. Most counties publish county avifaunas and produce annual bird reports (these are all listed in the *Birdwatcher's Yearbook*). In addition, many reserves, local bird clubs and observatories publish their own reports (usually annually), and most reserves produce free leaflets. Almost all organisations and many reserves now have their own websites, and these are perhaps the best way to get up-to-date information on sites.

A sound knowledge of status and distribution will help you get more out of your birding and a key reference for any serious birder is *The Status of Birds in Britain and Ireland* by David Parkin and Alan Knox, published in collaboration with the British Ornithologists' Union. The latest edition (2010) is both authoritative and up to date, containing a wealth of information on every British species and subspecies. For more detailed historical information on rarities and vagrancy patterns, the recently published *Rare Birds Where and When: An Analysis of Status and Distribution in Britain and Ireland* by Russell Slack (2009) is an essential reference. The first volume covers near-passerines and passerines, with a second volume covering non-passerines expected soon.

Finally, for those that seek current information about rarities there are various telephone hotlines, both for national and regional news, and also websites that give up-to-date bird news. If you must keep abreast of rare bird news as it happens, then a personal pager or 'smart' phone is the only solution. The digital age is already with us. Never before have we had so much information at our fingertips. Whether you read this book on paper or on screen, it is likely to be only one of a number of sources of information to help you get the most out of your birding.

ACCESS FOR THE DISABLED

We are sympathetic to the needs of disabled birders, and very much wanted to include information on disabled access to the sites in this book. Unfortunately, we have not been able to give *comprehensive* information on disabled access, due to the large number of sites included in the book and the range of disabilities that would have to be considered. A number of sites (notably bird reserves) offer productive birdwatching to those with limited mobility, but some sites are obviously not likely to be suitable for the disabled, requiring long walks over difficult terrain. Many more fall into a 'grey' area where some parts of a site or reserve have good disabled access while much of it remains inaccessible to those with limited mobility. Nature reserve websites may give more detailed information about disabled access to specific sites, but in all cases, where possible, we suggest that you contact the reserve warden or site owners to discuss your specific needs.

ASSESSING DISABILITY ACCESS TO BIRDING LOCATIONS

We welcome feedback from disabled readers about access to specific sites in this book, so that we may improve the information given in future editions. The following guidelines, produced with the assistance of the Disabled Birders Association, outline some of the considerations that need to be taken into account when assessing access possibilities for disabled birders.

Probably the most important point to note is that the vast majority of birders who have mobility problems will not be wheelchair users; just 0.1% of mobility disabled people are full-time wheelchair users and fewer than 1% use mobility scooters. So, 99% of birders with mobility problems can and do walk, but are restricted in the range they can cover or by the terrain. For the 0.1% of mobility disabled people who must use wheelchairs there are particular considerations (see below). For the majority the main problem areas can be remembered as the five 'P's – Permission, Priority, Parking, Paths and Perches.

Permission Some reserves and private sites have drivable tracks and many are happy to allow Blue Badge Holders to drive closer to the birds. If permission does not currently exist, but would be possible, please contact the Disabled Birders Association who will happily try and negotiate access.

Priority Some reserves have created special provision for disabled birders such as viewpoint lay-bys where disabled birders have priority over the able-bodied.

Parking Where there are parking facilities, many will include designated disabled parking spaces that are closer to hides and trails, and which allow extra space for wheelchair users to transfer from their vehicles to wheelchairs.

Paths This is perhaps the most crucial aspect to take note of (and for site managers making provision to address) when assessing disabled access. Note needs to be taken of the length, surface, width and incline of the paths. The distances involved are vital as many disabled people find it difficult to travel very far. Furthermore, most people cannot accurately judge distances and invariably under-estimate them. Surfaces are equally important. Some, such as sand or gravel, are very problematic for wheelchairs; others, such as sticky mud or wet grass, are treacherous for the unsteady. Camber or exposed tree roots, etc. can also impede progress of the least sure-footed or those unable to lift their feet very high. Boardwalks are terrific in principle but if they are bare wood they can be very slippery if wet (even for the able-bodied). Width of trails is also important. If paths are too narrow wheelchairs cannot access them, and boardwalks that are narrow are a frightening prospect if you are in fear of a wheel slipping over the edge. Incline is a problem both for the ambulant and wheelchair users. Walking up or down steep hills is difficult for those who are infirm and may be impossible for some. Anything above a 12% gradient is virtually impossible for a self-propelled wheelchair user.

Perches This is an aspect virtually ignored by most service providers or those assessing access. By 'perches' we mean either purpose-made seating of any sort or convenient perches such as fallen trees, large rocks, banks, etc. Making use of such perches can greatly increase the distance that a person who is 'hard of walking' can travel, as it allows them to rest and recover and go on again. Generally, 150 yards between each perch is ideal.

Wheelchair Access

As mentioned above, gradient is crucial for those using wheelchairs and this is particularly so for ramps into hides; if they are too steep they are as big a barrier as steps. Within hides, viewing slots need to be at a low enough level for viewing from a sitting position and the hide design needs to allow for the knees of a wheelchair user, so they can get close enough to view. Furthermore, if the viewing slots have shutters they must not be opened upwards as wheelchair users cannot open these.

KEY TO THE MAPS

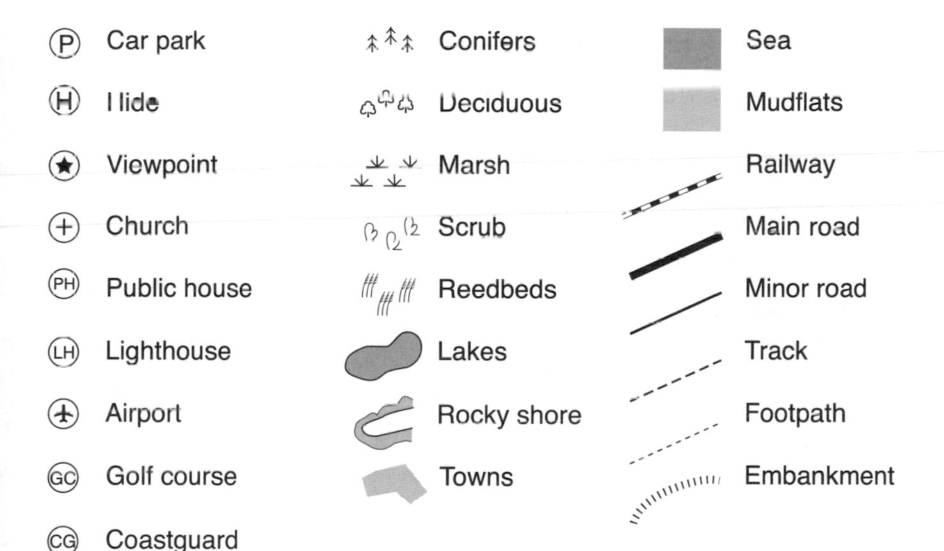

SOUTH-WEST ENGLAND

ISLES OF

1–6 Isles of Scilly

1 St Mary's
2 St Agnes
3 Tresco
4 Bryher
5 St Martin's
6 Uninhabited Islands
7 Pelagic Seabird Trips

CORNWALL

8–10 Land's End

8 Porthgwarra
9 Sennen Cove
10 St Just
11 Pendeen Watch
12 St Ives
13 Hayle Estuary
14 Drift Reservoir
15 Marazion
16 Mount's Bay
17 The Lizard
18 Stithians Reservoir
19 Fal Estuary
20 Gerrans Bay
21 Par Beach and Pool
22 Newquay
23 Camel Estuary
24 Davidstow Airfield
25 Crowdy Reservoir
26 Colliford Reservoir
27 Siblyback Reservoir
28 Rame Head
29 Tamar Estuary
30 Tamar Lakes

DEVON

31 Lundy
32 Taw–Torridge Estuary
33 Roadford Reservoir
34 Prawle Point
35 Start Point
36 Slapton Ley
37 Labrador Bay
38 Yarner Wood

39–43 Exe Estuary

39 Exminster Marshes
40 Bowling Green Marsh
41 Powderham
42 Dawlish Warren
43 Exmouth
44 Axe Estuary and Seaton Bay

SOMERSET

45 West Sedgemoor
46 Avalon Marshes
47 Bridgwater Bay
48 Cheddar Reservoir

AVON

49 Blagdon Lake
50 Chew Valley Lake

DORSET

51 Portland Harbour, Ferrybridge and The Fleet
52 Portland Bill
53 Radipole Lake
54 Lodmoor

55–60 Poole Harbour and Isle of Purbeck

55 Poole Harbour
56 Brownsea Island
57 Arne
58 Studland Heath
59 Shell and Studland Bays
60 Durlston Head–St Aldhelm's Head

61–63 Christchurch Harbour

61 Christchurch Harbour
62 Stanpit Marsh
63 Hengistbury Head

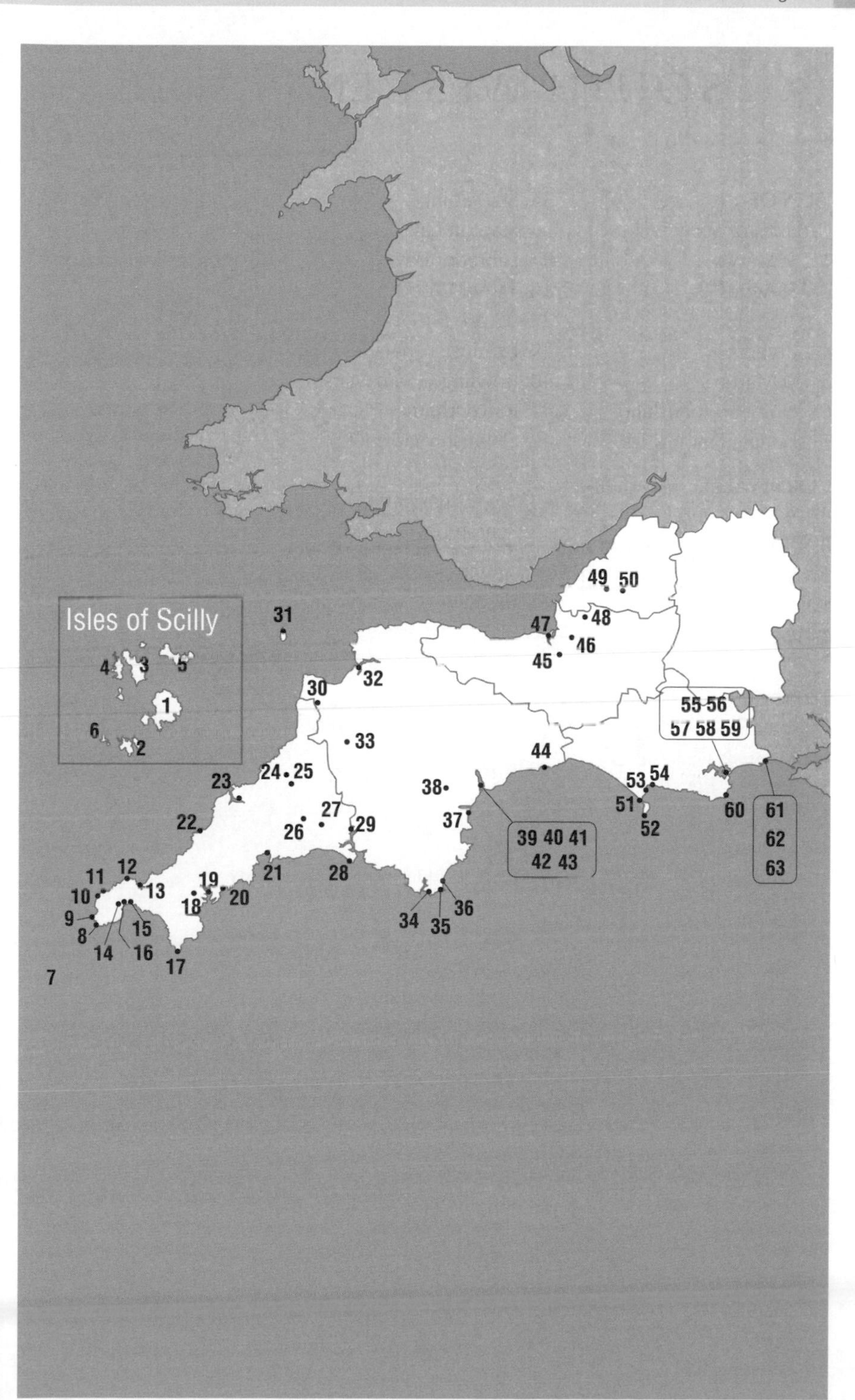
Isles of Scilly
1
2
3
4
5
6
7
8
9
10
11
12
13
14
15
16
17
18
19
20
21
22
23
24
25
26
27
28
29
30
31
32
33
34
35
36
37
38
39 40 41
42 43
44
45
46
47
48
49
50
51
52
53
54
55 56
57 58 59
60
61
62
63

1–6 ISLES OF SCILLY

OS Landranger 203

About 150 islands, of which only the five largest are inhabited, constitute the picturesque archipelago of Scilly. The entire group is only ten miles across and lies 28 miles off Land's End. The geographical position of Scilly ensures a wide variety of migrants in spring and autumn, and many seabirds breed in summer. The best times for migrants are mid-March to May and August–October; breeding seabirds are best seen May–July.

Over the last 40 years the Isles of Scilly have established themselves as the premier site in Britain for rarities, even overtaking the legendary Fair Isle. Although Nearctic species are particularly sought after and are now an annual feature, almost anything can and does turn up. October is the key month but late September and early November can also be outstanding. In addition to true vagrants, some of Britain's scarcer migrants, which occur only irregularly on the mainland, can almost be guaranteed on Scilly.

Weather has a major influence on birding on Scilly. The conditions bringing vagrants to the islands are complex and only partially understood. Light southerlies are best for falls of migrants

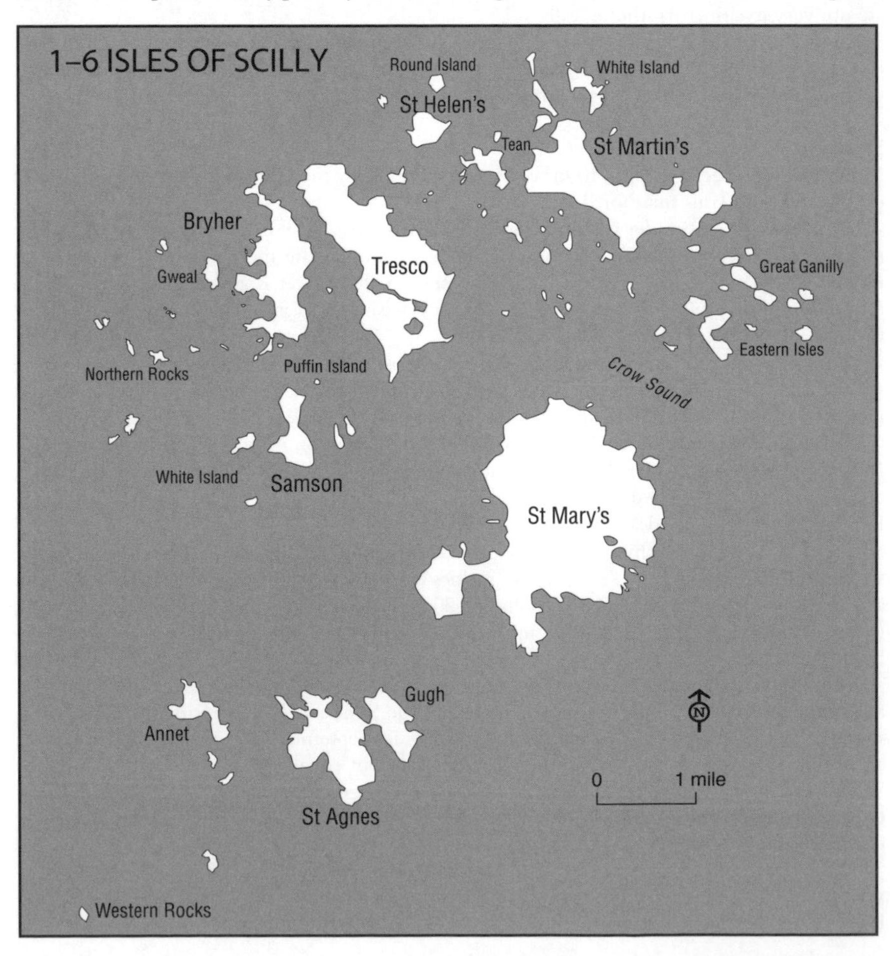

in spring. In autumn south-west winds prevail, and fast-moving wave depressions originating off North America are sometimes responsible for small falls of American landbirds, chiefly in late September to late October. American waders are less affected by such weather systems and tend to arrive somewhat earlier. South-east winds in autumn may bring Asiatic birds but sometimes rarities appear in apparently unsuitable weather conditions. Strong south-west to north-west winds can be good for seawatching but this pursuit is rarely well rewarded on Scilly, although occasionally there is a good passage of large shearwaters in August.

Scilly is exceedingly popular with traditional holidaymakers during summer but by October they are replaced by legions of birdwatchers. Strangely, the islands are still comparatively neglected by birders in spring.

Habitat

The smaller islands tend to be rocky and sparsely vegetated, while the larger ones, particularly those that are inhabited, possess a range of habitats. Around the coast are low granite cliffs, rocky shores and sandy beaches. Exposed areas with poor soil remain as heaths or grassy downs, and on St Mary's the golf course and airfield have created an important habitat. The few freshwater pools and associated marshland with dense clumps of sallows are especially attractive, but there is little natural woodland. Cornish Elms grow in more sheltered areas and stands of introduced pines occur in places. Tresco Abbey Gardens hold a wide variety of exotic trees and shrubs but are not very exciting ornithologically.

Large areas of the inhabited islands are farmed. This includes grazing land, but the dominant industry on Scilly, other than tourism, is flower farming. The tiny bulb fields are surrounded by tall hedges of introduced shrubs to protect them from the winds.

Species

More than 400 species of birds have been recorded on Scilly including a remarkable number of rarities and numerous firsts for Britain.

Spring passage gets under way in March and the main passage continues into April–May, but numbers are rarely impressive. Spring overshoots are annual, the most frequent species being Hoopoe in March–April. A Woodchat Shrike or something rarer is a distinct possibility, with this period often producing records of the scarcer British herons, and small numbers of Golden Orioles are regular in May.

Summer is essentially for breeding seabirds. Most are present April–July, though some species can be seen around the islands all year. Shag and Kittiwake are common and Fulmar has increased recently. Razorbill is the most numerous auk but only c.200 pairs breed. Common Guillemot is rather less numerous. Most of the Puffins breed on Annet, which is also home to Manx Shearwater and Storm Petrel. The latter two, however, are rarely seen around the islands during the day and are best seen from the *Scillonian*. Common Tern breeds on the islands, but Roseate Tern no longer does so.

Since 1997 deep-sea fishing boats have provided sightings of Wilson's Petrel and large shearwaters, with occasional Fea's Petrels. Mega rarities have included a Madeiran Petrel in July 2007 and a Black-browed Albatross in September 2009. Between July and September two boats (*Sapphire* and *Kingfisher*) regularly go up to 8 miles south or south-west off St Agnes (usually 5pm to dark), giving extremely close views of petrels and shearwaters (see below for details).

Autumn wader passage occurs July–October with a peak in September, but numbers are generally small. Commoner waders include Little Stint and Curlew Sandpiper. American vagrants (18 species recorded) are annual, mainly in September. American Golden Plover and Pectoral Sandpiper are among the more frequent but equally predictable is Buff-breasted Sandpiper, a grassland wader uncommon on the east coast of America but a near-annual visitor to the golf course and airfield in September. One or two Dotterel quite often share this habitat, occurring late August–October.

Seabird passage is rarely impressive and is usually best during or after south-west or north-west gales. The best seawatching points are Peninnis Head and Deep Point (St Mary's), Horse Point (St Agnes), and the north tip of Tresco. Seabirds are often better observed from the

Scillonian. Great and Cory's Shearwaters can be expected in late July and August in the right conditions, from southern viewpoints. Manx and Sooty Shearwaters and Great and Arctic Skuas are quite regular in September. Later in the autumn, gales may bring Leach's Petrel or Grey Phalarope.

Passerine migration is most evident in September and includes Common Redstart, Ring Ouzel, Whinchat and Pied Flycatcher. Many of the scarcer migrants are annual on Scilly and sometimes outnumber their more familiar counterparts. Among the earlier arrivals are Icterine and Melodious Warblers and Red-backed and Woodchat Shrikes, which occur from late August to October. Other species recorded annually in variable numbers, mainly September–October, are Wryneck, Richard's and Tawny Pipits, Bluethroat, Barred Warbler, Common Rosefinch, and Lapland, Snow and Ortolan Buntings. Yellow-browed Warbler, Firecrest and Red-breasted Flycatcher are regular between late September and early November, sometimes in comparatively large numbers, and can almost be guaranteed in October. A number of rarities are now annual, or almost so, on Scilly. These include Short-toed Lark, Rose-coloured Starling, and Rustic and Little Buntings. Late September to the end of October is the best time for American landbirds; perhaps the most frequent is Red-eyed Vireo, but Grey-cheeked Thrush, Blackpoll Warbler, Rose-breasted Grosbeak and Bobolink have all put in a number of appearances, and a host of other Nearctic passerines has been recorded.

Asiatic visitors tend to arrive slightly later, mostly early October to early November: the most regular in the last decade have included Olive-backed Pipit and Booted, Radde's, Dusky and Pallas's Warblers. But extreme vagrants from this sector also occur with some frequency, with Pechora Pipit, Siberian and White's Thrushes and Yellow-browed Bunting all found in recent years. Rarities from closer to home also make landfall; Red-throated Pipit and Subalpine, Greenish, Arctic and Western Bonelli's Warblers have each been recorded a number of times.

Raptors are not infrequent in autumn with Merlin and Peregrine seen almost daily from October. Rarer visitors include occasional Osprey, Honey-buzzard, harriers and Black Kite, and Scilly recently played host to Britain's first Short-toed Eagle. Late September and October usually produces a few Spotted Crakes and Corncrakes but their retiring habits cause them to be easily overlooked. By November, the birding scene is considerably quieter; migrant passerines include thrushes, finches, Blackcap, Chiffchaff and a few Firecrests, some of which winter in the archipelago. New arrivals include Woodcock, one or two Short-eared Owls, and perhaps a Long-eared Owl (the latter winters regularly on Tresco, but numbers fluctuate considerably). Black Redstarts sometimes arrive in considerable numbers at this time, frequenting the rocky coasts and bulb fields.

Winter is bleak and few birders visit the islands. Severe gales are frequent, though freezing conditions are rare. Several Great Northern Divers are usually present in the channels between the islands, Crow Sound being particularly favoured. Gannet and Kittiwake can usually be seen offshore and numbers of Purple Sandpipers frequent coasts. Geese are occasional visitors in very small numbers and a variety of duck winters on the Great Pool on Tresco. One or two seaducks and grebes (especially Slavonian) winter in The Roads. Visiting Water Rails greatly outnumber the few residents and a wintering Merlin could be seen anywhere. Rarities are possible even this late in the year – Britain's first Great Blue Heron was found on St Mary's in December 2007.

Access

By air British International Helicopters operates between Penzance and St Mary's and Penzance and Tresco daily (except Sunday). The journey takes 20 minutes, and the number of flights per day is considerably reduced November–March. Penzance Heliport lies 1 mile east of Penzance by the A30, and parking is available (fee). The Isles of Scilly Skybus operates frequent flights throughout the year from Land's End Airport, St Just, and from Newquay (except Sundays). A car park (fee) is available at St Just, and also a minibus connection to Penzance railway station. Between mid-February and October, Skybus also offers flights from Southampton, Bristol and Exeter. The eight-seater Skybus can also be chartered for urgent twitches.

By sea The *Scillonian III* sails from Penzance to St Mary's. The voyage takes 2.5 hours, but may be prolonged by rough weather. The usual departure times are 9.15am from Penzance, 4.30pm from St Mary's but these times vary on Saturday sailings in summer. There are daily sailings (except Sundays) mid-March to October, but no service November to mid-March inclusive. Always check departure times in advance.

There is a wide variety of accommodation on the islands, with the greatest choice on St Mary's. Information and accommodation lists are available from the Tourist Information Centre, and early booking of all transport and accommodation is essential, especially for October.

Once on the islands it is easy to walk anywhere, but there are taxis and buses on St Mary's. Passenger launches leave St Mary's quay daily to the off-islands March–October inclusive, normally at 10.15am and 2.15pm, with extra sailings on demand. 'Seabird Specials' are organised in season for locally breeding seabirds. Full details are posted on the quay. For those staying on other islands, inter-island transport is more problematic and, apart from the regular services to St Mary's, ad hoc arrangements must be made.

Roads, footpaths and nature trails access most areas on the islands. Birders should keep out of fields and other private areas. In addition, the airport may only be viewed from the perimeter and the golf course should not be traversed when play is in progress.

1 ST MARY'S (Isles of Scilly)

This is the largest island and contains a wide variety of habitats. In addition to the sites detailed, it is worth checking all sheltered bushes and trees for migrants and the ploughed and grassy

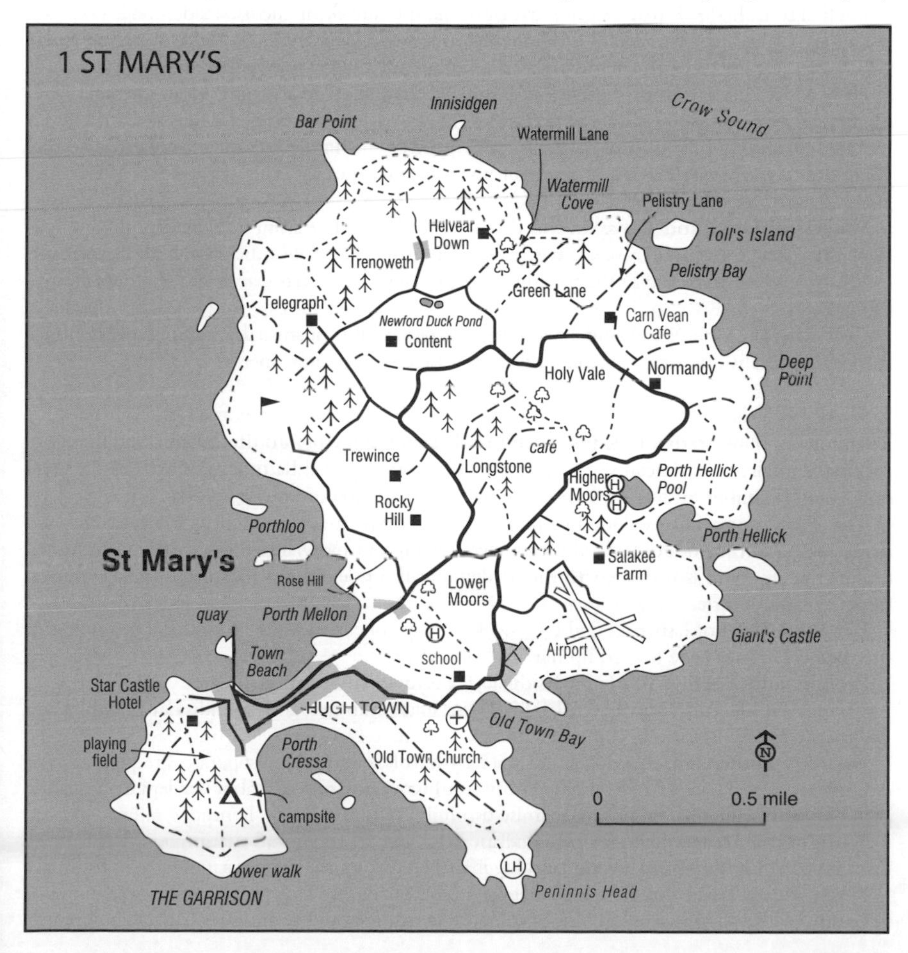

fields for larks, pipits and buntings.

The Garrison A choice of trails leads around this headland. Access is from Hugh Town, up a steep hill past Tregarthen's Hotel. There is a stand of pine trees next to the playing field (campsite nearby) and many sheltered spots for migrant passerines amid the more open areas. The Lower Walk, accessible via Sally Port, has attractive belts of elms and takes you to the main circular path.

Peninnis Head The fields and bushes are well worth checking on the way. Certain migrants favour the short turf around the headland and the Head itself is one of the best seawatching points on St Mary's. The elms in Old Town churchyard regularly attract migrants. There is a temporary trail in October.

Lower Moors A nature trail (from Rocky Hill via Rose Hill to Old Town Bay) crosses the marshes and reedbeds and there are two hides and a small scrape. Check the dense sallows carefully.

Airport Waders (including the occasional Dotterel and Buff-breasted Sandpiper), pipits and buntings occur occasionally. Check from the edges only. In October there is arranged access to an observation area by the windsock, accessed from the coastal path.

Salakee The fields around the farm have produced some outstanding rarities. Access from the airport is no longer allowed, but Salakee can still be reached from Porth Hellick or the road at Carn Friars Lane.

Porth Hellick Pool and Higher Moors A nature trail skirts the west side of the pool and continues through the marshland of Higher Moors. Two hides overlook the pool and reedbeds, which attract wildfowl and a few waders; migrant passerines favour the shallows.

Holy Vale This is one of the best areas and receives considerable attention. The trail from Higher Moors crosses the road and continues through sallows, brambles and marshland into a narrow trail through tall elms with dense undergrowth – a jungle by Scilly standards. The trail continues to Holy Vale Farm and the main road. Two temporary trails are opened in October at the discretion of the farmers.

Watermill Lane and Cove Newford Duck Pond, the elms alongside Watermill Lane, and the bushes near Watermill Cove are worth checking. From the Cove, a coastal path heads south to Pelistry and north to Bar Point, passing through bracken-covered slopes and scattered clumps of pines.

Golf course Dotterel and Buff-breasted Sandpiper are almost annual, and pipits and buntings turn up regularly. The surrounding fields may also hold the same species.

2 ST AGNES (Isles of Scilly)

The most south-westerly of the inhabited islands, facing directly into the Atlantic and therefore the first landfall for American vagrants. It is a little more than 1 mile in length with a very small population. The islet of Gugh is connected to St Agnes at low tide by a sandbar.

The Parsonage This secluded garden and small orchard surrounded by elms, hedges and dense vegetation is undoubtedly the best spot for migrant passerines and rarities on St Agnes. Unfortunately there is no access to the garden, but the trees of the Parsonage can be viewed from the lane outside.

Big Pool This pool, surrounded by sedges, usually appears devoid of birds but a number of unexpected species have turned up here over the years.

Periglis and Porth Killier The best bays for waders on St Agnes.

Chapel Fields and Troy Town Fields These small bulb fields, surrounded by brambles and hedges, are always worth checking in autumn.

Barnaby Lane A tree-lined lane favoured by warblers and Firecrest.

Covean A track leads to this sheltered bay from Covean Cottage in Higher Town. Tamarisks, brambles and hedges surround the bulb fields.

Wingletang Down This area of moorland with gorse bushes and granite boulders is often disappointing for birds but Lapland Bunting is fairly regular and several rarities have been found.

Horse Point This is the best seawatching site on St Agnes.

Gugh The small overgrown fields in the middle of the island sometimes harbour migrants.

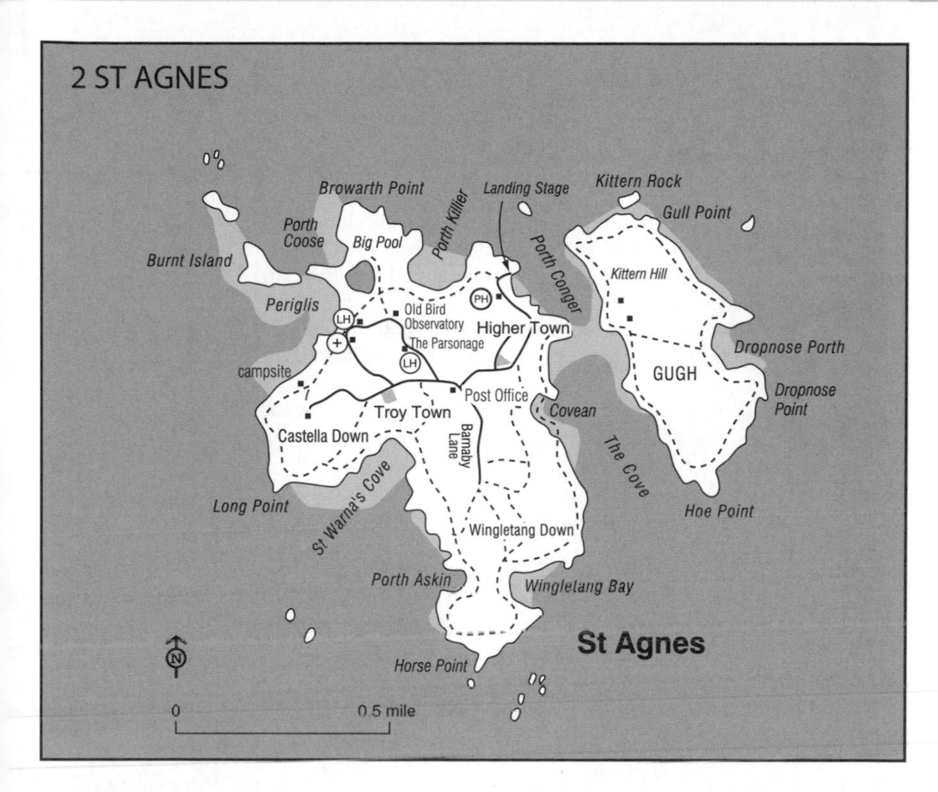

3 TRESCO (Isles of Scilly)

The island, dominated by Tresco Abbey and its gardens of exotic trees and plants, is a privately leased estate. Depending on the tides, boats land either at Carn Near or New Grimsby. Though private, access is generally unrestricted along footpaths and roads. Areas of woodland occur in many places but most are unproductive pine plantations. Tresco is a large island with a great potential for rarities. Most birders only make brief visits and to the same few areas – greater coverage would undoubtedly produce more.

Abbey Pool A small pool with sandy edges easily checked for waders.

Great Pool The largest area of fresh water on Scilly, surrounded by reeds and sallows. Not all of the pool is easily viewed; in particular try at each end and from certain points along the north side. Two hides provide good viewing opportunities for wildfowl and waders. Wildfowl are a notable feature and rarer species are found quite regularly. The sallows around the edges should be checked, and the fields on the north side often hold finches and buntings.

Borough Farm The fields and hedges around the farm have attracted several rarities in recent years. Access is limited to the main tracks through the farm.

4 BRYHER (Isles of Scilly)

Lying immediately west of Tresco, Bryher is surprisingly little visited by birders. The weedy fields in the centre of the island attract finches and buntings but there is little farming here now, and any area of bushes should be checked. The pool on the west side has produced some unusual waders, though the commoner species prefer the coastal bays of which Stinking Porth is one of the best.

5 ST MARTIN'S (Isles of Scilly)

This large island lies north of St Mary's. There are two widely separated landing points: Lower Town at high tide, Higher Town at other times. Much of the island is cultivated and the higher

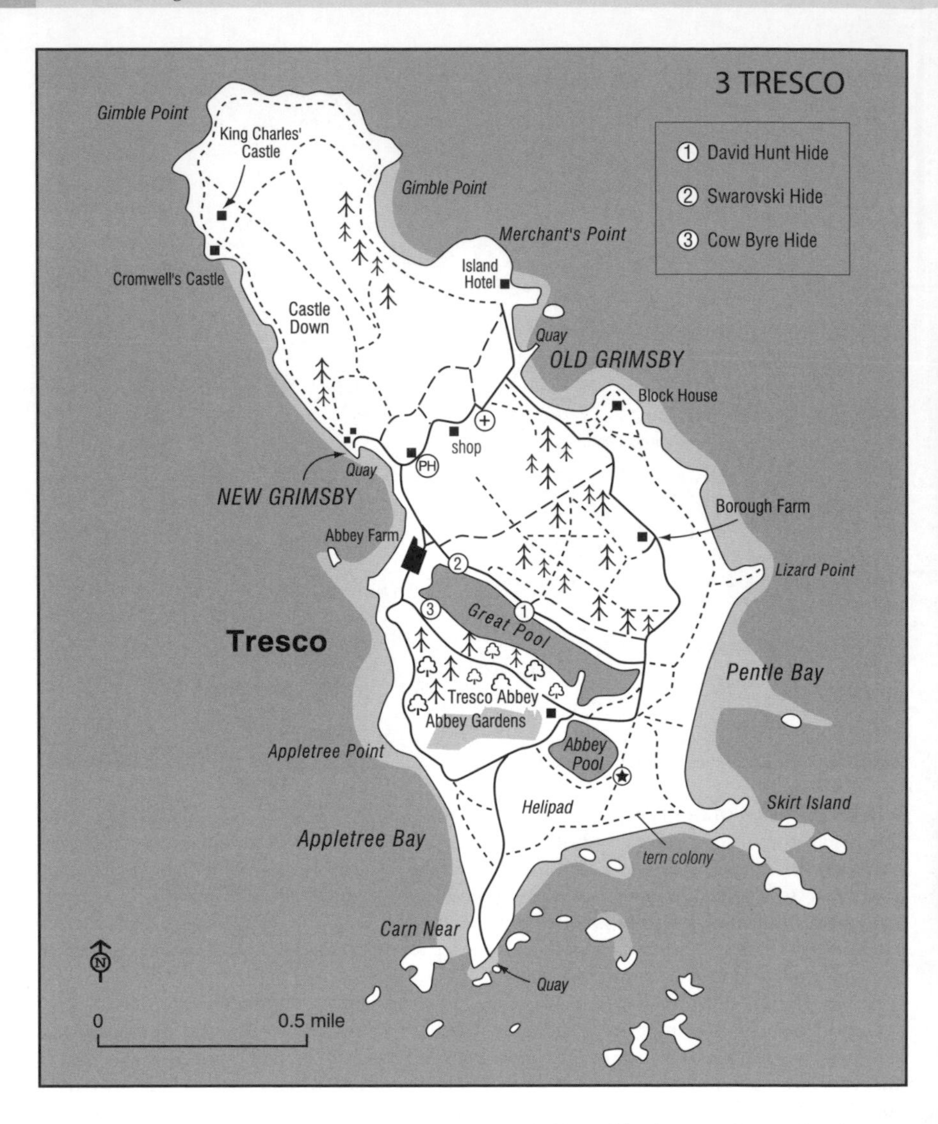

part is rather open. The most sheltered areas are the fields and hedges on the south side. One of the best areas is east of Higher Town: from the quay, walk towards Higher Town and bear right around the coast checking the bushes and trees on the left. After 0.5 miles turn left along a track to a farm where a short nature trail runs through fields; there are some mature elms here.

6 UNINHABITED ISLANDS (Isles of Scilly)

Annet is the second largest uninhabited island (after Samson) in the archipelago. It is famous for its breeding seabirds, notably Puffins. It also holds a significant population of Storm Petrels and Manx Shearwaters, but these are unlikely to be seen by day. Access is restricted in summer (March to August) to protect the breeding seabirds, but regular boat trips from St Mary's take visitors around the island, giving good views of the birds. There are three main groups of smaller uninhabited islands (Western Rocks, Northern Rocks and the Eastern Isles) which, together with Round Island and St Helens in the north of the archipelago, are also important for breeding seabirds.

FURTHER INFORMATION

Grid refs:
St Mary's: The Garrison (campsite) SV 898 102; Peninnis Head SV 911 093; Lower Moors SV 912 105; airport SV 917 105; Salakee SV 921 106; Porth Hellick Pool SV 923 107; Holy Vale SV 920 115; Watermill Lane SV 919 121; golf course SV 908 117
St Agnes: The Parsonage SV 880 082; Big Pool SV 878 085; Periglis SV 877 083; Barnaby Lane SV 882 081; Horse Point SV 882 071; Gugh SV 887 083
Tresco: Abbey Pool SV 897 140; Great Pool SV 891 148; Borough Farm SV 898 149
Bryher: Quay SV 880 149; pool SV 874 148
St Martin's: Lower Town SV 914 161; Higher Town SV 930 151

Accommodation:
Tourist Information Centre, Isles of Scilly: tel: 01720 422536; email: tic@scilly.gov.uk; web: www.simplyscilly.co.uk

Travel to islands:
British International Helicopters, Penzance: tel: 01736 363871. British International Helicopters, Scilly: tel: 01720 422646; web: www.islesofscillyhelicopter.co.uk
Isles of Scilly Travel (Isles of Scilly Steamship Group): tel: 0845 710 5555; web: www.islesofscilly-travel.co.uk (operates Skybus and *Scillonian*)

Other information:
Seabird boat trips and guided walks: Will Wagstaff: tel: 01720 422212
Pelagics off Scilly: email: admin@scillypelagics.com; web: www.scillypelagics.com
Joe Pender, skipper of *Sapphire*: tel: 01720 422751 or 07776 204631; email: joesapphire@aol.com
Alec Hicks, skipper of *Kingfisher*: tel: 07768 662229; web: www.scillyfishing.co.uk or www.scillyshortrangepelagics.co.uk
Contact Bob Flood with general enquiries at tubenose@tiscali.co.uk
Isles of Scilly Bird Group: web: www.scilly-birding.co.uk
Isles of Scilly Wildlife Trust: web: www.ios-wildlifetrust.org.uk
In October, a bird log is called each evening at the Scillonian Club in Hugh Town. This is an excellent source of information, and an enjoyable occasion for all who participate. Temporary membership of the Scillonian Club is required for all visiting birders. In addition, a blackboard with indispensable bird information is located outside the back door of the Pilot's Gig Restaurant, opposite the Mermaid Inn. It also contains notices of the Isles of Scilly Bird Group (ISBG), and now remains here all year. A similar, less extensive, noticeboard is located outside St Agnes Post Office.

7 PELAGIC SEABIRD TRIPS (English Channel)

Pelagic trips into the rich waters of the Western Approaches to the English Channel often yield good views of species unlikely to be seen well from the mainland. A much sought after speciality is Wilson's Petrel, which is recorded regularly on pelagic trips but very seldom seen from land. Charter trips take place in late summer and early autumn, this being the peak period for seabirds. Birds are most numerous around trawlers and are often drawn in towards the charter vessel by using 'chum', a foul-smelling mixture of fish oil and offal that floats on the surface.

Habitat

These deep and rich seas hold a great diversity of marine life, making them attractive feeding grounds for wide-ranging seabirds.

Species

Storm Petrel is common while Wilson's Petrel is reasonably regular, albeit usually in tiny numbers. Great Shearwaters can be fairly numerous but Cory's Shearwater is scarce. All of the commoner skuas occur with Long-tailed Skua being the rarest, and Sabine's Gull is another rarity that is seen regularly. There is always the possibility of an extreme rarity; for instance, a Fea's Petrel followed one boat for a considerable period in 2001, permitting all those on board to gain excellent views.

Access

See further information for booking trips. Interest in these trips is usually high and it is therefore best to book well in advance. Private vessels can be chartered from large ports such as Penzance or Falmouth by arrangement. For pelagics out from the Isles of Scilly, see page 33.

FURTHER INFORMATION

The Isles of Scilly ferry, the *Scillonian III*, runs one pelagic most years (usually in mid-August) sailing from Penzance (16 hours return trip). Contact the Isles of Scilly Steamship Company, Penzance (0845 710 5555 or 01736 362009/362124) or look for advertisements in birdwatching magazines.

8–10 LAND'S END (Cornwall)

OS Landranger 203

The Land's End area is famed for attracting migrants in spring and autumn and, in particular, rarities. Though overshadowed by Scilly, it has produced some outstanding rarities and is worthy of more extensive coverage, especially in late autumn. Seawatching off Porthgwarra can be rewarding especially in August–September.

Habitat

The exposed peninsula of Land's End comprises moorland and rough grazing fields surrounded by impressive granite cliffs. Trees and bushes only gain a foothold in the more sheltered valleys and hollows, and these are important refuges for migrants.

Species

Spring migration commences early, in March, and a variety of migrants pass through in April–May, though generally in small numbers. Typical species include Turtle Dove, Cuckoo, Common Redstart, Whinchat, Ring Ouzel, Grasshopper Warbler, and Spotted and Pied Flycatchers. An unusual spring migrant or a rarity is most likely in May, though March–April can produce unexpected rarities such as the white-phase Gyrfalcon that graced the cliffs near St Just in 2000.

Seabird passage in spring is light, peaking April–May. All three divers may be seen as well as small numbers of Manx Shearwaters, Common Scoters, Whimbrels and Sandwich and Common Terns, and the occasional skua. Numbers and diversity of seabirds increase from July. Balearic Shearwater and Great and Arctic Skuas are regular in small numbers. Much less frequent, and generally later in autumn, are Great and Sooty Shearwaters and Pomarine Skua. The best winds are usually south-east to south-west and not too strong. In recent years, Cory's Shearwater has regularly appeared off Porthgwarra in July–August, sometimes in large numbers. Fea's Petrel has also been claimed several times recently.

Passerine migrants reappear from August and a few scarce visitors, such as Melodious Warbler or Woodchat Shrike, are found annually. Later, light easterlies may bring Wryneck, Barred and Yellow-browed Warblers or Red-breasted Flycatcher. October–November sees movements of

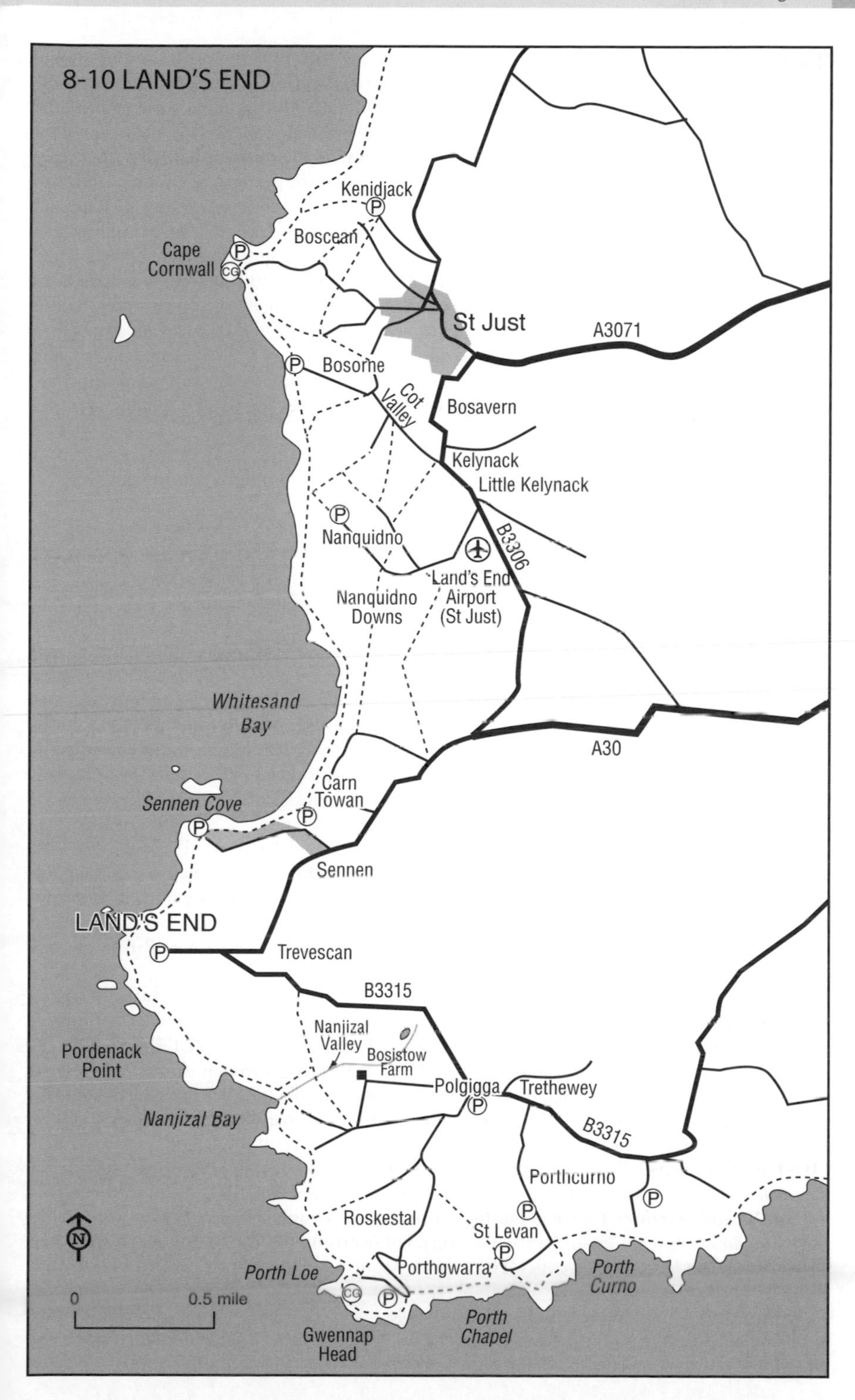
8-10 LAND'S END
Kenidjack
Boscean
Cape Cornwall
St Just
A3071
Bosorne
Cot Valley
Bosavern
Kelynack
Little Kelynack
B3306
Nanquidno
Land's End Airport (St Just)
Nanquidno Downs
Whitesand Bay
A30
Carn Towan
Sennen Cove
Sennen
LAND'S END
Trevescan
B3315
Nanjizal Valley
Bosistow Farm
Pordenack Point
Polgigga
Trethewey
Nanjizal Bay
B3315
Porthcurno
Roskestal
St Levan
Porth Loe
Porthgwarra
Porth Curno
0
0.5 mile
Porth Chapel
Gwennap Head

Skylarks, pipits, Robins, Goldcrests and finches, and a few Black Redstarts, Ring Ouzels and Firecrests. A sprinkling of raptors arrives at this time including Hen Harrier, Merlin, Peregrine and Short-eared Owl, but late autumn is most notable for rarities, including occasional Nearctic landbirds, especially after strong south-west winds. While the valleys are perhaps most famed for their attractiveness to North American migrants, just as on Scilly vagrants have appeared from most points of the compass. Be prepared (but never hope) for the unexpected. The short turf of St Just airfield attracts Golden Plovers and Lapwings in autumn, and species such as Ruff and Whimbrel may accompany them. Very small numbers of Dotterels are regular in autumn and occasionally also Buff-breasted Sandpiper. Richard's and Tawny Pipits, and Lapland and Snow Buntings occur infrequently, and most exceptionally both Blyth's and Pechora Pipits have been recorded. Some of the fields and grassy clifftops offer a similar habitat.

Winter is fairly quiet. A few seabirds such as Gannet can be seen and divers are sometimes offshore. Small numbers of Hen Harriers and other raptors winter on the moors and Chiffchaff and Firecrest in the sheltered valleys. Resident species are usually in evidence, including Common Buzzard, Sparrowhawk, Peregrine, Green Woodpecker, Rock Pipit and Raven.

Access

All of the following sites are close together and several may be combined in a single visit.

8 PORTHGWARRA (Cornwall)

To the south of Land's End, Porthgwarra well deserves its status as a rare-migrant trap, but Nanjizal is also worthy of exploration.

Porthgwarra The most famous valley in west Cornwall, its position at the south-west tip of the peninsula ensures a regular trickle of American landbirds. Leave Penzance west on the A30. At Catchall, turn left on the B3283 to St Buryan. Beyond St Buryan, join the B3315 and continue towards Land's End for 2 miles. At Polgigga, turn left to Porthgwarra. There is a car park in the village, just beyond the final hairpin at the end of the valley. Migrant passerines may use any suitable cover. In particular, check the dense bushes in the valley and gardens in the village. From the car park the road continues to the coastguard cottages and several paths lead to the cliffs and across the moorland to a small pool at the head of the valley. Seawatching is best from Gwennap Head, near the coastguard lookout. From Porthgwarra village a coastal footpath heads east to St Levan, which has more gardens and trees.

Nanjizal This scenic valley lying between Porthgwarra and Land's End is much less well known and less easy to reach. It can be accessed from Land's End, about a mile away along the coastal footpath or, more conveniently, from Polgigga on foot along a private track to Bosistow Farm. It is well worth checking the hedgerows and fields for migrants on the way. In the valley itself, check the bushes and pools for rarities, keeping to the footpaths for safety. Part of the valley can be viewed from the B3155, but the road is sometimes busy.

9 SENNEN COVE (Cornwall)

This village lies at the south end of Whitesand Bay, 1 mile north of Land's End. Divers, grebes and various seabirds may be seen offshore in season. From Land's End, turn left off the A30 beyond Sennen on a minor road to Sennen Cove. There is a car park overlooking the bay, which is used by surfers all year round. Walk the beach or seawatch from the coastguard lookout west of the village.

10 ST JUST (Cornwall)

There are several sites in the St Just area which are worthy of exploration:

Land's End Airport This small airfield 3 miles north-east of Land's End is not heavily used and attracts waders. From Land's End, turn left north on the B3306 towards St Just. The airfield is on the left after 1.5 miles. The best views are from the minor road to Nanquidno along its north side.

Nanquidno This sheltered, well-vegetated valley west of Land's End Airport is another excellent area for migrant passerines, and several American rarities have been found. Park at the end of the road and explore the dense bushes and trees in the valley bottom.

Cot Valley This valley lies north of Nanquidno and is also good for migrants. Leave St Just village on the road to Cape Cornwall and almost immediately turn left on a minor road signed Cot Valley. Parking is limited along the road. Check the gardens at the head of the valley and any other areas of cover. The road and a footpath continue alongside the valley to the coast.

Kenidjack This valley lies to the north of Cot Valley, and north-west of St Just. It is also good for migrants. Leave St Just northwards on the road to St Ives and take the first turning on the left at the bottom of the hill. There is limited parking here. Walk the valley towards the coast.

FURTHER INFORMATION

Grid refs: Porthgwarra SW 370 217; Nanjizal SW 365 239; Sennen Cove SW 355 263; St Just SW 370 314; Land's End Airport SW 370 297; Nanquidno SW 364 292; Cot Valley SW 355 309; Kenidjack SW 365 323

11 PENDEEN WATCH (Cornwall)

OS Landranger 203

Pendeen Watch lies at the extreme north-western tip of the Land's End peninsula and is an important seawatching site.

Habitat

The watchpoint is an open grassy clifftop, beside the lighthouse.

Species

Seawatching at Pendeen is rather similar to St Ives (see below), and indeed many of the same birds pass both sites within a short time of each other. The same weather conditions apply to both sites, although slightly more westerly winds can produce good numbers of birds at Pendeen. The passing seabirds can sometimes be a little distant compared to St Ives, but Pendeen is generally better for the larger shearwaters: Great, Cory's and Sooty are regular here in July–September. As for St Ives, it is vital to check the weather forecast to ensure a successful visit.

Access

The village of Pendeen lies on the B3306 between St Just and St Ives. From the village, a minor road is signed to the lighthouse. After 1 mile, park at the end of the road and descend the steps to the right of the lighthouse, to seek shelter from north-west winds at the base of the east-facing wall. This is better than watching from the car park.

FURTHER INFORMATION

Grid ref: SW 379 359

12 ST IVES (Cornwall)

OS Landranger 203

St Ives is well known as one of Britain's best seawatching points, if not the best, though in recent years it has been somewhat overshadowed by the advent of regular pelagic trips into the South-west Approaches and greater attention having been paid to Porthgwarra due to the discovery of regular (sometimes large) movements of Cory's Shearwater off the latter. In appropriate weather in autumn the seabird passage can be truly spectacular, but if the winds are wrong you could see nothing. In addition to large numbers of birds, good days offer a chance of seeing species rarely seen from land. The main attractions in winter are divers and perhaps one of the rarer gulls.

Habitat

St Ives lies on the west side of St Ives Bay and north of the town is The Island, which is actually a rocky headland.

Species

Weather is critical for large movements of seabirds. Ideal winds are west-north-west to north, preferably preceded by a south-west gale. Such conditions may only occur a few times a year, and rather than visit on the off chance you should ideally wait until the correct conditions are forecast. Good winds can produce seabirds at any season but the greatest number and diversity occur late August–November. In west winds seawatching can be better at Pendeen Watch (see page 37), c.10 miles west of St Ives, though the birds are generally more distant.

A few Great Northern Divers and Slavonian Grebes are regularly seen in St Ives Bay in winter. Black-throated Diver is less common and more likely in late winter or early spring. Seaduck (mainly Common Eider and Common Scoter) are not numerous. Gannet, Cormorant and Shag may occur offshore at any time. A few Purple Sandpipers winter on the headland. Gull flocks should always be checked; Glaucous is the most regular of the scarcer species and late winter is the best time. Iceland Gull is also frequent at this time.

In spring, Mediterranean and Little Gulls are occasional. There is a small passage of Sandwich Terns, with a few Common and Arctic. In appropriate weather there may be a movement of Manx Shearwaters and Kittiwakes, and perhaps the occasional Storm Petrel.

Early-autumn passage is dominated by Fulmar, Manx Shearwater, Arctic Skua, and Common and Sandwich Terns. Smaller numbers of Sooty and Balearic Shearwaters, Great Skuas, Little Gulls and Roseate, Little and Black Terns are seen regularly. Storm Petrel occurs but in very variable numbers. North-west winds in September are likely to bring in thousands of Gannets and Kittiwakes. Leach's Petrel and Sabine's Gull are two of St Ives' specialities and sometimes occur in quite large numbers though a few of each is more normal (beware of misidentifying young Kittiwake as Sabine's Gull – it happens frequently). Pomarine Skua is occasional and Long-tailed Skua rare. Auk passage almost exclusively consists of Razorbill and Common Guillemot, sometimes in huge numbers in late autumn.

Seabird passage peters out in November though Great Skua, Kittiwake and auks are still much in evidence. Two late species seen in variable numbers are Grey Phalarope and Little Auk. All three divers may be found, including Red-throated, normally the scarcest species in Cornwall. Rock Pipit occurs around the coast and one or two Black Redstarts or Snow Buntings may visit The Island.

Access

At the north end of the town follow signs to the car park on The Island and walk to the top of the headland. Seawatching is best from the perimeter of the coastguard lookout. For gulls, check the harbour and (from the car park) the sewage outfall. Other parts of the bay can be viewed from various points in the town and it is possible to walk east along the shore to search for terns on Porth Kidney Sands.

FURTHER INFORMATION

Grid ref: SW 520 410

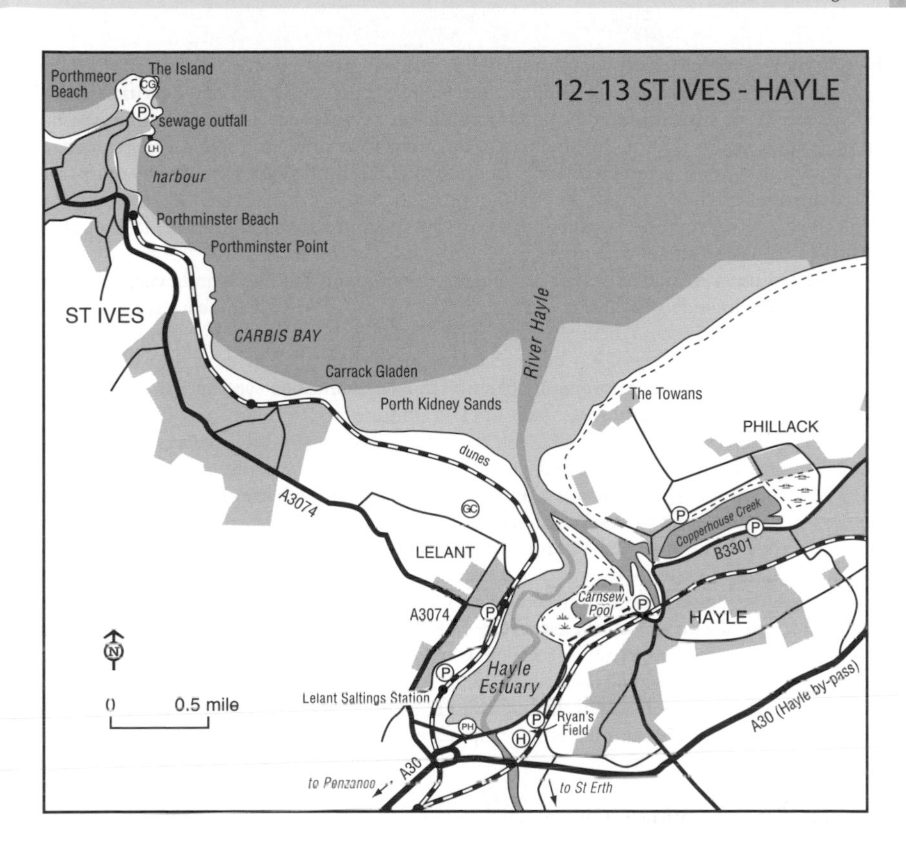

13 HAYLE ESTUARY (Cornwall)

OS Landranger 203

The Hayle Estuary's compact size and easy access permit close views of a good variety of wildfowl, waders and gulls. The RSPB bought the estuary in 1992, and added Ryan's Field in 1995. It is best visited in autumn and winter, and American vagrants are a regular feature.

Habitat

The estuary is surrounded by urban development and is effectively divided into two sections, the main bay and Copperhouse Creek; the latter lies within the town of Hayle itself. At low tide, most of the estuary is mudflats with an area of saltmarsh in the south-west corner. Carnsew Pool is separated from the main part of the estuary by a narrow embankment; it is tidal but retains some open water even at low tide. Ryan's Field (RSPB) lies across the road at the southern end of the estuary. This is now being managed, and a lagoon and four islands have been constructed.

Species

In winter Black-throated and Great Northern Divers and Slavonian Grebe are occasionally present on Carnsew Pool. A few Goldeneyes and Red-breasted Mergansers are regular. Most ducks on the estuary in winter are Teal and Wigeon with smaller numbers of Gadwalls. A few Knots, Spotted Redshanks, Greenshanks and Common Sandpipers usually winter among the commoner waders. Large numbers of gulls are present; a few Mediterranean, Little and Glaucous are regularly seen, especially in late winter, and Ring-billed and Iceland Gulls are annual. Search for the former at Copperhouse Creek, a regular site in winter where the birds are often attracted to the car park by birders throwing food down for them. Peregrine is seen irregularly through the winter and Little Egrets may be seen on Ryan's Field.

Spring passage brings small flocks of Whimbrels, and a few Little Ringed Plovers and Wood Sandpipers are seen annually. Small numbers of terns pass through, and Roseate, Little and Black Terns may feature in these movements.

A greater variety of waders occurs in autumn and although the numbers of each are usually small, they may include Little Stint or Curlew Sandpiper. An American rarity is found almost annually, the most frequent being White-rumped and Pectoral Sandpipers and Long-billed Dowitcher. American Wigeon has been recorded more than once, with at least one multiple occurrence. Dedicated gull-watchers are likely to find Mediterranean and Little Gulls, and small numbers of terns are regularly seen.

Access

The B3301 (formerly the A30) runs alongside the estuary offering good views from the roadside with several parking places. Waders are best seen within two hours of high tide. Carnsew Pool can be viewed from a public footpath which follows the embankment around the pool and also gives good views of the mudflats. Access is from the B3301 west of Hayle or from the town itself – through a small industrial area west of the viaduct. It may also be worth walking along the spit to the mouth of the estuary. Small numbers of waders roost on the spit and the Pool's embankments. Copperhouse Creek can be watched from the B3301 and the footpaths surrounding it. Ryan's Field lagoon has a hide overlooking it, and an adjacent car park. Waders roost on the lagoon at high tide. To the west of Hayle, Porth Kidney Sands is the best beach for terns in late summer and autumn, and for Sanderling; it can be reached from St Uny church in Lelant.

FURTHER INFORMATION

Grid refs: Carnsew Pool SW 555 372; Ryan's Field SW 549 362
RSPB Hayle Estuary (warden): tel: 01736 711682; email: hayle.estuary@rspb.org.uk; web: www.rspb.org.uk/hayleestuary

14 DRIFT RESERVOIR (Cornwall)

OS Landranger 203

This small reservoir lies just west of Penzance and is good for waders and gulls.

Habitat

Areas of mud are exposed around its edges when the water is low, usually in autumn. The surrounding area is largely fields, but there is some woodland at its north end.

Species

Drift is noted for attracting gulls, particularly in winter, from the nearby seafront and Glaucous, Iceland, Ring-billed and even Bonaparte's Gulls have been seen there, though Mediterranean is more likely. Other winter/spring attractions include occasional scarce grebes and seaducks. In recent years vagrant Lesser Scaup has occasionally been found among small groups of Scaup.

In autumn attention is focused on passage waders such as Ruff, Curlew Sandpiper and Little Stint. Transatlantic vagrants have included Lesser Yellowlegs and Spotted Sandpiper.

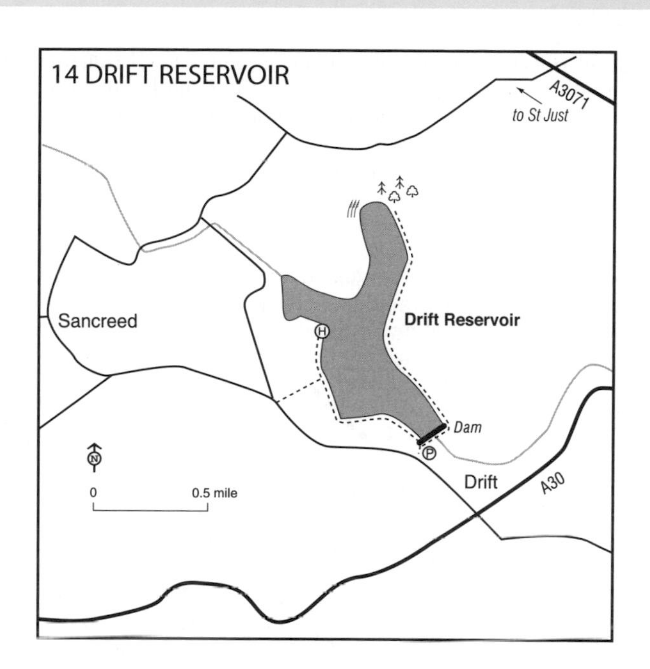

Access

Leave Penzance south-west on the A30 and at Lower Drift turn right on a minor road to Sancreed. After a few hundred yards the road passes close to the south end of the reservoir. Part can be viewed from here (there is an unlocked hide) and you can walk around the sides (no permit required). For views of the north end, continue along the road and fork right towards Sellan.

FURTHER INFORMATION

Grid ref: south end SW 436 288
Dave Parker: tel: 01736 710668; mobile: 07932 354711; email: damag@btopenworld.com

15 MARAZION (Cornwall)

OS Landranger 203

Marazion Marsh is on the coast, east of Penzance. It attracts a number of interesting species, including rarities, in spring and autumn.

Habitat

Most of the marsh is dense reeds and sedges (the largest reedbed in Cornwall) with scattered clumps of bushes. A few open pools with muddy margins attract waders. A road runs along the seaward side, a railway crosses the marsh, and housing flanks another side, but despite this the area remains relatively undisturbed. The 53-ha site is now an RSPB reserve.

Species

Cetti's Warbler is resident and with patience may be seen at any time. Wintering birds include small numbers of the commoner ducks, Water Rail (mostly heard), and occasionally Jack Snipe. Marazion also hosts a huge Starling roost in winter.

Spring sees a sprinkling of early passerine migrants and passage waders, and Garganey is regular in early spring. Later a scarce migrant or a rarity is almost inevitable; some of the more likely candidates are Spoonbill, Marsh Harrier and Hoopoe. Water Pipit is occasionally found on the drier edges.

Autumn waders include Little Stint, Curlew Sandpiper and Wood Sandpiper, and a North American wader such as Pectoral Sandpiper or Long-billed Dowitcher is found most years. Spotted Crake is fairly regular, especially in August–September. The passage of hirundines often attracts a Hobby in autumn and Little Gull and Black Tern are seen occasionally. A few Aquatic Warblers are seen each year, usually in mid–late August, and although this species is renowned for skulking in reedbeds Marazion probably offers the best chance of seeing one in Britain – but beware of young Sedge Warblers with crown-stripes.

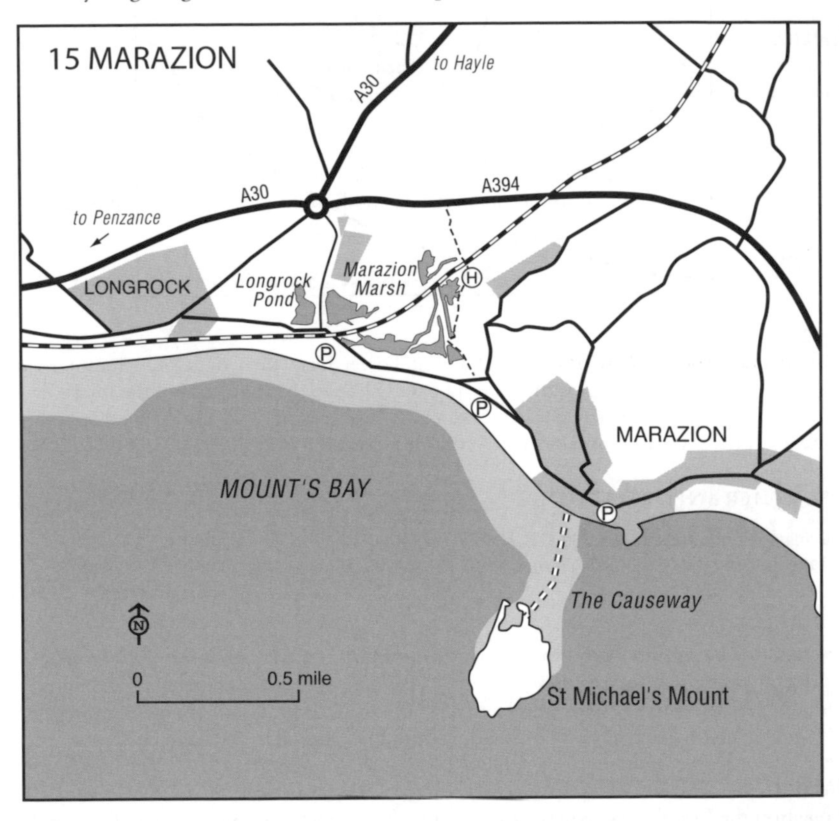

Access

Leaving Penzance on the A30 to Hayle, fork right at Longrock to follow the coast road to Marazion. The marsh is easily viewed from the road. There is a car park just beyond the railway bridge and another at the east end (Ropewalk car park) which is immediately opposite the reserve entrance. Trails (may not be wheelchair-accessible) lead to a hide and along the seaward side of the marsh.

FURTHER INFORMATION

Grid ref: SW 512 311
RSPB Marazion Marsh: tel: 01736 711682; email: marazion.marsh@rspb.org.uk; web: www.rspb.org.uk/marazionmarsh

16 MOUNT'S BAY (Cornwall)

OS Landranger 203

This large south-facing bay encompasses the towns of Penzance and Newlyn, as well as St Michael's Mount. It is an important site for seabirds in winter.

Habitat

The undisturbed parts of the shoreline comprise sandy beaches with rocky areas.

Species

Great Northern Diver is regular in winter and Black-throated Diver frequent but less common. Slavonian Grebe is also usually to be found. Sanderling is common on sandy beaches while Purple Sandpiper prefers rocky areas. Large numbers of gulls gather on the shore or at freshwater outflows when the tide is out. Scarce species such as Glaucous, Iceland, Mediterranean and Little Gulls are regular and rarities such as Ring-billed Gull, Baird's Sandpiper and King Eider have occurred. A small passage of terns occurs in spring and Kentish Plover is almost annual.

Access

Part of this huge bay can be viewed from the beach at Marazion. There is unrestricted access to other areas from the coast road but much of it is disturbed.

FURTHER INFORMATION

Grid ref: car park between Longrock and Marazion SW 506 311

17 THE LIZARD (Cornwall)

OS Landranger 203/204

The Lizard Peninsula on the south coast of Cornwall has few trees or other cover for passerines but its southerly aspect makes it attractive to numerous and varied migrants, including rarities, although the area is not as frequently visited as it perhaps deserves. It has now become famous for Choughs, following the return of breeding birds to Cornwall in 2001 after a 50-year absence.

Habitat

Habitat on the Lizard is extremely varied and includes cliffs and coves, gardens, fields and wet maritime heath, the latter being of European-level conservation importance.

Species

Winter is the dullest season though raptors may include Peregrine, Merlin and Hen Harrier.

Spring migration can be strong involving common arrivals such as Chiffchaff, Common Whitethroat, Wheatear and Whinchat. Scarce migrants typical of more open areas include Black Redstart and Ring Ouzel, and rarities have included Alpine Swift, Subalpine Warbler and Woodchat Shrike.

Early autumn is probably the best time for seawatching and may produce shearwaters and skuas. Early-autumn rarities recorded most years include Wryneck, Icterine and Melodious Warblers. Subsequently passerine migration increases and many species are commoner than in spring, e.g. Spotted and Pied Flycatchers are more numerous in autumn, as is Tree Pipit. Several extreme rarities have been found late in the year, most notably Upland Sandpiper and Little Bustard. More likely are Chaffinch and Brambling, or perhaps a Lapland or Snow Bunting.

As of January 2010 there were 20 Choughs in Cornwall, at least eight of which were in the Lizard area. They may be seen anywhere between the Lizard and as far west as Marazion, and are most likely to be found foraging in clifftop fields. The Chough watchpoint, staffed from April to June, is situated close to the Lizard Point car park. The birds nest to the west of the watchpoint and may be seen from the coastal footpath.

Access

There are seven major areas of interest to birders.

Lizard village and headland The peninsula is reached via the A3083 from Helston. Various public footpaths can be taken from Lizard village permitting coverage of the entire headland, and the village itself is a good area to search for migrants. Before entering the village a private toll road leads to Kynance Cove and within Lizard village a road on the left goes to Church Cove. Behind the church is a car park and the nearby sycamores have produced rarities such as Pallas's Warbler and Red-eyed Vireo.

Lizard Downs–Kynance Cove to Predannack Wollas This walk takes in cliffs and moorland as well as Kynance Valley, which can hold migrants. Another good spot to search for migrants is on the north route from Kynance NT car park, which passes Predannack airfield. Here there are bushes and fields of rough grass.

Windmill Farm nature reserve Bought in 2001, this reserve is jointly managed by the Cornwall Wildlife Trust and the Cornwall Bird Watching and Preservation Society (CBWPS).

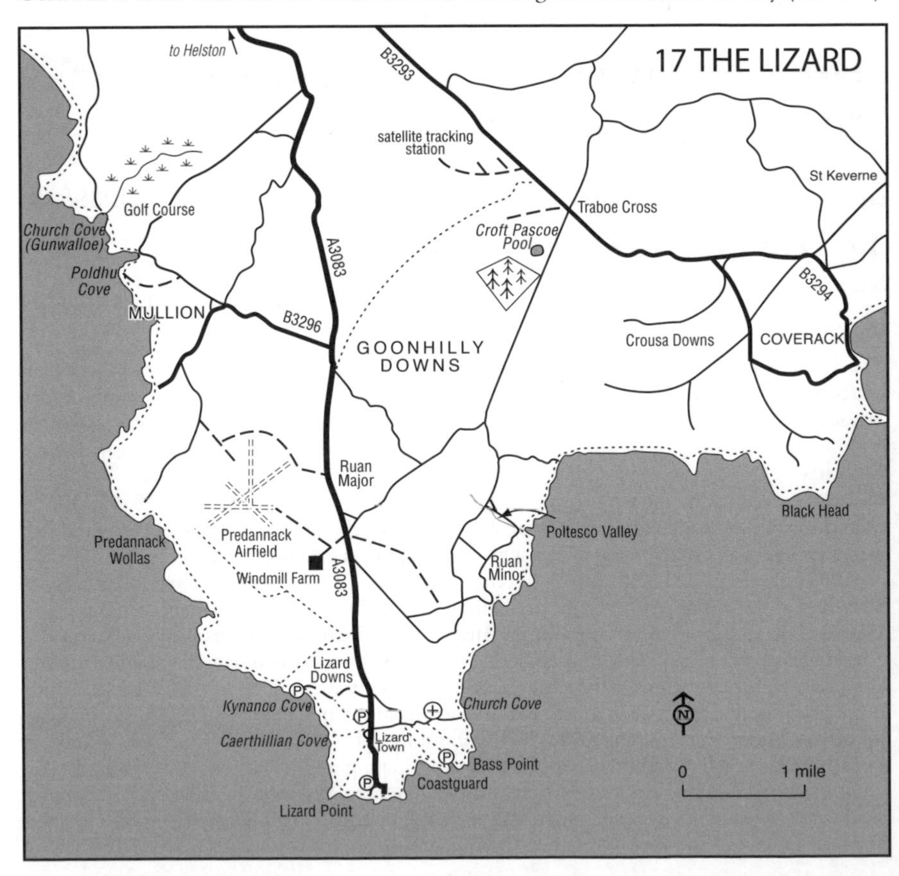

Considerable improvements to the farm have created some fine habitats providing some of the best birding on the Lizard. There are two hides overlooking Ruan Pool and further improvements are planned. From the A3083 southwards, turn right 1.75 miles after the B3296 turn-off to Mullion and follow the lane to the windmill.

Goonhilly Downs–Traboe Cross The moors can be viewed from Traboe crossroads on the B3293 and should be searched for raptors, while small pools in this area can produce interesting waders.

Crousa Downs and Coverack area Take the B3293 from the A3083 and continue until the B3294 to reach Coverack. Crousa Downs are accessed by taking a minor road from the B3293 at Zoar. Crousa Downs possess a variety of habitats that could hold migrants while the footpath east from the north end of Coverack village can also be worth checking.

Poltesco Valley This area is unusual on the Lizard in that it possesses extensive cover and tall trees, but it has not been intensively watched. To reach Poltesco take the road beside the church at Ruan Minor.

Gunwalloe and Poldhu Marsh Reeds in Gunwalloe Valley and around Poldhu Marsh attract unusual migrants such as Marsh Harrier and even rarities like Purple Heron and Aquatic Warbler. Poldhu is signed off the B3296 at Mullion village. From the NT car park walk the road towards the beach and turn left over a stone bridge on a public track at the valley mouth.

FURTHER INFORMATION

Grid refs: Lizard village SW 703 125; Windmill Farm SW 694 153
Cornwall Wildlife Trust: web: www.cornwallwildlifetrust.org.uk/nature_reserves/map

18 STITHIANS RESERVOIR (Cornwall)

OS Landranger 203

This well-known reservoir lies south of Redruth and is best in autumn and winter for wildfowl and waders. Indeed, it attracts an American wader almost every autumn.

Habitat

Much of the reservoir is shallow with natural banks and mud is frequently exposed when the water level drops, while marshy areas at either end provide variety. The surrounding area is largely open moorland. There is some disturbance by windsurfers at weekends.

Species

Waders are the prime interest. Spring passage is poor but autumn can be excellent. In 1994 a long-staying Pied-billed Grebe paired with a Little Grebe and produced several hybrid offspring. The most frequent species from July may include numbers of Little Stints, Curlews and Wood Sandpipers, and occasional Little Ringed Plovers. Later in autumn, numbers of Golden Plovers and Lapwings reach several thousand. The Golden Plover flock attracts an American Golden Plover almost every year, but the latter can be hard to find on the ground. Another American rarity of incredible regularity is Pectoral Sandpiper – sometimes several together. Other vagrants recorded several times are Long-billed Dowitcher and Lesser Yellowlegs. A variety of raptors has been seen, and in autumn Peregrine is not infrequent. A few Little Gulls and Black Terns occur each autumn. Ducks should not be ignored and Teal flocks should be checked for Garganey and Blue-winged Teal.

Most of the wildfowl arrive in November, a few Goldeneyes winter and Scaup is occasional. Few of the waders, other than Golden Plover and Lapwing, remain during the winter, though there are usually a few Ruffs.

Access

Leave Helston on the A394 towards Penryn. After 4.5 miles turn left (north) on a minor road signed Stithians. After Carnkie the road crosses the south tip of the reservoir. Waders can be seen well from the road here and there is a marshy area south of the road. Two hides are available (key required, see below) though they are not really necessary on a casual visit. Continue along the road to the north of the reservoir. Beyond the Golden Lion pub, the road again crosses an arm of the reservoir giving views over exposed mud and open water. Another hide overlooks the north-west corner of the reservoir. From here you can also walk along the west edge.

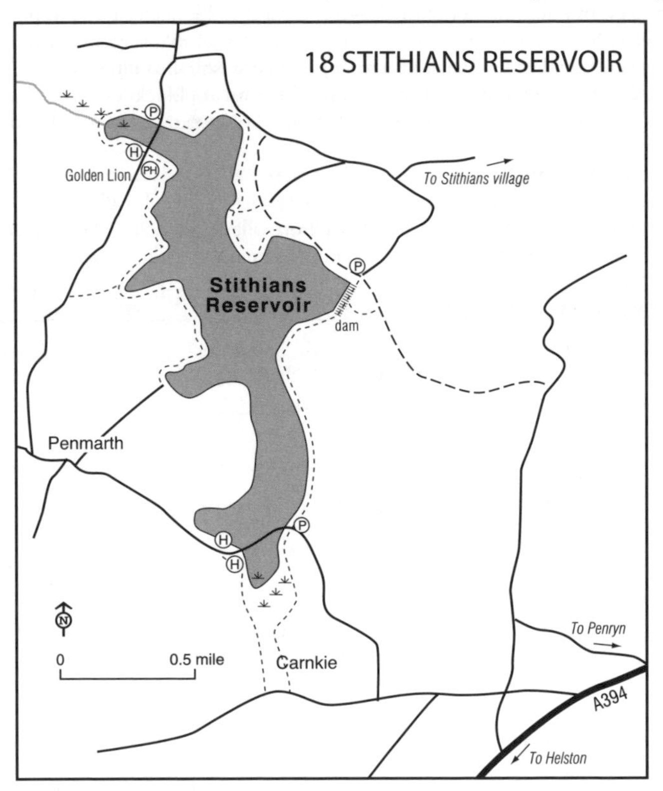

FURTHER INFORMATION

Grid refs: south end SW 712 350; Golden Lion SW 708 371
Warden: email: stithians_warden@cbwps.org.uk

19 FAL ESTUARY (Cornwall)

OS Landranger 204

The Carrick Roads waterway is formed by the confluence of the Truro and Fal Rivers between the headlands of St Anthony in the east and Pendennis in the west. Deep water makes this a particularly good area for diving birds in winter. On the southern edge of Falmouth Bay, the Rosemullion area has a small but growing reintroduced population of Cirl Buntings.

Habitat

Even at low tide the water is deep, making it ideal for diving birds. Creeks run away from both sides of the Roads but the largest of these, Restronguet, is heavily disturbed. Further north, to the east of Truro, a narrow tree-lined river, the Tresillian, is less disturbed.

Species

Diving ducks such as Goldeneye and Red-breasted Merganser are regular in the Roads in winter. Rarer species include all three divers and the scarcer grebes, with Black-necked Grebe regularly occurring in significant numbers (up to 40 prior to roost in late afternoon). Seaducks, including occasional Velvet Scoters, and auks may also be present – Black Guillemot is now annual.

Small numbers of passage waders including Spotted Redshank and Greenshank join Dunlins and resident Common Redshanks in spring and autumn. Shelduck, Grey Heron, Little Egret and Kingfisher are also resident; single Spoonbills often make extended visits in autumn/winter.

Access

There are four major areas of interest within the estuary.

Carrick Roads (see map) Drive to Mylor from the A39 at St Gluvias, and park in the village. Walk along the south side of the creek to Penarrow Point, which affords a view over the Roads. For Turnaware Point (best for Black-necked Grebe) take the A3078 Truro–St

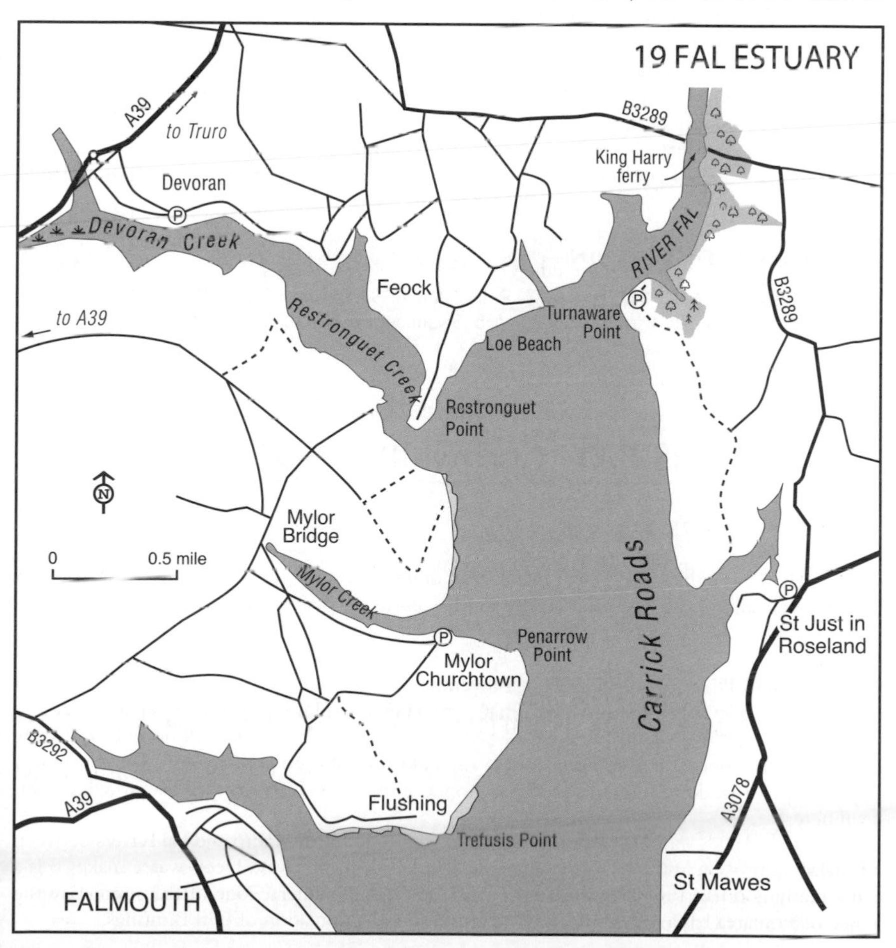

Mawes road, turning north onto the B3289 at St Just in Roseland then after 2 miles left onto a minor road which leads to an NT car park.

Restronguet Creek (see map) Devoran Creek, in the upper reaches of Restronguet Creek, is a good place to watch waders around high tide. It is reached by turning right off the A39 Truro road south of Carnon Downs onto Greenbank Road. Continue into Quay Road and park carefully on the side of the road after a very sharp bend. From here walk the road by the creek.

Ruan Lanihorne Situated at the head of the eastern arm of the estuary, this is another good place to watch waders on an incoming tide. From the A3078, take the minor road heading south-east at Tregony. Turn right in the village of Ruan Lanihorne and view the creek from the road.

Tresillian The northern arm of the estuary, south of the village of Tresillian, is also good for waders at low tide. After leaving Truro on the A390 towards St Austell, follow the sign to Pencalenick. Park carefully at the side of the road as close as possible to the public footpath at Pencalenick, and take the path that follows the river to St Clement.

To the south of Falmouth, two other areas are worth mentioning.

Swanpool This brackish lagoon on the outskirts of Falmouth frequently hosts divers or grebes in winter as well as a large number of gulls (check for rarer species). Wooded areas on the western side can harbour wintering Chiffchaff, Blackcap and Firecrest, and the bay can be scanned from Swanpool Beach for divers, grebes and seaducks. Park in the car park at the southern end of the pool; a footpath leads all the way round the pool.

Rosemullion This headland is situated on the west side of Falmouth Bay. From Falmouth, follow the Swanpool Beach road over the hilltop past the cafe, then fork left, about 2 miles by minor road. Continue beyond Maenporth for 2 miles to Mawnan village, parking near the old church. A footpath leads north-east to Rosemullion Head; seawatching from here can be profitable from late spring to autumn. Nearby, Cirl Buntings have been successfully reintroduced.

FURTHER INFORMATION

Grid refs: Mylor SW 820 352; Turnaware Point SW 836 382; Devoran Creek SW 797 390; Ruan Lanihorne SW 894 420; Tresillian SW 856 455; Swanpool SW 801 312; Rosemullion SW 787 272

20 GERRANS BAY (Cornwall)

OS Landranger 204

Gerrans Bay lies on the south Cornish coast east of the Fal estuary and is bounded by Portscatho and Nare Head. It is the best place in the south-west to see Black-throated Diver.

Species

Black-throated Diver normally occurs November–April, numbers peaking to around 50 in early April, and late-staying individuals may attain summer plumage prior to departing. A few Great Northern Divers and Slavonian Grebes are usually evident and small numbers of seaducks (including once a female King Eider) and auks may also be present. Nare Head is worth exploration for landbird migrants in the appropriate seasons – Red-rumped Swallow has been seen here.

Habitat

The shoreline is rocky and unprotected from heavy seas. The scenic Nare Head encompasses a range of habitats and dramatic views.

Access
Viewing is best from Pendower Beach within the innermost part of the bay. Access is from the A3078 south of Tregony along a minor road signed Pendower Beach and Hotel. Park by the road near the hotel and view from the clifftop. Alternatively, look out from the point of Pednavadan (north of Portscatho) on the west of the bay.
Nare Head Park in the NT car park to the south of Veryan. A clockwise circular route around the head is best, returning via Nare valley.

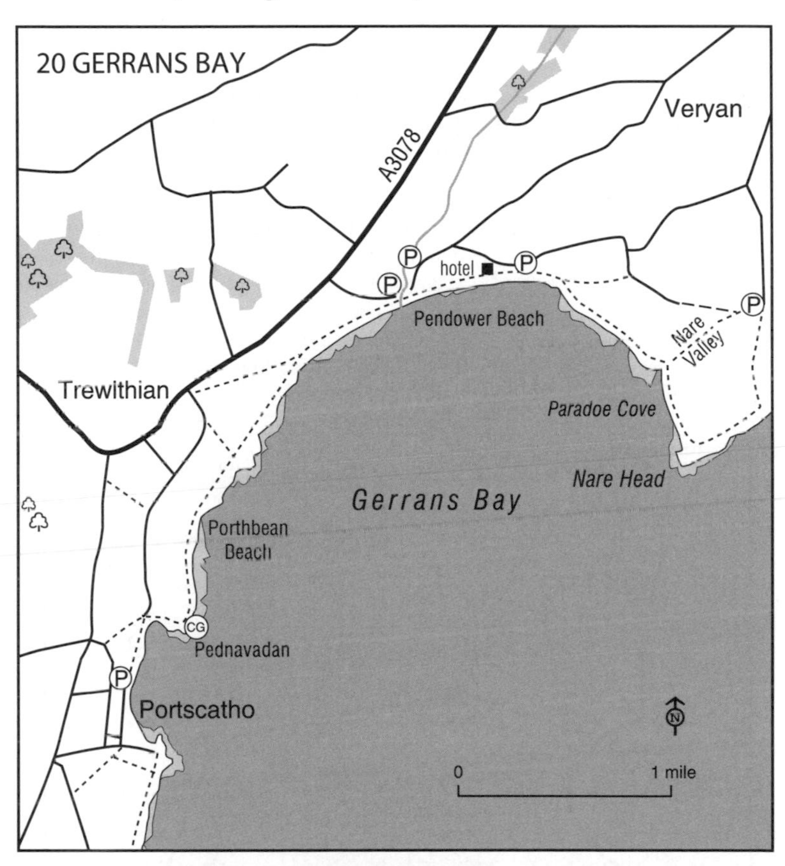

FURTHER INFORMATION

Grid refs: Pendower Beach SW 903 383; Nare Head SW 921 379

21 PAR BEACH AND POOL (Cornwall)

OS Landranger 200

Par Sands Beach is part of St Austell Bay and, despite the harbour waterfront alongside having been developed, the area is still rich in birds. Migrants occur around the pool with gulls congregating on the beach, and the sheltered bay provides refuge for divers, grebes and seaducks.

Habitat

Par Pool is a freshwater pool that covers approximately 1 ha. Though most of the banks are open there are some reed fringes. The sandy beach is broad and the water shallow, and offshore areas are partially sheltered.

Species

Migrants use the reedbeds around the pool and large numbers of Swallows and Yellow and Pied Wagtails gather there in autumn. Rare reedbed denizens passing through have included Bittern, Spotted Crake and Savi's Warbler.

In winter Reed Buntings and Starlings use the reeds. Large numbers of gulls gather on the beach and scarce species such as Mediterranean, Little, Glaucous or even Iceland may be located with skill, patience and luck – Ring-billed Gull is annual. Common waders include Oystercatcher and Turnstone, although Sanderling is present in surprisingly low numbers. Purple Sandpiper, Grey and Ringed Plovers, and Dunlin are also possible. There is a high-tide wader roost at Spit Beach. In the bay in winter Great Northern Diver may be joined by Red-throated and Black-throated Divers, Slavonian and Red-necked Grebes, Common Scoter and Common Eider.

Summer breeders and residents include Reed Warbler and at least one pair of Cetti's Warblers.

Access

From the A3082 take the minor road signed to Par Sands Holiday Park. Immediately west of the Ship Inn there is parking overlooking the beach, beside the pool.

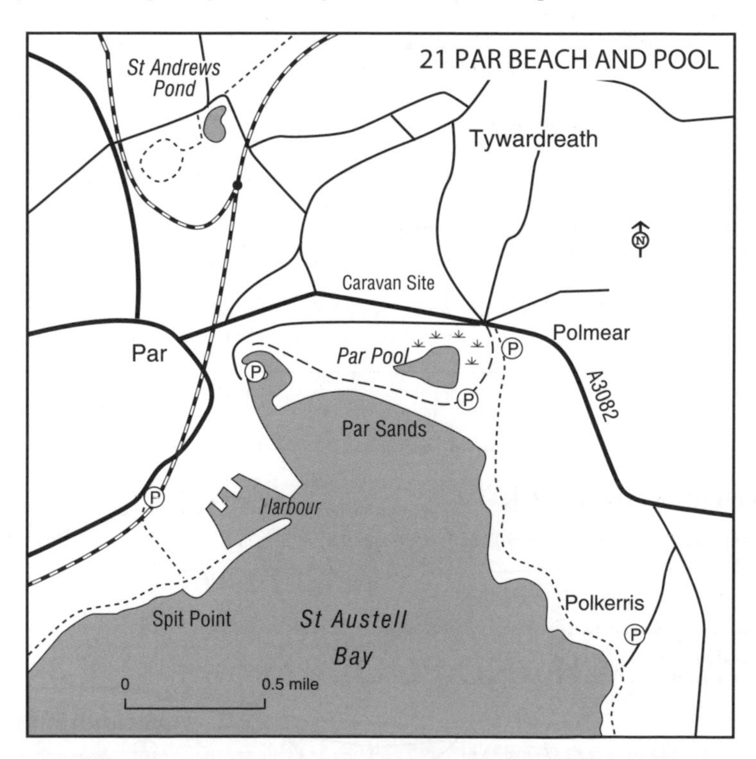

FURTHER INFORMATION

Grid ref: SX 087 532

22 NEWQUAY (Cornwall)

OS Landranger 200

Famous as a tourist resort, the town of Newquay and surrounding areas can yield migrants and good seawatching. Nearby, the Gannel Estuary should be checked for gulls and Porth Joke Valley attracts migrant passerines. To the east of St Columb Minor, Porth Reservoir attracts diving ducks in hard weather.

Habitat

Newquay is well known for its cliffs and beaches, but a feature of particular interest to bird-watchers is the sewage outfall, though ironically it is now less attractive to seabirds as the sewage is treated prior to release. This outfall is particularly well known for gulls, and in autumn Storm Petrel is regular on passage although scarcer than it was in previous years.

Species

Winter gull flocks can contain scarcities such as Mediterranean, Little, Glaucous and Iceland Gulls. All three divers occur, Red-throated being the commonest. Grebes, seaducks and auks are all possible.

Manx Shearwater, Sandwich Tern and Whimbrel are regular in spring while typical passerine migrants could include Grasshopper Warbler and Lesser Whitethroat, with scarcer species always possible. Unfortunately West Pentire headland may have lost its breeding Corn Buntings – this species is very scarce in Cornwall with Trevose Head in the north-west of the county being the only reliable site.

Late summer and early autumn sees the best seawatching. Storm Petrel is regular and other possibilities include Sooty Shearwater and Arctic and Pomarine Skuas. Later in autumn Leach's Petrel or Sabine's Gull could pass. Wheatears are common in autumn and rarities have included Firecrest and Red-breasted Flycatcher. From October, other scarce species that are possible are Grey Phalarope, Little Auk and Snow Bunting.

Access

Newquay is a popular tourist destination and is easily located. Follow signs north past the harbour towards Fistral Beach, turning right opposite Carnmarth Hotel for Towan Head. There is a small

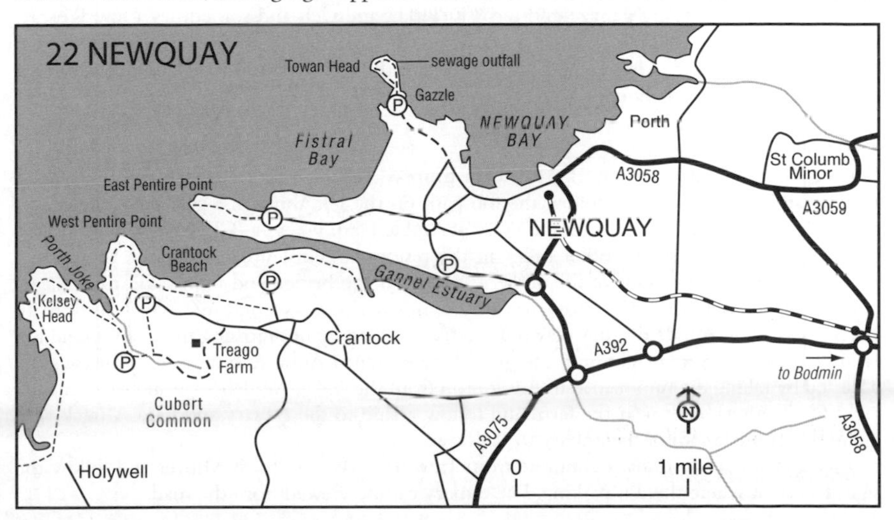

car park at the end, near the base of the head. The sewage outfall is directly in front of the head. For the Gannel Estuary take the A3075 towards Redruth and turn right at a mini-roundabout just beyond Trenance Park Lake (on the right). From the parking area 0.5 miles along this road (past the Medallion Court flats on the right) the river can be reached on foot. Porth Joke is reached by continuing from the mini-roundabout c.1 mile along the A3075 and taking the right turn to Crantock. From here follow signs to West Pentire and turn left on the minor road past Treago Farm. Continue beyond the farm, through a gate (please shut this behind you) and past the NT Cubert Common sign, keeping left on the sandy track to a car park at the end. Walk to the valley from here and check bushes and any flooded areas along the track. For Porth Reservoir, take the A392 eastwards towards Bodmin and follow the signs to the reservoir.

FURTHER INFORMATION

Grid refs: Towan Head SW 800 627; Gannel Estuary SW 809 608; Porth Joke SW 776 599; Porth Reservoir SW 862 618

23 CAMEL ESTUARY (Cornwall)

OS Landranger 200

Situated on the north Cornish coast, the Camel Estuary is productive for waders and waterfowl, including geese in winter.

Habitat

The upper estuary has marshy meadows and mudflats while the lower reaches are sandy.

Species

In winter up to 100 White-fronted Geese formerly occurred at Walmsley Sanctuary on the upper estuary, but these no longer arrive on a regular basis; small numbers are seen from time to time, while Bewick's and Whooper Swans also occur infrequently. Surface-feeding duck such as Wigeon and Teal are present in significant numbers as are Golden Plover and Lapwing. Divers and grebes are best looked for lower down the estuary.

Ruff appears on spring passage and terns include Sandwich and sometimes a few Roseate or Black.

Curlew and Wood Sandpipers and Little Stint pass through in autumn and a single Osprey is annual; Peregrines are frequently seen.

Access

For the upper estuary take the B3314 north from Wadebridge to Trewornan. From roadside parking at Trewornan Bridge follow the footpath on the left, through a gate into a field near the bridge. This path leads to Burniere Point at Amble Dam where a (CBWPS) hide is located (open access). To access two other hides, in the restored Walmsley Sanctuary, use the narrow pedestrian gate on the right of the verge just beyond the bridge and follow the path diagonally across the field to a stile. Cross this and the next field to a second stile, which leads to an open-access hide. A hide on stilts (key required) is reached by not crossing the second field but following the hedgerow right until the gated hide entrance. Another hide, at Tregunna, can be accessed by taking the minor road to Edmonton from the A39 at Wadebridge and then turning right to Tregunna. Park near the farm and follow a lane to the estuary and hide. A track runs along the entire length of the estuary to Padstow.

To reach the lower estuary continue from Trewornan Bridge to St Minver and follow the signs to Rock where there is parking. The estuary can be viewed from the road.

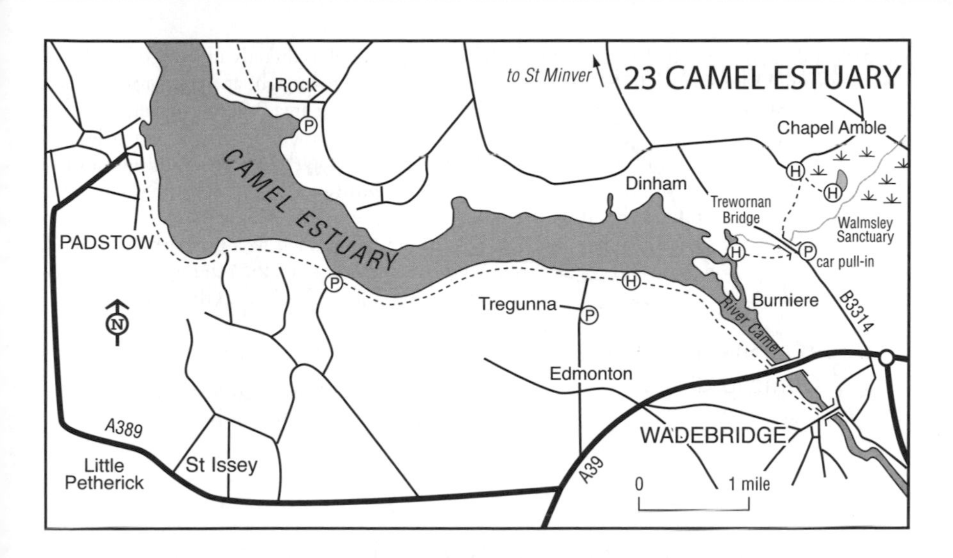

FURTHER INFORMATION

Grid refs: Treworan Bridge SW 987 743; Tregunna SW 965 735; Rock SW 940 757
Cornwall Bird Watching and Preservation Society. For access to the hides contact the current honorary secretary, Darrell Clegg: tel: 01752 844775; email: secretary@cbwps.org.uk; web: www.cbwps.org.uk

24 DAVIDSTOW AIRFIELD (Cornwall)

OS Landranger 201

A disused airfield on the north edge of Bodmin Moor, with the ruined control tower and runways still present. The short turf is grazed by sheep, and shallow pools occur after rain.

Species

Large flocks of Golden Plovers and Lapwings occur in autumn and winter and occasionally an American Golden Plover is with them. A few Ruffs are usually present and Buff-breasted Sandpiper now turns up annually. Very small numbers of Dotterels are reasonably regular in spring and autumn. Hen Harrier, Merlin and Peregrine are sometimes encountered in autumn and winter, Hobby in summer, and Goshawk has been reliably reported on a number of occasions.

Access

Leave Camelford north-east on the A39. After 2 miles turn right onto a minor road running through the airfield. Pull off to the left and drive along the old runways scanning for birds. Stay in your vehicle: this does not scare the birds and enables more ground to be covered. The control tower area is often best for waders.

FURTHER INFORMATION

Grid ref: SX 148 851

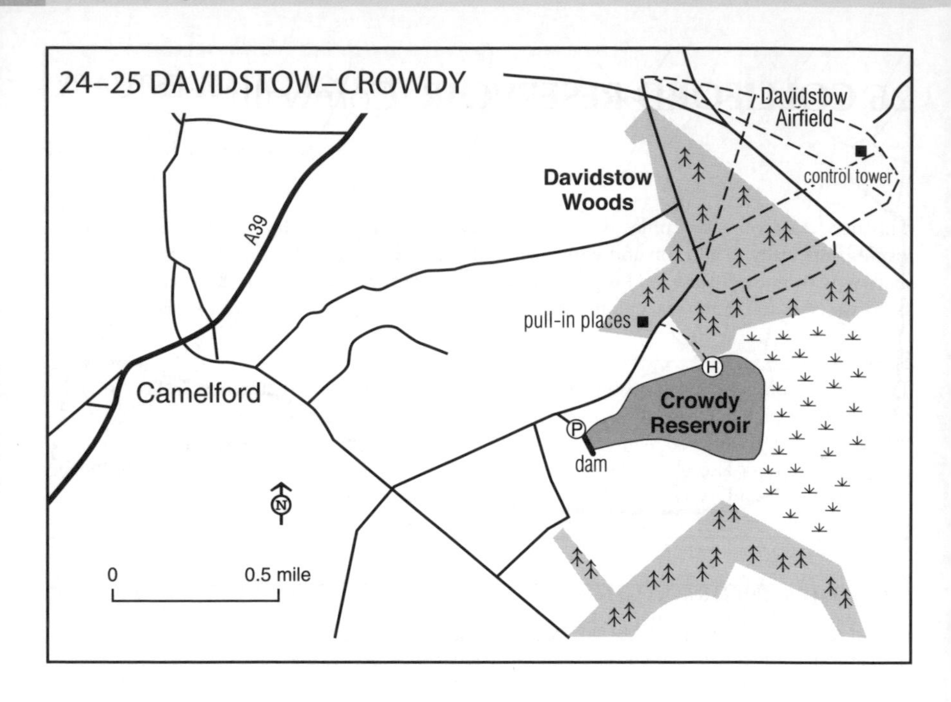

25 CROWDY RESERVOIR (Cornwall)

OS Landranger 201

This triangular reservoir, part of which is an SSSI, is very close to Davidstow Airfield.

Habitat

Mud is exposed when the water is low and there are marshy areas on the reservoir's east side that support a diverse mix of plant life.

Species

There are reasonable numbers of the commoner ducks in winter usually including a few Goldeneyes. Raptors may also be seen (as at Davidstow). A variety of waders frequents the muddy margins in autumn and occasionally includes a rarity. Very small numbers of Black Terns are recorded annually in spring and autumn.

Access

A signed road from the airfield leads to the reservoir. Alternatively it can be reached directly from Camelford: turn right (south-east) off the A39 at the north edge of the town and after 1 mile turn left to the airfield. The reservoir can be viewed from the road but for close views follow a track to the water's edge. A hide is sited at the north end.

FURTHER INFORMATION

Grid ref: SX 138 834
South West Lakes Trust: tel: 01566 771930; email: info@swlakestrust.org.uk; web: www.swlakestrust.org.uk

26 COLLIFORD RESERVOIR (Cornwall)

OS Landranger 201

This new reservoir a few miles west of Siblyback is the largest in the county and well worth visiting, especially in autumn and winter, for ducks and waders. The north-east part is relatively shallow with grassy islands, and is now a reserve. At present disturbance is minimal at Colliford as no water-based sports (except angling) take place, unlike at Siblyback.

Species

There is a significant autumn build-up of ducks, with a chance of Garganey, and the shallow margins have produced Pectoral and Buff-breasted Sandpipers and Spotted Crake as well as the commoner waders. In winter Smew and Goosander appear to be regular in very small numbers, and look also for the occasional Scaup. Towards dusk both Hen Harrier and Merlin roost around the reservoir, and Short-eared Owl winters in small numbers. Particularly good vantage points for these are at the north-west and north-east corners of the reservoir. Dippers inhabit the nearby Fowey Valley.

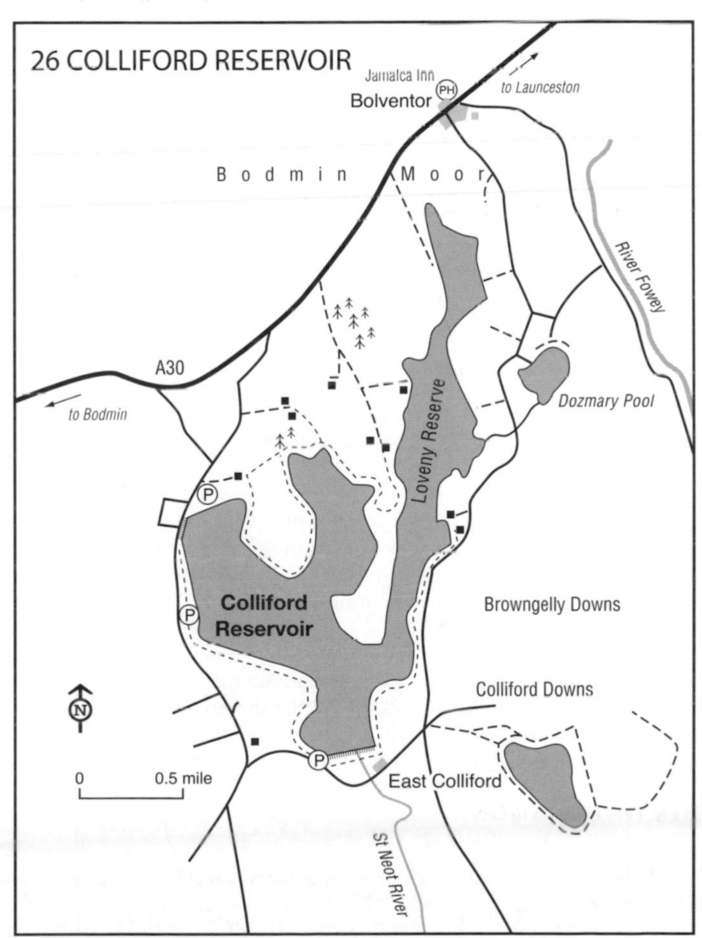

Access

Turn south off the A30 at Bolventor (virtually opposite Jamaica Inn) onto a road signed 'Dozmary Pool'. Follow this road for c.2 miles, passing Dozmary Pool to the left, and reasonable views may be had of the reservoir's north-east arm from the roadside. Continue south, turning west by the china-clay works, and pass below the dam (the dam area is poor for birds due to the deep water). Another road follows the reservoir's west side, affording reasonable views of the north-west arm before rejoining the A30. Footpaths exist around much of the reservoir, but a permit is required for the path covering the reserve area (available from the warden at Siblyback).

FURTHER INFORMATION

Grid refs: north-east arm SX 163 730; dam SW 176 709
David Conway: tel: 01208 77686

27 SIBLYBACK RESERVOIR (Cornwall)

OS Landranger 201

This reservoir lies on the south-east edge of Bodmin Moor, north of Liskeard.

Habitat

The reservoir is surrounded by moorland and grassland. Areas of mud are exposed on the north shore when the water is low.

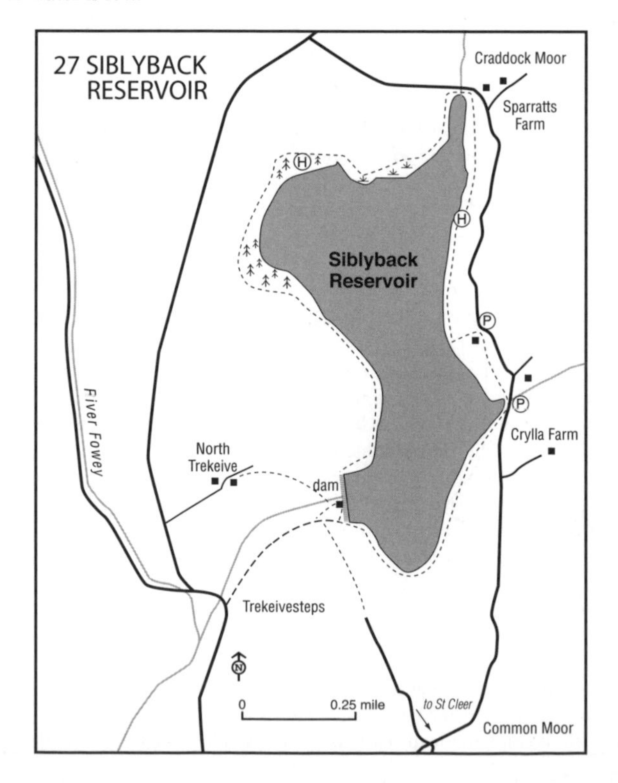

Species

There is a variety of ducks in winter, and a few Goosanders and Smews are annual visitors. Jack Snipe can sometimes be found in the marshy edges and Golden Plover and Lapwing are common in the fields. Hen Harrier, Merlin and Peregrine are seen occasionally in winter. Autumn wader passage includes most of the usual species and a rarity is found most years. Small numbers of Little Gulls and Black Terns occur each autumn.

Access

Leave Liskeard north on the B3254. After 1 mile fork left onto a minor road to St Cleer. Turn right immediately beyond the village and continue past the crossroads to the reservoir and car park overlooking it. You can walk in either direction but the north end is generally best and there is a hide here (permit from South West Water Authority).

FURTHER INFORMATION

Grid ref: SX 236 707
South West Water Leisure Services Department, Permits and keys to hides: tel: 01837 871565; email: info@swlakestrust.org.uk; web: www.swlakestrust.org.uk

28 RAME HEAD (Cornwall)

OS Landranger 201

Facing south on the south-east Cornish coast, Rame Head is a passerine-migrant trap and offers a good vantage point for seawatching.

Habitat

Cover for migrants can be found around Rame Church and Farm in the form of gardens, trees and hedgerows. The road to the old fortification at Penlee Point takes in a variety of habitats including a copse. Along the coast to the west is Whitsand Bay, and gorse and bracken slopes can be accessed in this direction, with fields on the tops.

Species

Passerines tend to move quickly from the exposed headland to the cover of the trees and hedgerows, and early morning is the best time to search for recent arrivals, feeding up before heading inland. Warblers are sometimes present in large numbers and scarcer migrants can include Pied and Spotted Flycatchers, Whinchat, Common Redstart and Black Redstart. In autumn rarities such as Pallas's Warbler have been found alongside common migrants like Goldcrest – major rarities have included Wilson's Warbler, Northern Parula and Red-flanked Bluetail. Fields should be checked for larks and pipits, while migrant raptors could also pass through.

Spring seawatching can produce divers, which are also present through the winter in Whitsand Bay. Scarce grebes, particularly Slavonian, and seaducks such as Common Scoter and Common Eider are often present offshore. Gannet and Manx Shearwater may pass in the hundreds with occasional skuas in spring and autumn. Terns, auks and waders also occur at both seasons though autumn seawatching is better when there is the possibility of scarcer species such as Sooty Shearwater.

Access

Take the A374 from Torpoint to Antony, then the B3247 to Tregantle and turn right on the coast road to Rame, via the village of Freathy. There is a car park at Rame Head and public footpaths around the coast. One leads east to Penlee while another good route heads in the opposite direction to Polhawn, permitting coverage of the bushes by the old fort (private) before looping back to Rame Farm and Church.

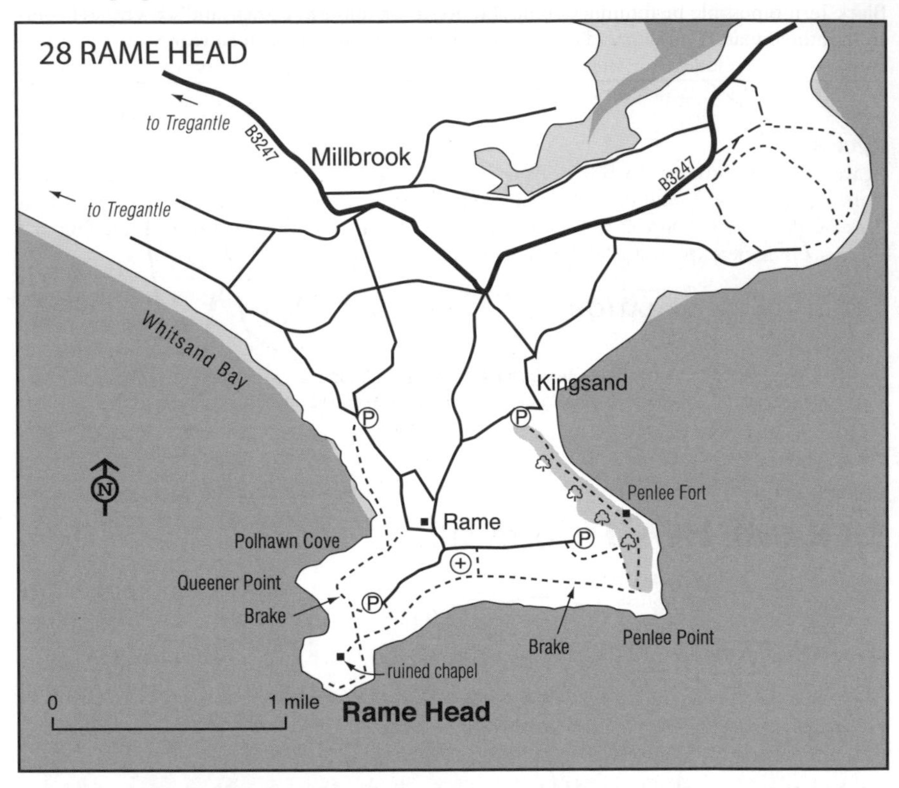

FURTHER INFORMATION

Grid ref: SX 420 488

29 TAMAR ESTUARY (Cornwall/Devon)

OS Landranger 201

Situated on the Devon/Cornwall border, the Tamar Estuary is best known as a wintering site for Avocet.

Habitat

Extensive mudflats are located between the Tamar road bridge and Kingsmill Lake. Further upstream the estuary narrows and is bordered by farmland and trees.

Species

Avocets arrive from early winter with 100+ present mid-season and peak numbers reach in excess of 200. Duck, often including Goldeneye, are present during winter and all three saw-bills have been recorded. Kingsmill Lake is actually a creek, which is good for waders such as

Common Redshank and Curlew on the incoming tide, and large numbers of feeding waders are often a feature of the area. Passage waders may include Curlew and Wood Sandpipers, while Black Tern is possible in autumn. Osprey also occurs on autumn passage and sometimes lingers around the estuary. Little Egret, Peregrine and Barn Owl may be seen all year.

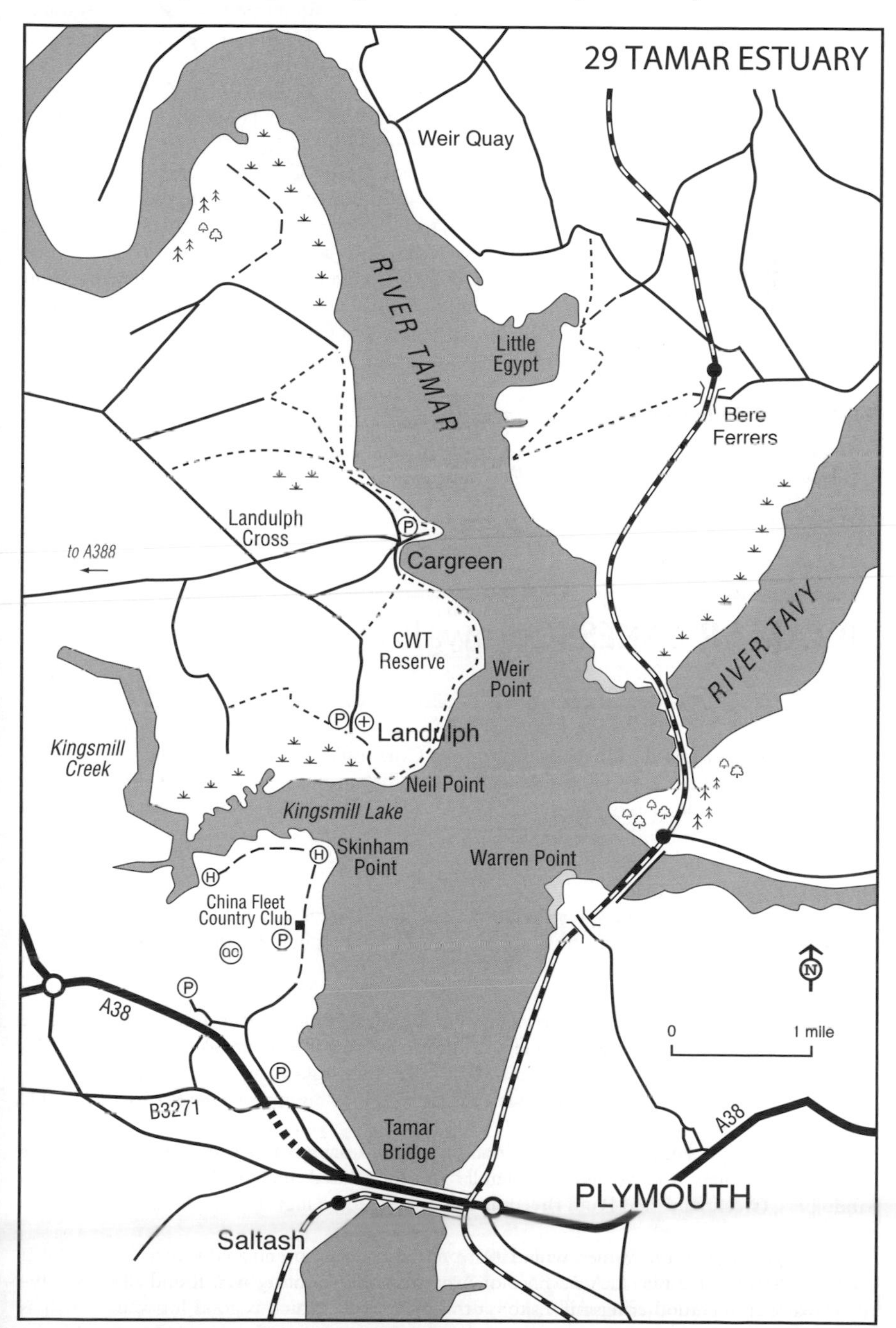

Access

The best areas are on the west side of the estuary.

Landulph Marsh This area, which is part of a Cornwall Wildlife Trust reserve, offers good birding throughout the year. Take the A388 northwards from Saltash and turn right to Landulph. Park near the church and follow the narrow road to the left, passing through a wooden gate and stile, and then crossing a meadow to a public path that leads to an embankment affording views of Kingsmill Lake. At low tide walk the beach and check the saltmarsh.

Cargreen The nearby village of Cargreen to the north is a regular Avocet feeding site and it is possible to observe the birds from a car if you park on the waterfront. Footpaths follow the edge of the estuary both northwards and southwards (to Landulph Marsh).

Kingsmill Creek The outer part of the creek, known as Kingsmill Lake, can be viewed from Landulph Marsh (see above) but Kingsmill Creek is more easily accessed from the south, where the light is better in the afternoons. From the Tamar Bridge at Saltash take the B3271, which runs parallel to the A38 road tunnel, and turn right onto a minor road signed Saltmill and North Pill. Follow signs to the China Fleet Country Club and park there. The club reception will give you the free key code for the two hides. A trail leads around the edge of the estuary to the hides. The first hide overlooks the Tamar estuary and the second gives views of roosting waders on Kingsmill Creek at high tide.

FURTHER INFORMATION

Grid refs: Landulph SX 431 614; Cargreen SX 436 627; China Fleet Country Club SX 428 604
Cornwall Wildlife Trust: tel: 01872 273939; web: www.cornwallwildlifetrust.org.uk/nature_reserves/map

30 TAMAR LAKES (Cornwall/Devon)

OS Landranger 190

These two reservoirs with natural banks lie on the Cornwall/Devon border, amidst open farmland. The 32-ha Upper Lake is used for sailing, and the smaller 16-ha Lower Lake is used for recreation only. The long-term future of Lower Lake and proposed height of water levels is in question owing to problems with the dam.

Habitat

The Tamar Lakes have wooded edges with shallow marshy areas. The natural habitats are attractive to birds, although there is some disturbance from fishing and recreational activities.

Species

In winter, a good selection of wildfowl may be found, although not in huge numbers. Rare American species have been found here in the past. Occasionally, small parties of Bewick's Swans or White-fronted Geese may visit, and the edges may hold a few Common Snipe and the odd Jack Snipe. The surrounding area hosts the usual wintering raptors and it may be possible to see a Hen Harrier or a Peregrine in the area, or perhaps even a Merlin.

The spring will bring a variety of migrants to the lakes, and they can be numerous. Large flocks of hirundines and Swifts are a regular feature. Waders include Common and Green Sandpipers, Greenshank and Ruff. Breeding species include Kingfisher and Willow Tit (try the trees by the Lower Lake).

Wader species are more numerous and more varied in autumn, and are likely to include Little Stint and Spotted Redshank. A number of American species have been found here over the years. Black Tern is another regular visitor at this time.

Access

The lakes lie to the east of the A39 Bude–Bideford road. From Bude, heading north on the A39, turn right at Kilkhampton onto the B3254 to Launceston, and after 0.5 miles turn left onto a minor road towards Bradworthy and Holsworthy. For the Upper Lake, turn left after 1 mile and left again to the car park. For the Lower Lake, continue towards Holsworthy and take the second left into the car park. The lakes are connected by a footpath. There is a hide overlooking the Lower Lake, part of which is maintained as a bird sanctuary from August to April.

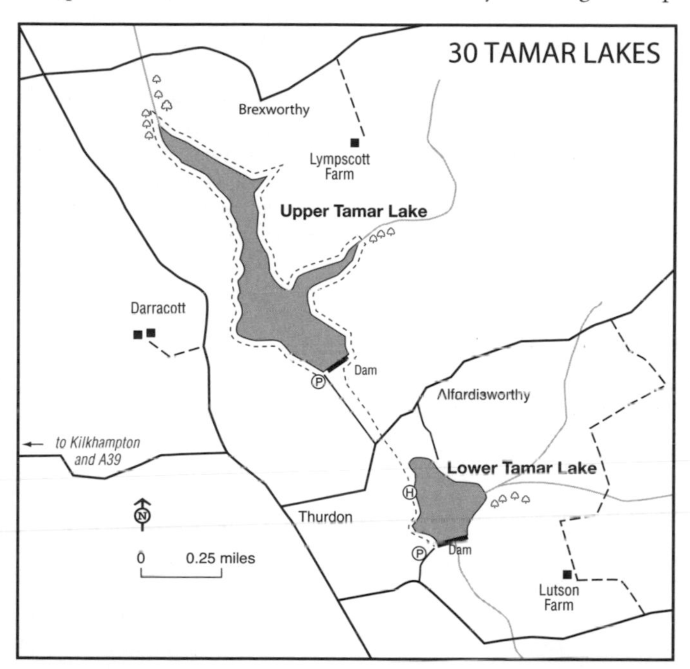

FURTHER INFORMATION

Grid ref: Lower Tamar Lake SS 293 107

31 LUNDY (Devon)

OS Landranger 180

The island of Lundy, c.3.5 miles long, lies in the Bristol Channel 12 miles north of Hartland Point. The cliffs hold good numbers of breeding seabirds but the island is best known as a migration watchpoint; a bird observatory operated from the Old Lighthouse from 1947 to 1973.

Habitat

The cliffs rise to over 300 feet and are most impressive on the west and north sides. Much of the island is barren moorland that is damp in places, with a few small areas of standing water, notably at Pondsbury. The sole habitation on the island is a small village at the south end where there are some fields. The Eastern Sidelands and sheltered areas such as Millcombe Valley contain bushes and a few small areas of woodland.

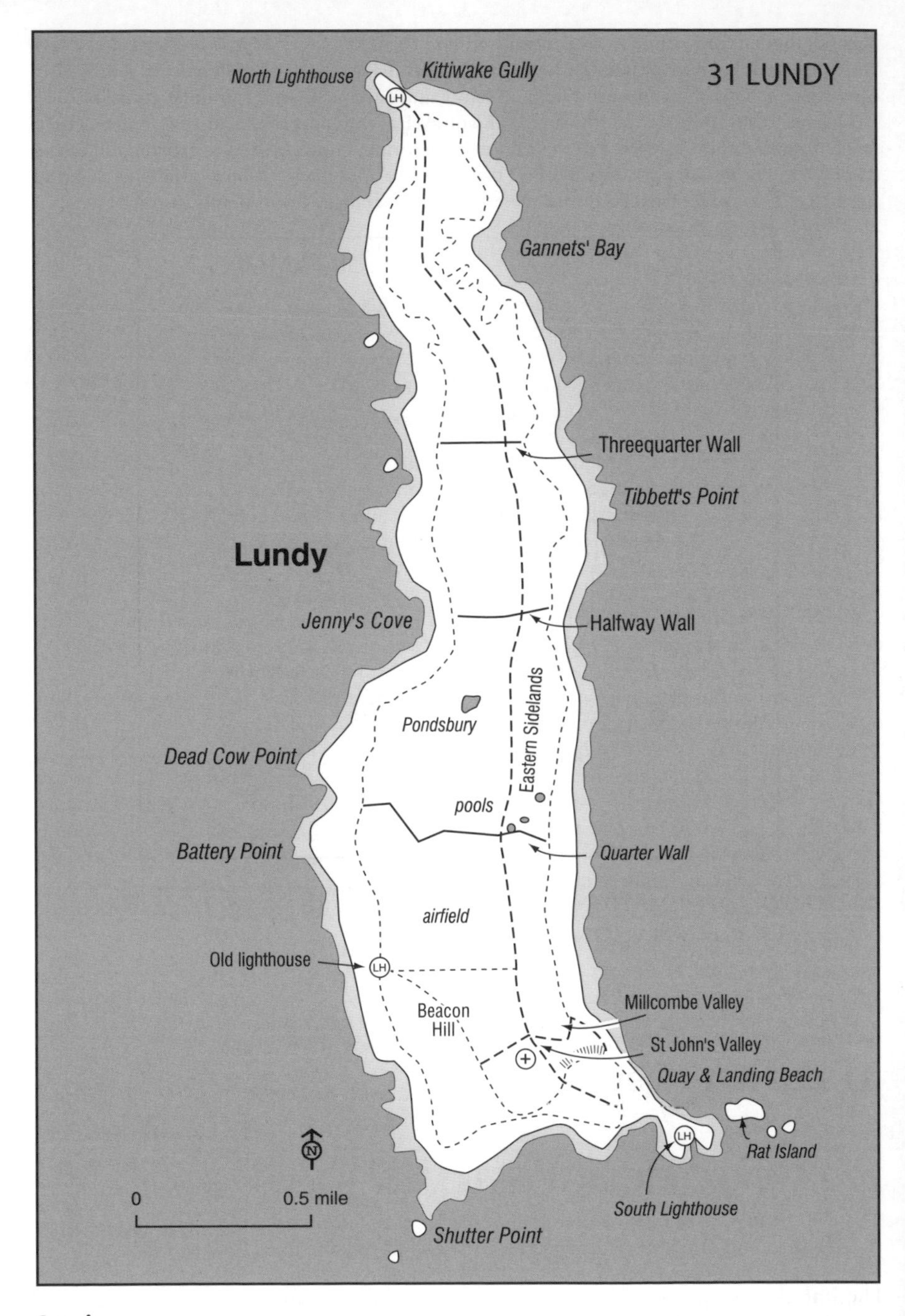

Species

Spring migration begins in March and usually includes a few Black Redstarts and Firecrests. A steady stream of commoner summer visitors passes in April–May. East or south-east winds are generally best for falls of migrants. Small numbers of Common Redstarts, Ring Ouzels and Pied Flycatchers are recorded annually, and scarce migrants are a regular feature, especially following light southerly winds. Hoopoe is generally among the first, in late March or April, and a few Dotterels are seen each year in late April or May. Both species favour short turf. Late spring often brings a Golden Oriole, and migrant raptors may include Merlin and Short-eared Owl.

Though many of the seabirds, such as Gannet, are visible offshore all year, the breeding species mostly return to their ledges or burrows in spring. Hundreds of Common Guillemots and Razorbills nest alongside over 1,000 pairs of Kittiwakes and smaller numbers of Fulmars and Shags. Up to 100 pairs of Puffins breed, mostly at Battery Point. Gulls (mostly Herring) are also numerous. Manx Shearwater and Storm Petrel breed in small numbers, largely on the west side, but only return to their burrows at night and the best chance of seeing them is on the boat crossing. Rock Pipit and Raven are resident.

Autumn wader passage is relatively insignificant due to scarcity of habitat, but several American waders have been recorded, usually after west gales; Pectoral Sandpiper is almost annual. A few returning Dotterels occur each year. Passerine migrants filter through in August–September, sometimes accompanied by a scarcer species such as Melodious or Barred Warbler. Ortolan Bunting is annual from late August and a few Lapland Buntings usually appear later. In late autumn there are large movements of larks, pipits, finches and buntings. Rarities can be found at any time but are most likely from late September to early November. Lundy has an astonishingly long list of vagrants, despite having been largely neglected by rarity-hunters. No fewer than nine firsts for Britain have been found here, and an Ancient Murrelet, first found in June 1990 among the breeding auks, was the first and so far the only Western Palearctic record.

Access

A regular supply ship operates from Ilfracombe and a steamer runs day trips in the summer (but these do not permit much time ashore). The *MV Oldenburg* sails to Lundy from Bideford; tickets should be obtained from Lundy Oldenburg Office, Bideford Quay, Devon (tel: 01237 470422). A variety of accommodation is available on Lundy but space is limited. For further information on transport and accommodation contact the Landmark Trust (see below).

The gardens and bushes of Millcombe Valley are perhaps the best area for migrants in spring and autumn. Other suitable areas for warblers and flycatchers are largely confined to the Eastern Sidelands, and a walk along the east coast checking the bushes and disused quarries may be profitable. The main seabird colonies are at North Lighthouse and Battery Point. A track from the village leads north across the moors to North Lighthouse. Immediately east of the lighthouse, a series of steps descends into Kittiwake Gully offering good views of the breeding seabirds.

FURTHER INFORMATION

Grid ref: SS 145 437
Lundy Field Society: email: lfssec@hotmail.co.uk; web: www.lundy.org.uk/lfs
The Landmark Trust, accommodation bookings and travel information: tel: 01628 825925; email: bookings@landmarktrust.org.uk; web: www.landmarktrust.org.uk

32 TAW–TORRIDGE ESTUARY (Devon)

OS Landranger 180

The extensive estuaries of the Taw and the Torridge, between Bideford and Barnstaple, are the largest in North Devon, supporting the region's largest wader population after the Exe and the Tamar.

Habitat

In addition to typical estuarine habitats, the area is fringed by sand dunes and rough grazing marsh. To the north lies the well-known Braunton Burrows, famous for its high dunes and rare plants, and to the south the smaller Northam Burrows. Areas of saltmarsh occur at Skern, Isley Marsh (now an RSPB reserve) and Penhill Point.

Access

There are a number of access points in the area, and the following is based on *Where to Watch Birds in Devon and Cornwall* (Norman & Tucker, 2008). The estuary is difficult to watch at low tide. An incoming tide is best, and a visit to a roost two hours up to high tide is recommended. There is a fair amount of disturbance at certain times from water sports, motorbike scrambling in the dunes and general public use, mostly at weekends and summer holidays.

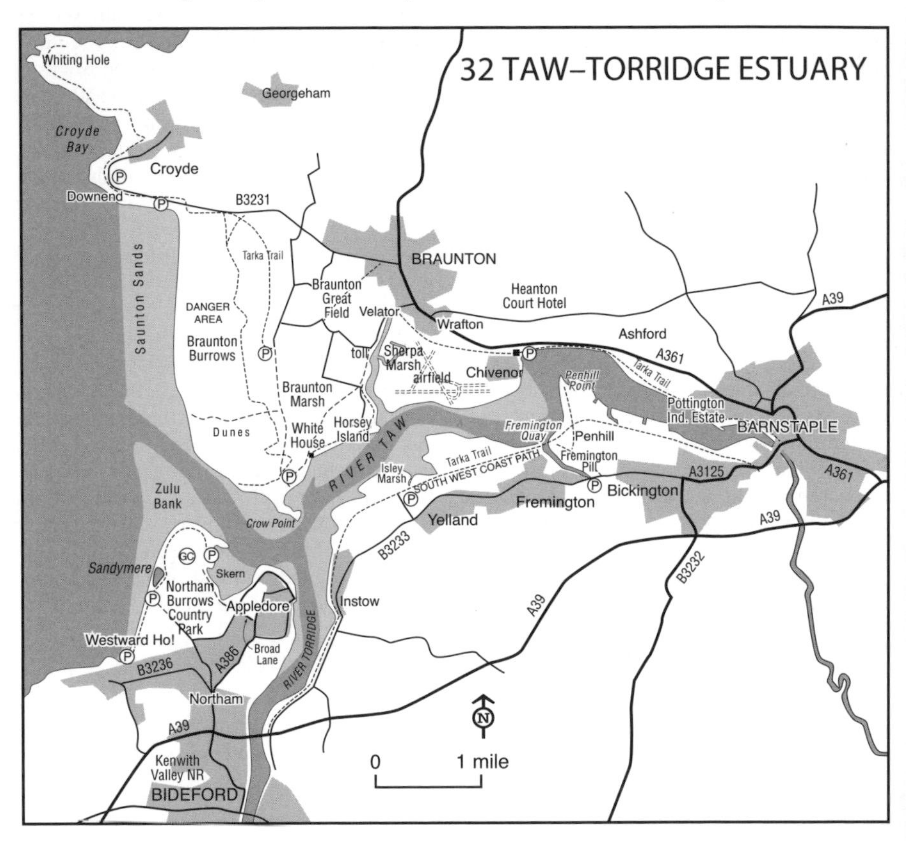

Species

Braunton Marshes hold large flocks of waders in winter, notably Lapwing and Golden Plover, with occasional Ruffs. A Peregrine or Merlin may hunt in the area. High tide roosts of waders at Crow Point or Northam Burrows hold all the usual shore species including Dunlin, Oystercatcher, Curlew, Common Redshank and Grey Plover, and smaller numbers of other species. Bewick's and Whooper Swans or White-fronted Geese may visit in cold weather, and up to 200 Brent Geese are now regular. A variety of the commoner ducks occurs in the estuary and on the marshes. Wigeon predominates, and Common Scoters occur in Bideford Bay. A regular non-breeding flock of some 20–30 Common Eiders in the estuary mouth is notable. Slavonian Grebes may occasionally be seen in the channel. Little Egrets are now present throughout the year, and Spoonbill is an occasional winter visitor, as is Greenshank and Spotted Redshank. Large numbers of gulls inhabit the estuary and rarer species such as Glaucous, Iceland and Ring-billed are almost annual.

Spring brings a notable passage of waders including Whimbrel and Common Sandpiper, as well as some warblers and other passerines. The Great Field at Braunton is a long established site for Quail although, as always, luck is required even to hear one in Britain. Unusually, Wheatears breed in the dunes around the estuary mouth.

Autumn wader passage may produce an American vagrant, but Little Stint or Curlew Sandpiper is more likely. Greenshanks may be seen in small flocks in the lower estuary and small flocks of Common, Sandwich and Little Terns frequent the mouth. Later, a Short-eared Owl or Hen Harrier may hunt over the dunes, and the area is notable for regular wintering Snow Buntings (especially along Northam pebble ridge). Twite, Shore Lark and Lapland Bunting have occurred, but are not annual. A Grey Phalarope may drop in after northerly gales in autumn.

NORTH SIDE

Braunton Burrows Leave Barnstaple on the A361 north-westwards towards Ilfracombe. In Braunton, turn left on the B3231 towards Saunton and Croyde, and after 1 mile turn left again onto a minor road to the Burrows. Park in car parks along the rough track to explore the dunes, or continue to the estuary mouth. After 2 miles, park and walk straight on to Crow Point for waders and views of the estuary, or right to the beach. To complete a circuit of the area, continue north-eastwards around Braunton Marsh. It is worth making a number of stops to check for birds in this rich area. Eventually you will reach a tarmac road and a tollgate at Velator, before returning to the main road.

Sherpa Marsh This area of grazing marshes lies across the pill (creek) from Braunton Marsh. From Wrafton village, a path follows the sea wall alongside the east side of the pill. It is worth checking the fields for geese and the pond for rails or possibly Bittern.

Downend Continuing beyond Braunton and Saunton on the B3231, the road reaches a rocky headland overlooking Croyde Bay. This is Downend. Park on the left at the sharp corner, and check the sea for Common Scoter and Common Eider. Purple Sandpipers may be found on the rocks here.

Heanton Court Hotel The car park at this hotel, 5 miles west of Barnstaple on the A361, gives great views over the central part of the Taw estuary. Penhill Point lies across the river, but views are distant. Check for duck flocks and perhaps a wintering Spoonbill. For closer views, walk 300 yards east where a footpath (opposite the road to West Ashford) leads down to the disused railway embankment. A stone lookout provides a good vantage.

Pottington Just west of Barnstaple, turn off the A361 (left) to Pottington Industrial Estate. At the end of the road turn left into Riverside Road until it bends away from the river. Park here and follow a signed path to check the inner estuary, with views to Penhill Point.

SOUTH SIDE

Penhill Point From Barnstaple head west along the B3233 (old A39). Between Bickington and Fremington, a rough track leads right towards the estuary. Park by the main road and walk down beside Fremington Pill (check for waders). Turn right along the estuary bank and after 1 mile you can view Penhill Point.

RSPB Isley Marsh In Lower Yelland, turn off the B3233 northwards for 300 yards towards the oil depot. This is a major access point to the Tarka Trail. Park at the end of the lane and walk right (east). The marsh can be viewed from the path, or from a coastal footpath around the western edge of the bay.

Northam Burrows Northam Burrows lies opposite Braunton Burrows. From Bideford, head north on the A386 to Northam. To view Skern saltmarsh, turn right to Appledore and second left into Broad Lane. Continue straight for 1 mile, following signs to the recycling centre, with views of marshes on the right and freshwater pools on the left. Walking across the dunes gives great views of the estuary. Alternatively, access the pebble ridge, on the west side, by following signs for Westward Ho! Turn right past the holiday camps and golf course to Sandymere Pool.

Kenwith Valley This small reserve on the outskirts of Bideford has produced the occasional scarce migrant passerine and even an overwintering Dusky Warbler. From Bideford town centre, head towards Westward Ho! and turn left at Raleigh Garage. Park after 400 yards and walk back to the reserve. The entrance is across a stile over a low wall. There are footpaths through the reserve and a public hide overlooking a lake.

FURTHER INFORMATION

Grid refs/postcode: Braunton Burrows SS 462 351; Sherpa Marsh SS 487 356; Downend SS 434 385; Heanton Court Hotel SS 513 349 / EX31 4AX; Penhill Point SS 516 331; Isley Marsh SS 491 325; Northam Burrows (Skern) SS 449 312; Northam Burrows (Sandymere) SS 438 305; Kenwith Valley SS 448 273
RSPB Isley Marsh: c/o RSPB South West Regional Office: tel: 01392 432691; web: www.rspb.org.uk/isleymarsh

33 ROADFORD RESERVOIR (Devon)

OS Landranger 190

At more than 283 ha when full, this reservoir is the largest in the region. Completed only in the early 1990s, it has still to reach its full potential.

Habitat

Roadford is the only reservoir in Devon that is not on acidic moorland. The two northern arms of the lake are designated as nature reserves.

Species

Good numbers of wildfowl use the reservoir in winter and, in addition to the usual dabbling ducks, the lake is particularly attractive to diving ducks. Goldeneye numbers can reach up to 100, and up to 50 Goosanders may gather in winter roosts (try the dam end). Smew turn up from time to time but this species is scarce in the south-west. Other rarities include occasional Scaup or Long-tailed Duck, or even an American vagrant – a Bufflehead wintered one year, but its origin was unproven. Single divers or one of the rarer grebes occur occasionally, and the reservoir attracts good numbers of gulls. Raptors are not seen regularly, but Hen Harrier, Short-eared Owl, Peregrine and Merlin are all possible. A wintering Kingfisher or Stonechat is more likely.

Waders are not significant at Roadford; their variety increases at passage times, but as yet the reservoir has not attracted many rarer species. Other passage birds may include Garganey, Green Sandpiper and Black Tern.

Resident breeding species are not especially notable. Lesser Redpolls and Willow Tits are still quite numerous in the surrounding areas, and a Barn Owl may be seen from time to time (dusk at the Southweek hide is perhaps the best bet). A few Grasshopper Warblers and Tree Pipits breed around the edge of the reservoir and a Hobby may hunt in the area in summer.

Access

From the A30 Launceston–Okehampton road, turn left (north) onto a minor road to Broadwoodwidger (the reservoir is signed). After 2 miles turn right to a car park (fee) just before the dam – there is a shop and cafe here (open Easter to October, 11am–5pm). Cross the dam and follow the footpath to Goodacre inlet and wood. On the east side, a network of trails follows the shoreline to Gaddacombe hide (public access). Further tracks lead to Wortha inlet where there is another car park. Alternatively, both these areas can be reached by road. Perhaps the best place for birds is the north-east arm of the reservoir, where there is another hide. Park on the left just before Southweek viaduct, and from here walk left through a gateway signed to the hide. The hide has open access and a logbook inside. After the viaduct, a 'no through road' to the left leads down to the north-west arm of the reservoir (Westweek). The west side can be checked from the car park at Grinacombe, but there are no paths on this side. Sailing and fishing activities can disturb the birds at Roadford at weekends.

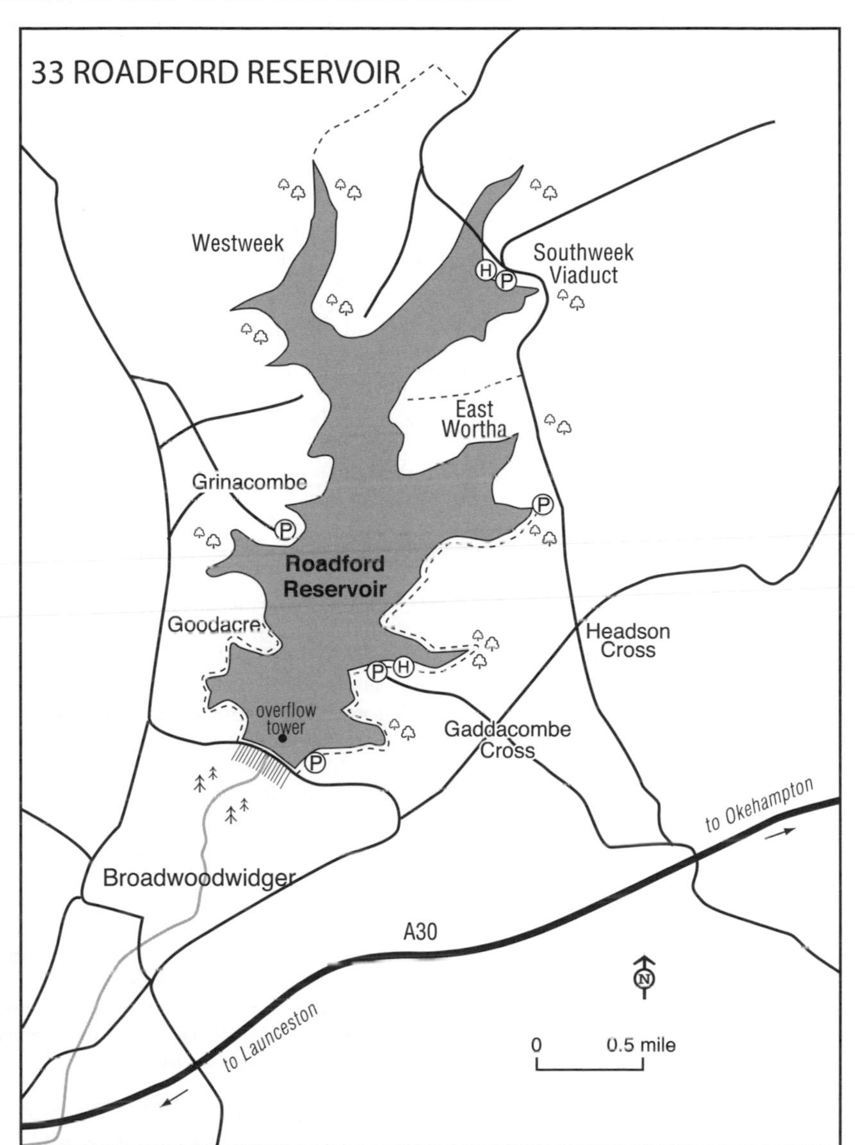

FURTHER INFORMATION

Grid refs: dam SX 423 898; Gaddacombe SX 428 905; Wortha SX 439 915; Southweek SX 435 930; Grinacombe SX 421 914

South West Water, Leisure Services Department: tel: 01837 871565; email: info@swlakestrust.org.uk; web: www.swlakestrust.org.uk

34 PRAWLE POINT (Devon)

OS Landranger 202

This outstanding migration watchpoint is the most southerly point on the south Devon coast, and lies south of Kingsbridge. Though best in spring and autumn there is usually something to see at any season. Prawle is also one of the most reliable places in Britain to see Cirl Bunting.

Habitat

The cliffs along this coast are quite spectacular. The tops are covered with grass, bracken, gorse and boulders, and behind lie fields surrounded by stone walls and hedges. Sheltered hollows and valleys contain dense vegetation and bushes attractive to migrants. There is a small wood near East Prawle.

Species

Among residents, Cirl Bunting is Prawle's speciality: up to 25 pairs breed in hedgerows around Prawle and in winter small flocks may be found in the bushes around the car park. Little Owl is frequently seen and other residents include Common Buzzard, Sparrowhawk, Rock Pipit and Raven.

In winter, a few divers occur offshore – Great Northern is the most frequent. Small numbers of Shags and Common Scoters are usually to be seen as well as many Gannets, gulls and auks. Flocks of Golden Plovers favour the short grassy fields and a few Purple Sandpipers winter below the cliffs. The bushes may hold the occasional wintering Chiffchaff or Firecrest, and more open areas should be checked for Black Redstart.

The first spring arrivals include Wheatear and Goldcrest. A few migrant Black Redstarts and Firecrests may be found from mid-March and Hoopoe is seen most years. A wide variety of migrant passerines occur, mostly in April–May, when a few raptors also pass through. Seawatching can be productive in spring: regular species include all three divers, Manx Shearwater, Fulmar, Common Eider, Common Scoter, Red-breasted Merganser, Whimbrel, Bar-tailed Godwit, skuas and terns; late April to early May is probably best. Scarce migrants and rarities are occasionally found, mostly in late spring.

Small numbers of Balearic Shearwaters tend to replace Manx from mid-July. They are sometimes accompanied by a Sooty Shearwater but the latter is more frequent in September. Arctic and Great Skuas are regular in autumn. Passerine migrants are mainly present August–September; the main species are Yellow Wagtail, Tree Pipit, Wheatear, Whinchat, Common Redstart, warblers and flycatchers. Late September and October bring large movements of hirundines, pipits and wagtails, and usually a few Ring Ouzels. A Merlin or Hobby is also often present at this time. Late autumn can produce the most exciting finds. A few Firecrests are usually present from late September into November, and Melodious, Icterine and Yellow-browed Warblers, and Red-breasted Flycatcher all occur most years, albeit in very small numbers. An American passerine is a possibility following west gales, with both Black-and-white and Chestnut-sided Warblers having been found in recent years, and an Asian vagrant such as Pallas's Warbler may turn up as late as November. Late autumn also brings a few Black Redstarts.

Access

Leave Kingsbridge east on the A379. Turn right at Frogmore or Chillington onto minor roads south to East Prawle. Continue through the village, down the no through road, to the NT car park within a sheltered hollow surrounded by bushes. These bushes often harbour migrants and, indeed, are one of the best places to look. Several other places should also be checked.

Prawle Point Continue to the Point from the car park, past the coastguard cottages, checking for migrants on the way. Seawatching is best from the memorial seat below the west side of the Point.

Eastern Fields From the car park walk east along the coastal path, checking bushes and gullies. Beyond Langerstone Point a small wood lies on the left. After checking this you can return to the car park along the lane, via the top fields.

Pig's Nose Valley This deep, lush valley lies about 1 mile west of the point. It can be reached either by walking north-west along the coastal path via Gammon Head or from East Prawle village by taking the lane past the duck pond.

Early mornings are generally best for observing migrant passerines. If a fall has occurred, all suitable cover should be checked thoroughly, especially in sheltered areas. There is disturbance by non-birders at weekends. Seawatching is often best in hazy conditions or after strong, onshore gales.

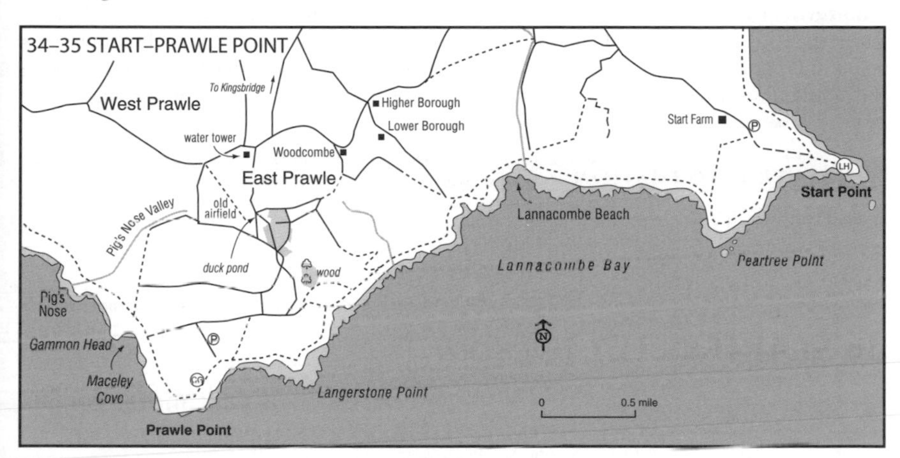

FURTHER INFORMATION

Grid ref: SX 774 355

35 START POINT (Devon)

OS Landranger 202

This migration watchpoint lies just east of Prawle. However, cover is limited and passerine migrants tend to move through quite quickly. Cirl Bunting is an equally prominent feature of the local avifauna.

Habitat

Start Point is a rocky promontory with open farmland lying immediately inland. Start Farm Valley, behind the point, is sheltered and has trees and bushes providing cover for migrants.

Species

In winter attention is best focused on the sea where Great Northern Diver and Common Eider can occur. Resident Raven, Peregrine and Rock Pipit should also be evident.

Chiffchaff, Grasshopper Warbler, Wheatear and Goldcrest are typical spring migrants, but more unusual species could include Firecrest, Black Redstart and Hoopoe. This season is also a good time to search for migrant raptors such as returning Hobby and Honey-buzzard or even rare vagrants like Black Kite, which has been recorded at nearby Prawle.

Summer is a good time to search for resident Cirl Buntings and Dartford Warblers as they are holding territory and singing. Skuas and shearwaters may be observed offshore from late summer, but the best time for seawatching is early autumn.

Early-autumn migrants can include Spotted and Pied Flycatchers, Yellow Wagtail and Tree Pipit. The chance of an extreme rarity increases later in the season and Booted and Arctic Warblers have been recorded at Start. Wryneck, Red-backed Shrike and Ortolan Bunting are possible drift migrants at this time. In late autumn passage is characterised by movements of larks, pipits, finches and buntings.

Access

From Stokenham village, south of Slapton village, turn sharp left and drive to the car park at the top of the Point (passing Start Farm valley on the right). A gate leads to Start Lighthouse access lane and a footpath right into the valley where there is some cover. For the Point continue along the lighthouse lane. A side valley immediately east of the Point is reached by turning off the coast path beyond the sandy cove.

FURTHER INFORMATION

Grid ref: SX 820 374

36 SLAPTON LEY (Devon)

OS Landranger 202

Slapton Ley is a nature reserve managed by the Field Studies Council. Its reedbeds and bushes attract a variety of migrants and some interesting breeding species, but the area is best known for wildfowl in winter.

Habitat

A series of freshwater pools lies behind a long shingle bank on the coast of Start Bay. The largest of these is Slapton Ley itself; 1.5 miles long and fringed by reeds, especially in Ireland Bay. Wintering ducks occur on the open water, often favouring the Lower Ley. Dense bushes along the seaward edge provide habitat for migrant passerines. Immediately north of Slapton Ley, and separated from it by a narrow strip of land, is Higher Ley. This is now mainly an extensive reed-bed with encroaching scrub. South of Torcross, the much smaller Beesands Ley consists of open water fringed with reeds. Further south still, Hallsands Ley is a semi-dry reedbed.

Species

Divers and grebes occur on the sea in winter, Great Northern Diver and Slavonian Grebe being the most regular. Shag, Common Eider and Common Scoter are also frequent offshore. Slapton is particularly attractive to diving ducks, and there are usually a few Gadwalls. Small numbers of Ruddy Ducks occasionally winter, and Long-tailed Duck turns up most years. In hard weather, Goosander and Smew may visit. There have been a number of records of Ring-necked and Ferruginous Ducks in recent years. Gull flocks on the beach at this season should be checked for rarer species such as Mediterranean or Glaucous Gulls. Several Water Rails inhabit the reedbeds, and in hard weather Bittern and Bearded Tit may visit. A few Chiffchaffs and Firecrests usually winter in the surrounding bushes and a Black Redstart may use the beach.

In spring all three divers occur offshore and passage terns are noted in small numbers. Garganey is a regular visitor in early spring. Migrant passerines may be found in the bushes and Wheatears favour the shingle bank. Scarcer migrants include the occasional Marsh

Harrier or Hoopoe.

In summer a few pairs of Great Crested Grebes breed and Sedge and Reed Warblers are common. Common Buzzard, Sparrowhawk and Raven are all resident and seen frequently, and the explosive song of Cetti's Warbler is a familiar sound year-round from the dense scrub and reeds. Cirl Bunting occurs in the sheltered valley leading inland; it often sings from high trees in summer, but in winter moves nearer the coast and favours hedgerows and low bushes. In late summer, light south winds can produce feeding flocks of Manx Shearwaters at sea as well as the more usual Gannets and Kittiwakes. Occasionally a Sooty Shearwater is seen.

Autumn sees large numbers of Swifts and hirundines gather over the lake and they are sometimes chased by a Hobby. Black Tern is regular and ringing has proven that the elusive Aquatic Warbler appears most years in small numbers following east winds, though the species is unlikely to be seen in the field. At sea Arctic Skuas chase the Sandwich Terns and, following severe gales in late autumn, oddities such as Grey Phalarope may shelter on the Ley.

Access

Slapton is c.6 miles south of Dartmouth on the coastal A379. There are two car parks along the road, which runs along the seaward side of the lake. Slapton Ley is easily viewed from the car parks, and the sea is worth checking at various points along the beach. Further exploration may

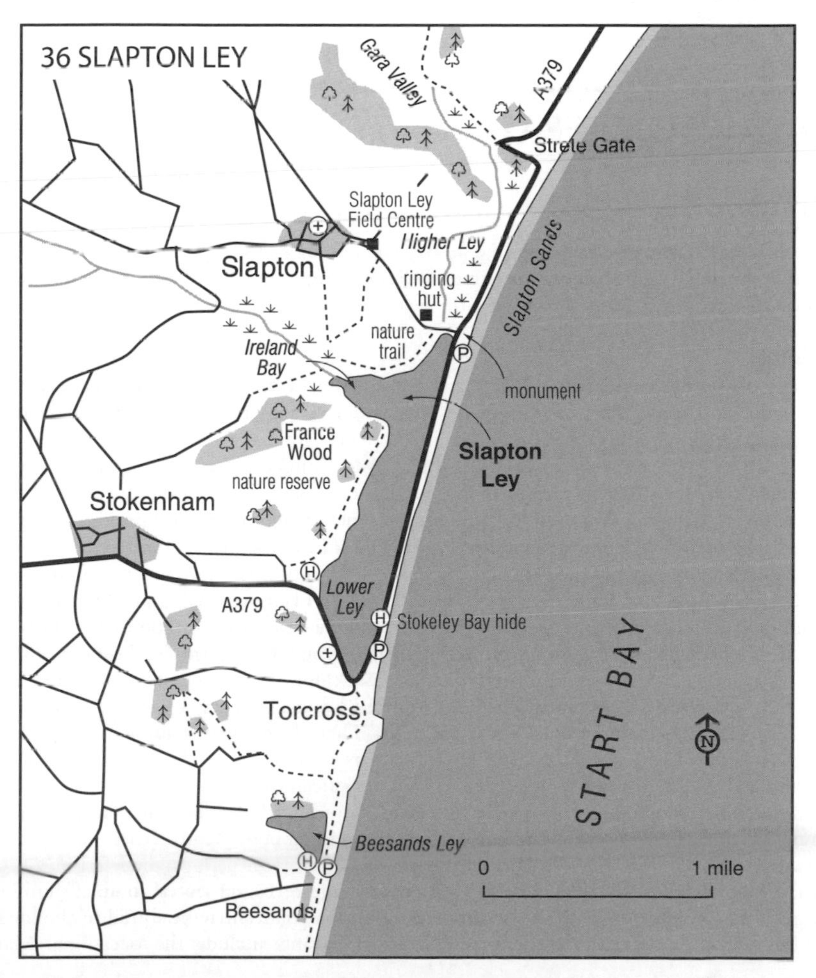

prove profitable. A ringing hut maintained by the Devon Birdwatching and Preservation Society is sited across the bridge from the monument between Slapton and Higher Leys, and a logbook is kept in a plastic ex-electric meter box on the left rear side of the hut, with key attached. Opposite here a trail leads through a gate and gives good views over Ireland Bay. Continue until a gated track leading to France Wood appears on the left (permit required); dense bushes and an overgrown quarry should be checked for wintering Chiffchaff and Firecrest. Higher Ley may be viewed from the coast road (e.g. at Strete Gate) or along a path which leads north from the ringing hut. A public hide (with wheelchair access) overlooks Lower Ley, close to the main car park. There is also a hide at Stokeley Bay (access details from Slapton Ley Field Centre). Beesands Ley can be reached by walking south along the coast from Torcross or by driving south from Stokenham (on the A379) on minor roads to Beesands village; the Ley is immediately north of the village. A public hide (with disabled access) is sited on the southern side of the Ley.

FURTHER INFORMATION

Grid refs: Field Centre SX 823 449; Lower Ley SX 823 423; Beesands Ley SX 820 407
Slapton Ley Field Centre: Slapton, Kingsbridge, Devon TQ7 2QP; tel: 01548 580466; email: enquiries.sl@field-studies-council.org; web: www.slnnr.org.uk

37 LABRADOR BAY (Devon)

OS Landranger 202

This new RSPB reserve overlooking Lyme Bay was purchased to help secure the future of Cirl Bunting. Situated near Shaldon, just south of Teignmouth, the area is already a well-known beauty spot, with free access throughout the year.

Habitat

The reserve comprises 1.25 miles of spectacular coastline, with habitats including coastal cliff tops, woodland, scrub and arable grassland. The reserve is part of a working farm where the fields are grazed by cattle and sheep, and some crops are grown.

Species

The reserve's star bird is Cirl Bunting which is present throughout the year. From early spring, pairs can be easily located from the distinctive rattling song of the male (although they may sing all year on sunny days), but they become harder to find in summer at the height of the breeding season. Autumn and winter are good times to see Cirl Buntings when small flocks may be found feeding on seeds in the specially managed stubble fields. At this time they can even be seen in the hedge around the car park. Yellowhammers are also present on the reserve. Other resident species include Common Buzzard and Skylark, and Peregrine is regularly seen around the cliffs. In summer, sheltered valleys hold summer migrants such as Common Whitethroat, Blackcap and Chiffchaff.

Access

Labrador Bay reserve is on the A379 coast road between Teignmouth and Torbay (not the B3199 as labelled on Google Maps). From Teignmouth, cross the Teign estuary into Shaldon and continue south for 1.5 miles. Park in the pay-and-display car park on the left. Entry to the reserve is free. The car park gives good views over the reserve, and several footpaths cross the site, including one that links to the South West Coast footpath. The terrain is fairly demanding and the footpaths are unsuitable for wheelchairs.

FURTHER INFORMATION

Grid ref: SX 930 703
RSPB Labrador Bay: c/o South West Regional Office, tel: 01392 432691; web: www.rspb.org.uk/labradorbay

38 YARNER WOOD (Devon)

OS Landranger 191, OS Outdoor Leisure 28

This small mixed woodland on the eastern edge of Dartmoor is an NNR, managed by Natural England. It is particularly rich in wildlife, harbouring a good variety of summer woodland birds and some unusual butterflies.

Habitat

Oak trees predominate in the woodland and there is a ground covering of bilberries.

Species

Yarner Wood is a great place to see woodland birds. In summer, there is a colony of over 50 pairs of Pied Flycatchers, most of which use the nest boxes provided. Some 20 pairs of Common Redstarts also breed, as well as Spotted Flycatcher, Blackcap, Chiffchaff, Willow and Wood Warblers, and the commoner tit species. Lesser Spotted Woodpecker is relatively easy to find here, particularly in early spring before the leaves appear. The surrounding heathland holds a few pairs of Nightjars most years and sometimes Grasshopper Warbler, as well as Stonechat, Whinchat and Tree Pipit. A few pairs of Dartford Warblers inhabit Trendlebere Down, to the north of Yarner Wood. Goosander and Mandarin have also nested in tree holes here. Overhead, Common Buzzard and Raven may be seen soaring, and a Hobby puts in an occasional appearance.

Access

From the A38 Exeter–Plymouth road, turn right (north) onto the A382 to Bovey Tracey (left goes to Newton Abbot). At Bovey Tracey, turn left onto the B3387 to Haytor and Widecombe. After 1 mile fork right towards Manaton, and after 1.5 miles turn left at a sharp right-hand bend, onto a tarmac lane into the wood. Park in the car park, beyond the warden's cottage. There are nature trails and a hide that overlooks several nest boxes. The wood is open from 8.30am to 8pm; access is free, but parties should book in advance.

FURTHER INFORMATION

Grid ref: SX 784 788
Warden: Yarner Wood, Bovey Tracey, Devon TW13 9LJ. Tel: 01626 832330.

39–43 EXE ESTUARY (Devon)

OS Landranger 192

The Exe, containing 6 miles of extensive tidal mudflats between Exeter and the sea, is the most important estuary for wildfowl and waders in the south-west. A variety of habitats ensures a strong diversity of birds all year though it is best during migration seasons and in winter. It is noted in particular for wintering Brent Geese, Avocets and Black-tailed Godwits.

Habitat

The estuary is tidal as far as Countess Wear on the outskirts of Exeter and is over 1 mile wide in places, with considerable areas of mud and sand at low tide. Two sand spits lie at its mouth: Dawlish Warren extends for 1 mile into the estuary from the west and contains a series of dunes, a golf course and a small reedbed. Behind lies an area of saltmarsh; to the east the spit at Exmouth is much smaller and there are high cliffs east of the town. Between Exminster Marshes and Topsham is a tidal reedbed in the estuary's narrower inner section.

Species

Red-throated Diver and Slavonian Grebe are regular in winter, usually off Dawlish Warren or Langstone Rock but sometimes also off Exmouth. Small parties of Common Eiders and Common Scoters are often on the sea and the latter may include a few Velvet Scoters. The regular wintering flock of Brent Geese numbers over 2,000 birds; the preferred feeding area is the bay north of Exmouth, though they are also seen off Dawlish Warren at high tide. Bewick's Swan and White-fronted Goose are only irregular at Exminster Marshes. Several thousand Wigeon winter, favouring the saltings behind Dawlish Warren or Exmouth. Other dabbling ducks include small numbers of Gadwalls and Pintails. Goldeneye and Red-breasted Merganser prefer the deeper channels and a Long-tailed Duck is found most winters. Scaup is irregular, off Turf Lock or on the sea. Peregrine is seen fairly regularly in winter while Common Buzzard and Sparrowhawk are resident and seen more frequently. Over 100 Little Egrets can now be seen on the Exe in winter, and occasionally a few Spoonbills. Thousands of waders winter, including Grey Plover and Turnstone, a few Knots and Sanderlings. Over 600 wintering Black-tailed Godwits are of national importance; they may be seen on the east shore or off Powderham and usually roost on Exminster Marshes (where they were briefly joined in winter 1981–82 by Britain's first Hudsonian Godwit), which also hold Golden Plover and several Ruffs. Common Redshanks roost in Powderham Park at high tide and should be checked for Spotted Redshank and Greenshank. The wintering Avocets, numbering 400–500 birds, form the third largest flock in Britain and are usually found off Turf Lock or Bowling Green Marsh. On the coast a few Purple Sandpipers inhabit the rocks below Orcombe Cliffs and less frequently at Langstone Rock. The large gull roost on the marshes consists mainly of Black-headed Gulls but should be checked for the occasional Mediterranean Gull. Short-eared Owl is often seen over Exminster Marshes or at the east end of Dawlish Warren and a Kingfisher may be found along the canal. A few Chiffchaffs winter in the bushes on Dawlish Warren, occasionally accompanied by a Firecrest. Cirl Bunting should be looked for in low vegetation on the seafront at Dawlish and Black Redstart occasionally winters. Water Rails and Cetti's Warblers can be heard along the canal at Exminster.

Spring is heralded by the arrival of Wheatears and sometimes a Garganey on Exminster Marshes. One year there was a Great Spotted Cuckoo at Dawlish Warren. Terns are regularly observed offshore and roost on the Warren; Sandwich is commonest with smaller numbers of Common and Little and a few Arctics, while one or two Roseates are regular in May. Passage waders include Whimbrel and Green Sandpiper, and one of the few British records of Semipalmated Plover was a bird found at this season, likewise Greater Sand Plover.

In summer a few Shelducks and Lapwings breed on Exminster Marshes and the reedbeds hold Sedge and Reed Warblers and Reed Bunting. There is a small heronry in Powderham Park. Cirl Bunting breeds in hedges near Langstone and around Exminster; they are present all year.

Autumn wader passage begins in July and occasionally a Little Stint or Curlew Sandpiper is found among the commoner species. Small numbers of Black Terns are regularly noted over the canal and Little Gull may also be seen. Osprey is almost annual in autumn and occasionally one stays several weeks. A Hobby sometimes chases the hirundines in late afternoon, particularly as they go to roost in the reedbeds. Gales in late autumn are likely to bring seabirds inshore. Gannet and Great and Arctic Skuas are often seen but seawatching is much better off more significant promontories such as Prawle. Scarce migrant passerines are found regularly at Dawlish Warren in autumn.

Access

The estuary can be observed from many points but the west side tends to be most profitable due to less development and restriction of shooting, though the light is inevitably unfavourable in the early morning. The waders can be very distant at low tide and are best seen 2–3 hours before high water. Alternatively, they may be observed at their high-tide roosts, notably

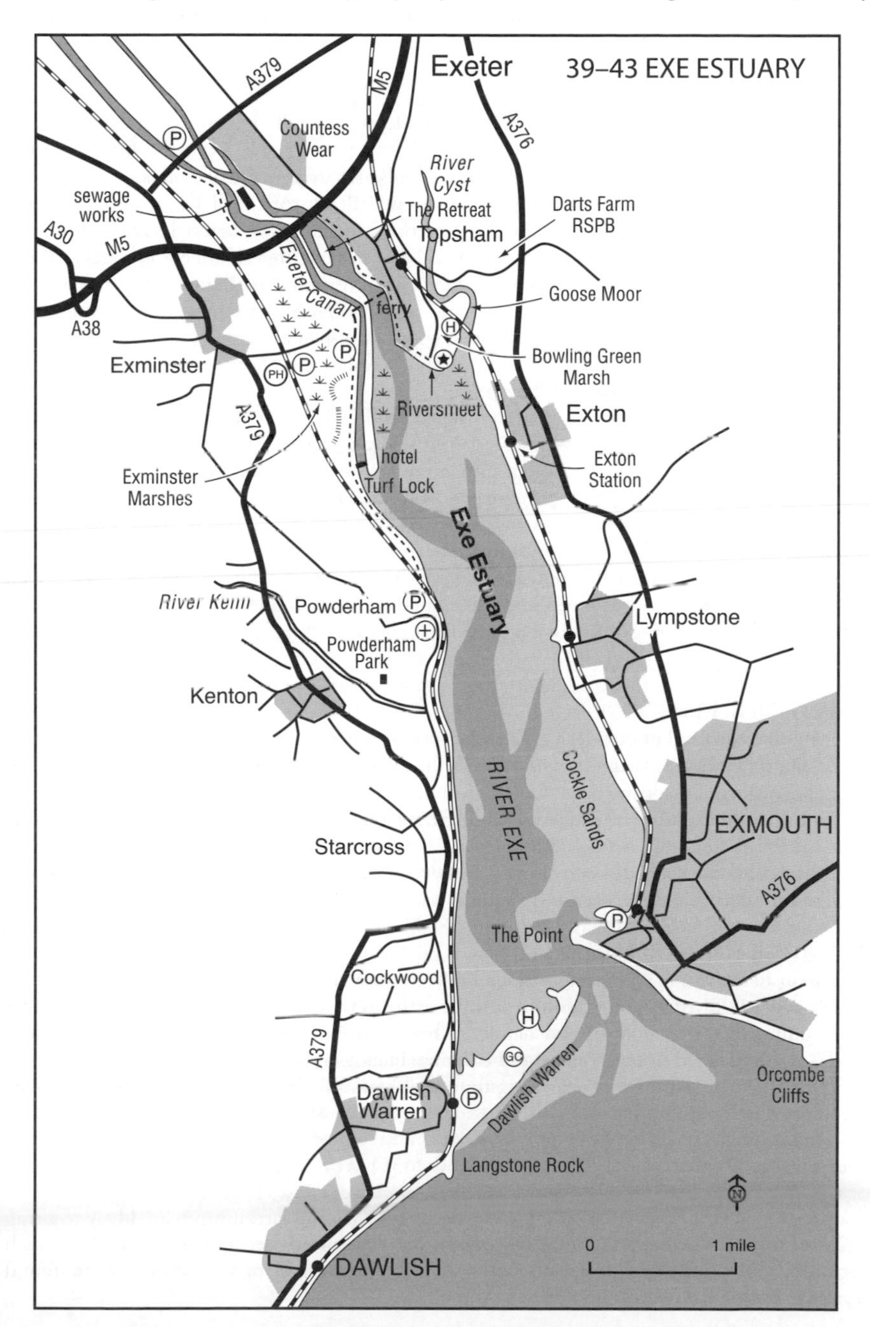

at Dawlish Warren, Exminster Marshes or Topsham. The sea is best watched from Langstone Rock (south-west of Dawlish Warren) or from Exmouth. The latter and Dawlish Warren are tourist havens particularly in summer; weekdays or early mornings are therefore best for birding. One of the best ways to see the birds is to take one of the regular birding cruises which operate throughout the year. The boats leave from Exmouth Dock or Topsham Quay. Further information and bookings can be made via the RSPB website or by contacting the RSPB South West Regional Office.

39 EXMINSTER MARSHES (Devon)

These freshwater marshes comprise rough grazing dissected by ditches and are subject to occasional flooding. They are situated on the estuary's upper west bank, separated from it by the disused Exeter Canal. The marshes form part of the RSPB's Exe Estuary reserve.

South of Exminster on the A379, park in the RSPB car park near the Swan's Nest Inn. Footpaths lead across the marshes to the estuary (where there is an occasional passenger ferry to Topsham). Continue south alongside the canal to Turf Lock (about 1.25 miles). This track affords good views over the marshes, the tidal reedbed and the estuary. At Turf Lock you can cross over the lock and walk a short way upstream to gain a better view of the estuary. Continuing south from Turf Lock, an embankment follows the edge of the estuary.

40 BOWLING GREEN MARSH (Devon)

The recent acquisition of Bowling Green Marsh by the RSPB is an important complement to their Exminster reserve on the other side of the estuary. Active management of the coastal grazing marsh is transforming the area into a valuable site for wildfowl and waders. A hide (wheelchair accessible) overlooking the marsh and a viewing platform overlooking the estuary are provided.

The reserve lies just south of Topsham, and is signposted from Holman Way car park in Topsham, close to the railway station (do not park in the lane by the reserve). Bowling Green Marsh is open at all times, but access is restricted to the public footpaths and hide. The Goatwalk, a path running alongside the estuary, is probably the best place to see the wintering Avocets at high tide.

41 POWDERHAM (Devon)

The wooded parkland of Powderham Park lies on the estuary's west bank immediately south of Exminster marshes. A small river runs through the park, occasionally flooding certain areas. The park and adjacent meadows to the north are used by roosting waders at high tide.

Continue south on the A379 from Exminster, turn left at the first crossroads to Powderham, and park by the church. A track north leads to an embankment, which gives good views of the estuary and reaches Turf Lock after c.1.5 miles. South from the church the road follows the edge of the estuary, and the park and its meres can be viewed from various points along here.

42 DAWLISH WARREN (Devon)

This is an LNR managed by Teignbridge District Council. The central area comprises dunes, bushes and a reedbed. A golf course lies on the north side and the beach is popular with tourists. A public hide overlooks the estuary and offers close views of roosting waders at high tide. Access is unrestricted except for the golf course, which is private.

South of Starcross (off the A379) turn left onto a minor road leading to Dawlish Warren. Cross the railway under the bridge and walk north-east across the dunes, past the reedbed. The estuary can be viewed from the north side of the Warren and the hide is on this shore beyond the golf course. From Dawlish Warren station walk south to reach the coast and Langstone Rock.

43 EXMOUTH (Devon)

The seafront gives good views of the sea and the mudflats. Head east along the seafront to reach Orcombe Cliffs. A sheltered bay north of the spit is favoured by Brent Goose; this can be viewed from the Imperial Road car park west of the station.

FURTHER INFORMATION

Grid refs: Exminster Marshes SX 954 871; Bowling Green Marsh SX 971 876; Powderham SX 972 844; Dawlish Warren SX 980 787; Exmouth SX 997 810
Dawlish Warren (warden): c/o Countryside Management Section, Teignbridge District Council: tel: 01626 215754. Visitor centre: tel: 01626 863980; postcode EX6 8RU; web: www.dawlishwarren.co.uk
RSPB Exe Estuary (reserve office): tel: 01392 824614; web: www.rspb.org.uk/exeestuary
RSPB South West Regional Office: tel: 01392 432691; web: www.rspb.org.uk

44 AXE ESTUARY AND SEATON BAY (Devon)

OS Landranger 192

Lying between Portland and the Exe estuary, the Axe estuary in east Devon flows into Seaton Bay at Seaton. The estuary contains two local nature reserves which, together with several good viewing points, provide good birding opportunities throughout the year. Nearby, Beer Head is a local migration watchpoint. The local nature reserves are managed by East Devon District Council (EDDC), which has plans to develop the estuary further. In recent years, dedicated coverage by a group of keen local birders has revealed the area to be of great interest ornithologically, with an impressive list of scarce migrants and national rarities.

Habitat

In addition to typical estuarine habitats on the estuary itself, there are tidal marshes at Colyford Common, grazing marshes at Seaton Marshes and reedbeds in the upper part of the estuary. EDDC has recently purchased Black Hole Marsh, lying between Seaton Marshes and Colyford Common, which has been converted into a shallow, tidally-affected lagoon with several islands. The open clifftops at Beer Head support some scrub and rough grazing.

Species

Red-throated and Great Northern Divers can be seen offshore in winter, but Black-throated is less regular. Seaducks include Common Scoter and occasional Common Eider, with small numbers of several other species most winters. The estuary attracts good numbers of wintering wildfowl, mainly Wigeon, Teal and Shelduck. Wintering waders include Lapwing, Common Snipe, Curlew and Black-tailed Godwit, and Little Egret is a familiar sight on the estuary. It is well worth checking gull roosts on the estuary; small numbers of Mediterranean are regular, while Glaucous and especially Iceland are recorded most winters. Vagrant American gulls have been found on several occasions. The reedbeds are home to several Water Rails, though they can be elusive. Kingfisher and Cetti's Warbler may be somewhat easier to find. Notable wintering passerines include several Water Pipits at Colyford Common, and coastal areas should be checked for Chiffchaff, Firecrest or Black Redstart, all of which winter in very small numbers.

Spring may bring a Garganey to the marshes, or a Spoonbill on the estuary. Passage waders include Little Ringed Plover, Bar-tailed Godwit and Whimbrel. Passerine migration is best observed at Beer Head, when Common Redstarts, Wheatears and a variety of warblers can be found. Seawatching in spring may be rewarded with Great and Arctic Skuas, and occasionally Pomarine. Summer is generally quiet, though breeding Peregrines and Ravens can usually be found.

Autumn is a good time for the Axe estuary. Returning waders regularly include Greenshank, Green and Wood Sandpipers, Little Stint, Curlew Sandpiper and Ruff, whilst an American rarity is always a possibility. Roosting gull flocks should be checked for Yellow-legged Gulls which are regularly reported in July and August (usually juveniles). Britain's fourth Audouin's Gull was found at Seaton Marshes in August 2007. Seawatching can be profitable in rough weather, although birds passing mainly westwards out of Lyme Bay tend to be rather distant. Numbers do not compare with major headlands such as Prawle or Portland, but Manx Shearwaters are regular from April, while Balearics are frequent in small numbers between July and September. Rarer species such as Sooty are noted occasionally, while Storm Petrel is more likely to be seen offshore. Arctic and Great Skuas are reasonably regular in autumn. Autumn raptors may include a long-staying Osprey or a Hobby. Passerine migration is best observed at Beer Head. As well as the usual Wheatears and warblers, less common species such as Common Redstart, Whinchat and Pied Flycatcher may be found in small numbers, with the distinct possibility of something rarer. Flocks of hirundines gather over the estuary and Yellow Wagtails may be seen in the meadows and especially at Beer Head, with peak numbers in late August. Later in the autumn, Ring Ouzels occasionally stop off at Beer Head, while the (sometimes considerable) westward passage of flocks of Woodpigeons, pipits and finches in October and November herald the onset of winter.

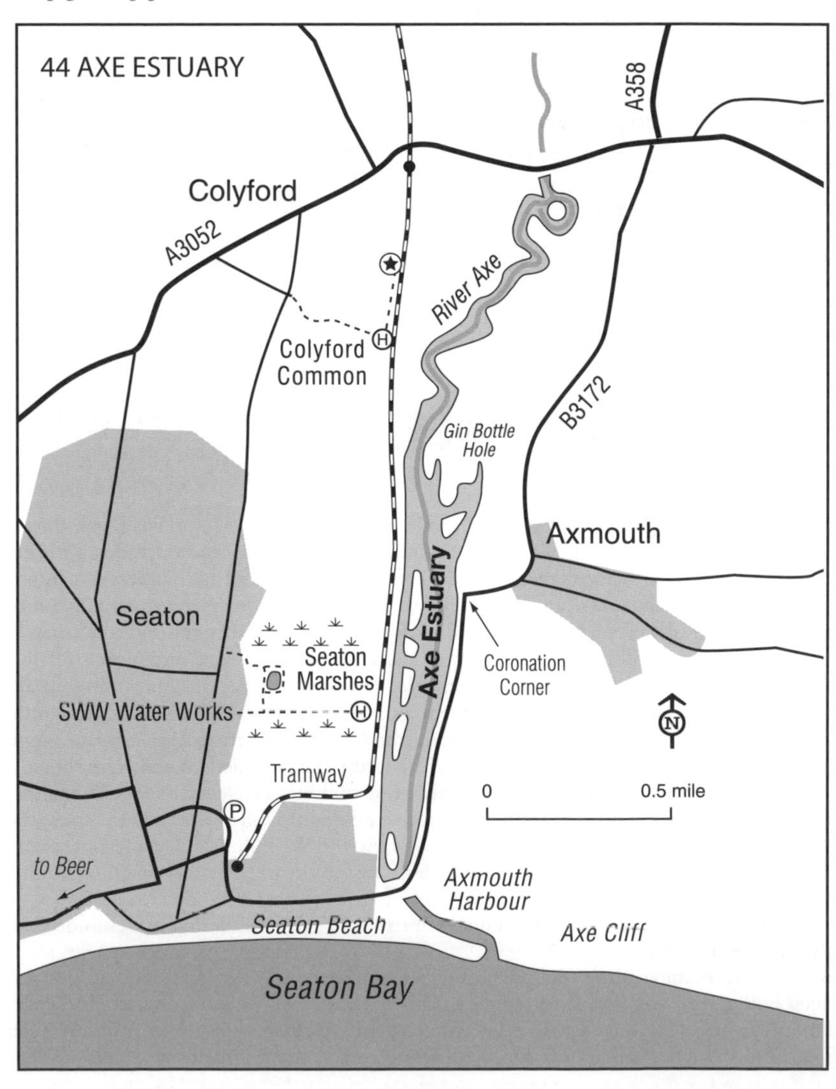

Access

The Axe estuary is small compared to the Exe but almost all of it is viewable from roadside vantage points (see map). One of the best viewing points is at Coronation Corner where there is pull-off for parking. The seafront at Seaton, or at nearby Branscombe, are good places to sea-watch from in suitable conditions in spring and autumn. The key sites in the area are as follows.

Seaton Marshes LNR This local nature reserve on the lower west side of the estuary has an excellent wheelchair-friendly hide overlooking the estuary and the marshes. It attracts good numbers of wintering wildfowl, especially Wigeon. The reserve includes Borrow Pit, an area of deep water that attracts a different range of species from the surrounding marshes.

Colyford Common LNR Situated on the upper west side of the estuary, this local nature reserve has a small hide and viewing platform providing two vantage points across the upper estuary marshes and two wader scrapes. This is one of the best sites in Devon to see Water Pipit.

Beer Head This chalk headland on the west side of Seaton Bay is a magnet for migrants, attracting good numbers of birds in the right conditions. Although not as impressive as Portland Bill, regular watching has turned up a number of rarities in recent years including Buff-breasted Sandpiper and Iberian Chiffchaff. The whole area is open to the public, with easy access; park in the large pay-and-display car park.

FURTHER INFORMATION

Grid refs: Seaton Marshes SY 252 908; Colyford Common SY 253 920; Coronation Corner SY 255 910; Beer Head SY 227 888
East Devon District Council: tel: 01395 516551; web: www.eastdevon.gov.uk/local_nature_reserves
Axe Estuary Birds (free email newsletter): contact David Walters: tel: 01297 552616; mobile: 07791 541744; email: davidwalters@eclipse.co.uk

45 WEST SEDGEMOOR (Somerset)

OS Landranger 193

The Somerset Levels are one of England's largest remaining wet meadow systems. They hold internationally important numbers of waterfowl in winter, with up to 70,000 birds in some years, and important numbers of breeding waders in summer. Access is generally difficult, and the RSPB reserve of West Sedgemoor is one of the best bets.

Habitat

The RSPB manages 534 ha of hay meadows and pasture, and 42 ha of woodland. The 'moorland' comprises flat, low-lying fields which are cut for hay and grazed by cattle to provide good conditions for wildlife; in addition, the water levels are controlled. In winter, the fields are waterlogged and occasionally flooded, attracting thousands of ducks and waders. To the south of the levels, Swell Wood has the largest heronry in south-west England, as well as a good range of woodland species.

Species

Wigeon is the most numerous duck on the levels, exceptionally numbering in excess of 20,000 birds, and Teal numbers can reach 10,000 if conditions are right. Other duck species are less numerous, but recent counts of 1,000 Pintails are significant. Lapwing (up to 40,000 birds) and Golden Plover also winter in large numbers. Other waders include several hundred Common Snipe and Dunlins, as well as a few Ruffs. Raptors are attracted to the area in winter, and Peregrine is frequently seen. Hen Harrier, Merlin and Short-eared Owl are also regular but less predictable.

West Sedgemoor is an important site for breeding waders in summer, mainly Lapwing, Common Snipe, Common Redshank and Curlew. A few pairs of Black-tailed Godwits may also breed. Hobbies are frequently seen in summer, and sometimes several at a time. Quail also occurs in the area, but is elusive as always. Amongst the passerines, Skylark, Yellow Wagtail, Whinchat, Sedge Warbler and Reed Bunting are the more typical species. Woodland birds include Common Buzzards and a few pairs of Nightingales.

Access

The reserve car park is signed just off the A378 Taunton–Langport road, 1 mile east of Fivehead. There is a woodland trail from the car park with a hide (wheelchair accessible) to view the herons (March to June is best). Another trail leads to the Moorland Hide, overlooking the levels. There is an information board here.

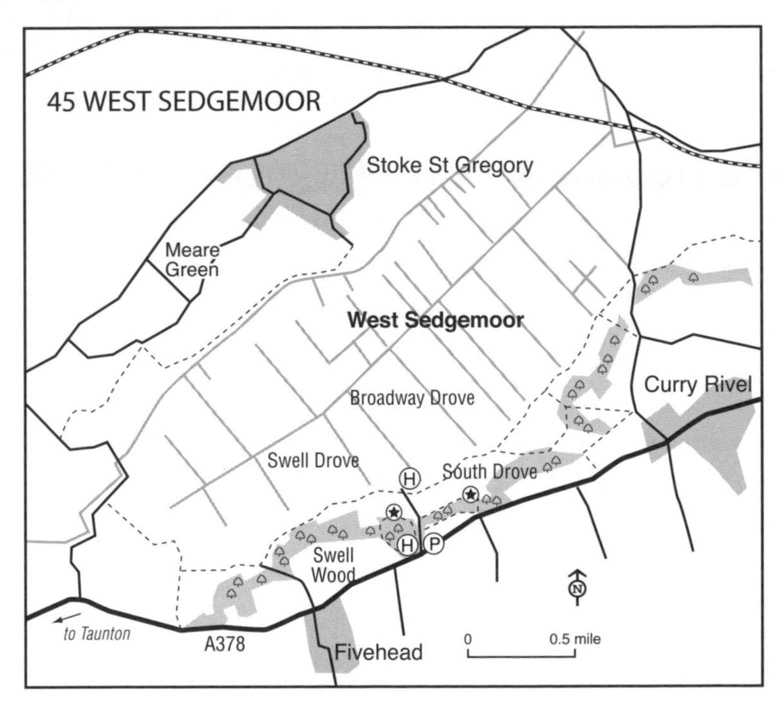

FURTHER INFORMATION

Grid ref: ST 361 238
RSPB West Sedgemoor: c/o South West Regional Office; tel: 01392 432691; web: www.rspb.org.uk

46 AVALON MARSHES (Somerset)

OS Landranger 182

This region of the Somerset Levels contains the remnants of once extensive raised bogs, a habitat much exploited for peat digging. There is also fenland, carr and flower meadows. Nowadays, much of the land is in the care of conservation bodies and there are four principal sites: Shapwick Heath NNR (run by Natural England), RSPB Ham Wall, Catcott Lows NR and Westhay Moor NNR (the latter two sites are run by the Somerset Wildlife Trust). Shapwick Heath is one of the best sites in England for Otters, and this most sought-after mammal may be seen at any time of day here, while Ham Wall is home to nesting Bitterns. The area has become well known for hosting one of the largest and most accessible winter Starling roosts in the country from November to March (but they don't always roost in the same place).

Habitat

The botanically rich Shapwick Heath contains fenland and bog myrtle pasture, grading to birch and alder carr. Nearby, RSPB Ham Wall is being developed as an extensive wetland reserve with reedbeds; further west Catcott Lows is restored fenland with areas of open water, and Westhay Moor has a mixture of reedbeds, open water and carr.

Species

Winter brings substantial flocks of Wigeon, Shovelers and Teal to the flooded meadows and peat workings, and diving ducks including Goosander and Goldeneye. Bitterns winter and in 2008 bred in the area for the first time since the 1960s. Marsh and Hen Harriers range across the area, and Peregrine and Merlin are also regular, especially at Catcott Lows. Although resident, Water Rail and Barn Owl may be easier to see in the winter months. One big draw for the local raptors is the huge and spectacular winter Starling roost in the reedbeds, which sometimes tops several million birds, and has become something of a tourist attraction in recent years. The Starlings begin arriving shortly before dusk (check the hotline for up-to-date information – see below).

In early spring a few Garganeys pass through – sightings are annual and breeding is probable. Spring wader passage includes Whimbrels and Ruffs, along with both godwits. It is hoped that the breeding Lapwings and Common Redshanks will be joined by Curlews and Common Snipe in due course. Interesting breeding passerines include many Cetti's Warblers, and also Bearded Tits. The numbers of Hobbies present in May are impressive and over 50 birds have been counted in most years recently. Britain's first Cattle Egrets bred in Somerset in 2008, and the species can frequently be seen in fields in the area, especially at Shapwick Heath or Catcott Lows. It is likely that they will be seen even more regularly in future.

Osprey and Spoonbill are among the more notable visitors during autumn passage, along with a good selection of the commoner waders including Little Stint.

Access

The minor road between the villages of Meare on the B3151 and Ashcott off the A39 provides access to Shapwick and Meare Heaths, and Ham Wall. Catcott Lows is on the minor road between Catcott and Burtle, and Westhay Moor is on the minor road between the B3151 and B3159, between Wedmore and Meare. Parking is limited at all four sites.

Shapwick Heath NNR A car park is available just south of the South Drain at Ashcott Corner – from here a disused railway line runs westwards to the east side of Shapwick and Meare Heaths. A footbridge over the South Drain 0.5 miles west of the car park leads to a hide overlooking Meare Heath to the north, and the path to the south leads to the Noah's Lake hide. Both hides are good for viewing Otters. Further west a minor road between Shapwick and Westhay runs through the middle of the Shapwick Heath reserve. There is limited parking on the roadside by the bridge over the South Drain, and this is the starting point for footpaths accessing both sides of the reserve. There are two more hides here.

RSPB Ham Wall From Ashcott Corner (the same car park as for Shapwick Heath), a wide track following a disused railway line leads east to the entrance of the Ham Wall reserve. A network of trails provides easy access (mostly accessible for wheelchairs), and there are several viewpoints, seats and viewing screens, some of which are roofed. The reserve is open at all times.

Catcott Lows NR Further west again, on the minor road between Burtle and Catcott, is an area of flooded fields which is the Catcott Lows reserve. Park at the end of a track off the minor road; two hides overlook the flooded fenland and the reserve has open access. This area is best from late autumn through to early summer.

Westhay Moor NNR To the north of Westhay village is Westhay Moor NNR. This is reached from the B3151 by turning right about 0.5 miles from the River Brue at Turnpike House and continuing for another mile until reaching the car park at the start of Daggs Lane Drove; this used to be the favoured site for the Starling roost, though in recent winters they have moved to Shapwick and Meare Heaths.

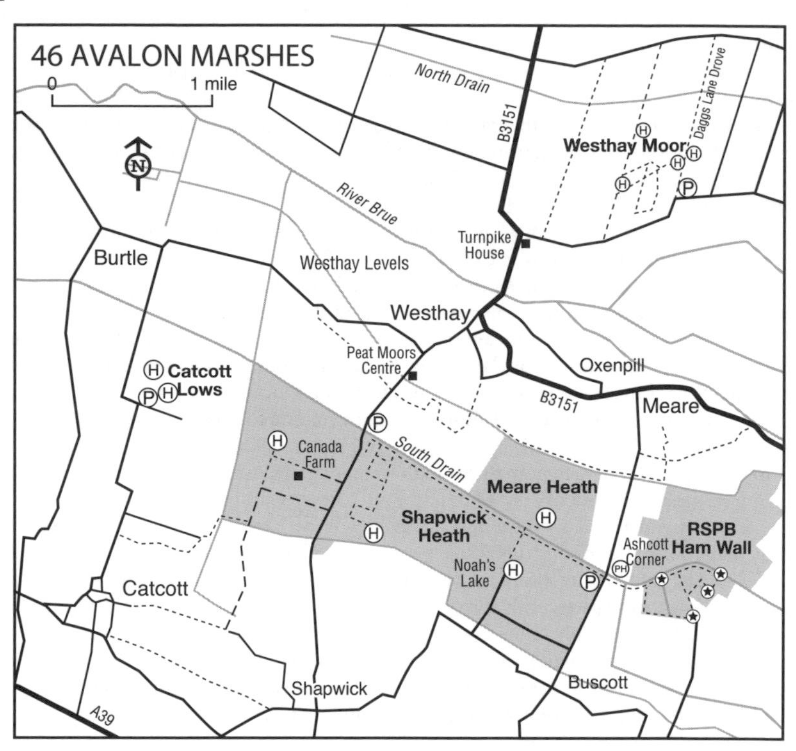

FURTHER INFORMATION

Grid refs: Shapwick Heath ST 448 396; Ham Wall ST 448 396; Catcott Lows ST 399 414; Westhay Moor ST 457 437
RSPB Ham Wall: tel: 01458 860494; email: ham.wall@rspb.org.uk
Natural England: tel: 01458 860120; web: www.naturalengland.org.uk
Somerset Wildlife Trust: tel: 01823 652400; web: www.somersetwildlife.org
Avalon Marshes Starling Hotline: 07866 554142

47 BRIDGWATER BAY (Somerset)

OS Landranger 182

The vast mudflats and saltmarshes of Bridgwater Bay hold important concentrations of waders and waterfowl in winter and on passage. The bay extends from Lilstock in the west to Burnham-on-Sea in the east, but the eastern part, around the estuary of the River Parrett, is the best area for birds. Some 2,559 ha are protected as a National Nature Reserve.

Habitat

The habitats of Bridgwater Bay range from extensive intertidal mudflats, saltmarsh and shingle shore to grazing marsh intersected by freshwater and brackish ditches. The River Parrett meanders north from Bridgwater across a low-lying plain, encircling Pawlett Hams before entering the bay between Stert and Berrow Flats. There are sand dunes at Berrow and Steart, and Stert Island, in the mouth of the Parrett, is mostly bare shingle and sand. There is a small reedbed west of Stert Point and several scrapes and pools have been excavated in front of the hides that overlook the estuary mouth.

Species

Over 10,000 Dunlins winter in the bay, with smaller numbers of Knots, Common Redshanks, Curlews, Grey Plovers and Oystercatchers. A few Spotted Redshanks also winter in the mouth of the Parrett, and occasionally Avocets. The majority of the wildfowl are Wigeon with some Mallards and Teal. In hard weather, Lapwing and Golden Plover are numerous, and a few Brent Geese are regular. Raptors are a feature of winter and include small numbers of Hen Harriers, Peregrines, Merlins and Short-eared Owls.

Numbers and variety of waders increase during spring passage, and include Bar-tailed Godwit and Whimbrel; there is a spring roost of the latter on Stert Island.

Summer is quiet, but Nightingales are still found in coastal copses, notably along the Hinkley Point nature trail. In late summer up to 2,500 Shelducks gather to moult, the major British site for them; smaller numbers are present throughout the year.

Autumn passage brings more waders, including Icelandic Black-tailed Godwits and occasional rarities. Yellow Wagtails are frequent and Aquatic Warblers have occasionally been found in the reedbed (but are more likely to be trapped than seen). Westerly gales may bring a wandering seabird to the estuary mouth.

Access

Junctions 23 and 24 on the M5 lead into Bridgwater. From here take the A39 westwards towards Minehead. In Cannington, turn right (north) onto a minor road to Combwich and Hinkley Point. About 0.5 miles after the turning to Combwich, turn right on a road signed to Otterhampton and Steart. After 1 mile turn right again to Steart itself. Park in the car park on the left in Steart and continue for 0.5 miles to a tower hide and two other hides. The hides have free access and are open daily. Waders roost at the point at high tide (best to arrive up to an hour

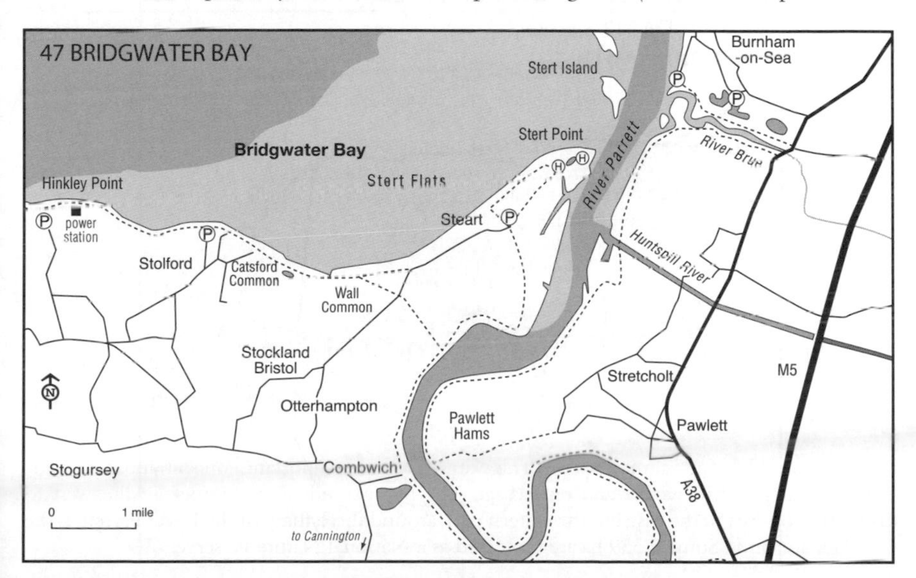

before), with the largest numbers just upriver on the west bank of the Parrett. The scrapes and pools can be viewed well from the tower hide.

Further west along the bay, waders can be seen for up to 2.5 hours after high tide at Wall Common – park by the road 1 mile west of Steart and walk west along the sea wall. Further west still, the bay can be viewed at Stolford and Hinkley Point. Park at Stolford and walk east to Catsford Common, or west to Hinkley Point. The nuclear power station's warm water outfall attracts gulls and terns (you can drive direct to the power station and park there).

It is essential to work Bridgwater Bay around high tide, as the mudflats are so extensive at low tide.

FURTHER INFORMATION

Grid refs: Steart ST 274 459; Stolford ST 228 456; Hinkley Point ST 206 458
Bridgwater Bay: Reserve manager, Robin Prowse: tel: 01278 652426
Natural England: web: www.naturalengland.org.uk

48 CHEDDAR RESERVOIR (Somerset)

OS Landranger 182

This reservoir, between Cheddar and Axbridge, has man-made banks but nevertheless holds a number of wildfowl in winter, which can include unusual species such as Smew. In autumn exposed mud attracts waders and there is a good passage of terns. Several rarities have been found in recent years. There are no access restrictions.

Habitat

Though Cheddar Reservoir has concrete banks and no marginal vegetation, muddy banks and gravel islands are exposed when water levels drop. Some small reed-fringed pools and willows can be overlooked from the south end. Disturbance from boating activities can be considerable.

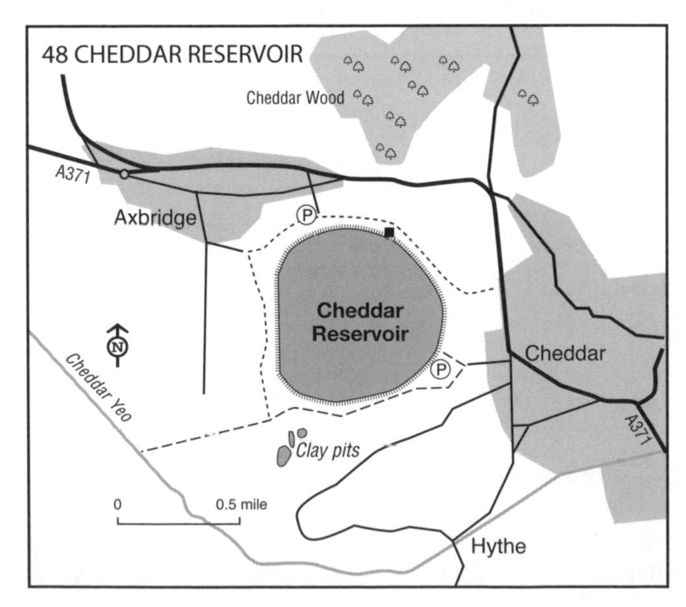

Species

In winter divers, grebes and seaducks can occasionally be found among more regular species on the open water, and the gull roost has produced Mediterranean, Yellow-legged, Ring-billed and Franklin's Gulls.

Common Sandpiper, Ringed and Little Ringed Plovers can occur in spring, as do Common, Arctic and Black Terns, and occasionally White Wagtail.

Waders and terns also pass through in autumn, when seabirds such as Grey Phalarope and Kittiwake have been recorded following strong gales.

Access

There are two main entrances to Cheddar Reservoir. The north entrance is reached by taking the A371 bypass and then the gated road 200 yards along the road to Axbridge. From the car park at the end a perimeter track is reached, but is subject to restrictions detailed on information boards at the entrance. Alternatively, take the A371 west of Cheddar and then the B3151 to Wedmore. Turn right on Sharpham Road after the bridge over the disused railway and in 300 yards take the middle road at a three-way fork and continue to the car park.

FURTHER INFORMATION

Grid ref: ST 447 534

49 BLAGDON LAKE (Avon)

OS Landranger 172 and 182

This 178-ha reservoir lies west of Chew and though generally less productive it can produce some interesting birds. The south and east sides of the lake are generally the best areas. It is possible to view from the road at the dam and at Rugmoor Bay, while along the south shore there are two hides (permits from Bristol Water at Woodford Lodge, as with Chew Valley Lake) and a footpath. At the north end of the dam there is public access to the reservoir's wooded north arm.

Habitat

There are some conifer plantations around the reservoir's shores as well as gardens and woodland near the dam. Mud is exposed when water levels drop, the best areas being at the east end.

Species

In winter commoner ducks such as Wigeon, Gadwall, Goldeneye and Goosander may be joined by scarcer species such as Smew or Scaup. Divers and scarce grebes are also possible (indeed, Britain's first Pied-billed Grebe was here for some years in the 1960s), while shoreline areas should be checked for Stonechat, Lesser Redpoll and Siskin.

Spring migrants include Common, Arctic and Black Terns as well as more regular species such as Common Sandpiper.

Autumn wader passage is more interesting and could yield Spotted Redshank, Curlew Sandpiper, Little Stint and Ruff, while rare waders also occur, with Pectoral and Buff-breasted Sandpipers among those recorded. Autumn is also the best time to look for Hobby, and Osprey may also pass through on migration.

Access

Blagdon Lake is situated between the villages of Blagdon and Ubley, north of the A368. A permit is required to access the reservoir enclosure (see below), but public roads allow reasonable coverage. For the west end follow signs for Butcombe from Blagdon village (Station Road) to the dam. A minor road at the south end of the dam affords views over the lake. A gated private road leads from where this road leaves the water and can be used by permit holders. Take the private road and turn left past the first bay, through trees to Home Bay hide. The road continues to the east end and another hide. For alternative access to the east end take the A368 and the left fork to Ubley village. Turn left at the church and left again at the T-junction 0.5 miles further on. Entry to the reservoir is possible from the road where it crosses the River Yeo. There is a hide right of the track after c.0.5 miles.

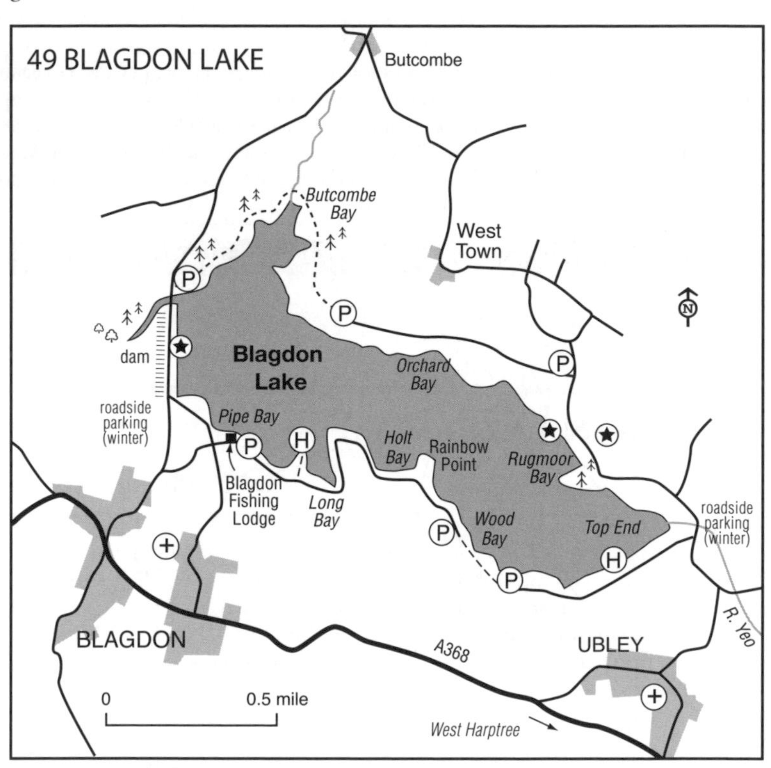

FURTHER INFORMATION

Grid refs: dam ST 504 600; fishing lodge ST 507 595; east end ST 530 590
Permits: Bristol Water, Recreation Dept, Woodford Lodge, Chew Stoke, Bristol BS18 8XH: tel: 01275 332339; web: www.bristolwater.co.uk

50 CHEW VALLEY LAKE (Avon)

OS Landranger 172 and 182

This reservoir south of Bristol covers c. 485 ha. Created in 1953, it has become one of the most important in Britain, noted for wintering wildfowl but also attracting an impressive passage of waders and other migrants. It is worth a visit at any time of year but is least productive in midsummer.

Habitat

Most of the lake is open water, part of which is set aside as a sailing area. The natural banks are attractive to waders, and in addition to exposed mud there are reedbeds, bushes and trees around the edge. A nature reserve has been established at Herriott's Bridge.

Species

Bewick's Swan is regular in winter as well as a variety of ducks including Gadwall and Pintail. Small numbers of Goldeneyes, Smews and Goosanders are regular but the lake is perhaps best known for its Ruddy Ducks – this was the species' first breeding site in Britain. They are most numerous at the south end of the lake, although DEFRA-organised culling since the late 1990s has reduced their numbers. Divers and the rarer grebes are not infrequent and many Dunlins remain through the winter. Water Rail and Bearded Tit are occasionally seen in the reeds and Golden Plover is found on the surrounding fields. There is a large gull roost in winter, mainly

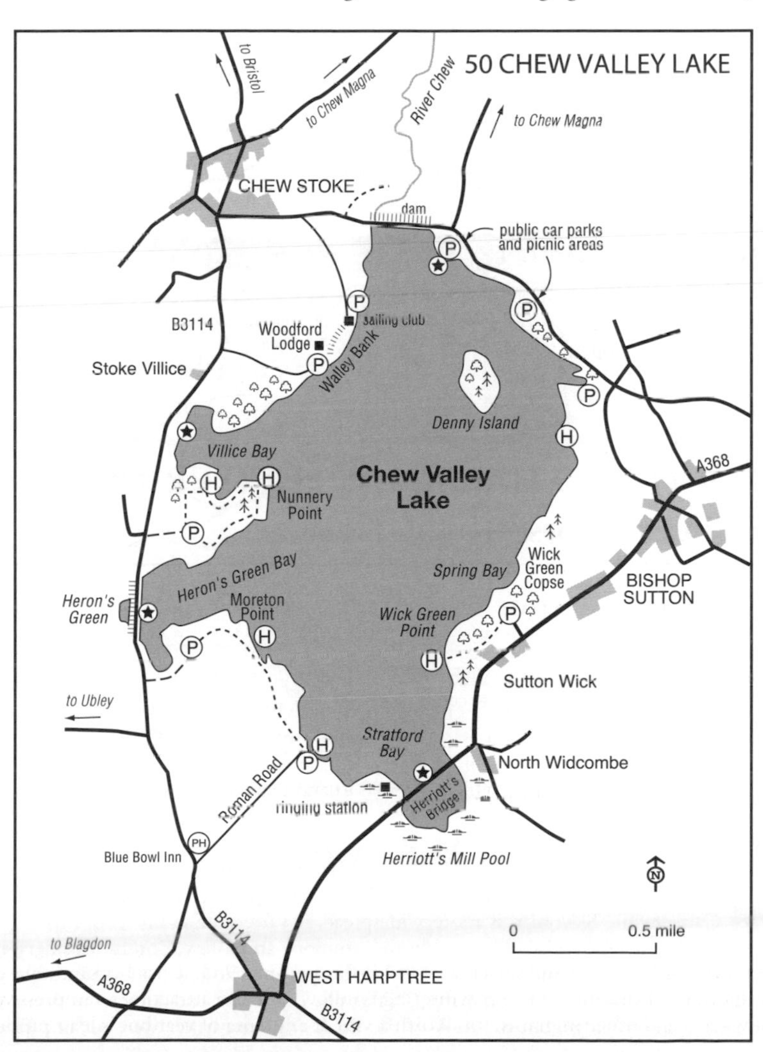

Common and Black-headed but a rarity such as Ring-billed is noted occasionally. A few Water Pipits are regular in winter.

A wide variety of waders occurs on passage, mainly in autumn, likely species including Curlew Sandpiper, Black-tailed Godwit, Spotted Redshank and Wood Sandpiper. There is a small passage of Black Terns in spring and autumn and a vagrant White-winged Black Tern is almost annual.

Summer is quiet. Garganey breeds but is only infrequently seen. Other breeding species include Gadwall, Ruddy Duck, and Reed and Sedge Warblers.

Access

The surrounding roads offer good views at the dam east of Chew Stoke, at Villice and Heron's Green Bays, and at Herriott's Bridge; the latter two are generally best. From Chew Magna on the B3130, take the B3114 to Chew Stoke. Continue on this road through the village and after 1 mile it runs alongside Villice Bay. Heron's Green is a little further on. Continue to West Harptree and join the A368 towards Bath. Herriott's Bridge is on this road c.1 mile north-east of the village. At Bishop Sutton turn left onto a minor road which completes the circuit back to Chew Stoke via the dam.

For a more thorough investigation six hides are sited around the edge with car parks off the roads. The hide at Nunnery Point is the best place for viewing gulls in the evening. Access to the shore is not allowed and a permit is required for the hides. An annual permit (£12) can be obtained by post or personal application from Bristol Water at Woodford Lodge, which is north of Villice Bay (open 8.45am to 4.45pm on weekdays, also at weekends from early April to mid-October). Day permits (£2) may be obtained from the refreshment area by the dam and bird wardens. Permits cover access to Blagdon Lake and other local waters.

FURTHER INFORMATION

Grid refs: Woodford Lodge ST 565 607; Heron's Green ST 554 593; Herriott's Bridge ST 571 581
Permits: Bristol Water, Recreation Dept, Woodford Lodge, Chew Stoke, Bristol BS18 8XH: tel: 01275 332339; web: www.bristolwater.co.uk

51 PORTLAND HARBOUR, FERRYBRIDGE AND THE FLEET (Dorset)

OS Landranger 194

These areas are best in autumn and winter for a variety of waterfowl, waders and gulls, and can be conveniently combined with a visit to Radipole Lake and Portland Bill.

Habitat

Portland Harbour is a large, artificially enclosed harbour lying between Weymouth and the Isle of Portland – not actually an island as it is connected to the mainland by the unique Chesil Beach, a great 18 mile long wall of shingle. Between the Beach and the coast is a narrow strip of tidal water known as The Fleet, which enters Portland Harbour at Ferrybridge.

Species

The harbour is best in calm weather, November–March. All three divers are regular in small numbers, Red-throated being least frequent. Black-necked Grebe haunts the vicinity of Sandsfoot Castle while Slavonian is more widespread. A variety of seaducks occurs including Red-breasted Merganser and Goldeneye (both common) and a few Common Eiders, Long-tailed Ducks and Common and Velvet Scoters. The bushes and gardens around Sandsfoot Castle usually hold small numbers of wintering Chiffchaffs, Blackcaps and Firecrests. Brent Geese are frequent at Ferrybridge, but the bulk of the flock winters further west on The Fleet. An

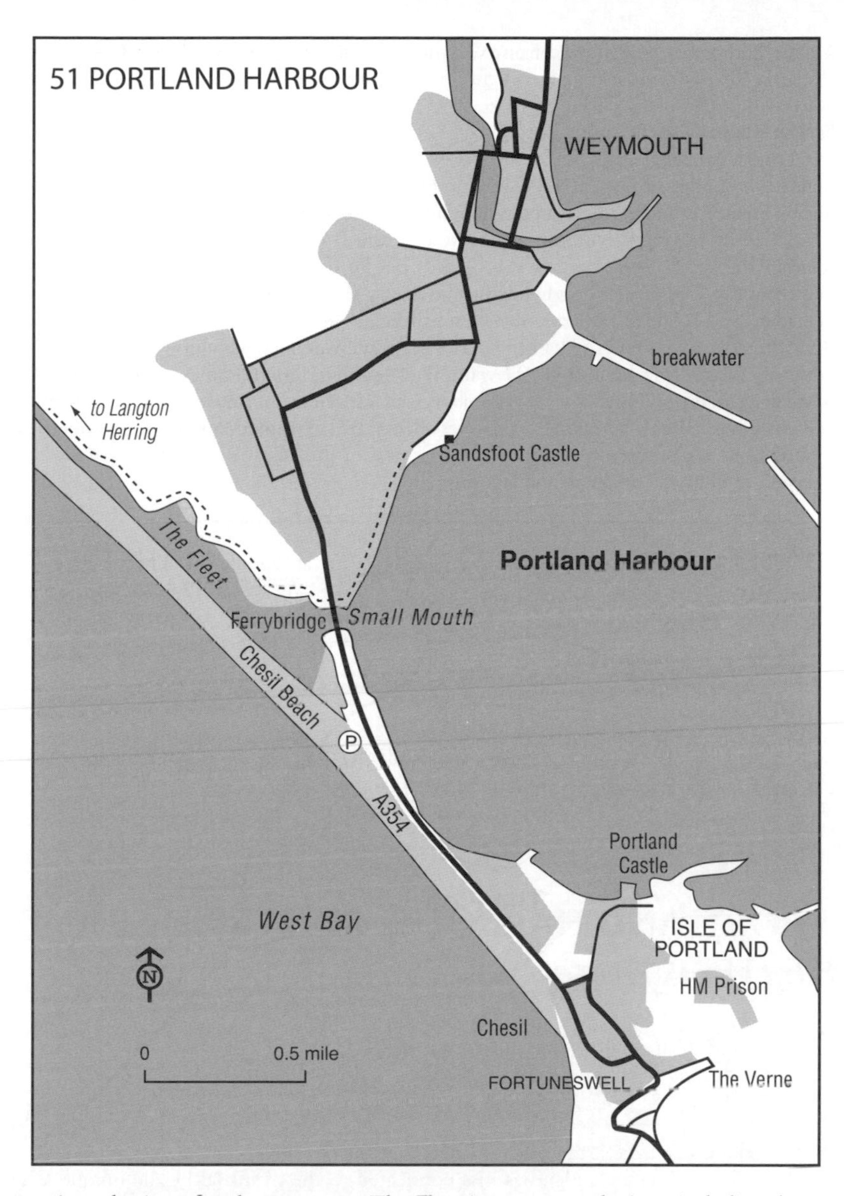

interesting selection of waders occurs on The Fleet in autumn and winter including the occasional rarity such as Kentish Plover. Gulls and terns should always be checked as several of the rarer species have made appearances, and Mediterranean and Little Gulls are regular.

In spring and autumn the Subtropical Gardens and the bushes and reedbeds along the beach at Abbotsbury hold a good selection of migrants. In winter the lagoon here is a haven for large numbers of wildfowl, and Bearded Tit can frequently be seen in the reeds.

Abbotsbury Swannery hit the headlines in early 2008 after a number of dead swans were found to be infected with bird flu. The area has since been given the all-clear.

Access

There are four main points of access.

1. The harbour is best viewed from Weymouth at the ruins of Sandsfoot Castle on Old Castle Road. It can also be seen from Ferrybridge but this area is often much disturbed by windsurfers.
2. The mudflats at the south-east end of The Fleet can be checked from Ferrybridge and there is a car park nearby which overlooks it. North of Ferrybridge a footpath follows the north-east edge of The Fleet to Langton Herring before turning inland.
3. The Fleet near Langton Herring is another good area for waders and wildfowl, and can be reached by leaving Weymouth north-west on the B3157. Turn left at a mini-roundabout near Chickerell, park at a gate c.100 yards before the Moonfleet Hotel, and follow the footpath to The Fleet. The bay north of Herbury Gore sometimes holds a good variety of waders, and Rodden Hive can also be rewarding.
4. At the north-west end of The Fleet is Abbotsbury Swannery, which is excellent for a wide variety of wildfowl besides Mute Swans. The Swannery lies 0.5 miles south of Abbotsbury village but is only open May–September. Outside these months the area can be viewed from Chesil Beach by taking a turning off the B3157 signed 'Subtropical Gardens' 100 yards west of Abbotsbury village. From the end of this road you can walk left along the beach to view the reedbeds and lagoons.

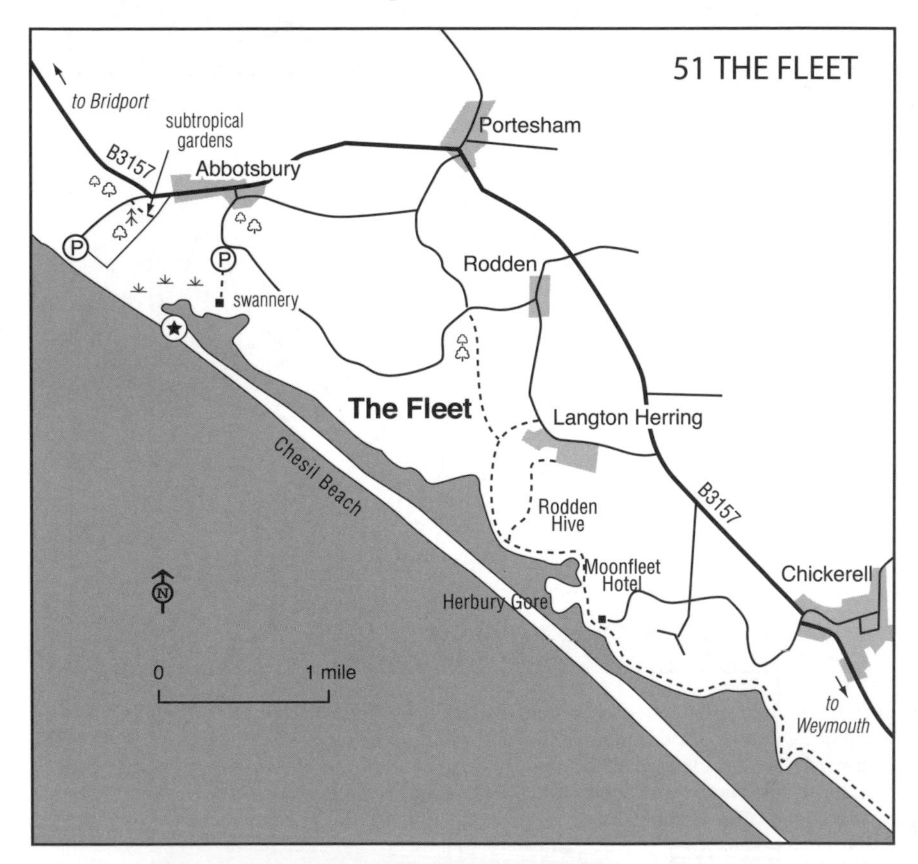

FURTHER INFORMATION

Grid refs: Sandsfoot Castle SY 674 773; Ferrybridge SY 668 754; Langton Herring SY 614 824; Abbotsbury Swannery SY 575 841

52 PORTLAND BILL (Dorset)

OS Landranger 194

The Isle of Portland protrudes c.6 miles into the English Channel from the Dorset coast, the south tip being known as Portland Bill. An active Bird Observatory is maintained at the Bill in recognition of its importance as a migration watchpoint for both landbirds and seabirds. Many extreme rarities have been found on Portland. The best times to visit are March–May and August–October.

Habitat

At first sight Portland appears uninviting being virtually treeless and heavily scarred by limestone quarries. Much of the north part of the island is settled or quarried but the south third is largely rural. This is the observatory's recording area, with the village of Southwell forming the north boundary. The various habitats include drystone-walled fields, clumps of bushes, hedgerows, grassy commons and disused quarries that have become overgrown with a rich variety of plants; all are attractive to migrants. The West Cliffs rise to over 200 feet and are used by breeding seabirds, but the East Cliffs are low and of little interest. There is virtually no wader habitat.

Species

Winter can be dismal but a variety of divers, seaducks and auks occur offshore. Gannets sometimes appear in large numbers in January and Purple Sandpipers winter on the rocks around the Bill. A few Black Redstarts may be present.

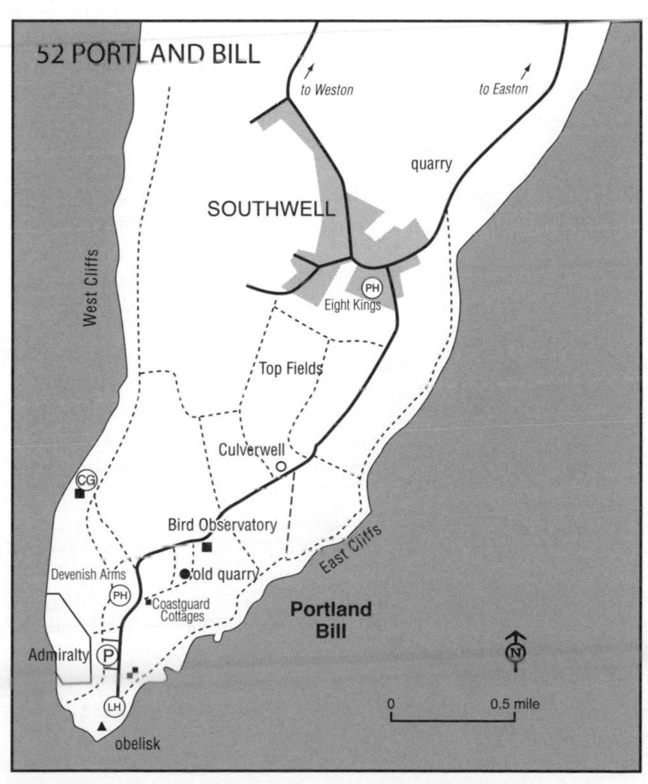

Spring seabird passage commences in March with divers, Gannet and Common Scoter heading east along the Channel. It continues throughout April into early May when Manx Shearwater, skuas and terns add to the variety. Many of the shearwaters are seen in the evenings. A small concentrated passage of Pomarine Skuas occurs in early May. Passerine migration also begins in March, and summer visitors arrive in force in April–May (Portland frequently hosts some of the earliest spring arrivals in Britain), frequently including a 'spring overshoot' such as Hoopoe; small numbers of Ring Ouzels and Firecrests are regular (as in autumn). Rarities are a feature of the late spring, with occurrences of Egyptian Nightjar and Lesser Short-toed Lark among the most outstanding.

Summer is fairly quiet but small numbers of Common Guillemots, Razorbills and Puffins breed alongside Fulmar, Kittiwake and a few Shags. Rock Pipit is resident and Little Owl inhabits the quarries.

Autumn seabird passage is less predictable than spring, and the more pelagic species are only noted occasionally, usually when onshore winds prevail. Balearic Shearwater is annual in July–October but numbers vary. Sooty Shearwater, Grey Phalarope and Little Gull are recorded irregularly at this time, but Great and Arctic Skuas are more frequent. The bulk of the landbird passage is August–September when chats, warblers and flycatchers can be much in evidence. Migrants also include Turtle Dove, Swallow, House Martin, Yellow Wagtail, Whinchat and Pied Flycatcher. The chance of finding one of the scarcer migrants adds spice to Portland at this season. Melodious Warbler and Woodchat Shrike may appear after south winds and Tawny Pipit is annual in September. Ortolan Bunting is regularly found in the Top Fields, and anticyclonic conditions may bring one of the eastern drift migrants such as Wryneck, Bluethroat, Icterine and Barred Warblers or Red-breasted Flycatcher. A few of each are recorded most years and Portland's list includes many outstanding rarities from all compass points, including Alpine Swift, Allen's Gallinule, Pechora Pipit, Cliff Swallow, Booted and Orphean Warblers and Northern Waterthrush.

In late autumn Portland witnesses large-scale movements of partial migrants such as Skylark, Meadow Pipit, Starling, Chaffinch, Goldfinch and Linnet. Spells of cold weather at the onset of winter frequently result in hard-weather movements of birds heading out to sea to the west or south. These typically comprise Lapwing, Redwing and Fieldfare, but Golden Plover and Skylark may also move on. Small numbers of some finches migrate throughout the winter.

Access

Leave Weymouth south on the A354 signed to Portland and continue through Fortuneswell and Easton to Southwell. Turn left in Southwell to the Bill. The Observatory is sited in the Old Lower Lighthouse beside the road c.0.5 miles before the Bill. There is parking at the Observatory for residents but casual visitors should use the large car park north of the new lighthouse (parking is not permitted beside the road between Southwell and the Bill).

Generally, the most productive birding areas are within the Observatory's recording area but much of this is common land and has become very popular with non-birders at weekends; on Bank Holidays it can be positively crowded. The best times to search for migrants are early morning and late afternoon when there is least human activity. Cover is fairly limited and any patch of bushes may hold migrants, but several areas deserve specific mention:

1. Observatory garden, containing a variety of cover and a small pond. Trapping and mist-netting are carried out here. (Residents and members only.)
2. The bushes around the Coastguard Cottages.
3. Culverwell, a dense area of bramble and elder.
4. Any of the overgrown quarries, especially the one behind the Eight Kings pub in Southwell (the latter is private property).

The Top Fields are noted for attracting migrants, especially finches and buntings (but be aware that much of this is private agricultural land – keep to footpaths). Certain species, such as pipits and wheatears, favour the short turf of the commons. Seawatching is best at the Bill beside the obelisk, which offers some protection to this hardy pursuit. In spring and summer the West Cliffs hold one of the south coast's few seabird colonies; a footpath runs along the clifftop. If time permits, the Verne, at the north end of Portland, should be checked for migrants. This scrub-covered plateau is situated above the dockyard and below Verne Prison (see Portland Harbour map, page 89).

Portland Bird Observatory is open all year. Comfortable self-catering accommodation is available for up to 20 in six bedrooms and for another four in a small self-contained flat, all at reasonable rates. Members of Portland Bird Observatory and Field Centre pay reduced charges – further details from the warden.

FURTHER INFORMATION

Grid refs: Bill SY 677 684; Bird Observatory SY 681 690
Portland Bill Bird Observatory: Warden (Martin Cade), Old Lower Light, Portland Bill, Dorset DT5 2JT; tel: 01305 820553; email: obs@btinternet.com; web: www.portlandbirdobs.org.uk

53–54 RADIPOLE LAKE AND LODMOOR (Dorset)

OS Landranger 194

Radipole Lake is an RSPB reserve of 78 ha in central Weymouth. Lodmoor is another RSPB reserve of 61 ha on the coast just east of town. Both are worth visiting at any season. Residents include Cetti's Warbler and Bearded Tit and both areas attract many migrants and wintering birds as well as regular rarities.

Habitat

Radipole Lake was formerly the estuary of the River Wey but since the completion of the Westham Bridge in 1921 has slowly become a freshwater lake. Extensive reedbeds cover much of the area but the remaining open water includes both shallow areas and deep channels. Dense scrub borders the trails and the perimeter while at the north end there are some water meadows. Lodmoor has pasture, shallow water, reedbeds and dykes, and is separated from Weymouth Bay by a sea wall.

Species

Cetti's Warbler and Bearded Tit are resident at both sites, as are Kingfisher and a few Water Rails. During summer many Reed and Sedge Warblers breed as well several pairs of Grasshopper Warblers and Lesser Whitethroats in the areas of scrub.

A wide variety of migrants has occurred in spring and autumn, and there is always a chance of something unusual. Garganey is regular in spring and is one of the first-returning migrants. Waders are generally present in small numbers depending on water levels, and possibilities include Little Stint, Black-tailed Godwit and Wood Sandpiper. Little Gull and Black Tern occur but cannot be guaranteed. Thousands of hirundines, hundreds of Pied Wagtails and up to 3,000 Yellow Wagtails use the reeds for roosting during passage. Spotted Crake is found most years but is easily overlooked. Cattle Egret has been seen several times recently at Lodmoor. Ringing at Radipole has demonstrated that small numbers of Aquatic Warblers pass through annually, but this denizen of dense reedbeds is rarely seen; one or two are found each year by a lucky few, usually in August or early September.

Winter sees the arrival of relatively small numbers of ducks including some Gadwalls, Scaup and Goldeneyes. Generally Lodmoor is more attractive to wildfowl than Radipole. A few Jack Snipe winter, and an influx of continental Water Rails increases the chance of seeing one. Cetti's Warbler is usually easier to see in winter and a few Water Pipits are regularly present at Lodmoor, often remaining until April by which time some will be in summer plumage. Gulls are prominent at both sites and an impressive list of rare species has been recorded; Mediterranean and Ring-billed Gulls are noted regularly November–May and Glaucous and Iceland are occasionally found in winter. The gulls are best seen at roost in the evening, close to the Information Centre at Radipole.

Rarities at Radipole have included Pied-billed Grebe, Squacco and Purple Herons, Little Bittern and Red-rumped Swallow.

Access

Both reserves are within easy reach of Weymouth.

53 RADIPOLE LAKE (Dorset)

The reserve lies close to Weymouth bus and railway stations. A large car park is available at the south end of the lake. A Visitor Centre sited here is open all year, 9am–5pm (4pm in winter). Several raised trails (wheelchair accessible) transect the reedbeds, affording good views over the lagoons and reeds. The reserve is open at all times but the North Hide is only open 8.30am –4.30pm, with a fee for non-members.

54 LODMOOR (Dorset)

Leave Weymouth northwards on the coastal A353 to Wareham. The entrance to the Lodmoor reserve is at the public car park by the Sea Life Centre along this road. The reserve is open at all times and viewing areas are provided off the perimeter footpath.

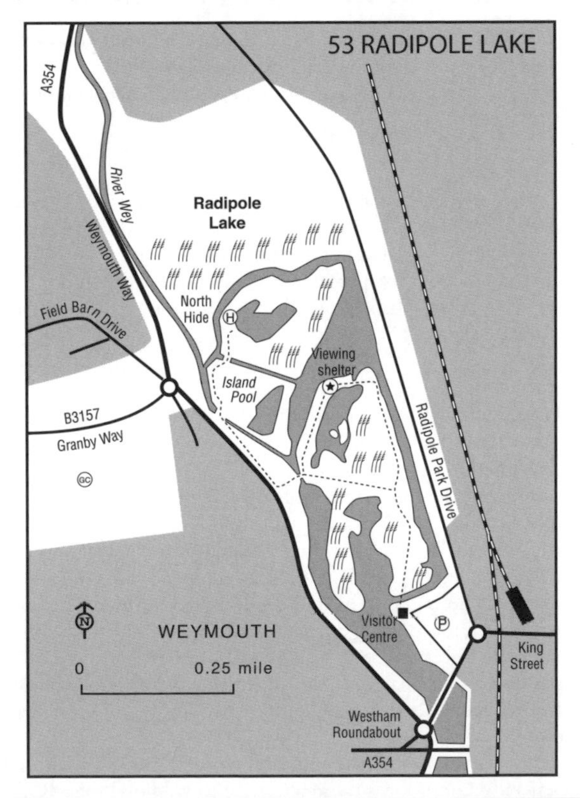

FURTHER INFORMATION

Grid refs: Radipole Lake SY 675 795; Lodmoor SY 688 809

RSPB Radipole Lake: Visitor Centre, The Swannery Car Park, Weymouth, Dorset DT4 7TZ: tel: 01305 778313; email: radipole.lake@rspb.org.uk; web: www.rspb.org.uk/radipolelake; www.rspb.org.uk/lodmoor

55–60 POOLE HARBOUR AND ISLE OF PURBECK

(Dorset) OS Landranger 195

Poole Harbour is the second largest natural harbour in the world. Its north and east shores have suffered considerable urbanisation at Poole and Bournemouth but the other sides are largely undeveloped. The harbour is good for wildfowl and waders in autumn and winter, and divers and grebes are often found in the coastal bays. To the south is the Isle of Purbeck (not actually an island), much of this being heathland and a stronghold for Dartford Warbler. Several areas are protected as nature reserves.

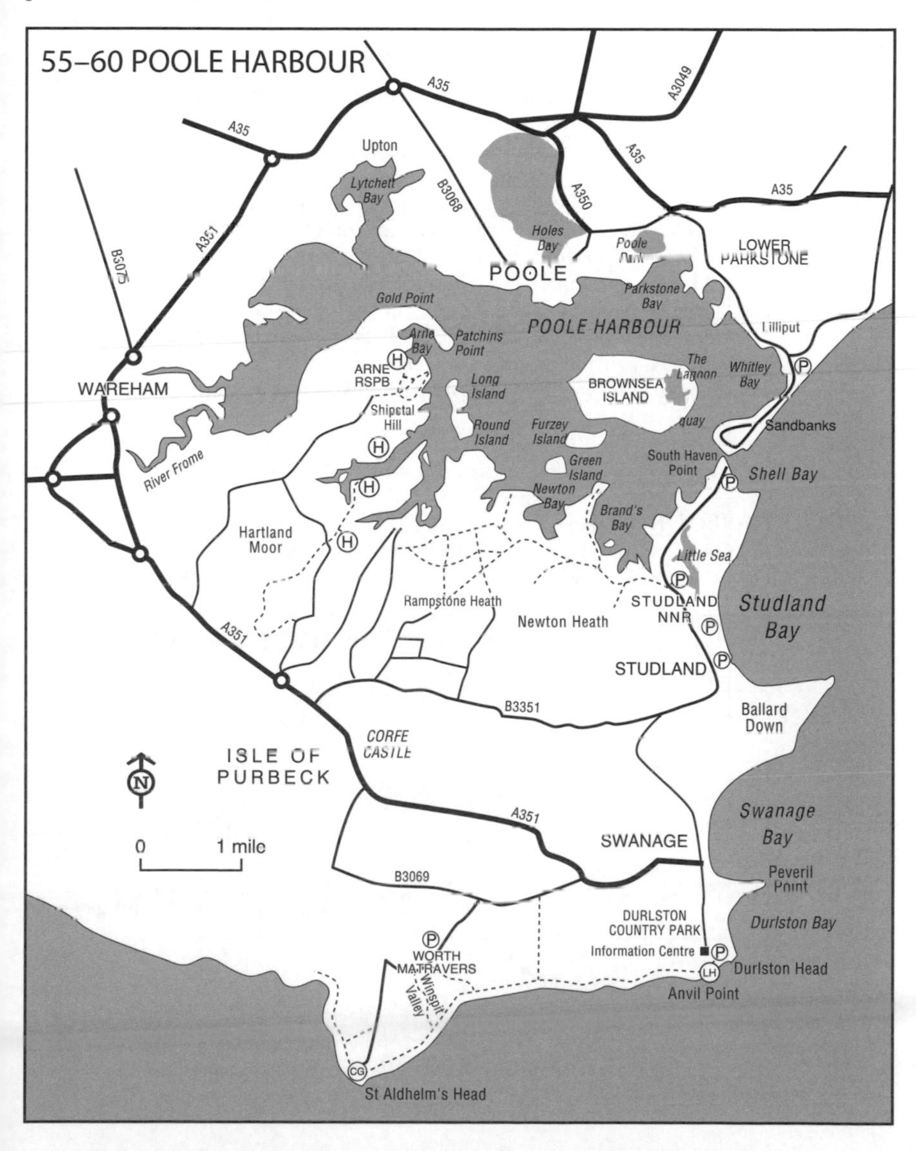

Habitat

Tidal mudflats and saltings surround the harbour, with the richest areas on the west and south shores. The open, dry heaths of the Isle of Purbeck are mainly heather and gorse with some bracken. There are small areas of both deciduous and coniferous woodland, carr and fresh marsh with reedbeds and small pools. The heaths have a particularly rich flora and fauna with many rare plants, all the British reptiles, and a host of insects.

Species

Winter is a good time to visit, and with luck you can see three species of diver and five species of grebe. Black-necked and Slavonian Grebes are regularly present, sometimes in the harbour but more frequently in the coastal bays. Seaduck also winter, mainly Red-breasted Merganser and Common Scoter. Cormorant, Shag and a few auks are also usually present on the sea. Wildfowl in the harbour include Brent Goose, Pintail and, in the channels, Goldeneye. Avocet, Black-tailed Godwit and Spotted Redshank winter (the former in good numbers), and Sanderling can be found on the shores of Studland Bay. Hen Harrier is occasional and Water Rail and Bearded Tit may be seen in the reedbeds at Arne or around Little Sea. Dartford Warbler is often silent and unobtrusive in winter. Black Redstart sometimes winters, usually on the coast, and in milder years one or two Firecrests and Chiffchaffs.

Spring and autumn bring a greater variety of waders to the harbour. Seawatching can be productive from the headlands, especially in spring, and the coastal bushes attract passerine migrants, including scarcities such as Wryneck, Icterine, Melodious and Yellow-browed Warblers, Red-breasted Flycatcher and Ortolan Bunting. Spoonbills are regular on Brownsea in autumn.

In summer Grey Heron, Little Egret and Common and Sandwich Terns nest on Brownsea Island, and Yellow-legged Gulls may also be present. Breeding birds of the cliffs include Fulmar, Shag, Kittiwake, Common Guillemot, Razorbill and Puffin. Dartford Warbler is more in evidence on the heath and Nightjar should be looked for at dusk. The small areas of woodland hold Sparrowhawk, Green and Great Spotted Woodpeckers and Marsh Tit.

Access

There are many places to birdwatch in this area. Six sites are detailed below:

55 POOLE HARBOUR (Dorset)

The harbour itself can be viewed from Arne and also at South Haven Point on Studland where, with the changing tides, duck, waders and grebes fly in from and out to sea. Brand's Bay, which

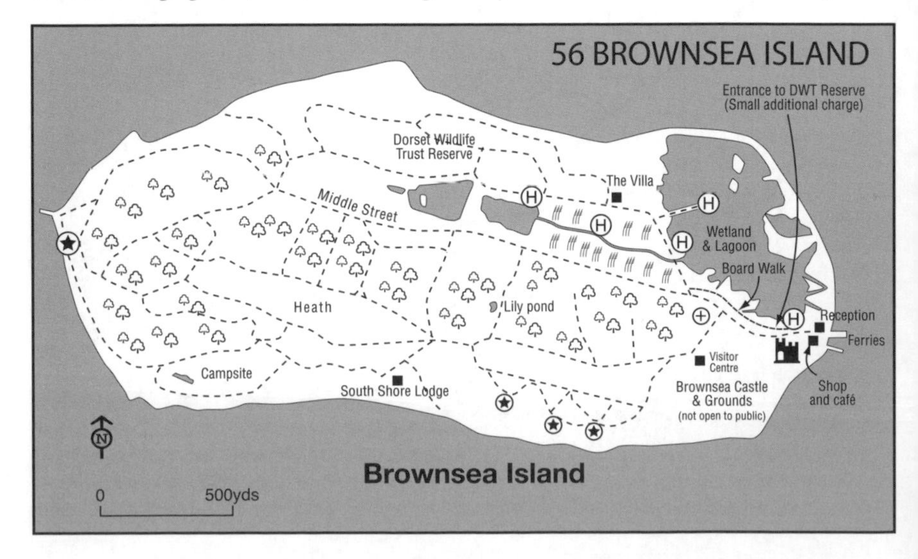

can also be viewed from Studland Heath, is probably the best area for wildfowl and waders. Gulls, waders and ducks frequent the north-east shore and the B3369 between Poole and Sandbanks can be excellent at low tide but it is very disturbed in summer.

56 BROWNSEA ISLAND (Dorset)

This 1.5 mile long island is the largest in Poole Harbour. It is owned by the National Trust, and the north part is managed as a reserve by Dorset Wildlife Trust. Its habitats include woodland, heath and freshwater pools and marshes. Several hides overlook the lagoon and marsh giving good views of the wildfowl, waders and terns. Large numbers of waders roost on Brownsea at high tide and there is a heronry (with Little Egrets as well as Grey Herons) and a population of Red Squirrels. Half-hourly boats to Brownsea Island go from Sandbanks (taking ten minutes) and Poole Quay (30 minutes); a landing fee is payable by non-NT members – a ferry fare is payable by all. The island is closed November–March.

57 ARNE (Dorset)

This peninsula protrudes into the south-west corner of Poole Harbour. The RSPB reserve of 529 ha consists mainly of heath but with some woodland and freshwater reedbeds and marsh. Leave Wareham south on the A351 and after 1 mile turn left at Stoborough; in c.2 miles the Reception Centre (open late May to early September) and car park are on the right. The reserve is open at

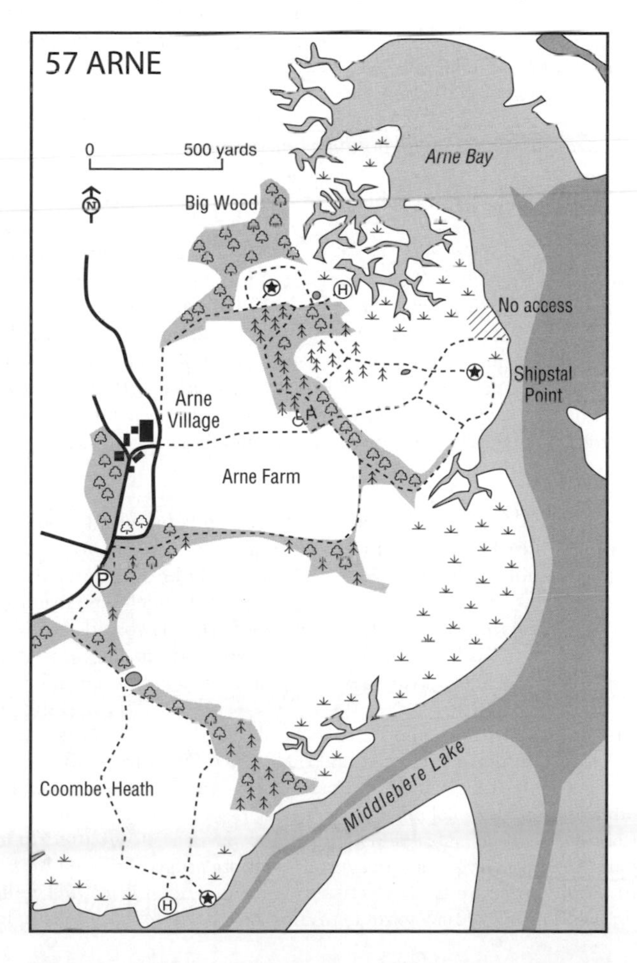

all times but visitors must keep to the footpaths, which are rather rough and rugged in places. There are two main trails: the Shipstal Trail and Coombe Heath trail. Between them they take in all the representative habitats, as well as giving fine views over the saltings and harbour; there is one hide on each trail. Both trails can be accessed from the RSPB car park or, alternatively, for the Shipstal trail you can continue beyond the Reception Centre and park in Arne village; the trail begins opposite Arne church. The reserve is one of the best places to see Dartford Warbler, which is present all year but often keeps hidden in the gorse, particularly in windy conditions.

58 STUDLAND HEATH (Dorset)

Much of this heath is an NNR managed by Natural England. The habitats are essentially as at Arne. Leave Wareham on the A351 and turn left on the B3351 at Corfe Castle. On reaching Studland, Ferry Road heads north to South Haven Point. A toll is payable on this road but there is a car park before it on the seaward side. The road continues through prime heath that is excellent Dartford Warbler country. The reserve lies on both sides of the road and access is unrestricted, but

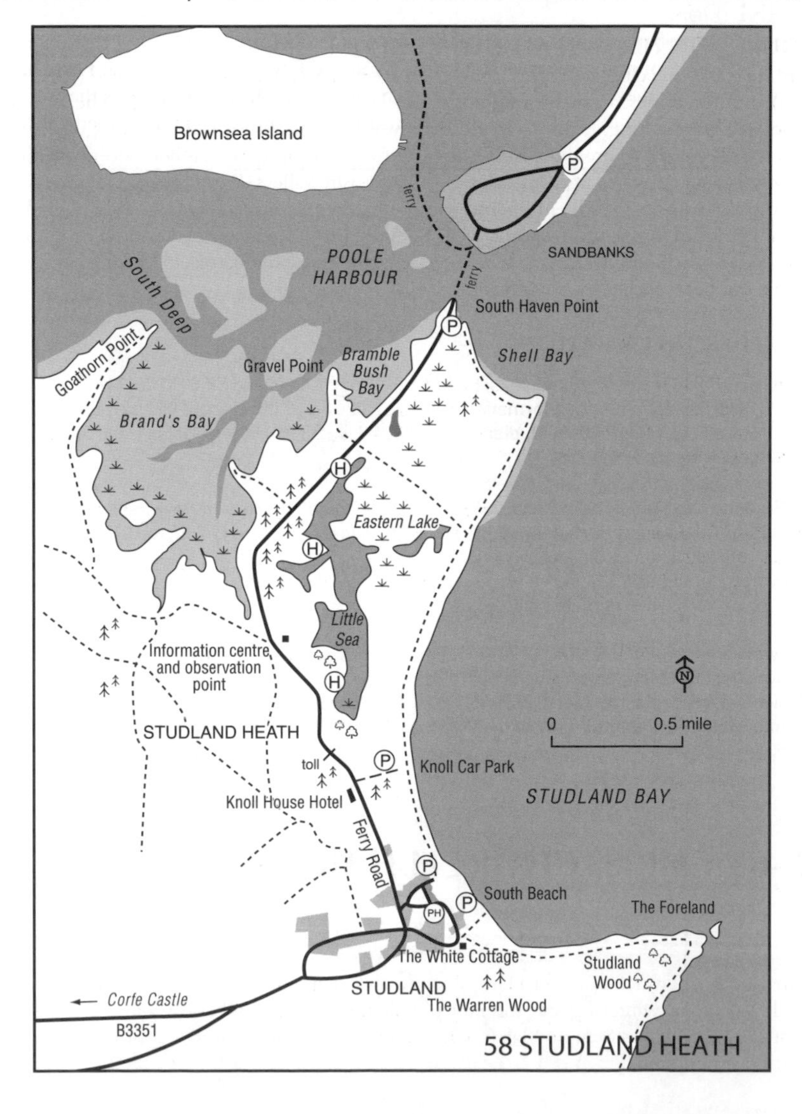

visitors should observe the notices at the reserve entrances. There are two nature trails: the Dune Trail is open year-round and the Woodland Trail operates April–September. A large freshwater pool (Little Sea) east of the road is attractive to ducks in winter. There is an Information Centre and an observation hut that offers a good view over the pool. A hide at the north end is always open. Alternatively, you can reach Poole from Bournemouth by taking the B3369 to Sandbanks and catching the ferry to South Haven Point (runs at 20–30 minute intervals).

59 SHELL AND STUDLAND BAYS (Dorset)

Both bays, north and east of Studland peninsula, are a favoured haunt of divers, grebes and sea-ducks in winter. Shell Bay is easily reached from the car park at South Haven Point. Studland Bay can be approached via the Knoll car park at the south end of Studland Bay or through Studland village; a minor road leads to the seafront where there is a car park. A bridleway from Studland village follows the clifftop to the Foreland. This gives good views over the bay and the two clumps of trees beside the path can harbour migrant passerines.

60 DURLSTON HEAD–ST ALDHELM'S HEAD (Dorset)

There are some fine cliffs along this stretch of the coast. A variety of seabirds breed (March–July is best), and passerines occur on migration, especially in the sheltered valleys. The Heads can be good for seawatching. A minor road from Swanage leads south to Durlston Country Park, which is well signed, where there is a Visitor Centre. Durlston Head has held several rarities and is heavily vegetated with large stands of trees and bushy valleys. From here you can walk west along the cliff path to St Aldhelm's Head (c.5 miles), the main interest of the walk being breeding seabirds including Puffin. Alternatively, St Aldhelm's Head may be reached on foot from Worth Matravers. Winspit Valley, a well-vegetated valley running between Worth Matravers and Winspit, can be excellent for migrants.

FURTHER INFORMATION

Grid refs: Poole Harbour (Sandbanks) SZ 044 876; Brownsea Island SZ 031 876; Arne SY 972 878; Studland Heath (Information centre) SZ 026 844; Little Sea SZ 027 849; Shell Bay/ South Haven Point SZ 035 861; Studland Bay SZ 034 834 / SZ 038 826; Durlston Head SZ 032 773; St Aldhelm's Head SY 960 756

RSPB Arne (Warden): tel: 01929 553360; email: arne@rspb.org.uk; web: www.rspb.org.uk/arne
Durlston: The Ranger, Durlston Country Park, Swanage, Dorset BH19 2JL: tel: 01929 424443; email: info@durlston.co.uk; web: www.durlston.co.uk
Brownsea Island (DWT Reserve Manager): tel: 01202 709445; email: dorsetwtisland @cix.co.uk; web: www.dorsetwildlife.co.uk

Brownsea Island (NT): tel: 01202 707744; email: brownseaisland@nationaltrust.org.uk; web: www.nationaltrust.org.uk/brownsea
Brownsea Island Ferries: tel: 01929 462383; web: www.brownseaislandferries.com
Studland Heath: Natural England: tel: 01929 450259; web: www.naturalengland.org.uk

61–63 CHRISTCHURCH HARBOUR (Dorset)

OS Landranger 195

Formed by the River Stour and Hampshire Avon, Christchurch Harbour is well known for its wide variety of wintering and migrant populations of wildfowl and waders. On the north side of the harbour lies Stanpit Marsh, while Hengistbury Head on the south side is a noted migration watchpoint.

Habitat

Much of the harbour is dominated by typical estuarine habitats such as mudflats and a shingle island, with pasture and marshy areas dissected by muddy channels, some of which have extensive reeds and sedges. A series of semi-permanent pools flank the landward edge of the marsh, while scrubby areas are dominated by gorse.

Hengistbury Head comprises a low-lying area of grassland and gorse scrub, a steeply sloping hill covered by open heath and densely vegetated areas dominated by birch and sallow merging into oak woodland at the base of the hill. The seaward side of the head consists of some unstable cliffs, a shingle beach and a low sandy spit at the point.

Species

In winter large numbers of wildfowl visit including Brent Goose, Wigeon and Teal, with smaller numbers of Gadwalls, Pintails, Shovelers and Goldeneyes. Commoner waders include Black-tailed Godwit, while up to 100 Common Snipe and a few Jack Snipe are present at Stanpit Marsh and Wick Hams. Purple Sandpiper frequents the breakwaters off Hengistbury Head. Large gull flocks host Mediterranean, Yellow-legged and Little Gulls on a regular basis. Reedbeds are home to Water Rail and the occasional Bittern. Good numbers of Little Egrets frequent the area throughout the year, while Rock Pipit, Stonechat and Cetti's and Dartford Warblers are also resident. Hengistbury Head may hold wintering Chiffchaff and Firecrest.

Migration periods bring numbers of waders in both spring and autumn. These include Little Ringed Plover, Ruff, Little Stint, Curlew Sandpiper, both godwits, Whimbrel, Spotted Redshank, Greenshank, and Green, Wood and Common Sandpipers. Kentish Plover is almost annual. Garganey sometimes joins the flocks of ducks. Raptor passage includes small numbers of Hobbies and occasional Marsh Harrier and Osprey. Seabird passage may occur during strong south-east to south-west winds and can include divers, Fulmar, Manx and Sooty Shearwaters (autumn), occasional Grey Phalarope, skuas, Little Gull, Black Tern and auks. Landbirds are most obvious at Hengistbury Head with Water Pipit, Black Redstart, Ring Ouzel, Pied Flycatcher,

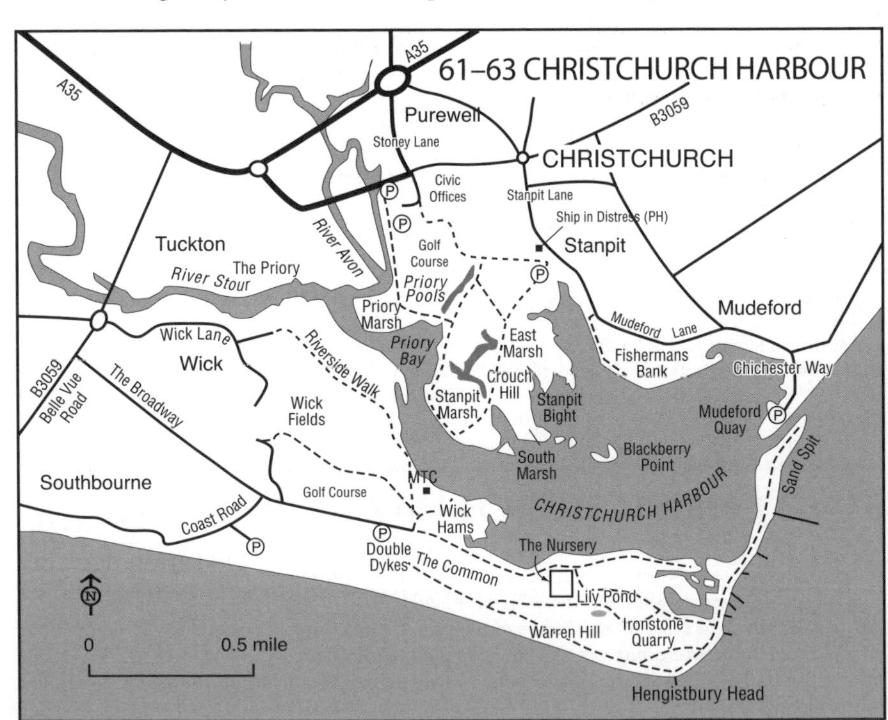

Firecrest and Wood Warbler all regular. Oddities include almost annual Spotted Crake, the occasional Spoonbill, Temminck's Stint, Hoopoe and Serin.

Breeding birds hold relatively little interest though good numbers of terns may be present and sometimes include Roseate. Reed and Sedge Warblers breed, while irregular offshore passage at sea in summer can include Manx Shearwater, Storm Petrel and Gannet.

Access

There are three main access points, with Stanpit and the north harbour both reached from Christchurch, and Hengistbury Head accessed via Southbourne.

61 CHRISTCHURCH HARBOUR (Dorset)

South of Christchurch, take Mudeford Lane and turn south into Chichester Way to reach Mudeford Quay. Alternatively take Argyle Road, which can be reached off Stanpit Lane to access Fisherman's Bank. Both will give access to the northern side of the harbour.

62 STANPIT MARSH (Dorset)

Accessed from the Recreation Ground car park, which is off Stanpit Lane south of the Ship in Distress pub. Alternatively park in the Christchurch Swimming Pool car park off Stony Lane South and follow the footpath south along the edge of the golf course to the marsh. The final option is to park in the Christchurch Civic Offices car park and follow the path south between the river and the golf course. Crouch Hill offers good views of the high-tide wader roost on Stanpit East Marsh.

63 HENGISTBURY HEAD (Dorset)

Hengistbury Head is reached from Southbourne. Park in the car park at the end of The Broadway. There is access to most of the headland. The woodland near the Old Nursery Garden is good for passerines. Seawatch from the beach huts or Warren Hill, and check the end of the spit opposite Mudeford for bird movements in and out of Christchurch Harbour. Wick Hams (a small marsh) and Wick Fields (rough pasture) are also worth checking.

FURTHER INFORMATION

Grid refs: Christchurch Harbour (Mudeford Quay) SZ 183 917; Stanpit Marsh SZ 170 923; Hengistbury Head SZ 162 911

Stanpit Marsh: Christchurch Countryside Service: tel: 01425 272479; email: countrysideservice@christchurch.gov.uk

Christchurch Harbour Ornithological Group: web: www.chog.org.uk

SOUTH-EAST ENGLAND

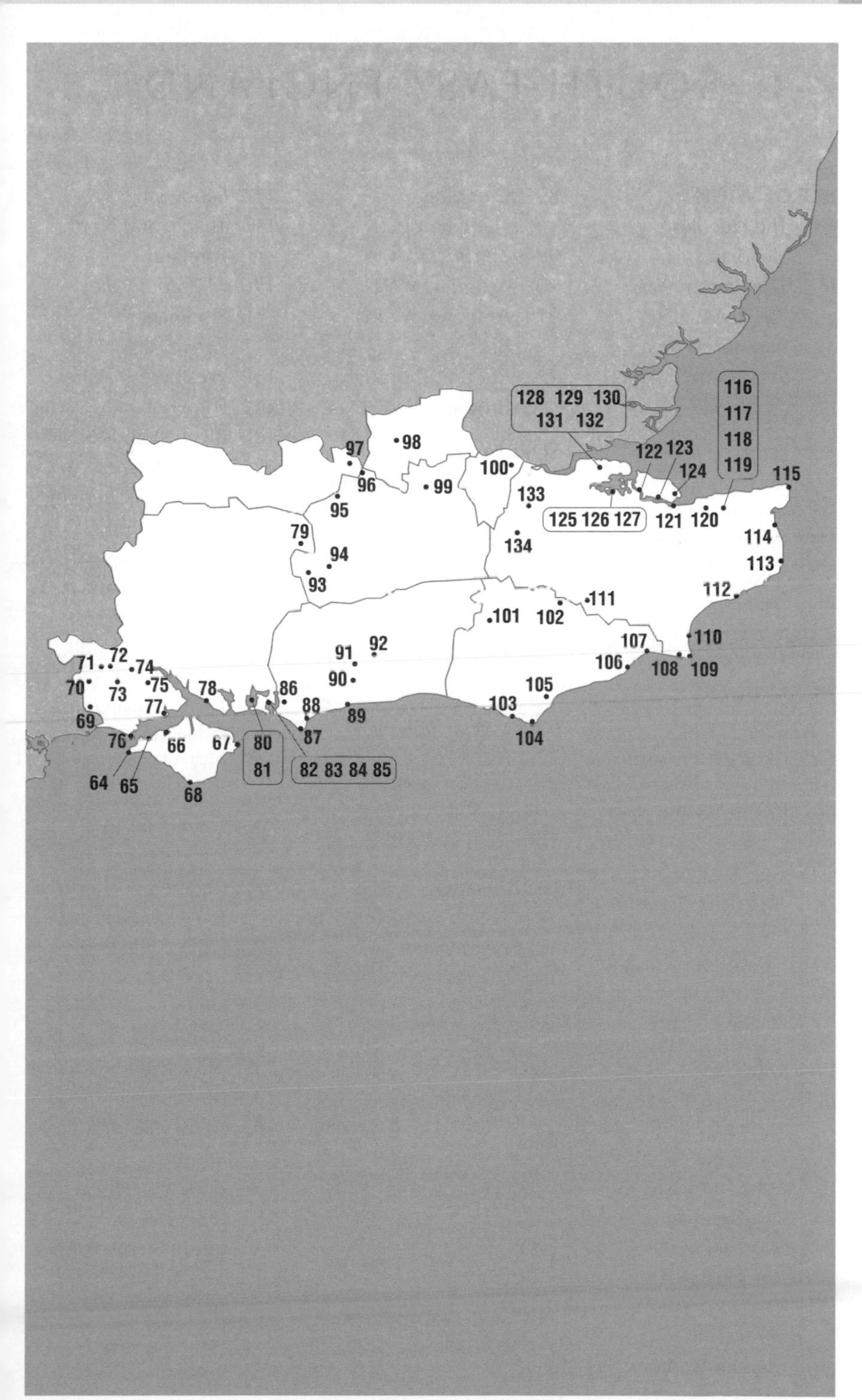

128 129 130
131 132
116
117
118
119
115
98
97
96
99
100
133
122 123
124
125 126 127
121 120
95
79
94
134
114
93
113
112
111
101
102
110
107
106
108
109
91
92
90
71
72
74
70
73
75
78
86
88
89
105
103
104
69
77
76
66
67
80
81
87
82 83 84 85
64
65
68

64 THE NEEDLES HEADLAND (Isle of Wight)

OS Landranger 196

The Needles – a famous series of chalk stacks that project into the Channel at the western tip of the Isle of Wight – mark the end-point of a scenic and biodiverse headland.

Habitat

From the chalk cliffs themselves to the scrub-filled gullies and heathy hilltops, this National Trust-owned area holds a variety of habitats with much to offer the birder. Seabirds breed on the high cliffs, and migrating landbirds are attracted to the sheltered, well-vegetated slopes. In autumn gales and in winter, seabirds may shelter in Alum Bay. The warm downland sward that covers much of the ridge of West High and Tennyson Downs is botanically rich and therefore attractive to insects, while further inland patches of woodland and farmland are also worthy of exploration.

Species

As spring migration gets under way, passerines such as Wheatear, Whinchat, Common Redstart, Yellow Wagtail and warblers begin to pass through, along with scarcer species such as Black Redstart and Ring Ouzel. Overshooting continental species are also found most years, Hoopoe being the most frequent, and Collared Flycatcher and Alpine Accentor the most notable recent rarities.

Breeding seabirds on the chalk cliffs include Fulmar, Cormorant, Shag, Common Guillemot and Herring and Great Black-backed Gulls. Other cliff breeders include Peregrine, Raven, Rock Pipit and the odd pair of Wheatears. Further inland, a few pairs of Stonechats and Dartford Warblers breed in the heathy areas.

Autumn landbird migration is more pronounced than in spring, with Pied Flycatchers and Firecrests regular in small numbers alongside more numerous Skylarks, Meadow Pipits, finches, numerous hirundines and other commoner species. Ring Ouzels are a particular feature, with day counts regularly reaching 20 birds. Rarities arriving at this time have included Hume's Warbler, and Yellow-browed Warbler is virtually annual.

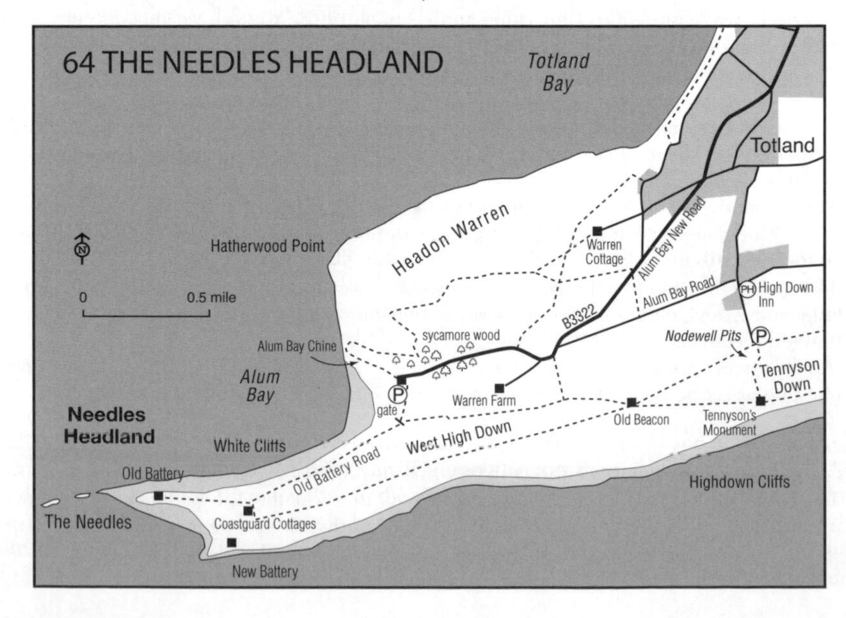

The Needles area is too elevated to offer good seawatching, but the occasional seabird may be seen passing the headland, while Alum and Freshwater Bay has offered shelter to Little Auk, Grey Phalarope and Leach's Petrel during spells of bad weather.

Winter is a quiet time but Black Redstart and Firecrest may be found, with the odd sea-duck offshore.

Access

The area is reached from the B3322, south-west from the junction of the A3054 and A3055 in Freshwater. From the large car park at Alum Bay, there are numerous footpaths from which the bay itself and the headland may be explored.

FURTHER INFORMATION

Grid ref: SZ 307 853

65 THE WESTERN YAR (Isle of Wight)

OS Landranger 196

This beautiful estuary on the north-west coast of the island is a great place for winter wildfowl-watching, with a good selection of breeding woodland and wetland birds.

Habitat

The estuary of the Western Yar, west of the town of Yarmouth, is designated an SSSI as well as being part of the Isle of Wight's Area of Outstanding Natural Beauty. The estuary mouth has extensive *Spartina* saltmarsh cut through with numerous muddy creeks, with areas of reedbed and marshland grading into damp woodland and scrub further inland.

Species

Winter wildfowl visiting the saltmarshes and channels of the Western Yar include substantial flocks of Brent Geese and Wigeon, along with Shelduck, Teal, Shoveler and Gadwall. Diving ducks include Pochard, Tufted Duck and, in harder weather, perhaps Goldeneye, Red-breasted Merganser, Scaup and Goosander. At low tide commoner waders such as Grey Plovers and Black-tailed Godwits forage on the exposed mud, along with a few Greenshanks and Spotted Redshanks. Little Egrets are present all year, and Mediterranean Gulls may join the commoner species.

Wader numbers build up through the spring passage period, with Whimbrel and Little Ringed Plover regular. Other migrants passing through at this time may include Garganey, Osprey, Yellow Wagtail and Whinchat.

In summer the woods and reedbeds hold a good selection of breeding warblers, including Cetti's, and many other common passerines. Breeding waterbirds include Little Grebe and Common Redshank.

Autumn sees another significant arrival of migrating waders, along with the chance of an interesting passerine migrant – Marsh Warbler and Golden Oriole are both possibilities.

Access

Its proximity to Yarmouth Harbour makes this site readily accessible to visitors from the mainland as well as residents and longer-term visitors on the Isle of Wight itself. From Yarmouth Harbour take the A3054 (River Road) south-east past the car park, then north-east a short distance before taking Mill Road south-east to Yarmouth Mill. From here, a footpath south along the sea wall offers excellent views across the estuary and marshland.

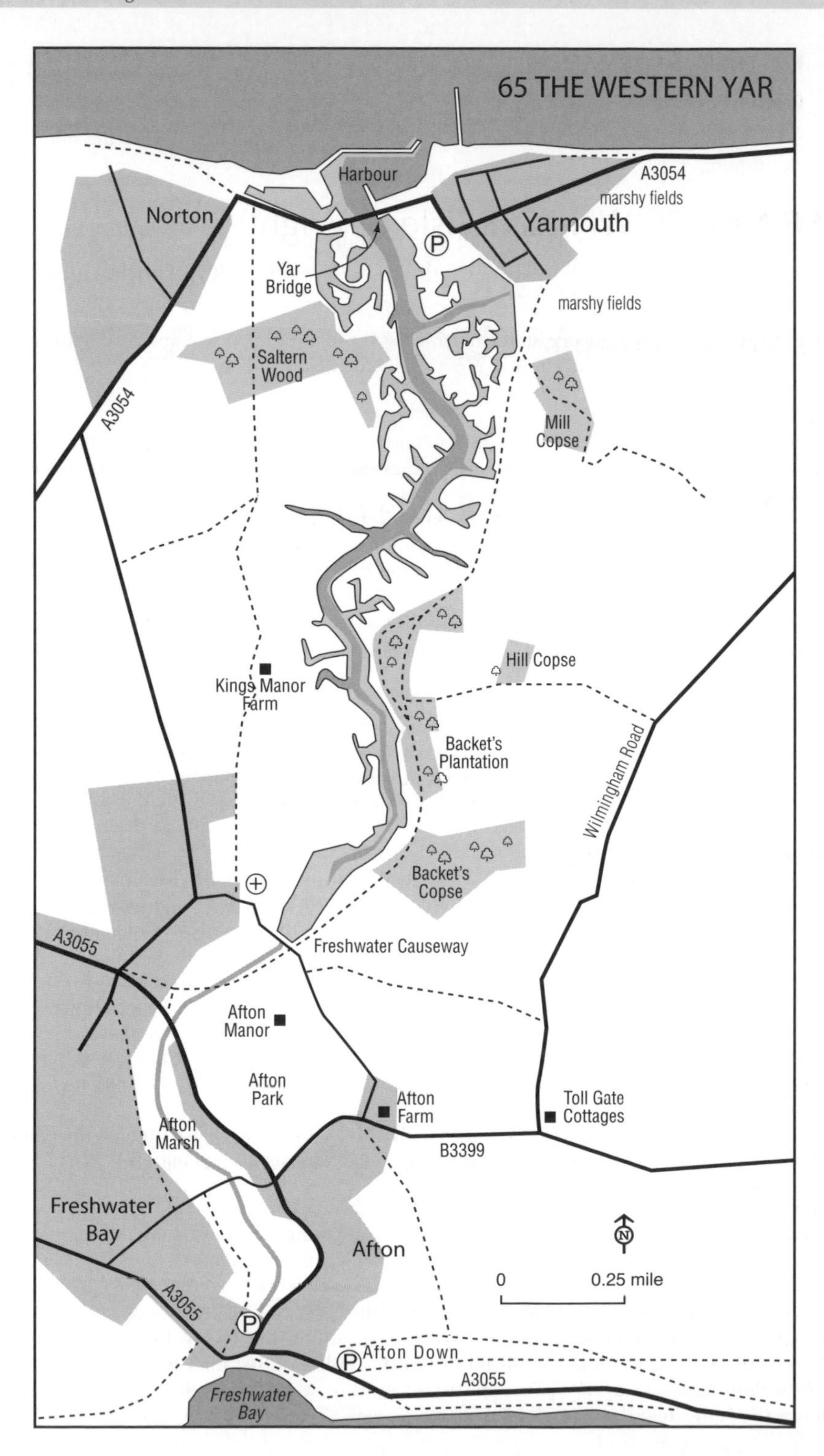
65 THE WESTERN YAR
Harbour
A3054
marshy fields
Norton
Yarmouth
Yar Bridge
marshy fields
Saltern Wood
A3054
Mill Copse
Hill Copse
Kings Manor Farm
Backet's Plantation
Wilmingham Road
Backet's Copse
A3055
Freshwater Causeway
Afton Manor
Afton Park
Afton Farm
Toll Gate Cottages
Afton Marsh
B3399
Freshwater Bay
Afton
0
0.25 mile
A3055
Afton Down
A3055
Freshwater Bay

FURTHER INFORMATION

Grid ref: SZ 353 895

66 NEWTOWN NNR (Isle of Wight)

OS Landranger 196

This large and complex estuary on the north coast is the most important and productive wetland site on the Isle of Wight. The area is owned by the National Trust.

Habitat

A breach of the sea wall in 1954 led to the flooding of a large area of grazing land – subsequently this area has reverted to *Spartina* saltmarsh and estuarine mudflats, with a network of broad channels along the course of the Newtown River and, to the east, the large Clamerkin Lake. The 'Main Marsh' at the centre of the complex is partially protected from the incoming sea by a low sea wall, which means it is exposed for longer spells and therefore more available for feeding birds. A man-made scrape just south of the Main Marsh offers more sheltered feeding and breeding opportunities.

A sand and shingle spit known as Hamstead Duver extends across the western side of the estuary entrance. Inland, 22.5 ha of farmland around Clamerkin Lake have recently been added to the reserve. Several hides overlook the marshes.

Species

As at the Western Yar, Brent Geese, Wigeon, Teal and Shelduck dominate the wintering wildfowl on the saltmarsh and surrounding fields along with smaller numbers of Shovelers and Pintails. Wild swans and White-fronted Geese may visit during colder spells. Diving ducks favour the deeper channels and Clamerkin Lake, especially in hard weather when seaducks may join the Goldeneyes and Red-breasted Mergansers. These waters may also attract divers and rarer grebes.

Wintering waders include numerous Golden Plovers, Knots and Black-tailed Godwits, as well as a few Bar-tailed Godwits, Greenshanks, Spotted Redshanks, Jack Snipe and occasionally Avocets. Little Egrets are now as much a feature of the area as the Grey Herons and Cormorants, and are present year-round.

Such concentrations of birds attract predators – Peregrines in particular are seen year-round, with Merlin and perhaps Hen Harrier and Short-eared Owl most likely in winter. Ospreys may stop off on their return migration, and Hobbies and Marsh Harriers also pass through at this time.

Passage periods see influxes of common waders with the chance of something scarcer, especially in autumn – Little Stint, Curlew Sandpiper, Green and Wood Sandpipers and Avocets are all found most years.

Wetland birds breeding at Newtown include Shelduck, Oystercatcher, Common Redshank and Common Tern, while the pockets of woodland have breeding Nightingales.

Access

Take the A3054 east of Shalfleet, and at the Shell garage turn left onto a minor road north-east towards Porchfield for 0.7 miles. Next take another minor road north (signed 'Newtown Old Town Hall'). In Newtown, take the first road on the left and continue past the reserve visitor centre and park in the designated area at the west end of the village. From here, a footpath leads towards Newtown Quay. Alternatively, continue along the minor road through Newtown village, and take the footpath that heads north 0.4 miles beyond the sharp right-hand bend.

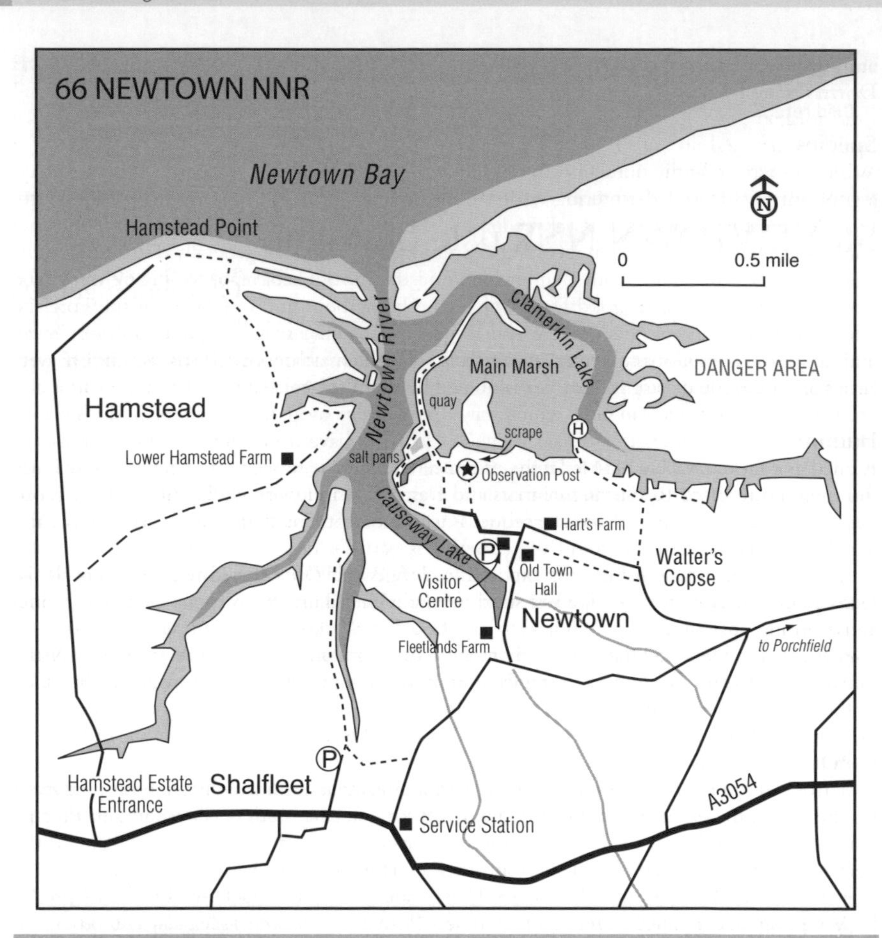

FURTHER INFORMATION

Grid ref: SZ 423 905

67 BEMBRIDGE AREA (Isle of Wight)

OS Landranger 196

The eastern corner of the Isle of Wight holds a number of interesting and diverse sites. This area is partly owned by the National Trust, and Brading Marshes is now an RSPB reserve; the remainder is in private ownership.

Habitat

Low tide at Bembridge Harbour reveals extensive mudflats and sandbanks, with saltmarshes and muddy creeks sheltered behind the sea wall. The Eastern Yar river flows into the harbour. North of the harbour lies Bembridge Sands, with Bembridge Pools, the larger Bembridge Pond and the rough wet pastures of RSPB Brading Marshes to the south. Overlooking the eastern approaches to the Solent is the Foreland, a low headland with seawatching potential, and offshore lie several rocky ledges extending south towards the high chalk ridge of Culver Down (National Trust).

Species

While not attracting the numbers found at Newtown NNR, the coastal parts of this area have a good variety of regular wintering waders – most notably Purple Sandpiper with 10–20 birds recorded most years on Black Rock Ledge just south of the Foreland. Long-billed Dowitcher and Lesser Yellowlegs have been recorded. Commoner wildfowl are generally distributed in the Bembridge area in winter, with Bembridge Pond the most reliable spot for diving ducks and Gadwall. This pond is also good for Water Rail. The harbour and foreshore off the Foreland attract large numbers of Brent Geese and Wigeon with other species such as Bewick's Swan and White-fronted Goose turning up in spells of cold weather. Grey Herons, Little Egrets, Shags and Cormorants are present here and elsewhere in the area all year. The Foreland is also the place for Mediterranean Gull, which may reach 100 or more in late winter – this is one of Britain's premier sites for this species. Ring-billed Gulls have been found in the large Black-headed and Common Gull roost off the entrance to Bembridge Harbour. Sheltered spots may hold wintering Black Redstarts, Firecrests and Dartford Warblers, while Bearded Tit occasionally winters – and has bred – in the reedbeds of Bembridge Pond. Kingfisher and Rock Pipit are also evident in winter.

Alongside the usual spring migrants, Little and Black Terns are sometimes seen, as are Marsh Harrier, Garganey and waders such as Sanderling, Ruff and Whimbrel. Spring rarities at Bembridge have included Night Heron, Woodchat Shrike and Little Swift. The Foreland has potential as a seawatching point – there are recent spring records of Pomarine Skua, Manx Shearwater and Storm Petrel, with single autumn records of Cory's and Sooty Shearwaters.

Interesting breeding birds in the area include Little Grebe, Tufted Duck and Pochard at Bembridge Pond, warblers including Cetti's at Brading Marshes, and Fulmars on the cliffs at Culver Down.

Culver Down is also a good place to observe autumn landbird migration, with numerous records of scarcities such as Wryneck, Tawny Pipit, Icterine Warbler, Common Rosefinch and Ortolan Bunting as well as good numbers of commoner species. The area also has the distinction of having hosted not one but two Desert Warblers, in 1988 and 1991.

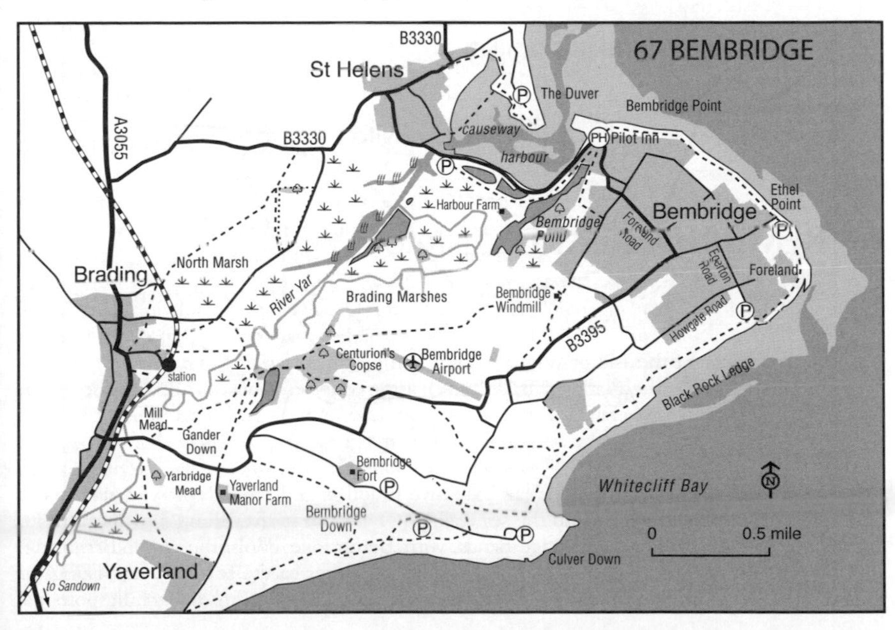

Access

The various sites in this area are accessed by footpaths off the B3395 and B3330. Bembridge Harbour and Bembridge Pools can be viewed from the roadside of the B3395 (Harbour Strand).

FURTHER INFORMATION

Grid refs: Bembridge Harbour and Pools SZ 635 884; The Duver SZ 636 891; Brading Marshes SZ 609 868; Culver Down SZ 636 855
RSPB Brading Marshes: tel: 01273 775333; web: www.rspb.org.uk/bradingmarshes

68 ST CATHERINE'S POINT (Isle of Wight)

OS Landranger 196

Yet another scenic spot in the National Trust's extensive portfolio, this headland at the southern tip of the island offers great seawatching and has an exceptionally interesting flora and fauna.

Habitat

The rocky shore and low, crumbling cliffs at St Catherine's Point are backed by a network of small, grassy fields, with dense scrubland further inland. North-west lie the high crags of Gore Cliff, with steep wooded gullies leading down from the crags to the shoreline.

Species

Spring sea passage begins in mid-March, and involves divers, Fulmars, Manx Shearwaters, Gannets, waders, terns and Common Guillemots, with regular reports of Arctic, Pomarine and Great Skuas, Velvet Scoters, Razorbills, Puffins and occasionally Balearic Shearwaters, primarily heading up-channel. Onshore southerly or easterly winds are most likely to produce good numbers.

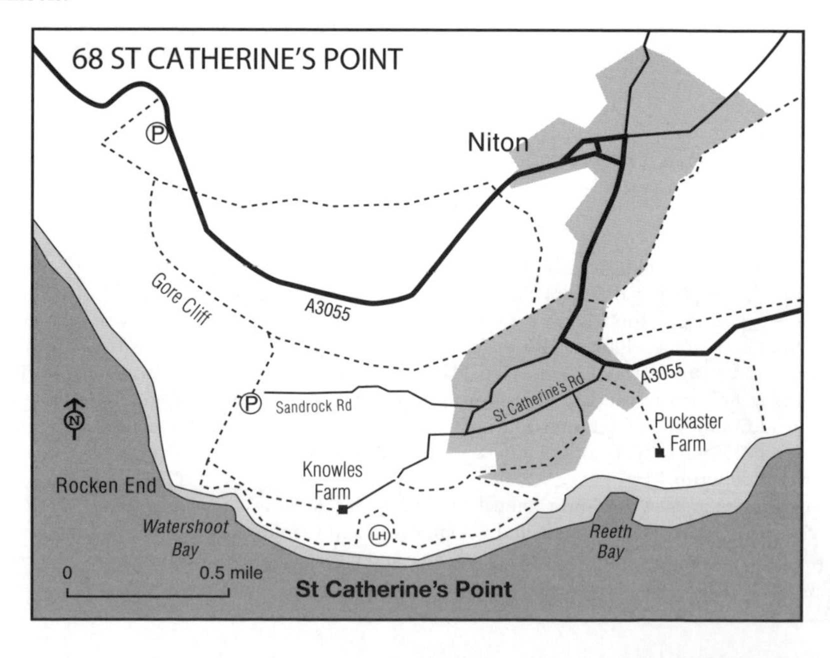

Landbird migration is also significant, with commoner species such as Yellow Wagtail, Whinchat, Wheatear, Common Redstart and various warblers sometimes joined by a scarcity or two – Serin is almost annual, and Alpine Swift, Bee-eater, Red-rumped Swallow and Melodious Warbler have all been recorded.

Typical coastal breeders such as Peregrine, Raven, Rock Pipit and Stonechat enliven an otherwise rather quiet birding scene in summer, while butterfly-watchers come here in search of the Glanville Fritillary – in the UK this species is restricted to southern undercliffs on the Isle of Wight.

Autumn seabird migration is less notable than in spring, but autumn landbird passage can be dramatic. Besides the common species, scarcities seen at this time have included Richard's Pipit, Bluethroat, Pallas's and Radde's Warblers and Red-breasted Flycatcher.

Winter is quiet but seawatching may produce divers, seaducks, Gannets and Kittiwakes.

Access

Take the A3055 (Undercliff Road) from Niton, then turn south-west down St Catherine's Road. At the end of the road, walk along the main track to the lighthouse. Alternatively, follow the road back north towards Niton and take Sandrock Road heading west to a small car park below Gore Cliff.

FURTHER INFORMATION

Grid ref: SZ 494 758

69 LOWER HAMPSHIRE AVON (Hampshire/Dorset)

OS Landranger 195

Much of this area is a designated SSSI that supports an excellent flora and fauna. Rising in the Vale of Pewsey the river derives much of its water from chalk aquifers in the Salisbury Plain, and the Hampshire Avon possesses a fine selection of riverine and wetland habitats.

Habitat

A broad flood plain is dominated by wet meadows that flood in winter. Marshy thickets of willows and alder, and in some places reedbeds, fringe the river. The Avon joins the Stour in Christchurch Harbour where there is an estuary.

Species

Good numbers of wildfowl occur during winter, mostly the commoner duck species, but perhaps also including Bewick's Swans and White-fronted Geese. Small flocks of the latter two species move around various sites in the valley but may be hard to locate. Flocks of Lapwings and Golden Plovers also occur in the area, and Common Snipe is frequent in the wetter areas. Raptors can include Hen Harrier, Peregrine and Short-eared Owl, and Barn Owls may be seen in the late afternoon hunting over the meadows. Water Pipits frequent the meadows near Christchurch and may remain until early spring, while almost any patch of alder and birch could produce flocks of redpolls and Siskins. Cetti's Warbler is resident in the riverside scrub.

In spring, Garganey is almost annual while migrant waders can include Ruff, Whimbrel, Black-tailed Godwit, Greenshank, and Green, Wood and Common Sandpipers. Common and Black Terns can also be found at this time. Scarce and rare species have been recorded here with some regularity.

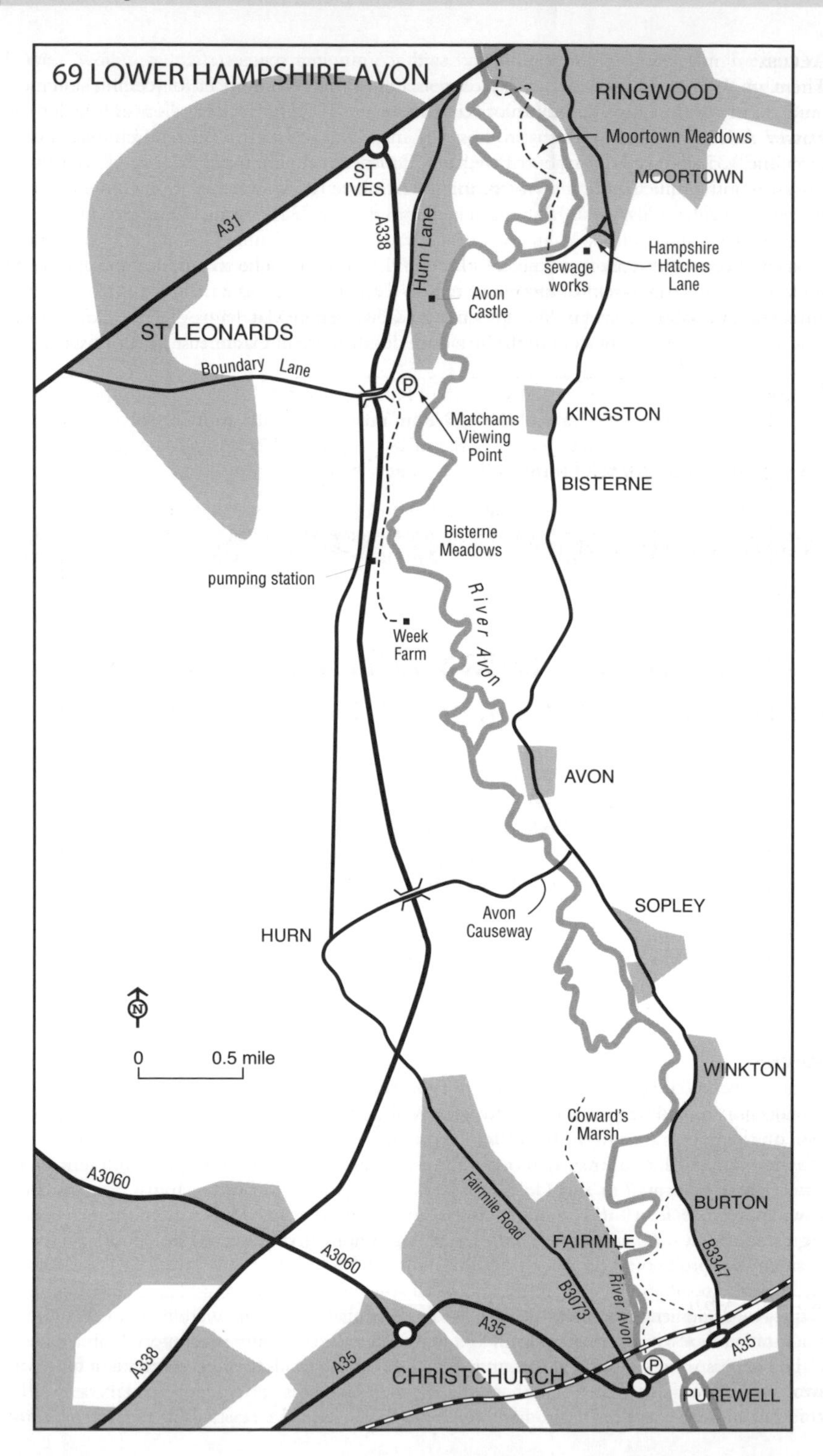
69 LOWER HAMPSHIRE AVON
RINGWOOD
Moortown Meadows
MOORTOWN
ST IVES
A31
A338
Hurn Lane
Hampshire Hatches Lane
sewage works
Avon Castle
ST LEONARDS
Boundary Lane
P
Matchams Viewing Point
KINGSTON
BISTERNE
Bisterne Meadows
pumping station
River Avon
Week Farm
AVON
SOPLEY
Avon Causeway
HURN
N
0
0.5 mile
WINKTON
Coward's Marsh
A3060
Fairmile Road
BURTON
FAIRMILE
B3347
A3060
B3073
River Avon
A35
A35
A338
A35
CHRISTCHURCH
P
PUREWELL

Access

There are several areas worthy of exploration, notably in the south close to Chichester, and immediately to the north and south of Ringwood.

Lower Avon Meadows Footpaths cross the meadows between Christchurch and Burton. From the A35 Purewell roundabout take the B3347 north. A path may be accessed from a small stream almost 1 mile north of the roundabout while a second footpath starts at a lay-by north of the railway bridge. Coward's Marsh is reached from the B3073 (Fairmile Road) in Christchurch by turning east into Suffolk Avenue, about halfway between Fairmile Hospital and the Jumpers Ilford roundabout and then proceeding through the housing estate to Marsh Lane. From here walk north along the track to the stile by Marsh Cottage.

Bisterne Meadows Bisterne Meadows can be viewed from Matchams Viewing Point, which is accessible via the southern end of Hurn Lane south of Avon Castle, just north of where the road crosses the A338.

Ringwood Meadows The meadows north of Ringwood can be observed from the small road to the electricity substation west of the A338, just north of the A31 roundabout.

FURTHER INFORMATION

Grid refs: Lower Avon Meadows SZ 160 947; Coward's Marsh SZ 150 948; Bisterne Meadows (Matchams Viewing Point) SU 134 021; Ringwood Meadows SU 147 056

70 BLASHFORD LAKES WILDLIFE RESERVE

(Hampshire) OS Landranger 195

This gravel pit complex in the Avon Valley, just 2 miles north of Ringwood, has been developed into a family-friendly reserve with good paths, a series of hides, and benches at regular intervals. Some of the lakes are used for other recreational activities.

Habitat

Blashford Lakes comprises a large series of flooded gravel pits that vary in age and support a broad variety of habitats from woodland and scrub to reedbeds and willow carr, and lakes with little vegetation and bare gravel shores. The 200-ha reserve is managed by the Hampshire and Isle of Wight Wildlife Trust. The nearby River Avon and adjacent meadows provide additional habitats.

Species

Winter wildfowl include small numbers of Bewick's Swans (best found near Ibsley Bridge) and numbers of commoner ducks such as Wigeon, Teal and Shoveler. Small numbers of Goosanders feed on the river and roost on Blashford Lakes, where Smew and Scaup may be present in cold weather. Riverside reeds may harbour a Bittern, while Little Egret can be found year-round. Large flocks of Lapwings and Golden Plovers move around the valley, and Common Snipe is widespread. A few Green Sandpipers favour Blashford Lakes. Wet conditions may attract a greater variety of waders, with Jack Snipe, Common Redshank and Black-tailed Godwit sometimes present.

Breeding species include small numbers of Lapwings, Common Redshanks and Common Snipe. Reed and Sedge Warblers occur in the reedy fringes, and Cetti's Warbler is well established at a number of locations. Grey Wagtail and Kingfisher prefer smaller streams and tributaries.

The area is interesting for migrants with wildfowl and waders more prominent when the area is wet. Blashford Lakes is perhaps the most reliable area for migrant waders, and Black Tern occurs with some regularity here. Garganey is frequent and a noticeable passage of raptors includes small numbers of Hobbies and annual Ospreys. Rarer raptors have included Montagu's and Marsh Harriers, and Red Kite. Passerine migrants are not a major feature, but Wheatear and Whinchat are usually conspicuous.

Access

From the roundabout on the A31 on the north-west side of Ringwood, take the A338 north towards Fordingbridge. After 2 miles turn right into Ellingham Drove, and right again to the Education Centre and car park. A network of relatively flat, well-surfaced paths provides access to six hides around Ivy Lake, Ibsley Water and Mockbeggar Lake. More paths and three new hides are planned in the future. The car park and hides are open daily, 09.00–16.30.

Kingfisher, Poulner and Linbrook Lakes From the A338, 0.5 mile north of the A31 roundabout at Ringwood, follow Hurst Road east through the housing estate to the end of the road. Walk north-east and east along the Avon Valley Footpath to view the lakes. For a longer walk, continue north alongside Snails, Blashford and Rockford Lakes (see below to access these lakes from the north).

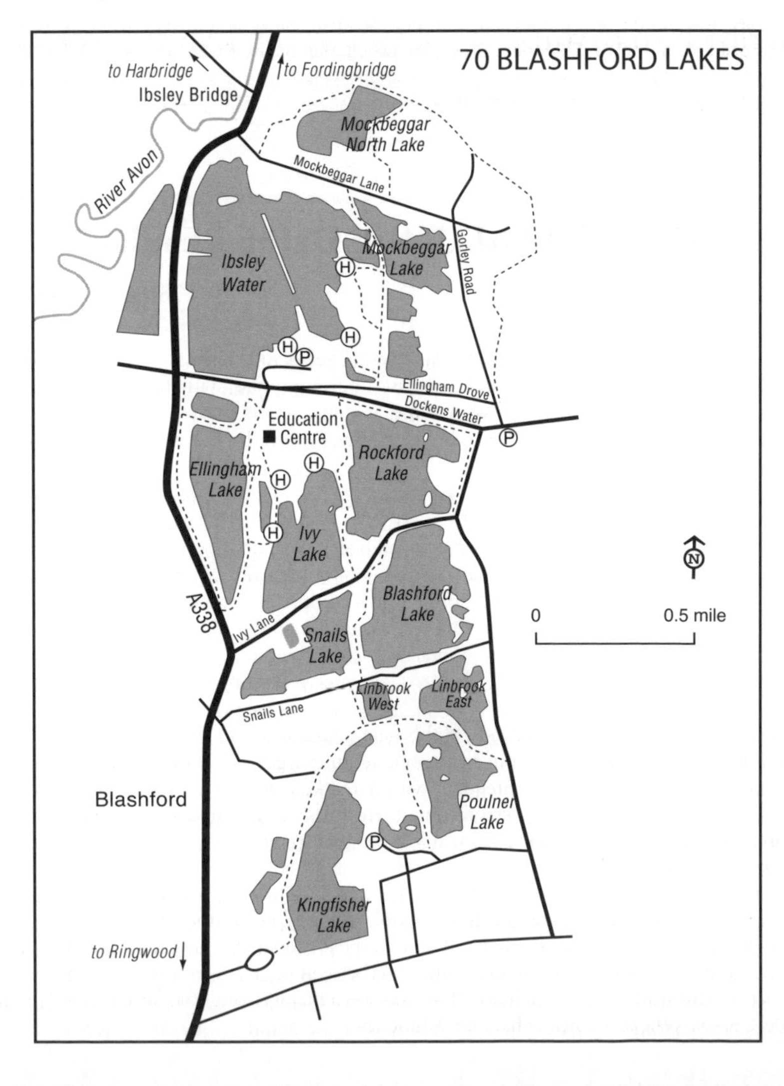

Snails, Blashford and Rockford Lakes These waters may be viewed at various points along Ivy Lane, which is signposted to Rockford from the A338. Snails, Blashford and Linbrook Lakes are also viewable from the Avon Valley Path; starting south of the junction of Ivy Lane and Gorley Road, follow the north shore of Blashford Lake south-west to Blashford Sailing Club. Continue south between Blashford and Snails Lakes to Snails Lane and then along the west shore of Linbrook Lake to join the footpath accessing Kingfisher and Poulner Lakes (see above).
Mockbeggar North Lake This can be reached from the A338 at Ibsley, taking Mockbeggar Lane east to Mockbeggar. A footpath on the left encircles the lake. Continue east along Mockbeggar Lane and south on Gorley Road for views of Mockbeggar Lake.
Ibsley Meadows and Bickton Mill: The river and meadows at Ibsley are best viewed from the minor road between the A338 at Ibsley Bridge and Harbridge Church. The minor road (Churchfield Lane) north from Harbridge Church to Harbridge Green offers further views of the valley.

FURTHER INFORMATION

Grid refs: Blashford Lakes (Education Centre) SU 151 080; Ibsley Bridge SU 150 096
Blashford Lakes Centre: Ellingham Drove, Ringwood, Hampshire BN24 3PJ: tel: 01425 472760; web: www.hwt.org.uk/reserve_detail.php/13/blashfordlakes
Hampshire and Isle of Wight Wildlife Trust: tel: 01489 774400; web: www.hwt.org.uk

71–75 NEW FOREST (Hampshire)

OS Landranger 195 and 196

This outstanding area of mixed woodland and heathland between Southampton and Bournemouth is managed by Forest Enterprise and is popular with tourists in summer. A number of scarce species breed (though some are difficult to locate) and a fine variety of common woodland and heath birds can be seen. May–June is the best time to visit, with early mornings most profitable.

Habitat

The extensive woods are a mix of mature oak and beech with conifer plantations (mainly Scots Pine). The conifers are generally less interesting ornithologically but can hold some unusual species, especially in the younger plantations. Heaths comprise heather, gorse and Bracken with areas of rough grassland and scrub; parts are waterlogged. Though much of the area is common land, some areas are permanently or temporarily enclosed with access forbidden or restricted.

Species

Common Buzzard, Sparrowhawk and Hobby are widespread and frequently seen. Honey-buzzard is a New Forest speciality but seeing one requires a great deal of luck. Goshawks breed in increasing numbers. Localised breeders with a marked preference for conifers are Siskin and Common Crossbill. Firecrest also favours these trees but not exclusively. The first British breeding records for Firecrest were from the New Forest, but it is very elusive and numbers fluctuate widely. Recently planted conifers or woodland edges are favoured by the localised Woodlark and commoner Tree Pipit; Woodcock and Nightjar are sometimes found here, as well as on heaths, and dusk is the best time to look for both. Deciduous woodland harbours all three woodpeckers, Common Redstart, Wood Warbler, Marsh Tit, Nuthatch and Treecreeper. Hawfinch is much less widespread and unlikely to be seen on a single visit. The star bird of the heath is Dartford Warbler for which the New Forest is a major stronghold; other heathland birds include Curlew, Wheatear, Stonechat and Whinchat.

In winter, Hen Harrier and Great Grey Shrike may be encountered on some of the less disturbed heaths, and with luck Merlin.

Access

Birds can be found almost anywhere and there are many good sites; it is usually best to seek less-disturbed areas. Good areas include Beaulieu Road/Bishop's Dyke/Denny Lodge in the east, Acres Down and Bolderwood in the central area, and Ashley Walk/Pitts Wood/Black Gutter/Hampton Ridge and Eyeworth Pond/Fritham in the north-west. Some of the best sites are detailed below and by visiting these most of the key species of the area can be found, but probably not all in a single visit.

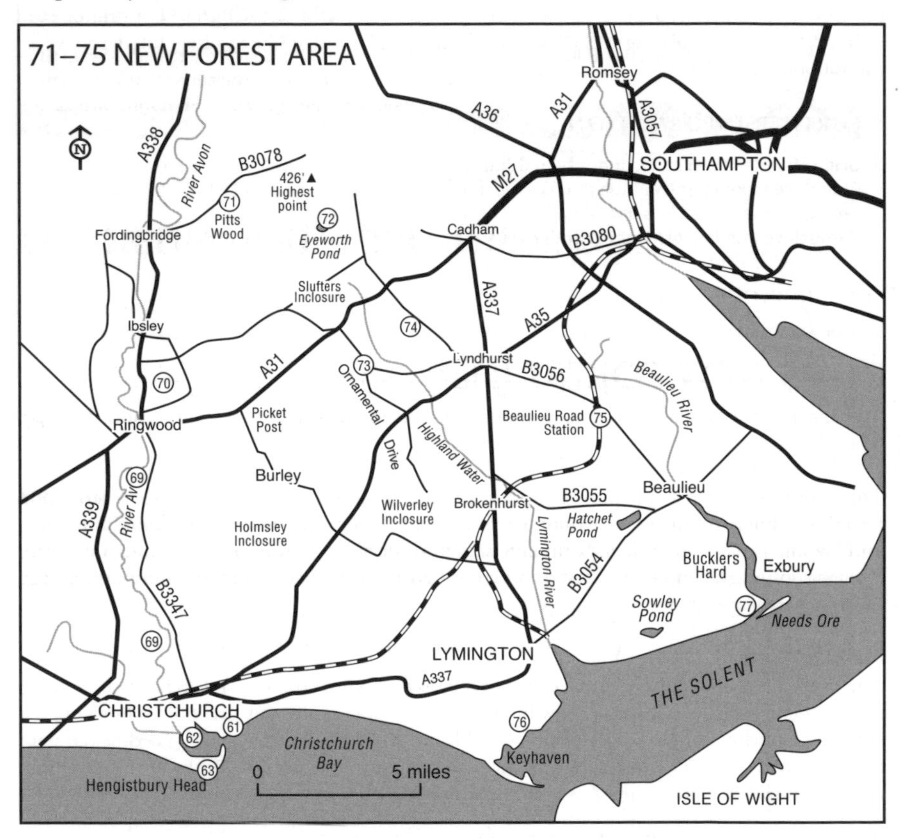

71 ASHLEY WALK AND HAMPTON RIDGE (Hampshire)

Several important sites are situated in the north-west of the forest. In the past, Hampton Ridge was an essential component of a visit to the New Forest, as it was one of the best places for Dartford Warbler and other interesting species. Nowadays, it has largely been superseded. Dartford Warblers are now commoner elsewhere in south England and may be found in many heathland areas. Ashley Walk is more convenient for Pitts Wood and, formerly, Black Gutter Bottom achieved fame as a site for a pair of Montagu's Harriers in summer.

Hampton Ridge Leave Fordingbridge east on the B3078. After 1 mile turn right to Blissford. Hampton Ridge is reached by tracks east from Blissford or from Abbots Well to the south. The main track across the ridge passes through excellent heath with gorse favoured by Dartford Warbler. It also offers fine views over the distant forest with a chance of raptors including Honey-buzzard. After almost 2 miles the track reaches woodland (Pitts Wood Inclosure and

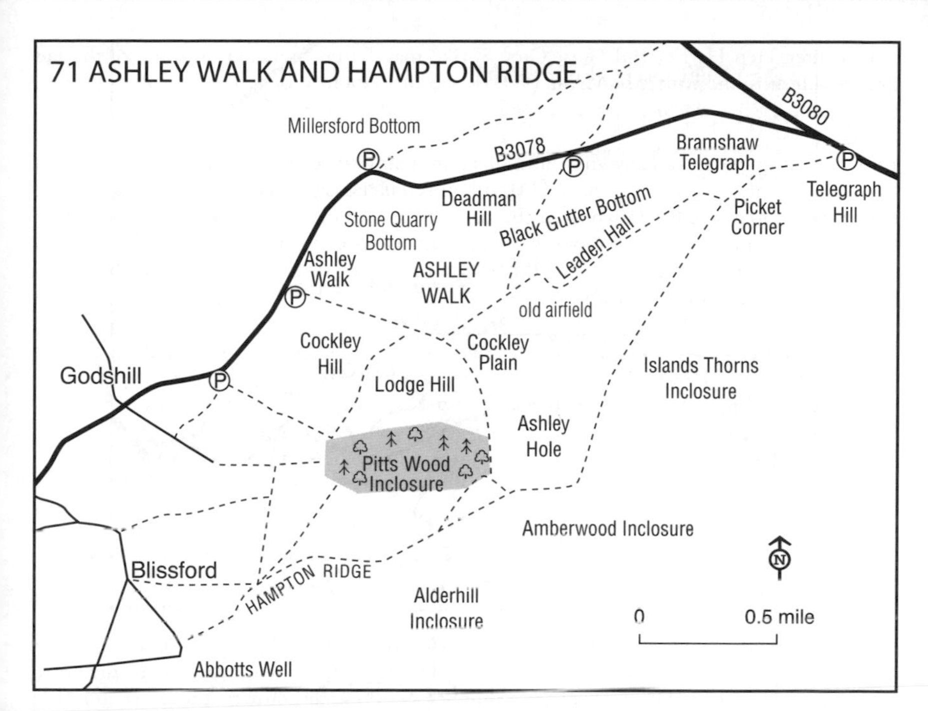

Amberwood Inclosure). The elusive Woodlark may be found here, as well as Common Redstart, Common Crossbill and Siskin.

Ashley Walk/Pitts Wood Continue east on the B3078, and beyond Godshill stop in the Ashley Walk car park. A track leads across the valley to Cockley Hill and Cockley Plain. Look for Hen Harrier in this area in winter. Fork right to Pitts Wood Inclosure. This is a good area for Hawfinch and other woodland species such as Common Crossbill and Siskin, as well as Tree Pipit in summer. Hawfinch is often elusive, but may be seen more easily in winter at the edge of the wood (right of the entrance gate) or in the holly bushes. Early mornings in spring are best to hear Woodlark.

Black Gutter Further east along the B3078 there is another car park at Black Gutter (now closed all year), set within a clump of Scots Pine. Dartford Warbler is common here and Hobby is a frequent sight in summer. Sadly, Red-backed Shrike is here no more – this was its last breeding haunt in the New Forest, in 1984. Montagu's Harrier also used to breed in the area. In winter look for Hen Harrier, and if you are very fortunate Merlin and Great Grey Shrike can sometimes be seen.

72 EYEWORTH POND AND FRITHAM (Hampshire)

Excellent woodland, holding Lesser Spotted Woodpecker, Hawfinch, Common Redstart and Wood Warbler, is situated in the vicinity of Fritham. Added variety is provided by Eyeworth Pond, a good site for Mandarin, which nests in the adjacent woodland. Fritham is located between the east carriageway of the A31 at Stoney Cross and the B3078 at Longcross Plain, and is reached by taking the road signed 'Fritham and Eyeworth only'. Continue through the village to the Royal Oak pub and from here head north-west on the metalled road to Eyeworth Pond or south-west on the gravel track to Fritham car park. Open heath areas should be searched for Nightjar, Woodlark and Dartford Warbler.

73 BOLDERWOOD (Hampshire)

Bolderwood provides forest walks and deer-viewing opportunities, and is thus popular with visitors. Some of the best woodland in the New Forest is found here. Bolderwood can be reached

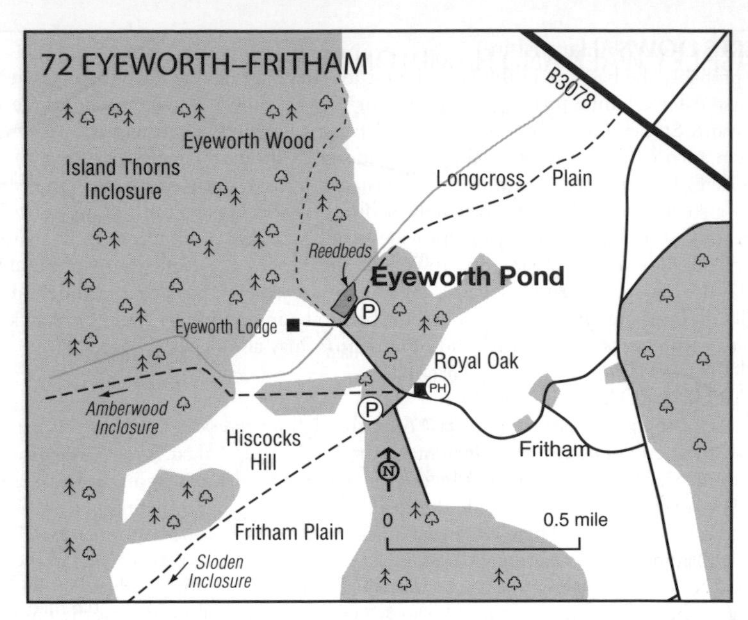

either along Bolderwood Ornamental Drive, c.2 miles south-west of Lyndhurst along the A35 or along a minor road west of the New Forest Inn at Emery Down. The car park lies by the north part of the Ornamental Drive. There are a number of good trails through the forest and most woodland species can be found in the area. Key species are Firecrest, Hawfinch and Common Crossbill. Firecrest can be found anywhere but the area between the car park and Bolderwood Cottage is often good; listen for their song from the treetops. Hawfinch sometimes favours the area around the Canadian Memorial, which is also good for Common Crossbill, or more reliably at Blackwater Arboretum (opposite Blackwater car park along Ornamental Drive at SU 243 086).

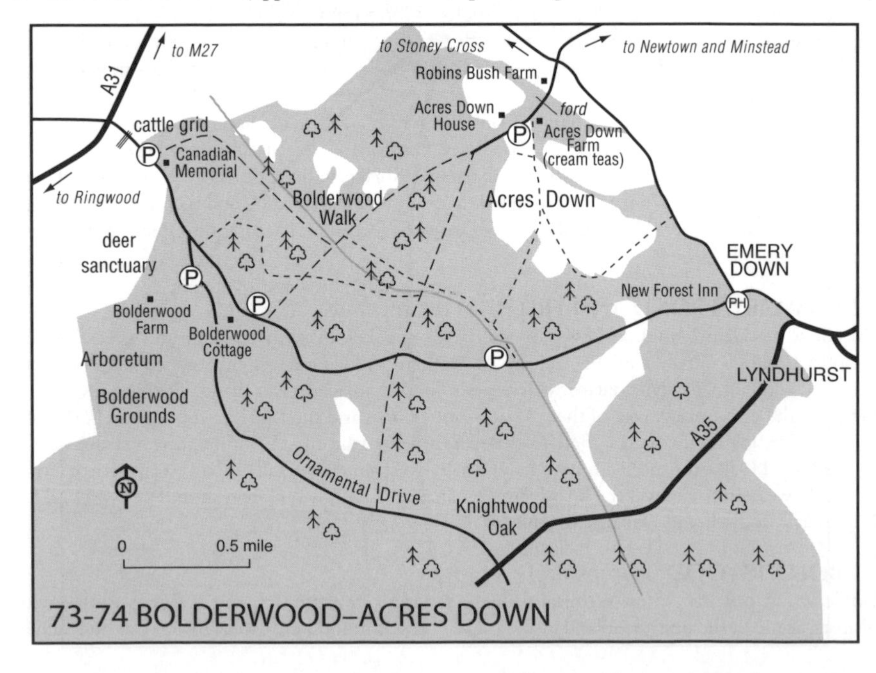

74 ACRES DOWN (Hampshire)

This is probably the best site in the New Forest for watching raptors. Take the A35 west from Lyndhurst. After c.1 mile turn right on a minor road to Emery Down and continue north-west towards Stoney Cross. Two miles beyond Emery Down, where there is a right turn to Newtown, turn left on a minor no through road signed 'cream teas'. After crossing a ford and beyond Acres Down Farm (cream teas), there is a small gravel car park on the left. From here walk up a steep small hillock immediately in front of you. At the top follow the track left through some scrubby trees. The ridge provides a good vantage point to overlook a large tract of woodland. Common Buzzard, Sparrowhawk and Hobby are regularly seen, but Honey-buzzard is the key species here. If a pair is breeding nearby, the birds may be seen here regularly, especially on mornings early in the season; it is best to be in position by 10am Goshawk is seen at this site with increasing regularity, and other raptors may also occur.

75 BEAULIEU ROAD (Hampshire)

This is one of the best areas in the New Forest throughout the year. Leave Lyndhurst east on the A35. At the end of the town turn right on the B3056 to Beaulieu. After 3 miles pull off to the right into Shatterford car park, just before the road crosses the railway line at Beaulieu Road station. The small patch of pines beside the road is a picnic site and sometimes holds Common Crossbill during irruptions. There are several tracks in the area, but the best route is a clockwise circuit via Bishop's Dyke and Denny Wood. Head south from the car park, parallel to the railway line. Woodlark occurs between the track and railway line, and Dartford Warbler can be seen on the heath to the right. In winter it may be worth taking a small detour to cross the railway over the old bridge to Stephill Bottom. Hen Harrier is regular at this season and with luck Great Grey Shrike may be present. Continuing south on the original track, cross Bishop's Dyke and a boggy

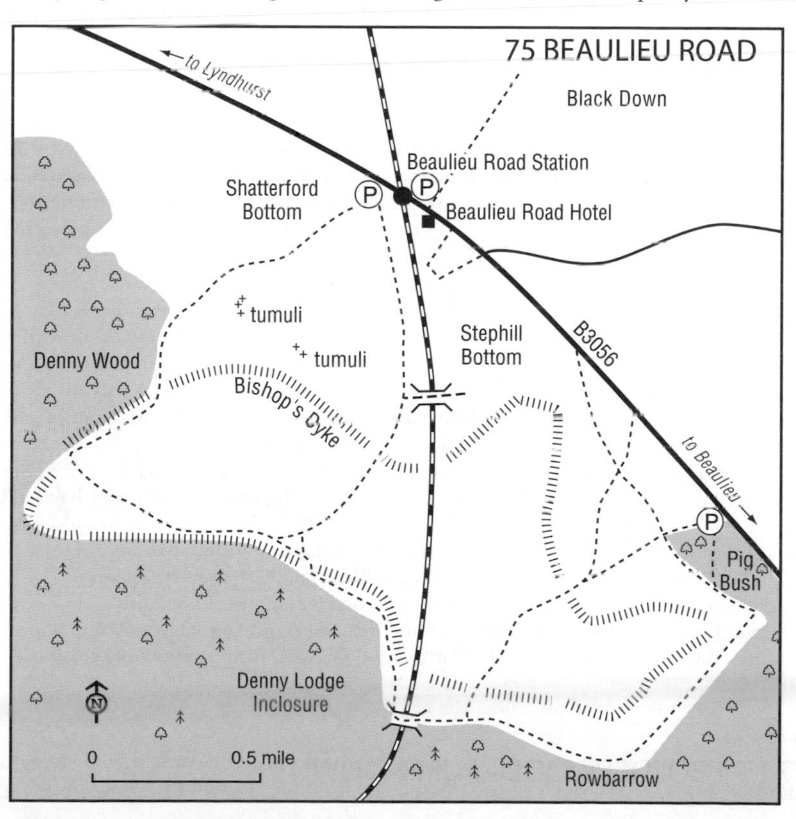

area. This is a favoured site for Hobby (especially in late afternoon) and Cuckoo, and it is worth scanning the wooded skyline (Denny Wood Inclosure) for Common Buzzard, Sparrowhawk and the occasional Honey-buzzard. Beyond the bog the track bears right to skirt Denny Lodge Inclosure. Tree Pipit occurs in summer and Woodcocks rode here on spring evenings. The track continues past Denny Wood, an excellent area of broadleaf woodland that holds all the expected woodland species, including Wood Warbler in summer and all three species of woodpecker. To complete the circuit, the track crosses the heath of Shatterford Bottom.

FURTHER INFORMATION

Grid refs: Ashley Walk SU 186 157; Eyeworth Pond SU 229 145; Bolderwood SU 241 091; Acres Down SU 267 097; Beaulieu Road SU 348 063
New Forest (Tourist Information): web: www.thenewforest.co.uk
Birds of the New Forest: A Visitor's Guide by A.M. Snook (1998)

76 KEYHAVEN AND PENNINGTON MARSHES (Hampshire)

OS Landranger 196

An extensive area of marshes noted for wildfowl, waders and raptors, particularly in winter. The area is a well-known wintering site for Short-eared Owl. It also has good potential for rarities during migration periods with aquatic species most likely, but occasional scarce passerines also show up.

Habitat

Keyhaven and Pennington comprise extensive marshes on either side of the Lymington River with muddy creeks and channels, shallow pools and natural saline lagoons separated from the saltings by a sea wall. The lagoons are of national importance due to their specialised flora and fauna. At the southern end, the marshes are protected by a long shingle spit leading to Hurst Castle. There are also some reedbeds alongside Avon Water. Around 728 ha are protected as a reserve and run by Hampshire County Council.

Species

Winter wildfowl numbers are impressive, with up to 4,000 Brent Geese accompanied by Shelduck, Gadwall, Wigeon, Teal, Shoveler and Pintail. Offshore scan for divers, grebes and seaducks. Waders include Black-tailed and Bar-tailed Godwits, and Ruff formerly overwintered in some numbers but are scarce now. Short-eared Owl, Peregrine, Merlin and Hen Harrier are sometimes present, and Snow Bunting is an occasional visitor.

Commoner spring migrants include Little Ringed Plover, Ruff and Greenshank, but Spotted Redshank and Whimbrel are also possible. Garganey, Marsh Harrier and Osprey have all been recorded at this season.

In summer, Little and Sandwich Terns breed on islands in the lagoons and on Hurst Beach. Breeding waders include Oystercatcher, Ringed Plover and Common Redshank.

Waders such as Knot, Sanderling, Curlew and Wood Sandpipers, Little Stint and Greenshank occur in autumn. Rarer species including Grey Phalarope and Spoonbill may also appear and sometimes migrant passerines, typically Ring Ouzel or Black Redstart, both of which use more open areas.

Access

There are several points of access to the marshes, which can all be reached via Lymington or Milford on Sea. The best views of the marshes are from the sea wall (the Solent Way), which runs all the way from Milford on Sea to Lymington, or from Hurst Spit. Circular routes of varying lengths take in the key areas of interest.

Keyhaven Marshes From Lymington, take the A337 south through Pennington and turn left at Everton onto the B3058 to Milford on Sea. Turn left in Milford to Keyhaven and park in the Keyhaven amenity car park opposite the pub at SZ 306 914 (pay and display). The Solent Way can be accessed from here and a loop route of 3 miles encircles Keyhaven Marshes. A longer walk will take in Pennington Marshes too.

Hurst Beach Continue from Keyhaven southwards along the coast and park at the base of Hurst Spit. The beach and spit can be worked from here. A ferry between Keyhaven Harbour and Hurst Castle operates regularly from April to October, and at weekends in winter.

Pennington Marshes South of Lymington leave the A337 in Pennington on a minor road south to Lower Pennington and continue to the coast. Park in the Fishtail Lagoon car park at SZ 318 927. The marshes can be viewed from the sea wall or take a circular route around them. From Creek Cottage, Lower Woodside, the Solent Way footpath follows the sea wall to Keyhaven Harbour.

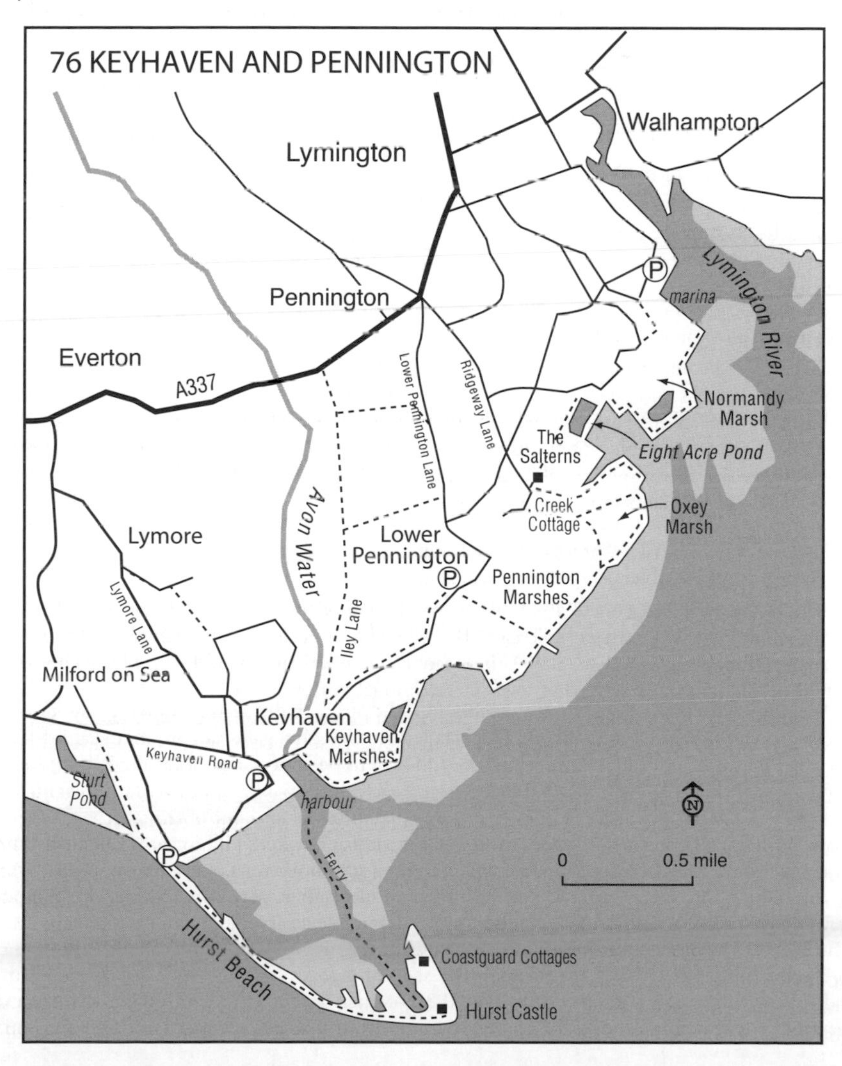

Normandy Lagoon Normandy Marsh and its lagoon lie to the east of Pennington Marshes, alongside the estuary of the Lymington River. From the east end of Lymington High Street take Captain's Row south, then turn left down Nelson Place, and turn right and follow Bath Road signposted to the Riverside Marinas south-east for 0.3 miles to the Bath Road amenity car park (fee payable March–October), at the junction with King's Saltern Road. Normandy Lagoon and The Salterns nearby can be accessed via a circular route from the car park to Creek Cottage, Lower Woodside, and back.

FURTHER INFORMATION

Grid refs: Keyhaven Marsh SZ 305 915; Hurst Beach SZ 299 908; Pennington Marshes SZ 318 927; Normandy Lagoon SZ 333 950
Hampshire County Council: tel: 0845 603 5630; web: www.hants.gov.uk

77 NEEDS ORE POINT (Hampshire)

OS Landranger 196

Needs Ore is part of the lower Beaulieu estuary and the saltmarsh and shore are particularly good in winter for wildfowl and waders, though spring and autumn migration can also provide exciting birding.

Habitat

The lower Beaulieu estuary incorporates estuary saltmarsh and Needs Ore Point on the coast, with attendant shoreline and offshore species.

Species

Wildfowl are conspicuous in winter and include Brent Goose and Shelduck. Red-breasted Merganser and Goldeneye also occur, and there have been recent reports of grey geese, mostly White-fronted Goose. Cold weather may produce Smew or Bewick's Swan. Black-necked and Slavonian Grebes are often present at the estuary mouth at this season. Good numbers of the commoner waders occur in winter, and Peregrine is regular in the area, as are Barn Owl and Merlin. Hen Harrier, Water Rail, Rock Pipit and Bearded Tit are other possibilities. Little Egrets and Cetti's Warblers are present year-round.

Among the commoner migrant waders, Avocet may occur in spring and other occasional passage visitors include Marsh Harrier, Garganey and Osprey. Little Gull and a variety of terns may also appear at this season. Rarities are occasionally noted, such as a Blue-cheeked Bee-eater in June 2009.

Autumn generally brings similar species to the area, and passage waders may include Greenshanks, Spotted Redshanks and Black-tailed Godwits, but gales can produce seabirds such as Grey Phalarope and Little Auk, which are unlikely to be seen at any other time.

Access

From Beaulieu take the B3055 south-west and turn left at Bunker's Hill towards Bucklers Hard. Continue south to St Leonards Grange and then turn left on Warren Lane. Park on Warren Lane by the hide (Shore Hide) or at the point by the sailing club. A permit is required for this site, available from the National Motor Museum, Beaulieu. The reserve comprises three main areas:
The Beach The beach is off-limits during the breeding season (March–July) to avoid disturbance to the birds. At other times, go through the second gate west of the hide on Warren Lane, and follow the rough path that runs past the Norman Pullen Hide, following the edge of the reedbed. This area is good for passerine migrants in autumn and look out for occasional

Bearded Tit in the reedbed. The beach holds the commoner wader species.

Blackwater Lagoon (North Solent NNR) A footpath opposite the parking area and hide on Warren Lane leads across a field and brook before reaching the lagoon. Four hides are provided. The Easterly Hide is best at high tide when waders roost directly in front of it (especially autumn). It is also good for wildfowl in winter. The Blackwater Hide is good for wildfowl and also for Cetti's Warbler. Water Rails can sometimes be seen at close range here. The NFOC Hide attracts wildfowl and waders, as does the James Venner Hide, which is also good for Little Egrets.

Needs Ore Point The Point gives good views into the mouth of the Beaulieu estuary, which in winter is good for grebes and waders. In the summer, you can see the gull and tern colony from here, but numbers are much lower than they used to be.

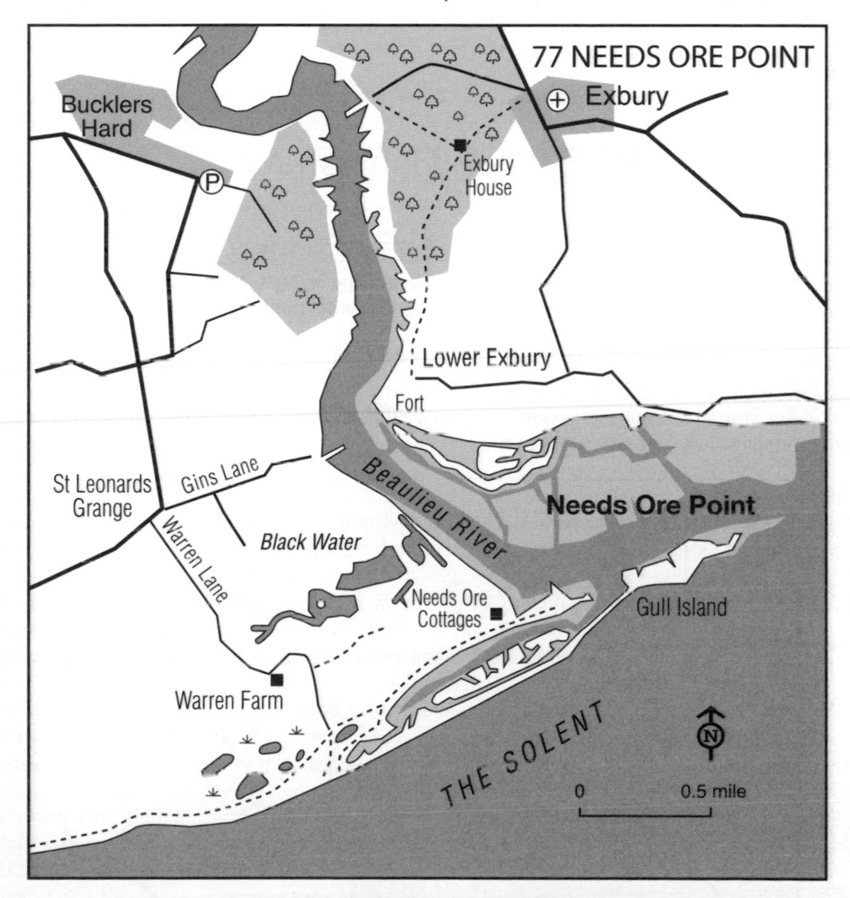

FURTHER INFORMATION

Grid refs: Warren Farm SZ 413 973; Needs Ore Cottages SZ 423 975
National Motor Museum, Beaulieu (permits): tel: 01590 612345
Natural England, Hampshire and Isle of Wight office: tel: 023 8028 6410

78 TITCHFIELD HAVEN (Hampshire)

OS Landranger 196

Titchfield Haven is a small, wetland NNR on the Solent, south-west of Fareham. It can provide good birding at any time of year, and is a stronghold for Cetti's Warblers.

Habitat

The 150-ha reserve comprises reedbeds, willow scrub, wet grazing meadows and freshwater scrapes.

Species

Large numbers of wildfowl visit the reserve in winter. Up to 1,500 Wigeon are present with good numbers of other species. Brent Geese feed in the surrounding fields, and occasionally grey geese also appear, most frequently White-fronted. Amongst many wintering waders, counts of over 1,000 Black-tailed Godwits are notable. Gulls are also in evidence, and may include Mediterranean. Bittern is a regular winter visitor, and Kingfishers are frequently seen.

Titchfield's most famous breeding inhabitant is Cetti's Warbler, and up to 50 singing males may be present. Bearded Tit also breeds in small numbers, as does Mediterranean Gull and Avocet.

Spring and autumn are good for passage waders, terns, and a wide variety of other species. Titchfield's list of scarce migrants and rarities is surprisingly good.

Access

Leave the M27 at junction 9 onto the A27 to Fareham. After 2.5 miles, turn right onto the B3334 to Stubbington. In Stubbington, turn right into Bells Lane (signed to Hill Head). Park in the car park on the left, adjacent to Hill Head Sailing Club. Haven House Visitor Centre is a large white

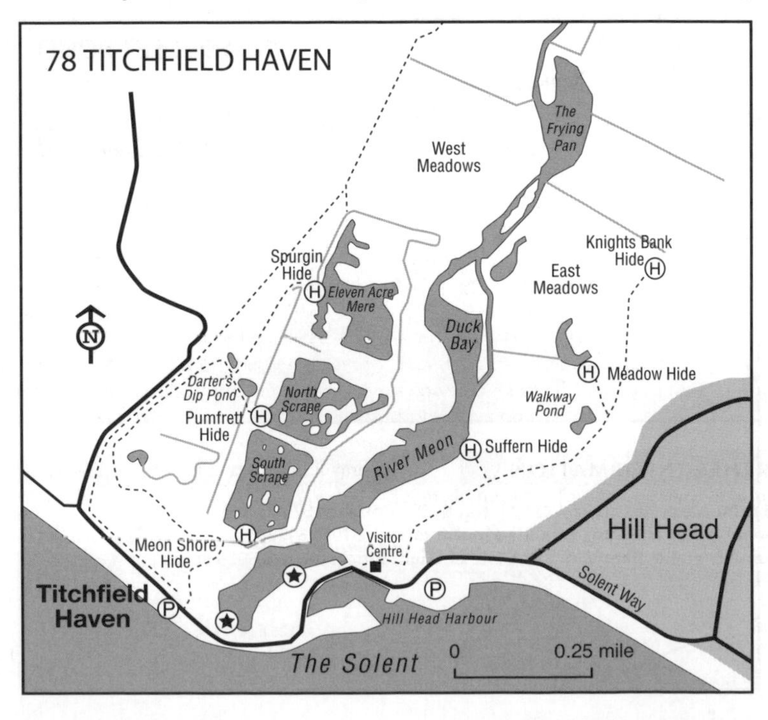

house across the road. Tickets need to be purchased from the Information Desk in the Visitor Centre. The reserve is open Wednesday–Sunday all year, plus Bank Holidays (except Christmas and Boxing Days), from 9.30am–5.30pm (9.30am–4pm in winter). There are six hides.

FURTHER INFORMATION

Grid ref: SU 534 023
Haven House Visitor Centre: Cliff Road, Hill Head, Fareham, Hants PO14 3JT; tel: 01329 662145; web: www.hants.gov.uk/titchfield

79 FLEET POND (Hampshire)

OS Landranger 186

Designated as an SSSI, Fleet Pond LNR (54 ha), located just north of Fleet town centre, is owned and managed by Hart District Council in the north-east corner of Hampshire. More than 200 species of birds have been recorded at this inland site, which remains the largest body of fresh water in Hampshire.

Habitat

The reserve offers a diverse range of habitats from open water and islands to extensive reedbeds and marshes, all encircled by mixed wet and dry woodland, dominated by alder and oak. The reserve includes several small wet and dry heaths that continue to be actively managed. At its eastern boundary the reserve adjoins several 'set-aside' fields. The precise origins of the pond are unknown, but it is believed to have been man-made in the 12th century through the damming

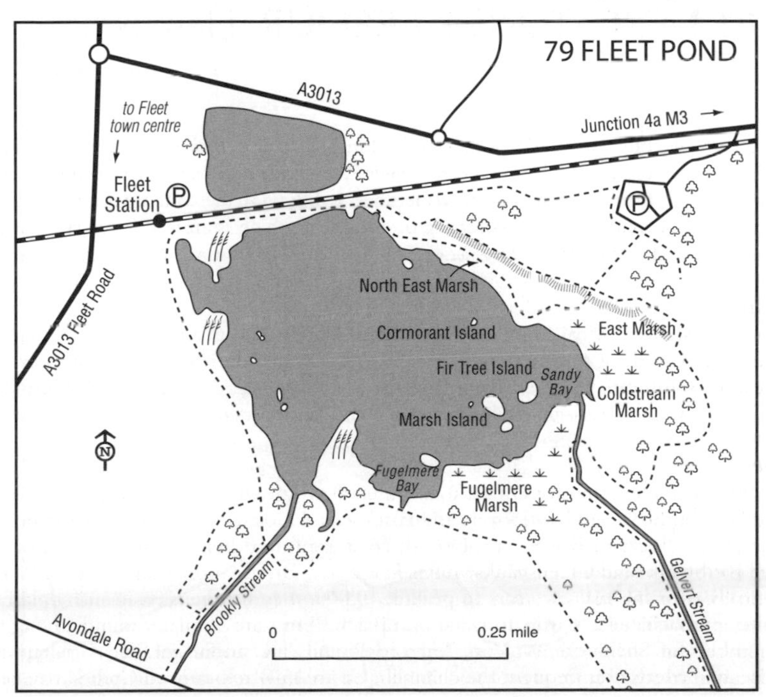

of local watercourses. The pond has suffered some degradation in recent decades from excessive silting as a result of run-off from adjoining MOD land, and the water depth rarely exceeds 1m.

Species

Breeding birds include Great Crested Grebe, Grey Heron, Sparrowhawk, Common Tern, Water Rail, Reed Warbler, Garden Warbler and Reed Bunting. Hobbies are present throughout the summer months, although they do not breed on the reserve. In winter the site regularly attracts Bittern as well as large flocks of Siskins, Lesser Redpolls and Bramblings. However, Fleet Pond is probably best known for its migrants and is a popular local site for visible migration, particularly in autumn. It lays claim to an impressive list of rarities including Little Bittern, several Purple Herons and Black Kites, both Whiskered and White-winged Black Terns, several Great Reed Warblers and a Citrine Wagtail.

Access

Fleet Pond can be accessed via the M3 (junction 4a). Take signs for Fleet and once on the A3013 look out for the reserve signpost to the left. Alternative access is possible via Fleet Station car park or at the end of either Chestnut Grove or Westover Road, the latter accessed from the A3013 via Avondale Road in Fleet. Maintained footpaths provide access throughout the reserve although there are surprisingly few vantage points offering full views of the pond itself, the station car park probably being the best.

FURTHER INFORMATION

Grid ref: SU 824 552
Fleet Pond Society: Tel: 01252 616531

80–81 LANGSTONE HARBOUR (Hampshire)

OS Landranger 196

The massive complex of tidal mudflats and saltings that also includes Chichester Harbour (82–85) is one of the most important estuaries on the south coast and is of both national and international significance for wintering waders and wildfowl. Langstone Harbour contains the largest expanse of mudflats in the complex, one-third of which is an RSPB reserve (no access, but can be viewed from surrounding areas). The best times to visit are autumn and winter.

Habitat

The estuary is largely tidal mudflats and saltings. Much of the surrounding area has been built upon, especially at the west end. There is some farmland on Hayling Island. Farlington was reclaimed from the sea several centuries ago and its grassland and scrub are controlled by grazing cattle. It also has some freshwater marsh and lagoons.

Species

Up to 90,000 waders (mostly Dunlins) and 15,000 wildfowl winter on the huge complex of mudflats. Numbers of Grey Plovers and Black-tailed Godwits are of national importance, and the largest flocks of Knots on the south coast are found here. Small numbers of Ruffs, Spotted Redshanks and Greenshanks winter in the area. Brent Goose numbers have increased dramatically over the last few years to peak at 10,000; they can be seen in many places and are quite approachable at Farlington, and odd Black Brants are regularly found among them. The numbers of Shelducks, Wigeon, Teal and Pintails are important, and Goldeneye and Red-breasted Merganser frequent the channels. Up to 15 Black-necked Grebes may be seen (one of the largest regular wintering flocks in Britain) and they favour the north-east part of Langstone Harbour. A few Slavonian Grebes may also be present. As many as 20,000 gulls roost in the harbour, mostly Black-headed. Sparrowhawk and Short-eared Owl are occasionally seen hunting over the marshes.

A wider variety of waders occurs in spring and autumn, and may include Little Stint and Wood Sandpiper. September–October is a good time to see passage waders.

Summer is the least interesting season ornithologically, but Sandwich, Common and Little Terns breed, as well as Oystercatcher, Ringed Plover and Common Redshank. A few pairs of Mediterranean Gulls also breed.

Access

The harbour can be reached from many points though certain areas have restricted access. The major wader roosts are on Farlington Marshes and its offshore islands, and west Hayling Island. The north shore can be watched from a coastal footpath (park at Broadmarsh, south-east of the A27/A3(M) junction). The mouth of the estuary can be viewed from the southwest corner of Hayling Island.

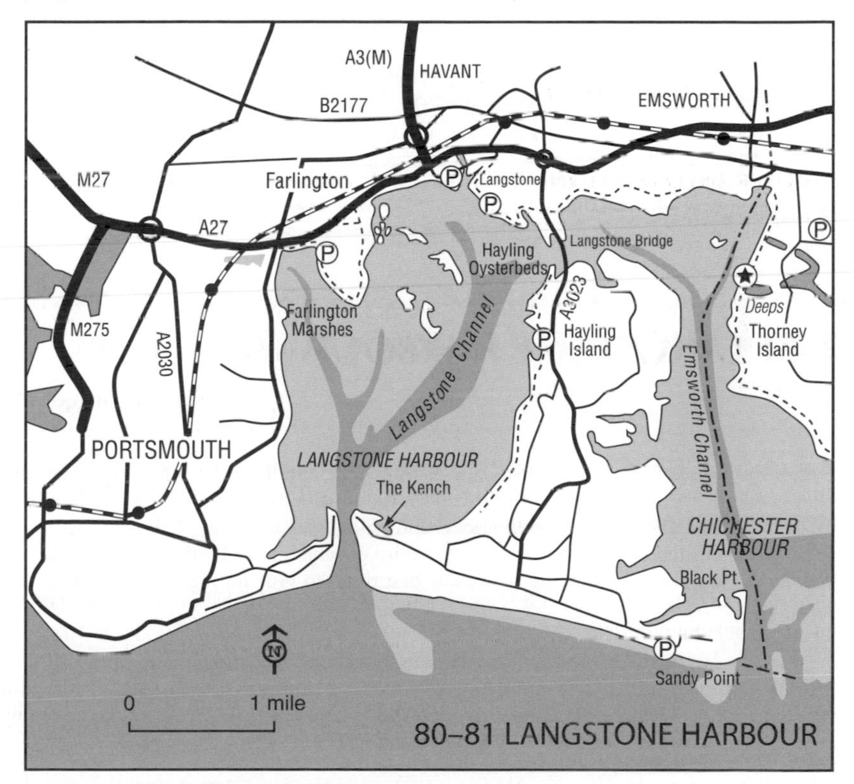

80 FARLINGTON MARSHES (Hampshire)

Leave the A27 Chichester–Fareham road at the intersection with the A2030 immediately north of Portsmouth. A track leads off the roundabout between the A27 east and A2030 south exits and runs east below and parallel to the A27 towards Farlington Marshes. Follow the coast on foot around the west side of the marshes. A freshwater lagoon and associated marshland are worth checking en route. Continue around the peninsula. Areas of scrub at the north end attract migrants. Alternatively, access from Broadmarsh car park and walk west along the coast to the reserve.

81 HAYLING ISLAND (Hampshire)

Take the A3023 south from Havant, and cross Langstone Bridge onto Hayling Island. A track around the west side of the island gives good views of the harbour. There are three prime areas to work:

Hayling Oysterbeds (West Hayling LNR) This lagoon on the north-west corner of Hayling is best accessed by parking behind the Esso garage on the Havant Road, about 0.5 miles onto the island from the bridge. Little Terns breed on the island in the lagoon and in winter the flock of Black-necked Grebes is best seen at high tide from the north-west corner of Hayling Island.

The Kench This small, protected inlet in south-west Hayling, close to the entrance to Langstone Harbour, is a local nature reserve run by Hampshire County Council. It is easily visible from Ferry Road and a path runs up the east side to the shore of the harbour. This is a good place to observe waders.

Sandy Point and **Black Point** These beaches lie at the south-east corner of Hayling, at the entrance of Chichester Harbour and can only be accessed on foot. Part of the area is a local nature reserve protecting South Hayling's original sandy heathland habitat. Black Point is a private beach on the other side of the peninsula, with access restricted to members of the sailing club.

FURTHER INFORMATION

Grid refs: Farlington Marshes SU 680 043; Hayling Oysterbeds SU 718 029; The Kench SZ 693 996; Sandy Point SZ 746 983
Hampshire and Isle of Wight Wildlife Trust: tel: 01489 774400; web: www.hwt.org.uk
RSPB South East Regional Office: tel: 01273 775333; web: www.rspb.org.uk

82–85 CHICHESTER HARBOUR (West Sussex)

OS Landranger 196

East of Langstone Harbour and part of the same estuary complex, Chichester Harbour has a similar mix of habitats. It comprises four main channels, Emsworth, Thorney, Bosham and Fishbourne, which separate the peninsulas of Thorney, Chidham and Bosham.

Habitat

The shoreline of Chichester Harbour is mostly undeveloped, unlike that of Langstone Harbour. Some of the surrounding area is farmed.

Species

A similar mix of species can be expected to those listed for Langstone Harbour (81–82), as the area is continuous and birds commute freely across the entire estuary complex. Additionally, in the mid-1990s the largest roost in Britain of Little Egrets was at Thorney Great Deeps, but the birds have sometimes relocated elsewhere in recent years.

Access

Chichester Harbour can be worked from four main areas, though access is generally more restricted than at Langstone Harbour. To view the north shore, walk east from the north side of Langstone Bridge along the coastal footpath towards Emsworth. The path continues around Thorney Island and all the way to Bosham. Emsworth and Thorney Channels are best viewed from Thorney Island. The mouth of the estuary can be viewed from East Head or the south-east corner of Hayling Island.

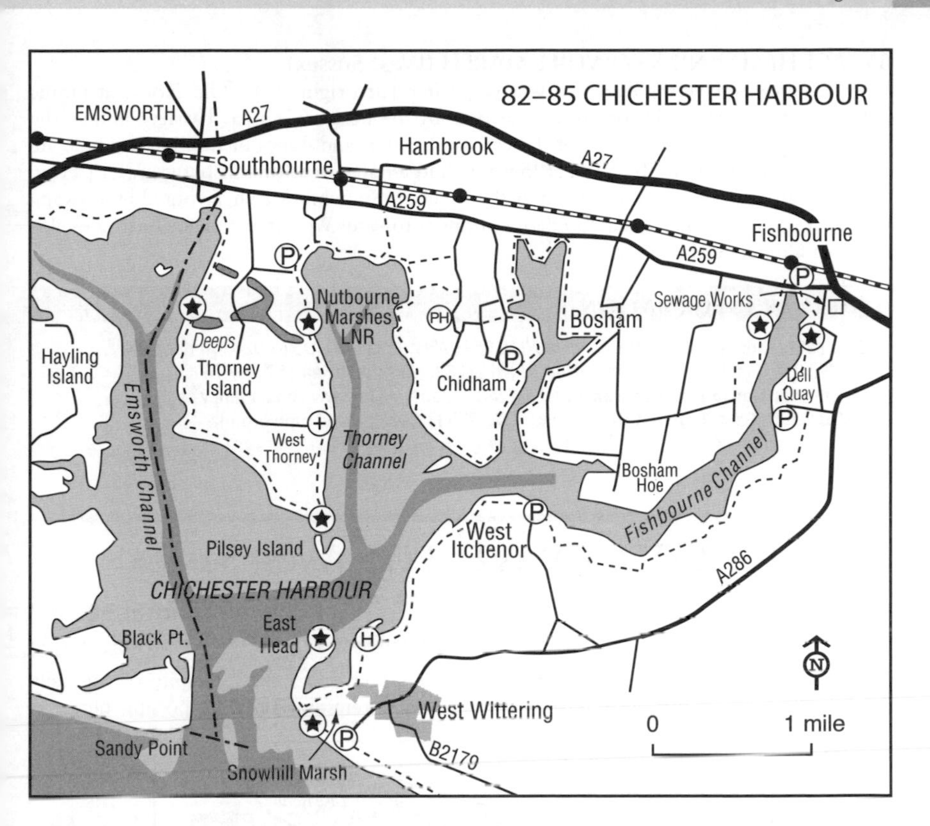

82 THORNEY ISLAND (West Sussex)

Leave the A259 at Southbourne on a minor road south to Prinsted. Park and walk south along the east shore of Thorney Island. Alternatively, from the A259 at Hermitage take the minor road to West Thorney and explore the east coast from here. Much of Thorney Island is MoD land with restricted access.

Pilsey Island This small island off the south tip of Thorney is an RSPB reserve comprising a wide range of coastal habitats. The reserve, together with the adjacent area of Pilsey Sand, is one of the most important roost sites for passage and wintering waders in the area at high tide. It can be viewed from the coastal path that runs around the Thorney Island MoD base.

83 CHIDHAM (West Sussex)

The Chidham peninsula lies to the east of Thorney Island and is much less well watched. From Chichester, take the A259 west towards Emsworth. After 3.5 miles (from the A27 roundabout) turn left in Nutbourne down either Chidham Lane or Cot Lane. Both lead to a car park south of Chidham village. From here a footpath leads east to the edge of Bosham Channel. You can follow the coast southwards to the tip of the peninsula, and on up the west side to have views of the mudflats of Nutbourne Marshes LNR.

84 FISHBOURNE CHANNEL (West Sussex)

From the Fishbourne roundabout on the A27, head west on the A259. After 1 mile turn left into Mill Lane and park at the end. Footpaths follow the sea wall on either side of the channel. Alternatively, leave the A27 further east at the Stockbridge roundabout and head south-west on the A286. After 1.5 miles turn right into Dell Quay Road. Park at the end near the Crown and Anchor pub. From here walk north along the sea wall to view the intertidal mudflats of the creek. There is a wader roost at the head of the creek (best on a rising tide).

85 EAST HEAD AND SNOWHILL MARSH (West Sussex)

From Chichester, take the A286 to West Wittering. Turn right at the Old House at Home pub in West Wittering and park in the car park by the beach (via a pay barrier). Follow the coast westwards from the far end of the car park to the sand dunes of East Head. There are extensive mudflats along the coast at low tide and an area of saltmarsh between East Head and the mainland. Behind the sea wall is the brackish Snowhill Marsh, favoured by roosting waders at high tide. The coastal footpath continues towards West Itchenor for further exploration of Chichester Harbour.

FURTHER INFORMATION

Grid refs: Thorney Island (Prinsted) SU 765 051; West Thorney SU 769 024; Chidham SU 793 034; Fishbourne Channel (Dell Quay) SU 835 028; East Head SZ 769 981
Chichester Harbour Conservancy: tel: 01243 512301; web: www.conservancy.co.uk
RSPB South East Regional Office: tel: 01273 775333; web: www.rspb.org.uk

86 CHICHESTER GRAVEL PITS (West Sussex)

OS Landranger 197

This is a series of gravel pits located by the A27 Chichester bypass. Despite occasionally intensive use of these pits for watersports and fishing, interesting birding is possible here.

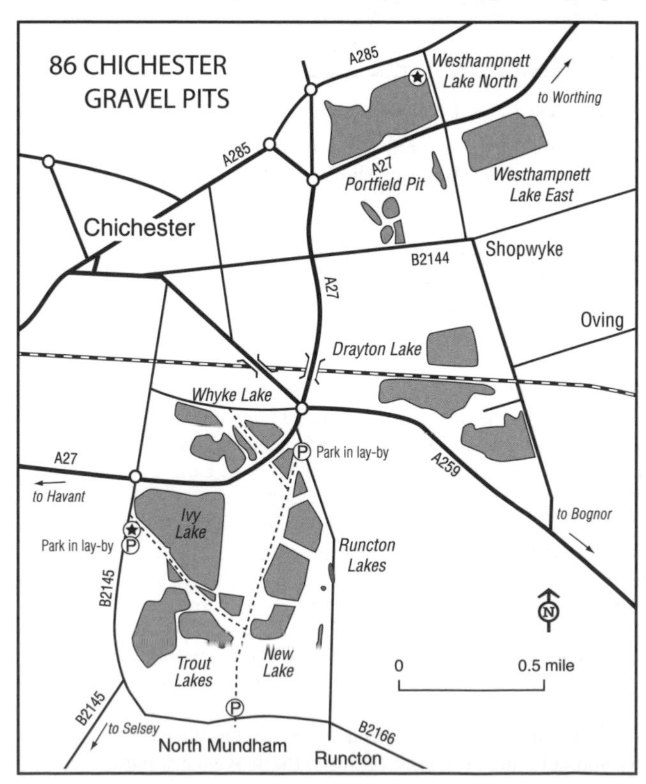

Habitat

A complex of gravel pits with little marginal reed growth, but in some places covered with willows, sallows and bramble scrub. There are few muddy fringes, but there are sand washings on the Portfield North Pit.

Species

In winter a few Scaup and Goldeneyes often occur among the rafts of more numerous Tufted Ducks and Pochards. Dabbling ducks are relatively scarce, though numbers of Shovelers and Teal occur. The pits are best in hard weather when Goosander, Smew and the occasional diver may appear. Cormorants roost on an island in Ivy Lake. Chiffchaffs winter in the waterside trees.

Easterly winds in spring can produce small numbers of Little Gulls and Black Terns over the pits, and Whimbrel and Bar-tailed Godwit sometimes pass overhead.

Breeders include Great Crested Grebe, Shelduck, Ruddy Duck, Pochard, Tufted Duck, Kingfisher and the commoner warblers. Common Tern breeds on rafts constructed by the Sussex Ornithological Society.

In autumn passage waders, including Greenshank, Little Stint and Ruff, frequent the muddy margins. Small numbers of Little Gulls and Black Terns are regular, and large numbers of hirundines roost at the pits.

Access

The pits are situated by the A27 Chichester bypass east and south of the town. Leave the A27 at the roundabout after Westhampnett and explore the northern pits. Most are viewable from roads. Rejoin the A27 southwards, and at the next roundabout take the minor road to Runcton. More pits are visible on the right and a footpath runs along their west side. Another pit can be seen from the B2145 immediately south of the A27. There have been many changes to the pits in recent years, with some being filled and others newly created.

FURTHER INFORMATION

Grid refs: Westhampnett SU 883 060; Runcton Lake SU 878 039; North Mundham SU 875 024; Ivy Lake SU 868 033

87 SELSEY BILL (West Sussex)

OS Landranger 197

Although not in the same league as Dungeness or Portland, Selsey is one of the most important seawatching sites on the south coast. The area is also attractive to passerine migrants in both spring and autumn. On the coast west of the headland lies Bracklesham Bay, a new RSPB reserve of wet grassland.

Habitat

Despite considerable development of the town of Selsey, the area still retains some large vegetated gardens and a small playing field, which are attractive to migrants. The old holiday camp was demolished in 1989 and subsequently areas of trees, scrub and patches of grass have developed. The shingle beach to the west of Selsey is bounded on the landward side by grassland, which gives way to arable farmland.

Species

In winter Red-throated Diver, Great Crested and Slavonian Grebes, Red-breasted Merganser, Common Eider and auks are all regular offshore. Black Redstart can be found feeding along the beach among the breakwaters. Gull flocks are joined by the occasional Mediterranean and Glaucous Gulls. The onset of freezing weather conditions may stimulate a westerly passage of Lapwing, Golden Plover, Skylark and winter thrushes.

In spring large numbers of Common Scoters migrate east, with the majority in late April, but Velvet Scoter is much less frequent. Other birds seen regularly with peak numbers in spring are divers, Common Eider, Red-breasted Merganser, Bar-tailed Godwit, terns and auks. A few Pomarine Skuas are recorded annually in early May, while Arctic Skua is not uncommon in both spring and autumn. Wheatears arrive in significant numbers during March, with a steady movement of other common spring migrants passing through.

Autumn witnesses regular arrivals of passerine migrants, with numbers usually exceeding those of spring. Warblers, Common Redstart and Pied Flycatcher frequent the scrubby areas, while Whinchat and Wheatear prefer the fields to the west. Easterly winds might produce something unusual, with Red-backed Shrike, Wryneck and Tawny Pipit possible. In October–November, the bushes may harbour a Firecrest among the larger numbers of Goldcrests, and perhaps a Ring Ouzel. Winter thrushes are a feature from late October. Visible migration can include Swifts and hirundines, Grey and Yellow Wagtails and Tree Pipits. Strong south-west winds may produce seabirds, with numbers of Manx and occasionally Sooty Shearwaters. Arctic, Pomarine and Great Skuas are also frequent in this period.

Bracklesham Bay has breeding Common Redshanks and Lapwings, with Short-eared and Barn Owls regular in winter.

Access

Selsey Bill To reach the Bill, continue south from Sidlesham Ferry on the B2145 and at Selsey turn left into East Street and right just before the Fisherman's Joy pub; continue to the sea. The best place to seawatch is a short distance east of the Bill. Peak seabird movements occur in spring, especially between late April and early May, the most favourable winds being light south-east.

Bracklesham Bay In Selsey turn right into Mill Lane and continue through West Sands Holiday Park. Park in a small parking area by a locked gate and continue westwards on foot. There is no access to the reserve but it can be viewed from the beach or from footpaths.

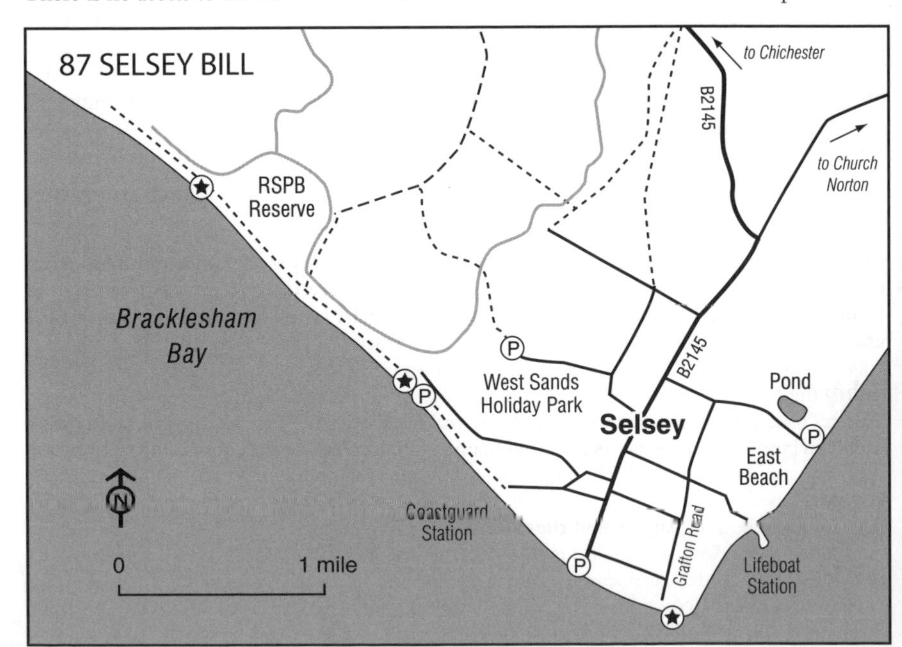

FURTHER INFORMATION

Grid refs: Selsey Bill SZ 850 923; Bracklesham Bay (West Sands Holiday Park) SZ 830 943

88 PAGHAM HARBOUR (West Sussex)

OS Landranger 197

This site, south of Chichester, is one of the best on the south coast. Part of the area is an LNR managed by West Sussex County Council. Although primarily a major wintering area for wildfowl and waders it is also noted for migrants and is an important breeding site for several species.

Habitat

More than 323 ha consist of relatively undisturbed tidal mudflats intersected by numerous channels, while shingle beaches flank either side of the harbour mouth. Patches of gorse and scrub occur around the perimeter alongside the harbour banks and sea walls, and are attractive to migrant passerines. Small areas of reeds grow around the non-tidal pools. The only woodland of note is at Church Norton and this can be productive for migrants. The harbour is surrounded by fields, which are frequently used by waders at high tide.

Species

In winter Red-throated Diver, Slavonian Grebe, Common Eider, Common Scoter and Red-breasted Merganser are regular offshore and best viewed from Church Norton. Common Guillemot and Razorbill are less frequent. In the harbour a few thousand Brent Geese winter as well as many dabbling ducks, including Pintail. Goldeneye favours the deeper channels and Smew is occasionally seen on Pagham Lagoon. Large numbers of waders use the harbour and a few Avocets, Ruffs and Black-tailed Godwits are often seen on the Ferry Pond. Water Rail, Kingfisher and Bearded Tit frequent The Severals in small numbers in winter. Short-eared Owl sometimes hunts over the saltings, which attract Rock Pipits.

In spring and (especially) autumn waders are more varied and regularly include Little Stint and Curlew Sandpiper, while Black-tailed Godwit sometimes occurs in large numbers. Sandwich and Common Terns are often present but Arctic and Black Terns are less frequent, and Little Gull only occasional. Red-necked Grebe is a rare autumn visitor, usually on the sea. Passerine migration is sometimes prominent, and typical species include Yellow Wagtail, Common Redstart, Whinchat, Wheatear, Spotted Flycatcher and a variety of warblers. Black Redstart and Pied Flycatcher are annual in small numbers. Firecrest is quite scarce, occurring mainly in spring, and Hoopoe is recorded most years. Wryneck and Melodious Warbler occur in autumn but are only occasional.

Summer brings breeding Oystercatchers and Ringed Plovers, breeding on the shingle beaches and shingle island within the harbour. A few pairs of Little Terns bred in 2007 after a decade-long absence. Shelduck nests in the adjoining fields and Reed and Sedge Warblers occupy the reedbeds. Barn Owl is often seen hunting in the Sidlesham Ferry area and breeds nearby. Though they may hunt by day, dawn and dusk offer the best chances of seeing one. Little Owl can sometimes be seen in the vicinity of Church Norton.

Access

There is a public footpath around the entire embankment enclosing the harbour, as well as a few other paths. Visitors should keep to these. Parts of the shingle are closed April–July to protect breeding birds. It is not practical to walk around the entire harbour but there are four good access points.

Sidlesham Ferry The B2145 from Chichester crosses the west arm of the harbour here. A

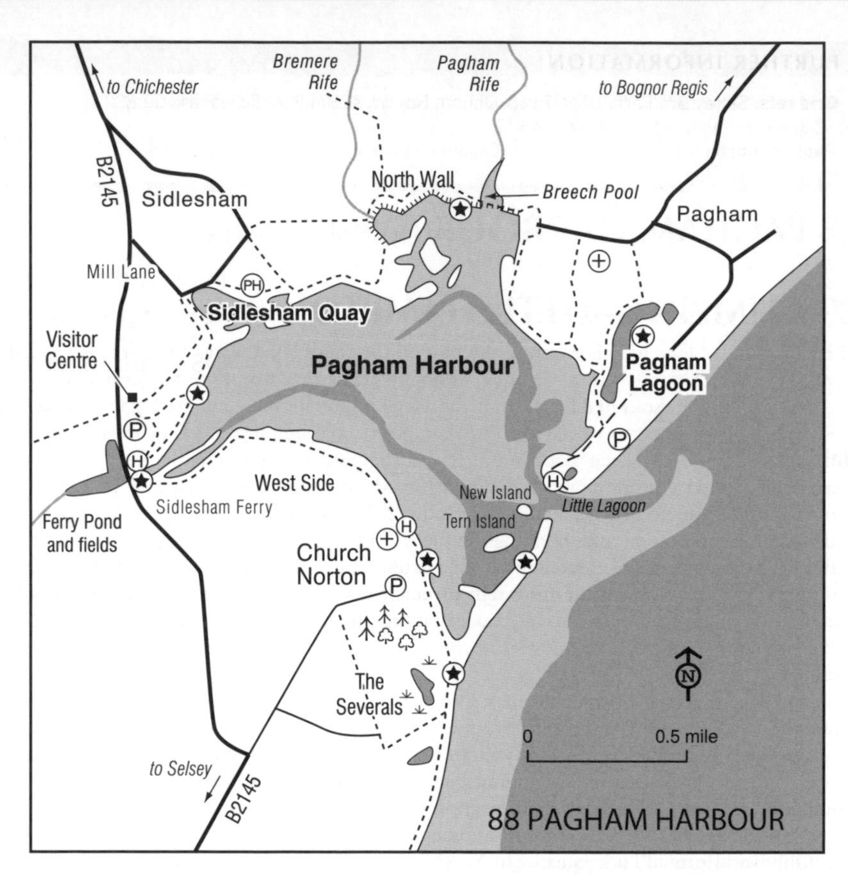

Visitor Centre, manned principally at weekends, and a small hide are sited beside the road on the north side. A nature trail starts here following the sea wall north-east towards Sidlesham Quay. Ferry Pond, west of the road and easily viewed from it, attracts many interesting waders in autumn and winter, and several rarities have been found. From Sidlesham Ferry, the walk along the sea wall around the south edge of the harbour to Church Norton is recommended. This stretch offers some of the best views over the harbour, although at low tide waders tend to be well scattered and distant.

Church Norton Continue south from Sidlesham Ferry on the B2145, turn left to Church Norton after 1.25 miles, and park at the end of the road by the church. A short track leads to the harbour. The woodland and hedges should be checked for migrants and the harbour mouth can be viewed from here. Turn right to reach the shingle shore. Seaducks and grebes are regularly seen offshore in winter. Just inland of the beach to the south there are some small freshwater pools and reedbeds (The Severals), which can hold interesting birds.

Sidlesham to Pagham The Crab and Lobster pub in Sidlesham Quay is at the edge of the harbour and from here it is possible to walk east around the sea wall towards Pagham village.

Pagham Lagoon From Chichester take the B2145 south. After 1.5 miles turn left onto the B2166 and in 2.5 miles turn right to Pagham village. The road west out of the village past the church leads to the harbour's north-east corner. Walking right eventually leads to Sidlesham and left takes you to the coast past the end of Pagham Lagoon. Alternatively, a road follows the coast past the lagoon to a car park on the spit. The end of the spit has a hide (open in winter only) and is a good place from which to see waders and ducks at the mouth of the harbour, especially at high tide. Pagham Lagoon can be quickly checked and is best in winter. The scrub and gardens near the lagoon may hold migrants.

FURTHER INFORMATION

Grid refs: Sidlesham Ferry SZ 857 965; Church Norton SZ 871 956; Sidlesham Quay SZ 861 972; Pagham Lagoon SZ 884 965
Pagham Harbour LNR: tel: 01243 641508; email: pagham.nr@westsussex.gov.uk; web: www.westsussex.gov.uk

89 CLIMPING–LITTLEHAMPTON (West Sussex)

OS Landranger 197

West Beach lies on the coast between Littlehampton and Climping and is of greatest interest during passage periods when good numbers of migrants pass through the area, including a number of rarities in recent years.

Habitat

Low sand dunes separate the hinterland from a sand and vegetated shingle beach and the sea. At low water the relatively wide area of beach comprises a mix of sand, shingle and stony ground. A golf course occupies the south-east corner, but much of the rest of the area is dominated by arable land. Inside the sea wall are areas of rough ground, hawthorn and blackthorn scrub, and small patches of private woodland.

Species

In winter Purple Sandpiper occurs at the river mouth at high tide, while other waders include Oystercatcher, Grey Plover and Sanderling. Fields support sizeable flocks of Lapwings, while Common Snipe and sometimes Jack Snipe frequent marshier areas. Offshore, Red-throated Diver, Great Crested Grebe, Common Scoter, Common Eider and Red-breasted Merganser feed. Gull flocks may include occasional Mediterranean, Glaucous or Iceland.

A good variety of migrants occurs in spring and autumn including small numbers of Common and Black Redstarts, Ring Ouzels, Firecrests and Pied Flycatchers. Hobby and Short-eared Owl are occasional and several rarities have been found in recent years. The golf course is particularly attractive to Wheatear and Whinchat, and has produced fairly regular records of Tawny Pipit.

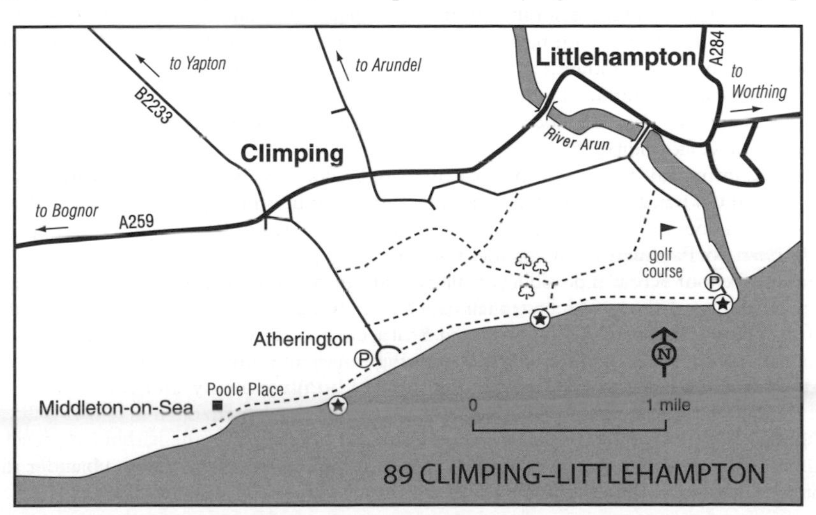

Breeding species are unremarkable, but include most of the commoner warblers, Yellowhammer and Corn Bunting.

Access

Leave Littlehampton west on the A259 and 0.5 miles after crossing the river turn sharp left to backtrack on the old road to Littlehampton. Continue for 1 mile and just before the footbridge turn right towards the mouth of the Arun. Park after 0.25 miles and walk south-west towards the coast along the north edge of the golf course, looking for migrants on the way. At the coast turn right and work the coastal bushes. Alternatively, a minor road leads to the coast from Climping, past the Black Horse pub. Check the trees and bushes in this area and continue east along the coast to the golf course.

FURTHER INFORMATION

Grid refs: Littlehampton TQ 026 012; Climping TQ 005 007

90–92 ARUN VALLEY (West Sussex)

OS Landranger 197

North of the A27 at Arundel, the beautiful valley of the River Arun is a winter home to 25,000–35,000 birds. Wildfowl are the predominant feature of this locality, but the patchwork of habitats offers some varied birding throughout the year.

Habitat

Though much of the valley has been drained, there is an interesting mosaic of lowland grazing meadows, patches of woodland and chalk escarpments. Arundel has one of the largest reedbeds in West Sussex. The RSPB reserve at Pulborough Brooks is characterised by grazing meadows managed with appropriate use of grazing animals and by varying water levels. It also features pasture, scrub and mixed woodland.

Species

Wintering wildfowl are the main attraction of the valley: significant numbers of the commoner ducks are usually present, with up to 100 Bewick's Swans, mainly at Waltham or Pulborough Brooks, and sometimes roosting at Arundel in cold weather. Geese, usually White-fronted, are only occasional, but Merlin, Hen Harrier and Short-eared Owl are often present. A small flock of Ruffs usually winters at Pulborough, while Arundel supports one or two wintering Green Sandpipers and Jack Snipe.

Both Waltham and Pulborough Brooks attract passage waders, including Little Ringed Plover, Black-tailed Godwit and Ruff, along with impressive numbers of Whimbrels.

Breeders include Garganey, Shoveler, Teal, Lapwing, Common Redshank, Common Snipe and Yellow Wagtail, with most of these at Pulborough. The reedbeds at Arundel support large numbers of Reed and Sedge Warblers, and Reed Bunting. Water Rails can be quite approachable at Arundel. The woodland between Rackham and Greatham has all three woodpeckers and Marsh Tits. Woods and heath close to Pulborough support Woodcock, Nightjar and Nightingale – the RSPB is restoring this area to improve its appeal to heathland birds. Barn Owls have bred in a nest box at the RSPB Pulborough visitor centre.

Access

A number of roads and footpaths permit fairly easy exploration of the valley. However three 'hotspots' provide the best birding opportunities.

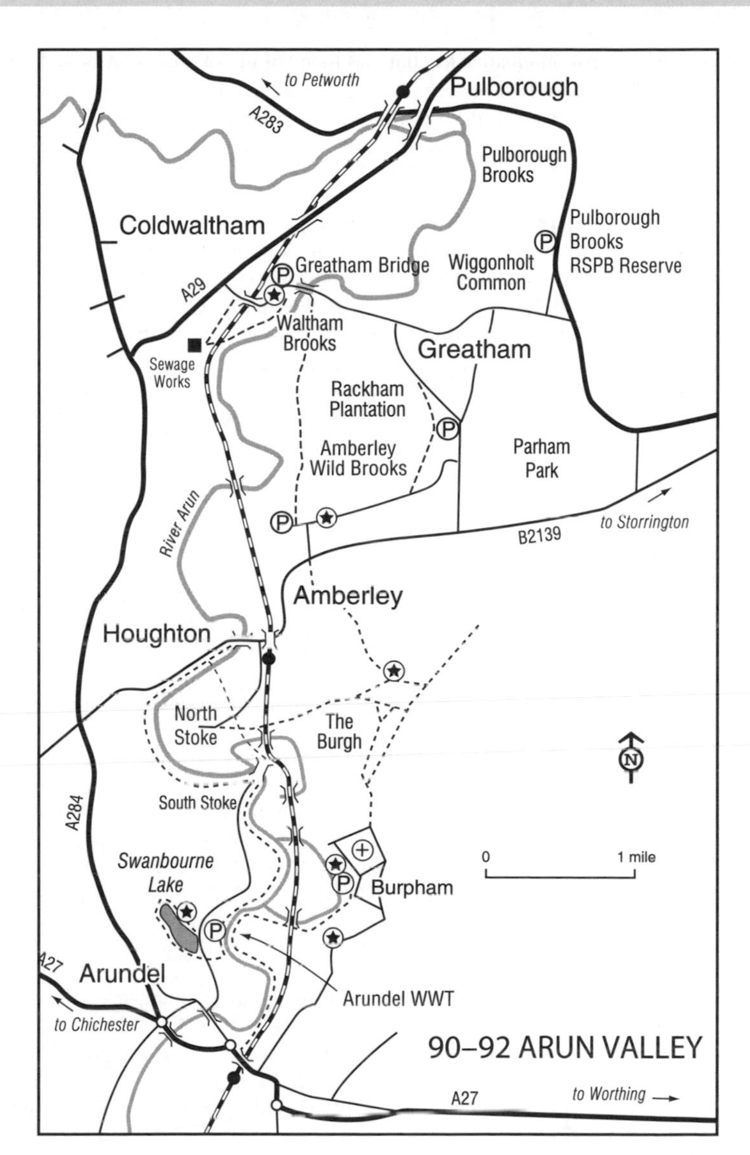

90 ARUNDEL WWT (West Sussex)

In the town of Arundel take the minor road just east of the castle towards Offham. This leads to the WWT wildfowl collection and nature reserve (open 9.30am–5.30pm in summer, to 4.30pm in winter; closed Christmas Day), where there is a large car park. Trails and hides around the reserve are wheelchair-friendly. Access to the bank of the River Arun is by walking south 200 yards from the WWT car park and then taking the path on the left bank of the mill stream for a further 200 yards. On reaching the river it is possible to walk the 1.5 miles to South Stoke, but it will be necessary to retrace your steps.

91 AMBERLEY WILD BROOKS (West Sussex)

These marshes in the valley of the River Arun are situated immediately north of Amberley, c.8 miles north of Littlehampton. From Arundel take the A284 north and after 2.5 miles turn right onto the B2139. Continue for 2.5 miles and turn left into Amberley. A track leads north across the marshes from the village and it is possible to walk to Greatham Bridge, 2 miles distant. Alternatively, drive to Greatham Bridge on minor roads around the east side of Amberley Wild Brooks, via Rackham and Greatham. Waltham Brooks lie just south of Greatham Bridge and are viewable from the car park.

92 RSPB PULBOROUGH BROOKS (West Sussex)

Situated 2 miles south-east of the village of Pulborough, the entrance to the RSPB reserve is on the left side of the A283, c.2 miles north of Storrington village. The reserve is open 9am–9pm (or sunset if earlier), closed Christmas Day, and there is a visitor centre (10am–5pm), tea room (closes earlier) and four well-positioned hides along a circular 2-mile nature trail.

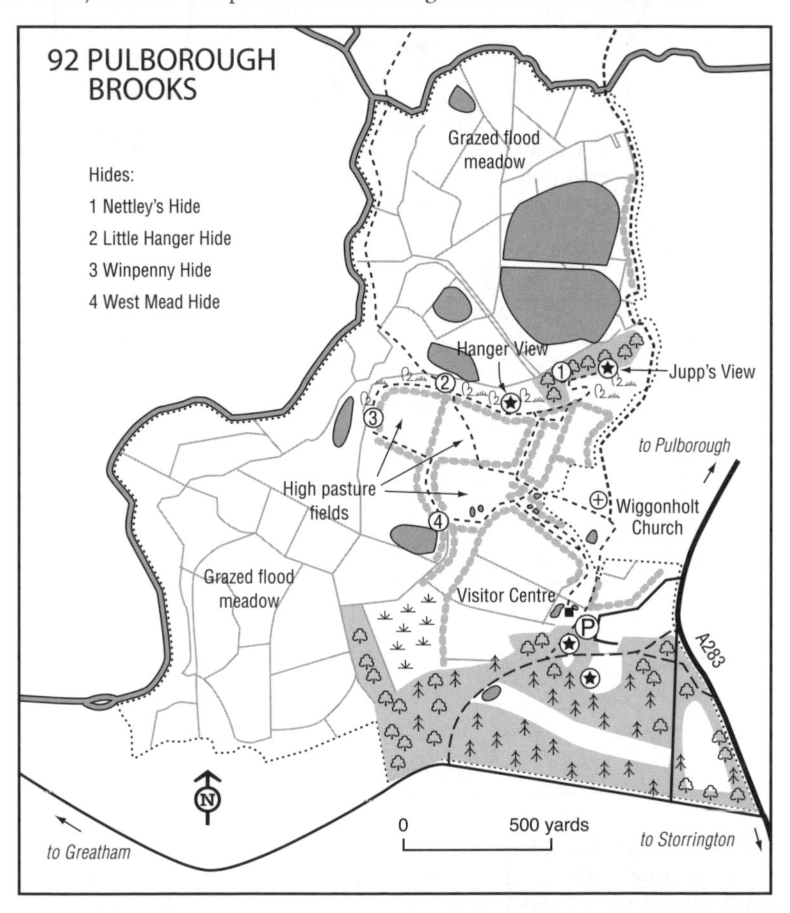

FURTHER INFORMATION

Grid refs: Arundel WWT TQ 020 080; Amberley TQ 030 133; Rackham TQ 050 143; Greatham Bridge TQ 030 161; RSPB Pulborough Brooks TQ 058 164

Arundel WWT: tel: 01903 883355; email: wwt.arundel@virgin.org.uk; web: www.wwt.org.uk/arundel

RSPB Pulborough Brooks: Upperton's Barn, Wiggonhalt, Pulborough, West Sussex RH20 2EL: tel: 01798 875851; email: pulborough.brooks@rspb.org.uk; web: www.rspb.org.uk/pulboroughbrooks

93 FRENSHAM COMMON (Surrey)

OS Landranger 186

This area of heath and woodland, a few miles west of Thursley and straddling the A287, is owned and managed by Waverley Borough Council and the National Trust.

Habitat

An area of undulating heath, woodland and artificial ponds, created by the damming of several streams, with shallow shores and reeds at one end.

Species

In winter small numbers of wildfowl use the area, with Smew occasionally present. Bittern may winter in the reeds, while the area is a noted regular haunt for Great Grey Shrike.

Breeding birds include Common Buzzard, Hobby, all three woodpeckers, Woodlark, Tree Pipit, Nightingale, Common Redstart, Stonechat and Dartford Warbler. Willow Tit (now extremely scarce in the south-east) may hang on here in small numbers.

Access

Take the A287 south from Farnham and then the signed turning on the right to reach the car park (locked 9pm–9am) at the north end of Frensham Great Pond. Here there is an information room, toilets and refreshments. Many paths cross the area but a walk through the woods and across the common to the Little Pond is probably the most profitable route.

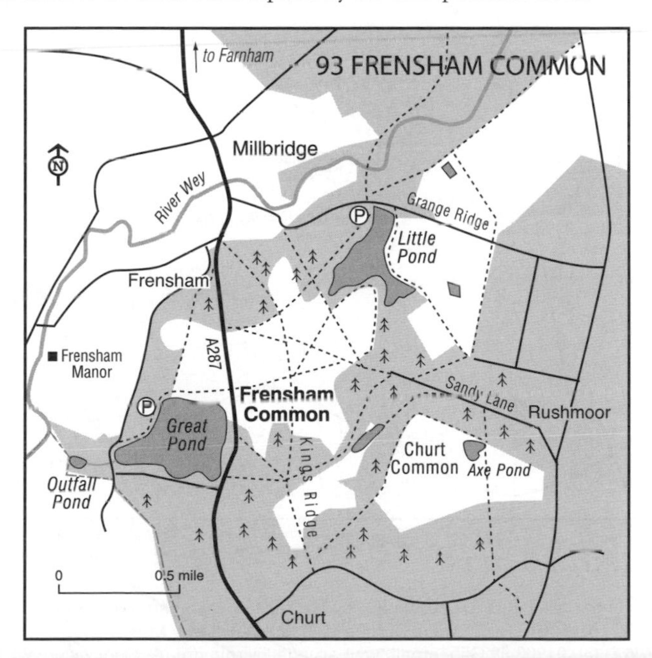

FURTHER INFORMATION

Grid refs: Great Pond SU 843 405; Little Pond SU 857 418
Ranger's Office: tel: 01252 792416

94 THURSLEY COMMON (Surrey)

OS Landranger 186

This 325-ha NNR is an important part of the west Surrey heaths. Several specialities breed, notably Hobby Woodlark and Dartford Warbler.

Habitat

The principal habitats are wet and dry heath, woodland and bog. The dry heath is dominated by heather with some gorse, and invading birch scrub is cleared to maintain this habitat. Water levels are regulated to prevent the bogs and pools drying out. Fire is a great hazard – controlled burning is carried out to assist the regeneration of heather, but in the summer of 2006, an arson attack devastated large areas of the heathland. The impact of this upon Thursley's wildlife seems

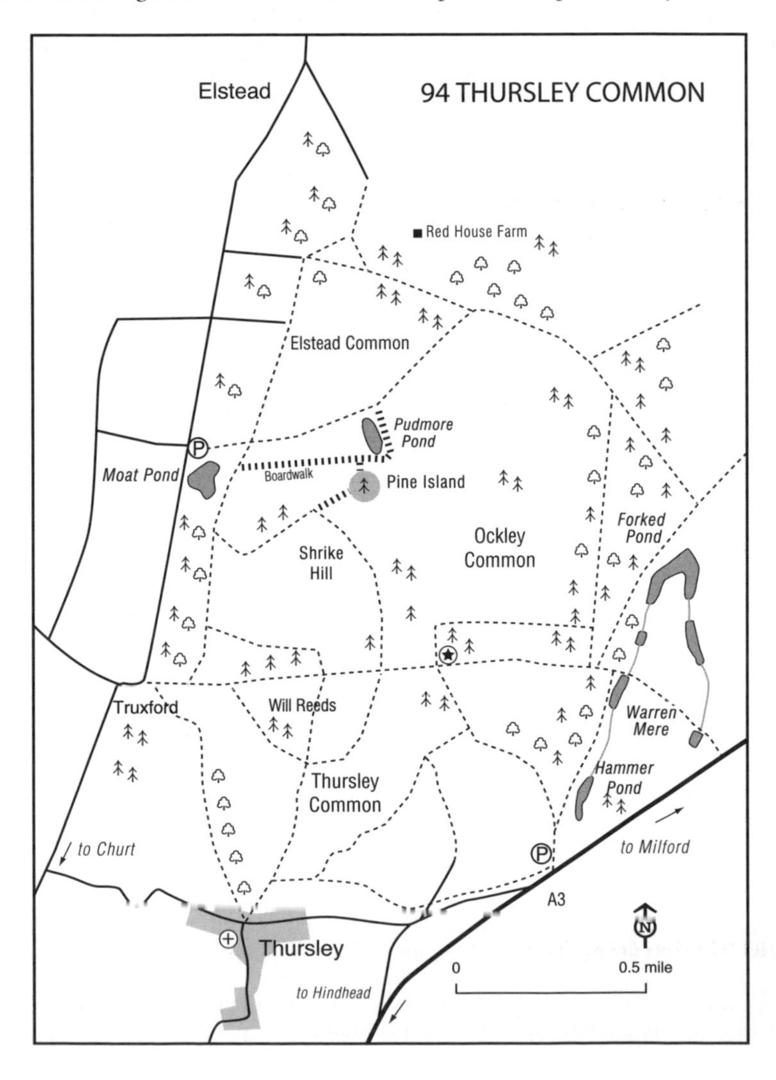

not to have been so severe as first feared; regeneration of this kind of habitat is quite rapid and many of the area's specialities can still be found quite easily.

Species

This is one of the best places to see Hobby. A good time to look is late afternoon during summer as they hawk for insects or chase hirundines; find a good vantage point to watch from. Sparrowhawk also breeds and is regularly seen. Common Snipe and a few pairs of Curlews nest, the latter a scarce breeding bird in south England. The fluctuating population of Dartford Warblers was hit hard by the fire of 2006 but some birds do remain – they favour areas of gorse and mature heather. Woodlark is another speciality and may be found on the scrubby heath or woodland edge. Though resident, the Woodlarks are often easiest to see when performing song flights in early spring. Commoner breeding species include all three woodpeckers, Tree Pipit, Common Redstart and Stonechat. Nightingale may also be seen but requires slightly more effort. Nightjars hawk for insects at dusk.

In winter, Hen Harrier and Great Grey Shrike are occasionally recorded.

Access

The Common is adjacent to the A3, c.8 miles south-west of Guildford. Access is from Thursley village, the road along the common's south edge, or the Churt–Elstead road on the west side. There is a car park on the latter road. The Common is open to the public free of charge but you should keep to the paths.

FURTHER INFORMATION

Grid refs: Thursley SU 901 397; Moat Pond SU 899 416
Natural England: tel: 01428 685878; web: www.naturalengland.org.uk

95 VIRGINIA WATER (Surrey/Berkshire)

OS Landranger 175

Lying at the south end of Windsor Great Park, a few miles south-west of Staines, this is a well-known haunt of Mandarin Duck and Hawfinch throughout the year.

Habitat

An area of sloping ground with stands of oak, sweet chestnut, beech, ash, pine and rhododendron surrounding the lake. There are some small patches of heath and wide lawns.

Species

Mandarin Duck can be found anywhere on the lake: the best areas are north of the car park and at the east end; in autumn flocks congregate on the Obelisk Pond. Common woodland species include all three woodpeckers, Nuthatch and Marsh Tit. Hawfinches can be hard to find here – numbers have fallen recently. Siskin, Lesser Redpoll and Brambling occur in winter. Ring-necked Parakeets are regularly seen in the area.

Access

Leave Staines on the A30 and beyond the Wheatsheaf Hotel turn right onto the A329, parking after 1 mile in the public car park on the right at Blacknest Gate. Walk towards Virginia Water, bearing left to cross an arm of the lake over a small stone bridge. There are some Hornbeams in this area, which are favoured by Hawfinches, though it is best to look for them in the early morning before the area becomes disturbed.

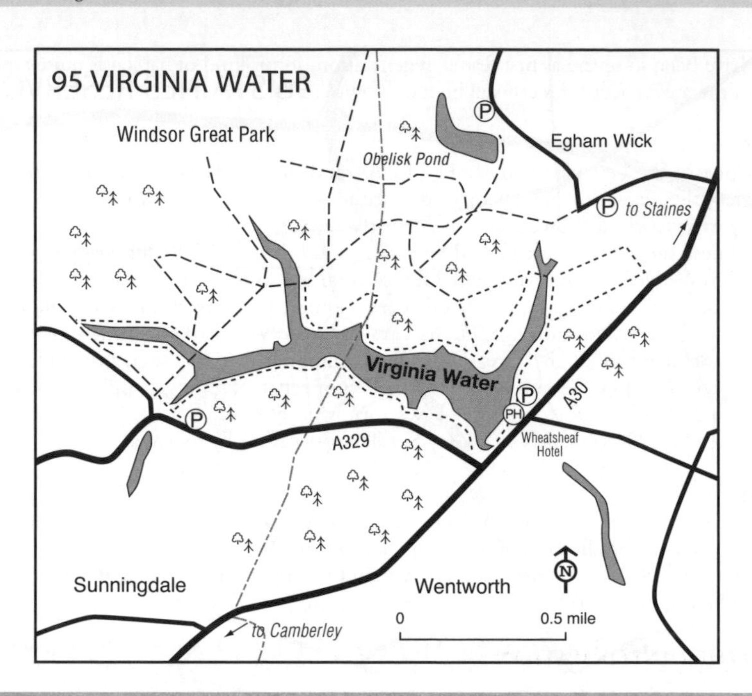

FURTHER INFORMATION

Grid ref: Blacknest Gate SU 960 686

96 STAINES RESERVOIRS (Surrey)

OS Landranger 176

Of the many reservoirs in the London area, Staines is undoubtedly the best known and most accessible. It provides sanctuary for many wintering waterfowl and attracts a variety of migrants.

Habitat

Staines' 170 ha are divided by a narrow causeway. The edges are of sloping concrete but nevertheless attract a few waders during migration seasons. Occasionally the two sections have been drained (separately) for many months at a time, and the resulting mudflats and pools have offered a haven to waders, including several rarities.

Species

Large concentrations of waterfowl are present in winter. Black-necked Grebe is regular in small numbers and can be seen almost year-round, though rarely in winter; peak counts are usually March–April and August–October. Although almost annual, Slavonian and Red-necked Grebes should be regarded as unusual. The majority of ducks are Tufted Duck and Pochard, but small numbers of Goldeneyes and Goosanders occur in winter. Smew is nowadays seen only infrequently.

A wide variety of migrants has occurred in spring and autumn. Common, Arctic and Black Terns are regular, as is Little Gull. Black Terns in autumn should be checked for a vagrant White-winged Black Tern (now almost annual in the London area). A sprinkling of waders use

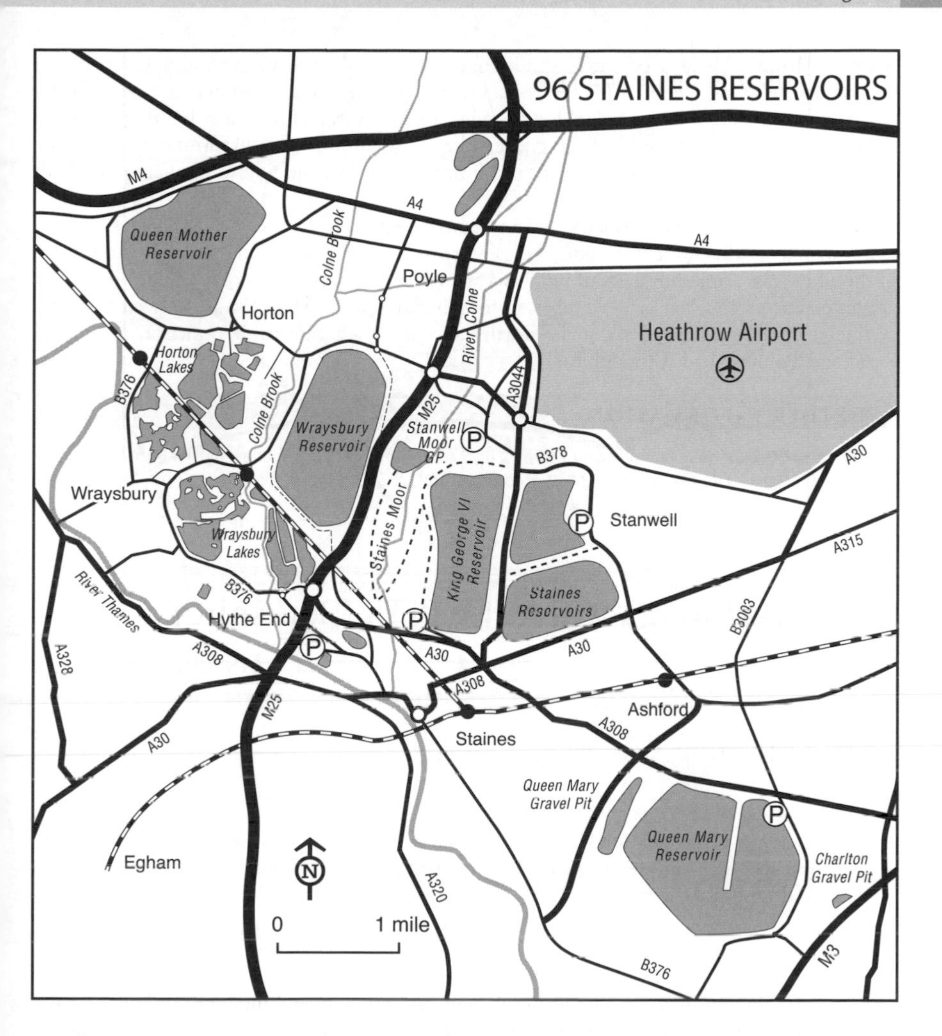

the concrete edges, though numbers and variety increase dramatically when the reservoir is drained. Some real rarities have been found among them, including five different Transatlantic vagrants. Migrant passerines often occur on the causeway, and include Wheatear, Yellow Wagtail and Whinchat. Hobby is regular in spring and autumn, often at dusk chasing hirundines.

Summer is usually quiet but some outstanding rarities have appeared in June–July, while Black-necked Grebe usually reappears from early July.

Access

Staines Reservoirs are more safely approached from the east side. Turn north into Town Lane from the A30 and park on the left beyond the causeway. Alternatively, the A3044 (Stanwell Moor Road) runs north between King George VI and Staines Reservoirs and almost halfway along there is a building on the left. Park on the grass verge and on the opposite side go through a kissing gate to the ramp leading to the causeway bisecting the reservoirs. The iron railings here are useful for resting telescopes if you have no tripod – a telescope is usually essential. There is no restriction on access to the causeway, and at weekends it can become almost crowded, mainly with birders and dog-walkers. There is no facility to walk around the reservoirs. Access to King George VI and Wraysbury Reservoirs is totally restricted without special permission.

Staines Moor This area of partly waterlogged meadow in the Colne Valley lies between Wraysbury and King George VI Reservoirs. Golden Plovers occur in winter, and occasionally Jack Snipe and Short-eared Owl. A long-staying Brown Shrike, the fourth for Britain, was a great celebrity in 2009. The southern end can be accessed from a lay-by on the A30, and the northern end from Stanwell Moor; park in Hithermoor Road and follow a concrete footpath to the moor.

Queen Mary Reservoir This large, deep reservoir lies to the south-east of Staines, and a long causeway almost divides it into two. The western side is used for sailing and the eastern is a wildfowl reserve. Large numbers of wildfowl winter and the reservoir holds a major gull roost in winter. Divers and the scarcer grebes are occasional. Queen Mary can be viewed from the perimeter or the causeway. Park at the north-east corner, adjacent to the junction of Ashford Road and Staines Road West (A308).

FURTHER INFORMATION

Grid refs: Staines Reservoir (Town Lane) TQ 055 734; Staines Moor (A30 lay-by) TQ 035 723; Stanwell Moor TQ 042 747; Queen Mary Reservoir TQ 079 703

97 WRAYSBURY GRAVEL PITS (Berkshire)

OS Landranger 176

These gravel pits lie west of Staines Reservoir in the Colne Valley and are a regular haunt of Smew in winter as well as a variety of other wildfowl.

Habitat

A complex of gravel pits surrounded by bushes and trees with adjacent arable areas. Parts of the site are protected as nature reserves, but a number of the lakes are used for other recreational activities and subject to disturbance.

Species

In winter, waterfowl include Great Crested Grebe and Cormorant and most of the commoner dabbling and diving ducks, including Goldeneye and Goosander. Smew favour Village Lake, immediately south of Wraysbury Station, and the pits west of Colne Brook. Bittern is occasional in winter. Wintering Chiffchaff and, less frequently, Blackcap may be present in the bushes. In summer, Common Tern is frequent and most of the commoner warblers can be found. Kingfisher and Ring-necked Parakeet are regularly seen in this area and are resident.

Access

Leave the A30 north-west of Staines on the B376 to Hythe End and Wraysbury. After 0.25 miles there is a stile on the right just before a bridge over Colne Brook, and this leads onto a footpath that parallels the river. The pits on the right are a favoured haunt of Smew, particularly the end pit. Silverwings Lake is easily viewed from the B376 between Hythe End and Wraysbury. Village Lake, immediately south of Wraysbury Station, is also good for Smew: leave Wraysbury towards the station and turn right to the station before the humpback bridge; the pit on the right is viewable from the road. South of Horton, Horton Gravel Pits can also be good; access from Park Lane. Wraysbury Reservoir lies east of the gravel pits and is an important gull roost, but there is no general access.

Queen Mother Reservoir To the north of Wraysbury, this large reservoir attracts grebes and the occasional diver. Access is limited to a small section of its north-east corner, which can be reached from the Colnbrook–Horton road.

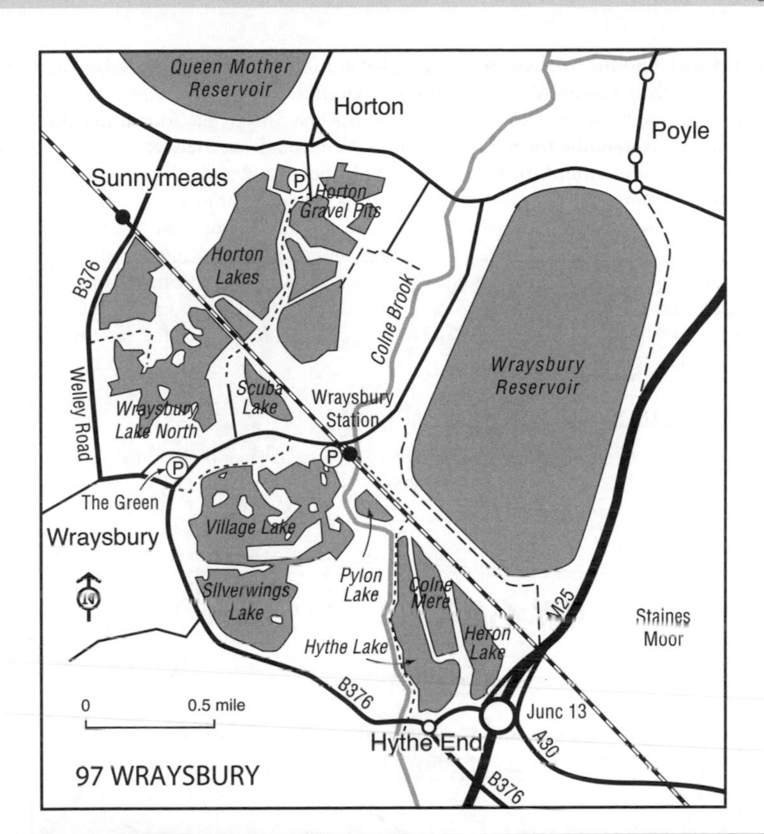

FURTHER INFORMATION

Grid refs: The Green TQ 004 740; Wraysbury Station TQ 013 741; Queen Mother Reservoir TQ 018 771

98 BRENT RESERVOIR (Greater London)

OS Landranger 176

Situated next to the North Circular Road in north-west London, Brent Reservoir is one of the best birding sites within the metropolis.

Habitat

This 52-ha reservoir is fairly shallow with natural, well-vegetated edges. There are areas of willows bordering the Eastern Marsh and several stretches of oak woodland. A finger of the reservoir extending northwards is known as the Northern Marsh and is bordered by a reedbed. The water level rarely fluctuates, but occasionally areas of mud are exposed, often at the eastern end.

Species

The commoner ducks are present in winter, but Smew is now rare. A few Water Rails inhabit the Eastern Marsh between September and April, and Common Snipe (and occasionally Jack

Snipe) are regular. Other waders are unpredictable. Large numbers of gulls are often present, sometimes including one of the less common species.

In summer, up to 50 pairs of Great Crested Grebes breed, as well as several pairs of Common Terns on rafts in front of the hides. Kingfishers occur throughout the year and the surrounding woodland holds a selection of the commoner warblers.

Spring and autumn offer the greatest variety of species and a number of unusual species have been found over the years.

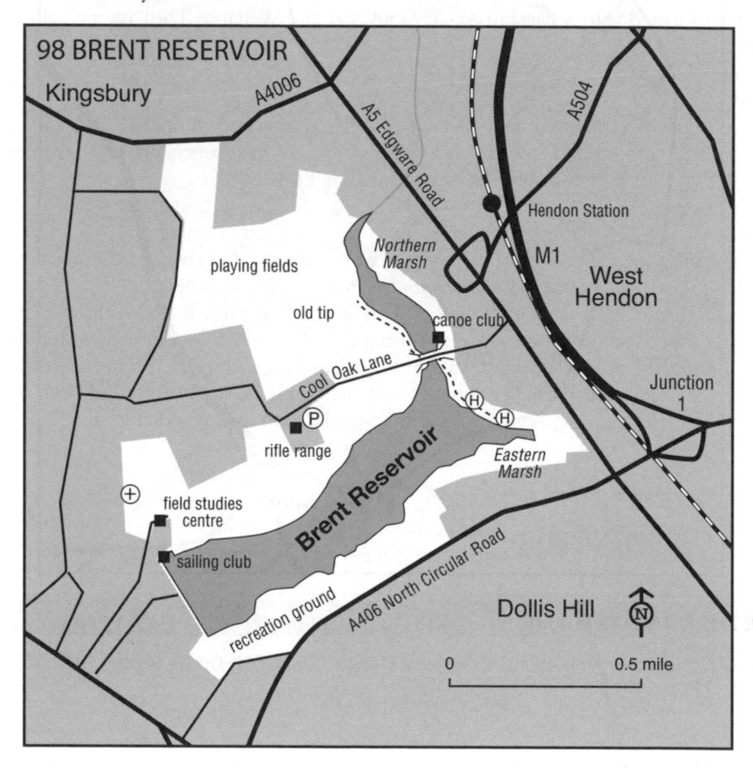

Access

The northern and eastern sections are recommended. Turn north off the A406 North Circular Road onto the A5 Edgware Road (signed to Edgware), and after 1 mile turn left into Cool Oak Lane. Cross the northern arm of the reservoir (Cool Oak Bridge) and after 0.5 miles park in the free car park. Walk back towards Cool Oak Bridge and turn left onto a track just before the bridge to explore the Northern Marsh. To visit the Eastern Marsh, cross the bridge and a footpath around the back of the canoe club car park leads to the reservoir bank and on to the Eastern Marsh.

There is no restriction on access to the reservoir except the dam wall at the western end, but much of the southern edge is not visible from the well-wooded paths. Two hides overlook the Eastern Marsh, but a key is required to use them (one-off fee, from Welsh Harp Conservation Group).

FURTHER INFORMATION

Grid ref: Cool Oak Bridge TQ 218 877
Welsh Harp Conservation Group: Tel: 020 8447 1810; email: barncol0181@aol.com; web: www.brentres.com

99 THE LONDON WETLAND CENTRE

(Greater London) OS Landranger 176

This former complex of four bleak, concrete reservoirs (known as Barn Elms) has been ambitiously transformed into a flagship 42-ha Wildfowl and Wetlands Trust reserve, close to the heart of London.

Habitat

The site consists of a mosaic of natural-edge ponds and lagoons with islands and some grazing marsh, reedbeds and scrub. There is also a plush visitor centre, observatory and seven hides. As at most WWT centres, a collection of captive wildfowl is kept here.

Species

Wintering wildfowl include nationally important numbers of Gadwalls and Shovelers. Water Rail is present in small numbers, Common Snipe in reasonable numbers and the occasional Jack Snipe occurs. Large numbers of gulls on the Main Lake infrequently include Yellow-legged Gull. Ring-necked Parakeets are common year-round, and Bittern is an occasional winter visitor.

Breeding species include Pochard, Lapwing, Little Ringed Plover, Common Redshank, the commoner warblers and Reed Bunting. Rarely, a singing Marsh Warbler has been found in spring. Hobbies are regularly seen hawking over the lakes in spring and autumn.

Access

The Wetland Centre is reached by taking the Rochampton exit to Barnes from the South Circular, and then turning right off Rocks Lane by the Red Lion pub. The site is open 9.30am–5pm during winter, and 9.30am–6pm in summer. There is a charge for non-WWT members.

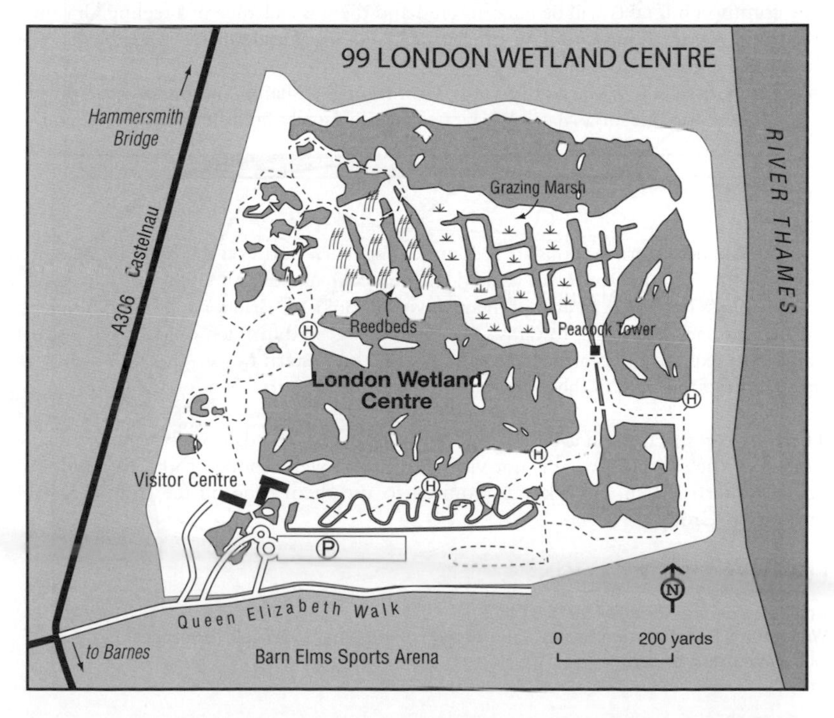

FURTHER INFORMATION

Grid ref: TQ 226 767
The London Wetland Centre: Queen Elizabeth Walk, Barnes, London SW13 9WT: tel: 020 8409 4400; email: info@wetlandcentre.org.uk; web: www.wwt.org.uk/london

100 RAINHAM MARSHES (Essex)

OS Landranger 177

The RSPB purchased Rainham Marshes, on the London–Essex border, in July 2000 and has since restored it to its former glory of lush wet grassland, reed-lined ditches and shallow pools. It is once again a magnet to birds and other wildlife and its accessibility is making it popular with London's urban birders. Nearly 270 species of birds, 31 species of butterflies and 21 species of dragonflies have been recorded, together with noisy Marsh Frogs and showy Water Voles.

Habitat

Classic lowland wet grassland, interspersed with reedbeds, the tidal Thames, rough grassland, and a small area of scrubby woodland with an adjacent landfill site (good for gulls!).

Species

Spring wader passage is often noteworthy with regular Whimbrel, Greenshank and Green Sandpiper as well as smaller numbers of Sanderlings, Knots and Bar-tailed Godwits plus occasional scarcities like Temminck's Stint, Little Stint, Curlew Sandpiper and even Stone-curlew. Spoonbill is annual and a Great White Egret usually drops in with the Little Egrets. All the common migrants will be encountered and there is a chance of a reeling Grasshopper Warbler or a passage Ring Ouzel, Whinchat or Garganey. Resident Cetti's Warblers are vocal at this time.

In summer, breeding warblers, Lapwings, Common Redshanks and ducks settle down and it is worth listening out for a Marsh Warbler as a male usually holds territory every year or so.

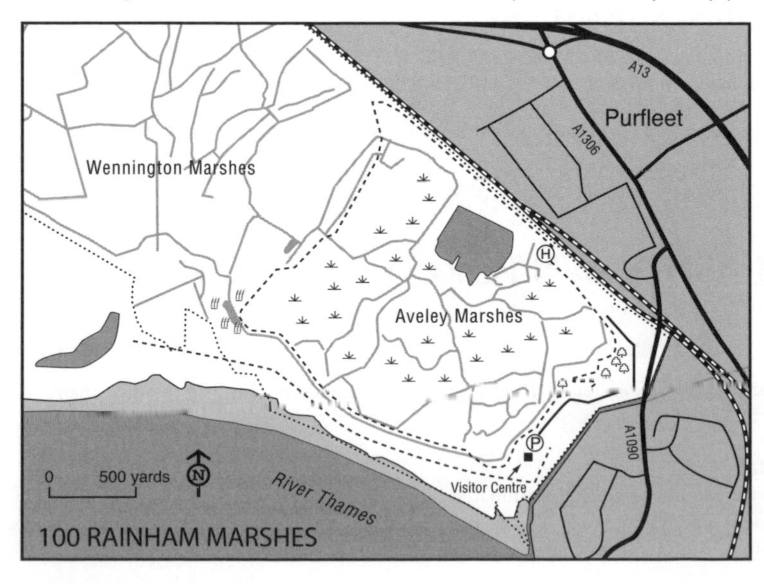

Hobbies start to use the site more as the dragonflies emerge and the likelihood of a Marsh Harrier is pretty good. Return wader passage depends on water levels and weather but is usually under way by late June and continues into late autumn occasionally including a rarity.

The site attracts most of the regular passage passerines in autumn with numerous Wheatears and Whinchats, and usually both Spotted and Pied Flycatchers, Common Redstart and Ring Ouzel passing through. Flocks of pipits, finches, thrushes and larks are on the move, and Woodlark and Tree Pipit are possible for those with their ears tuned in. Rarities such as Wryneck, Red-backed Shrike, Dartford and Yellow-browed Warblers, and Richard's and Red-throated Pipits have been recorded. October 2008 saw the unprecedented arrival of seven Serins that lingered with a huge finch flock until the New Year. Brambling, Twite and the scarcer buntings are sometimes found. It is worth keeping your eyes upwards as raptor passage has become prominent with all three harriers, Honey-buzzard, both kites, Osprey and Goshawk amongst the more expected species. Wintering Short-eared Owls usually arrive in late October.

Winter is a season for swirling flocks of ducks, Lapwings, Golden Plovers and Black-tailed Godwits with 40,000 gulls to search through for a Glaucous, Iceland, Yellow-legged or Caspian Gull. This is one of the most reliable sites in the country for the last-named species. The wintering Greylag flock occasionally attracts wild geese including White-fronted and Tundra Bean and in December 2007, 88 Barnacles touched down. Peregrines become a daily fixture. Since 2004, the reserve has been graced with wintering Penduline Tits, with up to six at a time from December to late March, and a Sociable Plover in December 2005 attracted a huge crowd. Stonechats are common and Bearded Tits winter in small numbers. Rock Pipits and Water Pipits (in the ratio of 10:1) spend the season on the Thames foreshore.

Access

From London, take the A13 eastwards; from the M25 clockwise, exit at junction 30 and head west on the A13 towards London. Exit at the A1306 to Wennington and follow signs for the A1306 east to Purfleet. (From the M25 anticlockwise, exit at junction 31 after leaving the Dartford Tunnel and follow the brown signs west on the A1306 towards Purfleet.) At the traffic lights turn right into New Tank Hill Road, go over the bridge and the reserve entrance is first on the right. The 2.5 miles of paths and boardwalks are suitable for wheelchairs, pushchairs and people with limited mobility. The Visitor Centre has a café and a shop. New facilities such as hides, paths and scrapes are being developed all the time.

FURTHER INFORMATION

Grid ref/postcode: TQ 546 788 / RM19 1SZ
RSPB Rainham Marshes: tel: 01708 899840; email: rainham.marshes@rspb.org.uk; web: www.rspb.org.uk/rainham
East London Birders Forum (local bird news): website: www.elbf.co.uk

101 ASHDOWN FOREST (East Sussex)

OS Landranger 198

The largest expanse of heathland in south-east England, this area is crossed by many paths and tracks that permit easy access.

Habitat

Extensive patches of heather-dominated heath have been invaded, in places, by Bracken, gorse, birch and pine. Large tracts of oak and Beech are present, but conifer plantations dominate other areas. Streams run off sandstone hills, cutting deep valleys and creating steep slopes where high humidity has encouraged the growth of woodland plant communities more typical of western Britain.

Species

In winter Hen Harrier may be seen quartering the heaths, and parties of redpolls and Siskins frequent streamside alders, but otherwise this is usually a very quiet season. Formerly regular, Great Grey Shrike is now very irregular in its appearances.

Spring and summer visits should prove most productive. Breeders include Common Buzzard, Woodcock, Nightjar, all three woodpeckers, Tree Pipit, Stonechat, Common Redstart, Dartford Warbler, Marsh Tit, commoner warblers (although sadly Wood Warbler no longer breeds here), and Hawfinch.

Access

The A22, A275 and B2026 are among the many roads which cross this large area. One of the best places to start is Ashdown Forest Centre situated 1 mile east of Wych Cross on the unclassified road to Colemans Hatch. The Information Centre is open 2pm–5pm on weekdays and 11am–5pm at weekends during the summer (1 Apr–30 Sep). In winter it is open at weekends only, 11am–4pm. Other large car parks are in the vicinity of Camp Hill, while many paths access the forest. Access is not permitted to the military training area near Pippingford Park.

Weir Wood Reservoir This flooded river valley lies on the north-west edge of Ashdown Forest, 2 miles south of East Grinstead. Its western end is a Local Nature Reserve, and is overlooked by a hide and a viewing platform. If water levels drop, the exposed mud attracts passage waders. From Wych Cross, take a minor road westwards signed to Hindleap Warren. At the crossroads continue straight on until the road reaches the reservoir, where there is a car park and hide.

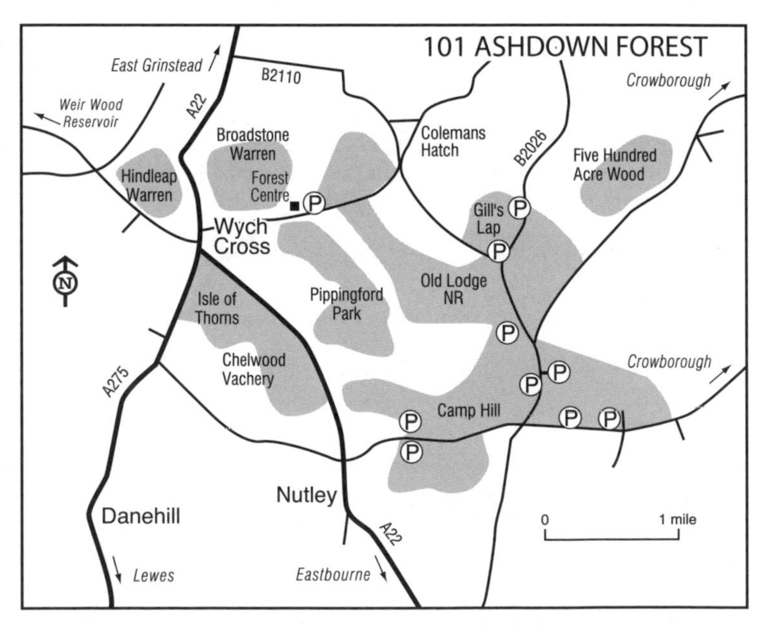

FURTHER INFORMATION

Grid refs: Ashdown Forest Centre TQ 430 325; Weir Wood Reservoir TQ 383 340
The Ashdown Forest Centre: Wych Cross, Forest Row, East Sussex RH18 5JP: tel: 01342 823583; web: www.ashdownforest.org
Friends of Weir Wood: web: www.weirwood.me.uk

102 BEWL WATER (East Sussex)

OS Landranger 199

Bewl Water is one of the largest reservoirs in southern England. It was completed in 1975 and covers over 300 ha. Though much of the area is used for watersports, part of the lake is a designated nature reserve (Sussex Wildlife Trust).

Habitat

This large lake has natural margins around most of its 15-mile circumference, though there is usually little exposed mud suitable for waders. Woodlands border some areas of the shoreline, as well as arable, scrub and orchards.

Species

In winter, moderate numbers of wildfowl use the reservoir, comprising the commoner dabbling species and including Wigeon. Smaller numbers of diving ducks occur, usually including a few Goldeneye, and occasionally Smew or Goosander. Great Crested Grebe is resident and sometimes one of the rarer grebes, such as Red-necked, may be present. More severe weather often produces a diver or rarer grebe.

Spring is also a good time to visit. The woods and bushes are full of singing passerines and migration is in progress. Most of the commoner species are present. There may also be a few terns or waders passing through.

By July return wader passage has commenced, and Green and Common Sandpipers, Little Ringed Plover and Greenshank are regular. Look also for Common and Black Terns. Hobby is quite frequent and Osprey annual. Indeed, this is one of the most regular sites for the latter in the south-east.

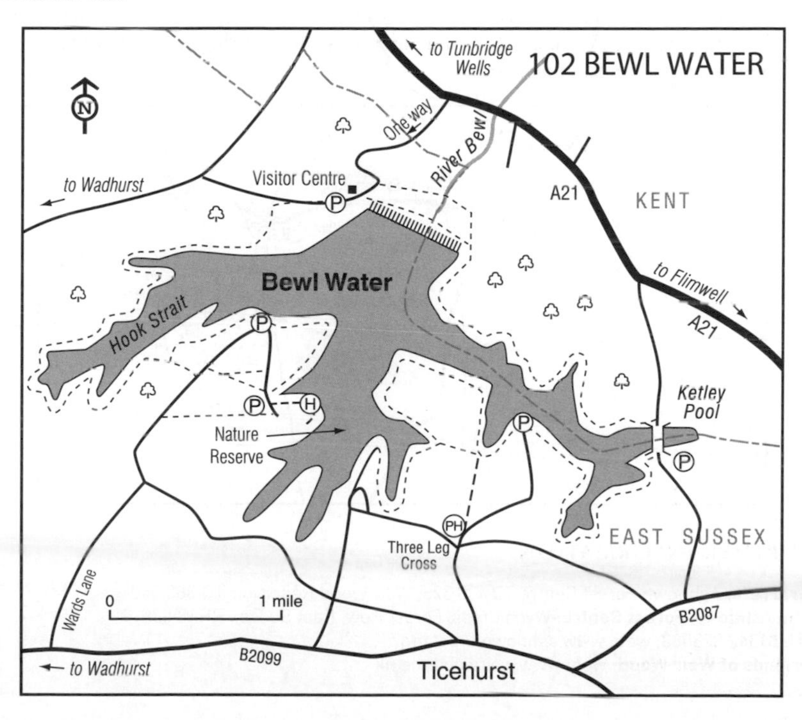

Access

The main visitor centre and car park (small charge) are only accessible from the A21 between Lamberhurst and Flimwell. From here a good footpath circuits the entire reservoir (13 miles). Several minor roads access other parts of the water; in particular at Three Leg Cross (near The Bull pub), and at Ketley Pool where the road crosses the east arm of the reservoir. There is a hide on the western shore of the central arm, overlooking the nature reserve. This is accessed by minor roads heading north from the B2099.

FURTHER INFORMATION

Grid refs: Visitor Centre TQ 675 339; Nature Reserve hide TQ 675 320
Sussex Wildlife Trust: tel: 01273 492630; email: sussexwt@cix.co.uk

103 CUCKMERE VALLEY (East Sussex)

OS Landranger 199

The valley of the Cuckmere River lies 3 miles west of East Dean, and is crossed by the A259. Seven Sisters Country Park lies immediately to the east (and reaches to Beachy Head), with Seaford Head to the west. Seaford is a well-known seawatching site.

Habitat

The lower reaches of the Cuckmere River comprise a series of grazing meadows and brackish marshes either side of the meanders of the old river, and the newer canalised river. A small scrape has been created close to the shore of Cuckmere Haven. Chalk downs lie to the east and west. An ambitious Environment Agency/Natural England/National Trust project to modify flood defences to restore a more natural estuary and potentially improve the area's potential for attracting wildlife is still in the planning stages.

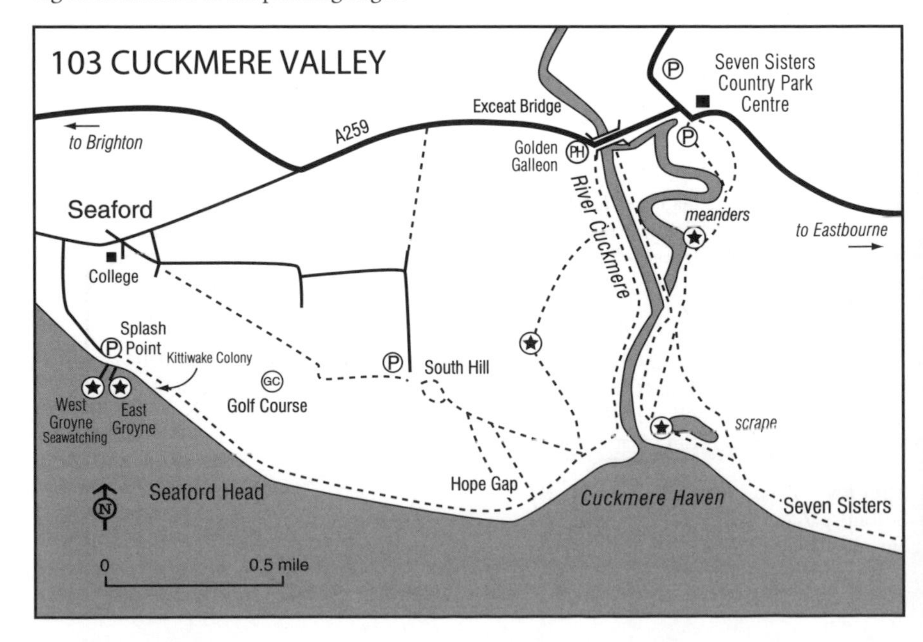

Species

The Cuckmere Valley attracts wintering wildfowl, notably several hundred Wigeon. These can be wary and are easily disturbed. Smaller numbers of other ducks may also be present, together with grebes and a few waders. Wildfowl numbers and variety increase in colder weather. Rock Pipit is regular by the river, and Twite occasionally overwinters. Kingfisher may be found (except summer) on the north oxbow and ditches west of the river. Peregrine and Little Egret are now regularly seen in the valley.

In spring and autumn migrants use the valley. Passerines are best seen in the bushes west of the river and waders favour the scrape. Check the smaller muddy pools (e.g. at Exceat Bridge) for Temminck's Stint in May. Rarities are occasionally found, especially in autumn.

Seaford Head attracts the usual downland species, while the cliffs have breeding Fulmars and Kittiwakes. Peregrines regularly patrol the cliffs, and Ravens are increasingly seen. Unsurprisingly, the seawatching at Splash Point is similar to Birling Gap, a few miles to the east, with Pomarine Skua in early to mid-May being the speciality.

Access

Cuckmere Haven Park in the large car park by the A259 (small charge). The Country Park's visitor centre north of the road has information and facilities. From here footpaths lead either side of the river to the Haven. A circular route east of the main river takes in both the old meanders and the scrape. To view the west side, walk along the road and cross the Exceat Bridge. From the Golden Galleon pub, a footpath follows the river directly to the Haven. Alternatively, another path traverses an area of bushes west of the meadows before reaching the shore.
Seaford Head to the west is a 2-mile walk, but the best place to seawatch is Splash Point, which lies on the east side of Seaford (and is more conveniently reached from Seaford itself).

FURTHER INFORMATION

Grid refs: Visitor Centre TV 518 994; South Hill TV 503 981; Splash Point TV 486 983
Seven Sisters Country Park: Exceat, Seaford, East Sussex BN25 4AD: tel: 01323 870280; web: www.sevensisters.org.uk

104 BEACHY HEAD (East Sussex)

OS Landranger 199

The cliffs of Beachy Head, immediately south-west of Eastbourne, jut south into the English Channel and this prominent position has produced one of the best migration watchpoints on the south coast, both for seabirds and landbirds. Generally, it is only worth visiting in spring and autumn, and even then can be very quiet indeed if the winds are unfavourable.

Habitat

A narrow belt of coastal downs extends for c.4 miles between Eastbourne and Birling Gap, south of the minor road that connects the two. Areas of gorse and scrub occur on the downland and there is a small wood at Belle Tout. North of the road is farmland. The eroding chalk cliffs vary from 50 feet at Birling Gap to over 500 feet opposite Beachy Head Hotel.

Species

Seabird passage is best in spring, between mid-March and mid-May. Light south-east winds are usually best. Good numbers of Red-throated and Black-throated Divers migrate east in early spring together with many Common Scoters. Brent Geese and a few Velvet Scoters can also be seen. From April, terns and Bar-tailed Godwit move in large numbers and skuas occur most days in the first half of May. Pomarine Skua is the key species here and is regularly seen in small groups, especially during south-east winds (from Birling Gap). A few Manx Shearwaters, Mediterranean Gulls, Roseate Terns and Puffins are seen each year and Little Gull and Black Tern are regular, sometimes in impressive numbers. Black Redstart and Firecrest occur from mid-March at Belle Tout and Birling Gap. The commoner migrants pass through in April–May and regularly include Nightingale, Common Redstart and Ring Ouzel. Grasshopper and Wood Warblers and Pied Flycatcher are seen most years, and scarce migrants such as Hoopoe and Serin are occasional. Migrant raptors can sometimes be seen at this time. Osprey, Marsh Harrier and Hobby are recorded annually and occasionally one of the rarer species.

In summer a few pairs of Fulmars nest and breeding passerines include Nightingale, Stonechat, Common Whitethroat and Lesser Whitethroat. Dartford Warbler formerly bred and is still occasionally seen in autumn.

There is little seabird passage in autumn. Migrant passerines, on the other hand, are usually present in good numbers, especially warblers and chats. August is the best month for the commoner species, notably *Sylvia* and Willow Warblers. In September, Chiffchaff and Blackcap predominate, with smaller numbers of Common Redstarts, Whinchats and flycatchers. Scarce migrants are found almost annually; some of the more likely candidates are Dotterel, Wryneck, Tawny Pipit, Barred, Icterine and Melodious Warblers, Red-backed Shrike and Ortolan Bunting. Perhaps the most regular is Tawny Pipit – check the stubble fields in the second half of September. In October, flocks of finches and pipits arrive, and species such as Black Redstart, Ring Ouzel and Firecrest are regular in late autumn. Yellow-browed Warbler is recorded annually and several rarities have occurred, one of the most frequent being Pallas's Warbler, which has been found several times in Belle Tout Wood in late October–early November.

There is little to see in winter, but movements of Gannet, auks or divers are always possible and Peregrines regularly hunt the cliffs. Ravens also breed locally. On the headland, resident passerines include Stonechat, Yellowhammer and Corn Bunting.

Access

There are three main areas to check, all accessible from the minor road which runs around the headland. The more energetic may wish to walk along the clifftop between each site in order to cover as much of the habitat as possible, particularly if it is a good day for migrants. Due to

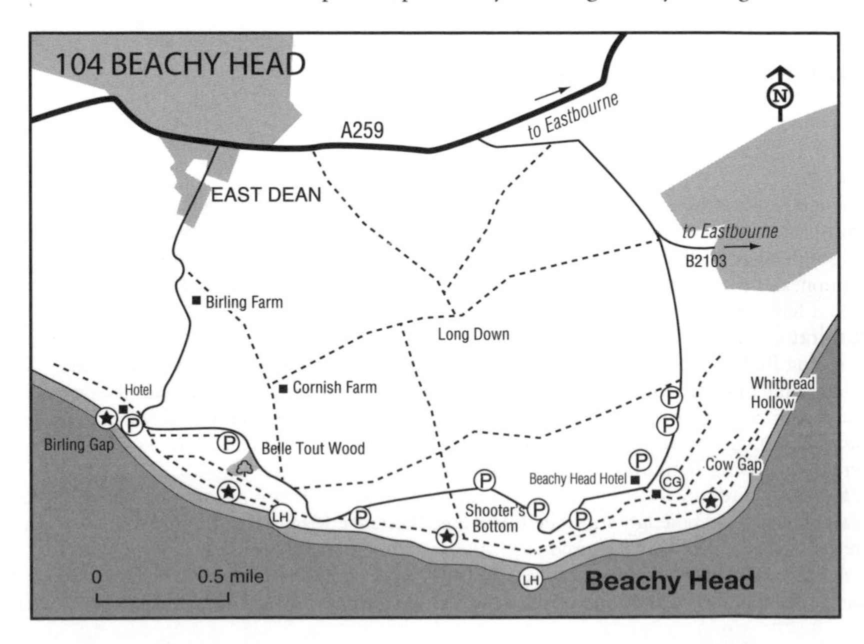

disturbance from non-birders, early mornings are always best.
Birling Gap From the car park, walk a short distance to the coast and turn right towards some beach huts. This is the best place to seawatch from. The area of bushes west of Birling is often good for migrants. Follow the track from the car park, which leads to the clifftop walk across the Seven Sisters. The bushes are just a few hundred yards from Birling.
Belle Tout Wood There is a small car park on the west side of the wood. Belle Tout Wood (also known as Horseshoe Plantation) is worth checking for passerine migrants, especially in autumn. Despite its small size, birds are often very hard to find here. The gorse and scrub on the slopes east of Belle Tout should also be explored for migrants.
Whitbread Hollow Park just east of the Beachy Head Hotel and walk along the cliff path to the top of the hollow, a large south-east-facing basin of scrub and woodland; the dense cover makes it a difficult area to work. Mist-netting is undertaken here by the Beachy Head Ringing Group (mainly at weekends, August–November). The cliff path and area between the hollow and the sea can also be rewarding.

FURTHER INFORMATION

Grid refs: Birling Gap TV 554 960; Belle Tout Wood TV 561 959; Whitbread Hollow TV 597 965

105 PEVENSEY LEVELS (East Sussex)

OS Landranger 199

Situated between Eastbourne and Bexhill, Pevensey Levels were formerly an important site for wintering wildfowl and waders. Although drainage and 'improvement' have taken their toll, some of the original habitat remains and interesting birds can still be found. Pevensey Levels NNR is part owned and managed by Natural England and the Sussex Wildlife Trust.

Habitat

Pevensey Levels originally comprised more than 4,000 ha of unimproved wet grassland, criss-crossed by a network of ditches. Although large areas were converted to arable, the area is now being managed to provide more wetland habitats. The smaller ditches hold some rare water plants and invertebrates.

Species

In winter, large flocks of Lapwings, Common Snipe and Golden Plovers are present. Other waders are less frequent, and wildfowl also occur, but in smaller numbers than previously. Wigeon and Teal are the most likely. Small numbers of swans include Bewick's and occasional Whooper. Hard weather often brings larger numbers of waders and wildfowl, and perhaps even a few geese. Raptors such as Hen Harrier, Merlin, Peregrine and Short-eared Owl are occasional. The coast at Pevensey Bay may be worth checking for gulls and waders (sometimes including Purple Sandpiper on rocks near Bexhill), and for seaducks and divers offshore.

In spring, migrants such as Yellow Wagtail are easily found, but Water Pipit and Garganey are much rarer. Several rarities have turned up over the years, including White Stork, Purple Heron, Great White Egret and Sociable Plover. Waders use the small pools on Pevensey Bridge Level, perhaps including Temminck's Stint or something rarer.

Breeding birds include Hobby, Lapwing, Common Redshank, Common Snipe, Yellow Wagtail, and Cetti's, Sedge and Reed Warblers. Recently, Marsh Harrier and Peregrine have bred on the levels for the first time.

Autumn migration is similar in extent and numbers to other coastal sites in the area, but a seawatch off Langney Point may be worthwhile. Skuas are regular and gulls favour the sewage outfall.

Access

Several minor roads cross the levels, providing good viewing (but parking is difficult). The area comprises nine levels, the best of which are:

Pevensey Bridge Level From the roundabout on the A259, just east of Pevensey, take the minor road to Normans Bay to view the south section of the levels.

Hooe Level Continue on the same road for a mile and cross the level crossing. Park on the shingle beach and walk back to gain access to the Hooe Level.

Horse Eye Level From the roundabout on the A259, take the minor road to Wartling and almost immediately turn left onto a minor road. After 2 miles you reach Rickney; take the next right and Horse Eye Level is on the left after 1 mile.

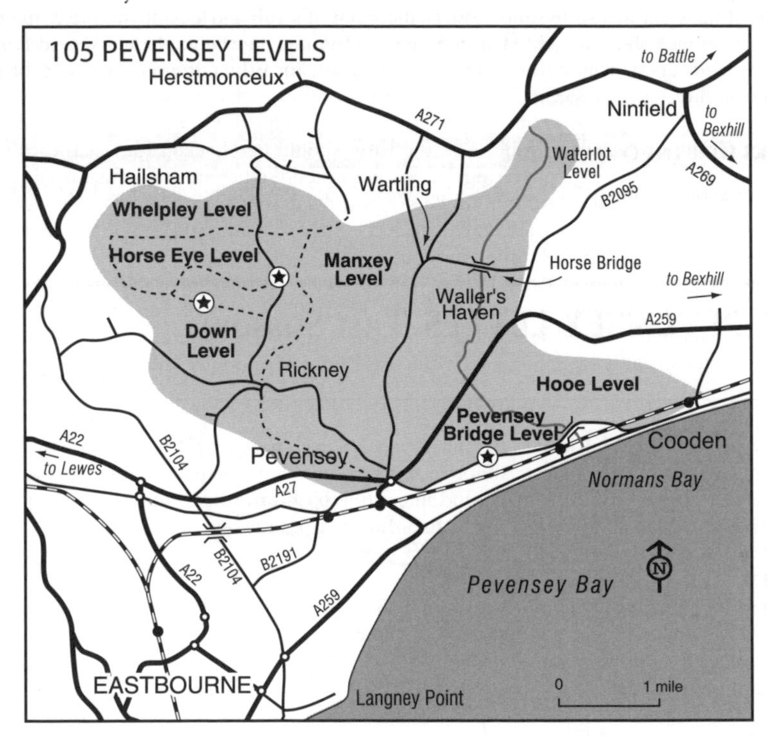

FURTHER INFORMATION

Grid refs: Pevensey Bridge Level TQ 671 056; Hooe Level TQ 697 060; Horse Eye Level TQ 630 086
Natural England: tel: 01273 476595; web: www.naturalengland.org.uk
Sussex Wildlife Trust: tel: 01273 492630; email: sussexwt@cix.co.uk; web: www.sussexwt.org.uk

106 PETT LEVEL (East Sussex)

OS Landranger 189

This series of pools surrounded by damp meadows lies beside the coast road between Winchelsea Beach and Fairlight. The pools are easily viewed from the road, and the nearest one is usually drained in mid-July to attract a variety of waders. As this site is close to Rye Harbour, the two are easily combined in a day (and both are close to Scotney and Dungeness).

Habitat

Pett Level is a large area of pasture, heavily grazed by sheep and crisscrossed by a network of narrow dykes. Pett Pools, a series of small lagoons (some with reedbeds) lie close to the road. More extensive reedbeds are found in the Pannel Valley, at the back of Pett Level, and two scrapes have been created there, with islands for breeding gulls and terns.

Species

In winter, Pett Level holds large numbers of wildfowl and waders. Lapwings, Golden Plovers, Curlews and Wigeon use the grassland, while a few ducks – perhaps including Goldeneye and Smew – visit the pools. At high tide, the commoner shoreline waders seek refuge on the levels, and it is worth checking any flocks of geese for White-fronted or Brent. Raptors are a possibility, such as Marsh or Hen Harriers, Merlin and Barn Owl. Bearded Tit can occasionally be seen in the reedbeds. Offshore, small groups of Common Eiders and larger flocks of Common Scoters can be seen; the latter should be checked for occasional Velvet Scoter (or even Surf). Large gatherings of Great Crested Grebes are frequent in winter, and Red-throated Diver and Common Guillemot also occur, but are perhaps better seen from the cliffs at Fairlight. Look out for Peregrine and Fulmar at Fairlight. The marshy areas in the Pannel Valley are good for Jack Snipe in winter.

In summer, the scrapes in Pannel Valley provide nesting habitat for waterbirds. Avocet and Mediterranean Gull may be present at this time. From July, Pett Pools can attract a variety of waders, which can be viewed at close range from the road. These often include Little Stint and Curlew Sandpiper, and sometimes a rarity such as Pectoral Sandpiper (September is best). The

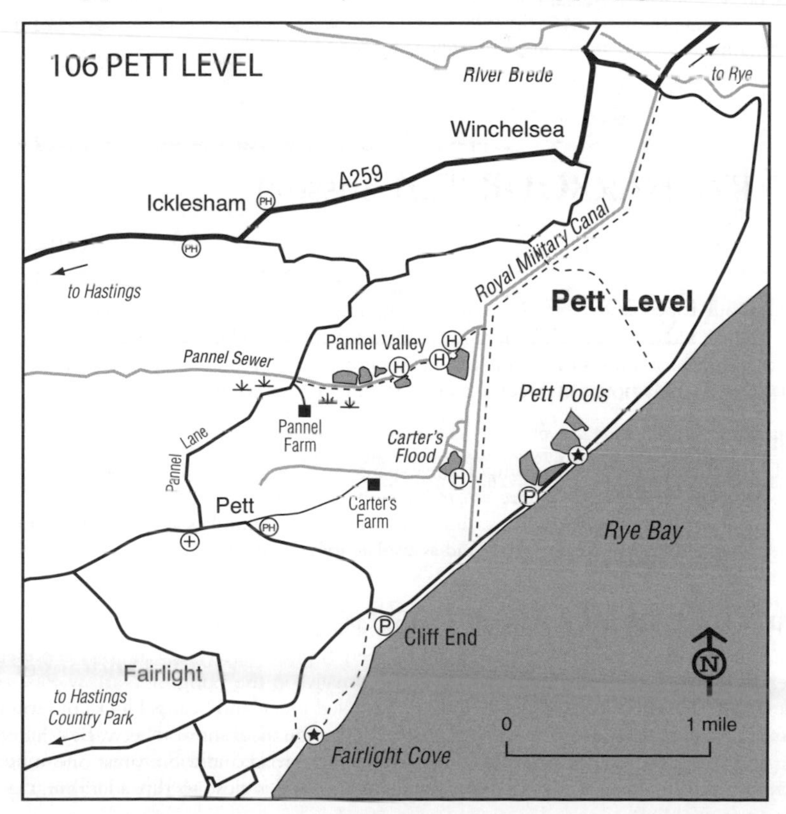

bushes at Hastings Country Park are good for migrant passerines, including the occasional rarity, and Dartford Warbler is regular.

Access

From Rye take the A259 to Hastings. After 2 miles (just before reaching Winchelsea), turn left beyond the River Brede onto a minor road to Winchelsea Beach. The road eventually follows the coast, behind a high sea wall, with the levels on the right. The pools are easily viewed from the road, and there is no access other than a footpath along the back of the levels (alongside the Royal Military Canal). Rye Bay should be checked in winter for seaducks, divers and grebes from the sea wall.

Pannel Valley Nature Reserve (see map) This private reserve to the west of the levels holds extensive reedbeds. Two pools are overlooked by three public hides. It can be accessed from the coast road at Pett Level, by walking north alongside the Royal Military Canal (check the hide overlooking Carter's Flood on the way). After 1 mile a footpath on the left follows the Pannel Valley westwards. Alternatively, this footpath can be reached from Pannel Bridge in Pannel Lane at TQ 881 151 (very limited roadside parking). It is essential to keep to public footpaths.

Fairlight Slightly further west, the cliffs at Fairlight provide a higher elevation for seawatching; a footpath from Cliff End at the western end of Pett Level leads up to the clifftops. To the west of Fairlight, Hastings Country Park, much favoured by dog-walkers, has some good areas of bushes and scrub that attract migrants.

FURTHER INFORMATION

Grid refs: Pett Pools TQ 900 143; Pannel Valley NR TQ 892 154; Fairlight cliffs TQ 883 124; Hastings Country Park TQ 860 116

Sussex Wildlife Trust: tel: 01273 492630; email: sussexwt@cix.co.uk; web: www.sussexwt.org.uk

RX Wildlife (local bird news): http://rxwildlife.org.uk

107 RYE HARBOUR (East Sussex)

OS Landranger 189

Rye Harbour is best known for its terns, notably Little Tern. Almost every species of tern on the British list has been recorded there (but most are very rare!). The reserve boasts a variety of habitats and its position on the south-east coast ensures good numbers of birds year-round. In recent years Rye Harbour has become a fairly reliable site for wintering Bitterns.

Habitat

This LNR chiefly comprises an important expanse of shingle, which hosts some rare plants. Past excavations of the shingle have created a series of pools that are now managed for birds. There are also saltings beside the River Rother and extensive mudflats at its estuary at low tide. Further inland, there are areas of grassland as well as arable and scrub.

Species

Rye Harbour is worth a visit at any time of year. Though particularly bleak in winter it is at this season that many interesting species can be seen. A good variety of wildfowl is present in winter, including Goldeneye and Smew (regular, usually on the Long or Narrow Pits, or on Castle Water). Less frequent visitors include Long-tailed Duck and Scaup. Flocks of Common Eiders and scoters, and sometimes divers, can often be found offshore in Rye Bay. All of the rarer grebes are recorded most years, particularly Black-necked and Slavonian. Numbers of waders use the mudflats at low tide, including many Sanderlings. Merlin and Peregrine are regular over the reserve, and Hen Harrier often winters. Short-eared and Long-eared Owls occasionally put in an appearance, while Barn Owl is resident (Castle Water is best). Winter passerines include flocks of Linnets and Greenfinches, plus smaller numbers of Reed and Corn Buntings, and perhaps Tree Sparrows. Rock Pipit can be seen along the Rother, but Twite and Shore Lark are less frequent. A winter visit should not miss Castle Water at dawn or dusk. There is a Little Egret roost here, and Water Rail is common but as always more often heard than seen. Several immigrant Bitterns have wintered here in most years since 2000, and when they are present sightings are almost guaranteed as they fly to their roost in the reedbeds at dusk. Conveniently, the reedbeds are overlooked by a viewpoint near the road. Bittern may sometimes be seen during the day from the hide at Castle Water.

Rye Harbour is a good area to find early spring migrants such as Wheatear, Sand Martin and Sandwich Tern. The commoner warblers and Yellow Wagtail soon follow, while occasional Arctic Skua and Gannet occur offshore. Hobby is regular in May. Particularly notable is a spring roost of Whimbrels in early May; up to 600 have been recorded, but you must arrive very early in the day to see them.

The most important breeding bird at Rye Harbour is Little Tern, although it has declined over the last 20 years. Since 2000, around 15–30 pairs have nested each year on the shingle, but although heavily guarded by volunteers, reproductive success has been low. Sadly, the birds failed

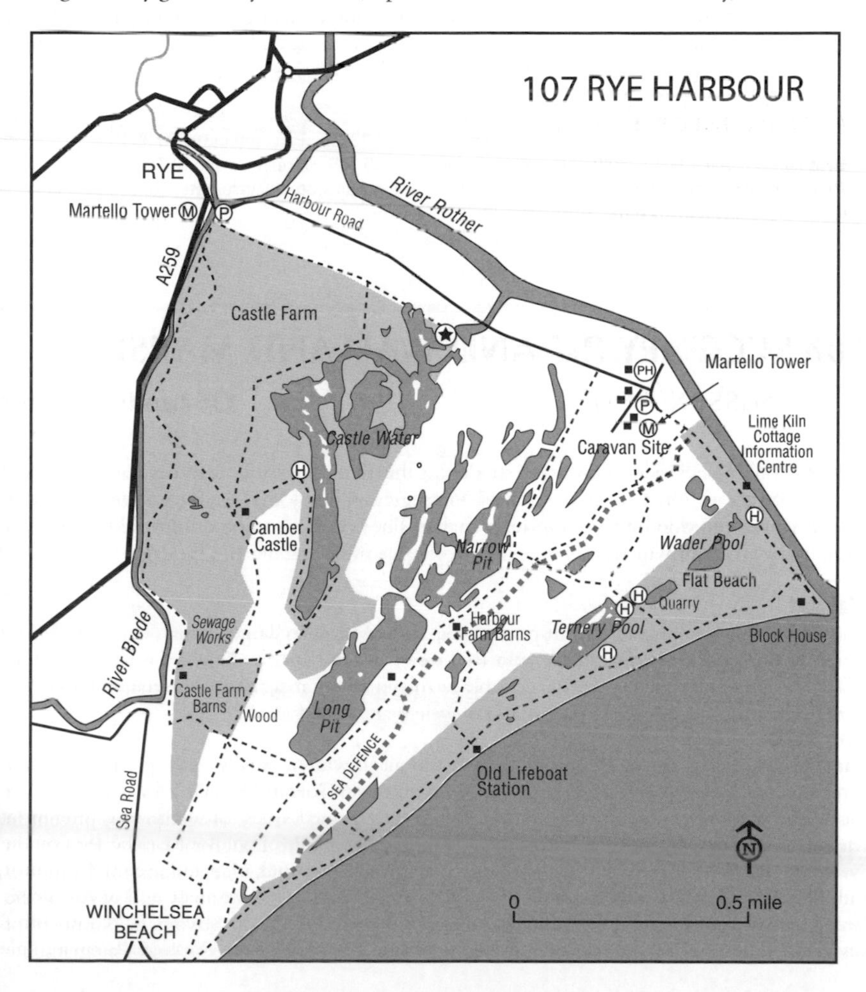

to return at all in 2008 and 2009. Common Tern breeds on the Ternery Pool, as does Sandwich Tern – numbers of the latter have reached several hundred pairs since 2000. Mediterranean Gull is a new and fast-increasing colonist, with at least 55 pairs nesting in 2008. Roseate Tern does not breed, but singles make occasional visits to the colony. Among passerines, there are several pairs of Wheatears on the reserve.

July–August is a good time to visit Rye Harbour with numbers of returning waders usually including Little Ringed Plover, Green, Wood and Curlew Sandpipers, and Little Stint. Rare American waders have appeared on a number of occasions, the most frequent being Pectoral Sandpiper. Passerine migration comprises the usual warblers plus Spotted and Pied Flycatchers, Common Redstart and Whinchat. Hobby is often evident at this time and Marsh Harrier sometimes passes through. Later in autumn, Firecrest and Black Redstart may occur.

Access

A minor road immediately south of Rye off the Winchelsea road (A259) follows the south side of the River Rother to Rye Harbour, with a car park and information point at its end. Continue on foot but keep strictly to the footpaths. The concrete road to the Rother estuary (no vehicles) leads to Lime Kiln Cottage where there is an information centre (open at peak times throughout the year). Continue to the scrape (overlooked by a hide) and the famous Ternery Pool (two hides). A new hide overlooks the Quarry area. A choice of paths will take you back via the caravan site or (for a longer walk) via the Long and Narrow Pits. For the more energetic, another path leads around Castle Water (where there is another hide).

FURTHER INFORMATION

Grid ref: Car park TQ 941 189; Lime Kiln Cottage TQ 945 185; Bittern viewpoint TQ 932 191
Rye Harbour Information Centre: tel: 01797 227784; email: yates@clara.net; web: http://rxwildlife.org.uk

108 SCOTNEY PIT AND WALLAND MARSH
(East Sussex/Kent) OS Landranger 189

Scotney is a large flooded gravel pit straddling the Kent/Sussex border between Lydd and Camber, on the southern edge of Walland Marsh. It can be viewed well from the road and is best in winter for wildfowl and the rarer grebes. The area is also noted for wild swans and a small harrier roost in winter.

Habitat

The pit has two small islands on the Sussex side, and its edges are largely open. There are MOD ranges to the south and other smaller pits to the east, but the surrounding area is largely pasture. Walland Marsh is an extensive area of arable farmland and grazing marsh dissected by dykes.

Species

Large flocks of Wigeon frequent the edges of the pit in winter, usually with good numbers of Greylags and Canada Geese, and often with Lapwings and Golden Plovers. A few feral Barnacle Geese sometimes occur and small numbers of White-fronted Geese are frequently present at the back of the pit (occasionally including a Bean). A good range of wildfowl can be seen on the water, and regularly includes Goldeneye and Smew. A sizeable flock of Scaup used to be annual, but the species is only seen irregularly now. Long-tailed Duck is less frequent, and once a Lesser Scaup was present for several weeks. Scotney is a favoured haunt for grebes (and sometimes divers) in winter, and all the regular species have been seen here – very occasionally on a single visit! Peregrine and Merlin can be found anywhere in the area in winter, and late afternoon is best for Hen and Marsh Harriers. The harriers can be seen at the back of Scotney in the late afternoon, but are best looked for at the Woolpack Inn. A flock of Bewick's Swans is present each winter on Walland Marsh, but they move around and could be on any of the pastures. A few Whoopers may also be found. Water Rails squeal from the reedbeds in winter, and Barn and Little Owls are frequently seen in the area. Look out for Tree Sparrows and Corn Buntings which can both still be seen here.

In spring and autumn, migrant waders and terns visit Scotney, and passerine migrants are likely to include Yellow Wagtail, Whinchat and Wheatear. Hobbies are frequently encountered in the area. Summer is quiet, apart from the nesting gulls on the islands in the pit.

Access

There are three areas worthy of exploration:

Scotney Pit From Lydd, take the minor road to Camber. After 1.2 miles, Scotney Pit is on the right. You can view the Kent side of the pit from the track to Scotney Court and further on there are two places where you can safely pull off on the right of the road, both on the Sussex side. These give good views over the pit.

The Midrips From Scotney, continue on the road to Camber for 1 mile until the road reaches the coast at Jury's Gap. Park on the left near the cottages, and walk eastwards along the coastal sea wall path. You can only do this if there is no firing on the ranges. Do not use this path if the red flags are flying. There are several small pools on the left of the sea wall, and the area can be good for waders and passage birds in spring and autumn.

Walland Marsh There is a well-known harrier roost near the Woolpack Inn near Brookland. From Rye take the A259 to Brenzett. The Woolpack is well signed on the right after about 7 miles (the pub is visible from the main road). Park near the pub, and take a track south for about 300 yards to a raised viewing point. The harriers come in shortly before dusk in winter. Wild swans may be anywhere on Walland Marsh.

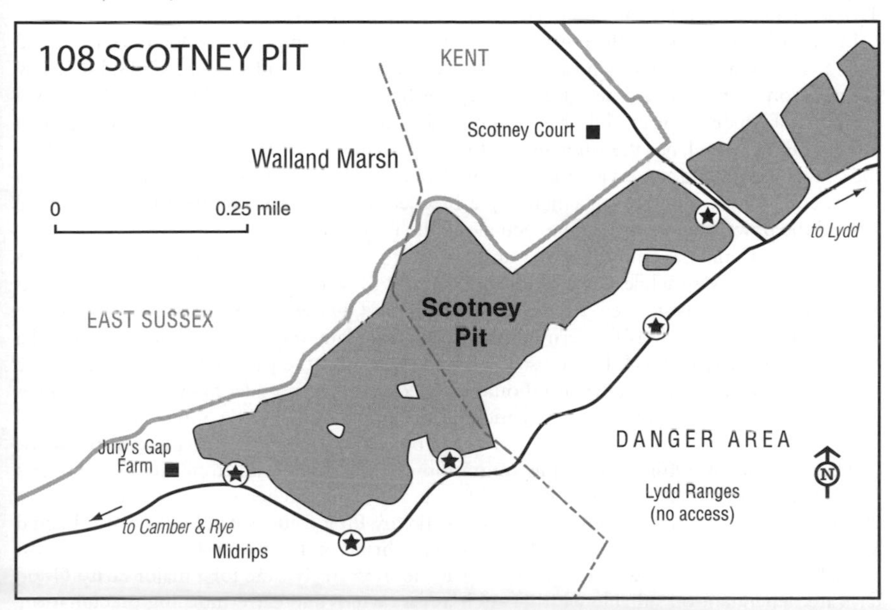

FURTHER INFORMATION

Grid refs: Scotney Pit TR 011 190; Jury's Gap TQ 990 180; Woolpack Inn TQ 978 244

109 DUNGENESS (Kent)

OS Landranger 189

This unique peninsula, between Hastings and Folkestone, is famed for migrants in spring and autumn, and has an active Bird Observatory. In addition, the RSPB maintains a bird reserve with important gull and tern colonies on the flooded gravel pits.

Habitat

The area is a flat expanse of shingle with ridges and hollows. The flora is generally sparse, though in the deeper hollows and other sheltered areas a denser vegetation including gorse and broom has developed. The only natural fresh water on Dungeness is the Open Pits and these are surrounded by dense reeds, sedges, brambles and sallows. Recently, gravel workings have created more open water with islands, considerably improving the diversity of birds.

Species

Spring migration gets under way in mid-March and the main passage commences in April and continues to mid-May. Most of the common migrants are well represented, while Ring Ouzel and Firecrest are regular. South or south-east winds in late May and June are likely to produce a rarity, Hoopoe being one of the more likely, and Wryneck, Icterine Warbler, Red-backed Shrike and Serin are almost annual at this time. Seabird passage is particularly good in spring with the peak occurring April to mid-May. The bulk comprises Common Scoter, Bar-tailed Godwit and Common Tern, together with a broad variety of divers, grebes, ducks, waders, skuas, gulls and terns. A concentrated passage of Pomarine Skuas is noted each spring, usually in the first half of May. Little Gull is sometimes numerous on the Patch, while Mediterranean Gull and Roseate Tern are regular in very small numbers.

In summer the islands in Burrowes and the ARC Pits hold impressive colonies of terns and gulls: mainly Common and Sandwich Terns and Black-headed and Herring Gulls, but a few pairs of Common Gulls also nest (the only regular breeding site in south England). Recently, a few pairs of Mediterranean Gulls have attempted to breed and for a few years a single pair of Roseate Terns nested, but the latter are no longer present. A few pairs of Little Terns breed on the coastal shingle and also on a specially created island on Burrowes Pit. Unfortunately, Kentish Plover and Stone-curlew are but a memory as breeders. Up to six pairs of Black Redstarts breed around the power station and can be seen along the perimeter fence.

In August, Gannet, Common Scoter and Sandwich Tern are likely to be the most obvious seabirds moving west, while Great Skua and Balearic Shearwater are occasional. The Patch is probably now at its best; a few Arctic Terns and Little Gulls are usually present amongst the commoner species, while Black Tern numbers peak from late August to early September and a vagrant White-winged Black Tern is usually found – although the Patch is the most likely place to look for one, all the pits in the area should be checked. Return wader migration commences in July and the commoner species may include small numbers of Little Stints, and Wood and Curlew Sandpipers. A good variety of waders can be seen from the hides on the RSPB reserve.

The volume of autumn passerine migration is impressive and, from late August, east winds may bring Wryneck, Common Redstart and Pied Flycatcher. By September there is a chance of scarcer migrants from the continent. Tawny Pipit, Bluethroat, Icterine and Barred Warblers, Red-breasted Flycatcher, Red-backed Shrike and Ortolan Bunting are almost annual but usually very few of each. October is the most likely time for a major rarity. Good days are dependent on suitable weather such as east winds and early-morning precipitation; consequently, birdless days are not infrequent. By late September the bulk of migrants will be arriving from the continent to winter, including Skylark, Meadow Pipit, Robin, thrushes, finches and even tits, and these movements continue into November.

Winter is quieter, though large numbers of ducks use the areas of open water. Among the commoner dabbling and diving ducks, a few Goldeneyes, Smews and Goosanders may be found, and sometimes a Long-tailed Duck. Rarer grebes are frequent in winter, especially Slavonian, and Dungeness is one of the few places where you may see five species of grebes in one day, though probably not on a single pit. In recent years Bittern has become a regular winter visitor with sightings possible throughout the day (often in flight). Gull flocks at this time are worth checking for Caspian. Winter raptors include Hen and Marsh Harriers, Peregrine and Merlin. Offshore, a variety of divers, grebes and seaducks occurs.

Access

From New Romney, turn left on the B2075 to Lydd c.1 mile west of the town centre (also signed Dungeness 6.5). On entering Lydd, fork left beyond the airport and on reaching a roundabout take the first exit. A long road leads directly to Dungeness. Lydd can also be reached by a minor road from Rye via Camber. Dungeness is conveniently worked from three points.

RSPB Dungeness About 1.5 miles after the roundabout in Lydd, Dungeness RSPB reserve is signed to the right along a track, with Boulderwall Farm on the corner. Follow the track

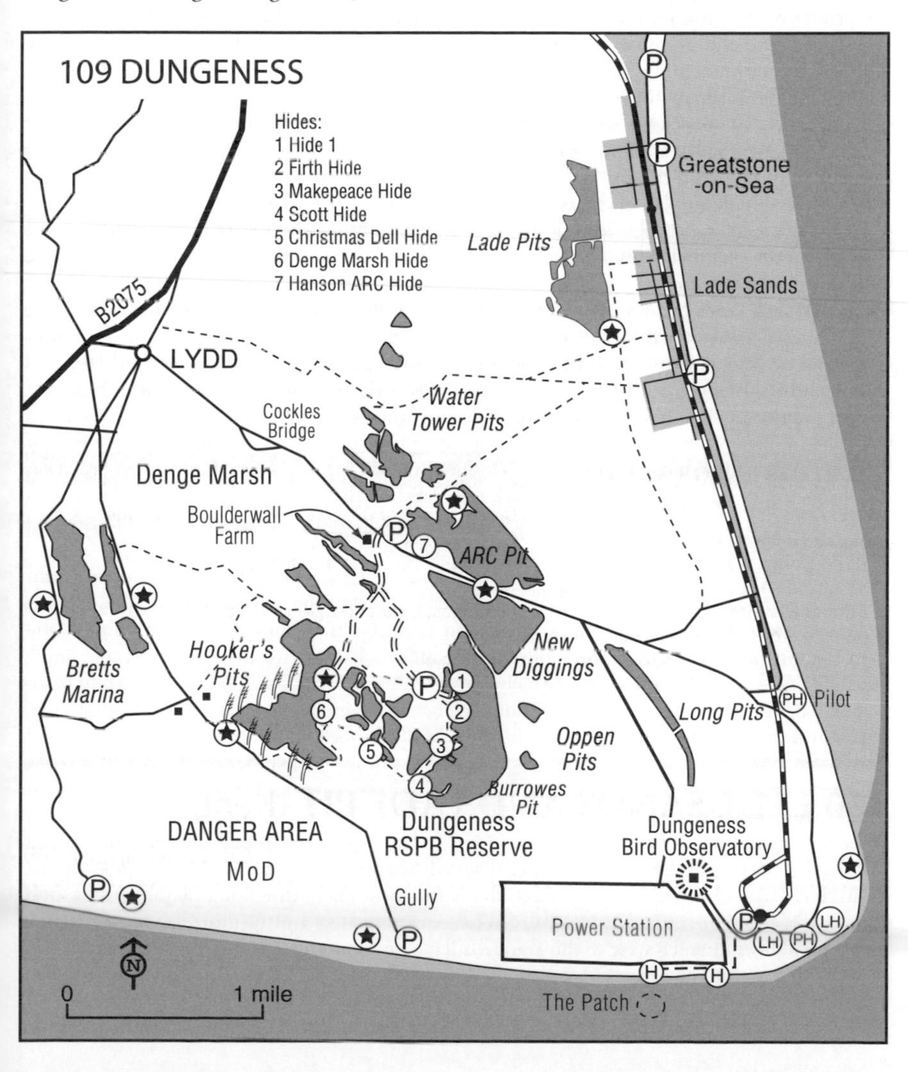

to the Visitor Centre (open 10am–5pm, until 4pm in November–February) and park. The modern visitor centre is sited at the edge of Burrowes Pit and there are six hides on the main nature trail affording good views of the pits and their islands. Most of the reserve's specialities can be seen from the nearer hides, but for the more energetic a trail of 1.5 miles will take in all six hides. If you only want to visit Denge Marsh hide overlooking Hooker's Pit, it is quicker to take the trail in an anticlockwise direction. The ARC pits on the other side of the main road are always worth a look. From Lydd, turn left instead of right at Boulderwall Farm and park in the small car park. A short nature trail leads to a hide overlooking the western end of the pit and its islands. Another trail leads to a viewing screen. You can also view the ARC pits from the main road, but parking is difficult and hazardous. The RSPB reserve is open daily, 9am–9pm (or sunset if earlier).

Dungeness Bird Observatory Continuing along the Lydd–Dungeness road, fork right c.0.5 miles after the ARC pits and again shortly after (do not take the first right fork to the power station). Continue around the headland to Dungeness. Shortly after the Britannia pub the road reaches the old (black) lighthouse. Park opposite (if visiting the observatory you can drive along the private road to it on the seaward side of the old lighthouse). Dungeness Bird Observatory (DBO) is at the end of a row of coastguard cottages and is surrounded by a bush-filled moat. This and the various clumps of bushes scattered across the shingle attract migrants in spring and autumn. Access is largely unrestricted.

Dungeness is famed for seawatching and the best point to watch from is opposite 'The Patch'. This disturbed area of water, caused by the power station outflow, regularly attracts large numbers of gulls and terns, often including some uncommon species. There is a seawatching hide on the shingle for Observatory residents and Friends of DBO.

Hostel-type self-catering accommodation is available at the observatory for up to ten people throughout the year (bring own bedding/sleeping bag and toiletries). Bookings should be made in advance with the warden.

Denge Marsh From Lydd, a minor road leads towards the power station, along the back of the RSPB reserve. In winter the fields on the left are worth checking for geese (especially White-fronts and occasional Beans, amongst the usual Canadas and Greylags), Whooper and Bewick's Swans, and Red-legged Partridge. Beyond the fields, a viewing point looks across to Hooker's pits on the reserve. The RSPB is working to restore reedbeds here.

FURTHER INFORMATION

Grid refs: RSPB Dungeness TR 067 184; Dungeness Bird Observatory TR 085 172; Denge Marsh TR 054 181

RSPB Dungeness: Boulderwall Farm, Dungeness Road, Lydd, Kent TN29 9PN: tel: 01797 320588; email: dungeness@rspb.org.uk

Dungeness Bird Observatory: 11 RNSSS Cottages, Dungeness, Kent TN29 9NA: tel: 01797 321309; email: dungeness.obs @tinyonline.co.uk

RX Wildlife (local bird news): http://rxwildlife.org.uk

110 LADE SANDS AND LADE PIT (Kent)

OS Landranger 189

Lade Sands is an area of sandy mudflats between Dungeness and Greatstone-on-Sea with a gravel pit a short distance west of the coast road.

Habitat

The extensive mudflats of Lade Sands are exposed at low tide. Lade Pit is a flooded gravel pit, similar in character to the many other pits in the Dungeness area, with shallow edges and vegetation such as sallows, osiers and reeds.

Species

Wintering waders on the shore include Oystercatcher, Ringed and Grey Plovers, Dunlin, Sanderling, Knot, Curlew and Bar-tailed Godwit. On the sea Red-throated Diver, Great Crested Grebe and possibly Common Scoter and Eider are often present. Check gull flocks for Mediterranean Gull. Lade Pit attracts some of the commoner dabbling and diving ducks, as well as a few Scaup, Goldeneyes and Smews. A diver or one of the rarer grebes is often present, while a Bittern may also take up residence. A Chiffchaff sometimes frequents the sallow scrub at the north end of the pit, while Stonechat, Black Redstart and, in some winters, a Dartford Warbler may all be present. Lade's biggest rarity was a Canvasback in 2001.

On passage, waders such as Green and Common Sandpipers are likely but rarer species such as Pectoral Sandpiper, Red-necked or Grey Phalaropes appear only occasionally. Black Tern and Little Gull may also be seen.

Access

Follow the coast road for 1.5 miles north of the Pilot pub. Turn left along Derville Road and park sensibly at the end of the road. From here walk across the shingle to view the pit. Lade Sands extend for c.4 miles north of the Pilot pub and may be viewed at a number of points from the coast road, the best being the Lade toilet block car park and the car park further north at Greatstone-on-Sea.

FURTHER INFORMATION

Grid refs: Lade Pit TR 078 215; Lade Sands (Greatstone-on-Sea) TR 082 228

111 BEDGEBURY FOREST (Kent)

OS Landranger 188

This large area of forest south of Goudhurst is noted as a regular haunt of Common Crossbill and Hawfinch.

Habitat

Extensive coniferous plantations are mixed with small areas of sweet chestnut coppice. The Pinetum contains an excellent variety of conifers planted on grassy slopes, and a small lake, as well as a variety of visitor facilities and attractions.

Species

Noted as a regular winter site for Hawfinch, though numbers have declined in recent years; Siskin, Lesser Redpoll and Brambling are also usually present. This is one of the best sites in the county for Common Crossbill during irruptions.

Breeders include Woodcock, Nightjar, Great Spotted and Green Woodpeckers, Nightingale, a variety of warblers, Goldcrest, Firecrest, Siskin and Lesser Redpoll.

Access

The Pinetum is open from 8am till dusk. From Goudhurst take the B2079 south towards Flimwell. A car park, complete with refreshments and toilet facilities, is reached after c.3 miles. After paying the entrance fee follow the path into the Pinetum and the cypress trees favoured by roosting finches are just beyond the sharp rise.

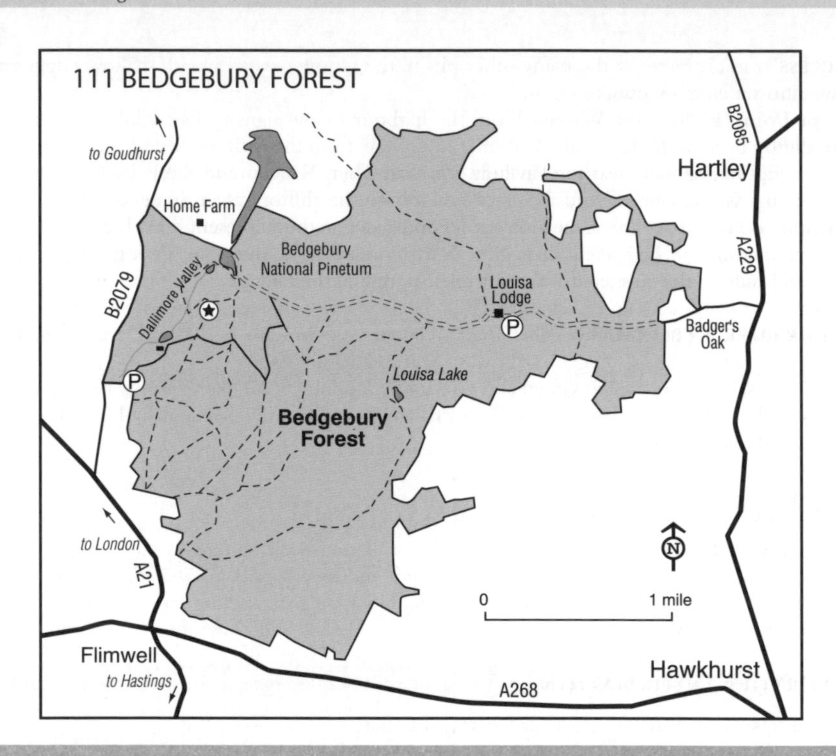

FURTHER INFORMATION

Grid ref: TQ 716 331

112 FOLKESTONE–DOVER (Kent)

OS Landranger 179

The sewage outfall at Copt Point, north of Folkestone Harbour, is a particularly good spot for Mediterranean Gull. Samphire Hoe, south-west of Dover, is a convenient place to see some species which are local in south-east England.

Habitat

The area is dominated by the famous 'white cliffs'. Samphire Hoe is a new site, largely created from material excavated from the Channel Tunnel.

Species

Mediterranean Gull is regular at the sewage outfall throughout the year, with the largest numbers occurring in autumn and winter. Some regularly frequent the clifftop fields where they may be attracted by bread to offer superb views. At high tide many gulls roost in Folkestone Harbour and Purple Sandpiper roosts on the harbour walls. The extensive scrub of Folkestone Warren attracts migrants in both spring and autumn, with Firecrest and Ring Ouzel regular. South-east winds may produce movements of seabirds including divers, shearwaters, skuas and terns.

Shag is sometimes seen at the west end of Samphire Hoe and check the sea wall and slopes for Stonechat, Black Redstart and Rock Pipit. Look out for Fulmar and Peregrine on the cliffs.

Access

Two sites are worth visiting:

Copt Point/Folkestone Warren From the harbour follow signs to East Cliffs and park on the clifftop beyond the East Cliff Pavilion. Either view from the clifftops or descend to the base of the cliffs by the path near the Pavilion, and walk along the shore to the Point. Walk east to explore the Warren, which may also be accessed from the clifftop café at Capel-le-Ferne.

Samphire Hoe From Dover, head towards Folkestone on the A20. Samphire Hoe is signed on the left, the minor road passing through a short tunnel to lead to the site. Park in the car park (fee) and walk to the west end of the complex alongside the railway.

FURTHER INFORMATION

Grid refs: Copt Point TR 240 365; Folkestone Warren TR 254 384; Samphire Hoe TR 294 390

113 ST MARGARET'S BAY (Kent)

OS Landranger 179

The coastline here is the closest point in Britain to France and consequently this can be an excellent place to observe visible passerine migration. Large falls of common migrants are not infrequent in autumn, especially October, and scarcer species are occasionally found. Weather conditions appear to be irrelevant and falls even occur in south-west winds when other migration watchpoints have few birds. An early-morning start is highly recommended as many birds move quickly through the area.

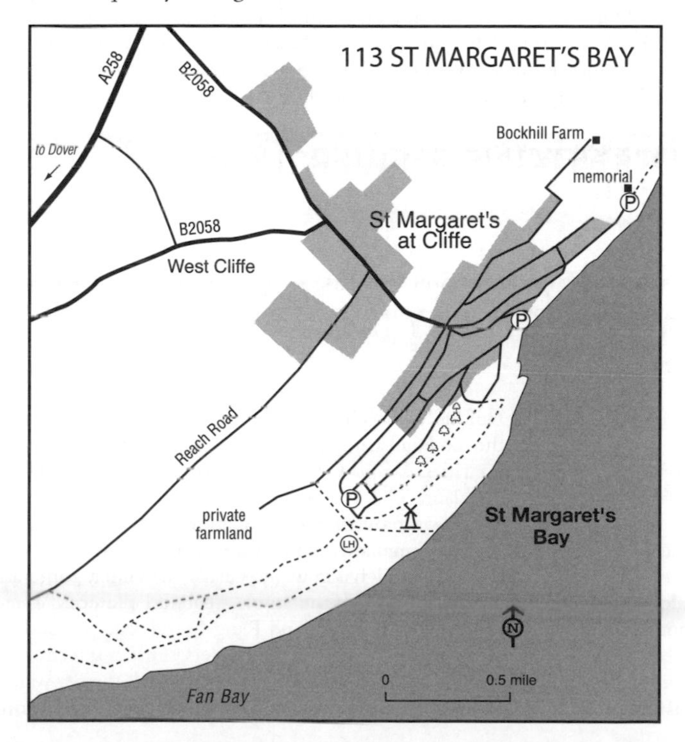

Habitat

While the coast faces south-east here, the main South Foreland valley runs north-east to south-west. Chalk cliffs reach a height of nearly 300 feet and have a rocky shoreline below. Scrub and copses are interspersed by grassy patches. North of the village is the NT-owned farm at Bockhill, which is surrounded by large trees and situated in a large area of arable land bisected by several paths and hedgerows.

Species

This area is at its best in spring (possibility of Marsh Warbler) and, in particular, autumn when large numbers of pipits, commoner thrushes, warblers and finches pass through. Falls of night migrants can be very impressive. Grasshopper Warbler is regularly flushed from clifftop grass, while other frequently occurring scarcer species include Ring Ouzel, Firecrest, Tawny and Richard's Pipits, Yellow-browed and Pallas's Warblers, Golden Oriole and Ortolan Bunting. Major rarities are found most years and have recently included Alpine Accentor, Red-flanked Bluetail, Booted and Radde's Warblers and Nutcracker. Dotterel has become a regular feature of late August–early September, with flocks often found in the fields south of the South Foreland lighthouse and in the vicinity of Langdon Bay. Passage raptors include numbers of Sparrowhawks, Marsh and Hen Harriers and Hobbies, with scarcer species such as Honey-buzzard, Rough-legged Buzzard and Osprey all recorded with some regularity. St Margaret's is not an ideal place to seawatch, but small numbers of skuas and shearwaters are recorded in most passage periods.

The cliffs are home to breeding colonies of Kittiwake and Fulmar. A few pairs of Rock Pipits also breed, but otherwise the area is rather quiet in summer.

Access

Leave Dover on the A258 to Deal and after c.2 miles turn right on the B2058 to St Margaret's at Cliffe. On entering the village turn right onto an unclassified road to the lighthouse and park before reaching it. Explore the valley in search of migrants. For Bockhill follow the road through the village and turn left onto Granville Road. Drive to the end and park by the monument. Visible migration can be observed here, and is at its best early morning. The farm is reached by walking along the path west from the clifftop.

FURTHER INFORMATION

Grid ref: TR 368 444

114 SANDWICH BAY (Kent)

OS Landranger 179

On the east coast of Kent south of Ramsgate, Sandwich is noted for migrants, including a sprinkling of scarce species and rarities. A Bird Observatory operates over an extensive area of privately owned dunes and marshland. The Observatory has recently undergone substantial redevelopment and now boasts much improved accommodation, a laboratory and educational facilities. The peak seasons are March–June and August–November. Pegwell Bay to the north and Sandwich Bay itself attract waders and wildfowl, especially in winter. Large areas of the mudflats, saltings and dunes around the mouth of the Stour estuary are protected as reserves and are maintained by the NT and Kent Wildlife Trust (KWT).

Habitat

The coast is dominated by a series of sand dunes, many of which have been converted into golf links. Inland, a large proportion of the area is farmed, and waders roost on these fields at high tide. Small areas of freshwater marsh, reedbeds, bushes and trees add diversity. To the north the River Stour enters the sea at Pegwell Bay and extensive areas of mud, sandflats and saltings surround the estuary.

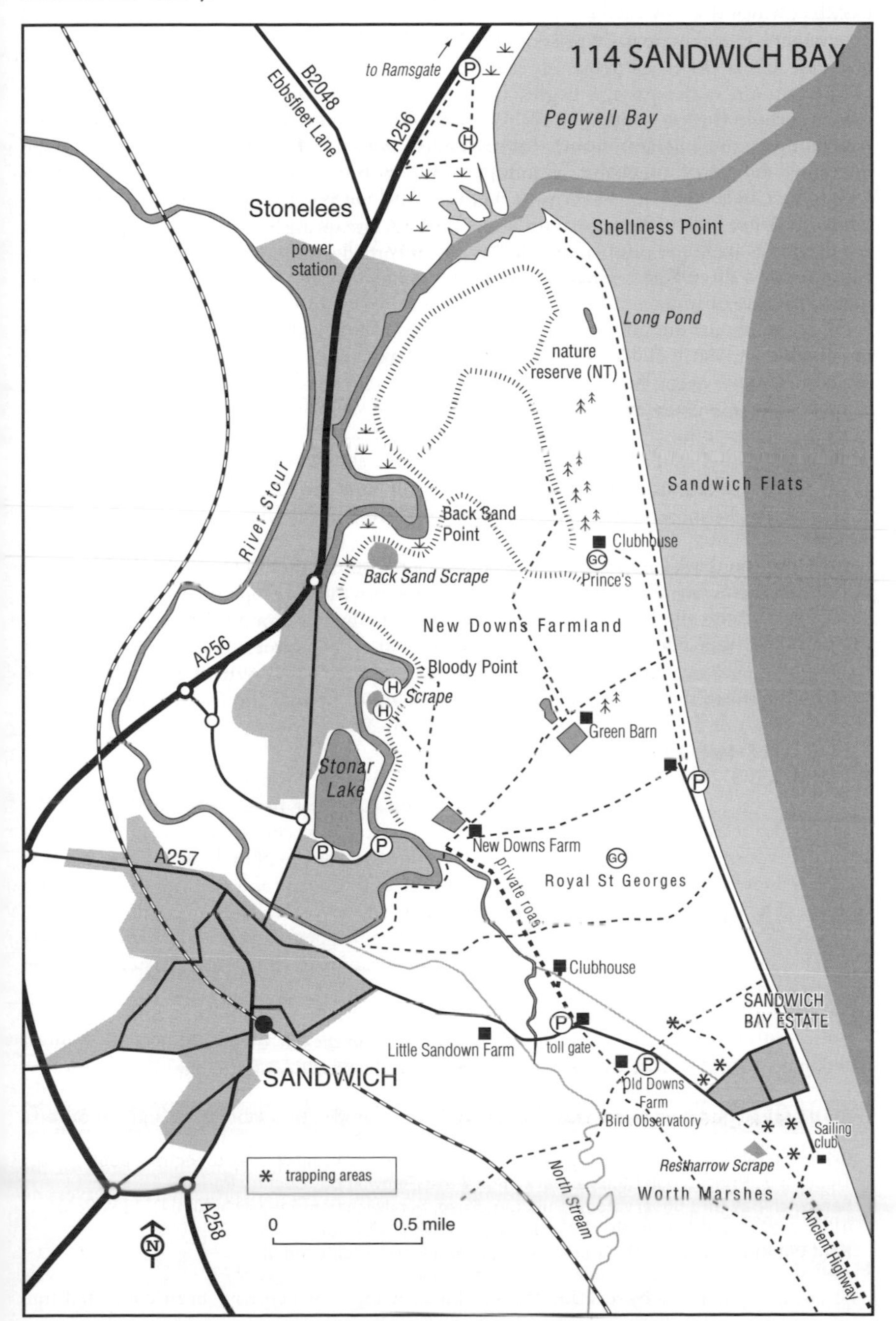

Species

The first spring arrivals are in March and include Garganey, Black Redstart and Firecrest, and by the end of the month Sandwich Terns begin to appear. The bulk of the passage is April–May, when a number of scarce migrants are regularly recorded, including Kentish Plover and Golden Oriole.

Summer is quiet but up to 30 pairs of Little Terns breed in the area, and Kittiwakes, from their breeding cliffs at Dover, are frequently seen.

The autumn wader passage begins in mid-July and continues into September. Typical species include Curlew Sandpiper, Black-tailed Godwit and Wood Sandpiper. August brings return of passerine migrants, mainly chats, warblers and flycatchers. East winds encourage the occasional fall, which inevitably includes a scarce species or two. By late autumn the bulk consists of Goldcrests, Robins, thrushes, Starlings and finches arriving from the continent to winter in Britain. Rarities can turn up at any time but late spring and particularly late autumn are the most likely periods. Sandwich Bay has recorded a number of outstanding species, and recently Pallas's Warblers have been almost annual in late October or early November, sometimes several at once.

Divers, grebes and seaducks may be seen offshore in winter and Hen Harrier and Short-eared Owl hunt the marshes. Golden Plover and Lapwing take up residence in the fields and a variety of the commoner waders inhabits the mudflats. Snow Bunting is regular on the shore and Twite can be found around the estuary.

Access

Leave Sandwich east along Sandown Road towards Sandwich Bay Estate. The observatory, at Old Downs Farm, is on the right shortly beyond the toll gate. Access to the main areas of interest is as follows.

Sandwich Bay Passerine migrants are mainly attracted to the bushes and gardens in the vicinity of the Estate. Access is from the toll road via the gate opposite the track to the Observatory. Cross the field to the stile and check the bushes around the Heligoland traps. There are no restrictions on visiting the Observatory but birders should not enter the enclosed trapping areas and should respect residents' privacy when watching birds in gardens.

Worth Marshes From the Observatory, continue south beyond the Estate on the Ancient Highway towards the Chequers pub. Footpaths cross Worth Marshes to North Stream and back to Old Downs Farm.

NT and KWT Reserves From the Observatory, continue through the Estate to the coast and drive north beside the beach to Prince's Golf Club car park. Walk north along the beach to the reserves. Access is largely unrestricted except to certain areas in the breeding season. It is possible either to return via the riverside footpath to New Downs Farm or cross the New Downs Farmland, checking the conifers at the north end of Prince's Golf Club en route. The entire area is privately owned and it is necessary to keep to roads and public footpaths at all times. Sandwich Bay Bird Observatory is open all year and provides hostel-type self-catering accommodation for up to 12 people. Further details are available from the honorary warden.

Pegwell Bay To view the mudflats of Pegwell Bay, continue north from Sandwich on the A256 towards Ramsgate. 0.7 miles after Ebbsfleet Lane on the left, turn right into the Pegwell Bay Country Park car park. A short trail leads to a hide overlooking the bay.

FURTHER INFORMATION

Grid refs: Bird Observatory (Old Downs Farm) TR 353 575; Worth Marshes (Chequers pub) TR 368 556; Prince's Golf Club car park TR 357 590; Pegwell Bay Country Park TR 342 634

Sandwich Bay Bird Observatory: Guildford Road, Sandwich Bay, Sandwich, Kent CT13 9PF; tel: 01304 617341; email: sbbot@talk21.com; web: www.sbbot.co.uk.

Kent Wildlife Trust: tel: 01622 662012; email: info@kentwildlife.org.uk

115 THANET AND NORTH FORELAND (Kent)

OS Landranger 179

Though heavily developed, the position of the Isle of Thanet in the extreme north-east corner of Kent, at the south end of the North Sea, makes it an obvious point for observing migration. It is a noted seawatching location and coastal scrub harbours a variety of passerine migrants.

Habitat

Chalk cliffs form the coastline from Ramsgate in the south to Margate. The rocky foreshore is interspersed by sandy bays. Grass covers most of the clifftop between Foreness Point and North Foreland, but more substantial cover occurs on the edge of North Foreland golf course and, 1 mile inland from Foreness Point, at the north end of Northdown Park.

Species

In winter you can expect to find Red-throated Diver, Great Crested Grebe, Common Eider and large auks (mostly Common Guillemot) offshore. The rocky shoreline holds a variety of wintering waders, most notably substantial numbers of Sanderlings and a regular group of Purple Sandpipers, which tend to roost in the sandy bays at White Ness. Small numbers of Rock Pipits and occasional Black Redstart are also present.

During passage substantial numbers of pipits, wagtails, chats, thrushes, warblers and finches pass through. Falls of the commoner night migrants are usually associated with east winds, with scarcer species such as Richard's Pipit, Icterine, Yellow-browed and Pallas's Warblers, and Red-breasted Flycatcher all regular in autumn. Autumn seawatching is most productive during north winds, with Gannet, Manx Shearwater, Arctic, Great and Pomarine Skuas, Kittiwake, terns and auks (including Little Auk in late October–November). Sparrowhawk, Merlin, Hen Harrier and Short-eared Owl all regularly arrive from the sea, with Osprey, Honey-buzzard, Common and Rough-legged Buzzards and Red Kite occasionally noted.

Breeding birds include colonies of Fulmar on the clifftops. Northdown Park has breeding Ring-necked Parakeet, which is easily found.

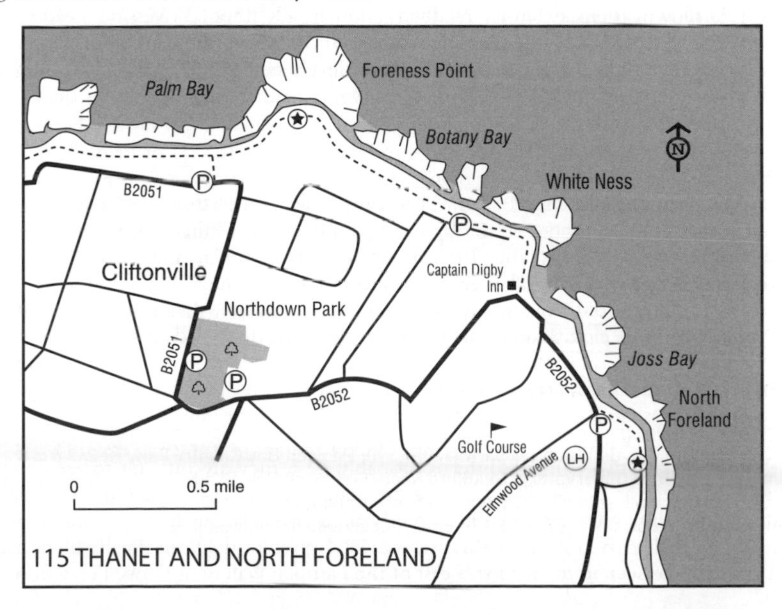

Access

North Foreland can be reached by following the B2052 north from Broadstairs. There is a car park, from where you can seawatch, overlooking Joss Bay, just north of the lighthouse. The clifftop between Foreness Point and Botany Bay may be accessed by following the B2052 west past the Captain Digby pub for c.0.5 miles. Turn right into St George's Road and take the second right, Kingsgate Avenue. Follow this to its end and park. Walk north-west along the coast to Foreness Point. It is possible to drive to Foreness Point by proceeding north along Queen Elizabeth Avenue, turning sharp right and then left into Princess Margaret Avenue until you reach the pitch-and-putt golf course. Park in the spaces provided and walk to the Coastguard Station, which offers some shelter for seawatching.

Northdown Park is bordered by the B2052 to the south and the B2051 (Queen Elizabeth Avenue) to the west, where you can park.

FURTHER INFORMATION

Grid refs: North Foreland TR 399 700; Foreness Point TR 391 710 / TR 379 712; Northdown Park TR 378 702

116–119 STOUR VALLEY (Kent)

OS Landranger 179

The Stour Valley, lying north-east of Canterbury, contains some of the finest wetlands in southern England. Stodmarsh NNR is the best known site, but other areas are also good. They are notable for breeding Cetti's and Savi's Warblers, but are also worth a visit during migration periods and regularly produce uncommon birds. Much of the area is managed by Natural England.

Habitat

The large shallow lagoons, extensive reedbeds and partially flooded riverside meadows have been created by a gradual subsidence of underground coal workings. Patches of woodland and scrub have developed in the drier areas. Water levels are carefully controlled and grazing is allowed on the damp meadows to produce suitable habitats; parts of the reedbed are cut annually to improve growth.

Species

Stodmarsh's position, in a major river valley only a few miles from the coast, ensures an impressive passage of migrants, especially in spring. It is one of the best places to see early arrivals before the main influx into Britain. A pair or two of Garganeys may breed on the reserve, but the best chance is of migrants in spring and autumn. The pools at Grove Ferry or the wet meadows south of the riverside path are favoured. Marsh Harrier and Hobby are regular, with concentrations of up to 50 of the latter recorded in May during recent years. Osprey is uncommon. Passage waders include Green, Wood and Common Sandpipers, Whimbrel, Ruff and Black-tailed Godwit, while Little Egret and, less regularly, Spoonbill are sometimes present. Recent autumns have seen regular appearances by Spotted Crake in August–September, with up to six noted on one day.

Although one or two pairs of Bitterns probably nest, they are difficult to see. In spring and early summer one is more likely to hear the male's characteristic booming. Savi's Warbler is another speciality; one or two pairs formerly bred annually, but in recent summers only single singing males have occurred, their stays brief. If present they can usually be heard and, with perseverance, seen in the reedbeds east of the Lampen Wall or at Grove Ferry from early

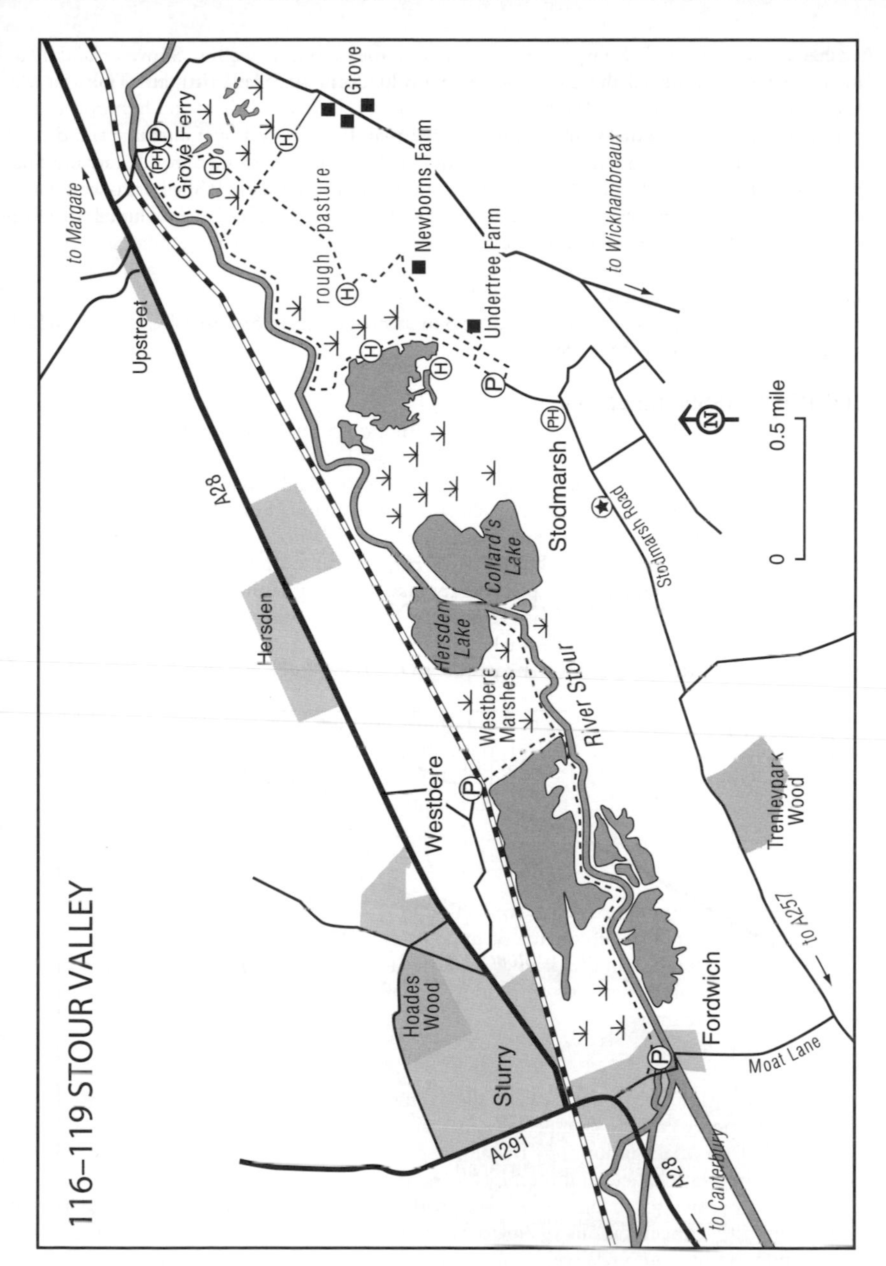

May. Savi's are usually most vocal at dawn and dusk, but may sing briefly during the middle of the day. Stodmarsh's other warbler speciality, Cetti's Warbler, is commoner, more vocal and resident. It is not unusual to hear up to ten singing explosively from the dense thickets and scrub between the car park and the river. With patience they can be seen but are often reluctant to leave cover. Bearded Tit is not uncommon and small parties can usually be found in the reeds, particularly on still days. Grasshopper Warbler has declined as a breeding species in recent summers, while Reed and Sedge Warblers and Reed Bunting are very common. Yellow Wagtail is frequently seen, but Water Rail is usually only heard. Lapwing, Common

Redshank and Common Snipe breed on the wet meadows, as well as Shoveler and Teal. Patches of woodland hold the usual species (woodpeckers, warblers, tits), but look out for Lesser Spotted Woodpecker at Westbere.

Large numbers of wildfowl winter on the reserve, mainly the commoner dabbling and diving ducks. Hen Harrier is frequently seen and Golden Plover can be found in the surrounding fields. A few Water Pipits favour the riverside meadows and a Great Grey Shrike is occasionally present, while Cetti's Warbler is often easiest to see in winter. Rarities have included Sociable Plover and Britain's first American Coot.

Access

Four sites in the Stour Valley are worthy of exploration. All except Stodmarsh are best reached from the A28 Canterbury–Margate road.

116 STODMARSH (Kent)

Leave Canterbury on the A257 towards Sandwich. After c.1.5 miles, just past the golf course, take an inconspicuous left turn onto an unsigned minor road. This leads to the village of Stodmarsh, c.4 miles away. In the village, turn left immediately after the Red Lion pub into a narrow lane; the reserve car park is on the right after a short distance. Various leaflets are on sale here from a small hut. Access to the NNR is along the Lampen Wall (a flood protection barrier), which provides an excellent vantage point. From the car park continue on foot to the end of the lane, turning sharp right and onto the Lampen Wall (or take the short nature trail through woodland). The path runs alongside woodland and scrub, then past open water and reedbeds into a drier, scrubby area, and finally after c.1 mile reaches the River Stour. It continues along the south bank of the river, with reeds, wet meadows and pasture to the right. You can walk all the way to Grove

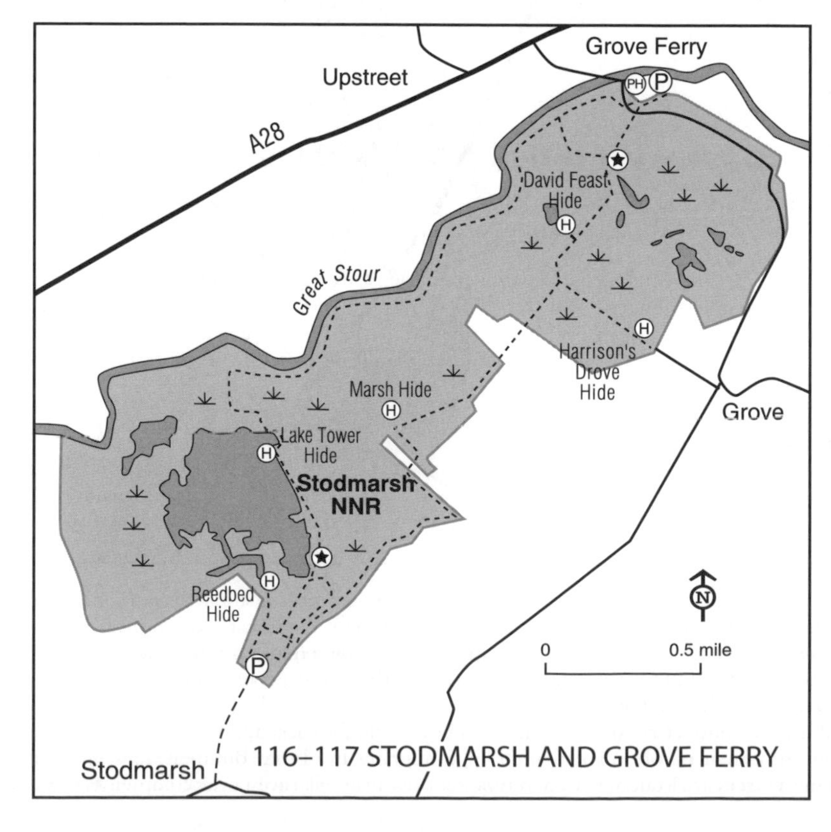

116–117 STODMARSH AND GROVE FERRY

Ferry (c.2 miles) and work the shallow pools and reedbeds in this area. A raised viewing area and hide overlooks this productive site. At Grove Ferry there is a pub, picnic site and café. You can return to Stodmarsh along the well-marked visitor trail via the Marsh Hide to the car park. If time is limited it is best to cover only a short stretch of the riverside path and then return to the car park.

117 GROVE FERRY (Kent)

An alternative approach to the area is from Grove Ferry itself. A right turn 3.5 miles after Sturry on the A28 leads to Grove Ferry down Grove Ferry Hill. Park in the car park by the pub and take the footpath leading to the recently created shallow pools and reedbeds. A riverside path to Stodmarsh offers alternative access.

118 WESTBERE MARSHES (Kent)

Leave the A28 west into the village of Westbere. Park at the west end of the village and cross the railway line. A footpath leads across the marshes to the river, which it then follows in both directions, alongside reedbeds and open water.

119 FORDWICH (Kent)

At Sturry, 2.5 miles north-east of Canterbury, a minor road leads south to Fordwich village. Park in the village and walk east along the river.

FURTHER INFORMATION

Grid refs: Stodmarsh TR 221 609; Grove Ferry TR 237 630; Westbere Marshes TR 196 610; Fordwich TR 180 598
Natural England: tel: 01233 812525; web: naturalengland.org.uk

120 BLEAN WOODS NNR (Kent)

OS Landranger 179

Blean Woods, situated close to Canterbury, is one of the largest deciduous woodland nature reserves in southern England. Much of the woodland has traditionally been coppiced and the entire area is now managed by a partnership of the RSPB, Natural England, The Woodland Trust and three local authorities. Apart from its woodland birds, the site is well known for the very rare Heath Fritillary butterfly in June.

Habitat

Although primarily deciduous woodland dominated by oak, hazel, birch and hornbeam, there are large areas of introduced sweet chestnut and small pockets of coniferous plantations. Open heathland areas add further variety, and these are managed to provide greater diversity.

Species

Blean holds the full range of resident woodland species, and in summer additional breeding visitors include Willow and Garden Warblers, Common Whitethroat, Chiffchaff, Blackcap and, notably, around 30 pairs of Nightingales (most easily found mid-April to mid-May). Nightjars are usually present in new chestnut coppice and in the heathland areas, and Woodcocks may be seen roding at dusk in the same habitat.

Winter is quiet at Blean, but at dusk Tawny Owls are vocal and Woodcock may be more easily seen, their numbers augmented by immigrants from the continent. This time of year, when the trees are leafless, is a good time to look for the elusive Lesser Spotted Woodpecker.

Access

Leave Canterbury on the A290 Whitstable road and after 1.5 miles turn left to Rough Common. After 0.3 miles, a signed track on the right leads directly to the car park. There are five trails of varying lengths, marked by coloured arrows, and the green trail is usually suitable for wheelchairs.

FURTHER INFORMATION

Grid ref: TR 124 594
RSPB Blean Woods (warden): tel: 01227 455972; email: blean.woods@rspb.org.uk; web: www.rspb.org.uk/bleanwoods

121 OARE MARSHES (Kent)

OS Landranger 178

The Oare Marshes LNR lies on the south bank of the Swale, immediately opposite Harty Ferry on the Isle of Sheppey, and is important for migratory, wintering and breeding birds.

Habitat

The reserve consists of 67 ha of grazing marsh with freshwater dykes, open water 'scrapes' and saltmarsh. Suitable habitat is achieved through manipulation of water levels and grazing.

Species

In winter the reserve hosts good numbers of wildfowl including Brent Goose and Wigeon, and many waders, notably Dunlin and Curlew. Like nearby Sheppey, raptors are often in evidence, and typical species include Hen Harrier, Merlin and Short-eared Owl. Other wintering species may include Bittern and Twite.

The diversity of waders increases during spring and autumn, and regular visitors include Little Stint, Curlew Sandpiper, Ruff, Black-tailed Godwit and Whimbrel. A rarity is a distinct possibility

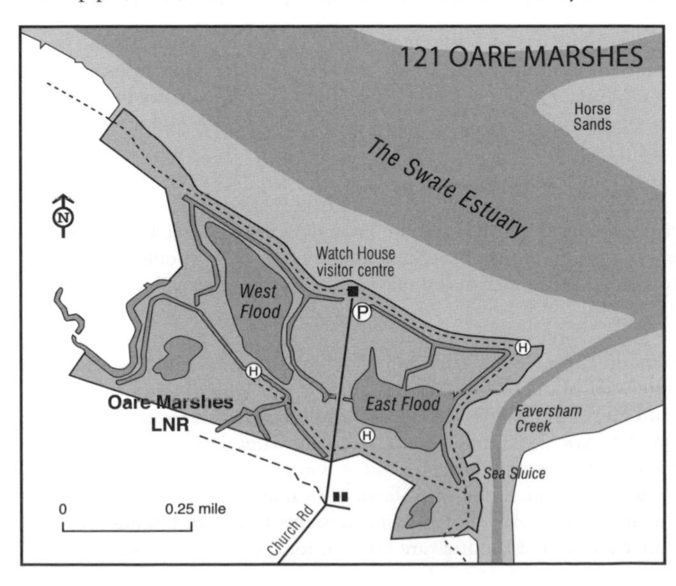

at the right time of year, and the reserve briefly hosted Britain's first Tufted Puffin in 2009.

Breeding birds at Oare include Garganey, Water Rail, Avocet, Lapwing, Common Snipe, Common Redshank, Common Tern and Bearded Tit.

Access

The village of Oare lies just north of Faversham. From Oare, head north for 1.5 miles and park opposite the Watch House, near the sea wall at the end of Harty Ferry Road. Access is restricted to the public footpath and nature trail to minimise disturbance to the birds. The whole reserve may be viewed from the trails and hides.

FURTHER INFORMATION

Grid ref: TR 013 647
Kent Wildlife Trust: tel: 01622 662012; email: info@kentwildlife.org.uk

122–124 ISLE OF SHEPPEY (Kent)

OS Landranger 178

The Isle of Sheppey lies on the south side of the Thames estuary, flanked by the Swale and Medway estuaries, and connected to the mainland by the Kingsferry Bridge. It is particularly productive in winter, harbouring significant numbers of wildfowl and waders in addition to several unusual species. A greater variety of waders, and sometimes impressive movements of seabirds, can be seen in autumn.

Habitat

The south half of the island is the most productive for birds. Much of this area comprises rough grazing and marshes intersected by dykes and creeks, as well as some arable, all protected from exceptionally high tides by long sea walls. At low tide, extensive mudflats are exposed on the Swale, bordered by saltings.

Species

Unless specifically mentioned for one site, the birds detailed below can be seen at any of the areas. Sheppey is particularly good in winter although it is often cold and windswept. Divers and grebes are regular off Shell Ness, the most frequent being Red-throated Diver and Great Crested Grebe; other species are occasional. Hen Harrier is a winter speciality, quartering the fields and marshes almost anywhere. Rough-legged Buzzard is an increasingly regular winter visitor – the Capel Fleet/Harty Marshes area is probably the best UK site for this species. Merlin and Peregrine are more frequent, though not guaranteed on a casual visit. White-fronted Goose is another winter speciality and all the fields should be checked. The Capel Fleet–Harty Ferry area is often favoured, and they sometimes roost at Elmley or the floods at Swale NNR. Small numbers of Bewick's Swans and the occasional Whooper sometimes frequent the Harty Ferry area while Brent Goose may be seen anywhere on the coast. Very large numbers of ducks winter, the commonest being Wigeon, Teal, Mallard, Pintail and Shoveler. Large rafts of ducks are often present offshore at Shell Ness and should be checked for Common Eider, Common Scoter, Goldeneye and Red-breasted Merganser. Velvet Scoter is less frequent. At low tide waders can be seen anywhere on the extensive mudflats and at high tide they roost at Shell Ness and Elmley. Knot and Dunlin are commonest, and other species include Sanderling. Short-eared Owl is present all year, though more numerous and most easily seen in winter. Lapland and Snow Buntings are sometimes seen on the beach at Shell Ness but Twite and Shore Lark are less frequent.

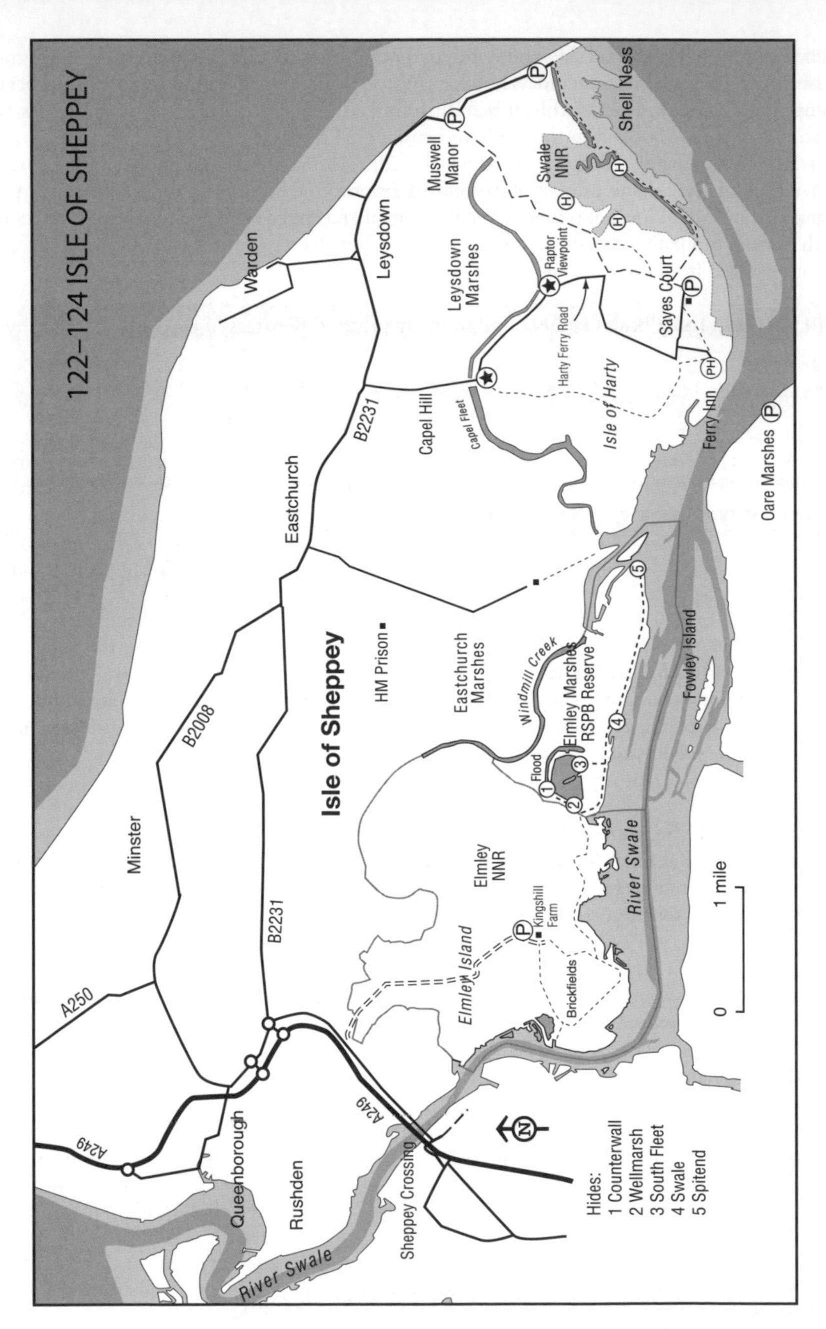

Species diversity increases in spring and autumn although the huge numbers of wildfowl and waders are absent. Marsh Harrier breeds and can be seen year-round. Garganey is regular at Elmley and the floods at Swale NNR in spring. The wide variety of waders may include Little Stint, Curlew Sandpiper and large numbers of Black-tailed Godwits, in addition to the wintering and breeding species, and rare waders are seen annually. Seawatching off Shell Ness can be productive in autumn, particularly after strong north winds. Gannet, Great and Arctic Skuas,

Kittiwake, and Common, Little and Sandwich Terns are the most frequent species. Migrant passerines include Wheatear and Whinchat and occasionally something rarer. A number of recent records of Aquatic Warbler (including four at Elmley in August 1995) suggests this species may be overlooked in the reeds and rush-filled ditches.

In summer a variety of commoner ducks breed, occasionally including Garganey. Lapwing and Common Redshank are common and Common Tern nests on some of the saltmarsh islands at Elmley – Avocet also breeds. Yellow Wagtail and Meadow Pipit are common breeding birds throughout.

Access

Although interesting birds can be found in many places, three areas in particular are worthy of investigation.

122 RSPB ELMLEY MARSHES (Kent)

Part of the area (the Spitend Marshes) has been successfully managed to become an important area for both breeding and wintering waterbirds. Three hides overlook the lagoons and marshes and two give views across the Swale. Follow the A249 north from junction 5 of the M2. Take the road signposted Iwade and Ridham Dock. At the roundabout take the second exit onto the old road bridge. On the Isle of Sheppey, after 1.25 miles, turn right following the RSPB sign. Follow the rough track for approximately 2 miles to the car park at Kingshill Farm. A walk of 1 mile is necessary to reach the hides. Keep to the main footpaths and walk below the sea wall to avoid disturbing the birds. The reserve is open daily (except Tuesdays) 9am–9pm (or sunset if earlier).

123 CAPEL FLEET AND HARTY FERRY (Kent)

From Leysdown return west along the B2231 for c.2 miles. Before reaching Eastchurch turn left on a minor road signed 'Ferry Inn' towards the Isle of Harty. This crosses Capel Fleet (often worth a look) and eventually ends at the Swale, beyond the inn. In winter this is a good area for geese, swans and raptors.

124 SHELL NESS (SWALE NNR) (Kent)

The small hamlet at Shell Ness lies at the east tip of Sheppey. Take the B2231 to Leysdown-on-Sea and continue along the seafront to Shell Ness. Park just before the hamlet and walk along the sea wall to the coast. It is worth checking the sea as well as the marshes and mudflats. Continuing along the beach, the bay beyond the pill box holds considerable numbers of wintering waders at high tide. After checking this area return along the sea wall a few hundred yards and turn left along the sea wall which runs approximately parallel to the coast. Walk for up to 2 miles to view the extensive shallow floods of Swale NNR, overlooked by three hides.

FURTHER INFORMATION

Grid refs: Elmley Marshes (Kingshill Farm) TQ 938 679; Capel Fleet TR 010 689; Ferry Inn TR 014 659; Shell Ness TR 051 683; Sayes Court TR 023 662
RSPB Elmley Marshes: tel: 01795 665969; web: www.rspb.org.uk/elmleymarshes

125–127 MEDWAY ESTUARY (Kent)

OS Landranger 178

The maze of tidal creeks and extensive mudflats that form the Medway Estuary are situated south of the Hoo Peninsula and immediately west of the Isle of Sheppey. Birds are similar to the estuaries of the Thames and Swale, and this area is at its best in autumn and winter.

Habitat

A large tidal basin with many saltmarsh islands and surrounded by mudflats at low tide. In the east the marshes of the Chetney peninsula are characterised by grazing marsh dissected by fleets and ditches. Further west, orchards, grazing and arable dominate the landward side of the sea walls around Ham Green. A small reedbed and extensive hawthorn and bramble scrub can be found at Riverside Country Park.

Species

The area is at its best in winter when huge numbers of wildfowl and waders use the estuary. Brent Goose, Shelduck, Wigeon, Teal, Shoveler, Pintail, Grey Plover, Dunlin, Knot, Common Redshank, Black-tailed Godwit, Curlew and Turnstone are all present in large numbers. Red-breasted Merganser and Goldeneye occur in the deeper channels, while seaducks and Scaup may move into the estuary in cold weather. One of the rarer grebes is usually present (often in Half Acre Creek or Otterham Creek) with all three sometimes present during hard weather. In most winters a diver is found, and there have been regular records of Great Northern Diver and Short-eared Owl. Fields around Funton and Chetney are frequented by flocks of Lapwings and Golden Plovers. Merlin, Marsh and Hen Harriers can be seen almost anywhere, but are perhaps most easily found on Chetney, as is a wintering Short-eared Owl. Look out for Rock Pipits at Motney Hill in winter.

Shelduck, Lapwing and Common Redshank breed on Chetney, while the islands of the river support large colonies of Black-headed Gulls, along with smaller numbers of Common, Sandwich and Little Terns. Mediterranean Gulls are also regular, breeding on Nor Marsh.

In autumn, the variety of waders increases. Chetney regularly attracts Little Ringed Plover, Ruff, Common, Green and the occasional Wood Sandpiper, and Little Stint. Spotted Redshanks regularly roost on the saltmarsh at Motney Hill, while Curlew Sandpiper is recorded most autumns. Small numbers of Garganeys also occur at Chetney in August–September. During south-east winds, Black Terns may appear at Motney Hill sewage outfall.

Access

125 RIVERSIDE COUNTRY PARK (NOR MARSH AND MOTNEY HILL) (Kent)

From Rainham take the B2004 north for 1 mile, turn left at the junction and travel for c.1 mile and park at the visitor centre car park. From here you can walk to Horrid Hill or west to Eastcourt Meadows. To access Motney Hill at the east end of the country park, you can use the small car park along Motney Hill Road (TQ 821 675), accessed from the B2004, 0.25 miles west of the junction. From here you can walk north along the sea wall to Motney Hill, viewing the estuary to the west and reedbed to the east. It is possible to walk along the shoreline and sewage works perimeter fence to view Rainham Creek and the saltmarsh that forms the RSPB reserves of Nor Marsh and Motney Hill. Mediterranean Gulls breeding on Nor Marsh can be viewed distantly from Horrid Hill. To view Otterham Creek walk to the sewage works gates and follow the path to the right, which runs along the edge of the field by the works fence. Berengrave Nature Reserve can be reached from a car park on the B2004 (TQ 820 675), or alternatively from Berengrave Lane itself.

126 HAM GREEN AND LOWER HALSTOW (Kent)

From Upchurch follow the road north-east for 1 mile, where there is limited space for parking. The narrow lane on the right leads to the sea wall. Walk right towards Lower Halstow in order to view the large expanse of mudflats in Twinney and Halstow Creeks. Walk left to view the deep Half Acre Creek.

127 FUNTON CREEK AND CHETNEY (Kent)

From Lower Halstow follow the road east for c.1 mile until the road runs alongside the estuary. There are several lay-bys where you can park and view the estuary. To reach Chetney follow the public footpath from the top of the hill across the fields to the sea wall. Please keep strictly to the public footpath at this sensitive site.

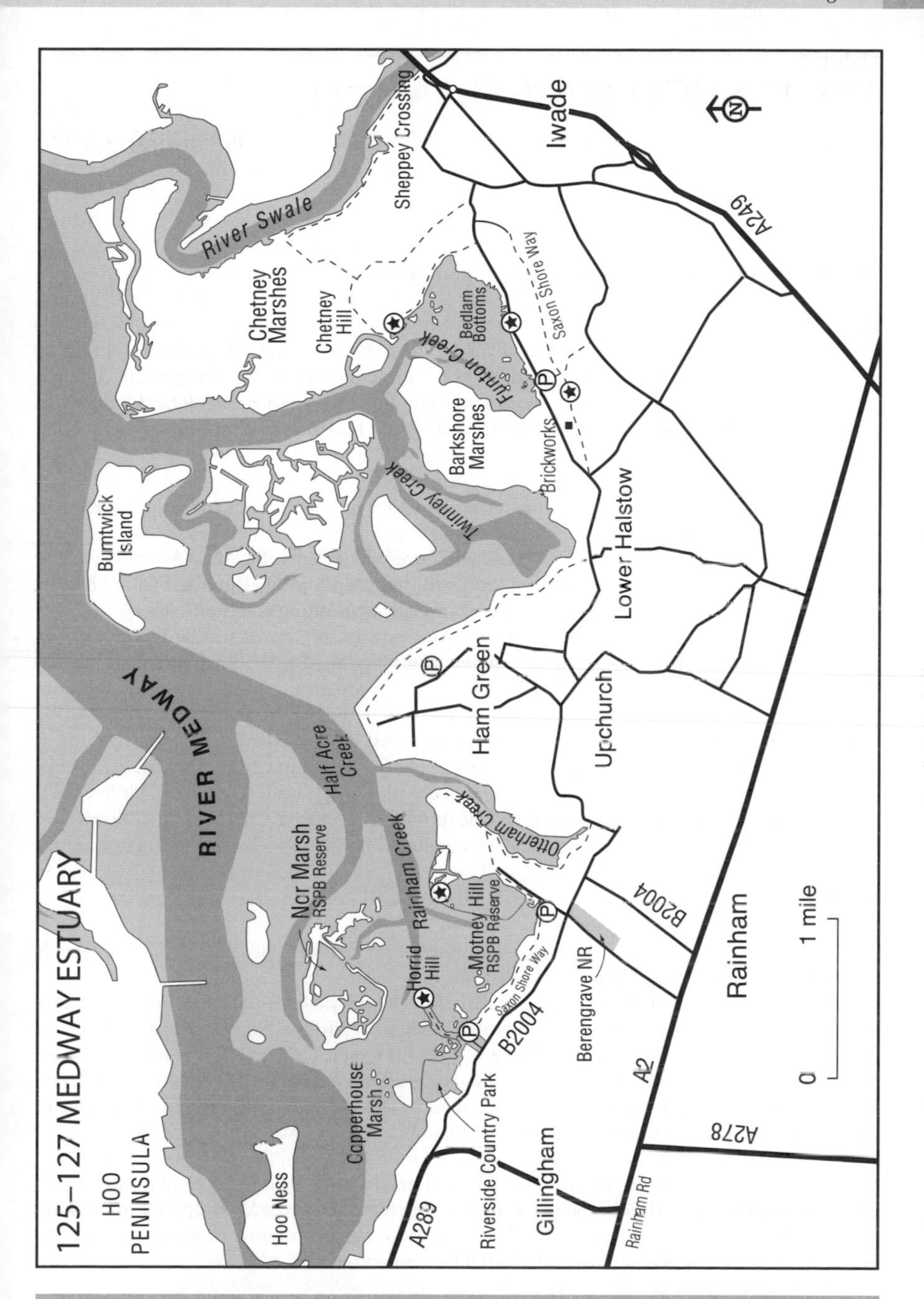

FURTHER INFORMATION

Grid refs: Riverside Country Park (Nor Marsh and Motney Hill) TQ 807 683; Ham Green TQ 847 688; Lower Halstow TQ 860 673; Funton Creek TQ 878 680; Chetney TQ 883 696
RSPB Nor Marsh and Motney Hill: tel: 01634 222480; web: www.rspb.org.uk/normarsh
Riverside Country Park: tel: 01634 378987; email: riversidecp@btinternet.com

128–132 HOO PENINSULA (Kent)

OS Landranger 178

These extensive marshes – often referred to as the North Kent Marshes – lie in the Thames Estuary immediately west of the Isle of Sheppey, bordered by the Rivers Thames and Medway. Similar to Sheppey, they are best in winter or during migration.

Habitat

Mainly grassland and saltmarsh dissected by dykes and sea walls, and surrounded by mudflats at low tide, much of the area has been drained for growing wheat. Flooded pits at Cliffe are the result of clay extraction and this area has recently been acquired by the RSPB. An area of deciduous woodland and scrub near High Halstow is the RSPB reserve of Northward Hill.

Species

Bewick's Swan and White-fronted Goose regularly winter on the marshes and large numbers of dabbling ducks include Pintail. Deeper pools attract diving ducks, sometimes including Smew and Scaup. Waders are abundant and typical wintering species include a few Avocets at Cliffe pools. Hen Harrier and Short-eared Owl can usually be seen hunting the marshes and Merlin is regular. Snow Bunting can be rather elusive but Lapland Bunting is more frequent.

Garganey may appear in small numbers in spring and Black Tern is regular in both spring and autumn, but the main interest is the passage of waders. Species that may be found include Little Stint, Curlew Sandpiper and Wood Sandpiper, and something more unusual is a distinct possibility.

Summer is quiet. Shelduck, Lapwing and Common Redshank breed on the marshes and at Northward Hill the heronry and woodland species are the main attractions (though the reserve is best visited April–June when it can be combined with passage waders at Cliffe). Little Owl is common in the area and quite easy to see, but though Long-eared Owl breeds at Northward Hill it is rarely seen. Other species on the reserve include all three woodpeckers and Nightingale.

Access

Several areas are worthy of exploration.

128 RSPB CLIFFE POOLS/RYE STREET (Kent)

At the west end of the marshes, these flooded clay pits are perhaps the most interesting area, attracting a variety of ducks in winter and many waders on passage. The water level determines the best pits for waders. This is one of the newest RSPB reserves in north Kent, but there are no visitor facilities (apart from six viewing mounds) as yet. Turn left off the B2000 at the north end of Cliffe village and on reaching the first pool turn right. This road passes around the perimeter of the pools (several of which can be seen well) and continues to a gate. A short walk beyond this leads to the sea wall. Going north to Lower Hope Point may be profitable; alternatively walk south back around the pits via Cliffe Creek. This is a long walk but if time is short much of the area can be viewed from the roads. In addition, the road to the Works passes other pools and a track runs through the pits to Cliffe Creek. Less than a mile east of the pools is RSPB Rye Street, an area of grazing marsh that attracts raptors in winter.

129 HALSTOW MARSHES (Kent)

In the centre of the North Kent Marshes, these are excellent for wildfowl and waders, especially at the coast. Egypt and St Mary's Bays are particularly notable for waders at high tide. From High Halstow continue east towards the A228. After 0.5 miles turn left (north) to Decoy Farm and park at Swigshole. Fork right across the marshes to St Mary's Bay. Continue along the sea wall west to Egypt Bay where another track leads back to Swigshole. Some of this area is now

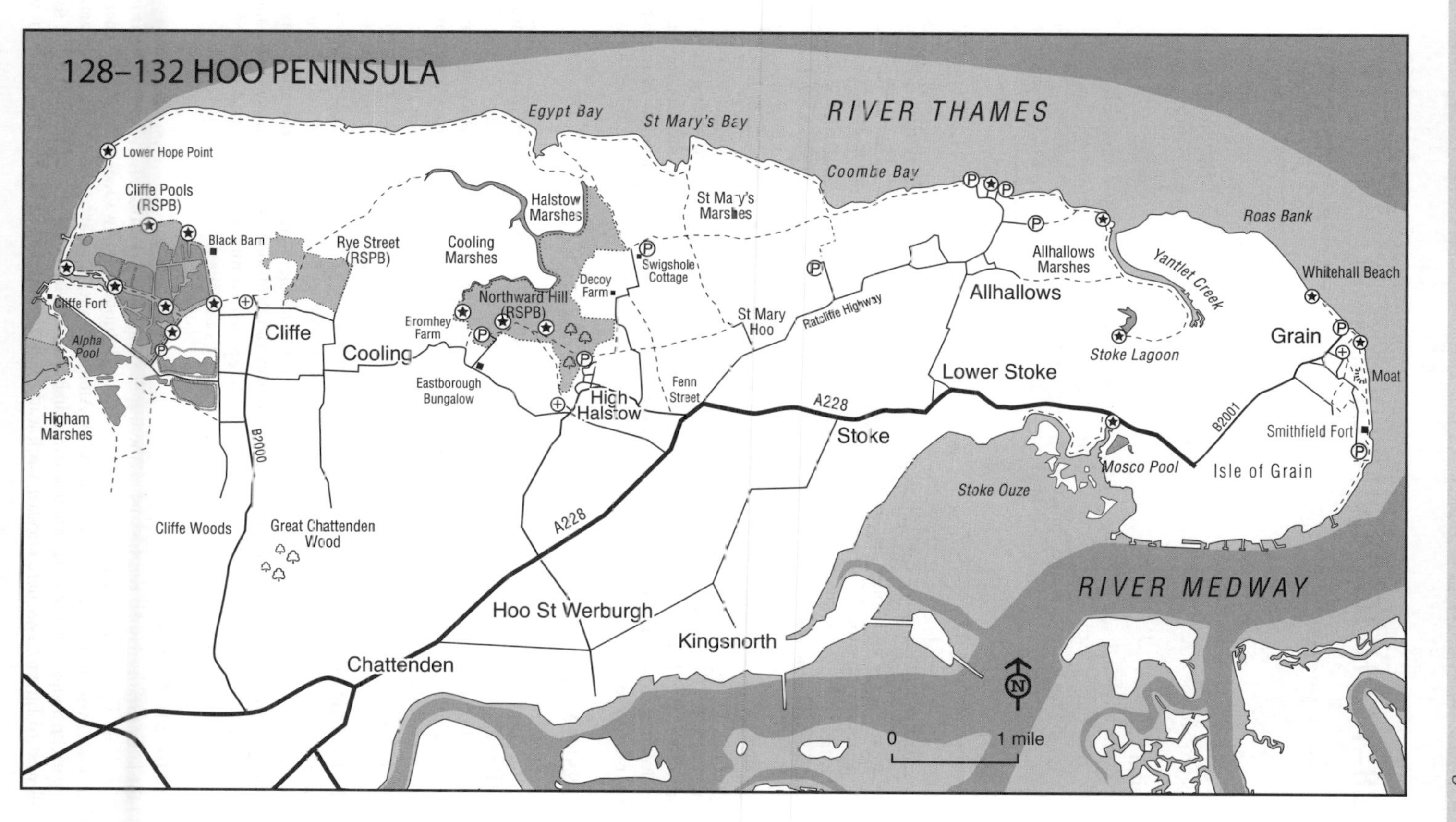
128–132 HOO PENINSULA
RIVER THAMES
Egypt Bay
St Mary's Bay
Coombe Bay
Lower Hope Point
Cliffe Pools (RSPB)
Black Barn
Cliffe Fort
Alpha Pool
Higham Marshes
Cliffe
Rye Street (RSPB)
Cooling Marshes
Halstow Marshes
Cooling
Eromhey Farm
Northward Hill (RSPB)
Decoy Farm
Swigshole Cottage
St Mary's Marshes
St Mary Hoo
Ratcliffe Highway
Eastborough Bungalow
High Halstow
Fenn Street
B2000
Cliffe Woods
Great Chattenden Wood
A228
Stoke
Allhallows
Allhallows Marshes
Lower Stoke
Roas Bank
Yantlet Creek
Stoke Lagoon
Whitehall Beach
Grain
Moat
B2001
Smithfield Fort
Mosco Pool
Isle of Grain
Stoke Ouze
RIVER MEDWAY
Hoo St Werburgh
Kingsnorth
Chattenden
0
1 mile

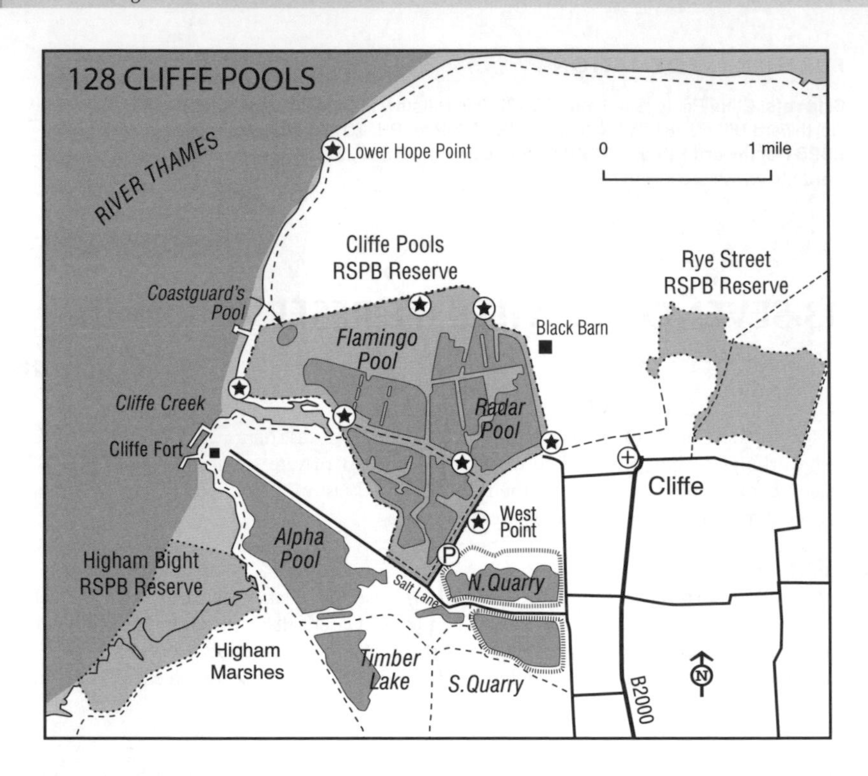

managed by the RSPB, and is best viewed from Bromhey Farm, reached by taking the road west from High Halstow towards Cooling for c.1 mile and then turning right after the sharp south bend. Continue to the farm buildings and park. The recently created flooded areas can be viewed from the nearby small hill.

130 RSPB NORTHWARD HILL (Kent)

This small reserve lies immediately north of High Halstow. The oaks hold Britain's largest heronry; over 150 pairs are present February–July, joined in recent years by a few pairs of Little Egrets. A range of woodland species can also be seen. High Halstow is north of the A228 in the centre of the North Kent Marshes. Parking is available by the village hall and the entrance to the reserve is north of Northwood Avenue. There is free access at all times (contact the reserve for advice on accessibility for disabled visitors) but the heronry may only be visited if escorted; apply in writing to the warden.

131 ALLHALLOWS (Kent)

Allhallows is reached by following the minor road off the A228 at Fenn Street for 3.5 miles. Turn right along Avery Way to the British Pilot pub. Park here and follow the track across the marsh to the sea wall. From here walk east to reach Yantlet Creek and then south and subsequently west for 2 miles to Stoke Lagoon. Keep below the sea wall to avoid disturbing the wildfowl and waders. Alternatively walk west to Coombe Bay.

132 ISLE OF GRAIN (Kent)

The Isle of Grain may be reached by following the A228 from Rochester. 1.5 miles beyond Lower Stoke you can park and view the mudflats and saltings of Stoke Ouze. The adjacent fleet attracts ducks. Follow the B2001 to Grain village, and follow Chapel Road until you are driving south and parallel to the sea wall. View the mudflats from the sea wall and work the bushes for migrants in autumn. The power station outflow can attract gulls and terns.

FURTHER INFORMATION

Grid refs: Cliffe Pools (Salt Lane) TQ 722 758; Halstow Marshes (Swigshole) TQ 788 775; Northward Hill TQ 781 757; Allhallows (British Pilot PH) TQ 844 783; Grain village TQ 888 769
RSPB Northward Hill and RSPB Cliffe Pools: tel: 01634 222480; web: www.rspb.org.uk/northwardhill; www.rspb.org.uk/cliffepools

133 SEVENOAKS WILDLIFE RESERVE (Kent)

OS Landranger 188

This site, in the Darent Valley in north-west Kent, has the distinction of being the first ever gravel workings to be converted to a nature reserve. In private ownership since the 1950s, management was fully taken over by the Kent Wildlife Trust in 2005. The Trust plans extensive improvements for this already wildlife-rich reserve.

Habitat

The reserve comprises deciduous woodland around a series of lakes, the largest of which (East Lake) has a collection of low muddy islands. The River Darent flows through the site. North of the East Lake are several small lakes fringed with reedbeds. Grebe Hide overlooks the smaller and deeper West Lake, and a busy feeding station beside this hide affords excellent close views of common woodland birds.

Species

A good range of common woodland and wetland birds can be found here. Visiting winter wildfowl include Wigeon and numbers of Shovelers, Teal and Gadwalls joining the resident common wildfowl, including large flocks of feral Canada and Greylag Geese. Lapwings and

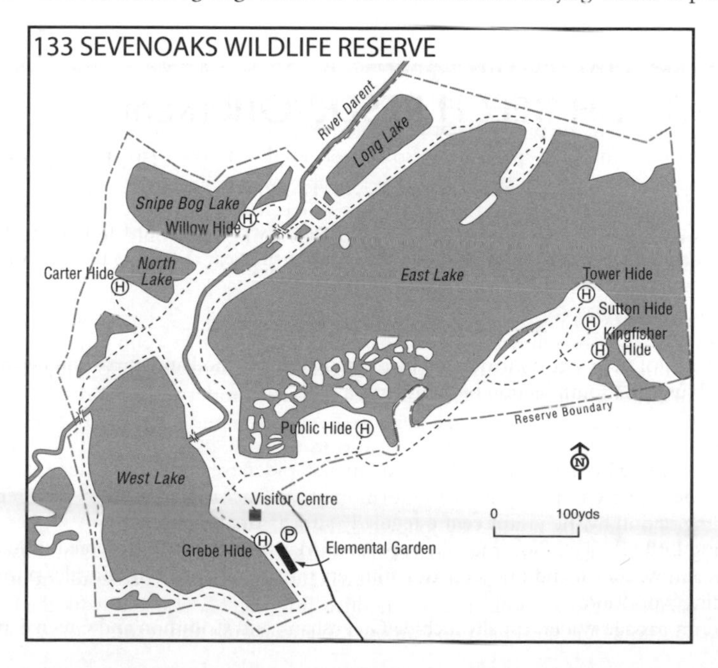

assorted gulls are also numerous in winter, and a few Common Snipe venture out of cover to feed on the islands of East Lake later in the day. The odd Goldeneye, Goosander or other scarcer diving duck visits from time to time, as may a Bittern. Siskins are common in winter and may visit the feeding station. A Firecrest is found most winters. Late winter and early spring is the best time to see Lesser Spotted Woodpecker, especially around the car park. Early spring visitors include Little Ringed Plovers and Common Sandpipers; one or two of each can usually be found on the main lake. Sand Martins feed over the lakes; a few pairs have nested in banks behind the trail down to Tyler Hide.

Cetti's Warbler is a recent addition to the reserve; listen for the rich explosive song from the reeds and scrub around Willow Hide. This is also a good place to find Reed, Sedge and Garden Warblers, Blackcaps and Reed Buntings. A few pairs of Egyptian Geese and Great Crested Grebes nest on the reserve, and Cormorants, Kingfishers and Ring-necked Parakeets are always around.

Unusual visitors have included Red Kite and Osprey in spring, and autumn brings a few passage waders such as Green Sandpiper, Greenshank and Whimbrel.

Access

The reserve is signposted along the A25 north of Sevenoaks, in between the junctions with the A224 and A225. A track leads down to a large car park. The reserve is open daily from dawn to dusk and entry is free. The visitor centre is open Wednesday, weekends and bank holidays 10am–5pm between the end of March and 31 July, 10am–5pm Saturday–Wednesday between August and October, and 10am–4pm Wednesday and weekends from November to the end of March. The centre has toilets and a café. The trails are mainly level and broad, though narrower and sometimes muddy at the northern edge of the reserve. There are seven hides.

FURTHER INFORMATION

Grid ref: TQ 519 566
Kent Wildlife Trust: tel: 01622 662012; email: info@kentwildlife.org.uk; web: www.kentwildlifetrust.org.uk

134 BOUGH BEECH RESERVOIR (Kent)

OS Landranger 188

The largest area of fresh water in Kent, this man-made reservoir was first flooded in 1969–70. Bough Beech attracts a variety of wildfowl throughout the year as well as passage waders, and it regularly produces rarities.

Habitat

A large stretch of fresh water surrounded by a patchwork of woodland, fields and hedgerows. A sailing club uses the south section of the reservoir.

Species

Wintering wildfowl include Pochard, Tufted Duck, Goldeneye and Goosander. Other waterbirds include Great Crested Grebe and Cormorant, and occasionally divers or rarer grebes. The feeding station by the visitor centre regularly attracts Bramblings.

In spring Little Ringed Plover usually appears in March and remains to breed in most years. Osprey is almost annual and one sometimes lingers for several days. Nightingales occur in the surrounding woodland.

In autumn passage waders usually include Greenshank and Common and Green Sandpipers.

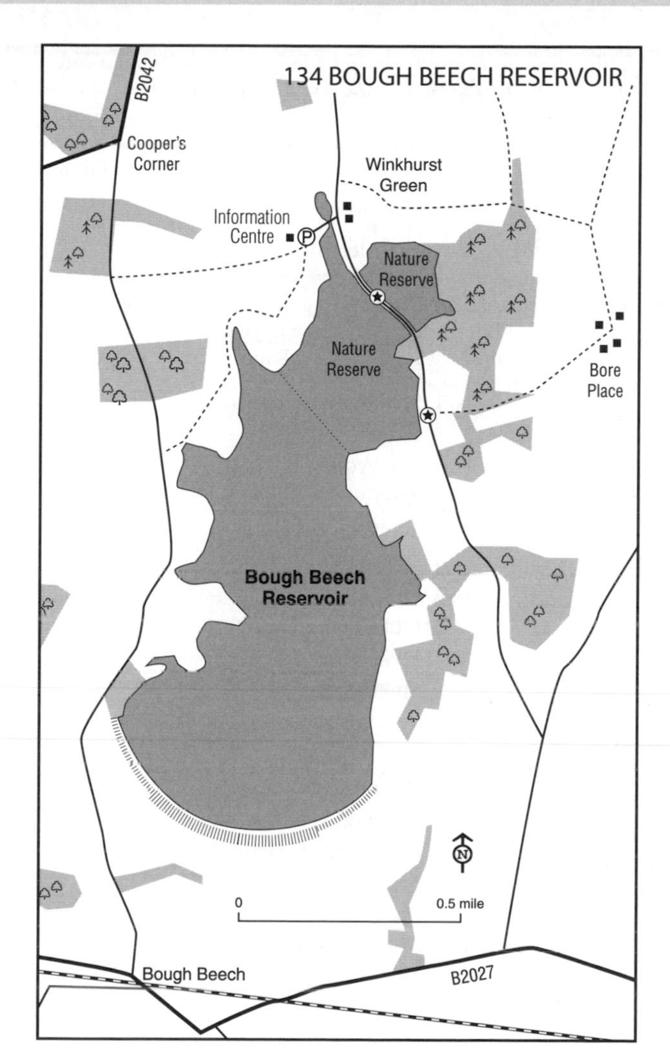

Garganey occurs in most years.

Breeders include Grey Heron, Shelduck, Mandarin, Hobby, Sparrowhawk, Kingfisher, the three woodpeckers, a variety of warblers, Marsh Tit, Nuthatch and Treecreeper.

Access

From Riverhead the B2042 runs south-west towards Ide Hill. One mile south of Ide Hill take the minor road east to Winkhurst Green. There is a Kent Wildlife Trust Information Centre north of the reserve with car parking and toilet facilities. The road running through the north end of the reservoir affords excellent views, while woodland birds may be found by taking the path towards Bore Place.

FURTHER INFORMATION

Grid ref: TQ 494 494
Kent Wildlife Trust: Tel: 01622 662012; email: info@kentwildlife.org.uk, web: www.kentwildlifetrust.org.uk

EAST ANGLIA

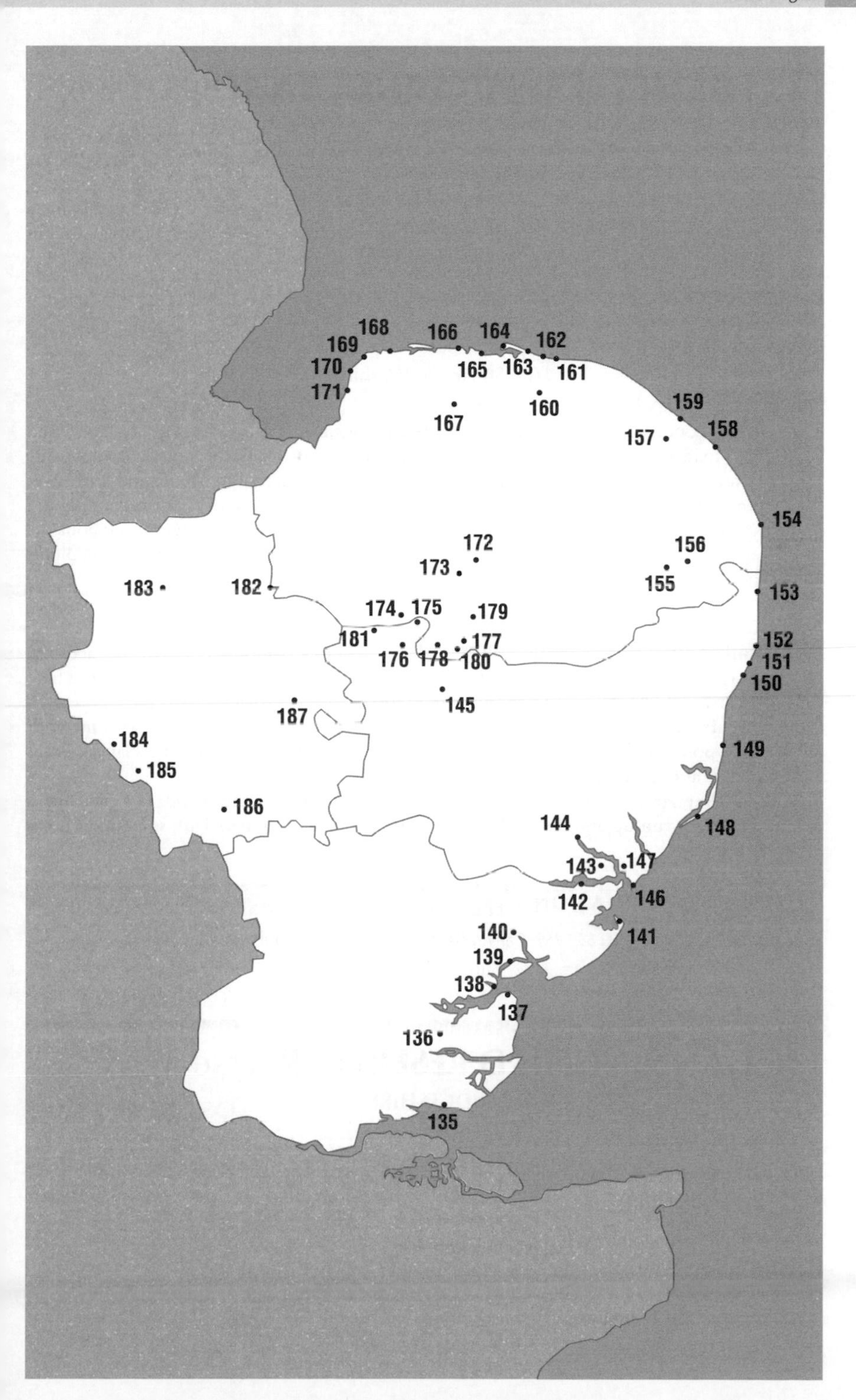

168
169
170
171
166
165
164
163
162
161
160
167
159
158
157
154
172
173
156
155
153
183
182
174
175
179
181
176
178
177
180
152
151
150
145
187
184
185
186
149
148
144
143
147
142
146
140
141
139
138
137
136
135

135 THAMES SEAWATCHING STATIONS (Essex)

OS Landranger 178

In late summer and autumn strong to gale-force winds with an easterly component (from north-easterly through easterly to south-easterly) push seabirds into the Thames estuary, and in such conditions the Essex shore can provide exciting seawatching.

Habitat

The sites look out over the Thames Estuary, and East Tilbury and Canvey Seafront also offer muddy foreshore attractive to waders.

Species

In the right conditions in autumn Fulmar, Manx Shearwater, Gannet, Arctic, Great and Pomarine Skuas, Little Gull, Kittiwake and good numbers of terns are recorded with some regularity, and Leach's Petrel, Sooty Shearwater, Long-tailed Skua, Sabine's Gull and Little Auk (late winter only) have been recorded on several occasions. In winter, a mix of divers, grebes and seaducks may be seen off Southend Pier and Mediterranean Gull is often seen throughout the area. There is a large wader roost at Canvey Point, and there are often a few Purple Sandpipers on Southend Pier.

Access

East Tilbury Seawatch from the sea wall immediately east of Coalhouse Fort, which is accessed via a minor road south from East Tilbury. There is a public car park nearby. The foreshore here is also attractive to waders.

Canvey Seafront This lies on Canvey Island between Labworth Cafe and Canvey Point, with suitable viewpoints at the old lifeguard station near the amusement arcades, and Canvey Point itself, which is also good for waders.

Southend Pier At 1.33 miles, this is the longest pleasure pier in the world. The pier opens at 8am and trains run up and down all day. In Southend itself Gunners Park sometimes holds passerine migrants.

FURTHER INFORMATION

Grid refs: East Tilbury TQ 689 769; Canvey Seafront TQ 801 824; Southend Pier TQ 884 851

136 HANNINGFIELD RESERVOIR (Essex)

OS Landranger 167

Lying 5 miles south of Chelmsford, this large reservoir attracts an excellent selection of water-birds throughout the year. Around 40 ha along the south shore are a reserve of the Essex Wildlife Trust (EWT).

Habitat

Flooding the valley of the Sandon Brook between West and South Hanningfield, the reservoir was opened in 1957 and covers 350 ha impounded behind the longest dam in Europe. It has a mixture of concrete and natural banks, with some wooded areas. Depending on water levels, variable amounts of mud may be exposed.

Species

Spring passage can be exciting, with movements of Arctic Terns in late April–May, and parties of Common Scoters occasionally pass through. Other migrants may include a variety of waders, and sometimes Osprey, Black Tern and Little Gull. On autumn passage a greater variety of waders may occur, including Greenshank, Ruff and Little Ringed Plover, and sometimes Little Stint, Curlew and Wood Sandpipers and Spotted Redshank.

In winter there is a sizeable gull roost, and Yellow-legged, Caspian, Mediterranean, Iceland and Glaucous Gulls can occasionally be found with diligence. Wintering wildfowl include Goldeneye, Shoveler and Pintail (the latter two peaking in autumn). Great Northern Diver and Red-necked, Slavonian and Black-necked Grebes are almost annual (and Black-necked Grebe and more rarely Red-necked Grebe may also turn up in summer and autumn).

Breeders include Great Crested and Little Grebes, Gadwall, Pochard, Common Tern and Yellow Wagtail. Hobby is often seen in summer, and there may be up to 80,000 Swifts, Swallows and martins feeding over the water during peak fly hatches.

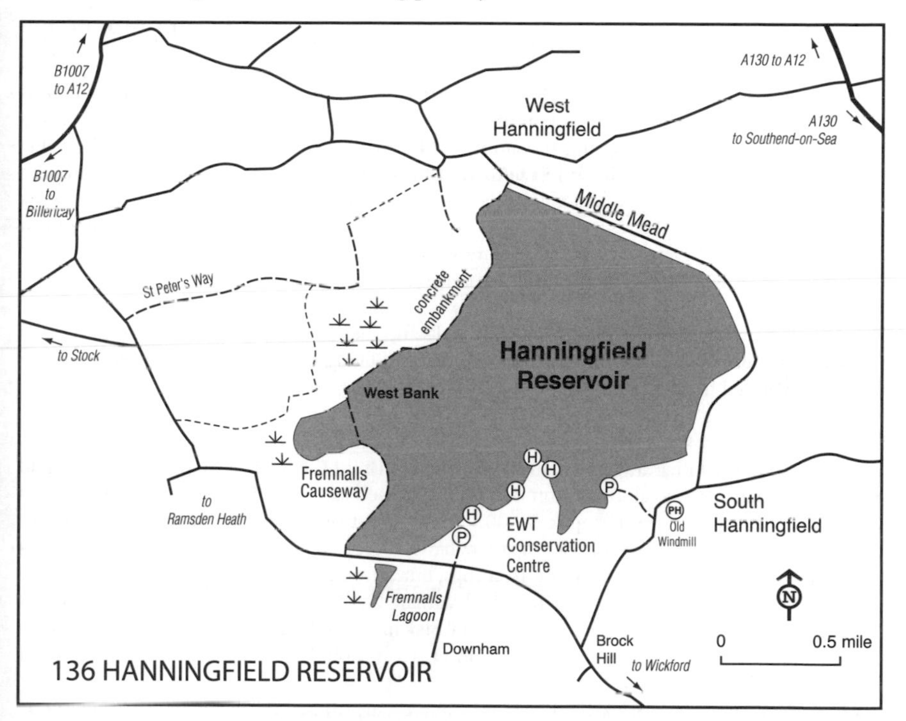

136 HANNINGFIELD RESERVOIR

Access

EWT Hanningfield Reservoir Visitor Centre Leave the A12 south on the A130 and, after 2.5 miles, turn west on minor roads to West Hanningfield. In the village turn left (south) along Middle Mead, which follows the northern, embanked, perimeter of the reservoir. After 2 miles, at the first junction, turn right, after 1 mile turn right again, and after a further 1 mile turn right following the reserve signs to the car park. Alternatively, turn off the B1007 (Billericay–Chelmsford road) on Downham Road and turn left on Hawkswood Road, with the visitor centre just beyond Fremnalls Causeway. The reserve and visitor centre is open daily from 9am–5pm (except Christmas Day and Boxing Day). The trail to the four hides passes through woodland and thus, as well as waterbirds, a variety of common woodland species can be seen.

Fishing Lodge When the reserve centre is closed, there is limited access to the reserve on weekdays from the Fishing Lodge car park on Giffords Lane (south of South Hanningfield

village, turn opposite the Old Windmill pub).
Fremnalls Causeway This lies at the south end of the reservoir, and is formed by the minor road between Stock and Wickford. From here views can be had north over the reservoir.

FURTHER INFORMATION

Grid refs/postcodes: Visitor Centre TQ 725 971 / CM11 1WT; Fishing Lodge TQ 736 975 / CM3 8HS; Fremnalls Causeway TQ 722 971
EWT Hanningfield Reservoir Visitor Centre: tel: 01268 711001; web: essexwt.org.uk

137 BRADWELL (Essex)

OS Landranger 168

Lying on the shore of the Dengie Peninsula near the mouth of the Blackwater Estuary, this bleakly attractive area is important for wintering wildfowl and waders, and attracts small numbers of passerine migrants in spring and autumn. The area around Sales Point (Bradwell Cockle Spit), a reserve covering 102 ha, is jointly managed by the EWT and the Essex Birdwatching Society, and the latter also operate Bradwell Bird Observatory in the grounds of Linnett's Cottage on the edge of the reserve. The mud and sandflats to the east, extending 1.5 miles from the shore, are part of Dengie National Nature Reserve.

Habitat

A mixture of arable farmland, bounded by sea walls and their associated borrow-dykes, bordered by extensive areas of saltmarsh and the huge mudflats of Dengie Flats. Near the bird observatory there is a wooded thicket.

Species

Wintering wildfowl include numbers of Dark-bellied Brent Geese, as well as Wigeon, Teal and Shoveler. Waders include all the common species, as well as Grey Plover, Knot and Bar-tailed Godwit; in autumn and winter up to 20,000 waders roost on the reserve at high tide. Raptors in the area can include Hen Harrier, Sparrowhawk, Peregrine, Merlin and sometimes Short-eared Owl. There may be a few Snow Buntings, but Shore Lark and Twite are now extremely scarce. Offshore, Red-throated Diver and Slavonian Grebe may be present, and sometimes Great Northern or Black-throated Divers and Slavonian or Red-necked Grebes. There may also be seaducks, with Red-breasted Merganser, Common Eider and Common and Velvet Scoters possible.

On passage, Wheatear, Black Redstart and Firecrest are annual in spring and autumn, while Whinchat, Common Redstart, Pied Flycatcher and Ring Ouzel are largely confined to autumn. Scarce migrants such as Wryneck and Barred and Yellow-browed Warblers have occurred on a few occasions, as have occasional rarities. Passage waders should include Sanderling (on the shell banks), and Green and Common Sandpipers in the dykes and saltmarsh gutters.

Breeders include Shelduck, small numbers of Little Terns (on Pewit Island), as well as Oystercatcher, Ringed Plover and Common Redshank, with Reed and Sedge Warblers in the dykes and Yellow Wagtail in some of the arable crops. In summer Marsh Harrier and Hobby are common visitors, and Peregrines have attempted to breed on the power station.

Access

Bradwell Leave Maldon south-east on the B1018 (and then B1010) for 5 miles to Latchingdon and, in the village where the main road turns sharp right by the church, continue straight ahead on the minor road to Bradwell-on-Sea, a further 8.5 miles. Approaching Bradwell-on-Sea this

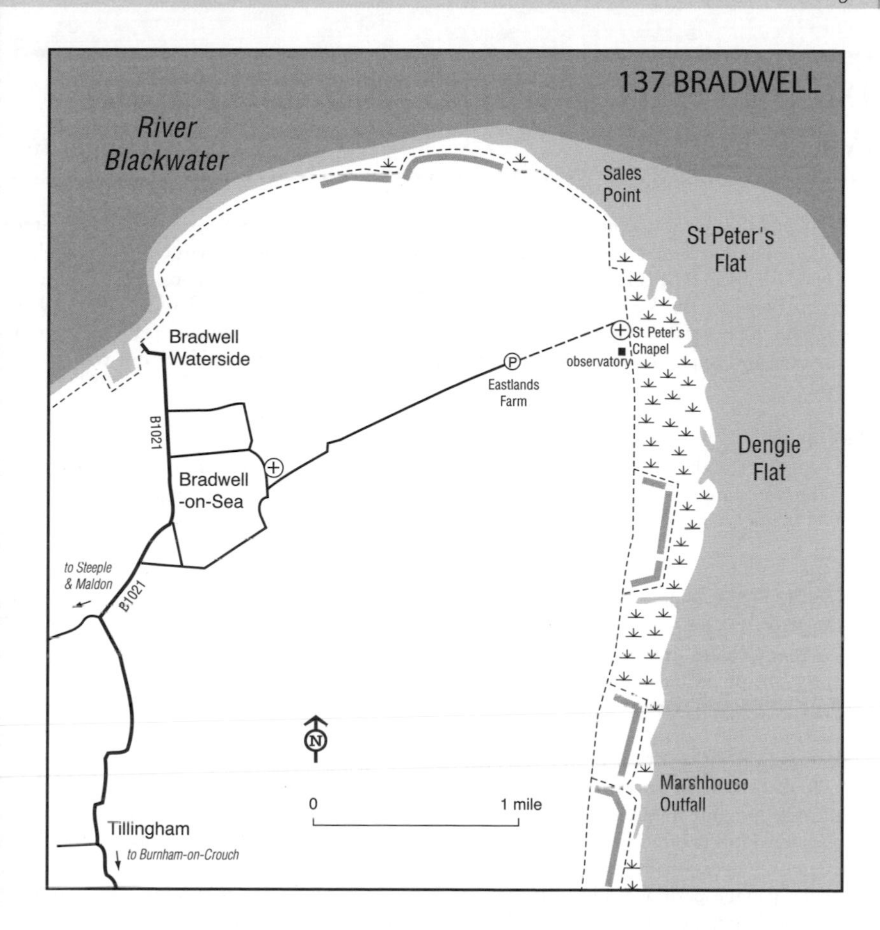

minor road meets the B1021. Turn left (north-east) and after 600 yards turn right (east) on a minor road to the village. Follow this and turn right again (immediately before the church) towards the Chapel of St Peter on the Wall, following the road for 1.3 miles to the car park at Eastlands Farm. From here walk along the track for 0.5 miles to the sea wall at St Peter's Chapel; the observatory lies 100 yards south of here and is manned every Sunday (it has dormitory accommodation for eight). From St Peter's Chapel you can either walk north for 0.5 miles to Sales Point, or south for 2.5 miles to Marshhouse Outfall. Both directions offer good views of roosting waders, with the period before high water being best.

Alternatively, from Bradwell-on-Sea, continue along the B1021 direct to Bradwell Waterside and follow the sea wall in either direction (west leads to St Lawrence Bay and Ramsey Marsh).

Dengie Peninsula It is possible to ride a mountain bike along the sea wall from Bradwell to the south end of the Dengie Peninsula near Holliwell Point. The dykes in this area are often good for migrant waders and wildfowl, and seabirds and sometimes passerine migrants are often commoner in this area. There is no access by car.

FURTHER INFORMATION

Grid refs: Bradwell TM 024 078; Dengie Peninsula TR 025 963
Bradwell Bird Observatory: Graham Smith, 48 The Meads, Ingatestone, Essex CM4 0AE; tel: 01277 354034
Essex Wildlife Trust: tel: 01621 86296; web: essexwt.org.uk

138 OLD HALL MARSHES AND TOLLESBURY WICK (Essex)

OS Landranger 168

Old Hall Marshes lie on the north shore of the Blackwater Estuary, 8 miles south of Colchester. They comprise the largest area of coastal marsh in Essex and the fourth largest in England. A total of 631 ha is an RSPB reserve, which includes Great and Little Cob Islands, and this in turn is part of the larger Blackwater Estuary NNR, which also includes the Tollesbury Wick EWT reserve which covers 243 ha. A variety of waders and wildfowl breed, but the primary interest is in winter and migration periods, when significant populations of wildfowl and waders are present.

Habitat

Three-quarters of the reserve is unimproved grazing marsh, dissected by numerous freshwater fleets, the rest is improved grassland, arable fields and saltmarsh, with a coastal lagoon and sizeable reedbed. Water levels are managed for the benefit of breeding and wintering wildfowl and waders.

Species

Up to 7,300 Dark-bellied Brent Geese winter in the Blackwater area, with up to 4,000 at Old Hall Marshes. They are often quite approachable, and it is worth checking at Old Hall for Pale-bellied Brent Geese and Black Brants among them. Ducks are numerous, especially Wigeon, Teal and Shelduck. Common Eider, Goldeneye, Red-breasted Merganser and occasionally Long-tailed Duck or Velvet Scoter may be found on the deeper channels. Great Crested, Little and Slavonian Grebes are all worth looking for, Red-throated Diver is common, and in recent winters up to five Great Northern Divers have been regular on the Blackwater; Black-throated Diver is an occasional visitor. Peregrine is regular, Hen Harrier, Merlin, and Barn and Short-eared Owls are quite frequent. The usual waders are numerous, with several thousand Lapwings and Golden Plovers often present, as well as small numbers of Ruffs. Sadly, Twite is nowadays recorded only occasionally.

During passage periods waders likely to be encountered include Little Stint, Green, Wood and Curlew Sandpipers, Ruff, Spotted Redshank, Greenshank, godwits and Whimbrel.

A range of species nest, including Shelduck, Pochard, Shoveler, Gadwall, Marsh Harrier, Avocet, Lapwing, Common Redshank, Yellow Wagtail and Bearded Tit. Little Terns sometimes breed and Common Terns may be viewed flying to and from Great Cob Island. Hobby and Garganey are not infrequently seen in summer.

Access

RSPB Old Hall Marshes From the A12 take the B1023, via Tiptree, to Tolleshunt D'Arcy. Turn left at the village maypole and then right into Chapel Road (the back road to Tollesbury). After 1 mile turn left into Old Hall Lane and carry on, over the speed bumps and through the iron gates, to the cattle grid, then follow the signs straight ahead to the car park. Parking permits should be obtained in advance from the reserve office. The reserve is open 9am–9pm or dusk if earlier. There are two trails following the public footpaths, one across the marshes which allows a circuit of 3 miles, the other following the sea wall around the perimeter of the marshes for 6.5 miles; two viewing screens overlook the saline lagoon at the eastern end of the reserve. Access on foot to the public rights of way is possible at all times (non-permit holders should use the car park at Woodrolfe Green, off Woodrolfe Road in Tollesbury, TL 964 107).

EWT Tollesbury Wick Marshes The whole of the north shore of the 10-mile-long Blackwater Estuary is an excellent birdwatching area in winter, and Slavonian Grebe is a speciality of this river. The grazing marshes and saltmarsh are a reserve of the EWT and lie immediately east of Tollesbury village. They can be viewed from the surrounding sea wall (which is continuous with that of Old Hall Marshes) and the birdlife is essentially the same. Enter Tollesbury on the

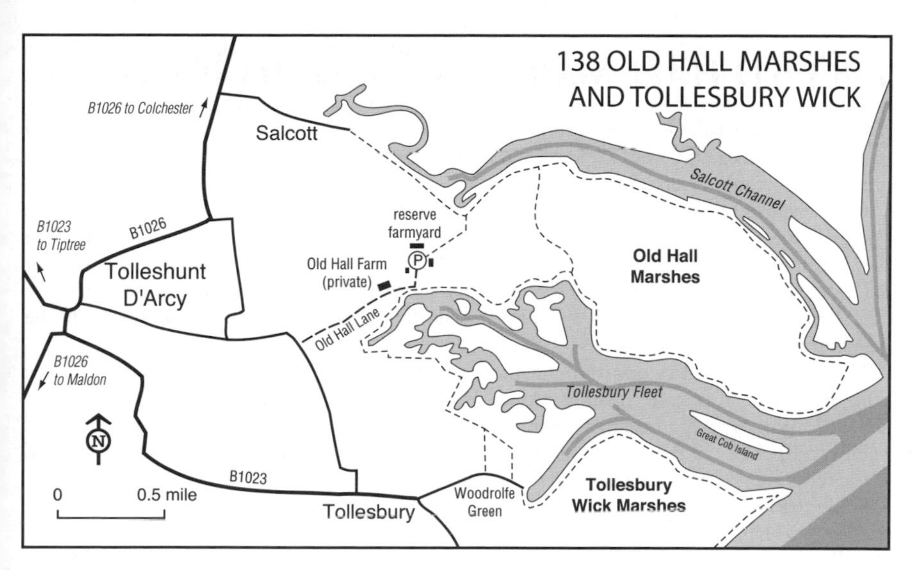

B1023 and fork left (north) in the village centre to park at Woodrolfe Green, and then walk past the marina to the sea wall. It is possible for the very energetic walker to continue along the sea wall to Goldhanger or even Heybridge Basin, and there are several public footpaths from the sea wall back to the B1023 near Tollesbury village.

FURTHER INFORMATION

Grid refs: RSPB Old Hall Marshes TL 959 122; EWT Tollesbury Wick Marshes TL 963 106
RSPB Old Hall Marshes: tel: 01621 869015; email: oldhallmarshes@rspb.org
Essex Wildlife Trust: tel: 01621 86296; web: essexwt.org.uk

139 ABBERTON RESERVOIR (Essex)

OS Landranger 168

Lying 5 miles south of Colchester, Abberton is the best-known reservoir in East Anglia. Noted for wildfowl, with nationally important concentrations of Mallard, Teal, Wigeon, Shoveler, Gadwall, Pochard, Tufted Duck and Goldeneye, its proximity to the coast also ensures a good variety of waders. A visit can be rewarding at any time of year.

Habitat

Covering 500 ha and 4 miles long, the reservoir is fed by Layer Brook at the west end, while the Roman River passes close to the dam at the north end. Most of the perimeter has concrete banks, but the west end has natural margins with lush vegetation and some bushes. Around 48 ha at the head of a sheltered bay adjacent to the B1026 is managed as a reserve by the EWT.

Species

Abberton attracts large numbers of waterfowl throughout the year and peak counts over the years have included 550 Great Crested Grebes, 40,000 Wigeons, 5,000 Pochards, 1,000 Goldeneyes, 113 Smews and almost 17,000 Coots. Indeed, Goldeneyes often number several hundred and Abberton is the most important inland site for this species in Britain.

In winter there are large numbers of the commoner wildfowl, especially Teal and Wigeon, as well as Goldeneye, Goosander and usually a few Smews. Bewick's Swan and White-fronted Goose occur irregularly. Large numbers of Canada and Greylag Geese are resident and, depending on water levels, waders, especially Dunlin, may winter (up to 3,000 Dunlins have been recorded), and large flocks of Golden Plovers are regular in the surrounding farmland. Other visitors can include divers, the rarer grebes, seaducks and occasionally Bittern.

In late summer there are large concentrations of moulting Mute Swans, Tufted Ducks, Pochards and Coots. Gadwall, Shoveler and Pintail are also commonest in autumn, although they occur throughout the winter. Red-crested Pochard may occur at any season but is most likely in autumn. The origin of such birds is perhaps suspect and some may be escapes. Water Rail, Kingfisher and Water Pipit are other species that may occasionally be seen outside the breeding season.

Many migrant waders pass through in spring and autumn, including Ringed and Little Ringed Plovers, Turnstone (mainly spring), Ruff, Black-tailed Godwit, Common Sandpiper and Little Stint (mainly autumn); Greenshanks and Spotted Redshanks are often present in good numbers, especially in the autumn. Marsh Harrier, Osprey, Garganey, Little Gull, and Arctic and Black Terns are recorded annually and there are sometimes huge flocks of Swifts and hirundines.

Since 1981 Cormorants have nested in willows between the two causeways and the colony peaked at 551 pairs in 1996. This is the largest tree-nesting colony in Britain, and most are of the subspecies *sinensis*, which may be a distinct species. The colony is easily viewable from the Layer Breton Causeway. Great Crested Grebe, Gadwall, Shoveler and probably Water Rail breed at the west end and Common Tern nests on specially constructed rafts. Other breeders include Yellow Wagtail, Nightingale and Cetti's, Reed and Sedge Warblers; Garganey, Ruddy Duck and Ringed Plover have bred.

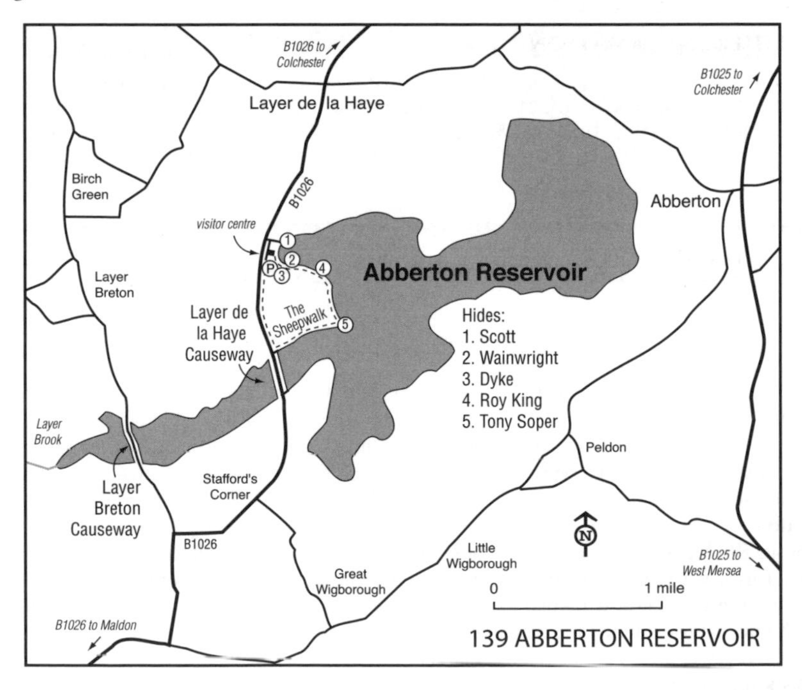

139 ABBERTON RESERVOIR

Access

Most species can be seen from the reserve and two causeways across the west arm; there is only very limited access (by permit only) to the remainder of the reservoir's concrete banks.

EWT reserve Leave Colchester south on the B1026 and after 4.5 miles the reserve and visitor centre lie east of the road. The centre is open 9am–5pm daily except Mondays (open Bank Holiday Mondays), Christmas Day and Boxing Day. There is a loop nature trail, with five hides, providing views over the reservoir and passing through farmland and woodland. Note that in 2010 Essex and Suffolk Water, owners of the reservoir, will start work to raise the levels of the reservoir, due to increased demand for water. This increase in water level means that the reserve and visitor centre will be moving to the peninsular field; up-to-date information is available from the current centre.

Layer-de-la-Haye Causeway Continuing south from the reserve on the B1026 for 0.5 miles the road reaches the Layer de la Haye Causeway.

Layer Breton Causeway Continue south on the B1026 for 1.25 miles to a T-junction and turn right onto a minor road north towards Layer Breton and Birch. After 0.5 miles the road reaches the causeway.

FURTHER INFORMATION

Grid refs/postcodes: EWT reserve TL 963 184; Layer de la Haye Causeway TL 963 173; Layer Breton Causeway TL 949 167

Abberton Reservoir Visitor Centre: Church Rd, Layer de la Haye, Colchester, CO2 0EU; tel: 01206 738172; email: admin@essexwt.org.uk

Essex Wildlife Trust: tel: 01621 86296; web: essexwt.org.uk

140 FINGRINGHOE WICK (Essex)

OS Landranger 168

Lying 5 miles south-east of Colchester, this EWT reserve encompasses a broad range of habitats in a relatively small area and is worth visiting throughout the year (although midsummer is quietest).

Habitat

Once gravel workings, 50 ha of undulating terrain have been transformed into a varied mosaic of habitats, including a freshwater lake, mature secondary woodland, planted conifers, scrub, stands of reeds and areas of heath. There is also a specially constructed scrape. Immediately to the east lies the River Colne with areas of saltmarsh and mudbanks along its margins, and to the south the extensive saltings of Geedon Marsh.

Species

Wintering wildfowl include good numbers (2,000+) of Dark-bellied Brent Geese around the estuary, with Cormorant, Red-breasted Merganser, Goldeneye and occasionally Long-tailed Duck on the river channel. The complex of lakes holds Shoveler, Teal, Gadwall and Pochard and Goldeneye, as well as Water Rail. Little Egret has become increasingly regular, and is most often seen from the hides overlooking the scrape or the saltings. Waders include Grey and Ringed Plovers, Turnstone, Curlew, Black-tailed Godwit, Dunlin and Common Redshank, and a flock of up to 700 Avocets is now regular along the Colne in autumn and winter, viewable from the hides. Winter raptors in the area may include Sparrowhawk, Marsh Harrier, Peregrine, Merlin, Short-eared Owl and occasionally Hen Harrier.

On autumn passage a variety of waders may occur, including Greenshank, Spotted Redshank, Green Sandpiper and Ruff.

Breeders include Little and Great Crested Grebes, Sparrowhawk, Barn Owl, Kingfisher, good numbers of Nightingales (up to 30–40 pairs), Reed Warbler, and Common and Lesser Whitethroats.

Access

Leave Colchester south on the B1025 towards Mersea, and after 4.5 miles cross the Roman River bridge and immediately turn east (signed to the reserve) on a minor road towards Fingringhoe. After a further 1.5 miles, turn right (east) on a minor road towards South Green, reaching the entrance to the reserve after a further 1.5 miles. Follow this track for 0.5 miles to the car park and Fingringhoe Wick Visitor Centre. There are two nature trails and 8 hides, and the visitor centre is open Tuesday–Sunday and Bank Holiday Mondays 9am–5pm; all visitors to the reserve are required to obtain a day permit. For waders on the estuary the period two hours either side of high water is best, but early morning can be problematic due to the east-facing aspect.

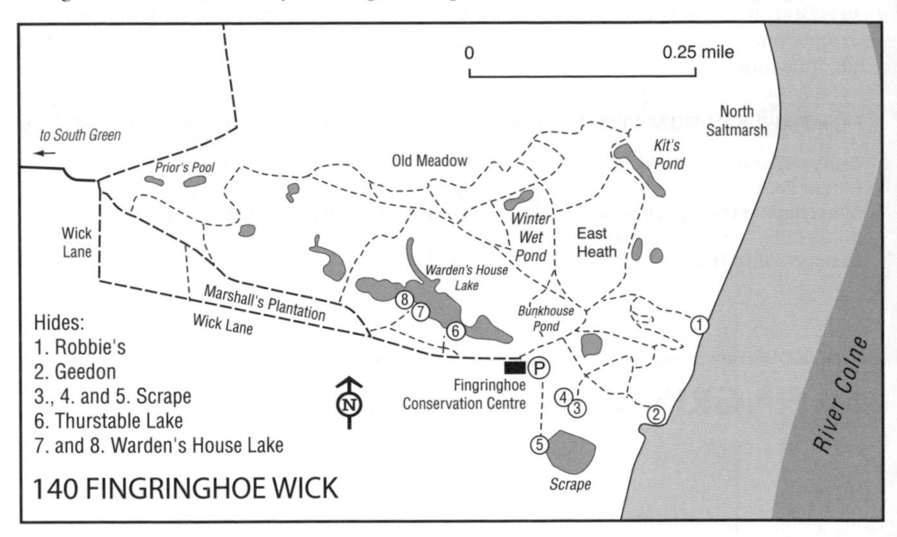

FURTHER INFORMATION

Grid ref/postcode: TM 047 192 / CO5 7DN
EWT Fingringhoe Wick Visitor Centre: tel: 01206 729678; email: louised@essexwt.org.uk; web www.essexwt.org.uk

141 WALTON-ON-THE-NAZE (Essex)

OS Landranger 169

Extending northwards from Walton-on-the-Naze, the Naze is a headland roughly 3 miles long by 1 mile wide that shelters Hamford Water from the sea. The whole area attracts wintering wildfowl and waders, while areas of cover along the coast offer shelter to a variety of migrants.

Habitat

At Walton the cliffs of soft Red Crag rise to 20 m, but they gradually fall in height northwards and from the tip of the Naze a mile long shingle beach extends north to Stone Point. Bushes along the cliff path attract migrant passerines and at the northern tip the John Weston EWT reserve comprises Blackthorn and bramble thickets, rough grassland and four ponds behind the sea wall, totalling 3.6 ha. In the lee of the Naze lie mudflats and saltmarshes with some rough grassland and pools.

Species

Several thousand Dark-bellied Brent Geese winter in Hamford Water and there is also a good chance of seeing Common Eider offshore, and Goldeneye and Red-breasted Merganser in the deeper waters off Stone Point or in Walton Channel. Other seaducks, including Long-tailed Duck, Velvet Scoter and Scaup, are occasional. Sanderlings and the odd Purple Sandpiper frequent the shore, and the common waders are well represented, with Avocet and Black-tailed Godwit in Walton Backwaters. Hen Harrier, Merlin and occasionally Short-eared Owl hunt the saltmarshes. Snow Bunting is also sometimes present along the shingle beach, while Rock Pipits favour the saltmarsh gutters.

In spring and autumn a variety of waders appear, including Greenshank, Spotted Redshank and, mostly in autumn, Curlew Sandpiper. Fulmar may be seen offshore in spring and summer, while in autumn passing Gannet and Arctic Skua are a possibility, and Manx Shearwater and Great Skua are sometimes seen in favourable seawatching conditions. The bushes on the clifftops are excellent for passerine migrants, with Black Redstart, Ring Ouzel and Firecrest possible in early spring. In autumn east winds during August to October are especially favourable. Ring

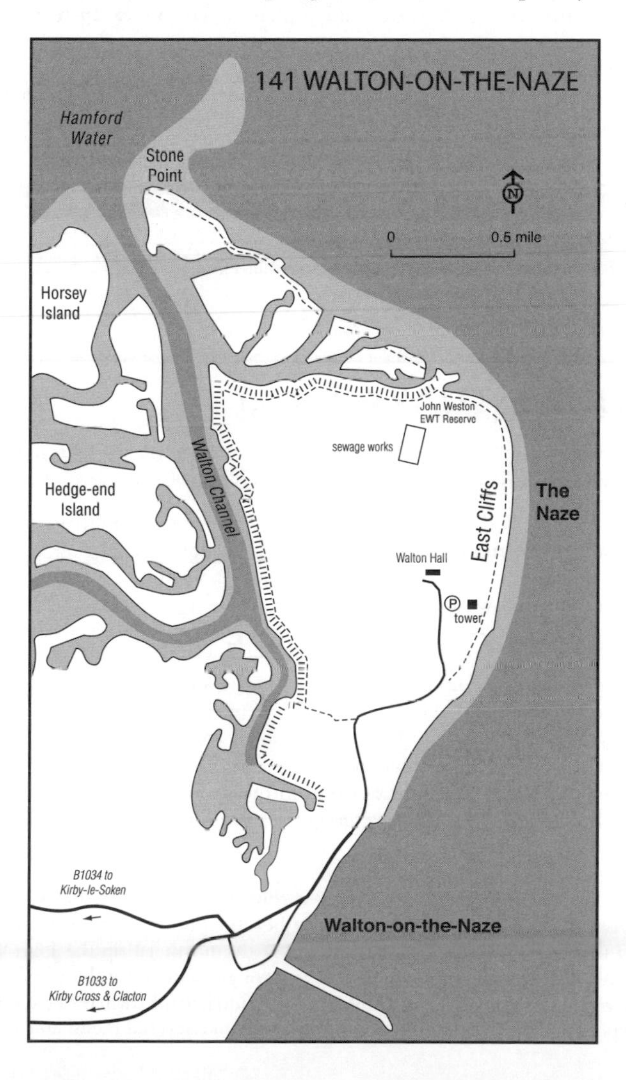

Ouzel, Common Redstart, Whinchat, Garden Warbler and Pied Flycatcher are regular, and scarcer species such as Wryneck and Red-backed Shrike are occasional, with Yellow-browed Warbler recorded annually in October.

Common and Lesser Whitethroats and occasionally Cetti's Warbler breed on the EWT reserve, with Sand Martins in the cliffs, while Shelduck, Oystercatcher, Ringed Plover and Little Tern nest along the beach towards Stone Point, but please respect signs restricting access to this area during the nesting period.

Access

On entering Walton, continue to the seafront and proceed north on Naze Park Road and then Hall Lane, turning right to the car park on the cliffs by the tower. Walk north along the cliff path, checking areas of cover for migrants in season; the area around the East Cliffs is the main birdwatching area at Walton, but it is possible (and worthwhile) to continue past the John Weston EWT reserve, crossing various creeks as necessary, as far as Stone Point: note that in recent years the beach has been built up, permitting access to the latter area at all states of the tide. The situation may change, however, and you are urged to seek up-to-date local advice, without which you should plan to be back at the sea wall at least three hours before high water. Alternatively, on reaching the sea wall by the John Weston reserve turn west for 0.75 miles towards Walton Channel. The sea wall follows the channel south and eventually back to the town; there is then a 0.5-mile walk north along the road back to the car park. The Naze is a public open space owned by Tendring District Council.

FURTHER INFORMATION

Grid ref: TM 264 234
Essex Wildlife Trust: tel: 01621 86296; web: essexwt.org.uk

142 STOUR ESTUARY (Essex)

OS Landranger 169

Lying on the south shore of the Stour Estuary, this RSPB reserve offers the unusual combination of both woodland and estuarine species, with the adjacent Copperas Wood EWT reserve adding more woodland interest. The woods are best in spring, while the estuary is most productive in autumn and winter.

Habitat

The RSPB reserve includes most of Copperas Bay on the south shore of the Stour Estuary, and extends east to near Parkeston. There are large areas of intertidal mud, with a little saltmarsh on the foreshore, and also small areas of estuarine reedbeds and bramble scrub. Stour Wood itself is mainly comprised of Sweet Chestnut and to the east, Copperas Wood has stands of Sweet Chestnut and Hornbeam; coppicing has been reinstated in both woods to provide a more varied habitat.

Species

The estuary attracts the usual range of wintering wildfowl and waders, including Dark-bellied Brent Goose, Shelduck, Wigeon, Teal, Pintail, Grey Plover, Curlew, Common Redshank, Knot and Dunlin, and good numbers of Black-tailed Godwits of the Icelandic race, with an average of 2,100 birds wintering on the Stour. Species present in smaller numbers include Turnstone, Oystercatcher and Ringed Plover, and Goldeneye may be seen on the river channel. On passage a greater variety of waders may be found, with Greenshank, Spotted Redshank and, in autumn,

Little Stint and Curlew Sandpiper possible, as well as Green and Common Sandpipers in the saltmarsh gutters, and there have been several records of Osprey.

Breeding birds in Stour Wood include Green and Great Spotted Woodpeckers, Nightingale, Garden Warbler and Blackcap, with Shelduck, Reed Warbler and Common and Lesser Whitethroats along the estuary shore.

Access

The RSPB reserve lies north of the B1352 between Harwich and Manningtree. The car park (signed) is north of the road 1 mile east of Wrabness, 600 yards after the wood appears on the left-hand side. Stour Wood is accessed via a circular walk of 1 mile (accessible by wheelchair when dry), while a circular walk of 2 miles leads from the car park to the estuary, returning via Copperas Wood. Two hides and a viewing point overlook the estuary, with the return trip from the car park to the furthest hide covering 4.3 miles. The reserve is open at all times, but the car park is locked from 7pm (or dusk, whichever is earlier) until 6am. For wildfowl and waders a rising tide, from 2 hours before high water until 2 hours after, is best. (NB the RSPB reserve also includes an area on the northern side of the river to the west of the A137 at Cattawade. This can only be viewed rather distantly from footpaths on the southern side of the river.)

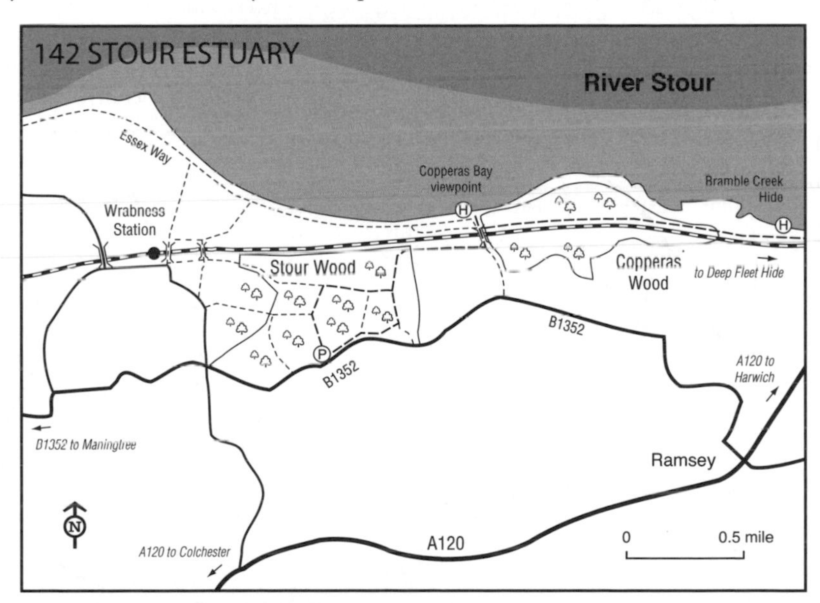

FURTHER INFORMATION

Grid ref: TM 190 310
RSPB Stour Estuary: tel: 01473 328006; email: stourestuary@rspb.org.uk

143 ALTON WATER (Suffolk)

OS Landranger 169

Just 5 miles south of Ipswich and close to the Stour Estuary, Alton Water is an important area for waterbirds, best in winter or during passage periods.

Habitat

The reservoir was opened in 1987, covers around 160 ha and has a largely natural shoreline, attractive to waders. It is owned and managed by Anglian Water. The surrounding area is farmed, with some woodland.

Species

The commoner ducks, such as Wigeon and Gadwall, reach several hundred in winter, with smaller numbers of Shovelers and Goldeneyes. Scaup, Long-tailed Duck, Smew and Goosander are regular, as are divers and the rarer grebes. Feral Greylag and Canada Geese are often present in surrounding fields. Jack Snipe is sometimes found in the lakeside vegetation and Bittern is another occasional visitor.

Passage brings a sprinkling of waders and terns of the more usual inland species (Coralline hide is often good), and Osprey and Black Tern are sometimes recorded.

In summer Alton Water is an important breeding site for Great Crested Grebe (65 pairs), with Common Terns on specially constructed rafts and good numbers of Nightingales in the surrounding woodland and scrub (try the area between the Larchwood hide and the ponds). Other breeding species include Oystercatcher, Barn Owl, Turtle Dove and Spotted Flycatcher.

Access

Alton Water: north arm To the south of Ipswich, leave the A14 dual carriageway southwards on the A137 towards Manningtree. After 2 miles the road runs alongside the north tip of the

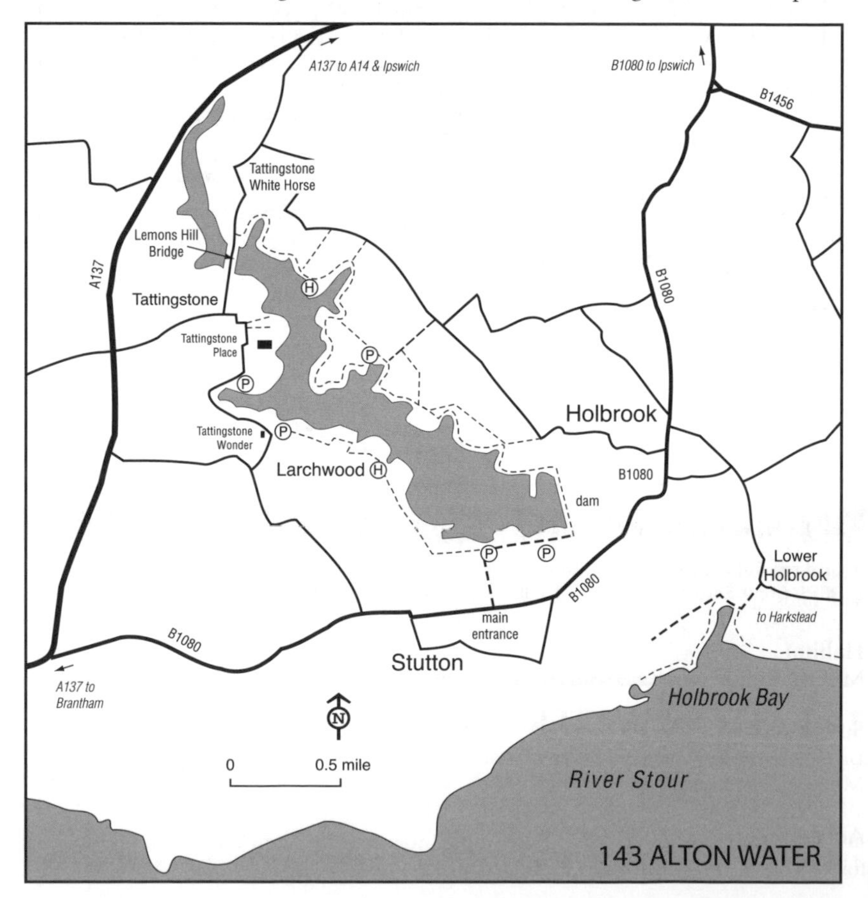

reservoir. This is probably the best area and is easily viewed from the road.

Alton Water: Tattingstone Continuing south on the A137, after 1 mile turn left (east) on a minor road to Tattingstone. In the village, footpaths lead to the reservoir banks (immediately north of Tattingstone Place), a good area for ducks and waders.

Alton Water: Lemons Hill Bridge Turning left (north) in Tattingstone, opposite the church, the road reaches this causeway across the north arm of the reservoir, affording further views, before continuing north to Tattingstone White Horse. It is possible to walk east or west from the causeway, on both the north and south banks (avoiding the marked conservation area on the north-west bank), and a hide overlooks the tern rafts.

Alton Water: central section, south shore From Tattingstone take the minor road south towards Stutton, and after 0.5 miles the road runs alongside the central section of the reservoir. At Tattingstone Wonder (a cottage that resembles a church), there is a car park and it is possible to walk east along the reservoir banks to a hide.

Alton Water: central section, north shore Turn off the A137 (1.25 miles south of the A14) on minor roads south to Tattingstone White Horse (also accessible from Tattingstone). Just north of the village take the minor road south-east towards Holbrook. The road makes two right-angle bends and, after 300 yards and 600 yards, footpaths lead to the shore and a hide. For further access, continue on the road for 0.7 miles and turn right (south-west) on a track to the reservoir shore.

Alton Water: dam end Turn north off the B1080 between Stutton and Holbrook at the official Anglia Water gate and continue towards the dam to the car park with toilets and a cafe. The deeper water near the dam is usually best for any wintering divers and a 3-mile nature trail on the south shore takes in woodland, meadow, ponds and a hide.

Holbrook Bay On the north shore of the Stour Estuary, south-east of Alton Water. Leave the B1080 at Holbrook on a minor road south-east to Harkstead. After 1 mile a track to the right (south) at the beginning of the hamlet of Lower Holbrook leads to the innermost part of the Bay. Walk in either direction along the sea wall. In winter Dark-bellied Brent Geese regularly congregate in the Bay and there is a large high-tide wader roost.

FURTHER INFORMATION

Grid refs: Alton Water: north arm TM 132 385; Tattingstone TM 137 370; Lemons Hill Bridge TM 136 375; central section, south shore TM 139 363; central section, north shore TM 145 378; dam end TM 156 353; Holbrook Bay TM 176 350
Anglian Water Visitor Centre: tel: 01473 328268

144 WOLVES WOOD (Suffolk)

OS Landranger 156

This small wood holds good numbers of Nightingales as well as many of the commoner woodland birds.

Habitat

Mixed broadleaf woodland with an area of coppiced scrub, surrounded by farmland.

Species

Breeders include Woodcock, Lesser Spotted Woodpecker, Nightingale, a variety of warblers, Marsh Tit and Hawfinch.

Access

The reserve lies just north of the A1071 between Ipswich and Hadleigh, 2 miles east of Hadleigh, and is signed from the road. Open at all times, with a circular nature trail of 1 mile, although the car park is locked from 6pm (or dusk if earlier) to 9am. The trail is not easily accessible by wheelchair as it is frequently muddy.

FURTHER INFORMATION

Grid ref: TM 054 437
RSPB Wolves Wood: tel: 01473 328006: email stourestuary@rspb.org

145 LACKFORD LAKES (Suffolk)

OS Landranger 155

This series of gravel pits lies in the Lark Valley between Mildenhall and Bury St Edmunds and attracts a variety of wildfowl throughout the year, as well as passage migrants, and has an excellent list of rarities to its credit. 121 ha form a reserve of the Suffolk Wildlife Trust (SWT).

Habitat

A complex of disused gravel pits, several of which have been specially landscaped to attract wildlife, while others are used for sailing and fishing; the site also includes areas of scrub, carr and meadow. To the north lies the King's Forest, a large area of conifer plantations (with Nightjar, Woodlark and Common Crossbill).

Species

Wintering wildfowl include Pochard, Tufted Duck, Teal and Gadwall, as well as small numbers of Goldeneyes, Goosanders and Shovelers; seaducks are recorded occasionally. Other waterbirds include Cormorant, Great Crested and Little Grebes and occasionally divers or the rarer grebes. Kingfisher is frequent (especially from the hides overlooking The Slough). The large winter gull roost on the Sailing Lake (peaking at 10,000+ birds) regularly holds Mediterranean and

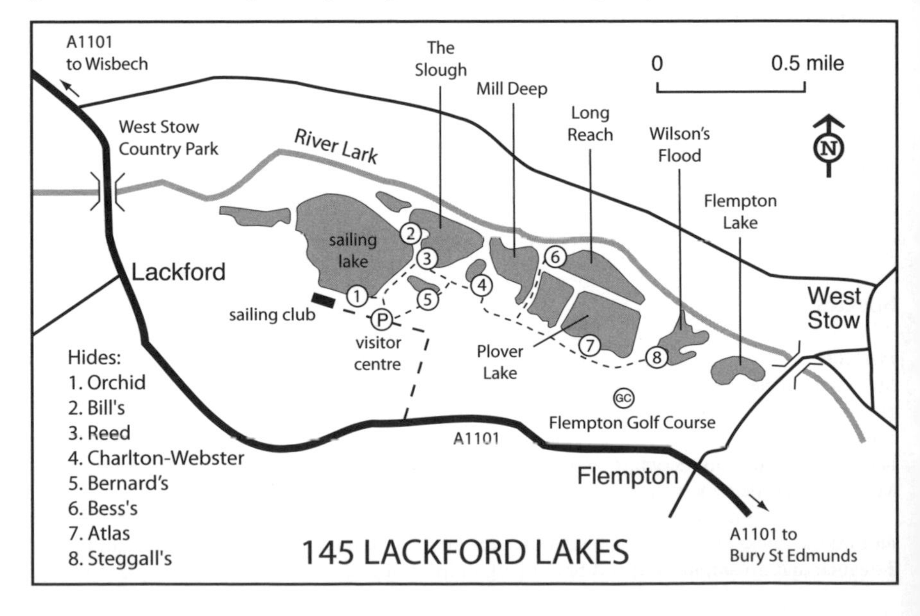

Yellow-legged Gulls, and the scarcer white-winged gulls are possible.

On passage Osprey is recorded annually, and Little Egret and a variety of waders may be seen, with Little and Temminck's Stints, Wood and Curlew Sandpipers, Black-tailed and Bar-tailed Godwits and Grey Plover joining more regular migrants. Terns may include Arctic and Black.

Breeders include Shelduck, Pochard, Water Rail, Common Redshank, Little Ringed Plover, Nightingale (the car park is a good area) and Reed and Sedge Warblers, while Hobby is a regular visitor, hunting dragonflies and hirundines, and Common Buzzard, Sparrowhawk and, much more rarely, Goshawk visit throughout the year. Common Crossbill and Siskin may fly over en route to and from nearby plantations.

Access

The reserve lies just north of the A1101 between Bury St Edmunds and Mildenhall, with the entrance track signed 1 mile west of the minor road to West Stow. From the car park a footpath leads to eight hides. The reserve and hides are open dawn to dusk every day, and the visitor centre is open Wednesday–Sunday 10am–4pm from 1 November to 31 March and 10am–5pm April to October.

FURTHER INFORMATION

Grid ref/postcode: TM 054 437 / IP28 6HX
Lackford Lakes SWT Visitor Centre: tel: 01284 728706; email: lackford.education@suffolkwildlifetrust.org
Suffolk Wildlife Trust: tel: 01473 890089; web: www.suffolkwildlifetrust.org

146 LANDGUARD POINT (Suffolk)

OS Landranger 169

This site has an outstanding reputation for attracting interesting migrants in spring and autumn, including many major rarities. A Bird Observatory was established here in 1983, housed in disused military buildings that formerly held gun batteries, and part of the point comprises an LNR, managed by the SWT.

Habitat

Landguard Point is a shingle spit lying immediately south of Felixstowe at the mouth of the River Orwell. Behind the shingle beach lies Landguard Common, an area of compacted sand and gravel covered by very short, rabbit-cropped turf, with some stands of scrub that are attractive to migrants. Landguard was used by the military for nearly 200 years and many old fortifications remain. All around is the dockland sprawl of Felixstowe and Harwich.

Species

The Point is most active in spring and autumn when hundreds of the commoner thrushes, warblers and finches pass through, with plenty of Wheatears on the short turf and sometimes falls of night migrants or arrivals of Redwing, Fieldfare and Starling. Up to 10,000 birds are ringed annually at the observatory. Some of the more regularly occurring scarce migrants include Wryneck, Bluethroat, Icterine and Barred Warblers, Firecrest, Red-breasted Flycatcher and Ortolan Bunting. Seawatching off Landguard is rarely notable but a few shearwaters and skuas are recorded each autumn, sometimes including Long-tailed Skua.

In summer, Ringed Plover and Little Tern breed on the shingle and Black Redstart nests around the docks and fort area, with Wheatear on the common. In winter Mediterranean Gull is regular and Snow Bunting and Shore Lark fairly frequent.

Access

Follow the A14 towards Felixstowe Docks and, at the roundabout where the dual carriageway ends, continue straight ahead into Felixstowe. Pass over the level crossing and turn right at the first set of traffic lights. The first turning on the left, Manor Terrace (signed Landguard), leads to a car park at the north end of Landguard Common. The second left turn (Viewpoint Road) leads to a car park at the south end of the common, close to Landguard Fort, where an old gun emplacement houses Landguard Bird Observatory (no accommodation). Interesting migrants can be found anywhere on the Common, to which there is free access at all times. The fenced-off bird observatory ringing area holds the best and most extensive areas of cover,

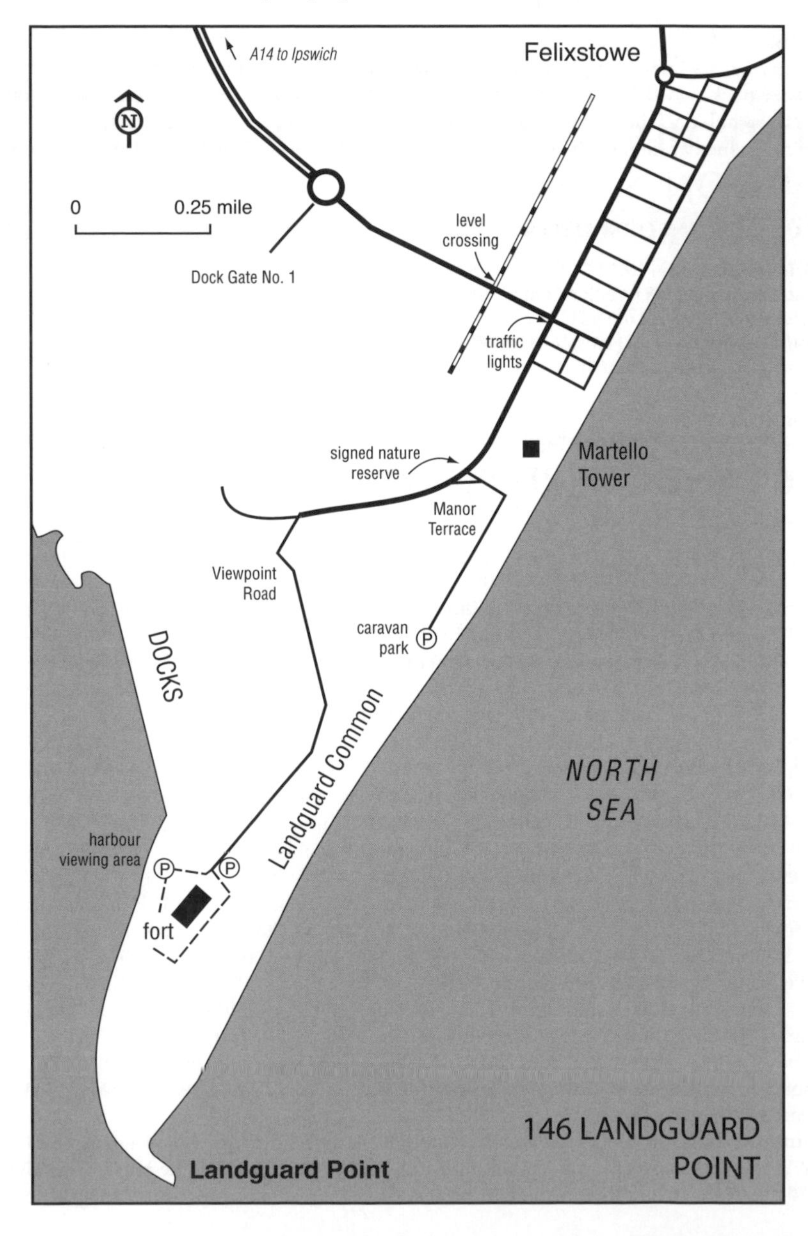

and though it is not open to the public, most of the birds can be seen from the footpath outside the fence – the west side of the compound can be accessed by walking around the large concrete building.

FURTHER INFORMATION

Grid refs/postcode: Observatory TM 054 437 / IP11 3TW; Landguard Common TM 285 315
Suffolk Wildlife Trust: tel: 01473 890089; web: www.suffolkwildlifetrust.org
Landguard Bird Observatory: tel: 01394 673782 (observation room); email: landguardbo @yahoo.co.uk; web: www.lbo.org.uk

147 TRIMLEY MARSHES AND LEVINGTON LAGOON (Suffolk)

OS Landranger 169

Immediately adjacent to Felixstowe Docks, Trimley Marshes SWT reserve comprises 84.4 ha and was created in 1990 to part compensate for the loss of intertidal mudflats caused by an extension to the port. A variety of wetland habitats, recreated from arable farmland, attract a broad range of species, with year-round interest. A little to the west, Levington Lagoon SWT reserve comprises saltmarsh and mudflats alongside the River Orwell.

Habitat

The area holds a range of wetland habitats, from open lagoons, saltmarsh and mudflats to shingle foreshore, with areas of scrub which attract passerines.

Species

Wintering wildfowl include Dark-bellied Brent Goose, Wigeon, Teal, Gadwall, Shoveler, Pintail, Pochard and Goldeneye. Waders include Common Snipe and Common Redshank, with Grey Plover, Curlew, godwits, Knot and Dunlin on the Orwell Estuary. Occasionally, Snow Bunting may be found on the shingle bund bordering the docks.

On passage a variety of waders has been recorded, unsurprisingly including some of the more uncommon species such as Spotted Redshank, Wood and Curlew Sandpipers, and Little and Temminck's Stints. Little Egret is regular, and several rarities have occurred.

Breeding birds include Avocet, Oystercatcher, Ringed Plover and Little and Common Terns on the specially constructed shingle islands, as well as Tufted Duck, Gadwall, Shoveler, Shelduck, Lapwing, Common Redshank, Black-tailed Godwit, Sand Martin and Reed and Sedge Warblers, with Yellow Wagtail and Corn Bunting in nearby fields.

Access

Trimley Marshes Leave the A14 dual carriageway immediately west of Felixstowe at the exit for Trimley villages. Follow signs into Trimley St Mary and from the main street turn south into Station Road (signed 'Nature Reserve'). Follow this over the level crossing and park on the right, just before the road ends at the farmyard. From here it is a 2-mile walk, following the signs, along a broad track to the river bank and lagoons. The reserve and five hides are open at all times, and the visitor centre at weekends, 10am–4pm.

Fagbury Cliffs Follow directions as for Trimley Marshes but after walking 0.5 miles carry straight on (rather than turning right towards the reserve). The trees and bushes here are favoured by passerine migrants.

Levington Lagoon Good for waders, although views tend to be a little distant. From the A12/A14 intersection take the A1156 south and, after 0.5 miles, follow the minor road towards Trimley St Martin and then turn right (south) to Levington. From the small car park on the minor road 0.25 miles east of the church follow the public footpath south along the eastern shore of Levington Creek to view. It is possible to continue to Trimley Marshes by continuing along the banks of the Orwell, past Levington Marina (check for interesting gulls) and Trimley Lake, following the sea wall for 2 miles to the reserve.

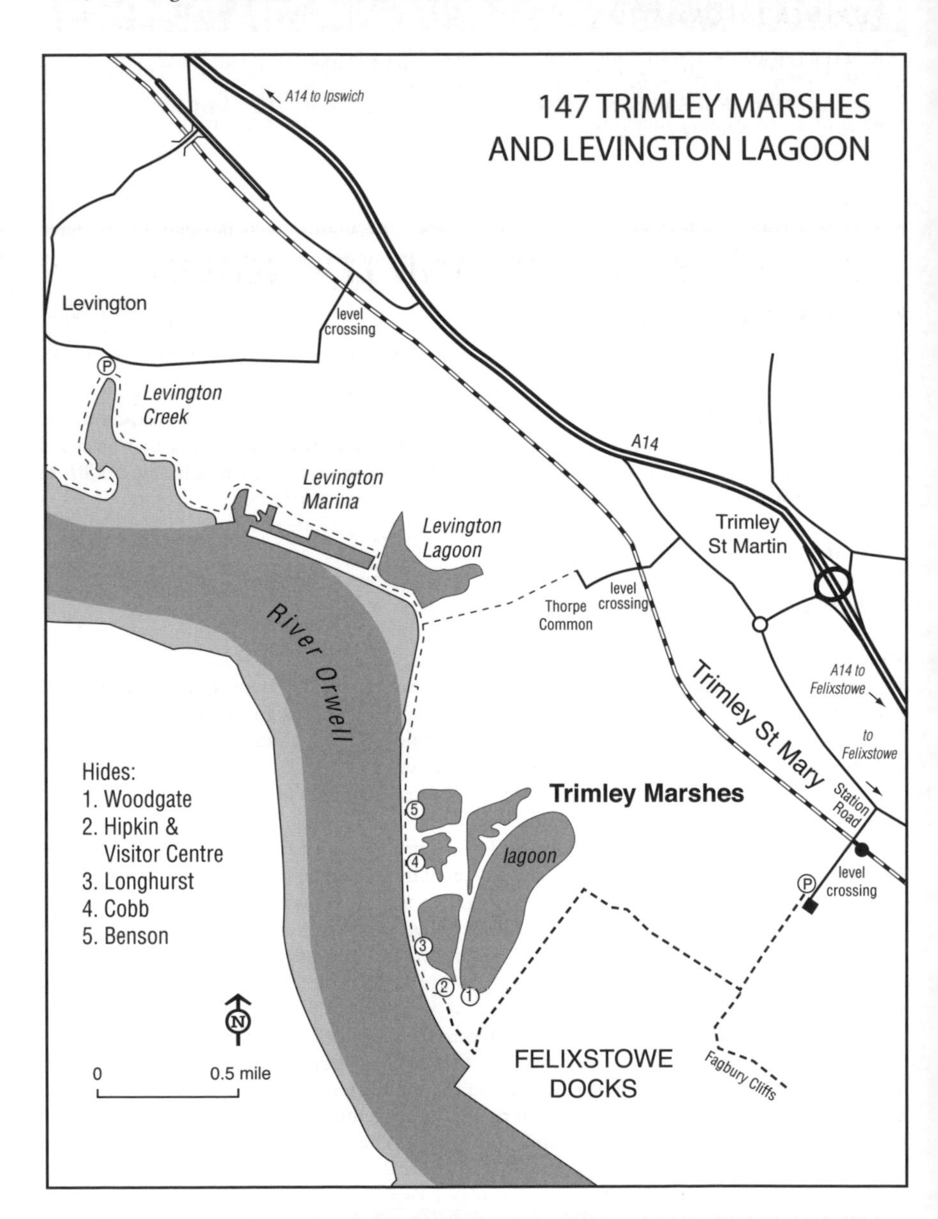

FURTHER INFORMATION

Grid ref: Trimley Marshes & Fagbury Cliffs TM 260 352; Levington Lagoon TM 238 389
Suffolk Wildlife Trust: tel: 01473 890089; web: www.suffolkwildlifetrust.org

148 HAVERGATE ISLAND AND BOYTON MARSHES (Suffolk)

OS Landranger 169

Flooded as part of wartime defence measures, Havergate Island was the jump-off point for the Avocet's recolonisation of Britain in 1947, and the colony now numbers up to 120 pairs, with several other interesting breeding species. On passage and in winter a variety of waders and wildfowl can be found.

Habitat

Havergate Island is a low embanked island in the River Ore, with some saline lagoons, complete with islands. It is bounded by shingle beaches and surrounded by saltmarsh. Boyton Marshes lie west of the River Ore, in the angle formed by its confluence with the River Butley. They comprise grazing marshes divided by a network of ditches and bordered by extensive areas of saltmarsh along the Butley.

Species

Breeding birds on Havergate Island include Shelduck, Oystercatcher, Ringed Plover, Common Redshank, Common and usually Sandwich Terns, a handful of Arctic Terns, Black-headed Gull and Short-eared Owl, as well as Avocet. Little Tern breeds on the shingle spit of Orford Ness, which separates the River Ore from the sea, and often visit Havergate. Boyton Marshes additionally support Gadwall, Shoveler, Lapwing, Reed and Sedge Warblers, and sometimes Grasshopper Warbler and Nightingale. Grey Heron, Little Egret and Barn Owl are resident in the area.

Many waders are present on passage, such as Black-tailed Godwit, Ruff and Greenshank in spring, with Turnstone, Little Stint and Curlew Sandpiper possible in autumn, and in recent years there have been flocks of more than 20 Spoonbills. Winter wildfowl include Wigeon, Teal, Pintail, Shoveler, Gadwall and occasionally Bewick's Swan, with Goldeneye and Red-breasted Merganser on the river channel. Grey Plover, Knot and Bar-tailed Godwit winter, as well as numbers of Avocets. Raptors may include Hen or Marsh Harriers and Short-eared Owl, and sometimes Rough-legged Buzzard.

Access

RSPB Havergate Island Only accessible by boat from Orford Quay (on the B1084), trips run April–August on the first and third weekends of the month and every Thursday, departing at 10am and returning at 3pm. In winter, September–March, boats run on the first Saturday of each month. Bookings can be made via the RSPB Minsmere reserve visitor centre, tel 01728 648281; there is a charge for both members and non-members of the RSPB. Facilities include a trail running for 1.5 miles along the length of the island, five hides, a viewing screen and toilets. The nature trail is not suitable for wheelchairs.

RSPB Boyton Marshes These overlook Havergate Island and share a similar range of wildfowl, waders and raptors. From the B1084 at Butley follow the minor road south to Capel St Andrew, then turn left towards Boyton village. Approximately 0.25 miles before the village, bear left down a concrete track on a sharp right-hand bend to the car park on the edge of the marshes. From here a footpath leads east to the river, where the coastal footpath follows the river bank in both directions. Paths may be difficult for the mobility-impaired.

FURTHER INFORMATION

Grid refs: RSPB Havergate Island TM 425 495; RSPB Boyton Marshes TM 387 475

RSPB Havergate and Boyton Marshes: tel: 01394 450732; email: havergate.island@rspb.org.uk

149 ALDE ESTUARY, NORTH WARREN AND SIZEWELL (Suffolk)

OS Landranger 156

This small estuary on the Suffolk coast attracts nationally important numbers of wintering Black-tailed Godwits and Avocets, and with the surrounding area also attracts wildfowl, a variety of waders and raptors. Nearby, the RSPB's North Warren and SWT's Sizewell Belts have a mosaic of habitats, including heathland, and further north still the outflow at Sizewell nuclear power station may attract gulls and terns.

Habitat

The Alde Estuary has mudflats and saltmarshes, and to the north and south lie grazing marshes, some of which flood in winter. To the north, the shingle beach at Thorpeness holds an important flora and occasionally attracts Shore Lark, and just inland the RSPB's North Warren reserve comprises an area of the Suffolk 'Sandlings', with grassy acid heath on sandy soils and areas of Gorse and birch, as well as areas of 'fresh' marsh. This coast faces east and regularly attracts scarce migrants wherever there is cover.

Species

Wintering wildfowl include numbers of Wigeon, Pintail, Shoveler, Gadwall and Teal, and sometimes Bewick's Swan, White-fronted and occasionally Tundra Bean Geese (especially on the grazing marshes around North Warren). Waders include Grey Plover and Turnstone, as well as several hundred Avocets and Black-tailed Godwits on the Alde. Small numbers of Ruffs usually overwinter at Church Farm Marshes. Raptors may include Hen Harrier and Sparrowhawk, and sometimes a Rough-legged Buzzard may take up residence for a season, while Red-throated Divers are usually present on the sea and can be seen in calm conditions.

Breeding birds in the area include Bittern, Grey Heron, Marsh Harrier, Shelduck, Gadwall, Teal, Shoveler, Oystercatcher, Avocet, Lapwing, Common Snipe, Common Redshank, Barn Owl, Kingfisher and Yellow Wagtail, with Bearded Tit and Grasshopper, Reed and Sedge Warblers in the reedbeds and marshes, Woodlark and Dartford Warbler on the heathland and Nightingale in woodland scrub, while Black Redstart breeds at Sizewell.

On passage Garganey and Ruff are often present in spring at Church Farm Marshes, making prolonged stays, and Spotted Redshank, Greenshank and Wood Sandpiper are fairly regular at both seasons. A variety of passerines may occur along the coast, especially in east or south-east winds.

Access

Alde Mudflats From the B1069 just south of Snape Maltings turn east on a minor road towards Iken, parking after a further 1 mile at Iken Cliff car park. From here walk east along the shore of the Alde for 0.5 miles to view the mudflats along the south shore of the estuary, a good area for Avocet and Black-tailed Godwit; an incoming tide is best. It is possible to follow the footpath further, as far as Iken church, and 123 ha of the tidal flats to the east of the church are an SWT reserve.

Hazelwood Marshes The SWT Hazelwood Marshes reserve comprises 62 ha of grazing marshes to the north of the estuary. From the A1094, just west of Aldeburgh golf club, turn north into the car park. From here follow the track past a reedbed, over an iron bridge and left over a stile. The right-hand fork leads to the Eric Hosking hide on the sea wall, giving views south over the estuary, while the left fork leads to grazing marshes. The reserve is open dawn to dusk.

Aldeburgh Marshes Lying immediately south-west of Aldeburgh, these may attract wintering raptors (sometimes including Rough-legged Buzzard). Follow the A1094 to its end at Fort Green, and then take the track south to Slaughden Quay, from where a public footpath follows the sea wall to circumnavigate the marshes.

The Haven, Thorpeness This SWT reserve comprises the shingle beach and adjacent areas of scrub and marsh inland of the coast road. From Aldeburgh take the minor road north towards Thorpeness, parking carefully after 1.3 miles by the road to explore the beach.

RSPB Church Farm Marshes Under the remit of North Warren reserve, this area of coastal grazing marshes is prone to flood in winter, with some small reedbeds, the reserve attracts a range of wildfowl and waders in winter and on passage. As wildfowl, especially geese and swans, are liable to be mobile, the first step is a circuit of the Aldeburgh–Thorpeness–Aldringham triangle, via the coast roads, the B1353 and B1122. The area can be accessed on foot via the footpath which runs east to west midway between Thorpeness and Aldeburgh.

RSPB North Warren A typical range of heathland birds is present. Leave Aldeburgh north-west on the B1122 towards Leiston and, after 1 mile, turn right at the RSPB sign on the track between the houses to the reserve car park. Access is also possible from the pay-and-display car park at The Haven on the Aldeburgh-Thorpeness road. The reserve is crossed by several public footpaths and nature trails. Access at all times.

RSPB Aldringham Walks Also under the remit of North Warren, this is another area of 'Sandlings' heath, lying north of the B1353 Aldringham–Thorpeness road. Access is via a maze of public footpaths that crisscross the area.

Sizewell Belts This extensive (94.5-ha) area is managed as a reserve by the SWT on behalf of British Energy. The mosaic of habitats includes heath, grazing marsh, reedbeds, fen and woodland. Leave Leiston north on the B1122 and, after 0.75 mile turn right (east) on a minor road

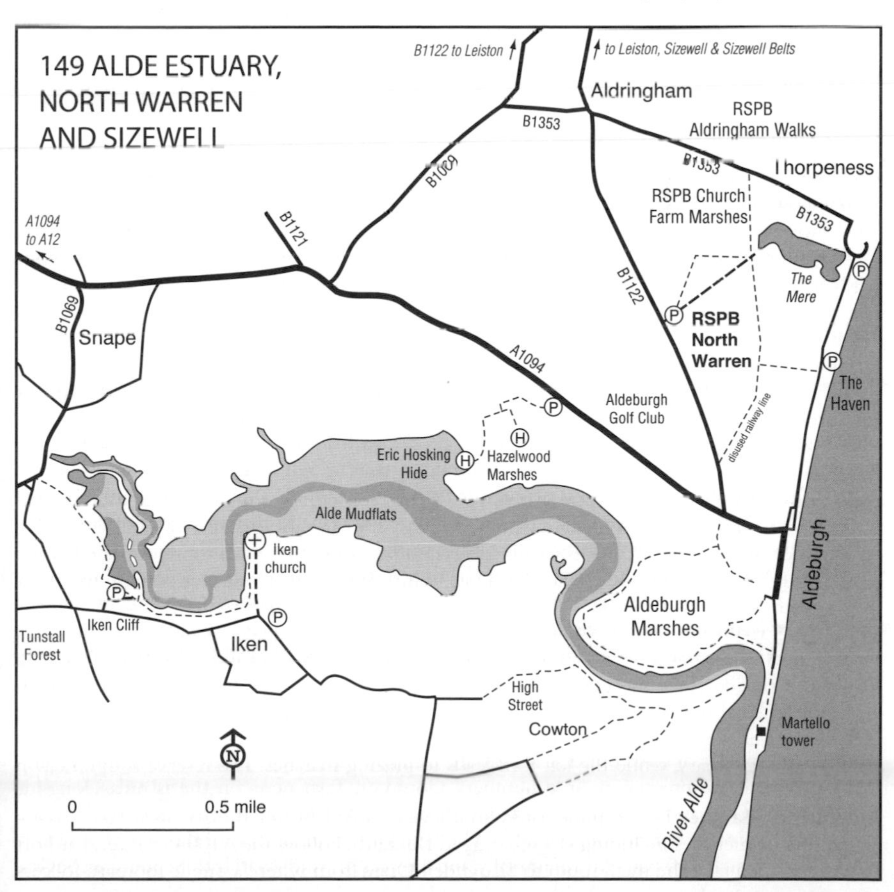

(Lover's Lane). The car park lies to the left after 0.5 miles, where the road bends south, and gives access to a network of nature trails.

Sizewell power stations The warm-water outflow offshore regularly attracts gulls and terns. A minor road east from Leiston leads directly to the power stations. From the seafront car park walk a short way north along the beach. Alternatively, you can walk 2 miles south along the beach from the sluice at Minsmere; this is most profitable in spring and autumn when passerine migrants may be present in the coastal bushes. Scarcer species occasionally include Wryneck, Barred Warbler and Red-backed Shrike. The power stations hold breeding Black Redstarts, which should be looked for in spring or summer around the perimeter fences. Notably, 200 pairs of Kittiwakes nest on the two offshore rigs (although the rigs may be dismantled as part of the decommissioning of the power station).

FURTHER INFORMATION

Grid refs: Alde Mudflats TM 400 562; Hazelwood Marshes TM 442 581; Aldeburgh Marshes TM 464 559; The Haven, Thorpeness TM 468 587; RSPB Church Farm Marshes TM 467 582; RSPB North Warren TM 467 576; Sizewell Belts TM 453 638; Sizewell power stations TM 474 628
RSPB North Warren: tel: 01728 648281; email: minsmere@rspb.org.uk
Suffolk Wildlife Trust: tel: 01473 890089; web: www.suffolkwildlifetrust.org

150 MINSMERE (Suffolk)

OS Landranger 156

Minsmere is one of Britain's finest reserves. Situated on the Suffolk coast between Southwold and Aldeburgh, it is owned by the RSPB. The diversity of habitats within its 930 ha ensures a wide variety of birds at all seasons, although it is perhaps best in May when over 100 species can be seen in a day. Several rare breeding birds occur and many interesting migrants are possible in spring and autumn.

Habitat

Originally rough coastal grazing, Minsmere was flooded in the Second World War as part of the local coastal defences. The resultant habitat proved attractive to Avocet, and this initiated the site's career as a reserve. There are now extensive areas of freshwater marshes, with a large reed-bed and several open meres. Management has produced Minsmere's most famous habitat, the Scrape, an area of shallow, brackish water and mud, with scattered small islands. Water levels and salinity are controlled by sluices to maintain optimum conditions at any given season, but the Scrape is best April–September. The reserve is bordered to seaward by a small area of dunes, and although the beach is rather disturbed, the bushes by the main sluice can be attractive to migrant passerines. Over 240 ha of the reserve comprise mixed woodland; the most accessible area is the South Belt, a narrow strip bordering the reedbeds. There is also extensive heathland, including a large area of farmland bought with the specific purpose of reverting it to heath. To the north of the reserve, Dunwich and Westleton Heaths hold a similar range of heathland species. Finally, to the south of the reserve, Minsmere Level is an area of rough grazing drained by the New Cut.

Species

Around 100 species of birds breed at Minsmere annually. The Scrape holds 20 breeding species, many of which are present in large numbers. Common Tern nests on the islands, alongside Black-headed Gull, and a few pairs of Mediterranean Gulls now breed, while Sandwich Tern is also regularly present but no longer breeds. The best-known inhabitant of the Scrape, however, is Avocet, of which 100 pairs breed annually and are present mid-March to September. Several of Minsmere's other specialities occur in the reeds. Several Bitterns breed but they are difficult to see; midsummer is the best time for a sighting, when they are feeding young and can be seen flying to and from the best feeding areas. On the other hand, Marsh Harrier is hard to miss in summer and is best seen from the Island Mere hide. Hobby also now breeds and up to 12 may be seen hunting over the reeds from West or Island Mere hides. Bearded Tit is often common and in autumn may number over 1,000 birds. Little Egrets are now resident in good numbers. Water Rail also breeds but this is another shy species, which is heard more than seen. Sedge and Reed Warblers are both common. Savi's Warbler has bred on a number of occasions (but has become increasingly erratic) and the song of the Grasshopper Warbler is a familiar sound of the drier edges of the reedbeds. Cetti's Warbler is resident, with good numbers throughout the reserve. Small numbers of Grey Herons nest in the reeds (despite there being no shortage of trees) and Kingfisher is present throughout the year around the meres. Little Terns nest on the beach (where an area is cordoned off in summer to protect them). The woodlands hold Green and Great Spotted Woodpeckers, Nightingale (an astonishing 40–60 pairs) and Common Redstart. Common birds of the heaths are Tree Pipit, Stonechat and Yellowhammer, as well as Woodlark and Dartford Warbler (which, following a long absence, returned to the Suffolk coast in 1997), and at dusk look for Woodcock and Nightjar, both of which are regularly seen in small numbers. Sand Martins breed in a purpose-built bank near the visitor centre.

A wide variety of migrants occurs in spring and autumn. Large numbers of waders visit the Scrape, the peak periods being May and July–August. Species regularly present include Knot, Little Stint, Curlew Sandpiper, Ruff, Black-tailed and Bar-tailed Godwits, Whimbrel, Spotted Redshank, Greenshank and Green, Wood and Common Sandpipers. Temminck's Stint is recorded annually, usually in spring, and rarities can appear at any time. Over 20 species of wader are regularly present in autumn. Spoonbill is a speciality of Minsmere and individuals may stay many weeks. Great White Egret and Purple Heron are less frequent but are amongst Minsmere's most regular rarities. A few Little Gulls often remain through the summer and there may be large post-breeding concentrations of this species in July, but Black Tern is more frequent in autumn. Falls of migrant passerines occur and it is worth checking any patch of cover on the coast.

In winter, the Scrape is quieter. Loafing gulls may include Caspian or Yellow-legged and a few waders remain, but wildfowl are more in evidence, with significant numbers of the commoner species. Goldeneye is regular on Island Mere and Goosander and Smew may visit, especially during cold snaps. Offshore, small parties of Common Scoters and Common Eiders are not unusual and large numbers of Red-throated Divers are present (e.g. 1,200 in January 1999). Small numbers of Bewick's Swans and wild geese sometimes graze the surrounding fields and these may roost on the Scrape. Greylag and White-fronted Geese are most likely but Bean and Barnacle Geese have also occurred. Marsh Harrier overwinters and is joined by Hen Harrier; both species roost in the reedbeds. Up to 20 Water Pipits winter and Snow Bunting may occur on the beach. Siskin is partial to the alders of the South Belt. Scarcer and irregular winter visitors include Peregrine, Rough-legged Buzzard, Jack Snipe, Great Grey Shrike, Waxwing and Shore Lark.

Access

RSPB Minsmere From A12 at Yoxford (if coming from the south) or Blythburgh (from the north) follow signs to Westleton. From Westleton take the Dunwich road, then take the first right, following the brown tourist signs. Turn left at the crossroads, then follow the reserve entrance track, turning left at Scotts Hall to reach the large car park and visitor centre after 0.5 miles. The reserve is open daily (except Christmas and Boxing Days), 9am–9pm or dusk when earlier, and the visitor centre is open 9am–5pm (to 4pm from 1 November–31 January). There are two circular nature trails, each 1.5 miles long, accessing six hides and a viewing platform. Most trails and hides are suitable for wheelchair users.

The beach and public hide From Westleton take a minor road east towards Dunwich. After 1.25 miles turn right, signed Dunwich Heath and Minsmere. Park in the NT Dunwich Cliffs car park at the end of the road, overlooking the reserve. A footpath leads down the cliffs to the beach. Continue south towards the sluice bushes and a track to the right goes to the visitor

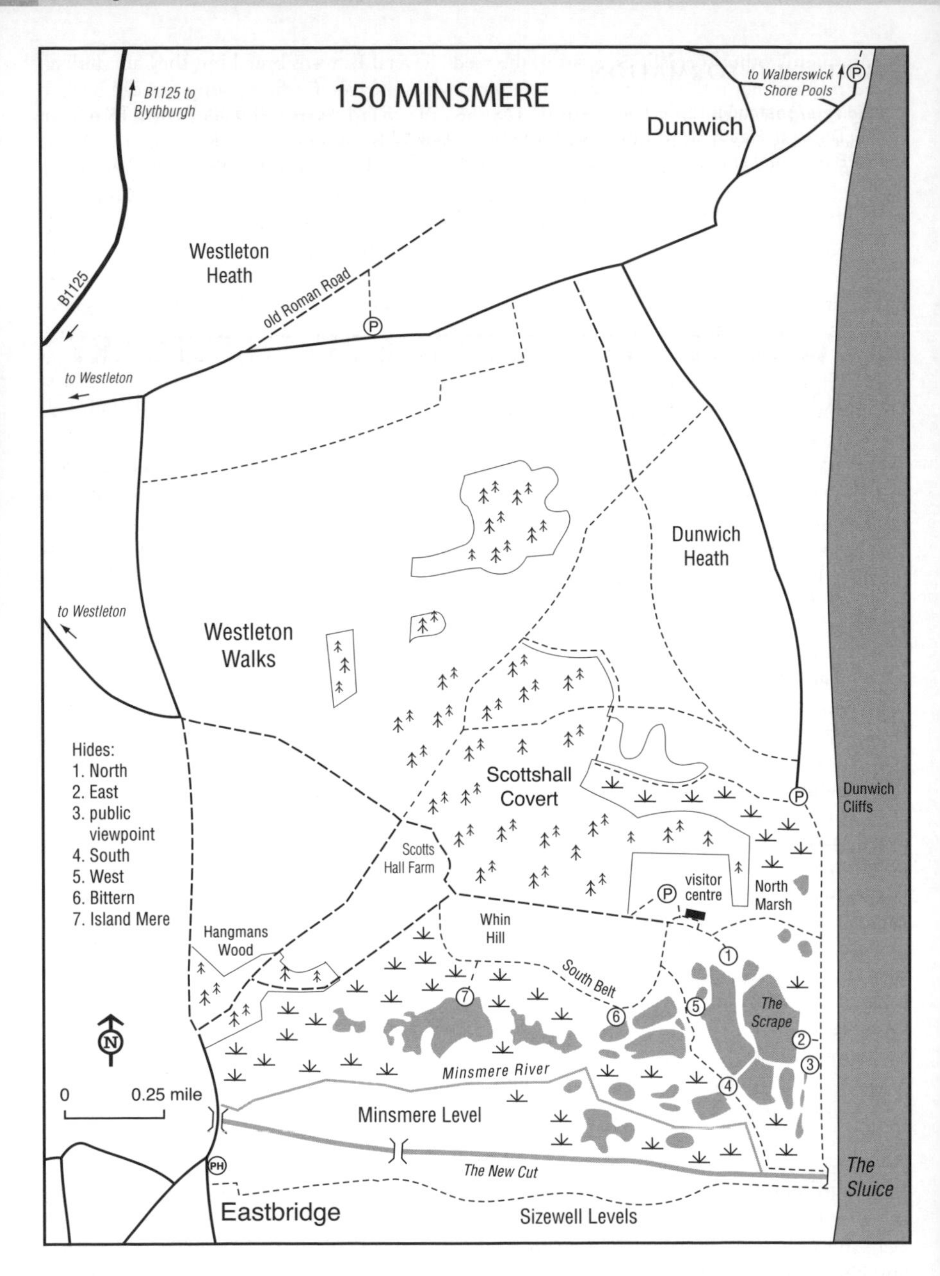

centre. Further along the beach there is a public hide on the sea wall giving excellent views over the Scrape.

Dunwich Heath Holds a good range of heathland species. Access is from the NT car park at Dunwich Cliffs; a walk west from here should, in season, produce singing Dartford Warblers.

Westleton Heath This NNR is owned by Natural England and covers 47 ha straddling the Westleton–Dunwich road. It holds a range of heathland birds. Only the area north of the road, as far as the old Roman Road, is open to the public (see map), with a car park at the south-east corner; otherwise access is limited to public footpaths.

FURTHER INFORMATION

Grid refs/postcode: RSPB Minsmere TM 473 672 / IP17 3BY; the beach and public hide TM 477 665; Dunwich Heath TM 477 677; Westleton Heath TM 453 695
RSPB Minsmere: tel: 01728 648281; email: minsmere@rspb.org.uk; web: www.rspb.org.uk/minsmere
Natural England, Suffolk: tel: 01284 762218; email: enquiries.east@naturalengland.org.uk

151 WALBERSWICK AND THE BLYTH ESTUARY (Suffolk)

OS Landranger 156

The Suffolk Coast (Walberswick) NNR covers 1,340 ha on the coast south of Southwold. It includes Westwood Marshes, which extend inland for 2.5 miles, and the mudflats and saltings on the south side of the Blyth Estuary. The variety of habitats guarantees an impressive list of breeding species and the coastal location ensures a variety of migrants. However, Walberswick is probably best in winter, particularly for raptors.

Habitat

The coast south of Walberswick village is bordered by shingle banks and a series of brackish pools. Further inland the 180 ha of Westwood Marshes form the largest single block of freshwater reeds in Britain. The marshes were reclaimed to form grazing meadows in the 18th century but, like Minsmere to the south, were flooded in the Second World War for defence purposes, and reverted to marshland. Now, invading scrub of sallow, alder and birch is controlled to protect the reedbeds while the reeds themselves must be checked to prevent their encroaching the open pools, and hay meadows have been recreated by summer-mowing areas of the reedbed. To the south, the 263 ha of Dingle Marshes (jointly owned by the SWT and RSPB) supports more reedbeds and grazing marsh. To the north of this complex of marshes lies one of the best-preserved remnants of the Suffolk 'Sandlings', heathland dominated by Heather and acid grassland, and to the south-west the shady expanse of Dunwich Forest. The River Blyth forms the north boundary of the reserve, and due to breaches in the river bank in 1921, 1926 and 1943 the river now broadens, 2 miles inland, into a large tidal estuary, important for wildfowl and waders.

Species

The Walberswick area attracts many of the species found at Minsmere and the list of species below is far from exhaustive. The main differences are that Walberswick lacks the Scrape and is not as well-watched.

The reedbeds hold breeding Bittern, Marsh Harrier, Water Rail and Bearded Tit, and Little Egrets are now resident in the area. Avocet breeds around the shore pools together with a few pairs of Little Ringed Plovers, with Common Snipe and Common Redshank on the grazing marshes. A few pairs of Common and Little Terns breed, as well as Gadwall, Shoveler and Teal, and Garganey occasionally does so. Barn Owl is resident, with Marsh Tit and smaller numbers of Willow Tits in the wetter woodlands. Reed and Sedge Warblers are common, and Nightingale and Grasshopper Warbler is not uncommon in summer. Areas of heath hold breeding Nightjar, Woodlark and Stonechat, and Dunwich Forest has Siskin and Common Crossbill. Hobby regularly visits in summer.

Waders are much in evidence in spring and autumn, with the greatest concentration on the Blyth Estuary. Species may include Grey Plover, Black-tailed and Bar-tailed Godwits, Whimbrel, Spotted Redshank and Greenshank. Other migrants can include Osprey, Black Tern and Little Gull. Migrant passerines can be interesting, especially in autumn, when falls of commoner species, such as Pied Flycatcher, Common Redstart, and Willow and Garden

Warblers can occur. Some of the more interesting scarce migrants, such as Wryneck, Icterine and Barred Warblers and Red-backed Shrike, turn up from time to time, often when there is no other obvious indication of migration. In recent years seawatching at Southwold has produced regular records of Long-tailed Skua in August–September, and Manx and Sooty Shearwaters, Sabine's Gull and Roseate Tern are annual.

In winter, Hen Harrier and Sparrowhawk are regularly seen, especially from Westwood Lodge, and Goshawk is also sometimes present. A handful of Marsh Harrier also usually over-winter and Rough-legged Buzzard, Peregrine and Merlin are not infrequent in some years, while Common Buzzard is possible at any time of year. Great Grey Shrike is now rare, but an individual may return to establish a winter territory for several years in the manner that made this a good spot for the species in the 1970s. At dusk Barn Owl may be seen around Westwood Lodge, while wintering Short-eared Owls prefer the shore area. The reeds and marshy pools hold wintering Water Rail and Bittern, and White-fronted Geese often winter in the area. The mudflats of the Blyth Estuary hold the usual common waders, as well as large numbers of Avocets, and there are often Black-tailed Godwit and a few wintering Spotted Redshanks. Red-throated Diver and small flocks of Common Scoters are regular offshore, sometimes with a few Velvet Scoters. Snow Buntings usually winter and should be looked for on the beach or in drier areas around Corporation Marshes, together with Twite, and Shore Lark is also occasionally seen.

Access

Access to the NNR is restricted to public rights of way, of which there are over 20 miles.

Walberswick Shore Pools and Dingle Marshes Enter Walberswick village on the B1387. Near the far end of the village turn right at the sharp bend to the beach car park. From here walk south along the beach, looking for birds on the sea as well as inland on Corporation Marshes. The shore pools are reached after 1.5 miles and are good for passage waders. Alternatively, park at Dunwich Beach car park and walk north.

Westwood Marshes Walking south along the beach as above, there are two banks close together on the right after 1.25 miles. Take the second one to the derelict wind pump; this trail offers good views over the reeds and lagoons, and the bushy slopes of Dingle Hill on the left can produce migrant passerines. Bearing right at the wind pump eventually brings you back to the car park. There are other tracks across the marshes for the energetic but it is essential to keep to the paths.

Tinker's Marshes In Walberswick village take the track to the Bailey Bridge over the River Blyth. Do not cross the bridge, rather walk west along the south bank of the river to view the fresh grazing marshes.

Southwold Town Marshes Proceed as above, but cross the Bailey Bridge and walk south-east along the north bank of the river, for views north over the marsh. The area is also accessible from Southwold by following the minor road south along the coast and then inland, along the north bank of the river, to the Harbour Inn. This route gives views over the marshes, as does the footpath running north-east from the inn to Southwold.

Southwold This is the best site for seawatching on the Suffolk coast. Watch from the seafront shelter, 200 yards south of the pier.

Southwold Boating Lake May hold odd seaducks or grebes, especially after bad weather.

Westwood Lodge From Walberswick village take the B1387 west. Shortly before leaving the village fork left on a minor road and follow it for 2 miles to Westwood Lodge. Park nearby. From here there is a magnificent view over the marshes. This is the best place to see raptors in winter, and late afternoon is the best time to look.

Blyth Estuary The A12 crosses the west end of the Blyth Estuary just north of Blythburgh. There is a lay-by on the west side of the road and the estuary can be viewed from the embank-ment opposite. In Blythburgh, the White Hart pub lies at the junction of the A12 and B1125. A footpath from here leads along the south edge of the estuary to a public hide (and continues for 1 mile to join the B1387). The estuary is best viewed on a rising tide (at low tide the waders may be distant but at high tide they will have departed).

Hen Reedbed SWT reserve (Norman Gwatkin reserve) Created in 1999, this reserve comprises 44 ha of reedbeds, fens, dykes and pools to the north of the Blyth Estuary. Access is

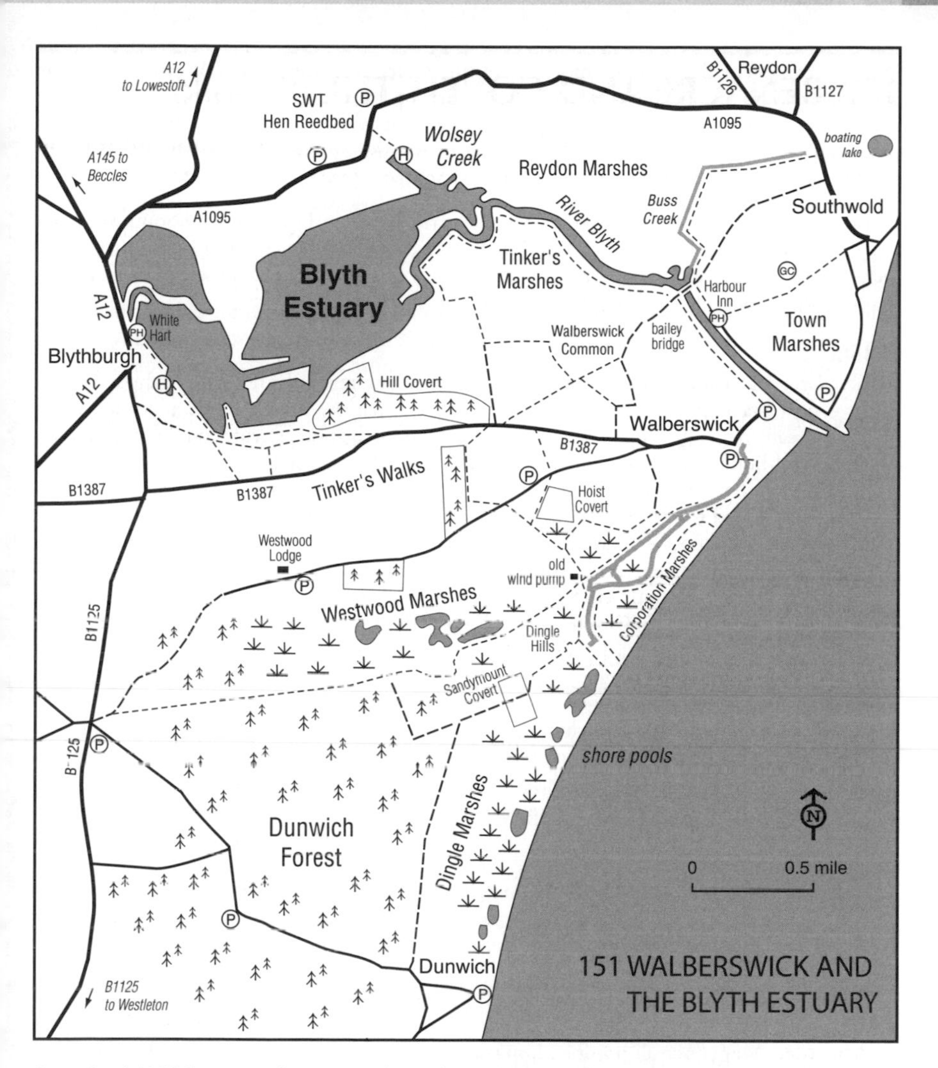

from the A1095 between the A12 and Southwold. The road passes through the reedbeds and the car park lies to the north of the road at the eastern end. For best views of the largest mere follow the way-marked trail through the reedbed to the viewing platform at Wolsey Creek Marshes. Breeding species include Bittern, Marsh Harrier and Bearded Tit; a heronry holding Grey Herons and Little Egrets can be seen in trees at the very western edge of the reedbed, viewable from a platform 100m west of the car park.

FURTHER INFORMATION

Grid refs: Walberswick Shore Pools and Dingle Marshes TM 499 745; Westwood Marshes and Tinker's Marshes TM 499 745; Southwold Town Marshes TM 499 749; Southwold TM 510 765; Southwold Boating Lake TM 512 769; Westwood Lodge TM 467 737; Blyth Estuary TM 451 753; Hen Reedbed SWT reserve (Norman Gwatkin reserve) TM 471 771
Natural England, Suffolk: tel: 01284 762218; email: enquiries.east@naturalengland.org.uk
RSPB East Anglian Office: tel: 01603 661662
Suffolk Wildlife Trust: tel: 01473 890089; web: www.suffolkwildlifetrust.org

152 BENACRE AND COVEHITHE (Suffolk)

OS Landranger 156

Benacre Broad is an interesting wetland on the coast between Lowestoft and Southwold and Benacre NNR comprises 393 ha, extending from Kessingland Sluice south to the Broad, and including Covehithe and Easton Broads. Although less extensive than Minsmere or Walberswick the area offers good birding, particularly in autumn and winter.

Habitat

Benacre Broad is a natural lake and although once 'fresh', more regular flooding by the sea means that it is now brackish. It is separated from the sea by a narrow shingle beach and water levels fluctuate greatly depending upon sea encroachment and irrigation activities on the surrounding estate. Several bunds have been constructed to protect the reedbeds from sea water, and five artificial lagoons have been created to compensate for the loss of habitat to the sea. To the north and south the Broad is bordered by mature mixed woodland and farmland, and a little further north there are some flooded gravel pits immediately inland of the beach, partially fringed with reeds and sallow bushes (and rapidly disappearing due to coastal erosion). Covehithe Cliffs to the south are being constantly eroded by the sea (which is moving landward faster here than almost anywhere in Britain). Covehithe and Easton Broads are smaller than Benacre and have deteriorated considerably, partially as a result of drainage, and now have little open water.

Species

A variety of waterbirds breed including Marsh Harrier, seven species of ducks (including Gadwall), Water Rail, Reed, Sedge and Grasshopper Warblers and Bearded Tit. Bitterns sometimes breed. Up to 70 pairs of Little Terns breed on the shingle beach, and also Common Tern. The woodlands hold most of the typical species including small numbers of Nightingales, as well as Common Redstart. Woodlark is found on the relict areas of heath, and Wheatear and Hobby sometimes breed.

Waders are a feature of Benacre Broad in autumn; Avocet is often present and several rarities have been found. Seabirds may be conspicuous offshore at this time and frequently include skuas and auks. A broad variety of migrant passerines may be seen, occasionally including scarce migrants such as Barred Warbler and Red-backed Shrike.

Winter is usually the best season at Benacre. Red-throated Diver is frequent offshore and there may be small numbers of seaducks, with Scaup, Common Eider, Common Scoter and Red-breasted Merganser all possible; Long-tailed Duck and Velvet Scoter are less frequent. Goldeneye is usually present on the Broad and Smew and Goosanders are regular in small numbers, while Red-necked and Slavonian Grebes may visit. Numbers of dabbling ducks are also present, including Wigeon, Teal and Gadwall. Marsh and Hen Harriers regularly hunt the reedbeds, and Common Buzzard, Goshawk, Sparrowhawk, Peregrine and Merlin are frequent. Several rarer raptors have been seen here, the most likely being Red Kite and Rough-legged Buzzard. Bewick's Swan and White-fronted, Tundra Bean and Egyptian Geese are sometimes found in the area, particularly on Kessingland Level, and Short-eared Owl may also be seen. With luck, Shore Lark and Snow Bunting may be encountered on the beach, and gulls occasionally include Glaucous, Iceland or Mediterranean.

Access

Access in the NNR is limited to rights of way and the concessionary path along the private beach. **Benacre Broad** Leave the A12 at Wrentham on a minor road east to Covehithe. Beyond the village the road terminates at the sea. Park and walk north along the top of the cliffs. After 0.25 miles the path descends to the beach and, after passing a small wood, Benacre Broad lies to the left. Much of the Broad can be seen from the beach while a track along the south side leads to

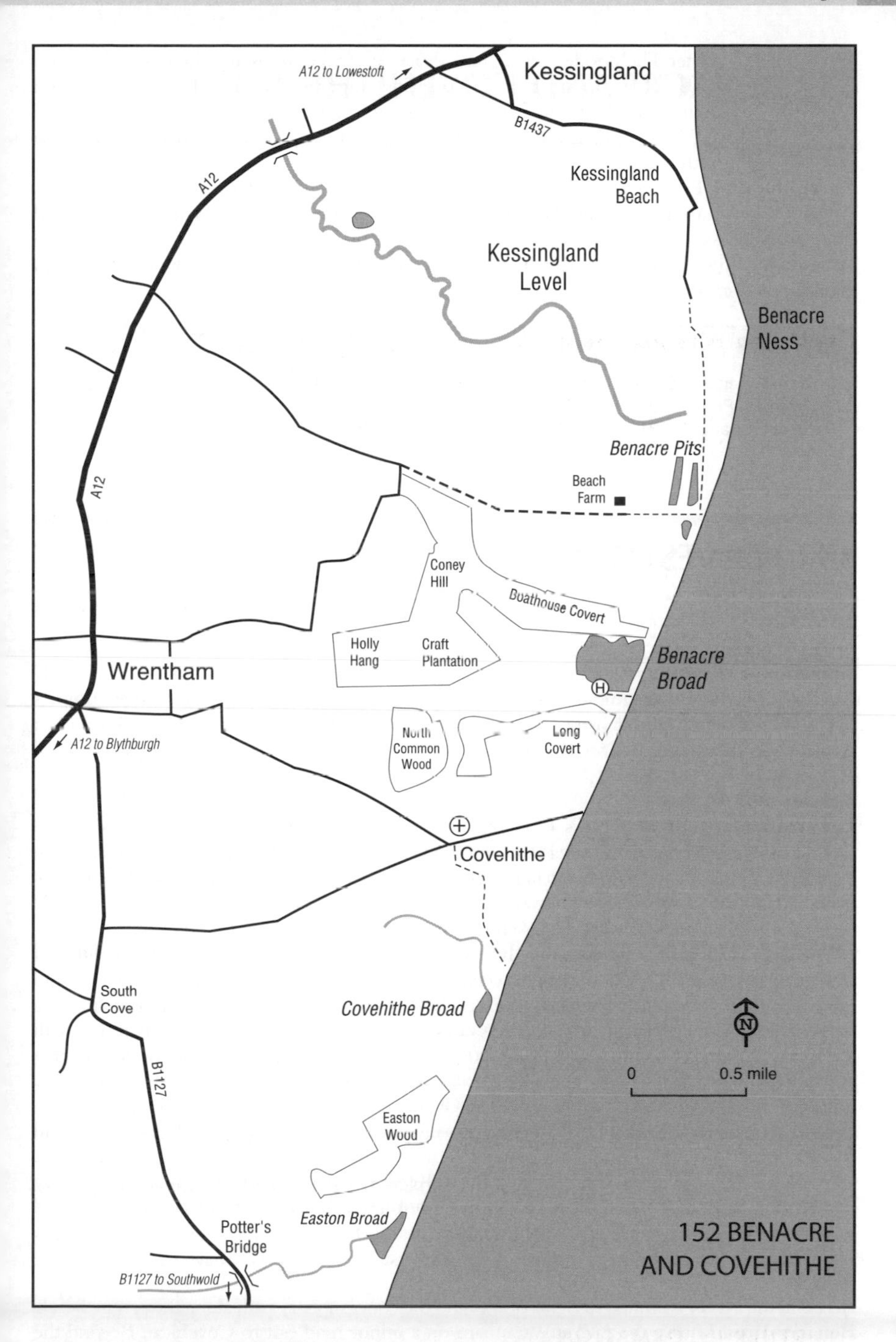

a hide (access unrestricted), which gives good views of the Broad.

Benacre Ness and Kessingland Level Continue north along the beach for 0.75 miles to the gravel pits and Benacre Ness. The latter is a potentially good place from which to seawatch in favourable weather. The flooded pits attract a variety of waterbirds and the bushes here and in the vicinity of Beach Farm are best for passerine migrants. The marshes of Kessingland Level may be worthy of exploration if time permits. Alternative access is from Beach Farm via Benacre village and from Kessingland Beach: in Kessingland village take the B1437 to the coast, park and walk south along the beach.

Covehithe and Easton Broads These lie 1 and 2 miles south of Covehithe respectively. A footpath leads south from Covehithe village to Covehithe Broad. Easton Broad can be reached along the beach.

Potter's Bridge The B1127 bisects the extensive reedbeds of the Easton Broad valley, with good views from Potter's Bridge.

FURTHER INFORMATION

Grid refs: Benacre Broad, Benacre Ness and Kessingland Level, Covehithe and Easton Broads TM 526 819; Potter's Bridge TM 508 791
Natural England, Suffolk: tel: 01284 762218; email: enquiries.east@naturalengland.org.uk

153 LOWESTOFT (Suffolk)

OS Landranger 134

Situated near the easternmost point in Britain, the gardens and parks in Lowestoft naturally attract a variety of passage migrants, while the fishing industry inevitably attracts large numbers of gulls in winter. Lowestoft also holds one of only two colonies of breeding Kittiwakes in East Anglia (the other being at Sizewell, see page 210).

Species and Access

Lowestoft Harbour and Ness Congregations of gulls in this area may include Glaucous, Iceland or Mediterranean. The area between the harbour and Lowestoft Ness is generally best. Purple Sandpiper is regular, favouring rocks or man-made structures (the broken concrete defences and sea wall between Ness Point and Lowestoft Harbour is the best site for Purple Sandpiper in East Anglia). A few Shags and Sanderlings are also usually found in winter.

The only colony of Kittiwakes on the east coast south of Yorkshire uses not sea cliffs but man-made structures, nesting on ledges on churches, the yacht club and other buildings around the harbour, and especially formerly the South Pier Pavilion. When this was demolished, a purpose-built artificial 'Kittiwake Cliff' was constructed on the north side of the harbour, with the result that the colony increased from 100 to 200 pairs; some remain throughout the winter. A few pairs of Black Redstarts also regularly breed, and are best looked for around the warehouses at Ness Point.

Lowestoft North Denes This is a grassy area north of the cricket ground. A camping site in summer, in spring and autumn it attracts gulls, waders, pipits and wagtails. The scrub and woodland on the cliff edge are worth checking for warblers, flycatchers and chats, as are the gardens of Belle Vue Park and Sparrows Nest Theatre. Look for Ring Ouzel on the cricket ground. Kensington Gardens, a small park and bowling green south of Clairmont Pier, are worth checking in autumn when the wind is from the north-east. Scarce migrants have been found here on several occasions.

FURTHER INFORMATION

Grid refs: Lowestoft Harbour and Ness TM 555 937; Lowestoft North Denes TM 552 945

154 GREAT YARMOUTH AND BREYDON WATER (Norfolk)

OS Landranger 134

Famous as a 'kiss-me-quick' seaside resort, immediately west of Great Yarmouth lies the vast expanse of Breydon Water, the landlocked estuary of the Rivers Yare and Waveney, part of which is a local nature reserve, while further west still lies the remote RSPB Berney Marshes reserve. An excellent variety of wildfowl and waders can be seen during both winter and passage periods. In Great Yarmouth itself, the cemetery regularly attracts rare migrants and the seafront can be productive, especially in winter.

Habitat

Breydon Water is the landlocked estuary of the Rivers Waveney and Yare, and at low tide extensive tidal mudflats are exposed. The Halvergate Marshes, an extensive area of low-lying grazing marshes, lie on either side of the estuary.

Species

Winter wildfowl include large numbers of Wigeon and Teal, as well as smaller numbers of Pintail, Gadwall and Shoveler, and sometimes Goldeneye and Scaup on the river channel. The grazing marshes hold variable numbers of Bewick's Swans and, more rarely, Whooper Swans, up to 200 White-fronted Geese and occasionally Tundra Bean Geese; recent winters have seen large numbers of Pink-footed Geese visiting east Norfolk, although they are very mobile and favour harvested beet fields. Hen and Marsh Harriers, Merlin, Peregrine, Barn and Short-eared Owls, Twite and Snow and Lapland Buntings are also found in the rough grassland, although the last may be very elusive. Waders include all the usual species, together with Black-tailed Godwits (which are present for most of the year), small numbers of Ruffs and up to 300 Avocets, while the gulls should be checked for Glaucous and Mediterranean. Small numbers of Rock Pipits winter on the saltmarsh.

Passage periods bring a wide variety of waders, which may include large numbers of Avocets as well as Ruff, Spotted Redshank, Black-tailed Godwit, Whimbrel and occasionally Temminck's Stint, while Broad-billed Sandpiper and Kentish Plover are almost annual in May. Other regular migrants include Spoonbill, Garganey, Black and Roseate Terns, and Mediterranean and Little Gulls; in late summer up to 130 Mediterranean Gulls join the roost on Breydon Water. Good numbers of Little Egrets are now resident in the area. Great Yarmouth cemetery regularly holds passerine migrants, and these have included some outstanding rarities.

Breeders include several pairs of Black Redstarts around the power station and in the town, Common Tern (on specially constructed rafts on Breydon Water, together with loafing Sandwich and occasionally also Roseate Terns too). There are nationally important numbers of Little Terns: the UK's largest colony breeds on Yarmouth North Denes beach, and from mid-May to late July RSPB wardens are on site. Berney Marshes holds good numbers of breeding Lapwings and Common Redshanks as well as Yellow Wagtail.

Access

Breydon Water Most of the waders on the estuary can be seen in the north-east corner 2 hours either side of high water as they gather on the last areas of mud to be exposed. A telescope is usually essential. Coming into Great Yarmouth on the A47 dual carriageway, follow signs to the railway station and drive past the station frontage into the large Asda supermarket car park by the river, parking in the south-west corner. Walk under Breydon road bridge to join the Weavers' Way footpath along the north shore of Breydon Water. After 200 m there is a rather tatty raised public hide, but waders can be viewed just as easily from the sea wall and it is possible to continue along the sea wall for several miles for more views of the estuary (although a more relaxed alternative is to take a boat trip out onto Breydon Water – to book phone 01603 715191).

Great Yarmouth seafront The harbour entrance can be productive for gulls and is accessed by following Marine Parade south into South Beach Road and then to the South Denes Industrial Area (breeding Black Redstart). Mediterranean Gull is, however, best looked for from the central beach area, while the large Little Tern colony can be accessed along the A149 north

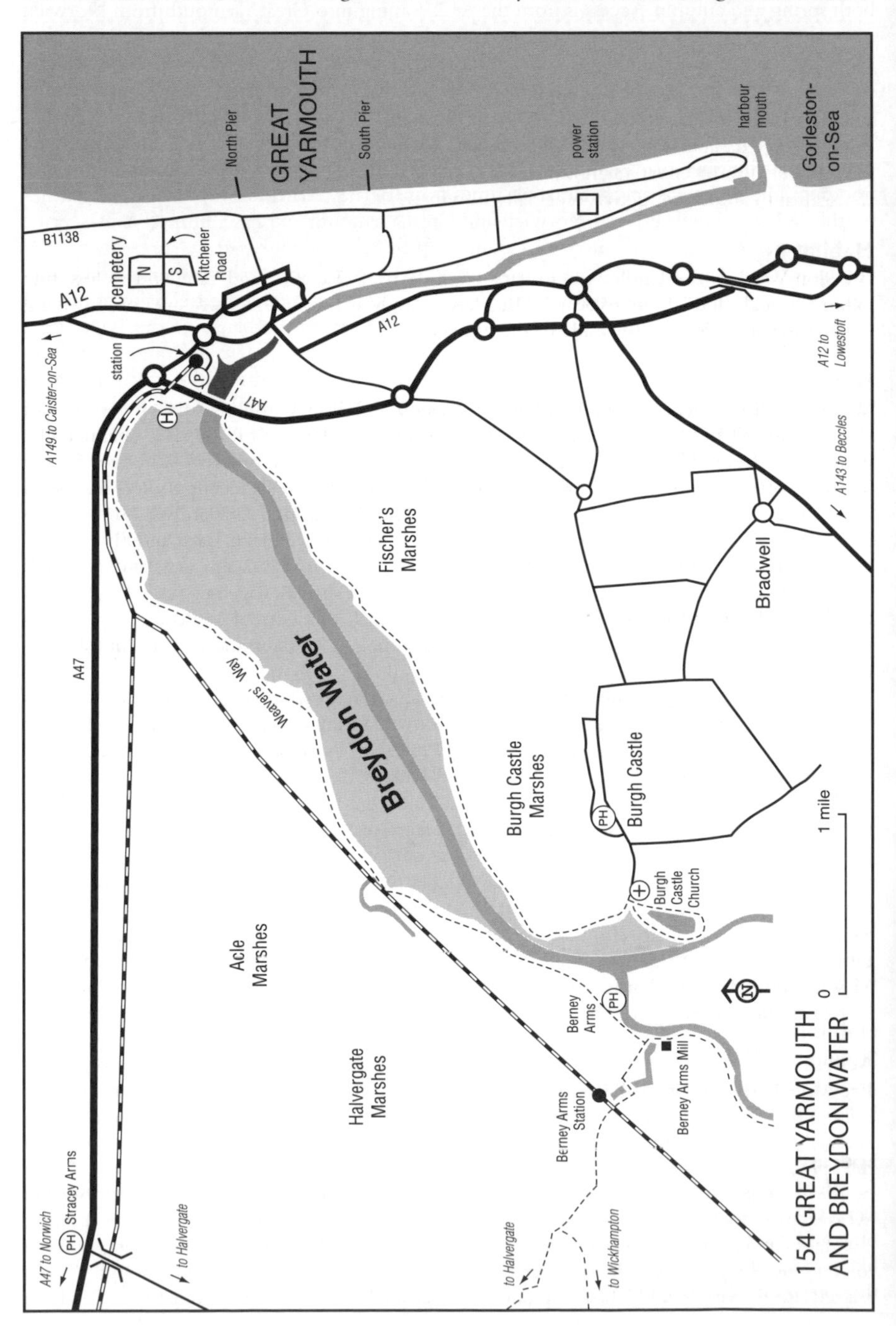

154 GREAT YARMOUTH AND BREYDON WATER

out of the town, turning right at the traffic lights into Jellicoe Road (signed to the racecourse), following this over the bridge and then turning left into North Drive. From here, it is a short walk to the beach.

Great Yarmouth cemetery The mature trees here consistently attract passerine migrants in both spring and autumn. Access is from the A12: coming into Great Yarmouth from Norwich go straight on at the first two roundabouts and then turn north (left) at the traffic lights into Northgate Street and right at the mini-roundabout into Kitchener Road (which divides the cemetery into its north and south sections). It is possible to park in Kitchener Road or in the pay-and-display car park found by driving to the end of the road and turning right.

RSPB Berney Marshes Covering 360 ha, this area forms part of the vast Halvergate grazing marshes, but management here has raised water levels to produce areas of shallow flooding. The reserve is accessible at all times from Berney Arms railway station, which lies on the 'Wherry Line' between Norwich and Great Yarmouth and has a limited daily service (most frequent on Sundays); note that trains stop by request only – make sure you tell the conductor you want to get off and give a clear signal to the approaching train when wanting to board. From the station follow the footpath south-east across the fields for about 0.3 miles to Berney Arms wind pump and the reserve office on the banks of the River Yare. Following the river downstream you soon reach Berney Arms public house and Breydon Water. The recommended tactic is to take the train to Berney Arms Station and then walk to Great Yarmouth Station, 4 miles away along the Weavers' Way long-distance footpath, following the north shore of Breydon Water. Swans, geese and birds of prey are the specialities, and Avocet has bred.

Burgh Castle and Fishers Marshes For the more energetic, these areas of rough grazing south of Breydon Water hold a similar selection of birds to Berney Marshes. Join the footpath along the south shore of the estuary at Burgh Castle church, and this route also affords the best views of the river channel for diving ducks and grebes.

FURTHER INFORMATION

Grid refs: Breydon Water TG 517 081; Great Yarmouth seafront TG 533 040; Great Yarmouth cemetery TG 526 083; RSPB Berney Marshes TG 464 049; Burgh Castle TG 476 050
RSPB Berney Marshes: tel: 01493 700645; email: berney.marshes@rspb.org.uk

155 STRUMPSHAW FEN (Norfolk)

OS Landranger 134

The marshes of the Yare Valley, east of Norwich, support some unusual breeding birds, with the RSPB reserves at Strumpshaw Fen and Surlingham being among the best areas.

Habitat

Strumpshaw Fen comprises reedbeds, alder and willow carr, damp woodland, open water and grazing marshes, with similar areas south of the River Yare at Rockland Broad and Surlingham Church Marsh.

Species

Resident species include Bittern, Marsh Harrier, Gadwall, Shoveler, Water Rail, Woodcock, Barn Owl, Kingfisher, Bearded Tit and Cetti's Warbler. Of these, Bitterns are usually most obvious in winter, Kingfishers are often seen from the Old Broad Hide, Bearded Tits are most likely to be seen on calm days in late summer and autumn, and Cetti's Warblers are vocal in the spring but always hard to see.

Summer visitors include Cuckoo, Reed, Sedge and a few pairs of Grasshopper Warblers, and other breeders include Pochard, Shoveler and Black-headed Gull (70 pairs), while Little Egret, Hobby and Common Tern are frequent visitors. Common birds of the grazing marshes are Lapwing, Common Redshank and Yellow Wagtail, and a good variety of woodland species include Green, Great and Lesser Spotted Woodpeckers and Marsh Tit.

On passage, look for Garganey (which has bred), while Osprey is a surprisingly frequent visitor to Strumpshaw. A few waders or Arctic and Black Terns may drop in, and Spotted Crake is occasionally seen (but more often heard on summer evenings).

In winter look for a variety of ducks, including Shoveler, Teal, Gadwall and Pochard, while the fen attracts roosting Cormorants and there can be flocks of Siskins and redpolls in the alders. Good numbers of Marsh Harriers roost on the reserve and these are occasionally joined by Hen Harriers, while other occasional visitors include Jack Snipe, Long-eared Owl and Water Pipit.

Access

RSPB Strumpshaw Fen At the roundabout on the A47 just east of Norwich take the exit to Brundall, and in the village follow signs towards Brundall Station and Strumpshaw; 400 yards after passing under the railway bridge on the outskirts of Brundall, bear right into Stone Road (signed to Hassingham and Cantley, and the RSPB's Strumpshaw reserve), and immediately right again into Low Road. The reserve car park is on the right after 0.5 miles. From the car park cross the railway line on foot to reach the reserve and information centre (open only in summer). There are two circular trails: the Fen Trail at 2.3 miles and the Woodland Trail at 1.6 miles, and these can be combined to make a 3-mile circuit. The trails are accessible by wheelchair in dry conditions – contact the reserve for advice prior to your visit. In addition,

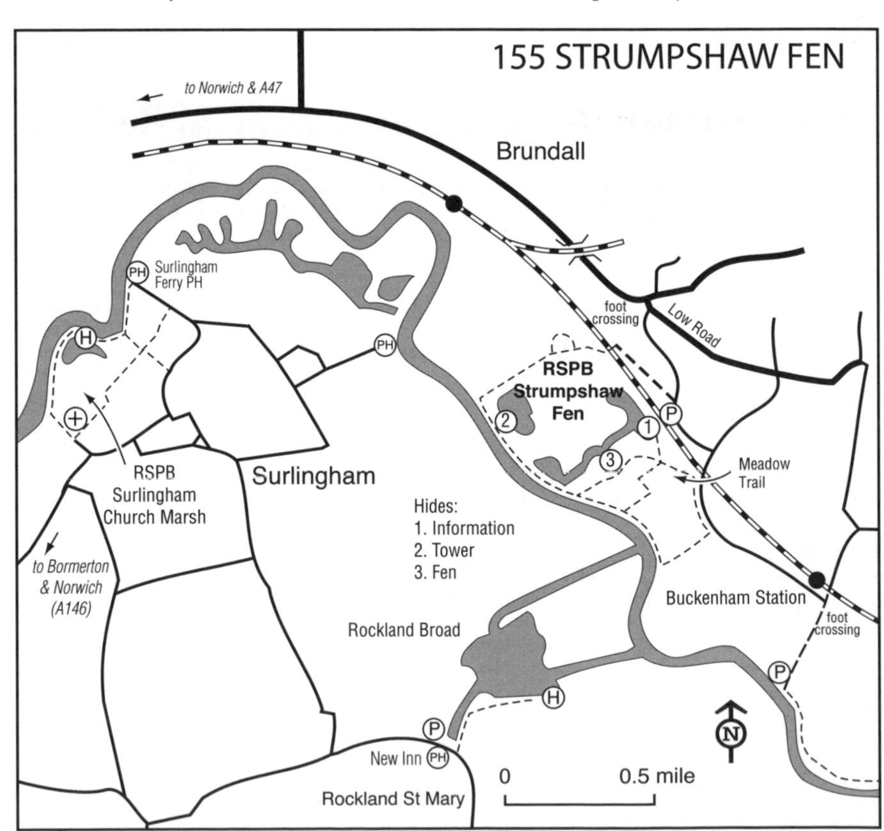

in summer the Meadow Trail makes a loop of 0.4 miles off the Woodland Trail. There is a hide at the reception centre and two additional hides, with the Tower Hide, in particular, affording excellent panoramic views. The reserve is open daily, dawn until dusk.

Rockland Broad Most of the RSPB Strumpshaw Fen reserve lies north of the River Yare, but an area of carr on the south bank extends as far as Rockland Broad. This is only accessible from Rockland St Mary, which is accessed from the A146 Norwich–Loddon road. A public footpath starts opposite the New Inn in Rockland St Mary and runs to a hide, giving good views over the reedbeds and Rockland Broad.

RSPB Surlingham Church Marsh Also south of the River Yare, the RSPB owns 68 ha at Surlingham Church Marsh with a similar range of birds. Park carefully by Surlingham Church, from where there is a 1-mile long circular walk around the reserve leading to a hide.

FURTHER INFORMATION

Grid refs: RSPB Strumpshaw Fen TG 341 066; Rockland Broad TG 328 045; RSPB Surlingham Church Marsh TG 305 064

RSPB Strumpshaw Fen: tel: 01603 715191; email: strumpshaw@rspb.org.uk; web: www.rspb.org.uk

156 BUCKENHAM AND CANTLEY (Norfolk)

OS Landranger 134

The marshes of the Yare Valley, east of Norwich, are important for wildfowl in winter and hold the only regular wintering flock of Taiga Bean Geese in England. Part of the area comprises an RSPB reserve.

Habitat

Buckenham and Cantley are rough grazing marshes adjacent to the River Yare. Recent management has aimed to raise the water table. At Cantley, the settling ponds in the sugar beet factory are attractive to passage waders.

Species

The flock of Bean Geese, the last to winter regularly in England, is of the subspecies *fabalis*, known as Taiga Bean Goose (often considered a separate species from Tundra Bean Goose *serrirostris*). These birds breed in central and southern Scandinavia (and the loss of many former wintering flocks in Britain is attributed to a decline in the breeding area and fatalities on migration in Denmark). Up to 150 Bean Geese are present between mid-November and mid-February, but in some seasons they may depart by early January. They move around the area, and although Buckenham Marshes have traditionally been favoured, in recent seasons they have mostly been at Cantley; they are very wary and should never be approached closely. A smaller flock of White-fronted Geese is also often present. Other birds in winter include up to 10,000 Wigeon, Golden Plover, Water Pipit, Fieldfare, Redwing and vast numbers of corvids – up to 20,000 Rooks and Jackdaws fly past Buckenham Station at dusk. Occasionally Peregrine, Merlin, Hen and Marsh Harriers or Short-eared Owl may visit the area.

On passage, depending on water levels on the settling ponds, an excellent selection of waders may be present at Cantley.

Breeding species in the area include Marsh Harrier, Avocet, Common Snipe, Common Redshank, Lapwing, Yellow Wagtail and a few pairs of Cetti's and Grasshopper Warblers, with Shelduck and Black-headed Gull at Cantley. Hobby, Barn Owl and Little Egret are regular visitors.

Access

At the roundabout on the A47 just east of Norwich take the exit to Brundall, and in the village follow signs towards Brundall Station and Strumpshaw; 400 yards after passing under the railway bridge on the outskirts of Brundall, bear right into Stone Road (signed to Hassingham and Cantley, and the RSPB's Strumpshaw reserve). There are now two options:

RSPB Buckenham Marshes Take the third turning to the right (also Stone Road), signed to Buckenham and, after 1 mile, turn right into Station Road and park at Buckenham Station. Cross the level crossing on foot (it is closed to vehicles) and follow the broad track for 0.5 miles to the River Yare. Following the river bank south-east for 0.5 miles takes you to the old windmill. This offers panoramic views of the grazing marshes. The Taiga Bean Geese sometimes feed in these fields but a telescope will be necessary to see them at all well (and views are often still poor); be sure not to attempt to approach the geese too closely as they will surely fly off. (Alternatively, if coming from the RSPB's Strumpshaw reserve, continue along Low Road from the Strumpshaw car park, turn right at the T-junction and cross over the level crossing.) The road continues along the south side of the railway line to Buckenham Station, where you should park and then walk to the River Yare. Access is also possible by train, but note that trains stop at Buckenham only at weekends and by request – make sure you tell the conductor you want to get off and give a clear signal to the approaching train when wanting to board.

Cantley Marshes Continue past the turn to Buckenham and turn right towards Cantley on the B1140. As you approach Cantley there is a sharp left-hand bend. Take Burnt House Road, which leads straight ahead, and park carefully beside the road at the end. Cross the railway on foot and proceed along the public footpath. Take a telescope and be content to view the geese at a distance, being very careful not to disturb them, even if you have right of way. As

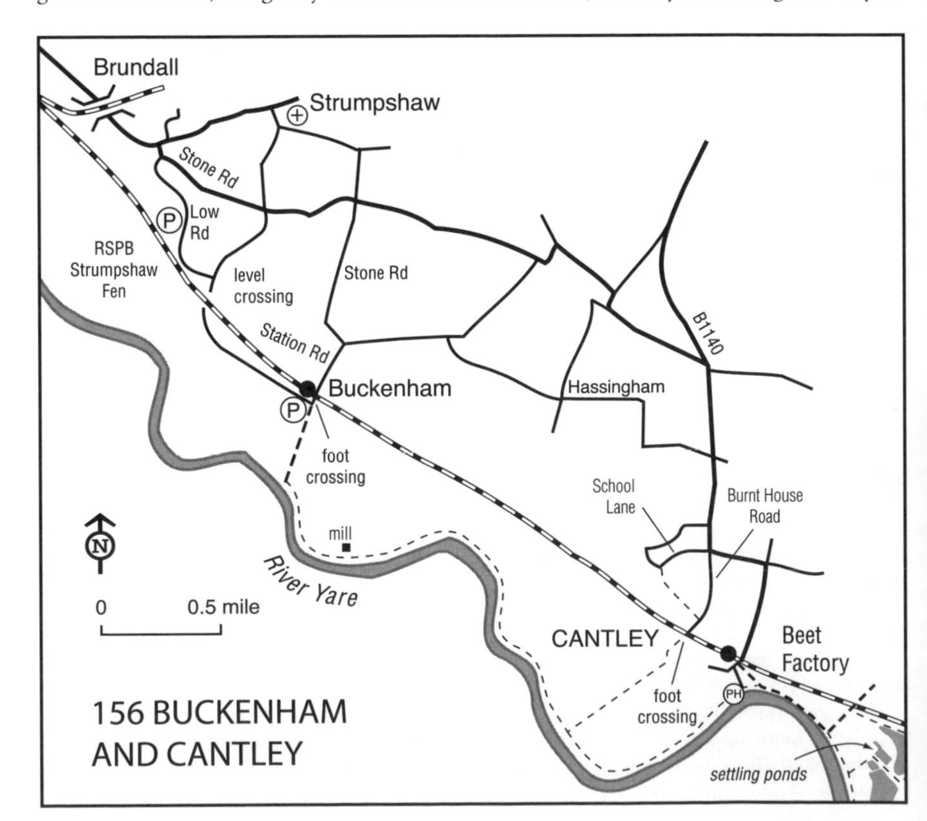

an alternative to entering the marshes and the possibility of disturbing the geese, the public footpath over the hill between Burnt House Road and School Lane provides a panoramic view of the area.

Cantley Sugar Beet Factory Follow the B1140 south over the railway and turn left into the British Sugar complex, following signs to the Reception. It is necessary to sign in and obtain a security pass to get decent views of the settling ponds, which attract passage waders, although there is a public footpath, marked on the road and pavement in yellow, which follows the main access track and then the north bank of the river, giving some limited views (and special access arrangements may be made for rarities).

FURTHER INFORMATION

Grid refs: RSPB Buckenham Marshes TG 350 056; Cantley Marshes TG 377 038; Cantley Sugar Beet Factory TG 381 036
RSPB Buckenham Marshes: tel: 01603 715191; email: buckenham.cantley@rspb.org.uk; web: rspb.org.uk

157 HICKLING BROAD (Norfolk)

OS Landranger 134 OS Explorer 40

Hickling Broad lies just north of Potter Heigham and only a few miles from the coast. An NNR covering 600 ha, owned and managed by the Norfolk Wildlife Trust (NWT), it holds a wide variety of breeding birds, augmented in spring and autumn by migrant waders, gulls and terns. Winter is often quiet, with the exception of the regular raptor roost and, especially in hard weather, a range of wildfowl.

Habitat

Lying in the upper reaches of the River Thurne, Hickling is the largest of the Norfolk Broads. Originally peat diggings, the broads flooded in the 14th century and now comprise areas of open water surrounded by extensive areas of reed and other fen vegetation, often invaded by scrub and wet carr woodland. Water pollution and neglect resulted in a serious degeneration of the wetland habitats at Hickling (as at many other broads), and the loss of Bittern as a breeder. Past management included the creation of a series of muddy scrapes, which have had a limited attraction for waders and ducks, and, more recently, significant efforts have been made to restore the reedbeds, resulting in the return of the Bittern after an absence of 20 years.

Species

Breeding specialities include Bittern, Marsh Harrier and Bearded Tit, and other breeders include Gadwall, Garganey, Ringed Plover, Common and Little Terns, Cetti's, Reed, Sedge and Grasshopper Warblers, and a selection of woodland birds such as Turtle Dove and Lesser Redpoll. Common Crane and Hobby breed in the area and are seen fairly frequently.

On passage a wide variety of waders occurs, including regular Avocet, Black-tailed Godwit, Ruff, Little Stint and, quite regularly, Temminck's Stint. Other migrants may include Spoonbill, Osprey, Black Tern and Yellow-legged and Little Gulls. Marsh Harrier is almost invariably present in the area.

Winter at Hickling is bleak, but wildfowl can include Gadwall, Goldeneye, Smew, Goosander, Pink-footed Goose and Whooper and Bewick's Swans, as well as Hen and Marsh Harriers, Sparrowhawk, Peregrine, Merlin, Barn and Short-eared Owls and Kingfisher; raptors can be elusive during the day, and are best seen from Stubb Mill towards dusk, and this is a regular spot for Common Crane. Occasional Ruffs may join the Golden Plover flocks on the surrounding pastures.

Access

Weavers' Way This long-distance footpath skirts the south edge of the broad and accesses a public hide overlooking Rush Hills Scrape (usually the most productive spot on the broad for migrant waders, gulls and terns). Park by Potter Heigham church and walk north along Church Lane for 150 yards; where the road bends sharp left take the footpath to the right for a short distance before turning left to cross a field, continuing over the stile and through the belt of trees to the Weavers' Way. Turn right for Rush Hills Scrape. The Weavers' Way is also accessible, at the

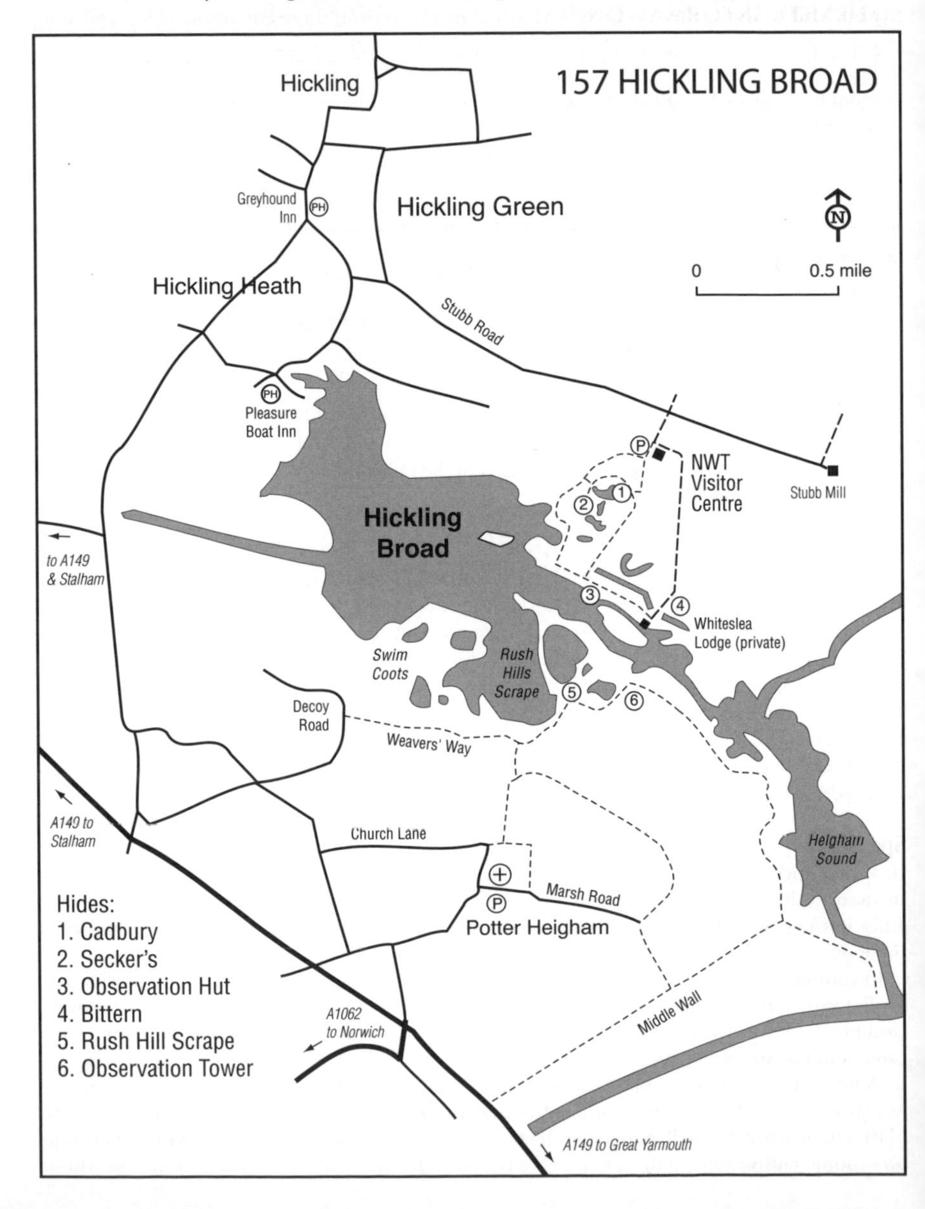

less interesting west end, from Decoy Road, but parking here is very limited.
Hickling NWT reserve Turn north off the A149 to Hickling Heath and continue to Hickling Green. At the Greyhound Inn turn east into Stubb Road and follow this for 1.5 miles, turning right at the NWT sign to the reserve car park. Several trails start from here, giving access to three hides and the observation hut, the latter offering panoramic views over the broad. The reserve is open all year and the visitor centre is open April to mid-September 10am–5pm daily.
Hickling Water Trail Mid-May to mid-September 2-hour boat trips from the NWT Visitor Centre explore the reserve via the Hickling Water Trail, which takes in a 60-foot Tree Tower giving panoramic views over the marshes, and hides overlooking the scrapes at Swim Coots or Rush Hills. Booking essential.
Stubb Mill In the winter up to 100 Marsh Harriers roost in the reeds around Horsey Mere, but they are best viewed from Stubb Mill to the west, with the light behind you (especially important on sunny evenings). A few Hen Harrier, Merlin, Sparrowhawk and Barn Owl are also regularly seen, and this is an excellent area to see Common Crane on winter evenings. Park at the NWT reserve car park and walk back along the access road, turning right at the crossroads and following the road for 0.5 miles to Stubb Mill (the lane is often rather wet). Immediately before the mill, a raised bank on the left is the designated viewing area. Marsh Harrier is often present during the day, but roosting Hen Harrier may not appear until 3pm. Arriving harriers may drop straight into the roost, offering only brief views, or they may quarter the area in search of a last meal.

FURTHER INFORMATION

Grid refs: Weavers' Way TG 419 199; Hickling NWT reserve, Hickling Water Trail and Stubb Mill TG 427 221
Norfolk Wildlife Trust: tel: 01692 598276; web: norfolkwildlifetrust.org.uk

158 WINTERTON (Norfolk)

OS Landranger 134

This extensive area of dunes on the coast at Winterton forms an NNR, and regularly attracts migrant passerines in both spring and autumn.

Habitat

A long line of dunes borders the coast, with some damp slacks in the hollows and, on the landward side, some patches of scrub and purpose-built small ponds.

Species

In spring and autumn a range of migrant landbirds may be found, with Ring Ouzel, Black Redstart and Firecrest possible in early spring, alongside common species such as Wheatear. Later in the season, in east or south-east winds, scarce migrants such as Bluethroat are sometimes present.

In autumn Winterton is one of the more regular sites on this coast for Wryneck, Red-backed Shrike and Barred Warbler, although all are elusive, and as autumn progresses almost anything could be found. Other birds to look for in late autumn include migrant Woodcocks and Long-eared and Short-eared Owls, and Little Gull can occur offshore. Divers and seaducks may be seen in winter and raptors are regularly recorded in winter and spring; the most frequent are Marsh and Hen Harriers, Sparrowhawk and Merlin.

Breeding species include Little Tern, Green Woodpecker, Nightjar, Stonechat and Grasshopper Warbler.

Access
Winterton-on-Sea lies just east of the B1159 and in the village follow signs to the beach car park (fee; closes 8pm in summer, 4pm in winter). From here it is possible to explore the dunes to the north and south, for a distance of 1 mile in both directions, concentrating on the areas of scrub on the west side of the dune system.

FURTHER INFORMATION
Grid ref: TG 498 197

159 WAXHAM AND HORSEY (Norfolk)

OS Landranger 134

This section of the east Norfolk coast can provide excellent birds in winter, with a range of wild swans and geese, and in spring and autumn, when it sometimes holds interesting migrants. At all times of year, the resident flock of Common Cranes is a star attraction.

Habitat
At Waxham the coast is bounded by a line of dunes and, immediately inland, a narrow band of dense woodland, attractive to migrants. Along the coast towards Horsey the landscape is completely flat, with a mixture of arable fields and pastures, interspersed by reed-lined ditches.

Species
In the winter herds of Bewick's and Whooper Swans are sometimes found in this area, together with up to 15,000 Pink-footed Geese, and White-fronted Geese and occasionally also Tundra Bean Geese are also seen. As well as the fields in the Waxham–Horsey area, it is worth exploring along the B1159 from East Somerton to Sea Palling and Stalham, and minor roads south from there to Hickling. The fields also hold flocks of Golden Plover and, when flooded, may also attract numbers of ducks such as Wigeon. Small numbers of raptors winter, including Barn and Short-eared Owls, Hen and Marsh Harriers and Merlin, but these are well scattered during the day and are best seen when arriving to roost around Stubb Mill, Hickling (see page 227).

Common Cranes are resident in the area and have been breeding since 1982 (the first successful breeding in Britain since around 1600), but they are very elusive in spring and summer and are best seen in autumn and winter, when up to 30 may be present. They can often (but not always) be seen from the roadside in fields of winter wheat or recently harvested potatoes, but for such large birds they can vanish into dead ground remarkably easily. Perhaps more reliably, they may be seen in winter in the late afternoon when they roost near Horsey Mere, and are then best viewed from Stubb Mill, Hickling.

In spring and autumn the stands of cover at Waxham and Horsey (and indeed, at any accessible point along the coast), may attract passerine migrants. As well as all the usual migrants, such as Garden and Willow Warblers, Pied Flycatcher and Common Redstart in autumn, Waxham regularly produces scarce migrants, and has been graced by several Pallas's and Greenish Warblers.

Access
The whole area is accessed from the B1159 between Sea Palling and West Somerton.

Waxham Turn east off the B1159, 1 mile south-east of Sea Palling, at the sharp right-hand bend. Follow the road past the church and park carefully at the end, around the T-junction. From here walk left, past Shangri-La cottage, to the dunes and beach. It is possible to explore the Sycamore woods that run north and south along the coast, immediately inland of the dunes. They are very dense, but with patience and perseverance, interesting birds can sometimes be found. Offshore, especially in late autumn and winter, small numbers of divers, grebes, seaducks and sometimes Little Gulls may be seen.

Brograve Farm to Horsey Corner This stretch of the B1159 is a favourite area for Common Crane and in winter sometimes also wild swans or geese. There are several spots where it is possible to pull off the road to scan the fields (those west of the road being favoured).

Horsey Gap At Horsey Corner take the track east to the beach car park. Scan the sea here for a variety of seaducks, grebes and Red-throated Diver. The dunes hold Stonechats and, in season, are worth checking for passerine migrants.

Horsey Mere This small broad is a NT reserve. Access is limited to a short track leading to the mere from the Horsey Windmill car park on the B1159. Hen and Marsh Harriers roost in the reeds around the mere, but are best viewed from Stubb Mill, Hickling (see page 227). Interesting wildfowl are sometimes seen on the Mere, with the best views from Horsey Mill in the early morning. The pull-off on the B1159, 0.6 miles south of Horsey Mill, is another favourite spot for Common Cranes and raptors.

FURTHER INFORMATION

Grid refs: Waxham TG 441 262; Brograve Farm to Horsey Corner TG 443 251; Horsey Gap TG 464 241; Horsey Mere TG 456 222

160 GREAT RYBURGH AND SWANTON NOVERS (Norfolk)

OS Landranger 133

These areas of parkland and ancient woodland between Holt and East Dereham are well-publicised sites for Honey-buzzard which, with luck, can be seen in late spring and summer.

Habitat

The raptor watchpoints give views across deciduous woodland and open grassland.

Species

Having bred for several years from 1989 onwards in the vicinity of Swanton Novers Great Wood, the Honey-buzzards have now moved a little further south to the Great Ryburgh area. They are, however, still regularly seen from the Swanton Novers raptor watchpoint, as well as at Great Ryburgh. Honey-buzzard is usually present late May–September, but becomes especially difficult to see in late August–early September. The optimum time for sightings is 10am–3pm on warm sunny days, with scattered cloud and only slight breezes, when they may be seen soaring or in wing-clapping display. They spend a great deal of time perched below the canopy, and a long wait may be necessary for a relatively brief and distant view. Views may sometimes be good, but equally on some days the birds are not seen at all. Other raptors in the area can include Hobby, Sparrowhawk, Common Buzzard, Marsh Harrier and Little Owl.

Access

The Wensum Valley Raptor Watchpoint at Great Ryburgh Turn off the A1067 2 miles north of Guist south-west onto the minor road to Great Ryburgh. After 1.5 miles turn south by the church in Great Ryburgh into Mill Road, bear left at the junction and park on the right after 1.25 miles, just past the disused railway bridge. The raptor watchpoint is well signed; follow the field border to the roped-off viewpoint on the rise and scan east towards the tower and Sennowe Park.

Swanton Novers Turn south-west off the A148 at the roundabout in Holt onto the B1110 towards Guist and East Dereham. After 7 miles (2 miles south of the staggered intersection with the B1354) turn west at the crossroads on the minor road signed Fulmodeston. After 0.5 miles the watchpoint car park lies north of the road.

FURTHER INFORMATION

Grid refs: The Wensum Valley Raptor Watchpoint at Great Ryburgh TF 970 256; Swanton Novers TG 010 302

161 SHERINGHAM (Norfolk)

OS Landranger 133

Sheringham is the premier seawatching station in East Anglia, noted especially for movements of Long-tailed Skua in September.

Habitat

The town lies on the coast close to the north-eastern corner of Norfolk, and thus is ideally placed for watching seabird passage from reasonably sheltered vantage points on the seafront.

Species

From late August to October strong north to north-west winds, coupled with poor visibility (at least overcast conditions, if not some mist out to sea) are likely to produce interesting movements of seabirds, particularly in early morning. Manx Shearwater, Fulmar, Gannet, Arctic Skua, Kittiwake and a variety of terns are regular, and Sooty Shearwater and Great Skua almost so. If conditions are just right, especially in September, there may be a passage of Long-tailed Skuas, but numbers are very variable and many days can be spent here without a sighting. In addition, they are hard to separate from Arctic Skua so care and practice are required. Nevertheless, Sheringham offers the best chance of seeing this species in south England. Pomarine Skua is also possible, especially as the season advances, and Leach's Petrel may sometimes be seen in small numbers. From late October, there is the chance of a passage of Little Auks, and other notable species from now through the winter include Red-necked and Slavonian Grebes, Red-throated and Great Northern Divers, and a variety of wildfowl and waders.

When seawatching is quiet, the clifftop fields between Weybourne and Sheringham often produce interesting migrants in spring and autumn. Notably, there are autumn concentrations of larks and finches, and Richard's Pipit and Lapland Bunting are recorded quite frequently. The best access is from Weybourne beach car park, walking east along the clifftop path.

Access

Follow signs from the A148 into Sheringham and, at the bottom of the hill near the town centre, turn west at the roundabout onto the A149 coast road. Take the first right over the railway bridge, and then first left (The Boulevard). At the end of the road go straight over the large roundabout to the coast. Park, go through the archway and down the steps to the lower tier of beach shelters (the best place from which to seawatch). In good seawatching weather, space is at a premium and latecomers must sit on the upper level (bring a camp chair) or find shelter elsewhere.

FURTHER INFORMATION

Grid ref: TG 155 434

162 SALTHOUSE–KELLING (Norfolk)

OS Landranger 132

Lying immediately east of the famous Cley (NWT) reserve, this area holds a similar range of birds through the year, but in a much more informal setting, without hides, visitor centres, etc. hough part is owned by the NT and part by the NWT. To the south, the heaths at Salthouse and Kelling are regular haunts of Nightjar.

Habitat

Most of the area north of the A149 coast road comprises rough grazing meadows intersected by numerous ditches, with some pools, while Kelling Water Meadows (also known as Kelling Quags) have been deliberately flooded to attract both breeding and migrant birds. This area is bounded to the north by the shingle bank which, after many years of being bulldozed into a high sea defence, is being allowed to re-profile naturally; it has already migrated inland for some distance at several places and who knows what will happen to this coastline should there be a really big storm! South of the coast road the land rises sharply to the Holt–Cromer Ridge, where sandy and gravelly soils support areas of heath at Salthouse and Kelling.

Species

On spring passage a wide variety of migrants may occur, including Garganey, Common and Green Sandpipers, Black Tern and Yellow Wagtail, and there are often good numbers of Whimbrels in the meadows; Ruff and Black-tailed Godwit may occur at almost any time of year.

Breeders include Shelduck, Avocet, Common Redshank, Oystercatcher, Ringed Plover and Sedge Warbler, and Sandwich, Common and Little Terns regularly patrol the beach; the former two species may spend time loafing on the pools. Little Egrets and Marsh Harriers are regular. Nightingales favour areas of dense blackthorn scrub on Salthouse Heath, and Nightjar can be found on the heaths, often joined on Kelling heath by Stonechat; Woodlark and Dartford Warbler have bred.

In autumn a wider variety of migrants may appear, and as well as the species mentioned for spring, waders such as Curlew Sandpiper and Little Stint may occur. The areas of bramble and other bushes around Little Eye, Gramborough Hill, Kelling Water Meadows, Kelling Quags and Muckleburgh Hill regularly attract interesting passerine migrants, with Wryneck and Barred Warbler being among the more regular scarce migrants. The rough pasture can also produce migrants, including Richard's Pipit and Lapland Bunting.

In winter the pools and flooded meadows hold numbers of Wigeons, Teals, Dark-bellied Brent Geese, Golden Plovers and Lapwings, and Barnacle and Tundra Bean Geese are very occasional visitors. The small pools behind the shingle ridge hold Common Redshank, Dunlin and Turnstone, and sometimes Shore Lark can be found around these wet areas (especially in the vicinity of Salthouse), while Snow Bunting is regular, if mobile, along the shingle bank. Red-throated Diver is regular offshore. Merlin, Hen Harrier and Short-eared Owl are irregular visitors, but Barn Owl is resident.

Access

Salthouse duck pond This lies just north of the A149 in Salthouse village, immediately west of the beach road, with space to pull off the road. There are often good numbers of gulls around the pond (or loafing in the fields to the north), which occasionally include Mediterranean or Yellow-legged Gulls.

Salthouse beach A metalled road leads to the beach car park from the A149 on the east side of Salthouse village. Snow Buntings are regularly seen in winter, especially on the wet areas west of the car park, and Shore Larks are occasional visitors, while Lapland Bunting can sometimes

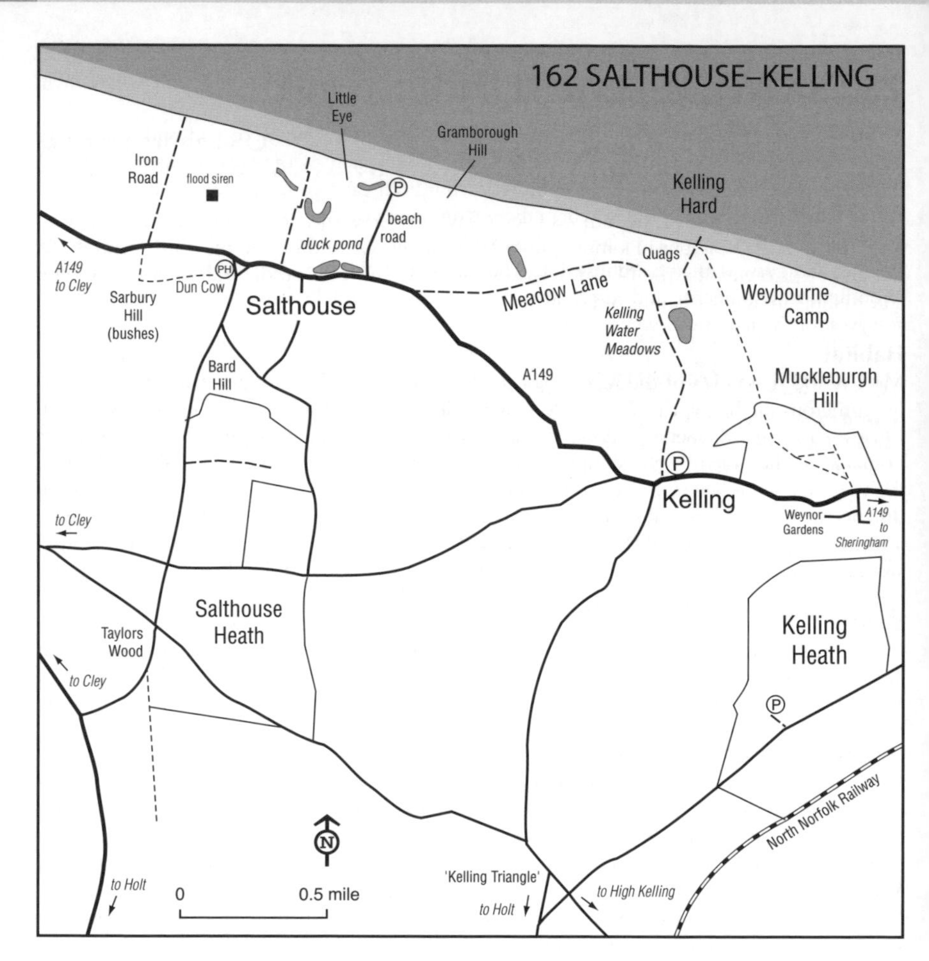

be found in the rough pastures. Two small hillocks, Little Eye and Gramborough Hill, lie west and east of the car park, and despite having very limited areas of cover (just a few small bushes), can hold interesting migrants in spring and autumn. It is possible to walk a circuit west from the car park to Little Eye, then south along a track to the A149, following the coast road east to the duck pond and then the Beach Road north back to the car park.

The Iron Road This track runs north from the A149 to the coast, west of Salthouse village. It can be good for waders in spring and autumn, depending on water levels. There is very limited parking – it is best to park by the village green (near the Dun Cow pub) and walk from there.

Kelling Quags and Water Meadows This area of grazing marshes lies immediately east of Salthouse Marshes, and includes several pools. In particular, the pool at Kelling Water Meadows, a 5.7-ha reserve of the Norfolk Ornithologists' Association (NOA), often holds interesting ducks and waders. About 0.5 miles east of Salthouse the A149 bends to the south, heading inland, and at this point a track (Meadow Lane) forks left. Park carefully at the beginning of this track and proceed on foot, checking the fields and marshes on the left. After 1 mile the track turns sharp right towards Kelling, passing alongside the Water Meadows pool. At this bend another track leads east to the beach at Kelling Hard, giving views of several small occasionally flooded areas on Kelling Quags. (Meadow Lane itself continues to Kelling village, and alternative access is to park on the side road south of the A149 in the village (opposite the gallery) and walk from there.)

Muckleburgh Hill This scrub and Bracken-covered hill lies immediately north of the A149 between Weybourne and Kelling and forms a northern extension of the heaths along the ridge.

It is attractive to migrants, although the extensive thick cover can be daunting to work. Turn south off the A149 to park in Weynor Gardens and cross the coast road to the hill, where there is a network of small paths.

Salthouse Heath A large area of heathland, much of which is now covered with thickets of gorse and birch, although management aims to restore much of the area to open heathland. There are breeding Nightjars and Nightingales, but both can be elusive; for Nightjar it is best to view from the road anywhere where there is an open vista.

Kelling Heath Another large area of heath; recent management has resulted in rather more heather and gorse than at Salthouse. The heath likewise holds breeding Nightjar and often also Woodlark and Stonechat, and can be accessed via a maze of tracks from the car park, which lies just west of the road from Weybourne to Holt.

FURTHER INFORMATION

Grid refs: Salthouse duck pond TG 079 438; Salthouse beach TG 081 442; The Iron Road TG 070 439; Kelling Quags and Water Meadows TG 082 437; Muckleburgh Hill TG 103 428; Salthouse Heath TG 071 424; Kelling Heath TG 098 417

163 CLEY (Norfolk)

OS Landranger 132

Cley is probably the best-known mainland birding site in Britain and over 360 species have been recorded in the National Grid 10-km square TG04, including a host of rarities and several firsts for Britain. Cley Marsh reserve covers 150 ha and is owned by the NWT, while adjacent Arnold's Marsh is owned by the NT. Birding at Cley is good at any time of year but the greatest diversity of species occurs in spring and autumn.

Habitat

Cley Marsh comprises extensive reedbeds, pools and grazing meadows. A series of scrapes have been excavated which are attractive to wildfowl and waders, and to the east Arnold's Marsh is a flooded area that is attractive to waders and terns. The entire area lies south of a shingle bank which (usually) protects it from the sea. Rising sea levels threaten the whole area, however, and the shingle bank is no longer maintained as a sea defence and has already migrated a considerable distance inland in one or two places.

Species

In early spring Wheatear can be seen along the beach, often in the vicinity of the Eye Field, and Firecrest and Ring Ouzel are occasionally found on Walsey Hills. Later in the season a variety of waders occurs, with Temminck's Stint being something of a speciality in May (although not guaranteed, the species is frequently present). Other regular spring migrants include Spoonbill, Garganey, Whimbrel, Little Gull and Black Tern.

Breeding birds at Cley include Bearded Tit, a reedbed speciality, but numbers fluctuate and it can be hard to see; calm days, when the birds are easier to hear and may be seen clambering to the top of reed stems, are best. Breeding waders include large numbers of vociferous Avocets, as well as Lapwing and Common Redshank, and several pairs of Marsh Harriers are resident but are most obvious late March–July; during the rest of the year they wander more widely (but may return in the late afternoon or evening to roost). Non-breeding Black-tailed Godwit and Ruff are present year-round, the former often in large flocks in the summer, while Sandwich, Common and Little Terns visit from the colonies on Blakeney Point. Roseate and Arctic Terns and Yellow-legged, Mediterranean and Little Gulls are less regular summer visitors. Reed and

Sedge Warblers are common in the reeds and Cetti's Warblers can be heard and sometimes seen, favouring the scrub by the A149 anywhere from Walsey Hills west to Cley village. Bitterns breed some years and are usually only seen in flight, flying to and from feeding sites; they are most active when feeding young in midsummer: choose a spot (such as the East or West Banks) with a panoramic view of the reedbeds and reed-filled dykes, wait and hope.

In autumn, wader passage commences with the arrival of Spotted Redshank and Green Sandpiper in late June, and numbers and variety increase to a peak in early to mid August. Typical species include Little Ringed Plover, Little Stint, Curlew Sandpiper, Whimbrel, Greenshank, and Wood and Common Sandpipers. Water Rail is more frequently seen in the autumn and occasionally Spotted Crake. One or two Wrynecks and Barred Warblers are found each year, usually in August–September; Walsey Hills is a favoured area, but any cover should be checked.

Terns are regularly present offshore in summer and autumn, and small numbers of Arctic Skuas are always present in the Cley–Blakeney area at this season, as are Gannets. From August onwards, north and north-west gales can produce a greater variety of seabirds, regularly including Manx Shearwater, while Long-tailed Skua is possible from mid-August, with the chance of Great and Pomarine Skuas increasing in September–October; Sooty Shearwater and Leach's Petrel tend to occur only in the 'best' conditions. Storms in late October–November can produce small 'wrecks' of Little Auks. Little Gull is regular offshore, especially in late autumn, and occasionally Woodcock and Long-eared and Short-eared Owls can be seen flying in off the sea at this time.

In winter Dark-bellied Brent Geese frequent the grassy fields alongside the flocks of feral Greylag and Canada Geese, and odd Pale-bellied Brent Goose or Black Brant are sometimes found in the flocks of Brent Geese. In recent years Pink-footed Geese have become much more regular, although they are usually only seen flying over. Golden Plover and Lapwing also favour the fields, with a few Water Pipits in some of the wetter areas. Barn Owl is resident and seen fairly frequently, but Merlin, Hen Harrier and Short-eared Owl are only occasional. Many dabbling ducks winter on the marsh, largely Wigeon, Gadwall, Teal, Pintail and Shoveler. Goldeneye is frequent on Arnold's Marsh and Red-throated Diver is almost always offshore (best seen in calm conditions); Black-throated and Great Northern Divers and Red-necked and Slavonian Grebes are very much less common. Snow Bunting flocks inhabit the beach anywhere between Blakeney Point and Weybourne, and are somewhat elusive, while the small and sporadic groups of Lapland Buntings prefer the roughest grazing marshes and are hard to find.

Access

Cley Marshes NWT reserve This lies just east of Cley next the Sea. The reserve is open daily, and access to non-members is by permit only, obtainable from the visitor centre, which lies south of the A149, 0.5 miles east of Cley village, and is open daily 10am–5pm (10am–4pm from 1 November to 1 March). A boardwalk leads from the visitor centre car park to Avocet, Daukes, Teal and Bishop Hides on the south side of the reserve (and also gives access west to the Beach Road and east to the East Bank). These hides are usually the best for wildfowl and waders, and the boardwalk out to Daukes Hide passes stands of reeds which can hold Bearded Tit (North Hide can be accessed from the beach car park; see below).

Cley coastguards and North Hide Just east of Cley village the Beach Road leads north to the beach car park (much reduced by the encroachment of the shingle bank). Immediately adjacent to the car park the Eye Pool sometimes hold a few waders, while the Eye Field is excellent for Dark-bellied Brent Goose and often also Golden Plover in winter. The beach car park is also the starting point for those walking to Blakeney Point (see site 164). The shingle bank by the car park is the favoured spot for seawatching at Cley, and the bank is favoured by mobile flocks of Snow Buntings in winter. From the car park it is possible to walk east along the base of the shingle bank, and after 500 yards a track leads south through a gate to the NWT's North Hide (permit required, although the North Scrape can be viewed more distantly with a telescope from the shingle ridge). The North Scrape is favoured by Cormorant, wildfowl, waders and gulls, but faces directly south, the light is poor for much of the middle of the day. Continuing along the shingle bank, the East Bank is reached after a further 500 yards.

The West Bank This parallels the Beach Road from the beach car park south towards the village,

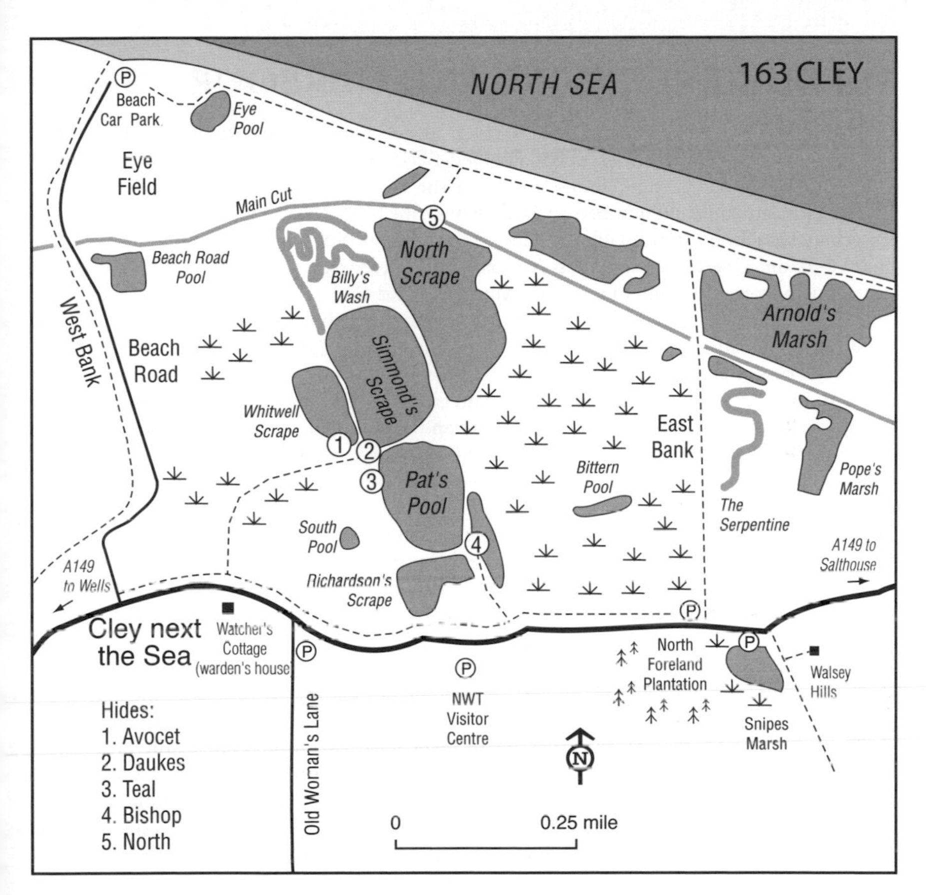

and continues, behind the windmill, to Cley Sluice. For its entire length, it affords panoramic views westwards over Blakeney Fresh Marshes and eastwards over Cley NWT reserve.

The East Bank This runs between the A149 (where there is a small car park) and the sea, and gives views west over the NWT reserve and east over the Serpentine and Arnold's Marsh. The extensive reedbeds west of the East Bank can be good for Bearded Tit. The wet fields east of the East Bank may hold Water Pipit in winter, while Arnold's Marsh usually has a selection of waders and is especially productive for loafing terns on summer evenings. A walk on the north side, inland of the shingle bank, will give closer views of birds at the back of Arnold's Marsh.

Walsey Hills and Snipes Marsh These lie south of the A149, 100 yards east of the East Bank car park (there is also space for a couple of cars to pull off the road at Walsey Hills). 1.2 ha of Walsey Hills is owned by the NOA, and the areas of cover on the hill and around the adjacent Snipes Marsh and North Foreland are attractive to migrants, while the summit of the hill gives views north over Popes Marsh and Salthouse Broad. If time permits, the complete circuit of the East Bank, shingle bank, Beach Road and the pathway that parallels the northern side of the A149, totalling 3 miles, can be profitable.

FURTHER INFORMATION

Grid refs/postcodes: Cley Marshes NWT reserve TG 054 440 / NR25 7SA; Cley coastguards and North Hide TG 048 452; The West Bank TG 043 437; The East Bank, Walsey Hills and Snipes Marsh TG 059 441

Norfolk Wildlife Trust, Cley Marshes: tel: 01263 740008; web: www.norfolkwildlifetrust.org.uk

164 BLAKENEY POINT AND HARBOUR (Norfolk)

OS Landranger 132

The unique 3.5-mile shingle spit of Blakeney Point is best known for attracting migrant passerines, sometimes in substantial falls, and for its large colony of terns. The Point is owned by the National Trust, has been a reserve since 1912 and is now an NNR. It is best visited in spring and autumn for migrants, and in summer for terns.

Habitat

Blakeney Point is the terminal portion of the long shingle bank, a product of long-shore drift that extends westwards from Weybourne. From Cley coastguards westwards the shingle bank departs from the land to shelter the mudflats and saltmarshes of Blakeney Harbour, while towards the tip of the point shingle is backed by extensive dunes. Thickets of Shrubby Sea-blite (often known by its generic name, *Suaeda*) occur on the southern, landward, edge of the spit, particularly at Halfway House, the Hood and the Long Hills, while the dunes have areas of Marram and other grasses and a few patches of denser vegetation; a tiny rectangular patch of stunted trees 200 yards east of the tea room is known as the Plantation. During migration periods, any area of cover can hold migrants.

Species

There is a large ternery at Blakeney Point, and although numbers are variable, in recent years there have been up to 3,500 pairs of Sandwich and 130 pairs of Common Terns, as well as 1,000 pairs of Black-headed Gulls and a handful of Common and Mediterranean Gulls; in recent seasons, one or two Roseate Terns have also summered in the area. There is also a nationally important population of Little Terns, with 60–215 pairs scattered in the roped-off areas between the Point and Cley coastguards. Other breeders include Shelduck, Oystercatcher, Ringed Plover and Common Redshank. Little Egret is now resident in the area and can be seen year-round, favouring the saltmarsh gutters.

In spring, migrants on the Point can include Yellow Wagtails and Wheatears in April, when Ring Ouzel, Firecrest or Black Redstart are also occasionally found. Autumn is, however, by far the best time to visit the Point. Falls can occur at any time of day, but are most frequent in the afternoon, and it should be quickly apparent as you walk out whether numbers of birds have arrived or not. In August–September the commonest species are Willow and Garden Warblers, Wheatear, Whinchat, Common Redstart and Pied Flycatcher. Sadly, due to the European-wide decline of common birds, substantial falls have become increasingly scarce. Scarcer migrants found annually or almost annually include Dotterel, Wryneck, Icterine and Barred Warblers, Red-breasted Flycatcher and Red-backed Shrike. Notably, Greenish Warbler is almost annual in late August and September; these and other rarities may occur alone, in the absence of a fall. In the late autumn migration is dominated by thrushes, Starlings, Goldcrests and Robins, and Woodcock, Short-eared Owl, Ring Ouzel and Black Redstart should all be looked for. Offshore, Arctic Skuas are invariably present from July to September, and north winds may induce an impressive seabird passage, typically Gannet, Kittiwake and a variety of terns, with a few Manx Shearwaters and Great Skuas. Leach's Petrel, Mediterranean and Sooty Shearwaters, Little Auk, Pomarine and Long-tailed Skuas and Sabine's Gull are only occasional, although Little Auks and Pomarine Skuas have occasionally been recorded in good numbers in late October and November.

In winter, Dark-bellied Brent Geese feed in Blakeney Harbour alongside many Wigeons and the commoner wintering waders. Goldeneye and Red-breasted Merganser frequent Blakeney Pit, and Common Guillemots, seaducks, divers (almost all Red-throated) and the rarer grebes are sometimes seen offshore. The fresh marshes also hold Brent Geese alongside Golden Plovers, and raptors, notably Marsh and Hen Harriers, Merlin and sometimes

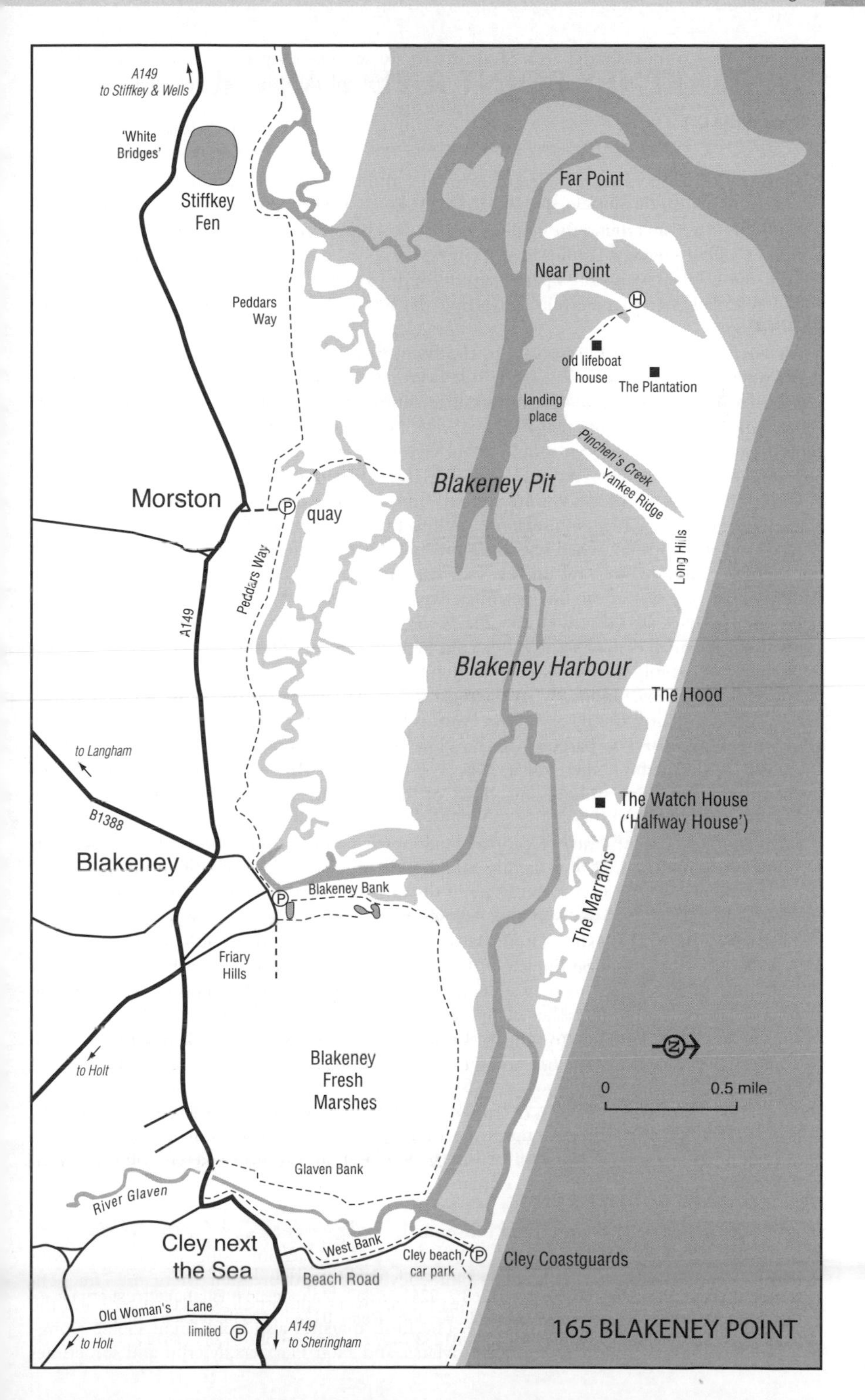

165 BLAKENEY POINT

Peregrine, may hunt over the area. Shore Larks are sometimes present and Snow Buntings occur along the beach but wander widely, and while Lapland Buntings are often somewhere in the area, they are usually very difficult to locate.

Access

Blakeney Point To view the breeding terns in summer the easiest option is to take a tripper boat from Morston Quay to view the seals on Blakeney Point; some of the seal trips also allow a limited time ashore around the landing stage at Pinchens Creek; these excursions usually also permit excellent views of Common and Grey Seals. The boats are, however, dependent on the state of the tide and can only operate around high water. They operate daily April–October and almost daily, weather permitting, during the rest of the year (booking always advisable July–August).

Alternatively, it is possible to walk to the Point from Cley beach car park. This is a long, tiring walk along the shingle but, if the tide is low, it is easier to walk along the beach where it is firmer underfoot. There is an information centre at the Point in the old Lifeboat House (open only in summer; note that the 'tern hide' no longer overlooks the ternery, which has moved to Far Point!). Visitors should avoid disturbing the tern colony; the main breeding areas are roped-off during summer and dogs are banned from most of the point April–August, but access is otherwise unrestricted, although it is best not to walk on the shingle ridge itself during the breeding season – stick to the beach or the fringe of the saltmarsh.

In spring and autumn, when searching for migrant passerines, it is best to walk in at least one direction along the landward side of the point (perhaps returning along the beach, or taking a boat one way if the tide permits). Areas of cover are scattered along the route (the best being perhaps at Halfway House, the Hood and Long Hills), and at the Point itself the Plantation, the small area of scrub by the old lifeboat station and the Elders all deserve careful scrutiny. Most migrants skulk deep in the *Suaeda*, and often only give brief, flight views, although there are exceptions. By far the most successful tactic involves teams of birdwatchers gently chivvying birds out of cover; the Point is hard work for a lone observer.

Blakeney Harbour The harbour can be viewed from the north, from Blakeney Point. To the south of the harbour the Peddars Way/Norfolk Coast Path skirts the southern edge of the saltmarsh, providing only distant views of the mudflats beyond. The best access can be had as follows:

1. Morston Quay. From the NT car park walk north, past the tripper boats, along an obvious but often muddy trail to a slightly raised, well-vegetated ridge, continue northwards to view the mudflats. You must be aware of the tide on this walk (and whenever you venture onto the saltmarshes) to avoid being caught by the rising tide.
2. Blakeney Bank. This extends from Blakeney Quay car park northwards along the sea wall with tidal saltmarshes on the left and the rough grassland of Blakeney Fresh Marshes on the right. At high tide, many waders roost on the saltmarshes. It is possible to follow the bank all the way round the marshes, returning via the Glaven Bank to the main A149 coast road at Cley sluice (on the A149 just west of Cley village). Blakeney Bank can be accessed at any state of the tide and is extremely popular with walkers during summer, and at weekends year-round.

Friary Hills, Blakeney This area of scattered bushes lies along the seaward edge of Blakeney village, and occasionally attracts interesting migrants (e.g. Ring Ouzel in spring); the scrub along the northern edge is good for Cetti's Warbler. Follow the road east for 200 yards from Blakeney Quay car park.

FURTHER INFORMATION

Grid refs: Morston Quay TG 006 442; Blakeney Bank and Friary Hills TG 028 441
National Trust, Blakeney office: tel: 01263 740241; email: blakeneypoint@nationaltrust.org.uk
Ferry operators: Graham Bean, 01263 740505; John Bean, 01263 740038; Bishop's Boats, 01263 740753, 0800 0740754; Roy Moreton, 01263 740792; Jim Temple, 01263 740791

165 STIFFKEY TO WARHAM GREENS (Norfolk)

OS Landranger 132

This stretch of coast holds a variety of raptors in the winter, while the area is backed by stands of trees and bushes which are attractive to migrants, especially in autumn. Management at Stiffkey has created a wetland area, Stiffkey Fen, which attracts a range of interesting birds throughout the year.

Habitat

The marshes along this stretch of coast form part of one of the biggest expanses of saltmarsh in Europe, and are included within the Holkham NNR. Immediately inland lie areas of mixed farmland, bisected by the Stiffkey River.

Species

In spring, and especially autumn, the area can be productive for passerine migrants, and has attracted a number of rarities, including several Pallas's Warblers. Such events are very weather dependent, however, with winds between north and east being best. Stiffkey Fen may hold Garganey and Mediterranean Gull.

In winter, visitors to the Stiffkey Fen area include Dark-bellied Brent Goose, Black-tailed Godwit, Ruff and Grey and Golden Plovers, and a variety of the commoner waders are found in the saltmarsh gutters as well as Little Egret and Rock Pipits. A few Hen Harriers roost on the saltings, and Short-eared and Barn Owls, Peregrine and Merlin may also occur; for raptors it is best to view in the late afternoon from the end of the Greenway at Stiffkey or from Warham Pits.

Summer is often quiet, but breeding birds at Stiffkey Fen include Avocet and Shelduck.

Access

Stiffkey Woods At the west end of Stiffkey village take the minor road (the Greenway) north towards the campsite and park at the end on the edge of the saltmarsh. Immediately to the east lies a narrow belt of trees, Stiffkey Wood, which can be attractive to migrants, while to the west are occasional areas of brambles and rough ground which may also hold a variety of migrants. If you walk directly north from the car park, you will come to a vegetated raised bank (Stiffkey Meals) and it is possible to continue north to the sea and view the tip of Blakeney Point. It is necessary to cross several shallow channels, however, and this should only be attempted at low

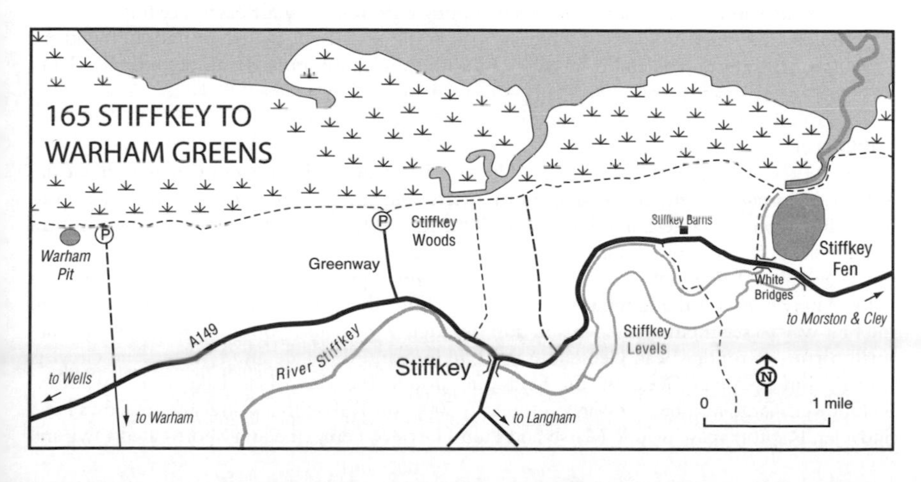

tide (it is essential to check accurate tidetables!).

Warham Greens Continuing west from Stiffkey Woods, the track eventually reaches an overgrown pit at Warham Greens, attractive to migrants. This can also be accessed by taking the track north off the A149 1 mile west of Stiffkey village (directly opposite the first turning to the south (to Warham) on the A149 west of Stiffkey village); although marked unsuitable for motors, there is a small car park at the end (do not take the track to the north 0.3 miles west of here, also marked unsuitable for motors, because it really is!).

Stiffkey Fen An area of 12 ha to the east of Stiffkey has been impounded to create a lagoon. The area can be viewed from the sea wall and the public footpath along the west bank of the River Stiffkey. Park at Morston Quay NT car park and walk west along the coast footpath for just over 1 mile to view.

FURTHER INFORMATION

Grid refs: Stiffkey Woods TF 964 438; Warham Greens TF 948 438; Stiffkey Fen TF 986 436

166 THE WELLS–HOLKHAM AREA (Norfolk)

OS Landranger 132

This stretch of the north Norfolk coast forms part of Holkham NNR, which extends from Burnham Norton to Blakeney and covers about 4,000 ha. Numbers of migrant passerines, sometimes including rarities, can occur in spring and autumn in appropriate weather conditions. The fields immediately inland of the coast regularly attract a variety of geese in winter, and hold breeding wildfowl and waders, while the sea holds divers, grebes and seaducks in winter.

Habitat

At low tide very extensive sand- and mudflats are exposed along the coast and bordering these is a long series of sand dunes, planted with a narrow belt of pines along a 3-mile stretch from Wells Woods in the east to Holkham Pines in the west (collectively known as Holkham Meals). The northern, exposed side of the pines does not support much vegetation but along the south edge dense areas of scrub and deciduous trees flourish (mainly birches and sallows), especially at the Wells end, where a particularly lush, overgrown hollow is known as the Dell. An area of fields lies between the pines and the A149 coast road, and management has raised the water table to create some flooded areas and greatly improve its suitability for both breeding and visiting wildfowl and waders. South of the coast road lie the extensive, walled grounds of Holkham Park, in which grassy parkland is interspersed with stands of mature deciduous woodland and there is a large artificial lake.

Species

Geese are the main attraction in winter. Up to 1,500 Dark-bellied Brent Geese frequent the areas of saltmarsh and the fields south of the pines, and the grazing marshes also attract up to 400 White-fronted Geese. Large numbers of Pink-footed Geese winter in the area, with counts of up to 70,000 in recent winters. In late autumn and late winter they favour the fresh marshes, but November–January/February most fly inland in impressive skeins to feed in sugar beet fields. Many roost on the sands at Wells/Warham, but others use Scolt Head Island, and often the best way to see them is at dawn and dusk on their way to and from their roosts. Sometimes small parties of Tundra Bean Geese join the Pinkfeet, and there are also resident feral Greylag, Canada and Egyptian Geese on the fields and around Holkham Hall Lake. Other wildfowl include very large numbers of Wigeon (up to 8,500) and Teal, with a scattering of Gadwall and Shoveler. Raptors may include Marsh and Hen Harriers, Peregrine and Short-eared Owl, and a few Merlins use the area, but are typically elusive; Merlin and Hen Harrier roost at East Hills in Wells Harbour.

The usual waders occur in the harbour and on the shore, notably good numbers of Sanderlings on the beach, with many Golden Plovers and occasional Ruffs in the fields. Divers, grebes and seaducks may also occur in the harbour channel, with occasionally a Shag or Red-necked Grebe near the quay. Twites also occur at roadside puddles along Wells beach road, and at Burnham Overy Staithe. In winter Holkham Park holds Goldeneye and Gadwall on the lake.

Spring migrants in the pines and dunes may include Ring Ouzel and Siskin, and occasionally also Firecrest or Black Redstart. However, autumn is usually more productive. Migrant numbers depend on the weather, however, with north-north-east winds being best. From early August, a variety of warblers, Pied and Spotted Flycatchers, Common Redstart, Wheatear and Whinchat occur, with Wryneck, Icterine and Barred Warblers and Red-backed Shrike annual in very small numbers; Greenish Warbler is one of the more regular early-autumn rarities. In mid-September to November the emphasis shifts to thrushes, Goldcrest, finches and buntings, with rather smaller numbers of warblers. Red-breasted Flycatcher and Yellow-browed Warbler are annual (as is Richard's Pipit on the grazing marshes) and Pallas's, Dusky and Radde's Warblers are among the more regular rarities. Passerine migrants may be found anywhere in the pines, although the east end, especially the Dell and Drinking Pool, are perhaps better in early autumn, with the west end being favoured in late autumn. In both spring and autumn, a wide variety of ducks, waders, terns (including Black), and Marsh Harrier may be at the fresh marshes south of the pines.

Breeders include several hundred pairs of Black-headed, Herring and Lesser Black-backed Gulls on the marshes at Wells, with Ringed Plover and nationally important numbers of Little Terns on the beach (keep out of the fenced enclosures). Siskin and Common Crossbills occasionally breed in the pines (and Parrot Crossbill bred near Wells beach car park in 1984 and 1985), with Grasshopper Warbler in the marshy scrub along the south fringes. The grazing marshes hold numbers of Lapwing, Common Snipe and Common Redshank. Marsh Harrier and Avocet breed and are usually present in summer, and the colony of tree-nesting Cormorants (75 nests in 2007) and the few nesting Grey Herons have been joined by Little Egrets. Bittern, Garganey and Bearded Tit have bred, and Spoonbill is an increasingly regular non-breeding visitor.

Access

Wells Harbour The harbour can be viewed from the A149 in Wells town (limited parking on the quay), and from here a minor road signed to the Beach and Pinewoods leads north to a large pay-and-display car park. To the left of this road, the grassy areas attract roosting waders and Dark-bellied Brent Geese. At the car park, the sea wall by the lifeboat station affords views of the outer part of Wells Harbour.

Wells Woods From the A149 in Wells town follow the beach road (signed Beach and Pinewoods) to the pay-and-display car park. From here go through the kissing gate and past the north end of Abraham's Bosom (the boating lake, which can occasionally hold a grebe or seaduck in winter). After the next gate, the left-hand path crosses an area of brambles and then follows the south edge of the pines. The Dell (a circular embanked area with birches concentrated along the seaward side) lies on the right after 400 yards, and a long, near-dry pit known as the Drinking Pool is hidden within a dense area of pines 400 yards further. Continuing, the path follows the landward fringe of the woods to Holkham Gap. The path to the right from the kissing gate heads towards the sea, passing the old toilet block.

Lady Ann's Drive and Holkham Gap This private road runs north from the village of Holkham on the A149 coast road towards Holkham Gap. Parking is available along the drive (toll). In winter, numbers of geese are present, often close to the drive. In Holkham Bay, as well as Red-throated Diver, small numbers of Slavonian and Red-necked Grebes may be found, especially in November. It is best to walk out across the saltmarsh and sands to the water's edge just west of the Gap at low water, but beware the rapid advance of the incoming tide. A few auks, Goldeneyes, Red-breasted Mergansers, Common Eiders or Velvet Scoters may join the

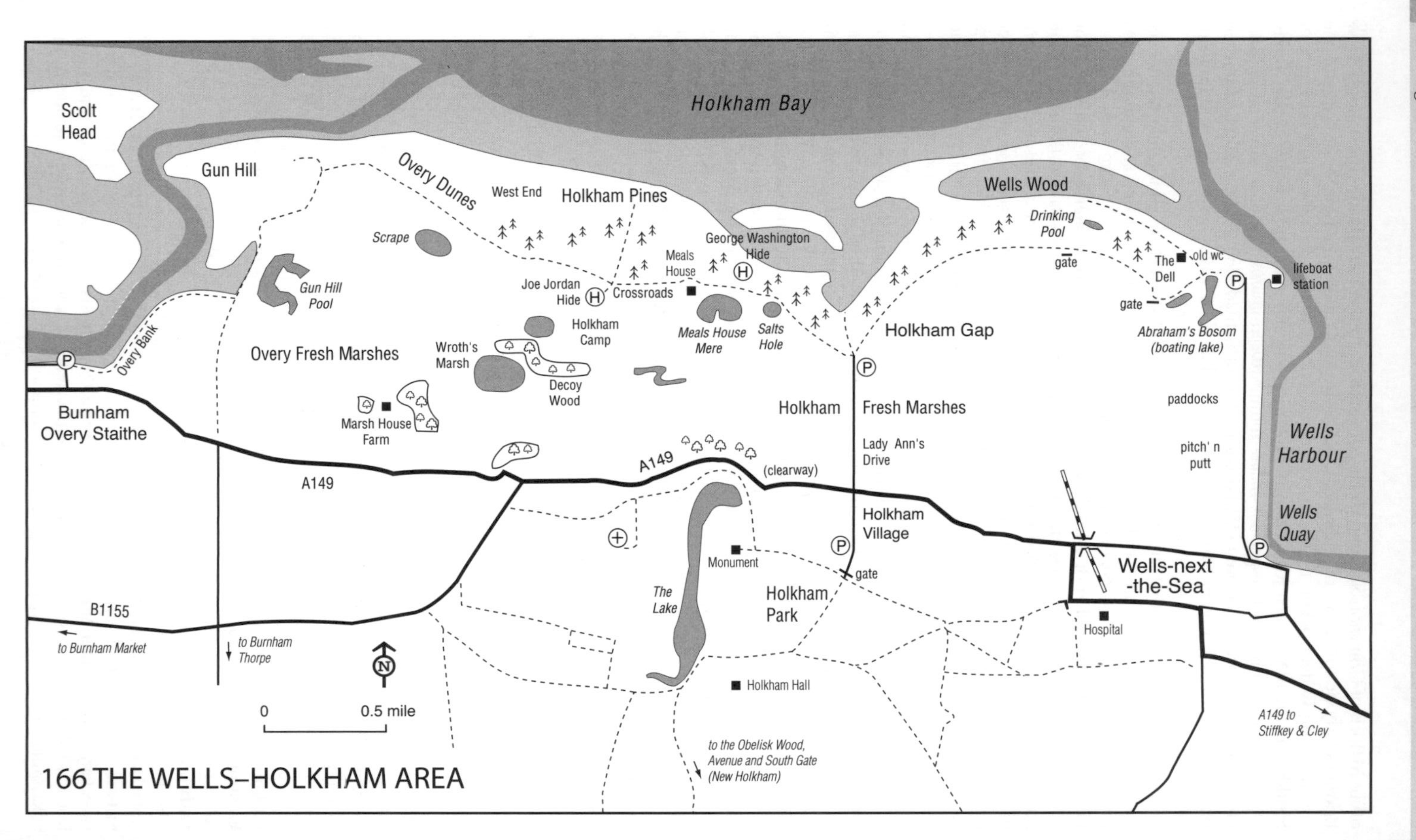

166 THE WELLS–HOLKHAM AREA

offshore flock of Common Scoters, which can number several thousand in some seasons. Snow Buntings may be found on the saltmarsh and shore, sometimes joined by Shore Larks or the fast-declining Twite.

Holkham Pines A track leads west from Holkham Gap along the south fringe of the woods, and this is good for migrants, especially in late autumn. Walking west, you pass a small pond (Salts Hole) and then reach the George Washington Hide which overlooks Meals House Mere, an artificial scrape which attracts a variety of ducks. The hide has good views south over Holkham Fresh Marshes for raptors and geese. Continue past Meals House to the Crossroads, where a short track leads from here to the Joe Jordan Hide, which has excellent panoramic views over the fresh marshes. Continuing west along the pines you reach the West End and emerge onto Overy Dunes. Migrants can occur at any point along this walk.

Holkham and Overy Fresh Marshes Traditionally a haunt of winter geese, management of this area of rough grazing inland of the pines has created areas of shallow flooding. Large numbers of ducks and waders now breed, and the pools attract migrant waders. Access to the marshes is limited, and although visible from the A149, views are very distant and the road is rather busy (the A149 is a clearway and the nearest parking is in Lady Ann's Drive); view the Fresh Marshes from Lady Ann's Drive, the George Washington and Joe Jordan Hides, or from Overy Dunes.

Gun Hill and Overy Dunes This area of dunes holds a few scattered pockets of bushes and although apparently rather bleak, has attracted some five-star vagrants. It can also be good for raptors in the winter. Access is along the sea wall (Overy Bank) from the quayside car park at Burnham Overy Staithe, or from the A149 at the Burnham Thorpe turning (though parking is difficult here) or from the west end of Holkham Pines.

Holkham Park Turn south off the A149 signed 'Holkham Hall, Pottery' and park on the right after 400 yards. The road continues to the main gate of the park and a small portal on the left permits access on foot. The mature woodland holds all three woodpeckers, Nuthatch, Treecreeper and Marsh Tit, and the sharp-eyed observer may spot a roosting Tawny Owl, especially in the evergreen oaks and the tall Cedar tree near the monument. The lake holds wildfowl and Cormorant. In the south part of the park, the Avenue (leading to the south gate) is a good area for geese and Brambling in winter.

FURTHER INFORMATION

Grid refs: Wells Harbour TF 917 438; Wells Woods TF 912 454; Lady Ann's Drive and Holkham Gap TF 890 447; Gun Hill and Overy Dunes TF 845 443; Holkham Park TF 891 435
Natural England, Site Manager: tel: 01328 711183; email: enquiries.east@naturalengland.org.uk

167 SCULTHORPE MOOR (Norfolk)

OS Landranger 132

Sculthorpe Moor Community Nature Reserve was created by the Hawk and Owl Trust in the Wensum Valley, and is an excellent site for a variety of wetland birds.

Habitat

The reserve comprises marsh, fen and willow and alder carr.

Species

Breeding birds include Marsh Harrier, Water Rail, Barn Owl, Kingfisher, Great and Lesser Spotted Woodpeckers, Grasshopper Warbler and Marsh Tit, and Hobbies are frequent visitors in summer.

In winter the hides are excellent places to see Water Rail, Siskin and Lesser and Common Redpolls, and Golden Pheasant may be seen by stalking quietly along the boardwalk through the woods.

Access

Turn south off the A148 at the sign for the reserve 1 mile west of Fakenham. A rough track leads to the car park and visitor centre, from where a well-marked trail leads to two hides and a viewing platform overlooking the River Wensum. The reserve is open every day except Monday and Christmas Day, 8am–6pm in summer, 8am–4pm October–March.

FURTHER INFORMATION

Grid ref/postcode: TF 900 304 / NR21 9GN
Hawk and Owl Trust, Sculthorpe Moor: tel: 01328 856788; web: www.hawkandowl.org/sculthorpehome

168 TITCHWELL (Norfolk)

OS Landranger 132

Titchwell is one of the RSPB's premier reserves (indeed, it is now their most visited) and, as a result of effective management, it has become one of the best birding sites on the Norfolk coast. Interesting birds can be found at any time of year, but the greatest diversity is in spring and autumn.

Habitat

The reserve contains fresh, brackish and saltmarshes, tidal and freshwater reedbeds, sand dunes and a sandy beach. Around the visitor centre and car park there is an area of scrub, willows and poplars.

Species

Breeders include Marsh Harrier, Bittern, Water Rail, Avocet, Cetti's, Reed and Sedge Warblers and Bearded Tit. On the beach there are a few pairs of Ringed Plovers and Oystercatchers. The Fen Trail winds through a scrubby area inhabited by a variety of warblers, sometimes including Grasshopper Warbler, which also attracts passerine migrants.

Spring comes late to the north Norfolk coast, but in mid-May and June Titchwell can be an exciting place. As well as the commoner waders, gulls and terns, Spoonbill, Garganey, Little Gull and Black Tern are regular migrants. Indeed, Spoonbills may occur at any time of year, and Little Egrets are now common year-round.

In autumn, from August to mid-October, a wide variety of passage waders is recorded, including Common and Green Sandpipers. Among scarcer species, Little Stint, Curlew Sandpiper and Black-tailed Godwit are regular. On higher tides, from late summer into winter, large numbers of waders may roost on the saltmarsh, including Oystercatcher, Grey Plover, Knot, Dunlin and Bar-tailed Godwit, and in late summer many of these will still have some summer plumage. Water Rails may be more conspicuous in autumn as their numbers are augmented by migrants. Large numbers of Swallows and martins roost in the reeds, and these may attract a Hobby. Seabird movements can occur offshore after a strong north wind, Gannet and Kittiwake form the bulk with a scattering of Great and Arctic Skuas and the occasional shearwater.

For divers, grebes and seaducks calm conditions around high tide are best for viewing. In recent years there has been a large flock of Common Scoter offshore, and there are usually also a few Common Eiders and Velvet Scoters and sometimes also Long-tailed Duck. Red-throated Diver is also regular and look out for Black-throated Diver and Slavonian and Red-necked Grebes, which are also recorded on odd dates through the winter. Numbers of Dark-bellied Brent Geese winter in the area, and Pink-footed Geese may fly over. The water on the marsh is deeper in winter, and Goldeneye and Red-breasted Merganser are often seen, together with large numbers of dabbling ducks, and Cormorant also roosts on the reserve. Waders include large numbers of Golden Plovers, and there may also be a handful of wintering Spotted Redshanks and also significant numbers of Ruffs. A few Purple Sandpipers are sometimes present on the rocky outcrops along the beach. Barn Owl is resident in the area and Merlin, Hen Harrier and Short-eared Owl occasionally hunt over the marshes, with small numbers of Hen Harriers roosting in the reedbeds. Snow Buntings are regular on the beach, where they are occasionally joined by Shore Larks. Rock Pipit can be found in the saltmarsh gutters, and there may also be a few wintering Water Pipits (which favour more freshwater habitats).

Access

RSPB Titchwell Approximately 0.3 miles west of the village of Titchwell turn north off the A149 on to the signed track to the reserve; the car park lies to the right after 200 yards. The visitor centre, shop and cafe are open weekdays 9.30am–5pm (9.30am–4pm mid-November to mid-February; closed Christmas Day and Boxing Day). From the shop follow the track north along the embankment, which gives good views of the reedbeds and lagoons, as well as access to the two public hides. This is the best area for wildfowl, waders and gulls, as well as Marsh Harrier, Bearded Tit and, with a lot of luck, Bittern. The bank eventually reaches a viewing platform overlooking the beach, and this is a good spot from which to scan for divers, grebes and seaducks around high tide. Additionally, starting at the shop, the Fen Trail leads for 250 yards through an area of fen and scrub to the Fen Hide, which overlooks the reedbeds, before

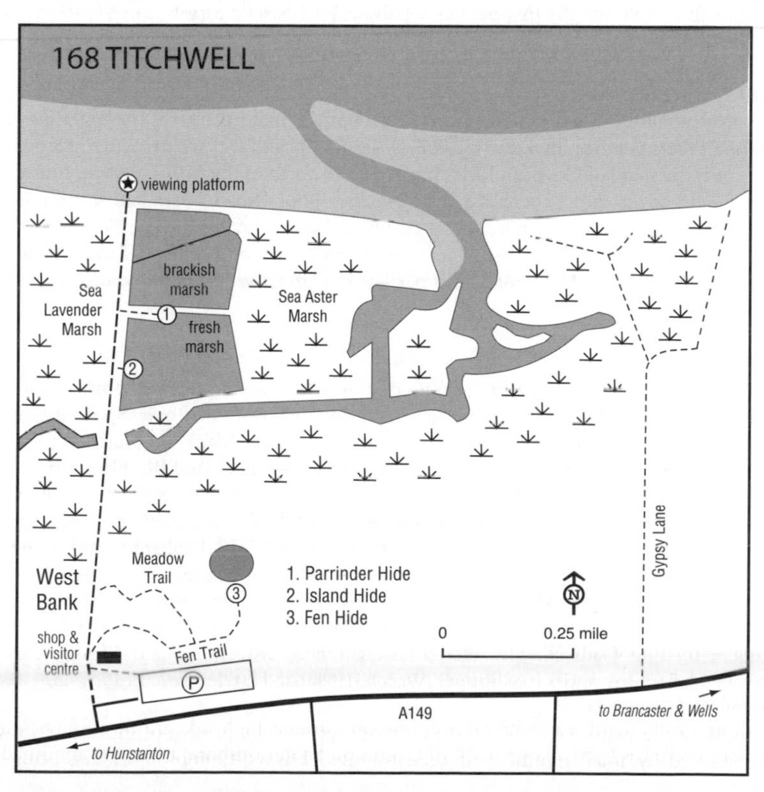

returning to the main embankment. All paths are firm and level, and all hides wheelchair-accessible. The reserve is always open, with access limited to the embankment, hides and beach. Note that as part of a programme of managed retreat implemented in 2009–11, sea water will be allowed into parts of the northern portion of the reserve while the banks to the south will be strengthened, changing the balance of habitats and possibly affecting access.

Gypsy Lane Continue east along the A149 through the village of Titchwell and, 1 mile east of the turning to the RSPB reserve, there is space to pull off the road by a belt of trees. From here a public footpath leads north along the west side of the trees and onto a bank, eventually reaching the beach. Raptors and Little Egret are frequently seen from this path.

FURTHER INFORMATION

Grid refs/postcode: RSPB Titchwell TF 750 437 / PE31 8BB; Gypsy Lane TF 764 437
RSPB Titchwell: tel: 01485 210779; email: titchwell@rspb.org.uk

169 HOLME (Norfolk)

OS Landranger 132

Situated on the north coast of Norfolk, at the point where an isolated stand of pines marks the entrance to the Wash, Holme is well placed to receive migrants and the varied habitats make a visit worthwhile at anytime of year. The NWT owns 240 ha (and leases a further 35 ha at Holme Marsh), while the NOA operates a private bird observatory from an enclave of 5 ha within this area.

Habitat

A small area of saltmarsh lies west of Gore Point but the main habitat is the extensive area of sand dunes. Thickets of sea buckthorn stabilise these in places and are particularly attractive to migrants, as is the stand of Corsican Pines immediately north of the NWT reserve centre at The Firs. Just inland of the dunes there is a large brackish pool, Broad Water, and some ponds and scrapes surrounded by reeds that are favoured by freshwater-loving waders. Between the coast and the village of Holme lies Holme Marsh, an extensive area of rough grassland. Management work to raise the water table has made this particularly attractive to winter wildfowl and breeding ducks and waders.

Species

Migrant passerines are dependent on suitable weather, north or east winds being best. Wheatear, Whinchat, Common Redstart, a variety of warblers, and Spotted and Pied Flycatchers are all frequent in spring and autumn, while Black Redstart, Ring Ouzel, Firecrest and Red-backed Shrike are also possible, with Wryneck and Bluethroat much more irregular visitors. Hoopoe and Nightingale are occasional in spring, while in most autumns Barred and Icterine Warblers are noted; Yellow-browed Warbler and Red-breasted Flycatcher are recorded most years, and Long-eared Owl, Woodcock and Lapland Bunting are quite frequent in late autumn. A wide variety of waders occurs on passage, including Black-tailed Godwit. Offshore, Gannet, Manx Shearwater, and Great and Arctic Skuas are regular from midsummer, together with the occasional Sooty Shearwater or Pomarine Skua.

Breeders include Gadwall, Shoveler, Oystercatcher, Avocet, Ringed Plover, Common Redshank, Little Tern and Grasshopper Warbler; Marsh Harrier and Hobby are frequent visitors in summer.

Throughout the year up to 2,000 Common Scoters may be present offshore, and in winter they are joined by small numbers of Red-throated Divers, Common Eiders, Long-tailed

Ducks (off Gore Point at high tide) and Red-breasted Mergansers. Careful searching may also turn up Great Northern or Black-throated Divers, Red-necked or Slavonian Grebes or Velvet Scoter. Up to 700 Dark-bellied Brent Geese and 1,500 Wigeons are regularly present on Holme Marsh, and Pink-footed Geese are regular visitors. Look too for Hen Harrier, Merlin and Barn Owl, with Kingfisher and Water Rail regular on the pools. Snow Buntings are often present on the dunes, beach and saltmarsh, with Twite towards Thornham Point at the eastern end of the area, but Rough-legged Buzzard and Shore Lark are only scarce and irregular visitors.

Access

Hunstanton Golf Club and The Paddocks At Holme next the Sea on the A149, take the westernmost road (signed NWT, NOA and Beach) to the sea. The beach car park (fee in summer) is reached after 1 mile and accesses the bushes and scrub on Hunstanton Golf Club (keep to the seaward edge, away from the greens and fairways; a footpath also skirts the landward edge of the course from the beach car park to Old Hunstanton). To the east, on the seaward side of a row of houses, lies The Paddocks, an excellent area of bushes attractive to migrants, which can be viewed from the public footpath along its northern edge.

Holme NWT Reserve, West End Turn right (east) off the beach road shortly before the beach car park onto a rough private track (toll in summer). After 0.5 miles a NWT car park lies to the left. From here you can explore the west end of the dunes and The Paddocks, walk out on the public footpath towards Gore Point, which can be good for seaducks in winter around high tide, or continue along the public footpath east towards The Firs.

Holme NWT Reserve, East End Continue along the track, through the gate and on to the second NWT car park, which lies to the left after 1 mile. The reserve is open daily, 10am–5pm

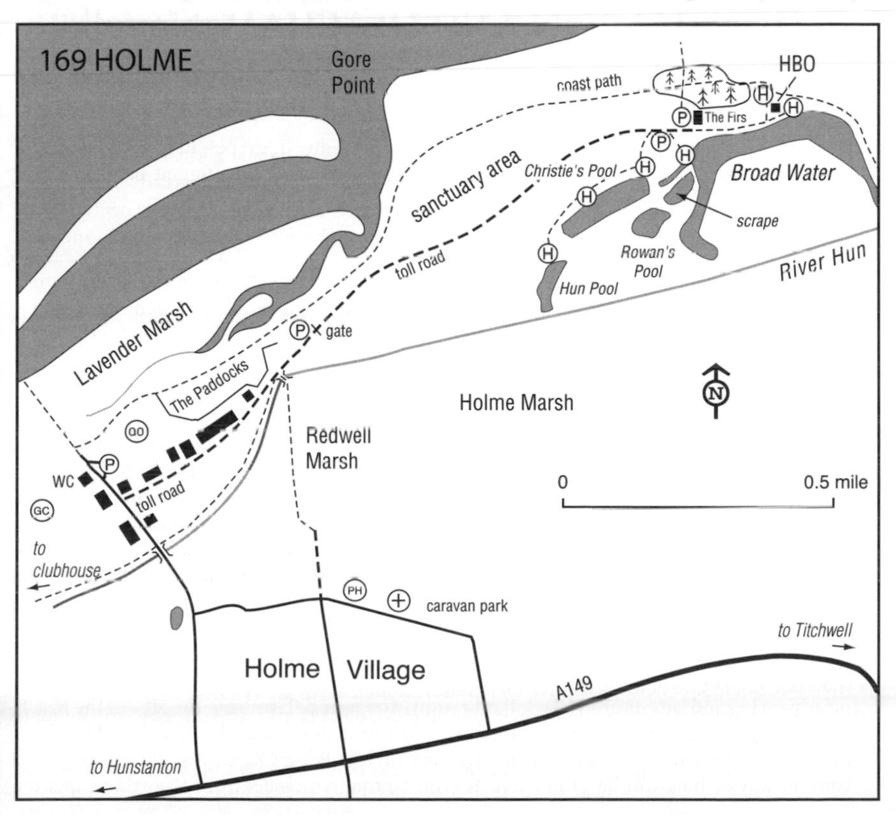

(outside these hours the gate is closed), and the NWT visitor centre at The Firs is open 10am–5pm April–October, weekends November–March. Three hides overlook Broad Water and the freshwater wader pools, while the pines and scrub on the dunes should be investigated for migrants (please keep to the marked paths). During strong onshore winds, seawatching is worthwhile around the period of high tide – watch from the top of the dunes where the footpath cuts through behind the car park at The Firs. Otherwise, the sea is always worth a look for divers, grebes and seaducks.

Holme NOA Reserve and Bird Observatory Follow directions as above for the NWT reserve, east end, but park in the NOA car park to the right (observatory visitors and NOA members do not need to buy a permit or car parking pass for the adjacent NWT reserve). Opening times are as for the NWT reserve. There are five hides overlooking a variety of habitats and winter feeding stations and a dedicated seawatching hide (exclusive to NOA members), while the Observatory Centre offers bird news and information.

FURTHER INFORMATION

Grid refs/postcodes: Hunstanton Golf Club and The Paddocks TF 697 438 / PE36 6JQ; Holme NWT Reserve, west end TF 702 442; east end, Holme NOA Reserve and Bird Observatory TF 714 449 / PE36 6LQ

NWT Holme: tel: 01485 525240; web: www.norfolkwildlifetrust.org.uk

NOA Warden: tel: 01485 525406; web: www.noa.org.uk

170 HUNSTANTON AND HEACHAM (Norfolk)

OS Landranger 132

The sea off Hunstanton regularly holds small numbers of divers, grebes and seaducks in the autumn and winter, and the former season can also be interesting for migrant passerines and seawatching in appropriate weather conditions.

Habitat

The coast at Hunstanton, a small seaside town, is bordered by low cliffs of chalk, surmounting a layer of carrstone, which extends to form the foundations of the beach and then offshore as a series of reefs. To the south the vast mudflats of the Wash border the coast at Heacham, with a series of old creeks, borrow pits and rough grazing to the landward side.

Species

In winter a few seaducks can be seen offshore at Hunstanton, including Goldeneye and Red-breasted Merganser, and other wildfowl include Dark-bellied Brent Geese, which favour the shoreline at low tide. Small numbers of divers and grebes are often present; Red-throated Diver and Great Crested Grebe are commonest, but Great Northern and Black-throated Divers and Red-necked and Slavonian Grebes are also possible with careful searching. Waders occasionally still include the odd Purple Sandpipers (formerly much more regular), which on a high tide favour the south end of Hunstanton promenade, but at low water may be found on the mussel beds offshore and the groynes towards Heacham. Other waders include Turnstone and, south along the coast into the Wash, numbers of Curlews, Bar-tailed Godwits, Knots, Dunlins, Common Redshanks and Grey and Ringed Plovers.

In early spring and late autumn Black Redstart may be found (favouring the area of the beach huts bordering the golf course on the north side of Hunstanton, and the chalets at Heacham South Beach). Other interesting passerine migrants, such as Ring Ouzel, may also occur.

Autumn seawatching can be productive at Hunstanton, with gales from a northern quarter being best (north-easterly gales may push numbers of birds into the Wash; some, like the skuas, may escape by moving overland, but many others will return past Hunstanton). Typical species include Fulmar, Gannet and Kittiwake, with Arctic Skua and, later in autumn, Great and sometimes Pomarine Skuas too. Long-tailed Skua and Sabine's Gull are rare, but Little Gull is regular in late autumn and winter (as often in calm conditions as in gales). Manx and Sooty Shearwaters and Leach's Petrel are also possible, together with a variety of grebes, divers, wildfowl and waders. In late autumn and early winter, such conditions may also produce Little Auk among the regular Common Guillemot and Razorbill.

Breeding birds include small numbers of Fulmars on the cliffs.

Access

Hunstanton lighthouse The B1161 north of Hunstanton passes close to the coast along the top of Hunstanton cliffs, before joining the A149. There are several shelters along the clifftop, and these are the best spots from which to seawatch. As the B1161 turns east to head inland, a short road leads to the lighthouse and a car park. In season, follow the footpath north-west from here to check for migrant passerines along the edge of the golf course and around Old Hunstanton.

Heacham North Beach Leave Hunstanton south on the A149 and 1.5 miles south of the roundabout turn west (opposite the B1454) into Heacham. Follow this road (ignoring side turns) for 1 mile and then fork right into Station Road to the North Beach. From the car park, explore the beach north towards Hunstanton, looking for Purple Sandpiper on the old groynes. This section of coast can also be accessed from the south end of Hunstanton promenade.

Heacham South Beach Leave the A149 at the southernmost turning to Heacham (signed Heacham Beaches) and, after 1 mile, turn left, following the road to the car park at South Beach. Walk south from here along the coast. At low water a vast expanse of mud is exposed, with a selection of common waders and large numbers of Common and Black-headed Gulls present. At high water grebes or seaducks may be present. There are a number of chalets with some areas of cover that are attractive to migrants, and immediately inland Heacham Harbour (a land-locked creek) can hold interesting ducks. It is possible to follow the coast south to Snettisham Coastal Park (see page 253).

FURTHER INFORMATION

Grid refs: Hunstanton lighthouse TF 676 420; Heacham North Beach TF 663 374; Heacham South Beach TF 662 368

171 SNETTISHAM (Norfolk)

OS Landranger 132

The Wash is internationally recognised as a major site for wintering wildfowl and waders and the RSPB reserve at Snettisham, on the east side of the Wash 8 miles north of King's Lynn, provides access to the wealth and diversity of its birds. The period from late summer into winter is the best time to visit.

Habitat

The RSPB reserve covers 1,820 ha, of which 1,670 ha are intertidal sand and mudflats (extending north towards Heacham South Beach), and 120 ha are saltmarsh. A series of brackish flooded lagoons (created by shingle extraction) and a shingle beach provide further habitats. The south pits have been managed to make them more attractive, and islands have been created which provide roosts for waders and nesting sites for terns.

Species

Waders peak during autumn and winter when up to 120,000 Knot have been counted. At low tide the waders are well scattered over the sand and mudflats, and large numbers are best seen at or just before high tide. Oystercatcher, Knot, Dunlin, Bar-tailed Godwit and Common Redshank are the commonest species, but Turnstone and Grey Plover are also present. Large numbers roost on the south pits, especially on the fortnightly higher 'spring' tides. Then, from the hides, the spectacle may include 70,000 Knots, 8,500 Dunlins and 6,000 Oystercatchers competing for space, all allowing close views. It is best to visit 2–3 hours each side of high water, and on a 'spring' tide (which coincide with the new and full moons), as these relatively high tides usually flood the mudflats and force the waders onto higher ground, including the pits at Snettisham, to roost. High water on 'spring' tides usually occurs late evening and early morning, which is fine in spring and autumn but in the winter months 'spring' tides peak during darkness and a late-morning high tide is the best compromise. (Contrariwise, the intervening neap tides leave the mudflats uncovered, and fewer waders move onto the pits to roost – the RSPB produces bird-watchers' tide tables to indicate the best dates each year – available from the Snettisham reserve office or the RSPB Titchwell shop). Wildfowl are also numerous, with especially large numbers of Shelducks, and up to 50,000 Pink-footed Geese roost on the mudflats November–February (best seen flighting from the roost at dawn). By day they can frequently be seen feeding on the fields east of the pits and sometimes they are joined by small numbers of Bewick's and Whooper Swans. Large numbers of Dark-bellied Brent Geese are present offshore, while small rafts of Common Scoters and Common Eiders may occasionally be seen on the open water at high

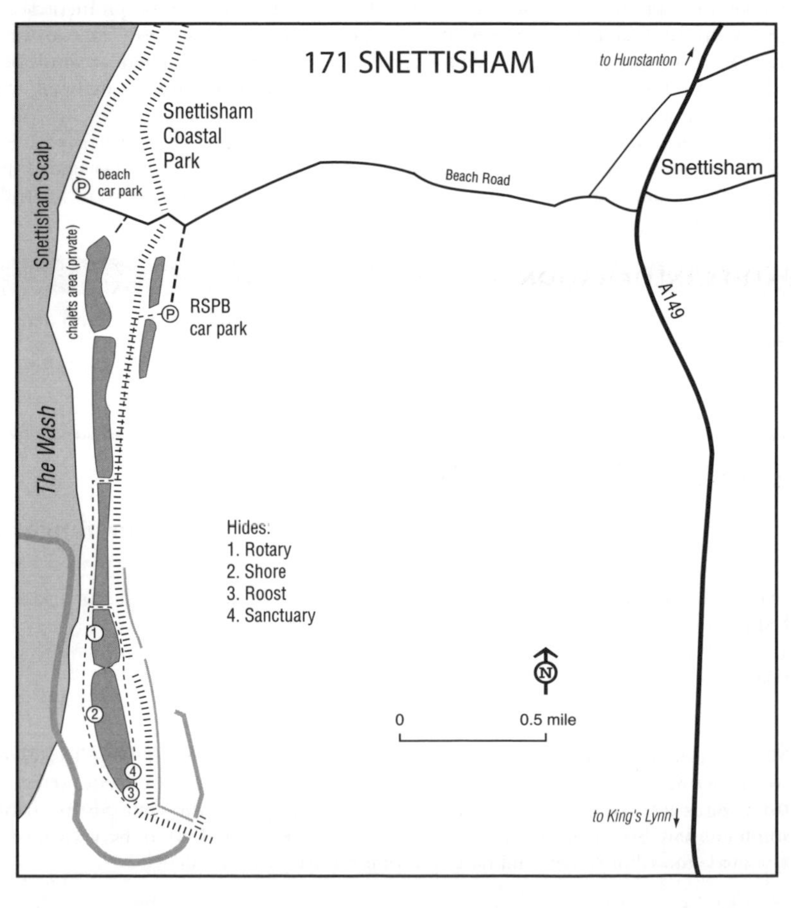

tide. Many ducks and waders use the pits in winter. Goldeneye and Red-breasted Merganser are invariably present, together with the commoner dabbling and diving ducks. Scaup is also quite regular but Smew and Long-tailed Duck are only occasional. Sometimes Slavonian or Red-necked Grebes or a diver will also visit the pits, or perhaps a Little Auk, and there is a roost of Cormorants and Little Egrets. On the beach Snow Bunting is sometimes present.

The variety of waders increases during migration and although there is little habitat here for freshwater species, Black-tailed Godwit, Greenshank, Spotted Redshank, Curlew Sandpiper and Little Stint may sometimes occur on the shore or in the wader roost on the pits. From July wader numbers begin to build up, and at this time large numbers of Knots and Bar-tailed Godwits may be seen with vestiges of their summer plumage. Among other migrants, Arctic Skua is frequent offshore in August–September, and Gannet and Kittiwake may be seen in strong onshore winds, but the area is not noted for good seawatching. Passerine migrants should be searched for in the bushes along the shore. In particular, the area north of the beach car park is graced by the occasional Ring Ouzel.

Midsummer is quiet. Common Terns and Black-headed Gulls breed on the pits together with a few Mediterranean Gulls and Avocets, with Oystercatchers and Ringed Plovers on the beach.

Access

RSPB Snettisham Leave the A149 signed to Snettisham Beach and follow the minor road for 1.5 miles, where the reserve car park is signed to the left (there is a 2.2 m height restriction). A well-marked footpath leads from the car park to the beach, through open countryside and scrub on the inland side of the pits. Disabled visitors may by arrangement drive to the first hide; two hides are wheelchair-accessible. The southernmost pits (with four public hides) are the best and lie 1.25 miles south of the car park. A circular trail accesses all the hides. The reserve is open at all times and a permit is not required. (Note that the properties at Snettisham beach are private and there is no public access to that area.)

Snettisham Coastal Park Instead of turning into the RSPB car park, follow the Beach Road to its end at the public beach car park. The Coastal Park lies to the north and is an area of sparse sandy grassland with areas of scrub and a series of small reed-fringed pools. It can be good for migrants in spring and, to a lesser extent, autumn.

FURTHER INFORMATION

Grid refs: RSPB Snettisham TF 651 329; Snettisham Coastal Park TF 647 335
RSPB Snettisham: tel: 01485 542689; email: snettisham@rspb.org.uk

172 WAYLAND WOOD (Norfolk)

OS Landranger 144

This stand of ancient woodland lies in south-west Norfolk, between Thetford and East Dereham, and is a site for Golden Pheasant. It is an NWT reserve.

Habitat

This is a block of mature oak, Ash and Hazel woodland, surrounded by open farmland.

Species

The speciality is Golden Pheasant. This introduced species can be hard to find as it prefers areas of dense cover. Winter or early spring, when males are calling, are the best periods to look and, surprisingly, mid-afternoon is often a good time to see a calling male. Otherwise, early morning and dusk are favoured times for a sighting. Other species include Woodcock.

Access

Leave Watton south on the A1075 and, after 1 mile, Wayland Wood is the first stretch of woodland alongside the road. The entrance to the small car park lies to the left of the road, 100 yards beyond the beginning of the wood. A network of paths allows access to the wood.

FURTHER INFORMATION

Grid ref: TL 923 996
Norfolk Wildlife Trust: tel. 01603 625540; web: www.norfolkwildlifetrust.org.uk

173–180 BRECKLAND (Norfolk/Suffolk)

OS Landranger 143 and 144

Breckland is a unique area of sandy heathland and conifer plantations centred around Thetford. Famed for several scarce breeding species, it is best in spring or early summer.

Habitat

Breckland (or The Brecks) lies on an area of light, sandy soil between the Fens to the west (now almost entirely drained) and heavier boulder clays to the east. The Little Ouse River bisects the area, but there is little standing water away from the mysterious ephemeral Breckland meres. Cleared for agriculture in prehistoric times, The Brecks quickly developed heathland on the poorer soils, intensively grazed by sheep and rabbits. Periodically, however, when demand was high, areas of heath would be ploughed and sown for a season; these were the original 'Breck'. Following the First World War, however, vast areas were planted with conifers, mainly Corsican or Scots Pines, and now only small relicts of heath remain. Forestry is now the dominant land-use, and as the plantations have matured, been felled and then replanted, they have provided habitats for a range of interesting species.

Species

Breckland birds can be divided into two groups: the traditional heathland specialities and birds of the conifer plantations. Of the first group, Stone-curlew is the most important. It is not, however, confined to relict grassy heaths, for it has adapted to arable fields, where it can breed successfully only with the support of conservation-minded farmers. Stone-curlews can be seen reliably at the NWT Weeting Heath reserve, and visiting birders need look no further for the species. They are in residence mid-March to September. A few pairs of Wheatears remain on the sandy heath (the species was once relatively common) and areas of heather and gorse may possibly also hold Nightjar.

The very extensive area of conifer plantations has now matured and is being felled and replanted, and each stage in their growth attracts a particular suite of species. The mature stands hold breeding Goshawks, with four or five pairs in Breckland. They are, however, often elusive, and best seen when displaying on clear sunny days in late February–April, usually at 9.30am–11am and 2pm–3pm. The key to success is to choose the right weather and a locality which offers a panoramic view over a wide area. The plantations also hold Hobby and large numbers of Sparrowhawks, but Breckland Goshawks, descended from imported falconers' birds, are usually rather large and pale (indicating a north European origin). While they are relatively easy to distinguish from Sparrowhawk, accurate identification calls for care and caution. Mature plantations also attract Common Crossbill, but the population of this species fluctuates dramatically and while they can be everywhere in some seasons, in other years they may be impossible to find. Listen for the dry *chip, chip* flight calls and in dry weather any water source, such as pools and puddles, is worth checking, as crossbills are thirsty birds. Siskin is a recent colonist, and the mature plantations also hold small numbers of Long-eared Owls, but they are seldom seen. The best chance is to listen for the calls at night in June/July when the juveniles are particularly vocal. Another recent colonist is Firecrest, with around 85 pairs in Breckland in 2008. It can be regularly seen at Brandon Country Park, High Lodge and around Lynford Arboretum. Stands of Douglas Fir are favoured and the birds' presence is given away by their distinctive ascending song, which can be heard throughout the day from April to July.

Once clear-felled, the large blocks of very young conifers attract Woodlarks; these increased tremendously up until 1999 with around 600 pairs but they have subsequently decreased to around 400 pairs. They prefer felled or recently planted areas of conifers and are easiest to see in late March–late April when in song flight; Woodlarks are largely absent in midwinter. Nightjars also favour young plantations, with around 350 territories recorded in 2004, as well as heaths, but are late-arriving summer visitors, best looked for at dusk on warm, still, summer evenings. The areas of clear-fell also attract Turtle Dove, Cuckoo, Tree Pipit, Stonechat (another recent colonist), Willow Warbler, Yellowhammer and occasionally also Grasshopper Warbler (which also favours the few damper heaths). Hawfinch prefers hornbeams, if available, and although widespread and resident, can be very unobtrusive; it is easiest to find in winter when it forms loose flocks. Some of the other interesting breeding species include Woodcock, Curlew, Lesser Spotted Woodpecker, and a few pairs of Nightingales and Common Redstarts, most of the latter favour deciduous woodland, which is usually found along watercourses and along the edges of conifer plantations. Recent colonists moving into the area include Little Egret, Goosander (the lakes and gravel pits are also good for this species in winter) and Cetti's Warbler, but Golden Oriole has almost vanished.

Breckland is generally quiet in winter, although Goshawk and Common Crossbill may still be found, together with redpolls, and there is a chance of a Hen Harrier or Great Grey Shrike on areas of heath or clear-fell, but spring comes early, with both Goshawk and Woodlark best looked for from March onwards.

Access

There are many interesting areas worthy of exploration, and many species are most easily found by looking for conifer plantations in the right stage of growth (with many of the best areas in the triangle formed by Mundford, Brandon and Thetford). The following sites are particularly notable.

173 LYNFORD ARBORETUM (Norfolk)

The arboretum is favoured by Hawfinch, especially in winter and early spring; try the trees at the west end of the arboretum or the large beech trees in the field beyond the lake. Common Crossbills are frequently present, and often also Firecrest. Turn east off the A1065, 0.3 miles north of the Mundford roundabout, on a minor road signed Lynford Hall and Arboretum. After 1 mile (just past the hotel) turn left (north) into the car park. Cross the road to the disabled car park, from where a number of footpaths permit access to the arboretum. The flooded gravel pits to the north of the road, although relatively disturbed (and thus best in the early morning), are attractive to wildfowl, including Goosander, and sometimes attract passage waders.

174 WEETING HEATH NNR (Norfolk)

Weeting Heath covers 138 ha and is owned and managed by the NWT. The reserve holds several pairs of Stone-curlews, which can be viewed from two hides; indeed, Weeting is *the* site for this elusive bird. Woodlarks are also present. Leave Brandon north on the A1065 towards Swaffham and, immediately after the level crossing, turn left on the minor road towards Weeting. After 1.25 miles, turn left (west) at the village green in Weeting, next to the phone box and post office, towards Hockwold cum Wilton. After a further mile the reserve car park lies to the left, concealed within a belt of pines. The reserve is open April–September, 7am to dusk, with access restricted to the tracks through the pine belt to the two hides. Stone-curlew is easiest to see in early spring; as the season progresses they may only be obvious in early morning and again in late afternoon and evening, and they can become very elusive when incubating in May. Late August and September can also be good times to see Stone-curlews; they gather to roost and often show close to the hides, and heat haze is less of an issue.

175 SANTON DOWNHAM (Suffolk)

This small village lies east of Brandon, off the B1107. The area of the village is good for Firecrest, while the dead trees between the bridge and the level crossing as well as the trees along the railway to west of the road are good for Lesser Spotted Woodpecker (probably the most reliable site in the Brecks). There are also redpolls in the winter.

176 MAYDAY FARM (Suffolk)

Leave Brandon south on the B1106 and, after 2 miles, turn right (west) into the FC car park at Mayday Farm. From here, walk directly south-west along the Goshawk Nature Trail, which follows the main ride (Shakers Road) for 0.75 miles until a large clearing comes into view to the left (south) of the track and a crossroads is reached. This is a good spot from which to scan for displaying Goshawks (as indeed is any area of clear-fell in Thetford Forest with a good view of the sky!). There are also Nightjars, Tree Pipits, Woodlarks and Common Crossbills in this area. It is possible to either return directly to the car park or to turn left, and then left and left again, to follow the nature trail on a circuit around the large clearing and rejoin Shakers Road 0.5 miles from the car park.

177 THETFORD (Norfolk)

At least one pair of Goosanders breed on the Little Ouse near Thetford and can often be seen on the river in Thetford itself, along with good numbers of Grey Wagtails. The woods along the river between Thetford and Brandon have a few pairs of Willow Tits and Lesser Spotted Woodpeckers – both scarce birds in Breckland. Thetford is the home of the British Trust for Ornithology (BTO) and the many active ringers based there mean that there are relatively large numbers of colour-ringed birds in the area. Details of any sightings can be reported to the BTO at www.bto.org.

178 THETFORD WARREN (Norfolk)

The extensive plantations in this area hold Goshawk, as well as Woodlark, Firecrest, Lesser Redpoll and sometimes Common Crossbill. The car park is at Thetford Warren Lodge, south of the B1107, 1 mile north-west of the A11 Thetford bypass, and 4 miles south-east of Brandon town centre. From the car park three marked trails permit exploration of Risbeth Wood.

179 EAST WRETHAM HEATH (Norfolk)

This NWT reserve comprises 143 ha, with areas of grassy heath, as well as some woodland, including hornbeam and ancient Scots Pine, and two meres: Langmere and Ringmere. Fed by groundwater, these fluctuate mysteriously and sometimes completely dry up; in the right conditions, there may be a few passage waders, such as Green Sandpipers. They may attract wildfowl and occasionally passage waders. Otherwise, the reserve holds one or two pairs of Common Redstarts, while wintering Hawfinch are occasionally seen in the mature hornbeams around Ringmere. Leave the A11 at the north-east end of the Thetford bypass on the A1075 towards Watton, and the reserve car park lies left (west) of the road after 2.25 miles. There is a signed nature trail and hide.

180 BARNHAM CROSS COMMON (Norfolk)

This is a site for wintering Hawfinch. Leaving Thetford southwards on the A134 towards Bury St Edmunds, park in a lay-by on the right about 150 yards after the left-hand turn to The Nunnery. Cross the road and head east straight across the common (short walk). Hawfinches favour the eastern side of the common – check the hawthorn hedges – and can be seen at any time of day, but are not present in the summer. The bench by the pumping station is a good place to view the Nunnery Flood, which can be good for passage waders.

FURTHER INFORMATION

Grid refs: Lynford Arboretum TL 823 943; Weeting Heath TL 758 879; Mayday Farm TL 795 834; Santon Downham TL 828 873; Thetford TL 873 826; Thetford Warren TL 842 841; East Wretham Heath TL 911 881; Barnham Cross Common TL 866 817
Norfolk Wildlife Trust: tel: 01603 625540; web: www.norfolkwildlifetrust.org.uk; NWT Weeting Heath: tel: 01842 827615.

181 LAKENHEATH FEN (Suffolk and Norfolk)

OS Landranger 143

In a bold step the RSPB created a completely new 200-ha wetland in the late 1990s on the very edge of the fenland plain, on what was then arable farmland (mostly carrot fields). The aim was to attract up to eight territory-holding male Bitterns and, although they have not yet bred, Bitterns are resident, while other star species are Golden Oriole and Common Crane.

Habitat

Lakenheath Fen comprises a mixture of wetland and woodland south of the Little Ouse. The wetland consists of a mosaic of washland (rough grazing that floods in the winter), wet reedbeds, ungrazed fen and wet grassland, while the poplar plantations beside the river have long been known for their breeding Golden Orioles.

Species

Breeding birds include several Marsh Harriers, Bearded Tit and several hundred pairs of Reed and Sedge Warblers and Reed Buntings. Two pairs of Common Cranes nested in 2007 and 2009 (successfully in the latter year – the first breeding in the Fens for 400 years!). Bitterns are seen throughout the year, although they are usually elusive, and Hobby is a regular visitor in summer. A handful of Golden Orioles breed in the remnant poplar woods, together with Blackcap and Garden Warbler. Early mornings and evenings are the best time for a sighting of Golden Oriole, which are present from mid-May until late summer, and the best tactic is to find a point where there is line of sight down a ride or gap in the trees and wait, scanning carefully, for orioles to fly across. Little Egret, Barn Owl, Kingfisher and Cetti's Warbler are resident in the area.

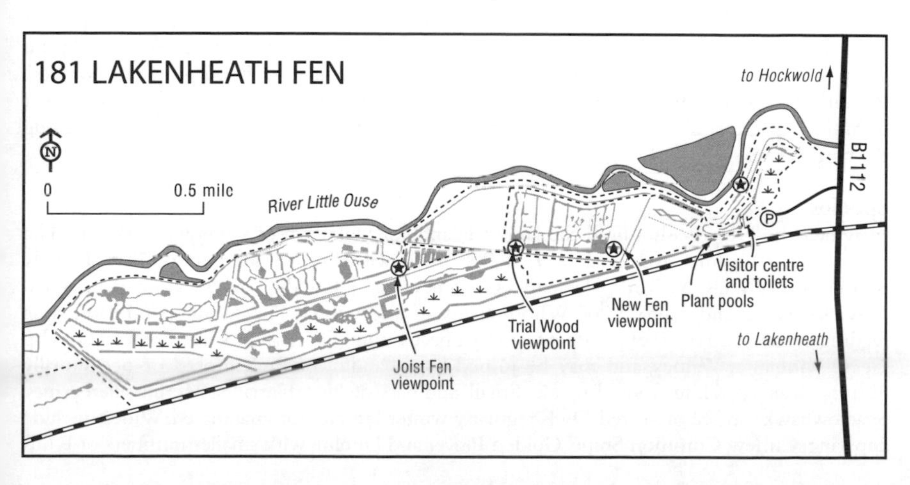

On passage small numbers of waders may be present, and in winter the fen attracts a variety of raptors, with Hen and Marsh Harriers, Peregrine, Merlin, Common Buzzard and Sparrowhawk all possible. Good numbers of wildfowl winter, including Teal, Gadwall, Wigeon, Shoveler and a few Pintails, with odd records of Whooper Swan, while gulls may include Yellow-legged. Water Pipits winter on the wet washes.

Access

Take the B1112 northwards out of Lakenheath towards Hockwold and, after 2 miles, cross the railway at the level crossing at Lakenheath Station. After a further 200 yards turn west into the reserve (coming from Hockwold the reserve lies 200 yards past the bridge over the Little Ouse). The reserve is open dawn–dusk all year, with the reserve centre open daily April–September, and at weekends in the winter. There are two trails, a 3.5-mile circular path and a 3.4-mile 'out-and-back' trail) giving access to three viewpoints, as well as a 0.7-mile circular path near the reserve entrance. Some paths are wheelchair-accessible.

FURTHER INFORMATION

Grid ref: TL 722 864
RSPB Lakenheath: tel: 01842 863400; email: lakenheath@rspb.org.uk

182 OUSE WASHES (Norfolk/Cambridgeshire)

OS Landranger 143

The unique wet meadows of the Ouse Washes, in the heart of the Fens, are renowned as a refuge for wildfowl in winter and for several rare breeding species in summer. Around three-quarters of the total area is now owned and managed by the RSPB, Wildlife Trust for Cambridgeshire and WWT. A visit is profitable at any time of year, but winter is the most spectacular season.

Habitat

The Ouse Washes lie between two straight, almost parallel artificial rivers (or drains), the Old and New Bedford Rivers. The Washes were created in the 17th century as part of the Earl of Bedford's ambitious scheme to drain large parts of the Fens, and were designed to flood as necessary in winter to control water levels elsewhere. Although little more than half a mile wide, the Washes extend for over 20 miles between Earith in Cambridgeshire and Denver Sluice near Downham Market in Norfolk. The area is permanent pasture, grazed in summer by cattle and sheep, and regularly flooded in winter; numerous ditches divide the flood plain into 'washes'. Recent management has created a series of permanently flooded lagoons and scrapes. Apart from the high banks alongside the drains, the area is entirely flat with a few trees growing along the barrier banks.

Species

In winter the Ouse Washes hold the largest inland concentration of wildfowl in Britain. The majority are Wigeons, averaging over 35,000 birds, with smaller numbers of Teals, Pintails, Shovelers, Gadwalls, Mallards, Pochards, Goldeneyes and Ruddy Ducks. More than 4,500 Bewick's Swans and up to 3,000 Whooper Swans also winter. The swans spend most time feeding in harvested fields of potatoes and sugar beet around the washes. A flock of feral Greylag Geese winters at Welney, and may be joined by a handful of Pink-footed or occasionally Tundra Bean or White-fronted Geese. Small numbers of Hen Harriers, Merlins, Peregrines, Sparrowhawks and Short-eared Owls regularly winter but are not guaranteed. Waders include Lapwing, Curlew, Common Snipe, Golden Plover and Dunlin, with smaller numbers of Ruffs and Common Redshanks. Jack Snipe and Water Rail winter but are usually elusive. The flooded washes also provide a roosting site for thousands of gulls, largely Black-headed, as well as Cormorants (which roost on electricity cables), and there are wintering Water and Rock Pipits, Stonechat, Brambling and occasionally Bearded Tit. The RSPB visitor centre has a feeding Station, which may attract Tree Sparrows.

A few pairs of Black-tailed Godwits breed. Their spectacular display flights are best observed in spring and early summer. Ruff breeds annually but in variable numbers. Males lek for a brief period in early spring, the majority moving on to leave the females to raise their young. As the season advances, the females become quite inconspicuous. Lapwing, Common Snipe and Common Redshank are common breeders, with a few pairs of Little Ringed Plovers and Avocets. Garganey is among the nine species of breeding ducks, but is unobtrusive in summer and best looked for in spring. Other breeding ducks include Shoveler and Gadwall. Common Terns breed, Marsh Harrier is frequently present in summer and a pair of Short-eared Owls usually nests. Black Tern has occasionally bred and Little Gull attempted to do so once. Spotted Crake is heard at night and probably breeds, but is unlikely to be seen. Reed Warbler and Yellow Wagtail breed in large numbers. Residents include Great Crested and Little Grebes, Stock and Turtle Doves, Barn and Little Owls, Kingfisher, Corn Bunting and Tree Sparrow.

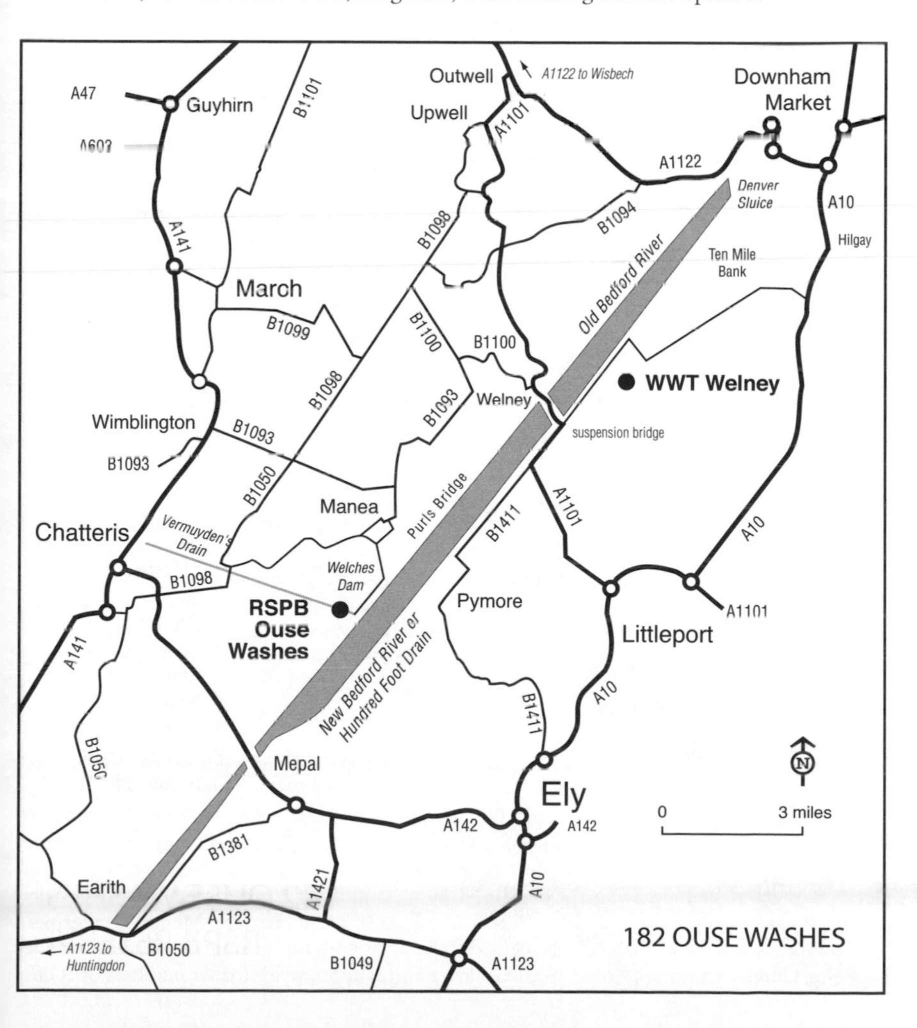

Spring can be a particularly rewarding time to visit. Several of the breeding species may be easier to see, their numbers being augmented by passage birds (e.g. Garganey, Ruff and Black-tailed Godwit, with up to 2,000 Icelandic Black-tailed Godwit in March–April). Marsh Harrier and Black Tern are regular, as well as a variety of commoner migrants. Waders are more in evidence in autumn, and most of the common species are present from July.

Access

Welney WWT lies on the east bank of the Washes, and is thus best in the morning (and again in the mid-afternoon/evening, when the swans are fed), while the RSPB Ouse Washes reserve faces east and is best in the afternoon. Note that when the Washes are fully flooded, the A1101 is closed where it crosses the Washes and thus a long detour is necessary if travelling between the two reserves.

RSPB Ouse Washes This reserve is jointly owned by the RSPB and the Wildlife Trust for Cambridgeshire. From the A141 at Chatteris take the B1098 signed Upwell and Downham Market. The road turns sharp left after 2 miles and after a few hundred yards crosses Vermuyden's Drain. Immediately after crossing the drain turn right to Manea (signed 4 miles). At the beginning of the village a right turn is signed Purls Bridge, Welches Dam and Nature Reserve. At Purls Bridge the road follows the Old Bedford River to Welches Dam, the headquarters of the reserve, where there is a car park. (Alternatively, turn off the A141 at

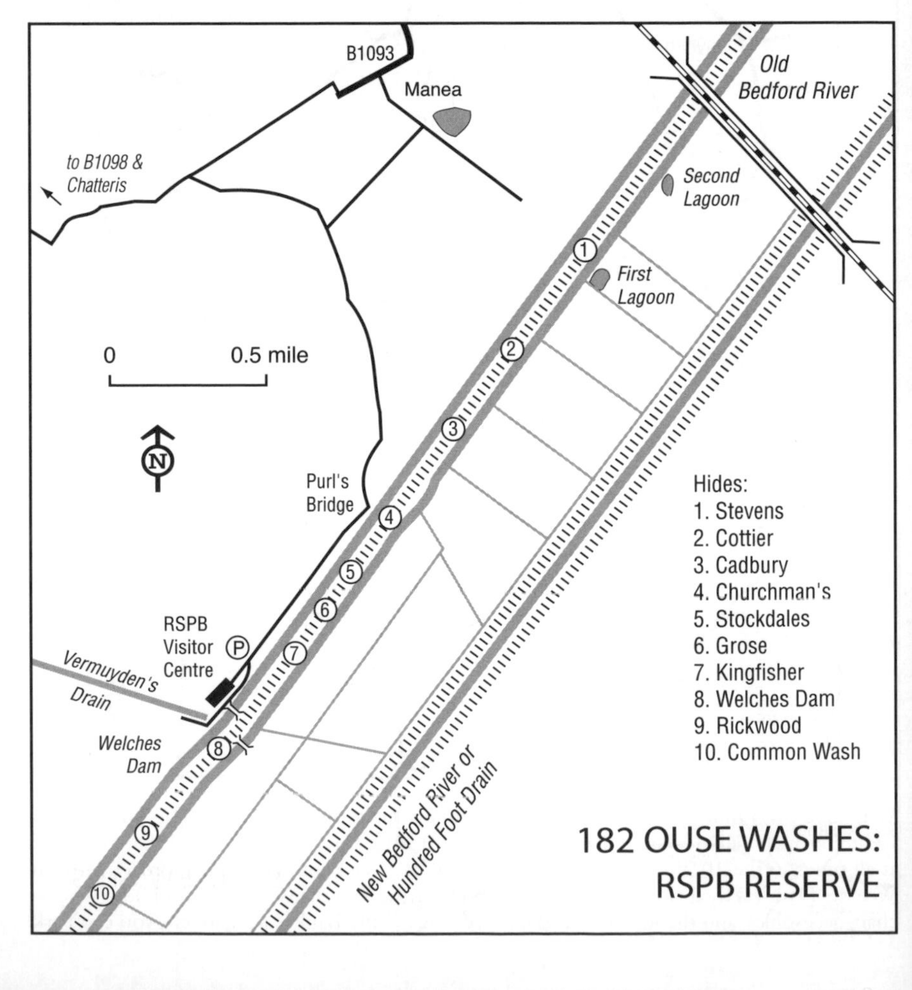

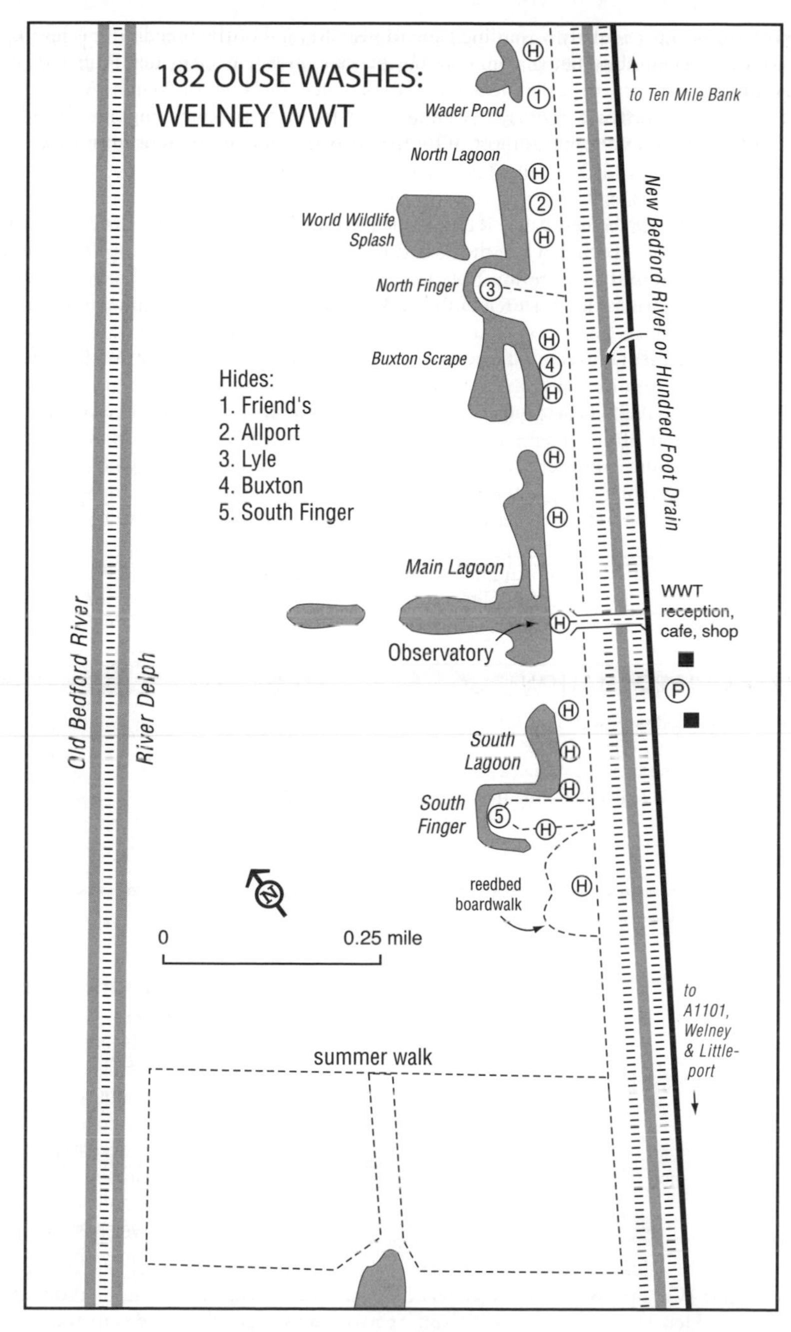

Wimblington, 4 miles north of Chatteris, and follow the B1093 to Manea). The Information Centre is open daily (except Christmas Day and Boxing Day) 9am–5pm, and there is access at all times to the 10 hides along 1.8 miles of the Barrier Bank. They are approached from Welches Dam via the bridge over the river and boardwalk to the Welches Dam Hide (wheelchair-accessible), and thence via the Haul Road below the Bank. Alternatively, you can park at Purls Bridge and use the floating bridge, a short distance north-east of the point where the road from Manea reaches the river, to access the northern hides, which are often the most productive.

WWT Welney Wildfowl Refuge Leaving the A10 in Littleport, take the A1101 west towards Outwell and Wisbech. After 4 miles the road reaches the New Bedford River and runs alongside it for a further mile. The main road then crosses the river over a suspension bridge to Welney village. Where the road turns north-west over the bridge, continue straight on along a minor road (signed to the Refuge and Ten Mile Bank) alongside the New Bedford River. After 1.5 miles the refuge's car park and reception centre lie to the right of the road; all visitors should report to reception on arrival. (Coming from the north, leave the A10 just south of Hilgay on minor roads to Ten Mile Bank and the refuge; both routes are signed from the A10.) The entrance to the observatory and hides is via a footbridge opposite the car park. The refuge is open daily (except Christmas Day) 10am–5pm in winter, 9.30am–5pm March–October. The observatory, a spacious hide overlooking a large lagoon, is heated in winter, and a wide variety of wildfowl can be seen at close quarters. In winter, the wild swans and ducks are fed daily in front of the observatory, with the main feed at 3.30pm. Numbers of Whooper Swans arrive to feed (making this the best place to see this species), but few, if any, Bewick's Swan come to feed, preferring to remain in the fields. The refuge remains open until 8pm, Thursday–Sunday, November–February, when the wild swans are fed by floodlight (phone or check the website for up-to-date opening and swan-feeding times). There are five other large hides situated along the bank north and south of the observatory, and also several small (2–6 seat) hides, but access to all but the observatory may be impossible at times, due to flooding. From May to August a 2-mile walk across the Washes is possible.

FURTHER INFORMATION

Grid refs/postcodes: RSPB Ouse Washes TL 471 860 / PE15 0NF; WWT Welney Wildfowl Refuge TL 546 944 / PE14 9TN
RSPB Ouse Washes: tel: 01354 680212; email: ouse.washes@rspb.org.uk
WWT Welney: tel: 01353 860711; email: info.welney@wwt.org.uk

183 NENE WASHES (Cambridgeshire)

OS Landranger 142

Lying just 6 miles east of Peterborough, this area of the Nene Valley attracts significant numbers of wintering wildfowl.

Habitat

The reserve comprises an extensive area of meadows alongside the River Nene that floods during periods of high water levels, usually in winter.

Species

Large numbers of wildfowl winter on the reserve when it is flooded in winter, including Bewick's and sometimes Whooper Swans, Wigeon, Teal, Shoveler and Pintail. Such concentrations of prey species attract Hen Harrier, Peregrine, Merlin, Sparrowhawk, Barn Owl and sometimes Short-eared Owl.

Breeders include Shoveler, Gadwall, Garganey, Shelduck, Lapwing, Black-tailed Godwit and Common Snipe (best when displaying on warm spring evenings) and sometimes Ruff. Spotted Crakes are heard calling at night most years and presumably breed, but are essentially impossible to see. Both Marsh Harrier and Hobby are regular visitors in summer.

Access

Access to Nene Washes is possible at two points. There are no visitor facilities.

The B1040 between Whittlesey and the A47 at Thorney This road crosses the Washes at the Dog-in-a-Doublet bridge. There are good views of the Washes from the bridge and just north of the bridge it is worth taking the minor road westwards for a short distance and checking the River Nene at the sluice. Continuing across the bridge, there is a track to the east by a row of trees at 90 degrees to the road. Follow this pot-holed track as it doubles back on itself. Parking at the main drove, it is possible to walk east along the drove for up to 3 miles.

Eldernell (South Barrier Bank) Leave the A60, immediately east of Coates, north along Eldernell Lane and park in the car park at the end of the road by the barrier bank. A public footpath runs both east and west from here along the bank (flooding permitting), affording good views, but it is best to walk west: the pool and reedbed to the left sometimes holds a wintering Bittern while the wood to the right contains a large heronry. As soon as you pass the wood, drop down out of sight below the bank to avoid disturbing the wildfowl (this is the best approach to the RSPB reserve in winter as walking down the main drive from the B1040 inevitably disturbs the ducks).

FURTHER INFORMATION

Grid ref: The B1040 between Whittlesey and the A47 at Thorney TL 274 993
RSPB Nene Washes: tel: 01733 205140

184 GRAFHAM WATER (Cambridgeshire)

OS Landranger 153

Two arms of this large reservoir comprise a 114 ha reserve, managed by the Wildlife Trust for Bedfordshire, Cambridgeshire, Northamptonshire and Peterborough. Large numbers of grebes and wildfowl winter and passage brings a variety of waders and terns.

Habitat

The reservoir was filled in 1964 and now covers 635 ha at the highest water levels, with nearly 10 miles of shoreline. There are a number of sheltered creeks and the reservoir is surrounded by small stands of deciduous woodland, an extensive mixed plantation, rough grassland with old hedgerows, and arable fields. Much of the open water is disturbed by sailing and fishing, but the sanctuary at the west end provides an undisturbed area.

Species

In winter up to 800 Great Crested Grebes are present, and are occasionally joined by one of the rarer grebes or divers. Wintering ducks include numbers of Teal, Wigeon, Goldeneye and Pochard. Goosanders use the reservoir as a roost and arrive late afternoon from the direction of the river and gravel pits to the east. Small numbers of Shovelers, Gadwalls and Pintails can also be seen, but Long-tailed Duck, Common and Velvet Scoters and Smew are rare visitors. A few Dunlins and Jack Snipes may winter, with Golden Plover on the surrounding farmland. The gull roost holds up to 30,000 birds, and Mediterranean, Iceland and Glaucous Gulls are all possible with diligent searching.

On spring passage Turnstone, Green and Common Sandpipers, Greenshank and Dunlin are possible, and in autumn these may be joined by Spotted Redshank, Ruff, Little Stint and Curlew and Wood Sandpipers. Common and Black Terns are regular on both passages, but Arctic Tern is less frequent, occurring mainly in spring. Small numbers of Little Gulls are also seen at these times, and Osprey is fairly regular.

Breeders include Ringed and Little Ringed Plovers, Lapwing, Common Redshank, Turtle Dove, Yellow Wagtail, Nightingale and Grasshopper, Reed and Sedge Warblers.

Access

Leave the A1 at the Buckden roundabout onto the B661, which passes the dam and then runs fairly close to the south shore into the village of West Perry. The reservoir can be viewed from two car parks along this road.

Plummers car park Lying to the east of Perry, this is a useful viewpoint.

Mander car park Situated on the west side of Perry, this accesses the Wildlife Trust reserve. A wildlife cabin here acts as an information centre and from the car park nature trails around the west shore of the reservoir lead to five hides.

North-east shore This can be viewed from a minor road that leaves the B661 before the dam and leads to Grafham village. There are two more car parks along this road: Marlow and the very small Hill Farm car park.

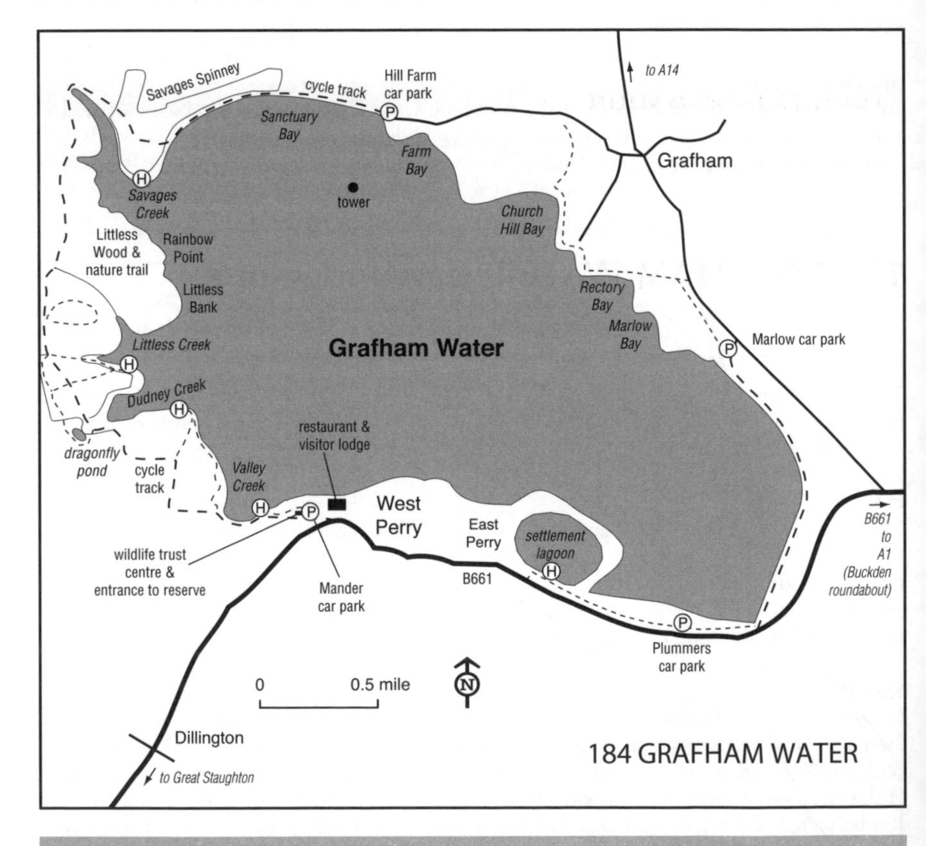

FURTHER INFORMATION

Grid refs: Plummers car park TL 162 664; Mander car park TL 142 672; north-east shore TL 165 682

Wildlife Trust for Bedfordshire, Cambridgeshire, Northamptonshire and Peterborough, Grafham/Buckden Education Centre: tel: 01480 811075; email: grafham@wildlifebcnp.org

185 PAXTON PITS (Cambridgeshire)

OS Landranger 153

Situated immediately north of St Neots and just off the A1, this site has a wide range of breeding species (including Cormorant and Nightingale), as well as a variety of wintering waterfowl and passage gulls, terns and (to a lesser extent) waders.

Habitat

The reserve is centred on a series of flooded, worked-out gravel pits, with areas of both deep and shallow water and several islands. The pits are surrounded by emergent vegetation, reedbeds and mature scrub, within an area of meadows, hedgerows, woodland and the former gardens of the now-demolished Wray House. To the north, working gravel pits still operate. The area is an LNR, part-owned and managed by Huntingdonshire District Council.

Species

Approximately 70 species breed annually, including Great Crested and Little Grebes, 100–180 pairs of Cormorants (best viewed from Hayden Hide), six to eight pairs of Grey Herons, Sparrowhawk, Ringed and Little Ringed Plovers, Common Redshank, Common Tern, Turtle Dove, Kingfisher, all three woodpeckers, Yellow Wagtail, Grasshopper, Reed, Sedge and Garden Warblers, Lesser Whitethroat and Corn Bunting. Hobby is a regular non-breeding visitor in

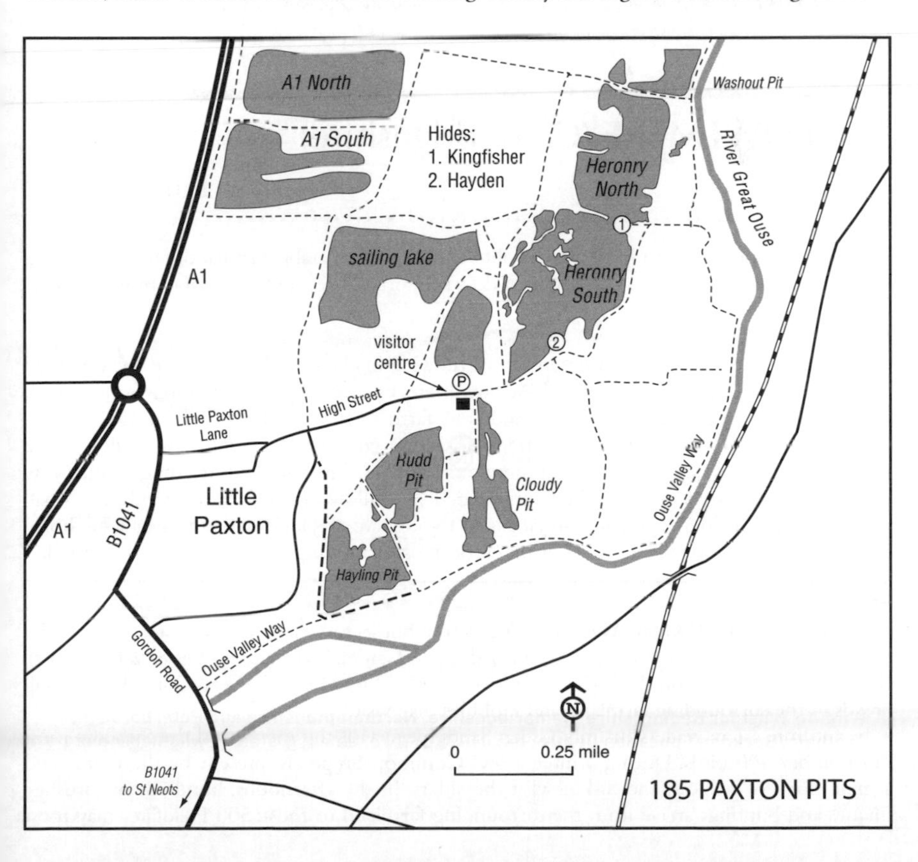

summer. Up to 25 pairs of Nightingales may be present, and the best time to hear them is at dawn or dusk, in scrub along the Haul Road or the Heronry Trail.

On passage Arctic Tern may be joined by Black Tern and Little Gull, and a variety of waders occurs (although in common with most gravel pit sites, there is only limited habitat for waders), including Dunlin, Ruff, and Common and Green Sandpipers. Other visitors can include Black-necked Grebe, Osprey, Common Buzzard and Red Kite.

In winter large numbers of Cormorants roost on the pits; around 400 are normally present, but there was a record-breaking total of 1,200 during cold weather in January 1997. Another notable species that roosts on the reserve is Stock Dove, with over 1,000 sometimes present. Wintering wildfowl include large numbers of Wigeons and Coots, as well as Teal, Gadwall, Pochard, Goldeneye, a handful of Goosanders and often also Smew, while the rarer grebes, especially Slavonian and Red-necked, may visit, as may Great Northern or Black-throated Divers.

Access

Leave the A1 on the B1041 into Little Paxton (2.75 miles south of the Buckden roundabout/5.25 miles north of the Black Cat roundabout). After 200 yards turn east into Little Paxton Lane and follow this, signed to the reserve, for 0.7 miles to the car park and visitor centre. The three trails and two other paths, together with the hides, in the reserve are open at all times, and the visitor centre is open most weekends.

FURTHER INFORMATION

Grid ref/postcode: TL 196 629 / PE19 6ET
Paxton Pits Nature Reserve, Visitor Centre: tel: 01480 406795; web: www.paxton-pits.org.uk

186 FOWLMERE (Cambridgeshire)

OS Landranger 154

This RSPB reserve forms a haven for wildlife within a sea of arable farmland. Though holding no concentrations of wildfowl or waders, it does support an excellent range of common species.

Habitat

Once part of the southernmost extension of the Cambridgeshire Fens, the area was drained long ago and given over to farmland. At Fowlmere, however, upwelling water from a chalk aquifer led to the establishment of a watercress farm in the late 19th century. Once cress-growing ceased in the 1970s, the wetter areas were colonised by reeds and willows. The RSPB has raised water levels, created meres and ditches, and removed invading scrub. Now the reserve is centred on a reedbed surrounded by willow and hawthorn scrub and small stands of ash and alder woodland. Adjacent to the reserve are areas of rough grassland, horse paddocks and a poplar plantation, but much of the surrounding land is intensively farmed.

Species

Breeders include Little Grebe, Water Rail (numerous but secretive), Turtle Dove, and nine species of warblers (including Grasshopper, Reed, Sedge and Garden Warblers and Lesser Whitethroat). Kingfisher rarely nests on the reserve, but is often seen throughout the year. Corn Bunting still breeds in the surrounding farmland and Hobbies may visit.

In autumn, a few waders (mainly Green Sandpipers) visit the meres, and the Swallow roost can number 400 birds. During winter, a few Common Snipe are present by the meres, and a small flock of redpolls and Siskins visit the alders. In the afternoons, hundreds of thrushes, finches and buntings arrive from the surrounding farmland to roost: 500 Fieldfares may roost in the bushes with fewer Redwings and other thrushes; Reed and Corn Buntings may both number 200 in the reeds, and Brambling often joins the Chaffinch roost in the scrub. This concentration of prey attracts raptors, especially Sparrowhawk, and also sometimes Merlin or Hen Harrier; Long-eared Owl may be seen at dusk. Bearded Tit and even Bittern are occasional winter visitors.

Access

Between Royston and the M11, leave the A10 south-east at Shepreth on the minor road to Fowlmere. After 1 mile turn right (south-west, signed) and the reserve entrance lies on the south side of the road after 0.6 miles. The reserve is open at all times, with access along the 1.9-mile marked trail to three hides, one of which (Drewer) is adapted for wheelchair users.

FURTHER INFORMATION

Grid ref: TL 406 461
RSPB Fowlmere: tel: 01763 208978; email: fowlmere@rspb.org.uk

187 WICKEN FEN (Cambridgeshire)

OS Landranger 154

Eleven miles north-east of Cambridge, Wicken Fen is among the few remaining examples of the once extensive fens of the Great Level. Its main interest is in winter for waterfowl and a Hen Harrier roost. Wicken Fen is the oldest nature reserve in Britain, and is owned by the National Trust. It is the lynchpin of the Trust's 'Wicken Fen Vision', a project to create a new nature reserve covering 5,600 ha (22 square miles) comprising wetlands, wet and dry grassland and woodland between Cambridge and Wicken Fen, and the Trust has already extended its land holdings to 930 ha.

Habitat

The reserve consists of four distinct areas: Sedge Fen, St Edmund's Fen, Adventurers' Fen and Baker's Fen. Sedge and St Edmund's Fens have never been drained or cultivated, and sedge and reed are still cut in rotation in the traditional manner, but on St Edmund's Fen natural succession has been permitted to proceed unchecked, and the area has developed into an area of carr with some mature trees and only a few small patches of reeds; similar habitats can also be found in parts of Sedge Fen. Adventurers' and Baker's Fens lie south of Wicken and Monk's Lodes and have undergone periods of drainage and cultivation (though currently wetlands again), but nevertheless hold much of the ornithological interest of Wicken, while the Mere is attractive to wintering wildfowl. Baker's Fen was purchased in 1993 and a programme to re-establish grazing marsh from arable farmland has resulted in an area that is particularly interesting in winter and passage periods. Recent management throughout Wicken Fen has concentrated on re-growing and restoring reedbeds in the hope of attracting Bitterns to breed again.

Species

Marsh Harrier, Water Rail, Bearded Tit and Cetti's Warbler are resident, and these are joined in summer by many Reed and Sedge Warblers and smaller numbers of Garden Warblers and Lesser Whitethroats while Grasshopper Warbler nests in the scrubbier sedge areas or reedbeds, especially on Sedge Fen. Areas of carr and woodland hold Woodcock (roding from early March), Turtle Dove and Willow Tit. Sporadically, Spotted Crake has been heard calling at night and Garganey is an irregular summer visitor. Other breeders include Lapwing, Common Snipe and Common Redshank. Hobby is a regular non-breeding summer visitor, and Barn

Owl is occasionally seen throughout the year.

In winter, the majority of waterfowl gather on or around the Mere and on Baker's Fen; mainly Wigeon and Mallard with smaller numbers of Gadwalls, Teals, Shovelers and Pochards, and occasionally Pintail or Goosander. Canada Goose is usually present on the drier parts of Adventurers' Fen and occasionally a few feral Greylag or White-fronted Geese are seen. Cormorants winter and roost on the island in the Mere. Bittern and Short-eared Owl are present most winters, usually just one or two of each, as well as variable numbers of Bearded Tits. All are best viewed from the Tower Hide or the hides on Adventurers' Fen. The greatest spectacle in winter is the Hen Harrier roost. Numbering up to ten birds, males frequently predominate. They begin to arrive an hour before dusk, usually from the west side of the Tower Hide, which provides an excellent vantage point. Sometimes Merlin, Peregrine or Sparrowhawk are also present. Lesser and Common Redpolls and Siskin may winter in the alders, and Golden Plover favours the surrounding farmland and Baker's/Adventurers' Fens.

On passage, reasonable numbers of waders may visit Baker's Fen, mostly Green and Common Sandpipers but sometimes including Black-tailed Godwit and Whimbrel, and Marsh Harrier is regularly recorded.

Access

Leave Cambridge north on the A10 and turn right onto the A1123 at Stretham. After 3.5 miles, at the beginning of Wicken village, turn right onto a minor road signed Wicken Fen. Park in the large car park on the left a few hundred yards along the track. Continue on foot to the William Thorpe Visitor Centre; permits (entrance free to NT members) and trail guides are sold here, and a list of recent bird sightings is available. The reserve is open daily 9am–5pm (9am–4.30pm November–December; closed all January and most Mondays in winter: check website for details). There is a choice of three routes. The Nature Trail around Sedge Fen takes in the Tower Hide, which affords a view over the entire reserve but is particularly good for watching

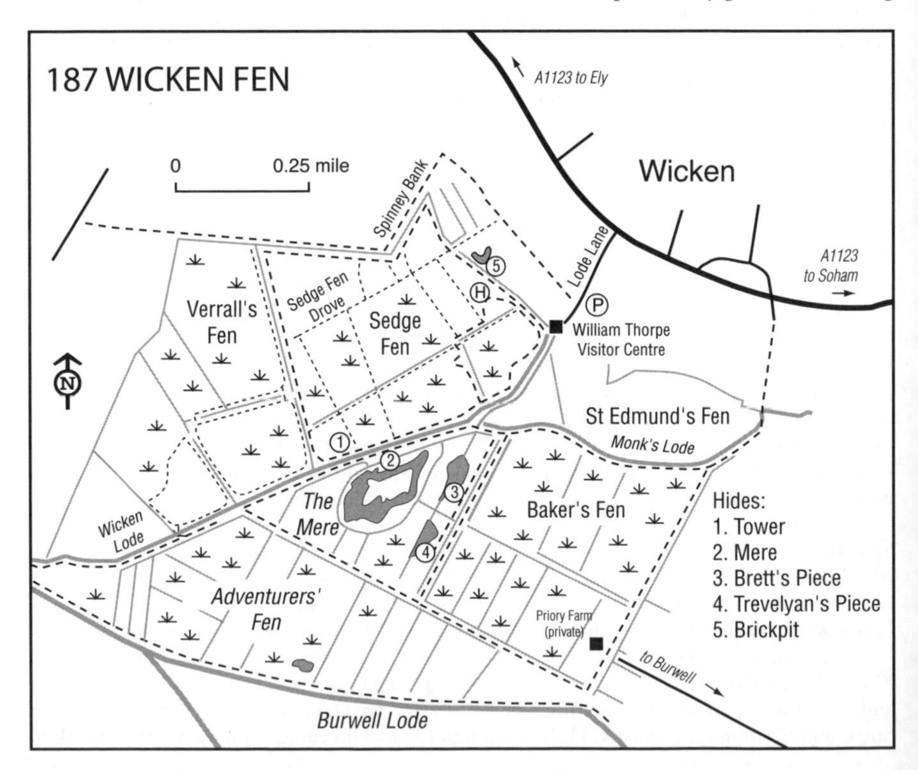

the Mere on Adventurers' Fen; various other paths cross the fen. Also on Sedge Fen there is a 0.75-mile wooden walkway permitting access to the less able-bodied, and visitors with families. Finally, there is a clockwise route, 2 miles in length, around part of Adventurers' Fen. Four hides overlook the various scrapes and the Mere. The flooded areas of Baker's Fen are also viewable at certain points along this route.

FURTHER INFORMATION

Grid ref/postcode: TL 563 705 / CB7 5XP
National Trust Wicken Fen: tel: 01353 720274; email: wickenfen@nationaltrust.org.uk; web: www.wicken.org.uk

CENTRAL ENGLAND

BEDFORDSHIRE
188 The Lodge
189 Harrold-Odell Country Park
190 Priory Country Park
191 Brogborough Lake

HERTFORDSHIRE
192 Rye Meads
193 Lee Valley Park
194 Tring Reservoirs

BUCKINGHAMSHIRE
195 Willen Lakes

OXFORDSHIRE
196 The Wessex Downs at Churn
197 Otmoor
198 Port Meadow, Oxford
199 Farmoor Reservoirs
200 Stanton Harcourt Gravel Pits

GLOUCESTERSHIRE
201 Cotswold Water Park
202 Slimbridge
203 Frampton
204 Highnam Woods
205 Forest of Dean
206 Symonds Yat

WORCESTERSHIRE
207 The Wyre Forest
208 Upton Warren
209 Bittell Reservoirs

WEST MIDLANDS
210 Sandwell Valley

WARWICKSHIRE
211 Tame Valley
212 Brandon Marsh
213 Draycote Water
214 Daventry Reservoir

NORTHAMPTONSHIRE
215 Earls Barton Gravel Pits
216 Ditchford Gravel Pits
217 Thrapston Gravel Pits
218 Pitsford Reservoir
219 Stanford Reservoir

LEICESTERSHIRE AND RUTLAND
220 Swithland Reservoir
221 Eyebrook Reservoir
222 Rutland Water

LINCOLNSHIRE
223 The Wash
224 Gibraltar Point
225 South Lincolnshire Coast: Chapel St Leonards to Huttoft
226 North Lincolnshire Coast: Mablethorpe to Grainthorpe Haven
227 Tetney Marshes
228 Covenham Reservoir
229 Killingholme Haven
230 The Inner Humber
231 Messingham Sand Quarry
232 Whisby Nature Park and North Hykeham Pits

NOTTINGHAMSHIRE
233 Lound
234 The Dukeries at Welbeck and Clumber Parks
235 Trent Valley Pits

DERBYSHIRE
236 Ogston Reservoir
237 The Upper Derwent Valley
238 Carsington Water
239 Staunton Harold Reservoir
240 Foremark Reservoir

STAFFORDSHIRE
241 Coombes and Churnet
242 Blithfield Reservoir
243 Belvide Reservoir

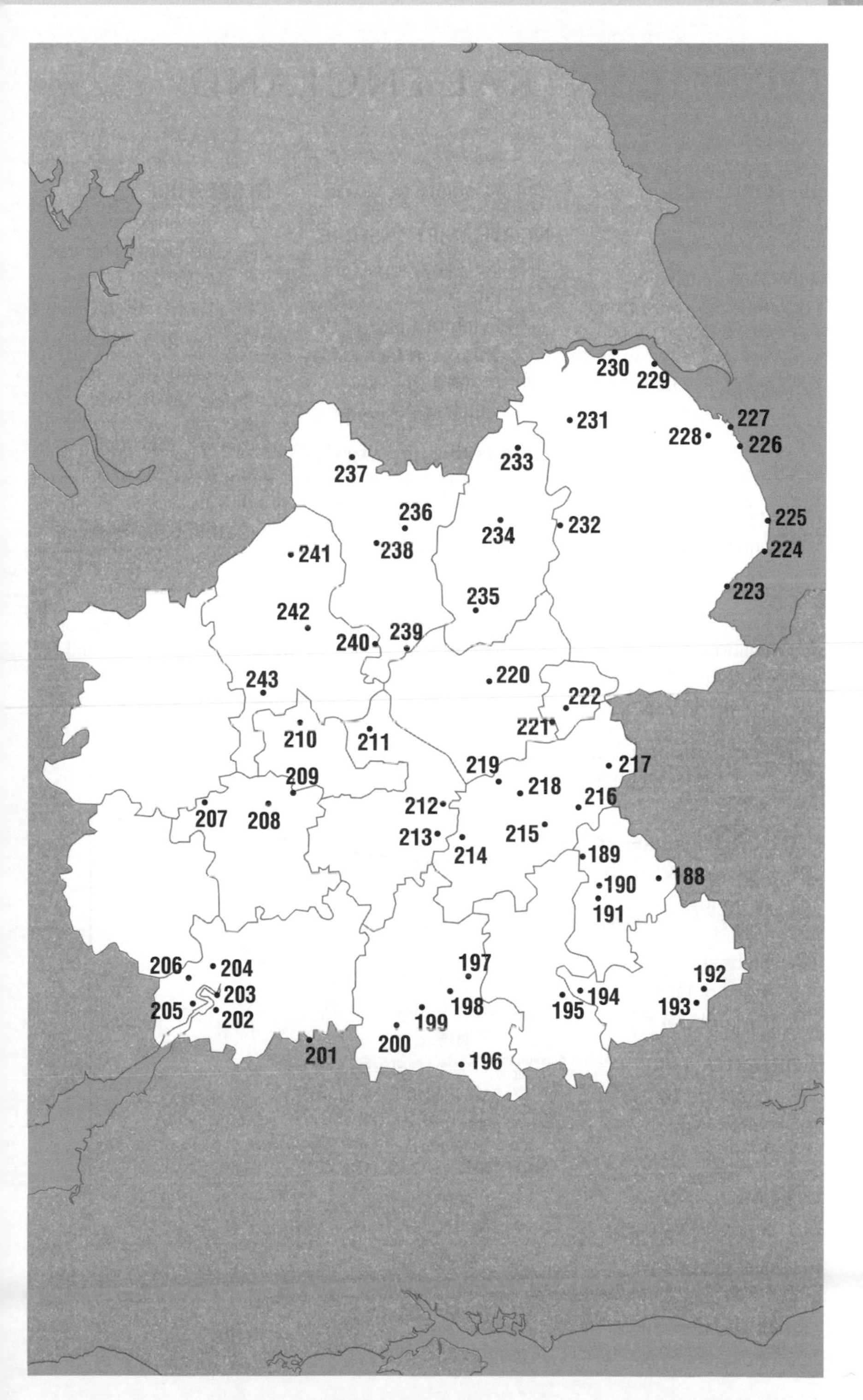

230
229
227
231
228
226
233
237
236
225
234
232
241
238
224
235
223
242
239
240
220
243
222
221
210
211
219
217
209
218
216
212
207
208
215
213
214
189
190
188
191
197
192
206
204
203
194
205
195
198
193
202
199
200
201
196

188 THE LODGE (Bedfordshire)

OS Landranger 153

The Lodge is the headquarters of the RSPB and is surrounded by 180 ha of partially wooded heathland with a good range of common birds.

Habitat

The Lodge is located on relict heathland, once part of Sandy Warren. Some is still heath, but much has developed into secondary woodland or is plantation, and it is planned to return a substantial area of this to heathland with the aim of attracting Nightjar, Woodlark and Dartford Warbler. Two ponds have been established, and The Lodge itself is surrounded by formal gardens.

Species

Resident species include Common Buzzard, Sparrowhawk, Woodcock, Tawny Owl, all three woodpeckers and Nuthatch. In summer these are joined by Hobby, Cuckoo, Turtle Dove, Tree Pipit, Garden Warbler, Lesser Whitethroat and Spotted Flycatcher. In winter Brambling, Siskin and Common and Lesser Redpolls may occur. Red Kite is a regular visitor, and Raven, Common Crossbill and Hawfinch are occasionally seen.

Access

Leave Sandy on the B1042 towards Cambridge and after 1.2 miles turn right at the top of the hill into the car park. There are around 4.5 miles of trails (including the Iron Age hill fort at Galley Hill) and a hide overlooking bird feeders and small pools. The reserve is open from 9am–9pm (or dusk if earlier), and the shop is open 9am–5pm on weekdays and 10am–5pm at weekends and Bank Holidays.

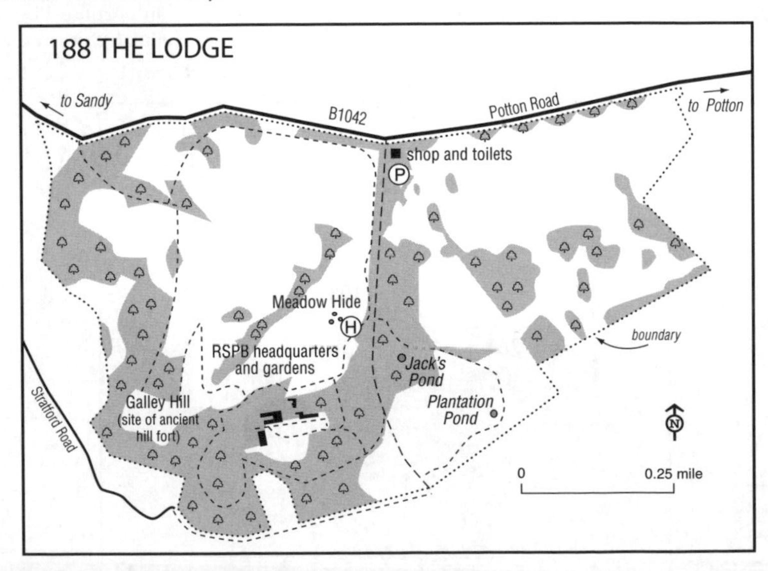

FURTHER INFORMATION

Grid ref/postcode: TL 191 485 / SG19 2DL
RSPB, The Lodge: tel: 01767 680541; email: thelodgereserve@rspb.org.uk

189 HARROLD-ODELL COUNTRY PARK

(Bedfordshire) OS Landranger 153

Around 7 miles north-west of Bedford, this series of gravel pits and associated habitats holds an interesting variety of species and is worth visiting year-round. It is owned and managed by Bedford Borough Council.

Habitat

Centred around a series of abandoned gravel pits beside the River Great Ouse, the 59-ha park includes riverside meadows and riparian willows and alders. At the east end of the main lake there is an interesting area of reeds and willow carr, while part of a specially created island is regularly cleared of vegetation to attract passage waders.

Species

Wintering wildfowl include up to 300 Wigeon, Teal, Shoveler, Gadwall, Pochard, Goldeneye, small numbers of Goosanders and occasionally Smew, as well as feral Canada, Egyptian and Greylag Geese. Other notable winter visitors include Water Rail, Common Snipe, Grey Wagtail, Siskin and redpolls in the riverside trees, and sometimes Stonechat. Bittern is occasionally recorded.

On passage small numbers of common waders are regular, including Little Ringed Plover, Dunlin and Common Sandpiper and, especially in May, occasionally Turnstone and Sanderling, with Curlew, Whimbrel, Greenshank, Green Sandpiper and Ruff more likely in autumn. Common, Arctic and Black Terns are also possible, especially in late April and May. Black-necked Grebes may pass through in late summer and autumn and Little Egret is now regular. Migrant passerines can include White Wagtail and Wheatear.

Breeding species include Little and Great Crested Grebes, Cormorant, Grey Heron, Sparrowhawk, Ringed Plover, Common Redshank, Common Tern (on purpose-built rafts), Grey and Yellow Wagtails and Reed and Sedge Warblers. Hobby is a regular visitor, especially on late summer evenings.

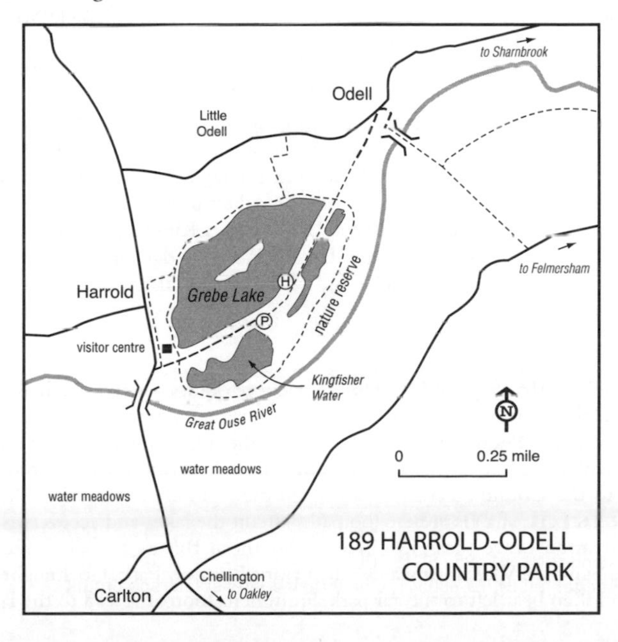

189 HARROLD-ODELL COUNTRY PARK

Access

Just north of Bedford leave the A6 west on minor roads to Oakley, and continue via Pavenham and Carlton towards Odell. Just south of Harrold the road crosses the River Great Ouse, and the entrance to the country park is on the right after 100 yards. There is a visitor centre at the car park and the main lake is circumnavigated by a footpath, with a hide on the south shore, and side paths to the river and reedbeds.

FURTHER INFORMATION

Grid ref/postcode: SP 956 566 / MK43 7DS
Harrold-Odell Country Park Visitor Centre: tel: 01234 720016; web: www.hocp.co.uk

190 PRIORY COUNTRY PARK (Bedfordshire)

OS Landranger 153

Close to Bedford town centre, this site attracts a range of wintering wildfowl and a variety of passage migrants. It is owned and managed by Bedford Borough Council.

Habitat

Comprising 146 ha and bordered to the south and east by the River Great Ouse, the Country Park contains a large (25-ha) disused gravel pit fringed by reeds and willows, while immediately to the north-east of the main lake are Finger Lakes, a pair of small overgrown pits. Elsewhere there are plantations of deciduous trees and areas of grassland, which at Fenlake and near the sewage works to the east may flood in winter.

Species

Wintering wildfowl include Wigeon, Teal, Gadwall, Shoveler, Pochard, small numbers of Goldeneyes and occasionally Shelduck, Pintail or sawbills. Other notable wintering species include Cormorant, Water Rail and Common Snipe.

On passage a variety of waders may pass through, but due to the lack of habitat, many simply fly over, with Little Ringed Plover, Greenshank and Common and Green Sandpipers among the more regular, with the best variety recorded in spring. Also, particularly in spring, Osprey occasionally occurs and parties of Common, Arctic and Black Terns and Little Gull may pass through. Other migrants include Grey, White and Yellow Wagtails, Whinchat, Wheatear, and occasional Shelduck, Jack Snipe, Kittiwake and Water Pipit in early spring.

Breeding species include Great Crested and Little Grebes, Kingfisher, Cuckoo, Lesser Spotted Woodpecker, Nightingale (around Finger Lakes), Reed, Sedge and Garden Warblers, and Spotted Flycatcher. Little Egret is regular and Hobbies may also visit, especially on summer evenings.

Access

Being so close to an urban area and used for a variety of outdoor activities, the Country Park is prone to disturbance, especially at weekends. Early mornings are generally best.

Main Entrance Leave Bedford town centre east on the A428 and, after 1 mile, turn south on the A5140 Newnham Avenue. After 0.5 miles, at the roundabout, turn east into Barkers Lane. The entrance to the Country Park car park is on the right after 0.5 miles. There is a visitor centre near the car park, and from here footpaths circuit the lakes and access two hides.

Southern Entrance Leave the A421 Bedford Southern Bypass at its intersection with the A603 and take Stannard Way northwards into Priory Business Park. Go straight over the first roundabout and then bear left to the car park. From here, footpaths lead to the lakes.

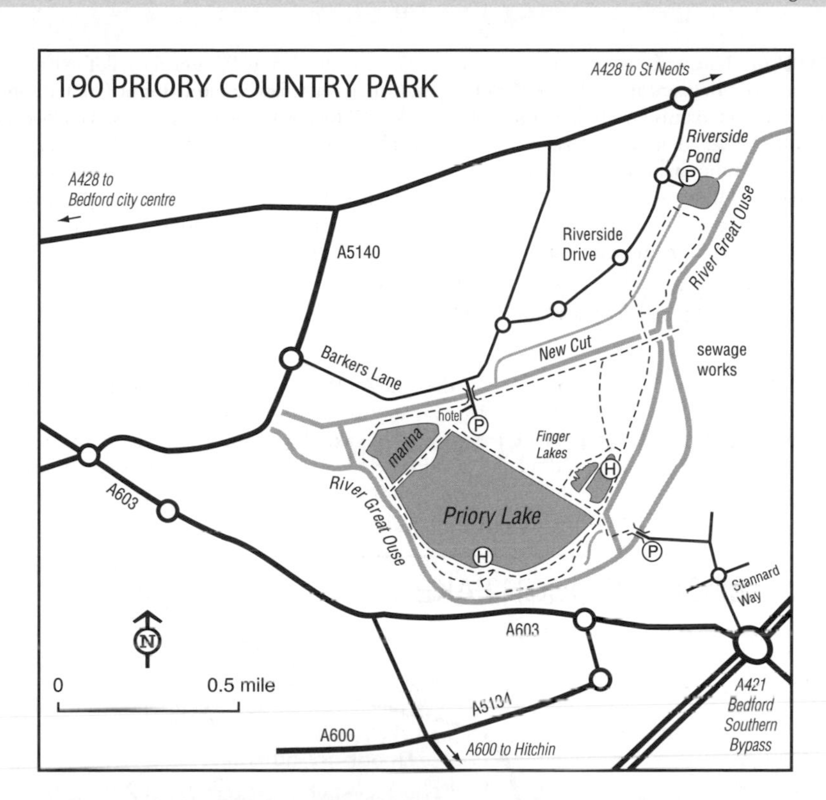

FURTHER INFORMATION

Grid ref/postcode: Main entrance TL 072 493 / MK41 9SH; Southern entrance TL 079 488
Priory Country Park: tel: 01234 211182; web: www.priorycountrypark.co.uk

191 BROGBOROUGH LAKE (Bedfordshire)

OS Landranger 153

This relatively small lake between Milton Keynes and Bedford holds a large winter gull roost, notable for being one of the few regular localities for 'white-winged' gulls in the Home Counties.

Habitat

This deep, largely steep-sided flooded clay pit is screened by a line of Poplars to the north, while elsewhere there are extensive areas of scrub with small stands of reeds in the south corner. The lake is disturbed by windsurfing and fishing.

Species

The principal attraction in winter is the large gull roost, which holds significant numbers of Black-headed, Common, Herring and Lesser Black-backed Gulls, together with Great Black-backed Gull. Among rarer species, Mediterranean and Yellow-legged Gulls are regular in very small numbers in autumn and winter, with a few Iceland and Glaucous Gulls in the latter period, and Kittiwake is possible in late winter and early spring. Wintering wildfowl include numbers of Pochards and up to 30 Goldeneyes. Other waterfowl include Great Crested Grebe and up to 30 Cormorants (which roost on the island). The rarer grebes (especially Red-necked), sawbills and seaducks are occasional. Other notable winter visitors include Water Rails, and a few Corn Buntings roost in the reeds.

On spring passage, Common Scoter, Arctic and Black Terns and Little Gull may pause briefly, and migrant passerines can include Yellow Wagtail and Grasshopper and Sedge Warblers. Terns also pass through in autumn, but the banks are too steep to attract many waders.

Breeding species include Shelduck, Ruddy Duck, Common Tern, Turtle Dove, Cuckoo, Kingfisher and Reed Warbler, and Hobby is a fairly frequent visitor.

Access

Leave the M1 north at junction 13 on the old A421 towards Brogborough and Marston Moretaine (not the new A421 dual carriageway towards Bedford). After 1.5 miles turn right on the minor road to Lidlington. There is access to the lake shore along a public footpath at one point along the south shore, and the gull roost is best viewed at the south-west end of the lake from this minor road, where there are convenient gaps in the hedge. The east end of the lake can be seen from the short footpath just south of the sailing clubhouse.

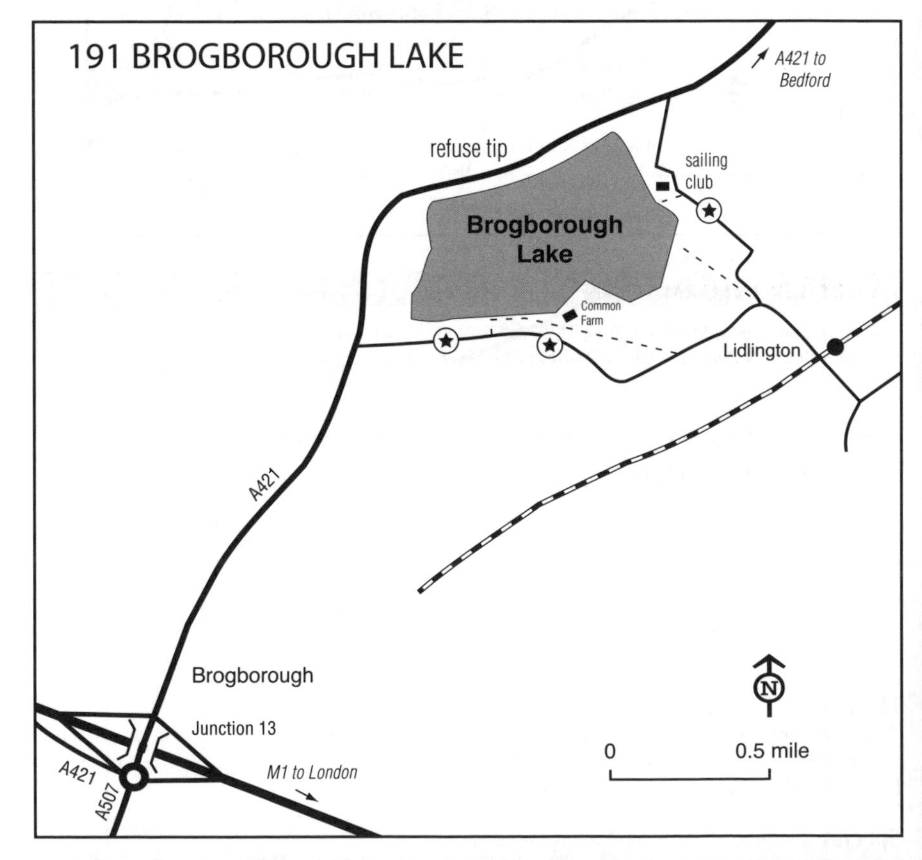

FURTHER INFORMATION

Grid ref: SP 974 392

192 RYE MEADS (Hertfordshire)

OS Landranger 166

Lying in the Lee valley just east of Hoddesdon, the Rye Meads area contains a variety of wetland habitats and is notable for regular wintering Bitterns and breeding Common Terns. The RSPB has a reserve at Rye Meads and jointly manages other areas with the Hertfordshire and Middlesex Wildlife Trust (HMWT).

Habitat

A complex of wet meadows, carr woodland and reedbeds has been supplemented with specially constructed scrapes, while a large part of the wider area is occupied by sewage treatment ponds.

Species

In winter small numbers of Water Rails are present – best looked for from the hides, as occasionally is Bittern. Both are very secretive, but with patience may show well. Water Pipit, Green Sandpiper and Common and Jack Snipes also winter, and may be found by careful scrutiny of the water's edge. Other winterers include Cormorant (on the North Lagoons), Pochard, Shoveler, Teal and Gadwall, with Golden Plover and Common Snipe on flooded meadows, Grey Wagtail, with Siskin and redpolls in alders. Stonechat and Bearded Tit are irregular visitors.

On passage small numbers of Greenshanks, Common and Green Sandpipers, Yellow Wagtails, Whinchats and Wheatears may pass through, with occasional Garganey and Pintail in late summer.

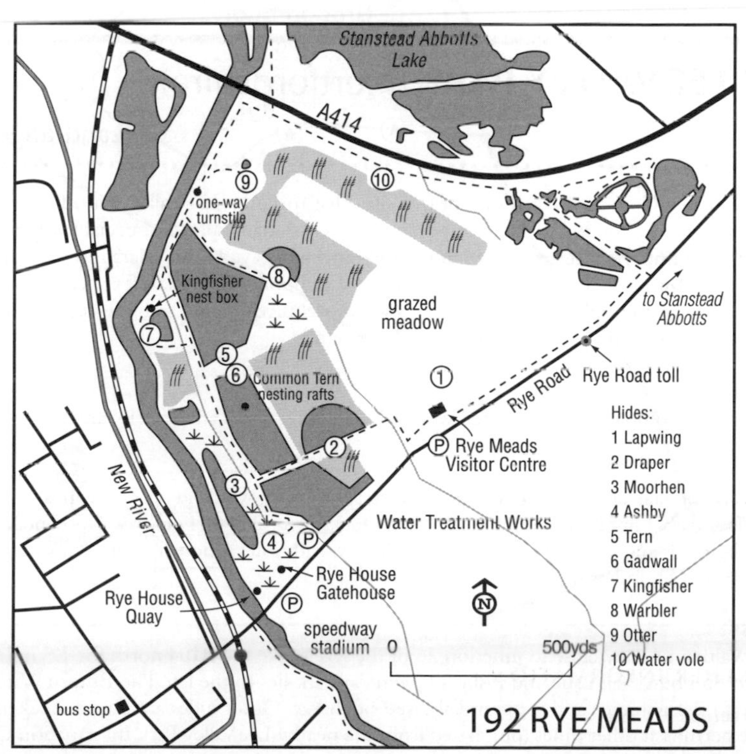

Breeding species include Gadwall, Little Ringed Plover, Common Redshank, Common Tern (breeding on specially constructed rafts), Turtle Dove, Kingfisher, Cuckoo and ten species of warblers including the resident Cetti's as well as Grasshopper, Sedge and Reed Warblers. Hobby is a regular visitor and Peregrine, Red Kite, Osprey and Marsh Harrier are possible.

Access

From the B181 north-east of Hoddesdon turn south-east on Rye Road towards Rye Park and Hoddesdon. Follow this past the toll (50p – 50p, 20p and 10p pieces accepted) and, after another 150 yards, look for the blue and yellow Rye Meads Visitor Centre to the right (brown duck sign). From the A10 take the Dinant Link road (a dual carriageway) into Hoddesdon and turn left (north) at the first roundabout into the A1170 Amwell Street. At the next large roundabout, continue straight on into Ware Road and then, at the mini-roundabout, turn right into Middlefield Road. This leads to a staggered crossroads controlled by two mini-roundabouts; go right and then left into Rye Road and follow this for about 0.5 miles past Rye House Station until you see the blue and yellow Rye Meads Visitor Centre to the left.

The reserve and visitor centre are open daily 10am–5pm (or dusk if earlier; the gates are locked when the reserve is closed; closed Christmas and Boxing Days). There are three trails: the Moorhen Trail (0.3 miles), Kingfisher Trail (0.5 miles) and Otter Trail (1 mile) and ten hides. The majority of hides and trails are wheelchair-accessible.

FURTHER INFORMATION

Grid ref/postcode: TL 389 103 / SG12 8JS
RSPB Rye Meads: tel: 01992 708383; email: rye.meads2@rspb.org.uk
Herefordshire and Middlesex Wildlife Trust: web: www.hertswildlifetrust.org.uk

193 LEE VALLEY PARK (Hertfordshire)

OS Landranger 166

The River Lee Country Park is one of five sites that make up Lee Valley Park. These flooded gravel pits close to London form an important refuge for many species, especially wildfowl, and are part of an important series of wetlands which includes Rye House Marsh to the north. The park's main claim to fame is, however, its Bittern Watchpoint. In the last few years, this site has become the most reliable place in Britain to see this elusive species, and not just fleeting flight views either. Here, you can see Bitterns on the ground.

Habitat

The park is a complex of open water, wooded islands, reedbeds and marshy areas.

Species

Bittern is the star attraction, but a good variety of other wetland species can also be seen here. Nationally important numbers of Tufted Ducks (over 2,500), Pochards, Goosanders, Great Crested Grebes and Coots winter in the Lee Valley, with internationally important numbers of Gadwalls and Shovelers.

Access

From Waltham Abbey, close to junction 26 of the M25, take the B194 north for 1.5 miles. Turn left into Stubbins Hall Lane and Fishers Green car park lies to the left. The Bittern Watchpoint is close to the entrance and is open daily, free of charge. Other hides are free at weekends, but require permits at other times (purchased from the nearby Lee Valley Park Information Centre).

FURTHER INFORMATION

Grid ref: TL 377 026
Lee Valley Park Information Centre: Abbey Gardens, Waltham Abbey, EN9 1XQ; tel: 01992 702200; web: www.leevalleypark.org.uk

194 TRING RESERVOIRS (Hertfordshire)

OS Landranger 165

Lying north-west of Tring, this complex of four reservoirs has a long ornithological history and, with a fine range of habitats, attracts interesting birds throughout the year.

Habitat

The reservoirs were built in the 19th century as canal feeders. Wilstone is the largest and has concrete banks to the north and natural banks to the south, with substantial stands of woodland along the water's edge and a marsh and damp meadow at the extreme south tip. To the east, the rather smaller Tringford has natural banks bordered by meadows and, in the south sector, woodland. Stantop's End and Marsworth also have largely natural banks, with a substantial reedbed at Marsworth, while to the east of these is a sewage farm which can be interesting at times. The reservoirs are managed by Friends of Tring Reservoirs.

Species

Wintering wildfowl include small numbers of Wigeons, Shovelers, Gadwalls, Teals, Pochards, Goldeneyes and Goosanders, occasionally also Smew and seaducks. There is a large gull roost, which mainly comprises Black-headed Gulls with a scattering of Lesser Black-backed Gulls. 'White-winged' gulls are rare, but Mediterranean Gull is sometimes recorded, and Kittiwake is possible, especially in late winter and early spring. Other notable wintering species include Corn Bunting, which roosts in the reeds, Siskin, redpolls, Chiffchaff, Blackcap, Common and Jack Snipes (especially at the sewage farm), and occasionally Peregrine, Merlin and Bearded Tit. Bitterns regularly overwinter at Marsworth Reservoir, with occasional sightings at Wilstone Reservoir and the Tring Water Treatment Works. The best tactic for a sighting is to wait on the causeway between Marsworth and Startops Reservoirs, especially early and late in the day.

On spring passage small numbers of waders pass through, including Curlew, Whimbrel, Common and Green Sandpipers, Common Redshank, Dunlin and Ringed and Little Ringed Plovers. Other migrants may include Shelduck, Common Scoter, flocks of Common, Arctic and Black Terns and Little Gulls, White and Yellow Wagtails, Whinchat and Wheatear. In autumn a greater variety of waders is possible, with Greenshank, Ruff, Curlew Sandpiper and Little Stint more likely than in spring. Black-necked Grebe, Garganey, Marsh Harrier and Osprey are occasionally recorded in both spring and autumn.

Breeding species include Great Crested and Little Grebes, Grey Heron, Gadwall, Pochard, Sparrowhawk, Water Rail, Common Tern (on specially constructed rafts at Wilstone), Little Owl, Kingfisher, all three woodpeckers (Lesser Spotted scarce) and Cetti's, Reed and Sedge Warblers. Little Egret is now regular and Hobbies often visit on spring and summer evenings. The reservoirs are famous as the site of England's first breeding Black-necked Grebes (1919) and Britain's first breeding Little Ringed Plovers (1938).

Access

Wilstone Reservoir Leave the A41 north at Aston Clinton on the B489 towards Dunstable, parking on the right after 2.5 miles in the car park by the reservoir embankment (note, the area has a bad reputation for car crime). Take the path onto the bank, walking right to reach

Drayton Bank and the hide (the footpath continues and it is possible to circuit the reservoir via the disused canal).

Startop's End and Marsworth Reservoirs Continue on the B489 and park on the right in the pay-and-display car park just before the Grand Union Canal. From here footpaths run along the north and west shores of the reservoirs and the causeway between them (hide) and it is possible to undertake a circular walk. For access to the Plover Hide at the Thames Water Sewage Farm Lagoon, contact Friends of Tring Reservoirs.

Tringford Reservoir A footpath runs along the west and south shores and accesses a small hide, accessible either from Little Tring Road or from the Tring Ford Road, the lane that separates Tringford and Startop's End (take the footpath through the wood to the hide).

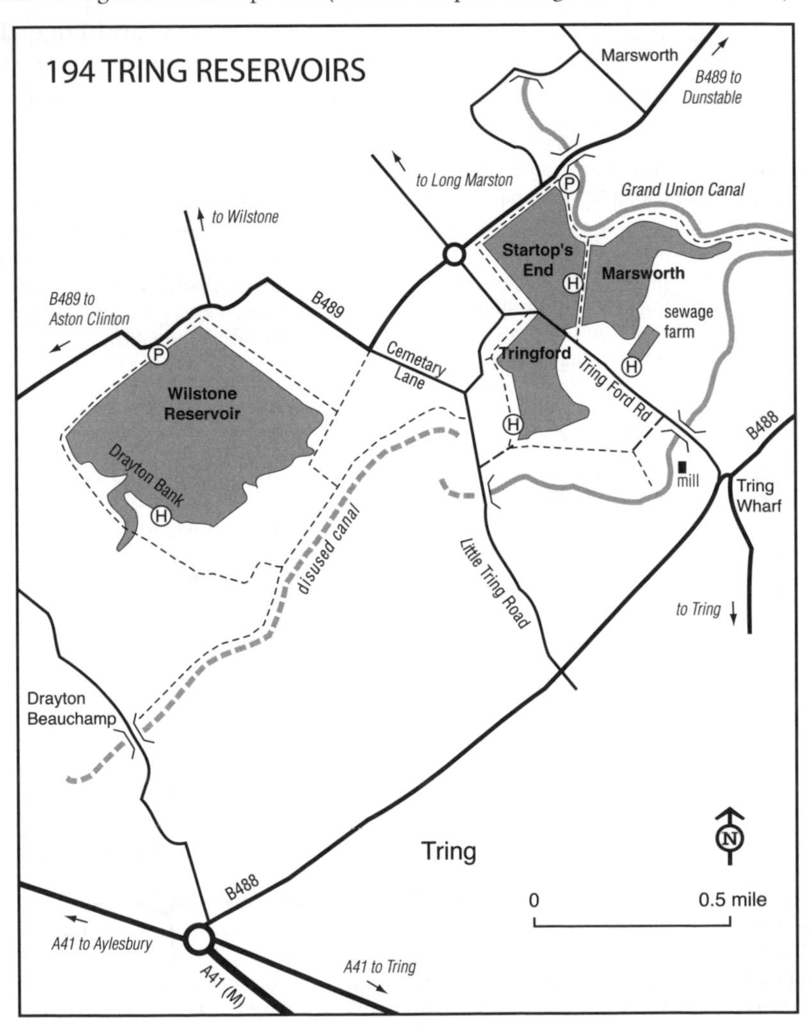

FURTHER INFORMATION

Grid refs: Wilstone Reservoir SP 904 134; Startop's End and Marsworth Reservoirs (Grand Union Canal) SP 919 140

Friends of Tring Reservoirs: web: www.fotr.org.uk

195 WILLEN LAKES (Buckinghamshire)

OS Landranger 152

Situated on the western outskirts of Milton Keynes, this site attracts a range of winter wildfowl, including Goosander, and a variety of passage waders.

Habitat

There are two lakes, separated by a road, the North Lake containing a large island that has a specially designed wader scrape on it. The surrounding area is open parkland with some shrubby areas.

Species

Wintering wildfowl include Wigeon, Teal, Shoveler, Pintail, Pochard and up to 30 Goldeneyes, while small numbers of Goosanders are regular. Other waterfowl include Cormorant, divers, rarer grebes, seaducks, and Whooper and Bewick's Swans are occasional visitors. There is a notable gull roost, and among the commoner species one or two Mediterranean Gulls are regular, and Iceland and Glaucous Gulls have been recorded. In late winter and early spring Kittiwake may also appear in the roost. Other notable winterers include Common Snipe, and occasionally Jack Snipe or Short-eared Owl.

In spring, flocks of Common, Arctic and Black Terns and Little Gulls may pass through. On both spring and autumn passage a variety of waders has been recorded, including the more usual Oystercatcher, Ruff, Dunlin, Green and Common Sandpipers and Ruff, while

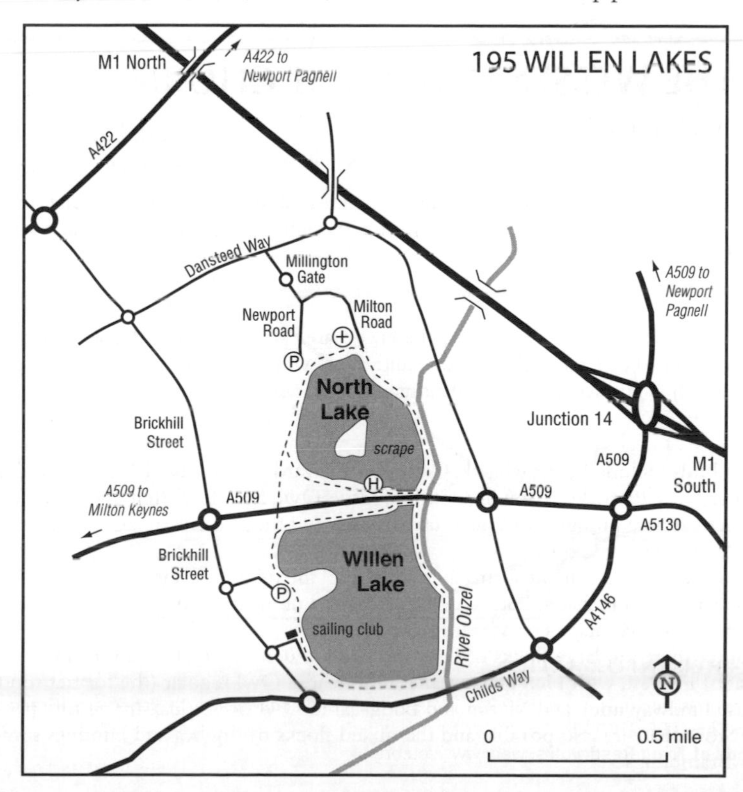

Grey Plover, Turnstone, Curlew, Whimbrel, Black-tailed Godwit, Knot, Sanderling, Curlew Sandpiper and Little Stint are also regular, especially in autumn and, unusually, some of these have been recorded in small flocks, while the scarce Temminck's Stint is almost annual. Among passerines, Rock and Water Pipits, Yellow and White Wagtails, Wheatear and Whinchat are regular. Scarce passage migrants include Osprey, Common Scoter, Garganey, Mandarin, Ring Ouzel, Common Redstart and Black Redstart.

Breeding birds include Little and Great Crested Grebes, Sparrowhawk, Ringed and Little Ringed Plovers, Common Redshank and Sedge Warbler. Little Egret is regular, and a Hobby occasionally visits, especially on late spring and summer evenings. Gadwall has bred, and Black-winged Stilt attempted to breed in 1993.

Access

Leave the M1 south-west at junction 14 on the A509 towards Milton Keynes, turning south after 1.5 miles at the third roundabout onto Brickhill Street and then turn left to the small car park by Willen Lake.

Alternatively, turn north at the third roundabout along Brickhill Street and turn right onto Dansteed Way. Turn right again into Millington Gate, go over the roundabout and turn right into Newport Road, proceeding to the small car park by the North Lake (or take Milton Road to join the footpath). Footpaths encircle both lakes, but the wader scrape is best viewed from the hide, and the North Lake is generally most productive.

FURTHER INFORMATION

Grid ref: Willen Lake SP 875 400

196 THE WESSEX DOWNS AT CHURN (Oxfordshire)

OS Landranger 174

The Wessex ('Berkshire') Downs comprise an area of open chalkland habitats in south Oxfordshire that can be interesting at any season, though a certain amount of luck is required and blank days are possible.

Habitat

Prior to the Second World War almost the entire area was unimproved grassland grazed by sheep and rabbits but large parts are now cultivated, with some areas of permanent grassland associated with racing stables and remnant stands of woodland.

Species

Breeding birds include Sparrowhawk, Common Buzzard, Hobby, Red-legged Partridge and the scarcer Grey Partridge, Lapwing, Curlew, Little Owl, Meadow Pipit and Corn Bunting. In summer variable numbers of Quails use the area and the occasional pair of Stone-curlews may persist.

On passage the escarpment forms a 'leading line' for migrating birds, attracting Wheatear, Ring Ouzel and Whinchat at both seasons, with good numbers of thrushes in autumn and, in May, sometimes a Montagu's or Marsh Harrier.

In winter there are large flocks of Lapwing and Golden Plover, but raptors probably hold the greatest interest, with Hen Harrier and Short-eared Owl regular (the former roosts near the disused railway line), and Merlin and Long-eared Owl occasional. Interesting passerines, such as Stonechat, are also possible and the mixed flocks of finches and buntings sometimes contain Bramblings.

Access

Exploration of public rights of way is recommended, notably the Fair Mile and Ridgeway, with three main access points off the A417 between Harwell and Streatley:

1. At the west edge of Blewbury leave the A417 south on Bohan's Road. Follow this concrete road through two left bends and park carefully on the grass verge near Churn pig farm.
2. Turn south off the A417 3 miles east of Blewbury (opposite the turn to Aston Upthorpe) and follow the track to park after 0.6 miles. From here walk south for 1 mile to the Fair Mile.
3. On the A417, 5 miles east of Blewbury, park at the end of the Fair Mile on Kingstanding Hill. The Fair Mile footpath can be followed west for several miles.

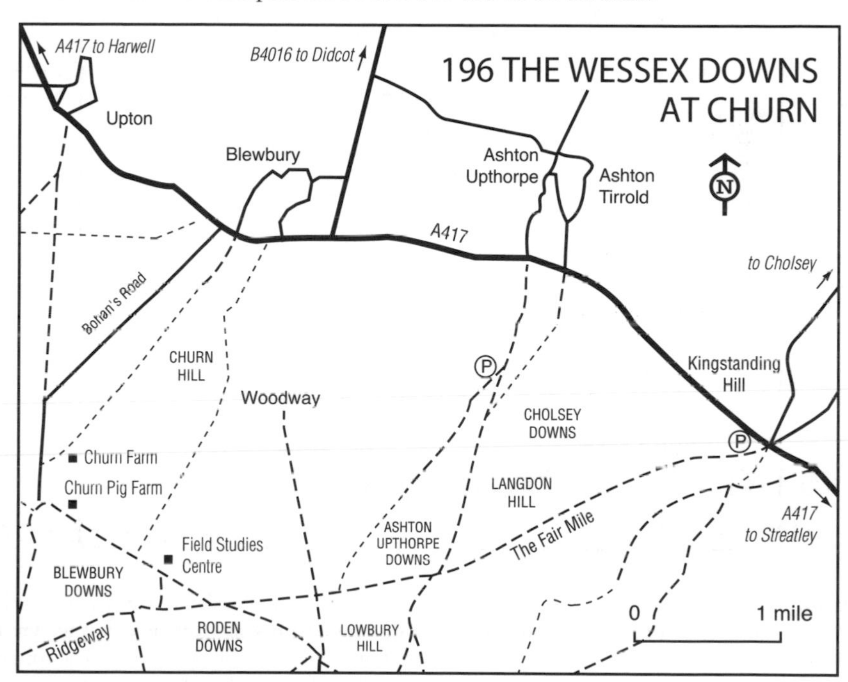

FURTHER INFORMATION

Grid refs: Bohan's Road SU 526 856; turn to Aston Upthorpe SU 552 854; Kingstanding Hill SU 573 838

197 OTMOOR (Oxfordshire)

OS Landranger 164

Just 7 miles north-east of Oxford, this area comprises the flood plain of the River Ray and in the 19th century held breeding Black Terns and Bitterns. Today, despite drainage, much is 'unimproved' and it still attracts wintering wildfowl and an interesting variety of breeding birds. Part of the area is an RSPB reserve covering 101 ha, and it is hoped that by recreating the once extensive reedbeds Bitterns will return as a breeding bird.

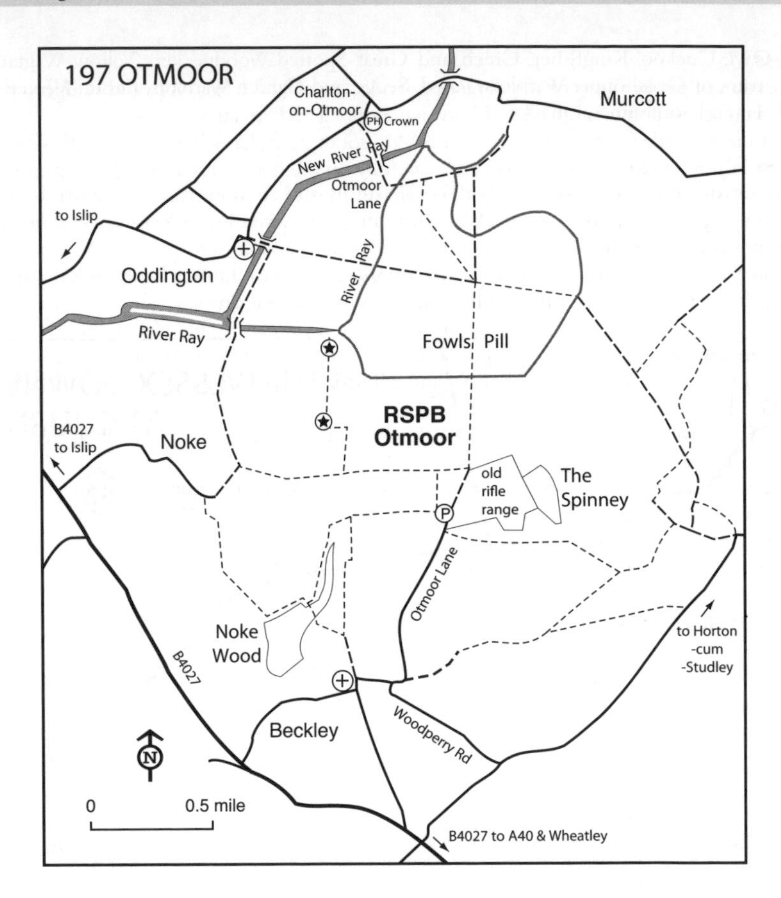

Habitat

The area comprises a low-lying plain, once highly prone to flooding, drained by the River Ray. Most has now been converted to agriculture but Fowls Pill still holds water year-round and can cover a large area in winter. The RSPB has created 22 ha of reedbeds. There are large areas of unimproved grassland with some scrub, and some mature woodland at Noke Wood and The Spinney.

Species

In winter the Fowls Pill area (and numerous ditches and dykes on Otmoor) attracts large numbers of Teals and Wigeons, and occasionally small numbers of Gadwalls, Pintails and Shovelers. Waders include Common Snipe, Golden Plover and, sometimes, Jack Snipe or Woodcock, and look too for Water Rail and Bittern – singletons of the latter have wintered in recent years. Sparrowhawk and Short-eared Owl are fairly frequent, and sometimes Hen Harrier, Peregrine or Merlin may be present. Look for Little Owl in the willows by the River Ray, and wintering passerines may include flocks of thrushes, finches, buntings and, sometimes, Stonechat.

On passage a variety of waders has been recorded, including Whimbrel, Greenshank, Green and Common Sandpipers, Ruff and Little Ringed and Ringed Plovers. Look for Garganey and Marsh Harriers in spring, while at both seasons Wheatear, Whinchat, Common Redstart and other passerine migrants may occur. Little Egrets are regular in summer and autumn.

Breeding species include Little Grebe, Shoveler, Pochard, Hobby, Water Rail, Lapwing, Common Redshank, Common Snipe, Curlew (most vocal at dawn and dusk), Turtle Dove,

Little Owl, Cuckoo, Kingfisher, Green and Great Spotted Woodpeckers, Yellow Wagtail, up to ten pairs of Grasshopper Warblers, Reed, Sedge and Garden Warblers, Lesser Whitethroat, Marsh Tit and, sometimes, Quail.

Access

RSPB Otmoor Leave the A34 on the B4027 to Islip and continue along the B4027 towards Wheatley. After 4 miles turn north-east on a minor road towards Horton-cum-Studley. After 0.5 miles turn left into Beckley on Woodperry Road, bear right into Roman Way, and at the T-junction turn right into High Street and then sharp left into Otmoor Lane. Follow this for 1 mile to the reserve, which is open daily, dawn to dusk. There is one trail, about 1.5 miles long, giving access to several viewpoints. Connecting with a variety of public footpaths also allows some long circular walks. NB some public footpaths are closed when the rifle range is active, which may be any day except Mondays and Thursdays. Wheelchair access is possible in dry conditions, though help may be needed.
Noke Wood Park in Beckley and follow the footpath past the church and then north-west through the wood.
Oddington Park at Oddington village green and walk east along Oddington Lane. On crossing the River Ray at the bridge either continue straight or turn south to follow the river.
Charlton-on-Otmoor Park by the Crown pub, then take Otmoor Lane south-east – this gives access to the Fowls Pill area and connects with the RSPB trails.

FURTHER INFORMATION

Grid refs: RSPB Otmoor SP 570 126; Noke Wood SP 561 115; Oddington SP 553 148; Charlton-on-Otmoor SP 562 158
RSPB Otmoor: tel: 01865 351163

198 PORT MEADOW, OXFORD (Oxfordshire)

OS Landranger 164

Within walking distance of Oxford city centre, this area holds a variety of winter wildfowl and waders, and a few interesting migrants on spring passage.

Habitat

Common land owned by the City Council, the area is largely unimproved grazing meadows between the River Thames and Oxford Canal, and is prone to flood in winter.

Species

Wintering wildfowl include up to 1,000 Wigeon as well as numbers of Teal, and smaller parties of Gadwall, Shoveler, Pintail and Shelduck. In cold weather Goldeneye and Goosander sometimes occur on the river. There is a large resident flock of Canada Geese and feral Greylag Geese and there may be up to 4,000 Golden Plovers, 2,000 Lapwings, 70 (exceptionally 1,000) Common Snipe, 40 Dunlins, up to 30 Black-tailed Godwits, ten Ruffs and small numbers of Common Redshanks. Many gulls use the area as a pre-roost. Other notable wintering species may include Water Rail, Jack Snipe, Grey Wagtail, Stonechat and Common and Lesser Redpolls.

On spring passage a few waders may occur, as well as Shelduck and occasionally Garganey. By autumn however, the meadow has largely dried out and is much less attractive, but Wheatear and Whinchat may pass through at either season.

Breeding species in the area include Sparrowhawk, Cuckoo, Barn Owl, Yellow Wagtail and Sedge and Grasshopper Warblers (Burgess Field), and Hobby may visit.

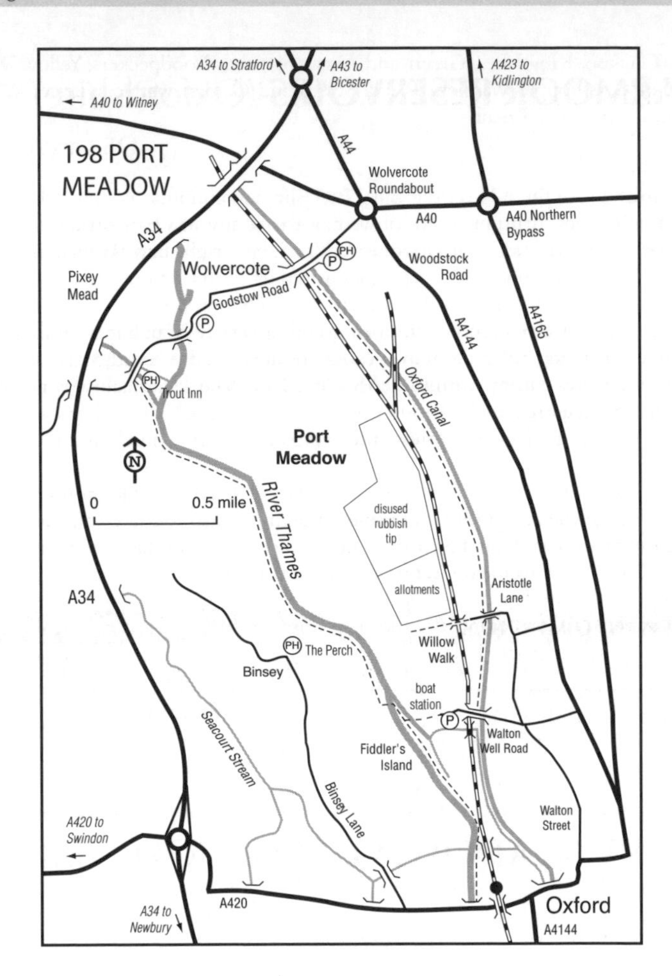

Access

Access to Port Meadow is unrestricted but the area is best worked early in the day before birds are pushed off by disturbance.

1. Leave Oxford city centre north on Walton Street and, after 0.5 miles, turn left into Walton Well Road and continue to the car park at the end. From here there are a couple of paths across the Meadow, one heading west towards the river and one heading north along a part-metalled track – the latter leads north to the main area of flooding and the end of Aristotle Lane (and the entrance gate into Burgess Field LNR – a reclaimed landfill site of about 35 ha; a circular path around the edge of the reserve goes through several small copses). Access to the area is possible via Aristotle Lane but parking is very limited.
2. Leave Oxford city centre north on the A4144 Woodstock Road, turning west at the Wolvercote roundabout (the junction with the A40 Northern Bypass and the A44) into Godstow Road. Park after 1 mile on the left by the River Thames. Continue on Godstow Road to the Trout Inn and take the towpath south along the west bank of the Thames, which affords good views in times of flood, and is often best for waders.

FURTHER INFORMATION

Grid refs: Walton Well Road SP 501 073; Godstow Road SP 487 094

199 FARMOOR RESERVOIRS (Oxfordshire)

OS Landranger 164

Completed in 1976 and lying just 5 miles west of the centre of Oxford, these two reservoirs adjacent to the River Thames are attractive to birds in winter and passage periods.

Habitat

A causeway separates the reservoirs and open water covers 153 ha, but both reservoirs have concrete banks, reducing their attractiveness to waders. To the west, 4 ha bounded by a meander of the Thames comprise Pinkhill Meadow NR, an area of shallow pools and willow scrub, while the 'Shrike Meadow' is another reserve on the banks of the Thames. Also adjacent to the Thames are flood meadows, shrubs and trees.

Species

Winter wildfowl include up to 60 Goldeneyes and 1,000 Wigeons, with smaller numbers of Goosanders, Pintails, Gadwalls and Shovelers, and at times oddities such as divers or seaducks. Indeed, Great Northern Diver and rarer grebes are reasonably regular. Notably, in hard weather the reservoirs may be the last of the local waters to freeze, thus attracting large numbers of wildfowl. Numbers of Cormorants use the reservoirs, roosting at Farmoor II, and Kingfisher is often present. The gull roost holds up to 5,000 Black-headed, 2,500 Lesser Black-backed,

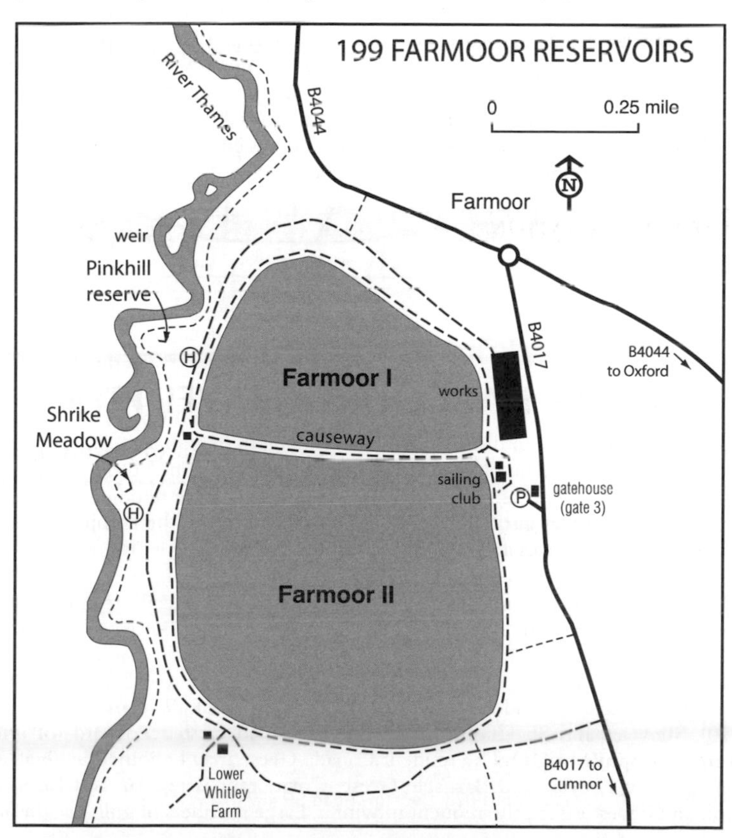

2,000 Herring and small numbers of Common and Great Black-backed Gulls. Up to 80 Yellow-legged Gulls may be present in late summer; Iceland, Glaucous and Mediterranean Gulls occasionally visit. The best place to watch the gull roost is the west end of the causeway (with the sun behind you).

On passage Black, Common and, sometimes, Arctic Terns and Little Gull occur (the latter two especially in spring). The most regular waders are Dunlin, Common Redshank, Common Sandpiper and Ringed and Little Ringed Plovers, but various other species are possible, including Ruff, Greenshank, Curlew Sandpiper, Little Stint and Sanderling. Migrant passerines may include Yellow Wagtail, Wheatear, Whinchat and, especially in early spring and late autumn, Rock Pipit. Large numbers of Swallows, martins and Swifts occur, especially in spring and late summer, when up to four Hobbies may be present, and Osprey is annual, mainly in spring. Common Buzzard is resident in the area.

Breeding birds include Black-headed Gull and Common Tern, which use specially constructed rafts, Cuckoo, Kingfisher and House Martin.

Access

Leave Oxford west on the B4044 and, upon entering Farmoor village, turn south on the B4017. The reservoirs' entrance and car park at Gate 3 are on the right after 0.5 miles. An individual permit to enter the reservoirs costs £10, valid for one year, and a one-day permit is available for £1. A key is needed to enter the Pinkhill Hide, available as a one-off purchase from the gatehouse. Parking at Gate 3 is free to permit holders. The gates are open from dawn to 30 minutes after dusk. Access is restricted to the embankment, and while a circuit of both reservoirs is worthwhile, the central causeway should be adequate for a short visit. Farmoor II is often disturbed by windsurfing and sailing, and though Farmoor I is generally disturbed only by fishermen, watersports are increasing here too (early mornings are perhaps best). Away from the reservoirs, there is access to public footpaths and the Thames towpath, which afford limited views of the reservoirs. To access Pinkhill Reserve follow the sealed road down the slope of the reservoir, to reach the Shrike Meadow Hide take the path directly from the western side of Farmoor II (please do not walk along the service road between the Reservoirs and the Shrike Meadow as this tends to flush all the birds from the reserve).

FURTHER INFORMATION

Grid ref: SP 451 060

200 STANTON HARCOURT GRAVEL PITS (Oxfordshire)

OS Landranger 164

Lying in the Windrush Valley just 6 miles west of Oxford city centre, this complex includes Dix Pit, one of the top birding sites in the county.

Habitat

The area comprises more than 30 sand and gravel pits.

Species

Winter wildfowl include Goldeneye, Goosander, Ruddy Duck, Wigeon, Gadwall, Pintail (up to 70), Shoveler and Shelduck, and occasionally Red-crested Pochard (of unknown but presumably captive origin). The resident Canada Geese may be joined by feral/escaped Bar-headed, Snow, Greylag and Barnacle Geese. Cormorant is regular, and Black-necked and Slavonian Grebes reasonably frequent in winter. Large numbers of gulls use the adjacent tip and many spend time loafing on Dix Pit; there is also a large gull roost with similar species to Farmoor Reservoirs, with which there is considerable interchange (see page 287). Yellow-legged Gulls are seen in most months of the year, with April and May the period when they are least likely. This has also become one of the best places in the country to find Caspian Gulls, although gull scaring in recent winters has decreased the once huge concentrations of birds. Glaucous, Iceland and Mediterranean Gulls are occasionally recorded. The surrounding fields and the dry pits hold large flocks of Lapwings, which are joined by a few Golden Plovers, and Common and Green Sandpipers occasionally overwinter. Raptors sometimes include Short-eared Owl or Merlin, and Little Egret is regular.

On passage a variety of waders is recorded, often including Greenshank and Ruff. Common, Arctic and Black Terns and Garganey also pass through, and Osprey is possible. Migrant passerines may include Wheatear, Whinchat and wagtails.

Breeding species include Great Crested and Little Grebes, Cormorant (there is a healthy colony on the islands), Grey Heron, Black-headed Gull, Common Tern, Turtle Dove, Little and Barn Owls, Kingfisher, Sand Martin, Nightingale and Corn Bunting, and in summer Hobby often visits.

Access

Many pits have restricted access and are heavily disturbed by watersports (though almost all can be viewed from a right of way, but note that these are sometimes re-routed). Four sites are best both for ease of access and ornithological interest.

Dix Pit The most important in the complex for wintering wildfowl. On the B4449 at Sutton village, keep straight on at a roundabout (signed Hardwick), then turn left (signed Waste Disposal Centre) then turn immediately right onto the Dix Pit approach road (Draw Road). The pit lies to the left and it is possible to view from the lay-by almost immediately on the left. Alternatively, keep straight on around the pit to the Waste Disposal Centre and park well off the road, to the left, just before the weighbridge, and view the pit (be sure to note the gate closing times – you may be locked in! If in doubt park outside the gate and walk).

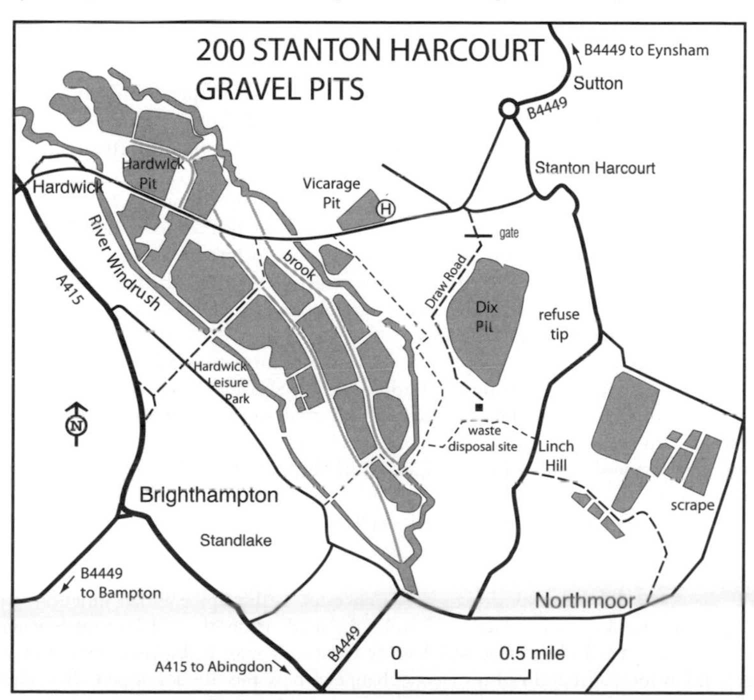

Linch Hill Pits Three pits managed by the ARC for fishing and water sports. The entrance is off a minor road 1.25 miles south of Stanton Harcourt. Entrance is free.
Northmoor Scrape A shallow pit, favoured by waders, Garganey and Hobby. View from the footpath running past the college to the pit, accessed via the gated minor road running north from Northmoor village.
Vicarage Pit View from the minor road between Stanton Harcourt and Hardwick (limited roadside parking) or from the hide in the south-east corner. This pit's ornithological interest has apparently declined in recent years.

FURTHER INFORMATION

Grid refs: Dix Pit SP 407 055; Linch Hill Pits SP 412 040; Northmoor Scrape SP 420 029; Vicarage Pit SP 402 055

201 COTSWOLD WATER PARK (Gloucestershire and Wiltshire) OS Landranger 163

Straddling the county border 4 miles south of Cirencester, this is a complex of over 140 gravel pits and counting – new lakes are created each winter and it is expected to become the largest man-made wetland area in Europe by 2050. The area attracts a good variety of birds throughout the year, although wildfowl and passage waders are the main draw.

Habitat

The extraction of gravel since 1920 has left a series of flooded pits situated amidst typical lowland farmland.

Species

On passage a good variety of waders occurs, the more regular species including Turnstone, Oystercatcher, Whimbrel, Ruff, Greenshank, Common and Green Sandpipers and Dunlin. Other migrants include Shelduck and a few Garganeys, while Common Terns are fairly regular migrants and odd Black and Arctic Terns may also occur.

Breeding birds include Great Crested and Little Grebes, Grey Heron, Sparrowhawk, Ringed and Little Ringed Plovers, Curlew, Common Redshank, a few pairs of Common Terns, Cuckoo, Kingfisher, Sand Martin, Reed and Sedge Warblers, Lesser Whitethroat, and small numbers of Nightingales, and Hobbies are regular in summer, attracted in part by the large flocks of feeding Swifts, Swallows and martins. Notable too are several pairs of feral Red-crested Pochards, and Dippers on some of the streams in the area.

Winter wildfowl within the complex include large numbers of Wigeons, Teals, Pochards and Tufted Ducks, with a scattering of Ruddy Ducks, Shovelers, Gadwalls and Goldeneyes and a few Goosanders. Divers, the rarer grebes, and small numbers of Smews may also occur, especially in hard weather. Other wintering species include Cormorant, Water Rail, large flocks of Lapwings and Golden Plovers, Common Snipe, a few Jack Snipe, and Siskin, while a gull roost at pit 16 near South Cerney attracts large numbers of Black-headed Gulls and several hundred Common and Lesser Black-backed Gulls, but the other large gulls are scarce.

Access

Turn off the A419 3.5 miles south-east of Cirencester at the Spine Road junction onto the B4696 signed to the Water Park. After 0.25 miles the road crosses the old Thames–Severn Canal and there is a car park and the Gateway Visitor Centre. A map is displayed here and at other car parks, and as footpaths and rights of way change as new pits are developed, these should be

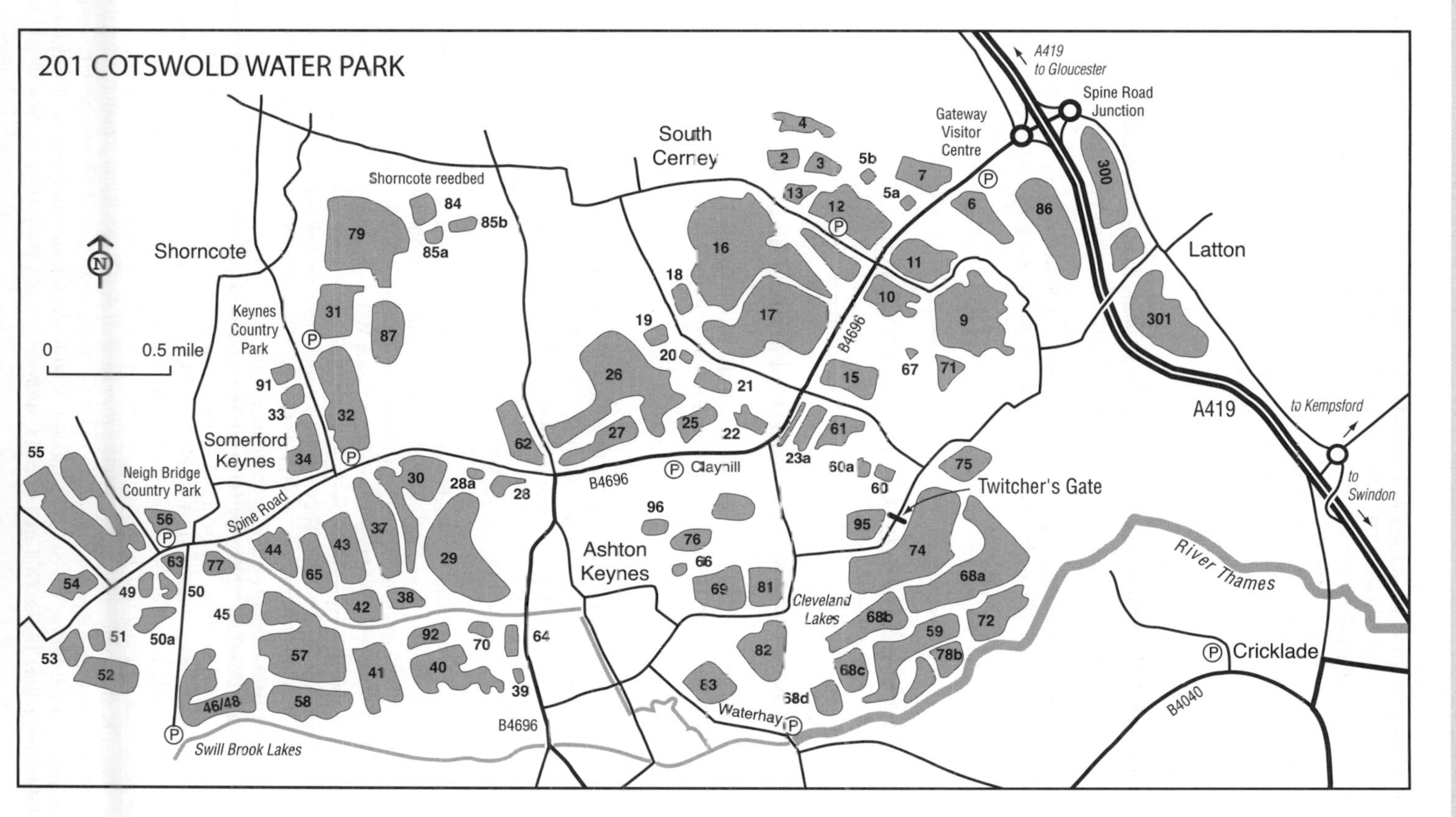
201 COTSWOLD WATER PARK
0
0.5 mile
Shorncote
Shorncote reedbed
Keynes Country Park
Somerford Keynes
Neigh Bridge Country Park
Spine Road
South Cerney
Gateway Visitor Centre
A419 to Gloucester
Spine Road Junction
Latton
A419
to Kempsford
to Swindon
Twitcher's Gate
B4696
Claynill
Ashton Keynes
Cleveland Lakes
Waterhay
River Thames
Cricklade
B4040
Swill Brook Lakes

consulted for the most up-to-date information. In such a big area it is hard to single out specific sites, but the Cleveland Lakes pits 74 and 68c (viewable from the 'Twitchers Gate') are often very productive for wildfowl, while the silt beds (pits 68c and 68d, accessed from the Waterhay car park) have been the best site in the county for waders for many years. Pit 44 is good for Smew, Swill Brook lakes (pit 46/48) for Nightingales, pits 57 and 41 for Hobbies, and 15, and 57 for wintering wildfowl. The Shorncote reedbed is being managed to attract Bitterns, and this area is also good for waders, especially pits 79, 87, 84, 85a and 85b, and these are accessed from the parking area by pits 31/32. Maps with the pits all numbered can be obtained free from the Cotswold Water Park Gateway Centre or at Keynes Country Park or may be downloaded from the website.

FURTHER INFORMATION

Grid refs/postcode: Gateway Visitor Centre SU 072 971 / GL7 5TL; 'Twitchers Gate' SU 065 946; Waterhay car park SU 060 933
Cotswold Water Park Society, Keynes Country Park: tel: 01285 862777; Biodiversity Officer (Gareth Harris): tel: 01285 861459; email: gareth.harris@waterpark.org; web: www.waterpark.org

202 SLIMBRIDGE (Gloucestershire)

OS Landranger 162

Slimbridge is the headquarters of the Wildfowl and Wetlands Trust (WWT), formed in 1946 by the late Sir Peter Scott. The area was chosen because of the largest flock of wintering White-fronted Geese in Britain, and these remain the major attraction for birdwatchers, together with a flock of Bewick's Swans. A visit is best in winter, when some 30,000 wild birds visit the reserve.

Habitat

Reclaimed meadows form an area of 400 ha known as the New Grounds, while 80 ha of grassy saltmarsh outside the sea wall are known as the Dumbles. Beyond this, large areas of mud are exposed alongside the River Severn at low water. Some shallow pools and scrapes have been created inside the sea wall, attracting waders, and these are visible from the Trust's hides.

Species

Once peaking at 6,700, White-fronted Geese have declined greatly at Slimbridge and nowadays only around 600 winter. Some White-fronts arrive in early October, but numbers remain low until late November, when they begin to build up rapidly, peaking in the New Year. Most will have departed by early March. One or two Lesser White-fronted Geese used to be found most winters, but are less regular now. Individuals of several other species of geese can usually be found with the White-fronts. They require a good telescope and a great deal of patience to find – not so the feral Barnacle, Canada and Greylag Geese. Several hundred Bewick's Swans winter; they feed on the fields and roost on the estuary, but also visit Swan Lake where they can be seen at very close range. Nine species of ducks are regularly present, including large numbers of Wigeon as well as Pintail, Gadwall and Shoveler, and many come into the collection to feed. Waders include several thousand Lapwings and Golden Plovers in the fields, sometimes together with a few Ruffs, as well as the common open-shore species. Common Buzzard and Peregrine are present all year and may be joined in winter by Merlin or Short-eared Owl. Little Egret and Water Rail are resident, and a Kingfisher may be seen from a special hide in the grounds.

Waders are more varied on passage, and can include Whimbrel, Black-tailed Godwit, Sanderling, Little Stint, Greenshank, Spotted Redshank, Green Sandpiper and Turnstone. Hobby and Yellow-legged Gull are regular in summer.

Access

Leave the M5 at junction 14 and drive north along the A38, turning left after 8 miles onto a minor road signed Slimbridge. The reserve is 2 miles from this turning. Coming from the north, the turning is 4 miles from junction 13 on the M5. The WWT's Slimbridge headquarters boasts a visitor centre with every facility imaginable. There are exhibitions, a cinema, an art gallery and a restaurant, as well as nature trails and a tropical house for hummingbirds. Visitors can view the wild geese from three towers and sixteen hides, accessible through the Trust's collection. This is open daily, 9.30am–5.30pm (to 5pm November–March), except Christmas Day. There is a charge for non-WWT members. As well as the wild geese, the collection of captive waterfowl has grown into the world's largest and best. A recent innovation has been the opening of a walk to a hide at Mid-Point throughout the spring and summer, allowing a close approach to the River Severn and views of waders.

FURTHER INFORMATION

Grid ref/postcode: SO 722 048 / GL2 7BT
The Wildfowl and Wetlands Trust, Slimbridge: tel: 01453 891900; web: www.wwt.org.uk

203 FRAMPTON (Gloucestershire)

OS Landranger 162

Flooded gravel pits on the outskirts of this picturesque village hold small numbers of wildfowl, while views of the Severn Estuary are possible from Frampton Breakwater where some good flashes attract passage waders.

Habitat

At Frampton the southern, larger pit is deep with steep banks and often disturbed by sailing, while the smaller northern pits are shallower, more heavily vegetated and less disturbed. Frampton Breakwater overlooks the vast Severn Estuary.

Species

Wintering wildfowl on the pits include Pochard, Teal, Shoveler, Gadwall, a handful of Goldeneye, and sometimes the odd Smew; there are also regularly numbers of Cormorants. Water Rails winter but are typically elusive, Siskins and redpolls may be found in the waterside alders, and Chiffchaffs may also winter, while the rarer grebes, divers or Bitterns are occasionally recorded. On the estuary, wildfowl and waders, as described under Slimbridge, may be seen in somewhat less artificial (and consequently less comfortable) conditions. The pools may hold Jack Snipe, and raptors such as Hen Harrier, Peregrine, Merlin or Short-eared Owl may occur.

On passage Garganeys are sometimes found, and small numbers of Common, Arctic and Black Terns and Little Gulls. Wader passage on the pits is not notable, and typically the longer-legged shanks and sandpipers are attracted.

Breeding birds include Ruddy Duck (it was from Slimbridge that British Ruddy Ducks originated), a resident flock of Mandarins (which may peak at over 50 in the autumn), and there are resident flocks of feral Greylag and Canada Geese, often joined by other geese of captive origin. Little Egret is resident and Water Rails sometimes breed, but Turtle Doves and Kingfishers are regular. Breeding passerines include Reed Warbler and several pairs of Nightingales. Hobbies may visit at any time in summer.

Access

Leave the M5 at junction 13 west onto the A419, and after 0.3 miles turn south at the roundabout onto the A38. After a further 0.5 miles turn west on the B4071 to Frampton on Severn. After 1.5 miles turn left into the village.

Frampton Pools Continue on to the village green in Frampton and park either in the small car park by the post office at the northern end of the green, or on the roadside at the far end of the green. A public footpath runs from the southern end of the green along a track to the yacht club, from which the Sailing Lake can be scanned, but the best access point for the Sailing Lake is from the top of Vicarage Lane. In recent years the field adjacent and to the south of the Sailing Lake has flooded and when wet attracts a good range of waders. Court Lake can similarly be viewed from public footpaths, with scrubby areas that are good for warblers and Nightingale.

Frampton Breakwater Leaving the village southwards, fork right at the industrial units and turn immediately left into the car park. To the north of the Splatt Bridge, which spans the Gloucester and Sharpness Canal, the Top Flashes can be viewed from the canal towpath – cross the bridge and walk north up the footpath, past the reedbed on the left and to view towards the river just beyond. This area may be productive in spring, but tends to be dry in summer. For the Hundred Acre, walk south down the towpath and go through a small gate to the right, just past a reedbed (there is a map just inside the gate); go straight across two fields, reaching Green Lane at the second gate; and walk left down the footpath (there is no access to the north) with pools to the right and then the Hundred Acre – view this only from the gate or viewing platform. Note that the gate just across Splatt Bridge is a good spot to watch for owls and raptors in winter and also the pre-roost gatherings of Starlings, while nearby, the small overgrown pool along Marsh Lane holds Reed and Sedge Warblers in spring and summer.

Nebrow Hill In 1998–99 150,000 trees were planted on 70 ha of arable farmland and in some years Hen Harrier and Short-eared Owl can be found in this area. Continue south past the turn to Splatt Bridge and past a row of cottages on the right to a five-bar gate; there is a car park immediately beyond to the right (information board with map). There is open access on foot and a few hundred yards further south along the track a large gravel mound (built to attract Sand Martins) gives good views of the whole area.

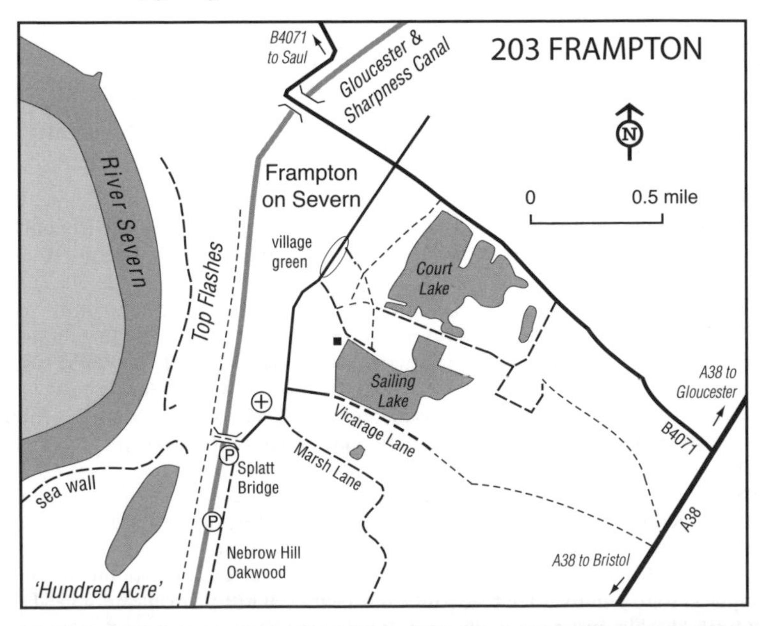

FURTHER INFORMATION

Grid refs: Frampton Pools SO 745 074; Splatt Bridge SO 742 067; Nebrow Hill SO 742 063

204 HIGHNAM WOODS (Gloucestershire)

OS Landranger 162

Lying around 4 miles west of Gloucester, this RSPB reserve was established to protect its population of Nightingales, which are best looked for from late April to early June.

Habitat

The wood is mixed, with a mixture of coppice and non-intervention high forest stands. Management aims to remove the non-native tree species.

Species

As well as around 20 pairs of Nightingales, breeding species include Woodcock, all three woodpeckers, Spotted Flycatcher, Nuthatch, Treecreeper and Marsh Tit.

Access

Leave Gloucester westwards on the A40 towards Ross. Go straight on at the roundabout after 1.25 miles and continue for 0.6 miles, where the wood is signed to the north of the road. The reserve is open at all times and there is a 1.25-mile nature trail (often muddy).

FURTHER INFORMATION

Grid ref: SO 778 190
RSPB Highnam Woods: tel: 01594 562852; email: highnam.woods@rspb.org.uk

205 FOREST OF DEAN (Gloucestershire)

OS Landranger 162

The Forest of Dean cloaks a large area between the confluence of the Rivers Wye and Severn and holds an excellent variety of woodland birds, including some sought-after specialities such as Nightjar, Pied Flycatcher and Hawfinch. May and June are the best months to visit.

Habitat

The Forest of Dean is a relict of the ancient wildwood that once covered Britain but, sadly, large areas have been felled and replaced with conifers and Pendunculate Oak. The RSPB's Nagshead reserve protects one of the largest and best stands of deciduous woodland, consisting largely of mature plantations of Pendunculate Oak, with some Beech and stands of conifers, while a fast-flowing stream runs from the Cannop Ponds through the oak woodland.

Species

Resident species in the forest as a whole include Common Buzzard, Sparrowhawk, Goshawk, Woodcock, all three woodpeckers, Dipper, Marsh and Willow Tits (the latter favouring dense stands of young conifers, reflecting their normal choice of habitat on the Continent), Nuthatch, Treecreeper, Raven, Common Crossbill and a few pairs of Siskins. Hawfinch is a speciality, and the key to finding the species in winter is to track down stands of Hornbeams, under which they can be seen quietly feeding.

Summer visitors include Nightjar (in the clearings formed by clear-felling and replanting conifers), Tree Pipit, Grey Wagtail, Common Redstart, a few pairs of Stonechats and Whinchats,

Grasshopper Warbler (in areas of young conifers), Wood and Garden Warblers, over 100 pairs of Pied Flycatchers, and a few pairs of Firecrests. Hobbies and Red Kites may visit.

In winter variable numbers of Bramblings arrive, and large numbers of Siskins and Lesser Redpolls augment the local breeding species.

Access

The following sites are recommended:

RSPB Nagshead Leave the B4234 westwards at Parkend on the minor road towards Coleford and, on the outskirts of Parkend, turn north at the RSPB sign onto a track, following this for 0.5 miles to the reserve car park. Access is unrestricted, and there are 1-mile and 2.25-mile waymarked trails, with two woodland hides overlooking ponds; an Information Centre is manned at the car park 10am–4pm at weekends, April–August. Around 50 pairs of Pied Flycatchers nest on the reserve, and Wood Warbler and Common Redstart are also common. Other species include Lesser Spotted Woodpecker and Hawfinch, while the plantations and clearings on the northern and western fringes of the reserve hold Tree Pipit, Whinchat, Lesser Redpoll and sometimes Common Crossbill.

New Fancy View Lying just east of Nagshead, this raptor watchpoint is manned by RSPB staff during the Goshawk display season, February–April. Leave the B4234 eastwards at the north end of Parkend on a minor road towards Upper Soudley and, after 1 mile, turn north to the car park at New Fancy View.

Speech House Hotel This lies to the south of the B4226 between Cinderford and Coleford, with a car park opposite (Common Redstarts and Pied Flycatchers breed around this car park). The arboretum and blocks of conifers to the south and east hold Willow Tit, Siskin and Common Crossbill.

Woodgreens Lake Hobbies may hunt over this large, rush-fringed pool, which attracts Common and Jack Snipes in the winter and has resident Willow Tits; the lake is a reserve of the Gloucestershire Wildlife Trust. The block of forest to the north, between the B4226 and A4136, holds Hawfinches, with Nightjars in the clearings.

Cannop Ponds These are signed off the B4234 north of Parkend. Look for Dippers on the stream, and explore the oak woodland to the south-east. The surrounding trees hold Siskins and redpolls in winter.

FURTHER INFORMATION

Grid refs: RSPB Nagshead SO 606 085; New Fancy View SO 628 094; Speech House Hotel SO 620 121; Woodgreens Lake SO 631 124; Cannop Ponds SO 607 108
RSPB Nagshead: Tel: 01594 562852
Gloucestershire Wildlife Trust: Tel: 01452 383333; web: www.gloucestershirewildlifetrust.co.uk

206 SYMONDS YAT (Gloucestershire)

OS Landranger 162

Since 1982 a pair of Peregrines has nested at Symonds Yat Rock, in the Wye Valley between Chepstow and Monmouth, with viewing facilities laid on by the RSPB. The Peregrines are resident, but most active June–August when feeding young.

Habitat

The lower Wye Valley is bounded by steep limestone cliffs, some of which rise to 300 feet, with stands of coniferous, deciduous and mixed forest in the valley.

Species

Breeding raptors in the area include Common Buzzard, Sparrowhawk, Goshawk, Hobby and Kestrel, in addition to Peregrine, while Red Kite and, less frequently, Osprey may pass by. The adjoining woodlands hold a range of typical forest species, including Wood Warbler, while Dipper and Grey Wagtail can be found on the River Wye.

Access

Leave the A4136 at Coleford northwards on the B4432, parking in the large car park on arrival at Symonds Yat. Alternative access is south off the A40 at Goodrich, via the B4229, crossing the Wye via a narrow minor road. From the car park a short trail leads over a footbridge to the viewpoint. The eyrie is located on the cliffs south of the river, a little upstream of the viewpoint. From the car park, a variety of trails leads into deciduous woodland to the south.

FURTHER INFORMATION

Grid ref: SO 564 158
RSPB Nagshead: tel: 01594 562852

207 THE WYRE FOREST (Worcestershire)

OS Landranger 138

Straddling the Shropshire-Worcestershire border to the west of Bewdley and covering around 2,500 ha, this area of woodland is not only scenic but also holds an excellent selection of woodland birds, including Common Redstart, Pied Flycatcher and Wood Warbler. Portions of the forest are protected as an NNR and as reserves of the Worcestershire Wildlife Trust (WWT) and West Midlands Bird Club.

Habitat

The forest contains some of the best stands of ancient woodland in Britain, predominantly oak, but just over half of the forest is actually conifer plantations, managed by the Forestry Commission. The Dowles Brook runs through one of the richest and most diverse areas of the forest, including meadows and abandoned orchards.

Species

Residents include Mandarin (along Dowles Brook, and sometimes also feral Wood Duck), Common Buzzard, Goshawk, Sparrowhawk, Woodcock, Kingfisher, all three woodpeckers, Grey Wagtail, Dipper, Marsh and Willow Tits, Nuthatch, Raven, Siskin and Hawfinch. In summer these are joined by Cuckoo, Turtle Dove, Tree Pipit (especially along the old railway line), Common Redstart, Wood and Garden Warblers and Spotted and Pied Flycatchers (the latter notably at Knowles Coppice and around Knowles Mill, although numbers have declined recently). Search also for Nightjar and Firecrest.

In autumn and winter there may be sizeable finch flocks, including Brambling, Siskin, Common and Lesser Redpolls and, sometimes, Common Crossbill (the Arboretum is often favoured, which is accessed from the Callow Hill visitor centre on the Buzzard Trail).

Access

Dowles Brook Leave Bewdley town centre westwards on the A456 towards Tenbury Wells, go up the hill and then turn right into The Lakes Road. From here, take the third turning on the left into Dry Mill Lane and follow this to the small car park (on the left just before the old railway bridge). From here take the footpath west along the old railway line, following

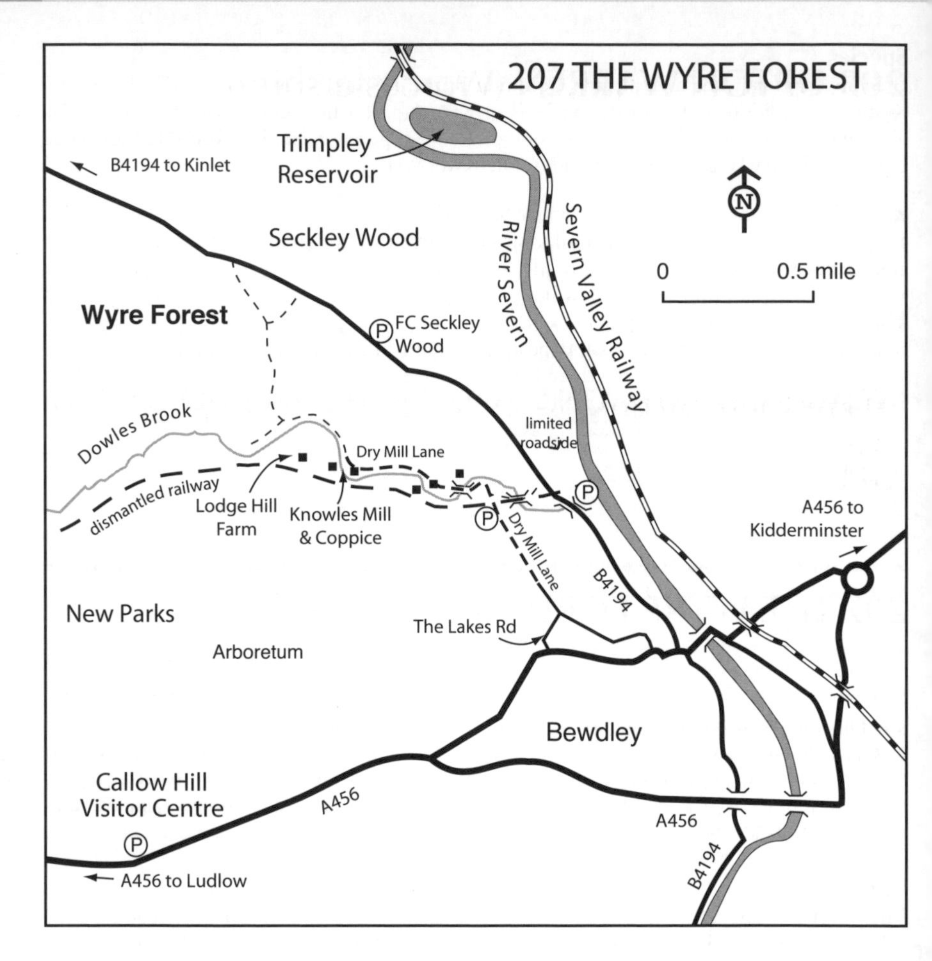

Dowles Brook, for about 0.5 miles to where a gate on the right gives access to Knowles Coppice WWT reserve and connects the old railway line to Dry Mill Lane, passing Knowles Mill and crossing the Dowles Brook via a footbridge (and a little further west, the area around Lodge Hill Farm can be good for Lesser Spotted Woodpecker and, in winter, for Hawfinch). In combination, Dry Mill Lane and the old railway make up a circular walk that should deliver most or all of the specialities.

Seckley Wood Leave Bewdley north-west on the B4194 Kinlet road and, 2.5 miles from Bewdley, park at the FC picnic site on the right. There are three marked trails and, after initially following the red trail, leave it on the path past Seckley Beech to a track that offers panoramic views over the River Severn and Trimpley Reservoir.

New Parks Leave Bewdley west on the A456 Ludlow road and, after 2.5 miles, park on the right at the FC Callow Hill Visitor Centre. From here follow the red or green marked trails through the principal area of coniferous forest. The red trail is 3.5 miles long and is connected by footpaths to Dowles Brook.

FURTHER INFORMATION

Grid refs: Dowles Brook SO 772 762; Seckley Wood SO 761 776; New Parks/Callow Hill SO 749 739

Worcestershire Wildlife Trust: tel: 01905 754919; www.worcswildlifetrust.co.uk

208 UPTON WARREN (Worcestershire)

OS Landranger 150

Also known as the Christopher Cadbury Wetland Reserve, this compact site (split into two halves) can produce a variety of birds, especially wildfowl and waders. The area is a WWT reserve covering 26 ha.

Habitat

The extraction of underground salt deposits has resulted in a number of subsidence pools, the Moors Pool to the north and the three Flashes (themselves saline) to the south. Between these is a relatively unproductive gravel pit. The pools are surrounded by rough grassland, with scrub bordering the Henbrook and alders the River Salwarpe.

Species

Winter brings small numbers of dabbling ducks, including Wigeon and Gadwall, up to 120 Shovelers, and also Ruddy Duck and the occasional Scaup, seaduck or sawbill. There are usually several Water Rails, Jack Snipes and Green Sandpipers. Cetti's Warbler is resident and in recent years there have been up to three wintering Bitterns. There are often Siskins and redpolls in the riverside alders.

It is during passage periods that Upton Warren, especially the Flashes, comes into its own. Regulars include a roost of up to 110 Curlews, and July–August gatherings of up to 25 Green Sandpipers and 2,000 Lapwings, as well as Common Sandpiper, Ringed Plover, Dunlin, Greenshank and Black-tailed Godwit. Scarcer species can include Ruff, Sanderling, Turnstone, Bar-tailed Godwit, Whimbrel and perhaps Wood Sandpiper, Spotted Redshank, Little Stint or Curlew Sandpiper. Indeed, almost any species of wader may turn up. Other migrants include Garganey, Osprey, Marsh Harrier, Common, Arctic and Black Terns, and a variety of passerines, notably Whinchat. The Black-headed Gull roost grows in size from late summer and often hosts Mediterranean Gulls with the occasional Little Gull.

Breeding species include Sparrowhawk, Water Rail, Oystercatcher, Little Ringed Plover, Lapwing, Common Redshank, Avocet (from 2003 onwards), Common Tern, Black-headed Gull, Kingfisher and Reed, Sedge and Cetti's Warblers. A feature of summer and early autumn is the regular evening forays of Hobbies among the Swallows and martins. Peregrine is present throughout the year, often perched on the radio masts to the south.

Access

The area lies east of the A38, 2 miles south-west of Bromsgrove, and can also be reached by leaving the M5 at junction 5, proceeding north along the A38 for 2 miles. Access is restricted to members of the WWT or other affiliates of the RSNC, and is otherwise by advance permit either from the Trust offices or from the Outdoor Education Centre adjacent to the reserve. There are seven hides.

Moors Pool Turn east off the A38 500 yards north of the Swan Inn, following a farm track for 200 yards and then parking on the left. There are three hides, one adjacent to the car park overlooking the North Moors Pool, the other two either side of Moors Pool. Alternatively, park at the Swan Inn and cross the A38, proceeding on foot through the gate opposite and following the path beside the River Salwarpe to the hides. Moors Pool is best for wildfowl, with Jack Snipe and Water Rail favouring the North Moors Pool, Bittern and Cetti's Warbler the surrounding reedbeds, and warblers and passerine migrants the adjacent vegetation.

The Flashes Park in the sailing centre car park adjacent to the A38 200 yards south of the Swan Inn (the turning is at the roundabout, and opposite Webbs Garden Centre) and follow the path beyond the sailing centre, around the south shore of the gravel pit and over the Henbrook to the hides. The Flashes are best for passage waders (an early-morning visit after overnight rain is recommended), and are usually quiet in winter, although they often hold numbers of Teal and the occasional Green Sandpiper at this season.

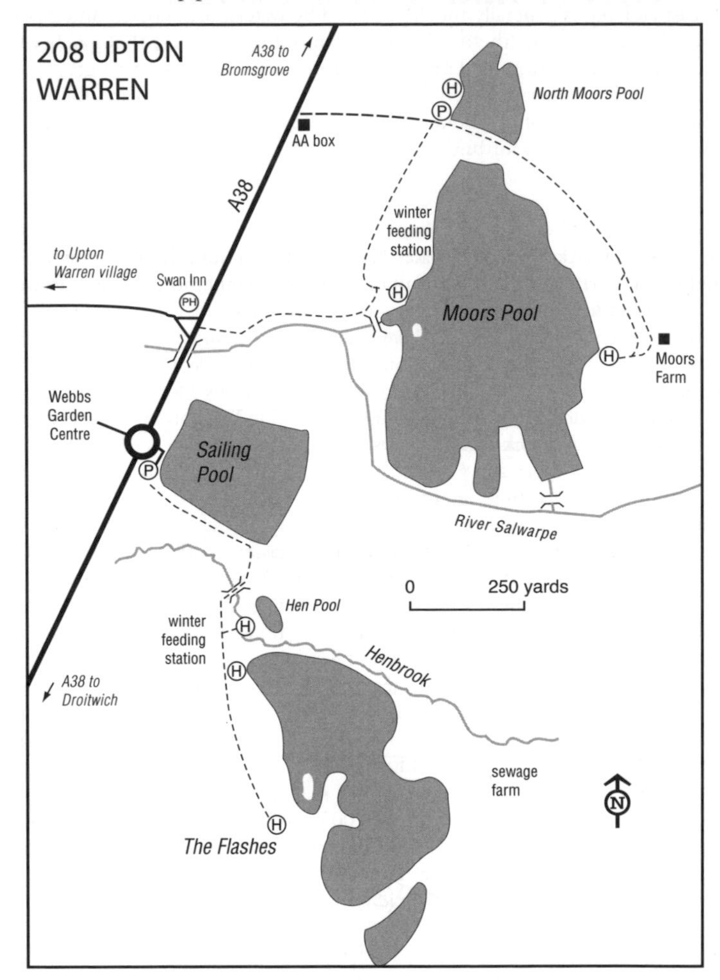

FURTHER INFORMATION

Grid refs: Moors Pool SO 935 677; The Flashes SO 931 671
Worcestershire Wildlife Trust: tel: 01905 754919; www.worcswildlifetrust.co.uk

209 BITTELL RESERVOIRS (Worcestershire)

OS Landranger 139

Just south of Birmingham, these reservoirs were built as canal-feeders and have a long ornithological history. A variety of wildfowl is usually present in winter, and small numbers of waders move through on passage.

Habitat

Upper Bittell is the larger water, covering 40 ha, with small areas of marsh and shallow water at the north-east end, but is generally much disturbed by watersports. Lower Bittell covers 23 ha and is bisected by a causeway (the northern portion of the reservoir is known as Mill Shrub). Much less disturbed, it is surrounded by woodland and extensive areas of rough grassland. Both reservoirs have natural banks. Between the two, stands of swampy woodland follow the course of a small stream.

Species

Winter is perhaps the most interesting season, with regular Teal, Wigeon, Pochard, Goldeneye and a few Goosanders, Shovelers and Ruddy Ducks, while other sawbills, seaducks and rarer grebes are occasional. Jack Snipe and Water Rail may be found in the wetter areas, with Grey Wagtail, redpolls and Siskin along the stream. There is a small gull roost, but it almost exclusively comprises Black-headed Gulls.

On passage small numbers of waders are recorded, including Ringed and Little Ringed Plovers, Ruff, Dunlin, Greenshank, Common Sandpiper and occasionally Turnstone, Little Stint or Curlew Sandpiper, as well as Black, Common and Arctic Terns. Hobby is fairly regular in late summer, and Rock Pipit may occur in late autumn.

Residents include Sparrowhawk, Kingfisher and Willow Tit.

Access

Lower Bittell The B4120 between Barnt Green and Alvechurch (accessed via the A441 from junction 2 of the M42) follows the south shore of the reservoir, giving limited views, though parking is difficult. From here turn north on Bittell Farm Road towards Hopwood; parking is easier and there are views of the open water over the hedge. Proceed to the sharp right-hand bend (limited parking) and then pass over the causeway between Lower Bittell and Mill Shrub, with good views from the road.

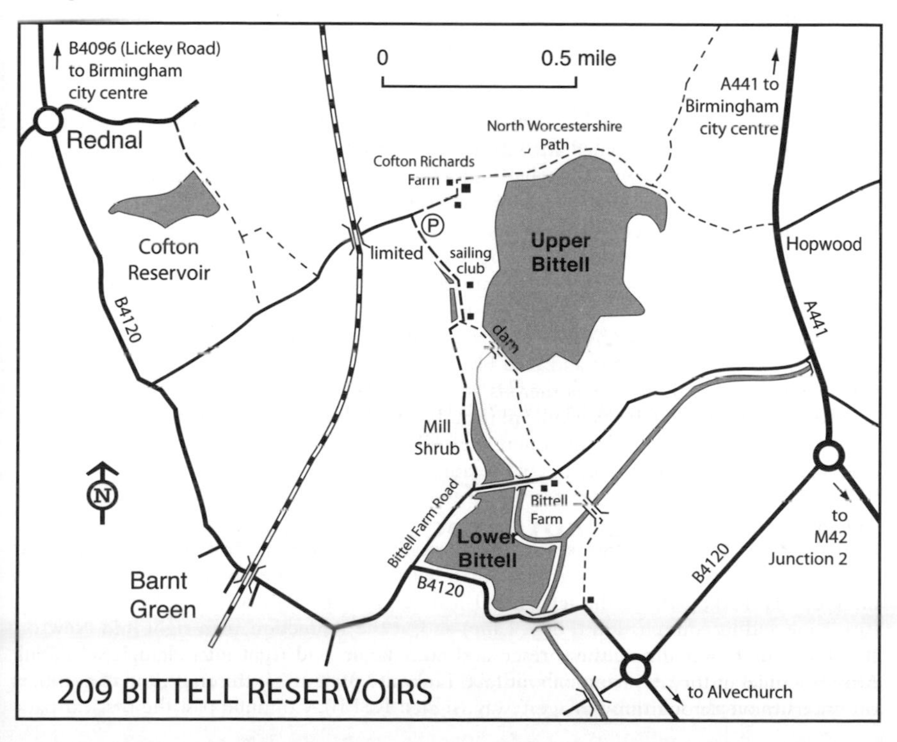

Upper Bittell From the right-angled bend at Lower Bittell causeway follow the lane north beside the stream for 0.5 miles to view Mill Shrub pool. At the two small ponds turn right over the stile by the old pump house on a public footpath along the top of Upper Bittell dam. Alternatively, proceed along the track past the two ponds and, after 0.3 miles, turn right to Cofton Richards Farm (limited parking at the corner). Follow the North Worcestershire Path over the fields until it parallels a short section of the north shore of the reservoir, giving limited views in a few places.

FURTHER INFORMATION

Grid refs: Lower Bittell SP 018 744; Upper Bittell (Cofton Richards Farm) SP 014 755

210 SANDWELL VALLEY (West Midlands)

OS Landranger 139

Sandwiched between Birmingham and West Bromwich, this area is an oasis of green amid the vast urban sprawl of the West Midlands, and attracts a remarkably rich variety of birds year-round.

Habitat

The River Tame meanders through the northern part of the valley, and a balancing lake (Forge Mill Lake) has been constructed next to the river, with the eastern end forming part of an RSPB reserve. Adjacent to the M5, Swan Pool is much more disturbed but can produce surprises. Otherwise the valley has four golf courses, active farmland and areas of rough grass, together with several small stands of mixed woodland, and part is managed as Sandwell Valley Country Park by Sandwell Metropolitan Borough Council.

Species

Residents include Sparrowhawk, Stock Dove, Green and Great Spotted Woodpeckers and Willow Tit, and these are joined in summer by Garden, Sedge and Reed Warblers, Lesser Whitethroat and, sometimes, Whinchat and Grasshopper Warbler. Lapwings breed, sometimes together with Oystercatcher, Common Redshank or Little Ringed Plover. Hobby and Mediterranean Gulls are occasional visitors.

In spring and autumn regular visitors include Common Redshank, Common Sandpiper, Little Ringed Plover and sometimes Oystercatcher, as well as Common, Arctic and occasionally Black Terns. Spotted Crake has appeared on several occasions in autumn. A variety of passerines is regular on passage, including White Wagtail (spring), Tree Pipit, Whinchat and Wheatear, and occasionally Common Redstart and Pied Flycatcher.

In winter Common Snipe is numerous and there may also be Jack Snipe and Water Rail. Look too for Stonechat, Grey Wagtail, Siskin and redpolls. Regular wildfowl include reasonable numbers of Pochards and Teals, and often a few Goldeneyes, Wigeons and Ruddy Ducks, and Goosanders may occur, especially in cold weather.

Access

From junction 1 on the M5 take the A41 towards Birmingham city centre and, beyond West Bromwich Albion football ground, turn left into Park Lane. To access Swan Pool and the Country Park, park on the left after 1.5 miles. For the RSPB reserve continue along Park Lane and its continuation (Forge Lane) and, at the T-junction, turn right into Newton Road, continue over the railway bridge and take the second right into Hampstead Road. After 0.7 miles at the mini-roundabout take Tanhouse Avenue on the right, and the reserve entrance lies on the left through a gateway (signed RSPB). Continue past the first car park

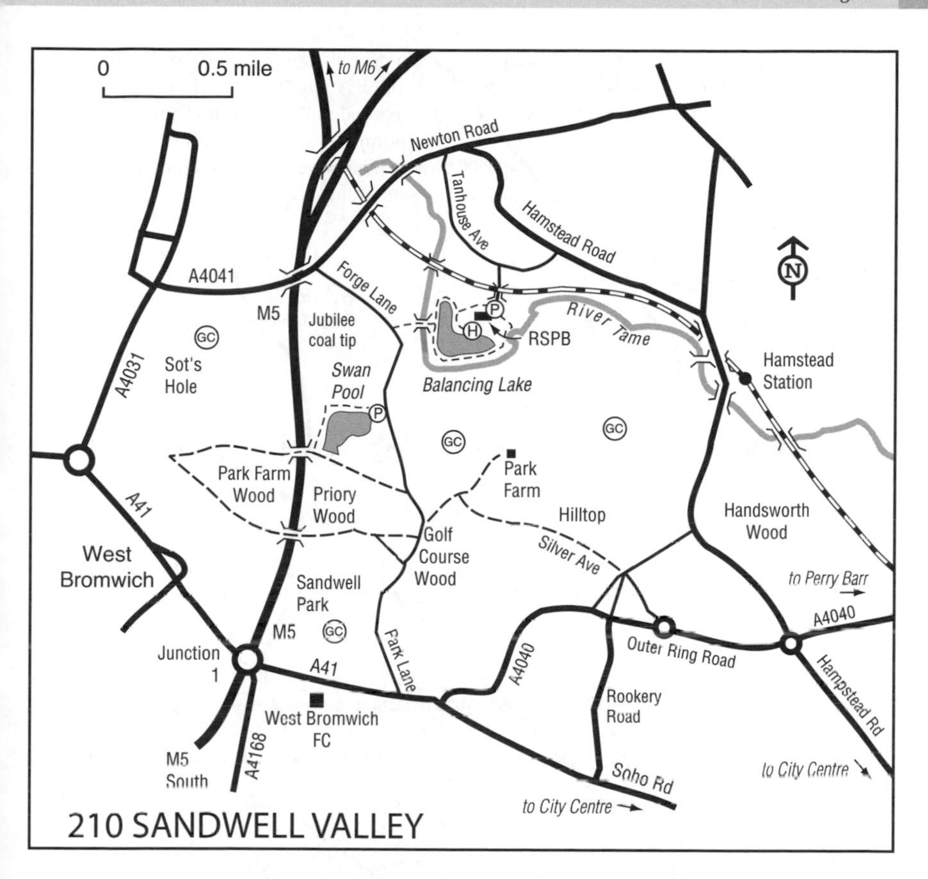

210 SANDWELL VALLEY

over the railway bridge to the RSPB car park outside the visitor centre. (Alternatively, the balancing lake on the RSPB reserve can be viewed from the pull-in on Park Lane.) The RSPB visitor centre offers views of the feeding stations and there is one hide (Lakeside Hide), a viewing platform and three viewing screens. All are wheelchair-accessible, though help may be required, especially in wetter weather. The reserve is open at all times; the hide 10.30am–3.30pm every day except Mondays, and the visitor centre Tuesday-Friday 9am–5pm; Saturday and Sunday 10am–5pm (or dusk in winter). In much of the area it is best to avoid weekends and Bank Holidays, as disturbance is especially heavy at these times.

FURTHER INFORMATION

Grid ref/postcode: SP 035 928 / B43 5AG
RSPB Sandwell Valley: tel: 0121 357 7395; email: sandwellvalley@rspb.org.uk

211 TAME VALLEY (Warwickshire)

OS Landranger 139

The Tame Valley between Tamworth and east Birmingham has a complex of gravel workings that have become one of the premier birdwatching localities in the Midlands.

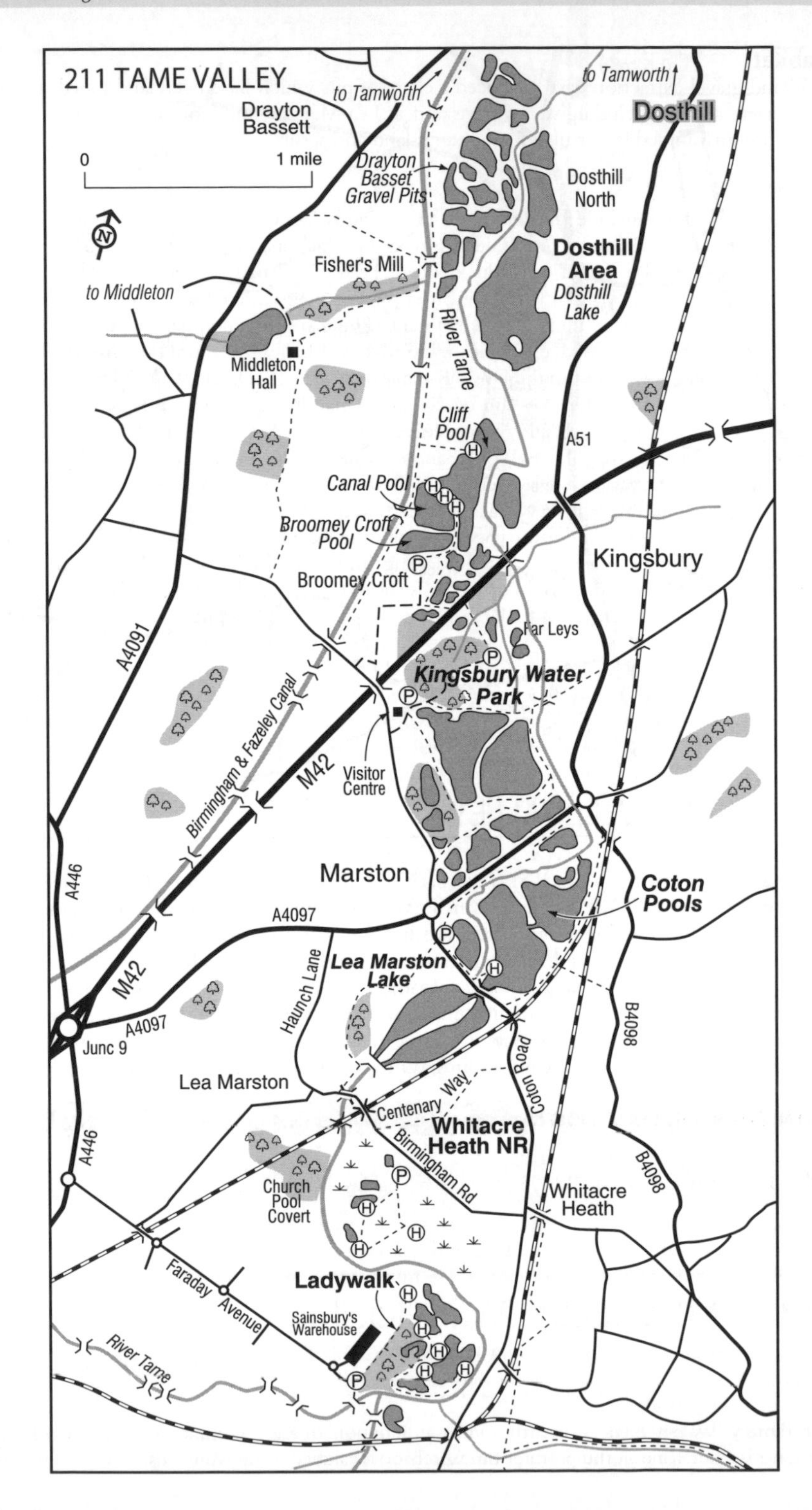
211 TAME VALLEY
to Tamworth
to Tamworth
Drayton Bassett
Dosthill
0
1 mile
Drayton Basset Gravel Pits
Dosthill North
Dosthill Area
Dosthill Lake
Fisher's Mill
to Middleton
River Tame
Middleton Hall
Cliff Pool
A51
Canal Pool
Broomey Croft Pool
Kingsbury
Broomey Croft
Far Leys
A4091
Kingsbury Water Park
Birmingham & Fazeley Canal
M42
Visitor Centre
A446
Marston
Coton Pools
A4097
Lea Marston Lake
Haunch Lane
M42
A4097
Junc 9
B4098
Coton Road
Lea Marston
Centenary Way
Whitacre Heath NR
A446
Birmingham Rd
B4098
Church Pool Covert
Whitacre Heath
Faraday Avenue
Ladywalk
Sainsbury's Warehouse
River Tame

Habitat
Sand and gravel extraction have produced a chain of pits which have been landscaped and put to a variety of uses, including water sports and, at Lea Marston and Coton, water purification. These are surrounded by farmland, rough grassland and scrub.

Species
Wintering wildfowl include large numbers of Pochards, Wigeons, Shovelers and Teals, as well as Goldeneye, Gadwall, and a few Ruddy Ducks, Goosanders, Pintails and sometimes Smew. Occasional divers, rare grebes or seaducks occur. Water Rail may be seen, especially in cold weather, and in recent years up to four Bitterns have overwintered at Ladywalk, while this reserve also holds a roost of up to 200 Cormorants. There are large numbers of gulls, sometimes including Yellow-legged or Mediterranean. Up to 1,000 Lapwings and 3,000 Golden Plovers frequent the fields; they roost in the Dosthill-Kingsbury area. Rock and Water Pipits may occur in late autumn, and the latter may also overwinter. A handful of Common Redshanks, Green Sandpipers and, sometimes, Ruffs may occur at the same season and, quite extraordinarily, even single Spotted Redshank and Wood Sandpiper have wintered. The concentration of birds attracts raptors, and Sparrowhawk, Merlin and Short-eared Owl are possible, while Peregrines favour the pylons and warehouses adjacent to Ladywalk.

A variety of waders occurs on passage, including Dunlin, Ruff, Whimbrel, Green and Common Sandpipers, Turnstone and Sanderling, and Temminck's Stint is almost annual in spring. Parties of Common, Arctic or Black Terns may pass through, augmenting the breeding Common Terns. Osprey and Marsh Harrier visit most years and Hobbies are frequent in summer.

Breeding species include Great Crested and Little Grebes, Common Buzzard, Shelduck, Gadwall, Water Rail, Oystercatcher, Little Ringed and Ringed Plovers, Common Redshank, Common Tern (usually on the Canal Pool at Kingsbury), Black-headed Gull, Turtle Dove, Barn and Little Owls and Kingfisher. Garganey, Teal and Shoveler have occasionally bred. Breeding passerines include ten species of warbler including a few Grasshopper and Cetti's (the latter two species favouring Ladywalk), Willow Tit and still a few Corn Buntings and Tree Sparrows, while Ravens breed nearby and visit the area.

Access
Ladywalk Reserve This 50-ha reserve of the West Midland Bird Club comprises floods and woodland lying within a loop of the River Tame at the eastern end of the former Hams Hall power station site. Leave the M42 at junction 9 and take the A446 south for 1 mile and at the roundabout take Faraday Avenue into the Hams Hall National Distribution Centre, turning right at the third roundabout (by the Sainsbury's warehouse) to the car park. From here public footpaths lead to two public hides (one, to the north, on the path to Church Pool Covert, the other, in the south, overlooking Whitacre Pool, on the path to Whitacre village); access is otherwise by permit only, issued by the Bird Club, giving access to four other hides overlooking the marsh. Bittern favour the River Walk Hide and Hide B and thus a permit is necessary to search for this species.
Whitacre Heath This 44-ha reserve of the Warwickshire Wildlife Trust has three hides overlooking a series of old gravel workings. Access is from a car park off the Lee Marston to Whitacre Heath road, and access is restricted to Trust members only.
Coton Pools and Lea Marston Balancing Lakes Leave the M42 at junction 9 and take the A4097 east towards Kingsbury. After 1.5 miles turn right at the first roundabout onto a minor road (Coton Road) and park on the left after 0.3 miles. Lea Marston Lakes lie to the south-west of the road, while Coton Pools lie to the north-east; both can be viewed from the causeway, and further views of Coton Pools are possible by following the signed path to the public hide or by walking east along the road and following the public footpath northwards from the railway bridge.
Kingsbury Water Park Leave the M42 at junction 9 and take the A4097 east towards Kingsbury. After 1.5 miles turn left at the roundabout (signed to the Water Park) and follow the

minor road for 0.5 miles to the entrance. Follow signs to Far Leys car park. It is usually best, however, to continue over the M42 and turn right immediately after into a small lane, then turn left at the sign to Broomey Croft. Follow this road, forking right to leave the caravan park to the left, through the barriers to the car park. A path leads from here north to several hides. Throughout the park there is a network of marked trails, and Broomey Croft car park is most convenient for the best areas. (Closed on Christmas Day.)

Dosthill Gravel Pits Follow directions to Broomey Croft car park (under Kingsbury Water Park). Walk north along the canal towpath for 1.5 miles to Fischer's Mill Bridge. Leave the canal here and follow the footpath eastwards to view the southern end of Drayton Bassett Gravel Pits. View from the canal side, the bridge or the public footpath to the A4091. In recent years this has been the best area in winter within the Tame Valley, attracting Red-necked and Black-necked Grebes and Smew.

RSPB Middleton Lakes An area of 160 ha of worked-out gravel pits in the Dosthill complex, adjacent to Middleton Hall and between the River Tame and the Birmingham and Fazeley Canal, was purchased by the RSPB in 2007. This new reserve is due to open to the public in 2010.

FURTHER INFORMATION

Grid refs/postcode: Ladywalk Reserve SP 208 915; Whitacre Heath SP 209 931 / B46 2ET; Coton Pools and Lea Marston Balancing Lakes SP 211 944; Kingsbury Water Park SP 203 958; Dosthill Gravel Pits/Broomey Croft SP 203 969

RSPB Middleton Lakes: tel: 01827 259454; email: middletonlakes@rspb.org.uk

West Midland Bird Club: web: www.westmidlandbirdclub.com

Warwickshire Wildlife Trust: tel: 024 7630 2912; web: www.warwickshire-wildlife-trust.org.uk

212 BRANDON MARSH (Warwickshire)

OS Landranger 140

Situated just south-east of Coventry, this 85-ha wetland attracts an excellent variety of breeding, wintering and passage birds. It is managed by Warwickshire Wildlife Trust.

Habitat

Centred around old colliery subsidence pools in the Avon Valley and many gravel pits, the area also has marsh, reedbeds, grassland, willow scrub and small stands of woodland.

Species

Wintering wildfowl usually include up to 300 Teals and smaller numbers of Wigeons, Gadwalls, Shovelers and Pochards, along with a few Goldeneyes. Wintering waders include Common Snipe, with Jack Snipe in late autumn and early spring, and the occasional Woodcock, Dunlin and Green Sandpiper. Raptors include Sparrowhawk, and sometimes Long-eared and Short-eared Owls. Up to three Bitterns have been recorded in recent winters but Bearded Tit, once regular, is now rather rare.

On passage Black-necked Grebe, Garganey and Marsh Harrier are occasional in spring, and waders may include Ringed and Little Ringed Plovers, Curlew, Black-tailed Godwit, Ruff, Greenshank, Common and Green Sandpipers, Dunlin and occasionally Sanderling (in May). Small numbers of Black, Common and Arctic Terns and Little Gull may visit. Little Egrets can now be seen at almost any time of year and Peregrine, Red Kite or Marsh Harrier are increasingly frequent.

Breeding species include Little and Great Crested Grebes, Canada and Greylag Geese,

Common Buzzard, Sparrowhawk, Common Snipe, Common Redshank, Little Ringed Plover, Lapwing, and occasionally Garganey, Gadwall or Shelduck. Kingfisher is frequent, several pairs of Water Rails breed and so do up to 50 pairs of Reed Warblers, together with a few Grasshopper and Cetti's Warblers. The woodland holds all three woodpeckers (Lesser Spotted is scarce), Turtle Dove and Willow Tit, and Hobby is a regular visitor in summer.

Access

From the Toll Barr Island, the point where the A46 intersects with the A46 Coventry Eastern Bypass, head south-east on the A45 dual carriageway and, 200 yards beyond the roundabout and just beyond the Texaco filling station, turn left (north-eastwards) into Brandon Lane (if approaching from the south-east, continue past Brandon Lane towards Coventry and perform a U-turn at the roundabout). Proceed along Brandon Lane and turn right after 1 mile to the nature centre car park. There are seven hides and a nature trail. The reserve is open 9am–5pm on weekdays and 10am–4pm at weekends (members of the Wildlife Trusts get in free and can also visit outside opening hours).

212 BRANDON MARSH

FURTHER INFORMATION

Grid ref/postcode: SP 386 762 / CV3 3GW
Warwickshire Wildlife Trust, Brandon Marsh Nature Centre: tel: 024 7630 8999; web: www.warwickshire-wildlife-trust.org.uk

213 DRAYCOTE WATER (Warwickshire)

OS Landranger 140 and 151

Draycote Water lies just south-west of Rugby. Though heavily disturbed by sailing and fishing it attracts a variety of ducks in winter, often including a few seaducks, as well as divers and grebes. The site is owned by Severn Trent Water.

Habitat

The reservoir covers 280 ha and the shoreline is partly natural and partly concrete embankments. There is a small marsh at the north-east end, overlooked by a hide.

Species

In winter divers are regular, especially Great Northern, and often make long stays, and Slavonian Grebes are occasionally recorded. There may be 250 Great Crested Grebes, 1,000 Coots and 100 Cormorants. 700-1,000 Wigeon occur, together with much small numbers of Gadwalls and Shovelers. Diving ducks are more numerous, with up to 1,200 Tufted Ducks, 1,000 Pochards, 200 Goldeneyes and up to 70 Goosanders. Common and Velvet Scoters, Red-breasted Merganser, Scaup and Common Eider have all been recorded and several of these may be present simultaneously, together with Smew. There is a huge gull roost (50,000+), comprising mainly Black-headed Gulls, but Glaucous is fairly frequent, sometimes together with Iceland or Caspian Gulls. There are several winter records of Mediterranean and Little Gulls, but Kittiwake is more frequent at this season, especially after severe coastal gales. Common Buzzard and Little Owl are resident in the area and Peregrine is fairly regular.

Wader passage in spring is usually poor due to high water levels, Common Sandpiper and Dunlin being the most frequent species. In autumn a greater number and variety of waders may be present, such as Ringed and Little Ringed Plovers, Ruff and Greenshank and perhaps also Oystercatcher, Green Sandpiper and Little Stint. Black, Common and Arctic Terns and sometimes Little Gull also occur on passage, sometimes in good numbers. Ospreys occur most years and Hobbies are regular in summer, while there is a notable late-summer build-up of Yellow-legged Gulls and Swifts, Swallows and martins.

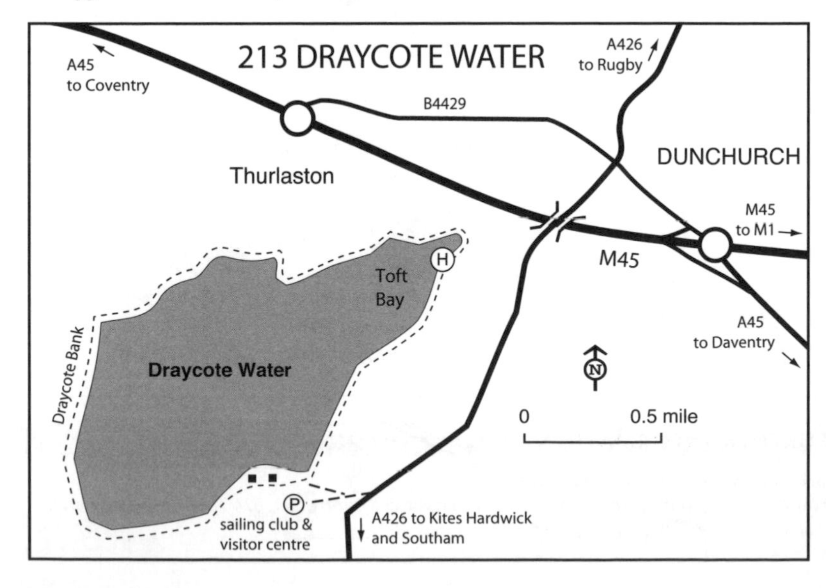

Access

Draycote Water Country Park covers 9 ha on the southern side of Draycote Water, and the entrance is off the A426 2 miles south of Dunchurch. The Country Park is open daily, 8.30am to dusk, and there is free access on foot to the 5-mile long track around the reservoir. The best areas for birds are Toft Bay and around Draycote Bank.

FURTHER INFORMATION

Grid ref/postcode: SP 466 691 / CV23 8AB
Draycote Water Country Park, Draycote Water Visitor Centre: tel: 01788 811107; web: www.moretoexperience.co.uk

214 DAVENTRY RESERVOIR (Northamptonshire)

OS Landranger 152

This small reservoir on the north-eastern outskirts of Daventry attracts a variety of winter wildfowl and passage waders.

Habitat

Built in 1804 as a canal feeder, Daventry Reservoir covers 52 ha with gently shelving natural banks, and is encircled by a narrow ring of mature deciduous woodland, broken only at the dam. The reservoir and surrounding area are a Local Nature Reserve and operated as a country

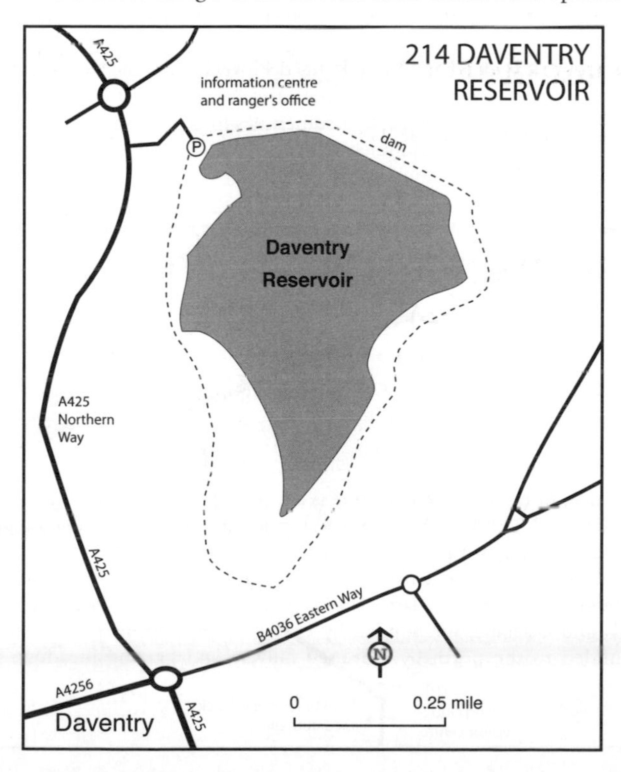

park by Daventry District Council: facilities include an information centre, adventure playground and a café along the track to the dam, while coarse fishing is permitted mid-June to mid-March.

Species

Wintering wildfowl include Teal and Pochard with smaller numbers of Wigeons, Shovelers, Goldeneyes and Goosanders, as well as Cormorant. Divers, the rarer grebes, Shag and a variety of seaducks have also been recorded, with Scaup being the most regular of the latter. Golden Plover and Common Snipe occur around the water's edge. There is a small gull roost, mainly comprising Black-headed Gulls; Mediterranean Gulls are annual (usually juveniles in August–September and adults in spring), as are Yellow-legged Gulls, usually in autumn. The roost is best watched from the north-west shore.

On spring passage Arctic and Black Terns and Little Gull may occur, especially after east winds, and Osprey and Common Scoter sometimes pass through. Of note in April and in late summer and autumn is the occasional occurrence of Black-necked Grebe, and on autumn passage a variety of waders is possible, including Ringed and Little Ringed Plovers, Dunlin, Greenshank, Common and Green Sandpipers and Ruff, with Little Stint, Curlew Sandpiper and Spotted Redshank reasonably regular; the south and west banks are best for waders. At both seasons interesting passerines may occur, such as Common Redstart, Whinchat, Wheatear, Ring Ouzel (spring) and Rock Pipit (late autumn).

Breeding species include Great Crested Grebe, Common Tern (on specially constructed rafts), Green and Great Spotted Woodpeckers and, sometimes, Grasshopper Warbler, while Gadwall and Shoveler have bred.

Access

Leave Daventry north-east on the A425 Northern Way and Daventry Reservoir Country Park is signed to the right after 1 mile. The reserve is encircled by a 2-mile footpath.

FURTHER INFORMATION

Grid ref: SP 576 641
Daventry Country Park: tel: 01327 877193

215 EARLS BARTON GRAVEL PITS (Northamptonshire)

OS Landranger 152

Lying in the Nene Valley just south of Wellingborough, this complex of gravel pits attracts a wide range of waders and wildfowl throughout the year.

Habitat

Within this section of the Nene Valley there is a series of gravel pits stretching over 3.5 miles between Wollaston and Earls Barton. Both working and restored pits exhibit a typical succession of habitats. In the centre of the complex and surrounded by older pits lies Summer Leys LNR, managed by the Wildlife Trust for Bedfordshire, Cambridgeshire, Northamptonshire and Peterborough. It comprises a large landscaped pit with gently shelving banks, specially created islands and a scrape, surrounded by extensive plantations of deciduous trees. The reserve is bounded to the north by a disused railway and grazing meadows alongside the River Nene.

Species

Wintering wildfowl include up to 900 Wigeon as well good numbers of Teal, Gadwall, and Pochard, small numbers of Pintails and Shovelers, and up to 40 Goldeneyes. Parties of Bewick's Swans occasionally visit, and small numbers of Smews regularly winter. Other notable wintering species include Water Rail, Golden Plover, Dunlin and Green Sandpiper. Small numbers of gulls are present, regularly including a few Common and Lesser Black-backed Gulls, and occasionally Mediterranean Gull. The feeding station attracts parties of Tree Sparrow throughout the day.

On spring passage Garganey, a variety of waders including Turnstone, Ruff, Dunlin, Sanderling, Bar-tailed Godwit, Whimbrel, Green Sandpiper and Greenshank, Little Gull, and Arctic and Black Terns occur, while migrant passerines may include White Wagtail, Whinchat and Wheatear. In autumn a similar variety of waders may occur, with the addition of Black-tailed Godwit and, from late August, sometimes Little Stint and Curlew and Wood Sandpipers. Scarcer migrants have included Black-necked Grebe, Osprey, Marsh Harrier and Red-crested Pochard. In late autumn there is a build-up of Golden Plovers in the area, sometimes topping 1,000 birds.

Breeding birds on the reserve and surrounding pits include Great Crested and Little Grebes, Cormorant, Grey Heron, Greylag and Canada Geese, Gadwall, Sparrowhawk, Oystercatcher, Ringed and Little Ringed Plovers, Lapwing, Common Redshank, Black-headed Gull, good numbers of Common Terns, Cuckoo, Sand Martin, Yellow Wagtail, Reed and Sedge Warblers, Tree Sparrow and, sometimes, Shoveler and Shelduck. Little Egret, Hobby, Garganey and Kingfisher are regular visitors.

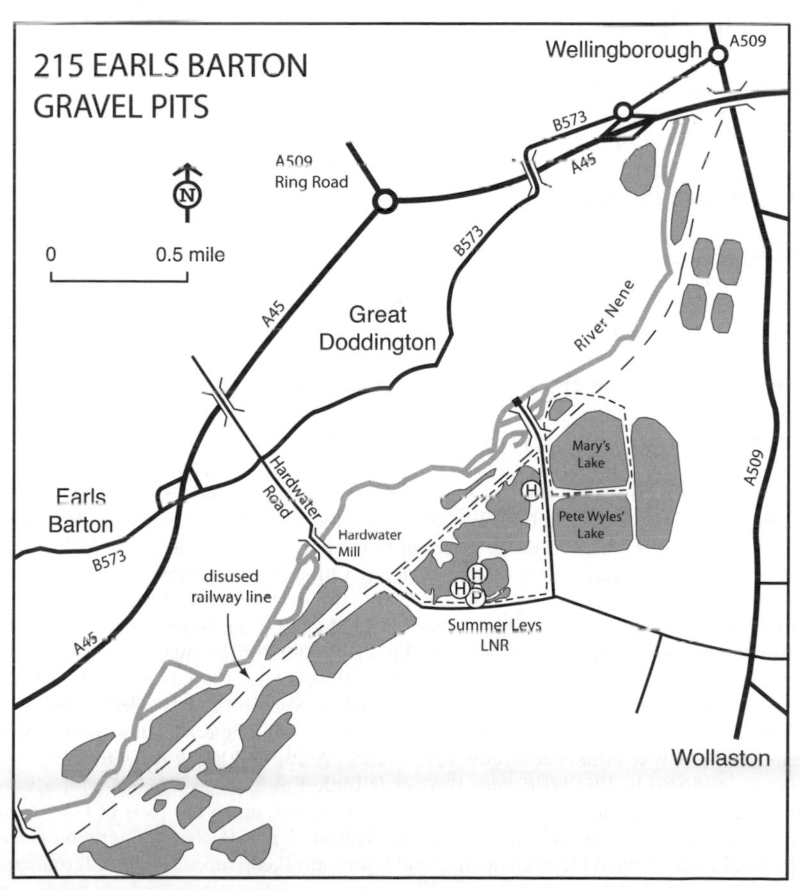

Access

Summer Leys LNR Leave the A45 5 miles east of Northampton on the B573 to Great Doddington. After 0.25 miles (just west of Great Doddington), turn right (south) on Hardwater Road towards Wollaston and follow this road over the River Nene by Hardwater Mill. The reserve car park is on the left after a further 0.25 miles. A 2-mile trail circles the reserve, with three hides near the car park. The reserve is open at all times.

Mary's Lake and Pete Wyles' Lakes These two pits lie east of Summer Leys and can be viewed from the gates along the single-track road that runs north to the river, 0.3 miles west of the reserve car park. Pits further west and newer ones to the east can also be explored by walking along the disused railway track in either direction.

FURTHER INFORMATION

Grid refs: Summer Leys LNR SP 886 633; Mary's Lake and Pete Wyles' Lake SP 890 640
Wildlife Trust for Bedfordshire, Cambridgeshire, Northamptonshire and Peterborough: tel: 01604 405285 (Northamptonshire office); web: www.wildlifebcnp.org

216 DITCHFORD GRAVEL PITS (Northamptonshire)

OS Landranger 152 and 153

Sandwiched between Wellingborough, Higham Ferrers and Rushden, this complex of gravel pits in the Nene Valley has a broad variety of habitats and attracts a wide range of species, as well as having an exciting list of rarities to its credit. The Wildlife Trust for Bedfordshire, Cambridgeshire, Northamptonshire and Peterborough has reserves at Higham Ferrers Pits, Wilson's Pits and Ditchford Lakes and Meadows.

Habitat

The gravel pits, varying in age, demonstrate the succession from bare sand and gravel with shallow pools to stands of mature willows. One of the most attractive to birds is also the largest, and lies immediately west of Higham Ferrers.

Species

Wintering wildfowl include good numbers of Wigeons, Teals, Gadwalls and Tufted Ducks, with smaller numbers of Goosanders, Pochards and Shovelers and a few Goldeneyes. Smew, Scaup, Common Scoter and Red-crested Pochard are occasional. The largest lake immediately west of Higham Ferrers usually has the best variety of wildfowl. Cormorant is regular and there are large pre-roost gatherings of gulls in late afternoon, including many Common, Herring and Lesser and Great Black-backed Gulls, but most gulls appear to roost elsewhere. Yellow-legged and Mediterranean Gulls are occasionally recorded. Other winter visitors include Water Rail, Stonechat and up to six Water Pipits, which favour the wet grassland along the River Nene, with Green Sandpiper and Grey Wagtail also possible by the river. Bitterns have wintered and look for Peregrines perched on the pylons at the western end of the area.

Passage waders favour the west pits, with Dunlin, Ringed and Little Ringed Plovers, Green and Common Sandpipers, Ruff and Greenshank all regular, and Black-tailed and Bar-tailed Godwits, Little Stint and Curlew and Wood Sandpipers possible, especially in autumn. A variety of scarce and rare species has also occurred. Common, Arctic and Black Terns are fairly regular on passage, and Shelduck and Garganey also occur. Little Egret is now a regular visitor.

Breeding species include Little and Great Crested Grebes, up to ten pairs of Grey Herons, Greylag and Canada Geese, Shoveler, Shelduck, Gadwall, a few Pochards, Common Buzzard, Ringed and Little Ringed Plovers, Lapwing and Common Redshank. Common Tern formerly bred and is usually present in summer, while Wigeon, Garganey, Teal and Oystercatcher may also oversummer. Reed and Sedge Warblers are common and join the resident Cetti's Warblers. Sparrowhawk, Kingfisher and Willow Tit nest in the area and Hobby is frequent in summer.

Access

The Nene Way long-distance footpath passes through the area, connecting many of the pits.
Higham Ferrers Pits From the A5028 in Higham Ferrers turn west at the Queen's Head pub into Wharf Road, parking in the small car park at the foot of the hill. Cross the bridge and bear left for 100 yards to cross the A45 dual carriageway via the footbridge, and there is then largely free access to the pits.
Wilson's Pits At the roundabout where the A45 meets the B465 and Crown Way, turn north on to a minor road and park at the reserve entrance on the right.
Ditchford Lakes and Meadows Leave the A45 on the A5001 (Rushden) and turn north at the roundabout onto Ditchford Road. The reserve entrance and car park lie on the right after 500 yards. (If coming from the north on the A45 leave the A45 northern junction with the A5001 and follow the A5001 Northampton Road towards Irthlingborough until you meet with the A45 again, where you cross over the A45 onto the Ditchford Road.)
New Pits To view the newest pits leave the A45 north on Ditchford Lane (see above). After crossing the river take the Nene Way to the left.

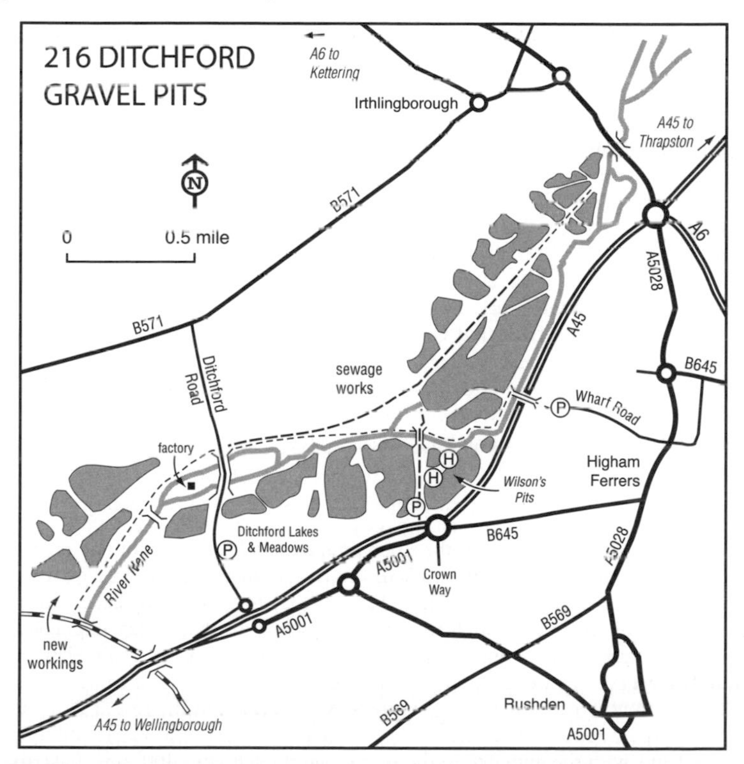

FURTHER INFORMATION

Grid refs: Higham Ferrers Pits SP 952 686; Wilson's Pits SP 943 679; Ditchford Lakes and Meadows SP 929 677; New Pits SP 930 683
Wildlife Trust for Bedfordshire, Cambridgeshire, Northamptonshire and Peterborough: tel: 01604 405285 (Northamptonshire office); web: www.wildlifebcnp.org

217 THRAPSTON GRAVEL PITS (Northamptonshire)

OS Landranger 141

This complex of gravel pits occupies 3.5 miles of the Nene Valley immediately north of Thrapston. A wide range of habitats is present, and the area often holds wintering Bitterns.

Habitat

As is usual with gravel workings, the pits are of various ages and the more recently excavated pits lie at either end of the complex. Town Lake is the largest and deepest, with several small wooded islands at its east end: it is used for sailing and fishing. The older central area of the complex forms the 72-ha Titchmarsh Reserve of the Wildlife Trust for Bedfordshire, Cambridgeshire, Northamptonshire and Peterborough, which includes an overgrown duck decoy now occupied by a heronry, while the largest pit on the reserve is surrounded by rough grassland, with stands of reeds at the west end and several islands that may attract nesting Common Tern, ducks and waders.

Species

Wintering wildfowl include large numbers of Wigeon (peaking at over 1,000), several hundred each of Pochard, Gadwall and Teal, and smaller numbers of Shovelers, Goldeneyes and Goosanders. Smew and Red-necked and Slavonian Grebes sometimes visit. Up to three Bitterns have wintered in some recent seasons, with the Heronry Hide being favoured. Cormorant, Water Rail and Kingfisher are fairly frequent, and Hawfinch is sometimes found. There is a substantial gull roost on Town Lake, and Yellow-legged Gull is regularly seen amongst the common species, especially in autumn.

On passage a variety of waders occurs, including Dunlin, Ruff, Greenshank, Common Sandpiper, Turnstone and Bar-tailed and Black-tailed Godwits. Common, Arctic and Black Terns and Little Gull may occur; all favour Town Lake. Other visitors are Shelduck, Garganey (occasional), and White Wagtail may occur in spring.

Breeding species include Little and Great Crested Grebes, Greylag and Canada Geese, and 50 pairs of Grey Herons. Several waders breed, notably Oystercatcher and Ringed and Little Ringed Plovers, as well as Common Tern, at Titchmarsh Reserve, and Common Redshank, Common Snipe and a few Curlews nest nearby. Other breeding species include Reed, Sedge and Grasshopper Warblers and a few Nightingales. Hobby is a frequent visitor, sometimes in some numbers, and Red Kites now breed in the area.

Access

Town Lake Leave the A14 north on the A605 and, after 0.25 miles, continue north over the roundabout. After a further 0.75 miles park in the lay-by on the left and follow the footpath beside the field to the disused railway line ('Town Walk') to view Town Lake. This path runs south into Thrapston, passing through mature trees and scrub that attract migrants and hold several species of warblers, sometimes Nightingale and, in winter, occasionally Hawfinch.

Titchmarsh Reserve Centred around Aldwincle Lake, the reserve also includes Heronry Lake and the heronry itself (no access). Turn west off the A605, 2 miles north-east of Thrapston, at the Fox Inn, to Thorpe Waterville. Take the minor road (past the pit) to Aldwincle and turn first left into Lowick Lane, with the entrance to the reserve on the left after 400 yards (take the bridge over Harpers Brook), with a small car park opposite. The Nene Way long-distance footpath runs along the west edge of the reserve, giving access to the East Midlands Electricity and Peter Scott Hides (overlooking the wader scrapes), and Kirby Hide (overlooking the main lake). Another footpath from the bridge follows Brancey Brook along the north side of the reserve, and the River Nene south along the eastern perimeter, giving access to the North (overlooking the main lake) and Heronry Hides. Entrance is possible at all times.

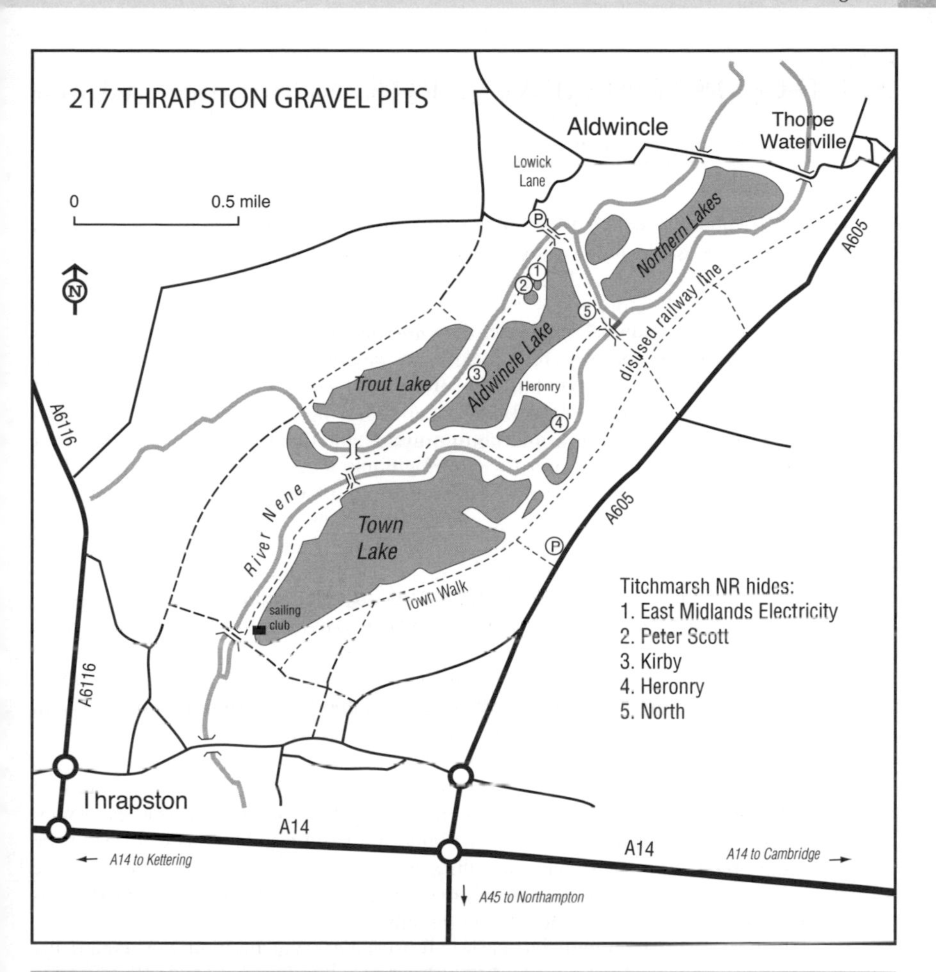

FURTHER INFORMATION

Grid refs: Town Lake TL 008 796; Titchmarsh Reserve TL 006 812
Wildlife Trust for Bedfordshire, Cambridgeshire, Northamptonshire and Peterborough: tel: 01604 405285 (Northamptonshire office); web: www.wildlifebcnp.org

218 PITSFORD RESERVOIR (Northamptonshire)

OS Landranger 141 and 152

Just 6 miles north of Northampton, Pitsford Reservoir holds a variety of ducks in winter and waders on passage, and is worth a visit during these periods. Indeed, with a species list of more than 240, it is the premier birdwatching site in Northamptonshire. The area to the north of the causeway is a reserve of the Wildlife Trust for Bedfordshire, Cambridgeshire, Northamptonshire and Peterborough, and the area north-west of the dam is part of Brixworth Country Park.

Habitat

The reservoir was filled in 1955 and covers 324 ha with natural, gently shelving banks. The perimeter of the north half has been extensively planted with conifers, and there is also a mature oak copse; otherwise the surrounding land is farmed.

Species

In late autumn and early winter there are often 100–200 Dunlins present if water levels are low. Winter wildfowl include up to 2,250 Coots, 1,700 Tufted Ducks, 1,200 Wigeons, 750 Teals and 420 Gadwalls, as well as smaller numbers of Pintails, Shovelers, Pochards and Goldeneyes and up to 50 Goosanders. Bewick's Swan, Red-crested Pochard, Smew, Red-breasted Merganser, Scaup and Common Scoter are annual, as are rarer grebes and divers (especially Great Northern). Numbers of Cormorants are present year-round. There is a large gull roost, sometimes holding 20,000 birds, largely Black-headed and Common Gulls, with smaller numbers of Lesser and Great Black-backed and Herring Gulls. Several Mediterranean and Yellow-legged Gulls are regular from August onwards, and Caspian, Glaucous and Iceland Gulls are recorded most winters. Wintering passerine flocks may include Brambling and Tree Sparrow; the winter feeding station in the Scaldwell arm is a favoured spot.

On passage there can be a good variety of the usual waders, with Golden and Ringed Plovers, Greenshank and Ruff among the commonest, and Little Stint, Curlew Sandpiper and Spotted Redshank are annual. Tern passage is also good: Common, Arctic and Black are regular, as is Little Gull, while Arctic Skua appears surprisingly frequently. Osprey is annual, especially in August, and Black-necked Grebe and Marsh Harrier are also reasonably regular in autumn.

Breeding species include Great Crested Grebe, Cormorant, Shoveler, Gadwall and Tufted Duck, while Common Terns breed on specially provided rafts. Eight species of warblers nest in the surrounding area, including Grasshopper, and Hobby is quite frequent.

Access

The reservoir lies just to the east of the A508 and slightly west of the A43. The reservoir is intensively used for sailing, trout fishing and walking, and there is public access to the entire shoreline south of the causeway (except the sailing club grounds) and it is encircled by a cycle track. Wigeon prefers more open areas of the south section, and the dam area can be good unless there are boats present. Access is via the following points.

1. The causeway on the minor road to Holcot from the A508 at Brixworth. There are car parks at the west end and at the Fishing Lodge at the east end of the causeway.

2. The dam end, accessed from the A508 or the village of Pitsford. There are car parks and toilets at either end of the dam, and also an information centre, café etc. at Brixworth Country Park at the north-west end.

3. The anglers' car park, outside the reservoir perimeter fence and just north of the sailing club, is a good place to observe the gull roost (although the reservoir is usually disturbed by boats until near dusk at weekends).

It is possible to view the northern portion of the reservoir from either end of the submerged road north of the causeway. Access to the northern basin is otherwise by permit only. Scaldwell and Walgrave Bays north of the causeway are best for waders, and the north half is also better for most ducks. Annual and day permits (fee) are issued by Anglian Water and are available from the Pitsford Water Fishing Lodge by the causeway or from the Wildlife Trust. There are eight hides on the reserve north of the causeway.

FURTHER INFORMATION

Grid refs: Causeway SP 780 701; Brixworth Country Park SP 753 693
Wildlife Trust for Bedfordshire, Cambridgeshire, Northamptonshire and Peterborough: tel: 01604 405285 (Northamptonshire office); web: www.wildlifebcnp.org
Pitsford Water Fishing Lodge: tel: 01604 781350

219 STANFORD RESERVOIR (Northamptonshire/ Leicestershire)

OS Landranger 140

Between Rugby and Husbands Bosworth, this small reservoir in the valley of the River Avon attracts good numbers of winter wildfowl and is also interesting on passage. Indeed, for its size, a relatively large number of scarce migrants and rarities has been recorded here. The reservoir is owned by Severn Trent Water and is managed by the Northamptonshire Wildlife Trust.

Habitat

A drinking-water reservoir set in arable farmland, open water covers 73 ha with a narrow belt of trees along the north-west shore, more extensive areas of woodland just to the north and south, and Hawthorn and willow scrub at Blower's Lodge Bay in the south-eastern corner of the reservoir, which is a reserve of the Wildlife Trust for Bedfordshire, Cambridgeshire, Northamptonshire and Peterborough. There are also some limited stands of reeds.

Species

Wintering wildfowl include large numbers of Pochards, Wigeons (up to 1,850) and Coots, as well as Teal, Shoveler, up to 90 Gadwalls, Goosander, and a few Pintails and Goldeneyes. Wild swans, rarer grebes (especially Black-necked) and divers are occasional. The large gull roost can hold as many as 2,000 Lesser Black-backed, and Mediterranean, Glaucous and Ring-billed Gulls have been recorded (the west hide offers the best vantage for gull watching, but is accessible by permit only). Check the dense Hawthorn scrub at Blower's Lodge Bay for the occasional wintering Long-eared Owl.

On passage a variety of waders may occur, including Little Ringed and Ringed Plovers, Dunlin, Ruff, Greenshank and Green and Common Sandpipers, with Little Stint, Curlew and Wood Sandpipers, Spotted Redshank, and Black-tailed and Bar-tailed Godwits possible, especially in autumn. Common, Arctic and Black Terns are relatively frequent.

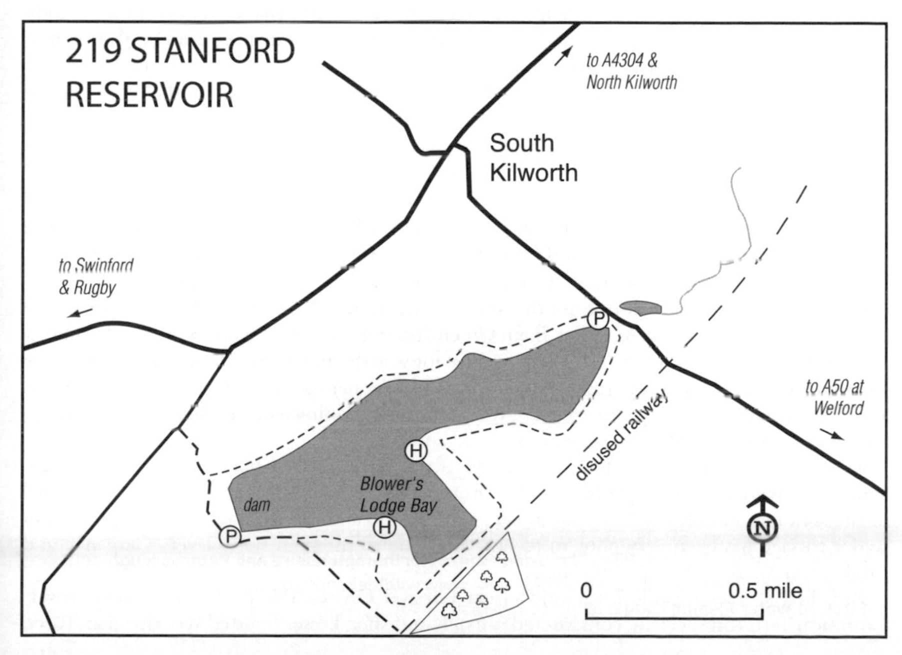

Breeding birds include Great Crested and Little Grebes, Common Tern (on purpose-built rafts), Turtle Dove, Grasshopper, Sedge and Reed Warblers, and Hobby is a regular visitor.

Access

The north-east tip of the reservoir, at the inlet stream, can be viewed from the minor road between South Kilworth and the A50 at Welford, with a convenient car park. Otherwise a permit, available from the Northamptonshire Trust, is required for access to the reservoir, including the car park by the dam (off the minor road between South Kilworth and Stanford on Avon); a footpath encircles the reservoir and there are two hides either side of the mouth of Blower's Lodge Bay.

FURTHER INFORMATION

Grid ref: SP 611 811
Wildlife Trust for Bedfordshire, Cambridgeshire, Northamptonshire and Peterborough: tel: 01604 405285 (Northamptonshire office); web: www.wildlifebcnp.org

220 SWITHLAND RESERVOIR (Leicestershire and Rutland)

OS Landranger 129

This relatively small reservoir holds a variety of wintering wildfowl and a notable gull roost that regularly includes Mediterranean Gull.

Habitat

A little over 1 mile from north to south and less than 0.5 miles wide, the reservoir is bisected by a viaduct carrying the Great Central Railway and is crossed at the south end by the minor road between Rothley and Swithland, the bridge having caused a shallow, reed-fringed lagoon to form at the inflow. Though water levels fluctuate considerably, exposing areas of shoreline, this is mostly stony and attracts few waders. The water is, however, relatively undisturbed. To the north, Buddon Wood has been reduced to a narrow strip of birches by the expansion of Mountsorrel Quarry.

Species

Wintering wildfowl include small numbers of Wigeons, Teals, Gadwalls, Shovelers, Pochards and Goldeneyes. Goosander, Smew and Pintail are occasionally recorded but divers, Red-necked Grebe and seaducks are all scarce and irregular. Up to 50 Cormorants gather in the late afternoon to roost. Other interesting wintering species include up to three Peregrines (one or two often sit in a prominent oak tree on the skyline of Buddon Wood in the late afternoon, and can be viewed from the dam), Water Rail, Green Sandpiper, Grey Wagtail, Siskin and redpolls. Among the prime attractions of Swithland in winter is the gull roost, which may hold up to 15,000 birds. Mediterranean Gull is regular from November to March but Caspian, Glaucous and Iceland Gulls are scarce. In March–April occasional Kittiwakes may appear and Yellow-legged Gulls are most likely in late summer and autumn.

Migrants in spring may include Arctic and Black Terns, Little Gull, Common Sandpipers, and occasionally Garganey or Osprey. On autumn passage Black-necked Grebe is near-annual in late summer, and Hobby is regular in late summer and autumn (Buddon Wood is good), but wader passage has been very poor in recent years due to the high water levels. Garganey and Black or Arctic Terns may pass through.

Breeding species include Little and Great Crested Grebes, Ruddy Duck, Sparrowhawk, Common Tern (on specially constructed rafts), Kingfisher, Lesser Spotted Woodpecker, Reed and Sedge Warblers, while Pochard, Gadwall and Shoveler have bred and Cormorant over-summers. Marsh and Willow Tits may still cling on in Buddon Wood.

Access

There is no access to the reservoir margins, but both sections are easily viewed from the road.
North and west shores Leave the A6 west on minor roads to Rothley and then take Woodgate and Westfield Lane towards Rothley Station. Turn north on The Ridings towards Rothley Plain and follow the road over the crossroads towards Swithland. The road passes under a railway bridge and immediately beyond this turn north into Kinchley Lane, which runs beside the south part of the reservoir before passing under the railway and continuing along the east and north shores. This is a very narrow lane and caution should be exercised, especially when parking. The gull roost is best watched from halfway along Kinchley Lane (but the light is only suitable on overcast evenings; on sunny days it faces the setting sun).
South shore The southern (inflow) end can be viewed from the minor road between Swithland and Rothley: park at the eastern end of the bridge at the south tip of the reservoir.

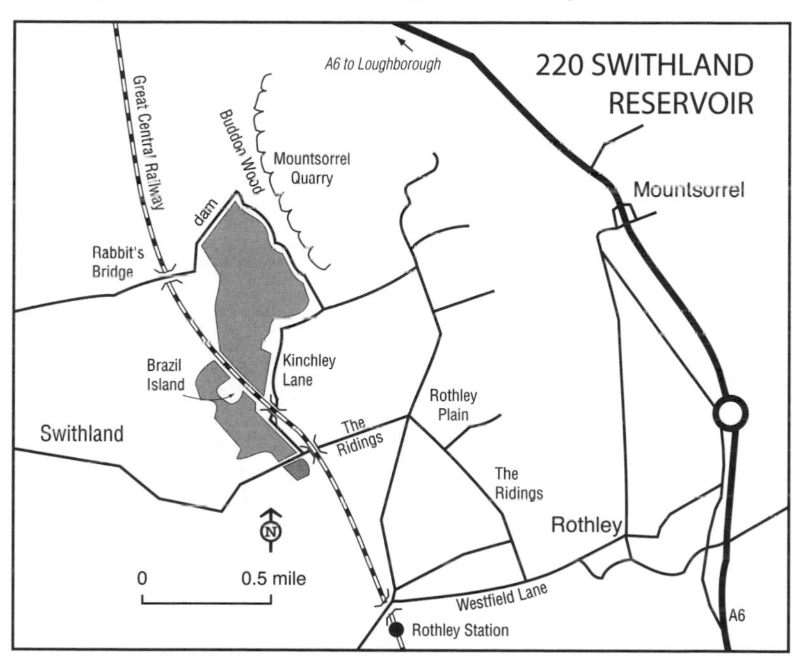

FURTHER INFORMATION

Grid refs: North and west shores SK 561 140; South shore SK 562 132

221 EYEBROOK RESERVOIR (Leicestershire and Rutland)

OS Landranger 141

Approximately 3 miles north of Corby, this reservoir holds a variety of winter wildfowl and passage waders, and in hard weather often attracts interesting birds. It is managed as a reserve.

Habitat

The reservoir was completed in 1940 and open water covers 162 ha with natural, gently shelving banks. There is an area of marsh near the inlet at the north end and this is usually the best area for passage waders, while the east shore has been extensively planted with conifers. The reservoir is used for trout fishing but is otherwise undisturbed.

Species

Wintering wildfowl include Teal and Wigeon, which may peak at 1,000, together with small numbers of Goldeneyes, Goosanders, Smews, Ruddy Ducks, Shovelers, Pintails and, sometimes, Red-crested Pochards. Hard weather can bring seaducks, especially Scaup, as well as Slavonian and Red-necked Grebes, and divers, particularly Great Northern. Bewick's Swan may occur, but the former high count of 80 in midwinter is now seldom approached. There are also feral Greylag Geese. Peregrine is sometimes seen, and wintering waders include up to 2,000 Golden Plovers, 150 Dunlins and the occasional Green Sandpiper or Jack Snipe. The large gull roost is comparatively easy to work, though a telescope is essential. It is best watched from the rod near the 'island'. Yellow-legged and Mediterranean Gulls are regular, while Caspian, Glaucous and Iceland Gulls are occasionally recorded. The feeding station may attract Tree Sparrows

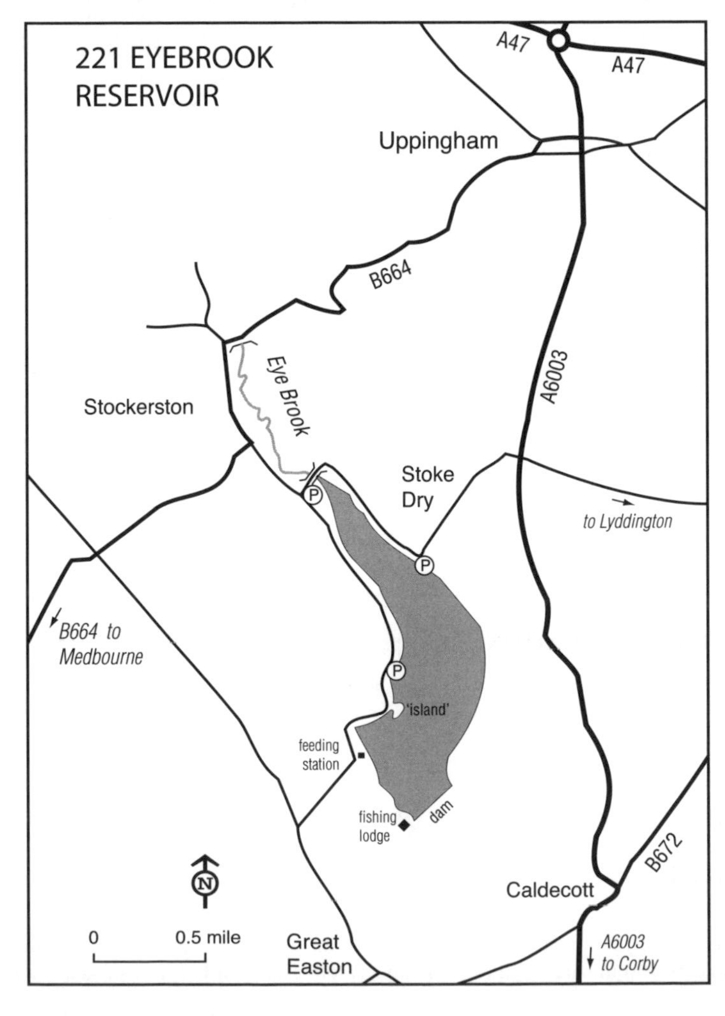

during the winter.

Common Scoter, White Wagtail and Water Pipit sometimes occur in early spring and Hobby is quite regular in spring and early summer (the north end of the reservoir is best). Black-necked Grebe is increasingly frequent on passage from late March and other spring migrants may include Osprey, Garganey, small numbers of waders depending on the water level, Common, Arctic and Black Terns, Kittiwake and Little Gull.

Osprey is also recorded almost annually in autumn and this season, when water levels have dropped, is also best for waders. Little Stint and Curlew Sandpiper are annual, and other species may include Ruff, Black-tailed Godwit, Spotted Redshank and Greenshank.

Breeding species include Great Crested Grebe, Common Buzzard, Gadwall and occasionally Shelduck or Teal, and Barn Owl, and look for Little Owl on the Stockerston Road.

Access

From the A6003 between Corby and Uppingham take the minor road west to Stoke Dry and continue through the village to the reservoir. The road runs around the north end and then parallels the west shore with limited roadside parking at three points. The bridge over Eye Brook at the north tip and the west shore are best for passage waders. Access off the road is only granted to members of the Leicestershire and Rutland Ornithological Society, but good views of the whole reservoir are possible from the road.

FURTHER INFORMATION

Grid ref: northern tip SP 843 970

222 RUTLAND WATER (Leicestershire and Rutland)

OS Landranger 141

Rutland Water is the largest man-made reservoir in Britain. Since its construction in 1975 it has become one of the most important wildfowl sanctuaries in Britain (it is a Special Protection Area and a Ramsar site), and is well worth visiting, especially during winter and passage periods. Nine miles of shoreline (comprising one-third of the total) along the west arms form two reserves of the Leicestershire and Rutland Wildlife Trust (LRWT), covering 281 ha, and the Anglian Water Birdwatching Centre at Egleton is home to the annual British Birdwatching Fair.

Habitat

As well as 1,255 ha of open water, at the west end of the reservoir there are purpose-built islands, shingle banks and three lagoons, with areas of scrub woodland at Gibbet Gorse and Gorse Close, and stands of mature deciduous woodland at Lax Hill. Other interesting habitats include ancient meadows and reedbeds.

Species

All three divers are near-annual in winter, with Great Northern the most likely. Red-necked and Slavonian Grebes are also almost regular in winter, while Great Crested Grebe has astonishingly peaked at over 1,000, with up to 800 regularly present. Other waterfowl include several hundred Cormorants. Wildfowl include feral Greylag and Snow Geese, up to 4,500 Wigeons, 1,500 Gadwalls, 2,500 Tufted Ducks (and an astonishing total of 8,487 moulting birds was recorded in October 2005), 2,000 Pochards, 400 Ruddy Ducks and 400 Goldeneyes, and peak autumn counts of 750 Shovelers and 1,800 Gadwalls are outstanding. Species occurring in smaller numbers include Pintail, Goosander (usually 50 but up to 130), Smew (including up to seven males each winter) and, notably, Red-crested Pochard (up to seven have occurred, mainly

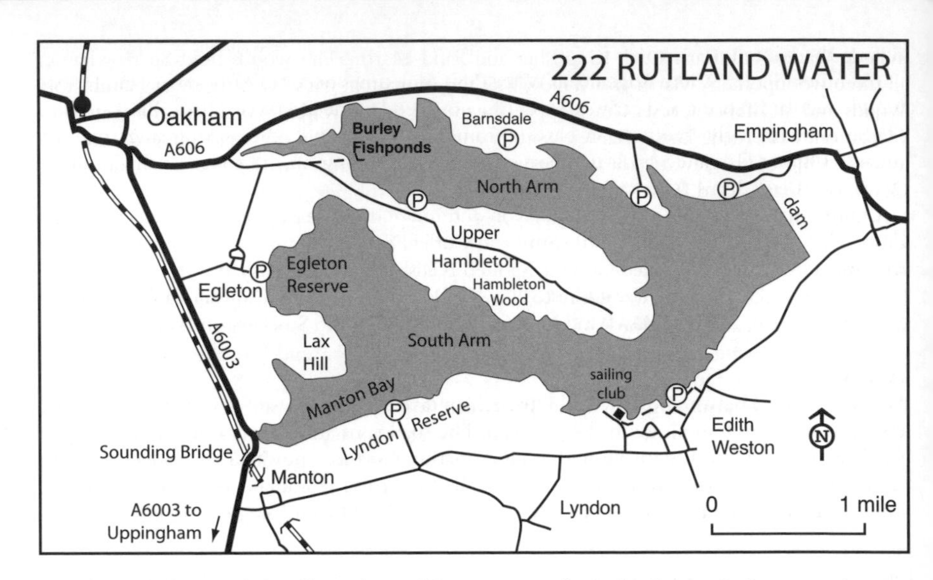

in autumn and winter). Small numbers of Scaup are regular and other seaducks appear in ones and twos.

Water Rail and Jack Snipe winter in marshy areas, and grassy areas, especially the slopes of the dam, often attract a large flock of Golden Plovers. Other waders include up to 20 Ruffs and 120 Dunlins as well as Black-tailed Godwit and Green Sandpiper. Numbers of gulls use the reservoir to roost. Largely Black-headed and Common Gulls, the roost also attracts Herring Gull and up to 400 Great Black-backed Gulls. Yellow-legged and Mediterranean Gulls are regular, and records of Glaucous and Iceland Gulls are increasing. The roost is rather scattered but is best watched from Gadwall or Goldeneye Hides on Egleton Reserve. Short-eared Owl is sometimes seen hunting over rough ground and occasionally Long-eared Owl is found roosting in the scrub, particularly at Lax Hill and Gibbet Gorse, though they are difficult to find. Peregrine is a regular visitor and occasionally Bearded Tit and Cetti's Warbler are recorded. Other notable species at this season include Tree Sparrow, especially at the feeders by the Birdwatching Centre. Rock Pipits are a possible in late autumn and late February–early April (try the rocky face of the dam) and look for Siskin and Common and Lesser Redpolls in the alders behind lagoons II and III and at the Lyndon Reserve. Brambling also visits in winter.

Little Egret is now regular in good numbers and Black-necked Grebe is quite frequently seen, especially in spring, late summer and autumn (usually on Lagoon III or the north arm), while Marsh Harrier, Garganey and Common Scoter are also regular on passage. Waders may include all of the usual species such as Ringed and Little Ringed Plovers, Greenshank, Common Redshank, Common Sandpiper and Dunlin, as well as Grey Plover, Turnstone, Black-tailed Godwit, Whimbrel and Spotted Redshank. Wood Sandpiper, Little Stint and Curlew Sandpiper are annual, usually in autumn, and Pectoral Sandpiper nearly so. The lagoons, Manton Bay and north arm are best for waders. Little Gull, Kittiwake, and Common, Arctic and Black Terns are regular migrants, favouring the main water. Migrant passerines include Wheatear, Whinchat and occasionally Common Redstart.

Great Crested and Little Grebes, Cormorant (65 pairs), Grey Heron, Greylag and Egyptian Geese, Shelduck, Pochard, Gadwall, Shoveler, Teal and Garganey breed, as do Oystercatcher, Lapwing, Little Ringed and Ringed Plovers, Common Snipe, Common Redshank and 60 pairs of Common Terns on specially constructed rafts. Astonishingly, an Avocet laid eggs in 1996, the first recorded breeding attempt at a freshwater site anywhere in the world. In 1996 young Osprey chicks were reintroduced into England at Rutland Water as part of a translocation programme. The project has been a success and Ospreys have now bred successfully at Rutland Water and can be seen from early April to mid-September. Other breeding birds on the reserve include Water Rail, Barn Owl, Kingfisher and Sand Martin. The woods hold Sparrowhawk, all three woodpeckers, Marsh and Willow Tits and Nuthatch, with Nightingale at Hambleton Woods and at Gibbet Gorse. Grasshopper, Reed and Sedge Warblers breed in the waterside vegetation, with Turtle Dove, Garden Warbler and Tree Sparrow in the scrub; indeed, Rutland Water is one of the few places in Britain where Tree Sparrows are increasing in numbers. Common Buzzard and Red Kite breed in the area and Hobbies are often seen from spring to early autumn, being most regular in the evening.

Access

Rutland Water lies 1 mile south-east of Oakham and is easily accessed from the A1 via the A606 west of Stamford, and from the A47 along the A6003 north of Uppingham. Note that major new works, involving the creation of new lagoons, are under way and may affect access and the location of hides and trails.

Egleton Reserve and Anglian Water Birdwatching Centre From the A6003 1 mile south of Oakham take the minor road to Egleton. The entrance to the reserve is at the southernmost point of the village and is well signed. It is open 9am–5pm daily (except Christmas Day and Boxing Day). A permit is required by non-members of the LRWT, and is available from the Anglian Water Birdwatching Centre (which has an upstairs viewing gallery giving panoramic

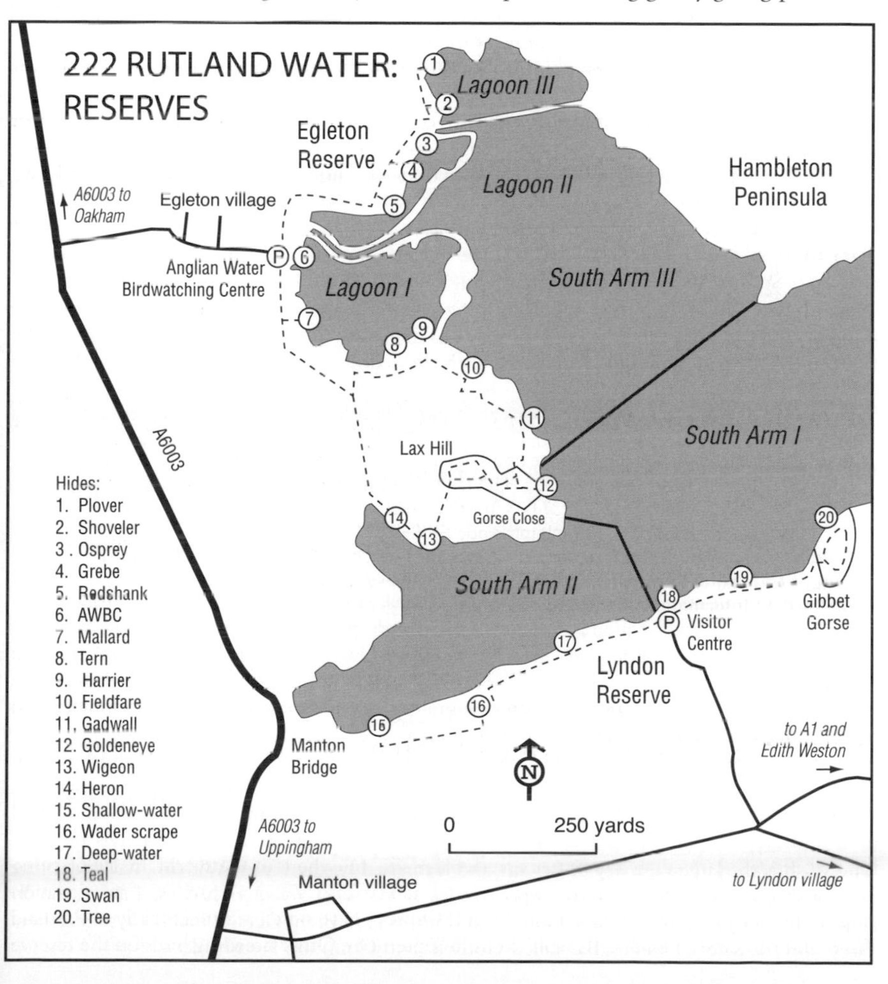

views over the lagoons). There are 16 hides and a woodland trail through Lax Hill and Gorse Close. The reserve contains a wide variety of habitats: open water, muddy edges, reedbed, scrub, woodland and grassland, and a visit here is likely to produce the greatest variety of species.

Lyndon Reserve and Lyndon Hill Visitor Centre Leave the A6003 eastwards, 2.5 miles south of Oakham, to Manton and take the first turning on the left after the village to the Visitor Centre. Access from the A1 is via the A606, following signs to Edith Weston and Manton. A nature trail leads from the Visitor Centre to seven hides overlooking the reservoir and a woodland walk; an eighth hide overlooks a pond in Gibbet Gorse. The reserve is open 10am–4pm daily (except Mondays) May–October, and on Saturdays, Sundays and Bank Holidays November–April. A permit is required by non-members of the Leicestershire and Rutland Wildlife Trust (LRWT), and is available from the Visitor Centre. The Manton Bridge area is one of the better areas for waders and can be viewed from the Shallow-water Hide, and in late summer Lyndon is also one of the best places to see Ospreys, which often sit on dead trees in Manton Bay.

Sounding Bridge Manton Bay can also be viewed from Sounding Bridge, parking in the lay-by on the opposite side of the road.

South arm The elevated cycle track east of the Lyndon Reserve Visitor Centre offers good views over the south arm, with signed car parks at Edith Weston.

Dam area Often holding congregations of ducks in winter, especially diving ducks, and also good for divers, this may be the last area to freeze in cold weather. Park at the end of Sykes Lane, signed off the A606, or on the grass verge at the southern end of the dam.

North shore Two signed access points to the shore off the A606 afford views of the reservoir for ducks, gull, and terns.

North arm and Burley Fishponds The fishponds can be viewed from the private road just off the Hambleton Road and the north arm from the end of this road or from the small car park north-west of Upper Hambleton. The fishponds may hold Goosander and Smew, and have a heronry and colony of Cormorants. The north arm is one of the best places to see Black-necked Grebe between April and September.

Barnsdale Wood The car park is signed off the A606 opposite Barnsdale Avenue. A good spot for Lesser Spotted Woodpecker and Brambling.

Hambleton Wood The most regular site for Nightingale. In Upper Hambleton, take the narrow road opposite the Finch's Arms towards the south arm and Old Hall and park at the bottom. Walk east for about 0.5 miles to reach the wood.

FURTHER INFORMATION

Grid refs/postcodes: Egleton Reserve and Anglian Water Birdwatching Centre SK 878 072 / LE15 8BT; Lyndon Reserve and Lyndon Hill Visitor Centre SK 894 054 / LE15 8RN; Sounding Bridge SK 877 052; dam area SK 937 083; north arm and Burley Fishpond SK 896 079; Barnsdale Wood SK 908 087; Hambleton Wood SK 897 071
Rutland Water Nature Reserve, Anglian Water Birdwatching Centre: tel: 01572 770651
Lyndon Hill Visitor Centre: tel: 01572 737378; web: www.rutlandwater.org.uk
Up-to-date information on the Ospreys: www.ospreys.org.uk

223 THE WASH (Lincolnshire)

OS Landranger 131

The Wash is the largest estuary in Britain, and is formed by the Rivers Witham, Welland, Nene and Great Ouse. It is also the most important for wildfowl and waders, holding internationally important concentrations of Pink-footed and Dark-bellied Brent Geese, Pintails, Oystercatchers, Grey Plovers, Knots, Dunlins, Bar-tailed Godwits and Common Redshanks.

Habitat

Vast areas of saltmarsh, tidal sand and mudflats are backed by artificial sea walls that serve to defend areas of reclaimed land, now largely used for arable farming. The excavation of clay to construct these sea banks has formed a series of borrow pits, many overgrown with reeds and sometimes willows. In this flat and featureless region, stands of trees and bushes are otherwise relatively scarce. Several areas are protected as reserves, notably Frampton Marsh, where both the RSPB and Lincolnshire Wildlife Trust (LWT) possess holdings, and Freiston Shore, where the RSPB manages 683 ha of saltmarsh and mudflats, with a 15-ha saline lagoon recently created in the 'borrow area' where silt had been removed for building up the sea banks; Freiston Shore is also the site of one of the UK's largest 'managed retreat' projects in which the old sea walls have been breached to convert 66 ha of coastal farmland into tidal saltmarsh.

Species

Wintering wildfowl include large numbers of Dark-bellied Brent Geese, with a maximum winter count of around 20,000, and several thousand remain into spring, especially around Frampton Marsh. Also present are Pink-footed Goose (at Holbeach Marsh), 10,000 Shelducks, Wigeon, Mallard and Pintail (especially at Nene Mouth). Seaducks include Red-breasted

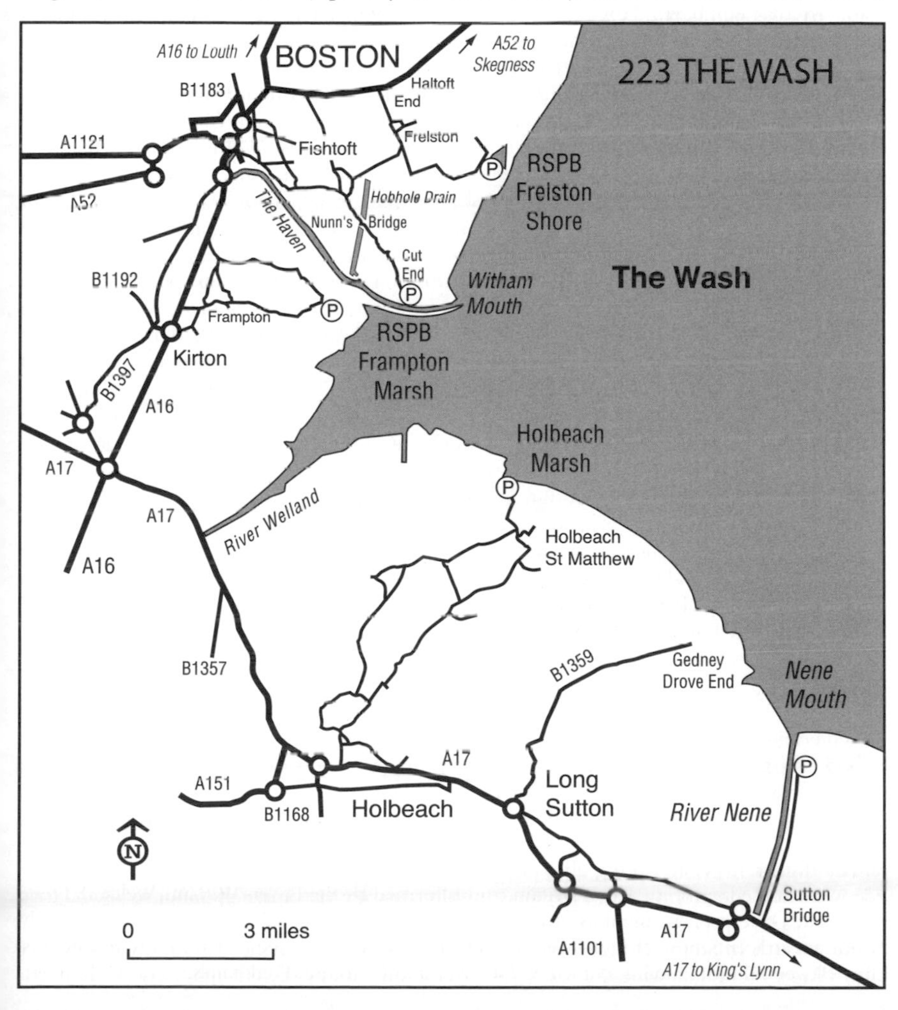

Merganser, Common Eider and Goldeneye, but other species are much rarer and less predictable. Witham Mouth is favoured by seaducks, and divers and rarer grebes are also occasionally recorded. Other waterfowl include Cormorant. Waders are abundant with huge numbers of Lapwings (46,500), Golden Plovers (22,000), Dunlins (36,600) and Bar-tailed Godwits (16,500), as well as Oystercatchers, Ringed and Grey Plovers, Knots, Sanderlings, Curlews, Common Redshanks and Turnstones. At low tide most are dispersed over the flats and often offer only very distant views, but at high tide they gather to roost, and on spring tides do so in the fields immediately inland of the sea wall. Scarcer waders include Black-tailed Godwit, which may winter at Holbeach Marsh. Large numbers of commoner gulls use the Wash, and Glaucous Gull is occasionally found in winter, with Mediterranean Gull also possible, especially in late summer. Wintering raptors include Hen Harrier, Sparrowhawk, Merlin, Peregrine and Short-eared Owl, with a Long-eared Owl roost at Nunn's Bridge. With luck, Kingfishers may be seen in the creeks. Wintering passerines include Rock Pipit on the saltmarshes; in late winter and early spring look for individuals of the Scandinavian race as they begin to assume breeding plumage. Corn and sometimes Snow Buntings occur on the sea walls, as may Twite and Lapland Bunting, which also favour rough ground and stubble fields (the latter species is likeliest at Frampton Marsh).

On passage numbers of Oystercatchers (15,600), Ringed Plovers (1,500), Grey Plovers (13,100), Knots (69,000), Sanderlings (3,500), Curlews (9,500), Common Redshanks (6,400) and Turnstones (900) are higher than in winter, and there may also be significant totals of Whimbrels, Black-tailed Godwits (7,000), Greenshanks and Spotted Redshanks. In autumn, a few Little Stints and Curlew Sandpipers may join the Dunlins and Knots on the flats and Common, Green and sometimes Wood Sandpipers favour the saltmarsh gutters. As with the other waders, high-tide roosts offer a chance to find the scarcer species, though identifying birds in the tightly packed flocks can be difficult. Interesting waders are more easily found on borrow pits if water levels have fallen in autumn, and on the scrapes at Frampton Marsh and Freiston. Terns are regular in summer, especially Common Tern, but also Little and Black Terns and Little Gull in autumn. A feature of autumn is wind-blown seabirds, which may be forced by north gales to shelter in the Wash. They follow the coast in a clockwise direction to leave the Wash at Gibraltar Point. In the appropriate conditions Fulmar, Manx Shearwater, Leach's Petrel, Gannet, Common Guillemot, Kittiwake and Arctic and Great Skuas (and sometimes Long-tailed and Pomarine Skuas) are possible. The best localities for seabirds are Witham Mouth, where there is a hide, and Holbeach Marsh. Some species do not leave the Wash via the sea, however, and skuas in particular are prone to move inland, either directly overland or along the rivers (especially the Nene), giving exceptionally good views. The Wash coast is relatively unsheltered but interesting passerines, such as Ring Ouzel, Wheatear, Whinchat and Stonechat, may be found on migration.

Breeding species include Marsh Harrier, now relatively numerous, and there are irregular sightings of Montagu's Harrier. Avocets now breed at Frampton Marsh and Freiston Shore, as do Ringed and Little Ringed Plovers (the latter at Frampton), Common Redshank, Barn Owl and Tree Sparrow (Freiston Shore); Short-eared Owl sometimes nests.

Access

The pace of birdwatching in this huge area is largely determined by the tides and, ideally, visits should be timed to coincide with high water to view roosting waders, seaducks and, in season, seabirds. In spring and autumn the fortnightly spring tides are ideal, as these usually flood the saltmarshes and force the waders onto drier ground to roost. High water on spring tides usually occurs in the late evening and early morning. In contrast, the intervening neap tides leave parts of the mudflats exposed and fewer waders move into view of the sea walls, while seaducks too remain distant. In winter spring tides peak during hours of darkness and a late-morning high tide is the best compromise. Access to the sea walls is usually via a maze of minor roads and thus the OS map is especially useful in this area.

Nene Mouth In Sutton Bridge leave the A17 northwards on a minor road immediately east of the River Nene, following this for 3 miles to the car park by the old lighthouse. Walk north

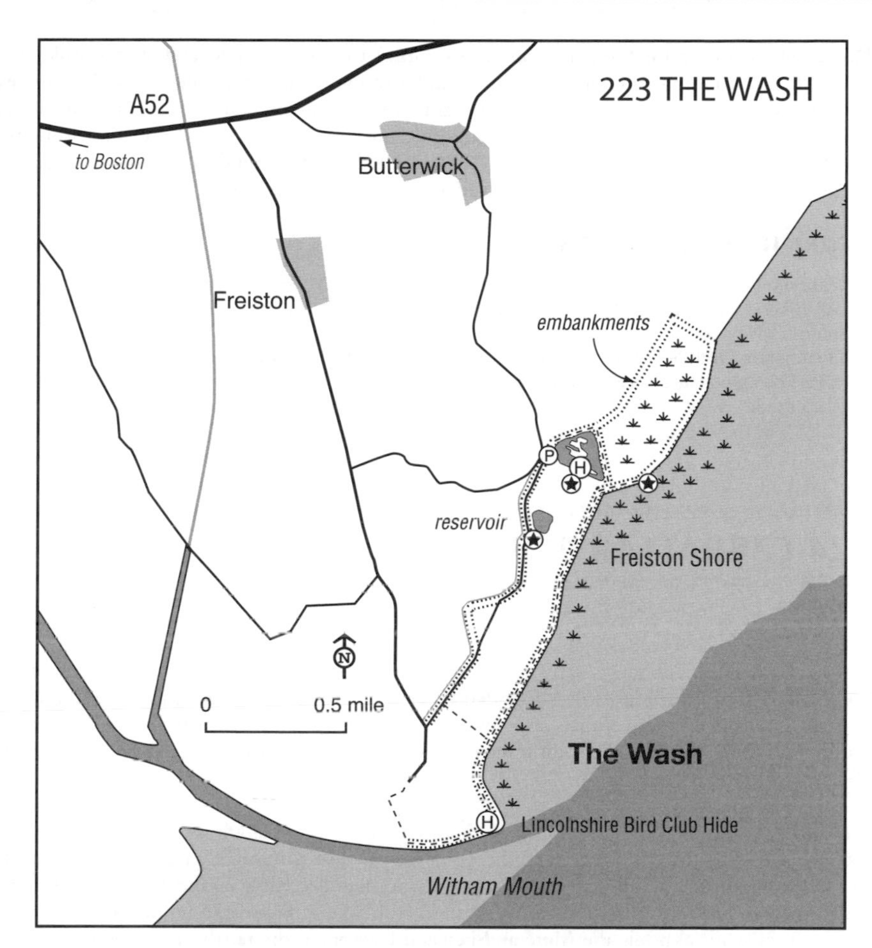

from here along the sea wall to the river mouth, which is a good area for seaducks. This is also the starting point of the Peter Scott Walk which follows the coast to King's Lynn.

Holbeach Marsh Leave the A17 at Holbeach and follow signs through the maze of minor roads north to Holbeach St Matthew. From the phone box in the village take the minor road north for 1 mile to the car park at the sea wall. From here walk east along the sea wall to view a vast area of saltmarsh, or north and then west for 2 miles with interesting borrow pits by the bank much of the way. Search for seaducks offshore around high water.

RSPB Frampton Marsh and LWT Reserves Leave the A16 east in Kirton on minor roads to Frampton village and then follow signs to the RSPB reserve. The visitor centre overlooks the reedbed reservoir and there are four hides overlooking freshwater scrapes. The reserve is open at all times and the visitor centre is open daily 10am–4pm. Follow the footpath along the south bank of The Haven (River Witham) – not suitable for wheelchairs.

Witham Mouth From central Boston follow signs to the docks and thereafter Fishtoft. In the village centre bear right by the pub to Nunn's Bridge. Continue for 2 miles to Cut End on the north bank of the River Witham, parking on the roadside. Walk east for 2 miles to the river mouth where there is a seawatching hide; this is a good site for seaducks.

Hobhole Drain Follow directions as for Witham Mouth, and as you approach Nunn's Bridge the road crosses Hobhole Drain. Immediately beyond the bridge turn sharp right (south) onto the bank of the Drain. A few Long-eared Owls roost in the Hawthorns on the opposite bank of the drain here.

RSPB Freiston Shore From Boston, take the A52 towards Skegness and after 2 miles, in Haltoft End, turn south and follow signs through Freiston to RSPB Freiston Shore. The reserve car park is accessed via a steep concrete ramp by the Plummers Guest House. The Wetland Trail is a circular route around the developing wet grassland with views over to the saltmarsh, while the Lagoon Trail gives access to the Lagoon Hide – trail and hide are both wheelchair-accessible. On big tides large numbers of waders may roost on the lagoon by the car park.

FURTHER INFORMATION

Grid refs: Nene Mouth TF 493 255; Holbeach Marsh TF 407 339; RSPB Frampton Marsh and LWT reserves TF 356 391; Witham Mouth TF 380 391; Hobhole Drain TF 367 415; RSPB Freiston Shore TF 397 424
Lincolnshire Wildlife Trust: tel: 01507 526667; web: www.lincstrust.org.uk
RSPB Frampton Marsh & Freiston Shore: tel: 01205 724678; email: lincolnshirewashreserves@rspb.org.uk

224 GIBRALTAR POINT (Lincolnshire)

OS Landranger 122

South of Skegness, Gibraltar Point marks the spot where the Lincolnshire coast turns into the Wash. A Bird Observatory was founded in 1949 to study the resultant concentrations of migrants, and the area also attracts large numbers of wintering birds. It is worth visiting year-round. An NNR covering 450 ha, it is managed by the LWT, and the area has been designated a Ramsar site and SPA.

Habitat

Two major ridges, the East Dunes and the older West Dunes, run roughly north–south parallel to the sea, supporting stands of Sea-buckthorn and some Elder; between these are areas of fresh and saltmarsh. To the south, an extensive sandbar (the Spit) attracts large numbers of roosting waders. Two artificial pools, the Mere and Fenland Lagoon, in the north of the reserve attract migrant ducks and waders. Extensive mudflats and saltmarsh border the Point.

Species

Spring passage is heralded in March by migrant Stonechats, Wheatears and Black Redstarts and Woodlark is possible in late February–late March, with occasionally up to six present, favouring the close-cropped dune ridges. By May, east or south-east winds with cloud, rain or poor visibility may produce falls of warblers, flycatchers and chats, sometimes including scarcer species such as Nightingale, Ring Ouzel, Wood Warbler, Pied Flycatcher and Firecrest. Marsh and, occasionally, Montagu's Harriers pass through, as may Osprey and, from mid-May, Hobby. Visible migration (best viewed from Mill Hill in moderate north-west winds) is a particular feature of Gibraltar Point; as at Spurn, the majority of diurnal migrants (comprising a wide range of species, notably pipits, wagtails, finches and buntings) fly south in spring as well as autumn. Other migrants may include Common, Arctic, Sandwich and Black Terns, Little Gull, and a variety of waders, including up to 8,000 Knots and 2,000 each of Sanderling and Grey Plover in their magnificent summer plumage.

Notable breeding birds include Shelduck, Avocet, Redshank, Lapwing, Oystercatcher, Ringed and Little Plovers, Black-headed Gull, Common Tern, a variable number of Little Terns, and Common and Lesser Whitethroats. Hobby and Marsh Harrier summer in the area. In autumn wildfowl, waders and gulls throng the shore. Large numbers of Common, Arctic and Sandwich Terns offshore may attract all four species of skuas, though Arctic Skua is by far the most regular. Black Terns may also pass through and Mediterranean and Yellow-legged Gulls can put in an appearance, while during onshore north-east winds Manx Shearwater is possible. The freshwater pools may attract Garganey, Water Rail, Avocet, Little Ringed Plover, Black-tailed Godwit, Little Stint, Curlew and Wood Sandpipers, and Spotted Redshank, in addition to the common waders. From mid-July large flocks of open-shore waders, often still in breeding plumage, roost on the Spit or the beach to the north. Up to 50,000 Knots, 20,000 Oystercatchers, 5,000 Bar-tailed Godwits, 5,000 Dunlins, 5,000 Grey Plovers and 2,500 Sanderlings have been counted. The largest totals are recorded on spring tides.

In autumn falls of night migrants occur, and these may be substantial with a chance of Wryneck, Icterine and Barred Warblers, and Red-backed Shrike from late August. By October large movements of thrushes, especially Blackbird and Fieldfare, become prominent, as are Goldcrest, tits, finches and buntings. Visible migration is most obvious in south or south-west winds. In late autumn, look for Water Rail, Woodcock, Little Auk (after strong north winds), Long-eared and Short-eared Owls, Richard's Pipit, Yellow-browed Warbler, Firecrest, Great Grey Shrike (which occasionally winters) and Lapland Bunting, and more erratically Rough-legged Buzzard and Waxwing.

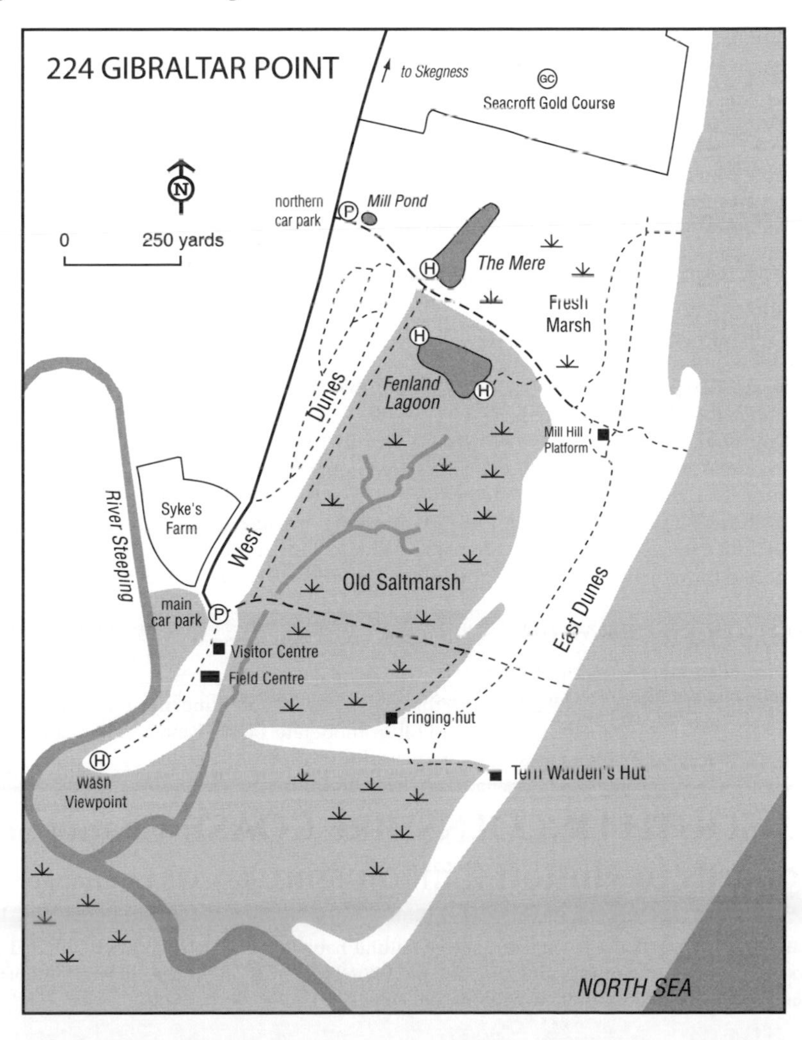

Winter brings good numbers of Red-throated Divers and auks offshore, and in some seasons up to 1,000 Common and 60 Velvet Scoters and a few Common Eiders and Red-breasted Mergansers; these are occasionally joined by Long-tailed Duck, one or two Black-throated and Great Northern Divers or a Slavonian Grebe. The reserve attracts up to 2,000 Dark-bellied Brent Geese and sometimes skeins of Pink-footed Geese pass over, especially in late winter. Other possibilities include a few Whooper or Bewick's Swans, or a Mediterranean, Iceland or Glaucous Gull, and increasing numbers of Little Egrets over-winter. There are large congregations of common open-shore waders, including up to 10,000 Knots, with Sanderling on the beaches. Sparrowhawk, Hen Harrier, Peregrine, Merlin and Barn and Short-eared Owls are regular. Rock and Water Pipits, finches (including Common and Lesser Redpolls and Brambling) and Snow and Corn Buntings frequent the saltmarsh, dunes and beach, although the Snow Bunting flock is mobile and can be hard to find; Shore Lark and Twite are erratic visitors, most likely in late autumn. Cold weather may bring movements of Lapwing, thrushes and finches.

Access

Leave Skegness seafront south on a minor road signed to Gibraltar Point. Parking is available at the beach car park and visitor centre (fee). No permit is required but visitors should keep to roads and marked paths, and not enter marked sanctuary areas. The visitor centre is open daily. Accommodation is available at the Wash Study Centre (contact the LWT for details).

The Mere and Tennyson's Sand Both a short walk from the beach car park, with three hides.

Fenland Lagoon There are two hides, at the east and west ends.

Jackson's Marsh Overlooked by a spacious hide accessible from the main car park.

East Dunes Accessed from the beach car park or via South Marsh Road from the main car park. Extensive areas of Buckthorn, though difficult to work, provide cover for migrants. The height of the dunes offers an overview in many places (notably at Mill Hill), and at the north end of the dunes Shoveler's Pond, a small reed-fringed pond surrounded by willows, is especially attractive to migrants.

Mill Hill Observation Platform At the north end of the East Dunes, this provides an all-round view and is ideal for observing movements of diurnal migrants in early morning.

The Wash Viewpoint South of the main car park, this offers panoramic views of the saltmarsh and mud and sand flats of the Wash.

The Spit Wildfowl and waders are best seen as they fly to the Spit to roost at high tide. Largest numbers are in spring and autumn, arriving up to 2 hours before high tide, and a north-west to north-east wind produces the highest counts. View from the Shorebird Warden's Hut.

Seawatching This can be a problem in early morning, because of the light, and at low tide. The best place to watch from is the Shorebird Warden's Hut at the base of the Spit, though in northbound movements a position further north may be better.

FURTHER INFORMATION

Grid ref/postcode: TF 556 580 / PE24 4SU
Lincolnshire Wildlife Trust: tel: 01507 526667; web: www.lincstrust.org.uk

225 SOUTH LINCOLNSHIRE COAST: Chapel St Leonards to Huttoft (Lincolnshire) OS Landranger 122

This stretch of coast has a restricted range of natural habitats but is rather under-watched and offers the chance to find interesting migrant and wintering birds on the sea, the various borrow pits and in the limited areas of cover along the coast.

Habitat

On this section of coast the foreshore is rather narrow and mostly bounded by artificial sea walls, and the extraction of clay to construct these sea defences has formed a series of borrow pits just inland of the sea wall. Comprising deep water fringed by reeds and willow scrub, most of the pits are managed as reserves by the LWT. Inland are extensive areas of arable farmland, and some sand dunes with stands of Sea-buckthorn between Huttoft and Chapel St Leonards. An important habitat for migrant passerines is formed by gardens along the coast, especially around Anderby Creek and Chapel St Leonards.

Species

In early spring, passage migrants include Wheatear and sometimes Ring Ouzel, Stonechat, Black Redstart and Firecrest. Later in the season, Yellow Wagtail, a variety of warblers and Whinchat are regular, and scarcer species include Common Redstart, Wood Warbler and Pied Flycatcher. There is little habitat for waders but in both spring and autumn Black-tailed Godwit, Ruff and Whimbrel can be found on areas of short grass, notably at Sandilands Golf Club.

Breeding species include Sparrowhawk, Water Rail, Barn Owl and Grasshopper, Reed and Sedge Warblers, and Bearded Tit may summer.

Autumn passage sees arrivals of the 'usual' east coast species such as Common Redstart, Willow and Garden Warblers and Pied Flycatcher, and occasionally also Wryneck, Icterine and Barred Warblers, Red-breasted Flycatcher or Red-backed Shrike. In late autumn Richard's Pipit (on rough grassland south of Anderby Creek and Sandilands Golf Club), Black Redstart, Yellow-browed and Pallas's Warblers, Firecrest and Great Grey Shrike are some of the scarce migrants to search for, and more predictable late-autumn migrants are thrushes, Long-eared and Short-eared Owls and Woodcock.

Early-autumn gatherings of Common, Sandwich and Little Terns offshore attract Arctic Skuas, and Mediterranean Gulls can also be found with diligence. Low water levels in late summer on the borrow pits (especially Huttoft Bank Pits) attract passage waders, including Common and Green Sandpipers, Greenshank and sometimes Spotted Redshank, Wood and Curlew Sandpipers and Little Stint. Strong onshore winds, especially prolonged north–north-east, may prompt a passage of seabirds, including Gannet, Fulmar, Kittiwake, terns (including Black) and Arctic Skua, and occasionally Manx and Sooty Shearwaters. In late autumn seaducks, Dark-bellied Brent Goose, Little Gull, Great and Pomarine Skuas, Common Guillemot and, sometimes, Little Auk are possible.

Offshore in winter look for Red-throated Diver, Common Scoter and small numbers of Common Eiders. Following strong north winds other seaducks are more likely, including Red-breasted Merganser, Long-tailed Duck, Velvet Scoter and Scaup, as well as Common Guillemot and a few Razorbills. Winter gull flocks may harbour Glaucous or Iceland Gulls, and flocks of Little Gulls are often present off Huttoft Bank (best searched for in calm weather). Due to the limited foreshore there are few waders but Sanderling and occasionally Purple Sandpiper may occur, as well as Rock Pipit, Twite and Snow Bunting. Inland of the sea walls, the borrow pits may hold occasional seaducks (offering much better views than birds out to sea), Smew, and perhaps also divers or Red-necked and Slavonian Grebes. Further inland still, check the flocks of Lapwing and Golden Plover on the fields for Ruff. The farmland also attracts Peregrine, Merlin, Hen Harrier and Short-eared Owl, although these are well-scattered and hard to find, and occasionally also wild swans or geese.

Access

All points are accessed from the minor road between Sutton on Sea and Chapel St Leonards.

Sandilands Pit Turn east off the A52 at the south end of Sutton on Sea on the minor road signed to Sandilands. After 0.3 miles the road bends sharply south to parallel the sea wall, and after a further 0.25 miles the pit forms a 1.5-ha LWT reserve. There is limited access, and cars may be parked in the access track (next to Stymie Cottage), but do not obstruct the field access.

Huttoft Bank Pits After a further 0.75 miles there is a pull-in on the inland side of the road, opposite the public footpath sign on the sea bank, with room for a couple of cars (there is also parking 0.5 miles south at the Huttoft Car Terrace). Cross the footbridge over the dyke and take the field-side track to the hide overlooking the largest pit, a 4-ha LWT reserve. (Following the road south for a further 1.5 miles there are more borrow pits beside the road, just beyond the turning to the dunes car park at Moggs Eye.)

Huttoft Car Terrace A further 0.5 miles south of the Huttoft Pits pull-in, there is a minor crossroads with the west turn signed to Huttoft, while the road to the east leads to the sea. The raised sea wall and the car park (fee in summer) offers a vantage for seawatching, best around

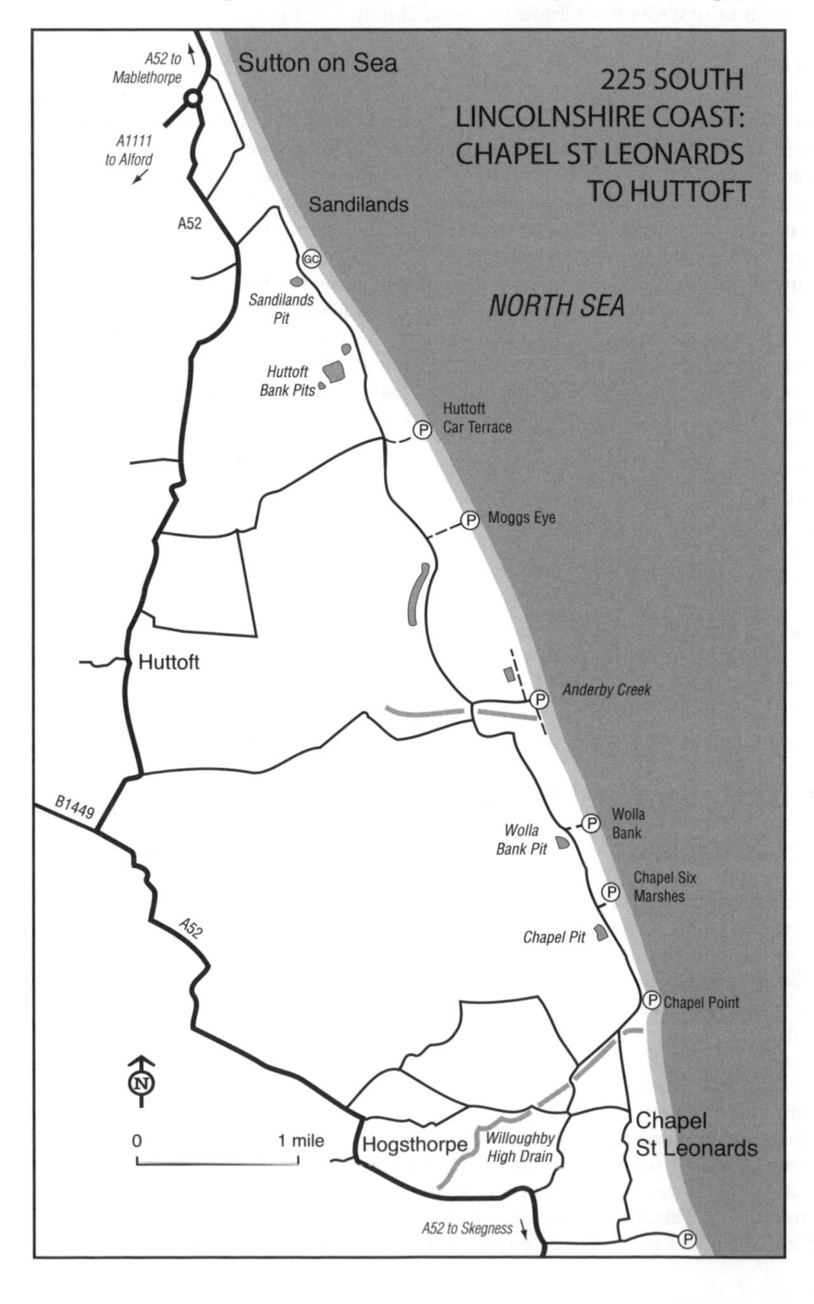

high tide, and the car can be used as a shelter in poor weather. Gulls roost offshore here.
Anderby Creek Turn east off the coast road 3 miles south of Sandilands to Anderby Creek and park at the end of the road by the dunes. A track runs north, passing some gardens, then a stand of young trees by the open-air sports complex (with a path leading into the trees), and to another patch of trees. The gardens hold migrants but discretion is necessary when peering into them. The small pit among the houses, just inland of the track, attracts occasional seaducks. Another path leads south along the coast from the car park, past the mouth of Anderby Creek, to a patch of scrub with some mature trees. Thirdly, take the track over the dunes to a small shelter which, though affording limited views, can be useful for seawatching.
Wolla Bank and Chapel Six Marshes Continue south on the coast road and tracks lead east to car parks in the dunes at Wolla Bank and Chapel Six Marshes. There are some extensive areas of scrub around these, and a small stand of conifers at Chapel Six Marshes, while the area between the car parks comprises grazed grassland.
Wolla Bank Pit Lies inland of the coast road opposite the turning to Wolla Bank. A 3.9-ha LWT reserve, it can be viewed from a short path off the road.
Chapel Pit Lies inland of the coast road 200 yards south of the Chapel Six Marshes turning. A 3.2-ha LWT reserve, it can be viewed from a short path off the road.
Chapel Point Turn east off the A52 6 miles north of Skegness to Chapel St Leonards and drive through the village for 1.5 miles to the northern outskirts at Chapel Point. There is a car park at the coastguards' lookout on the seaward side of the road, with scrub around the car park and on the other side of the road. Chapel Point offers reasonable seawatching from the car park and it is possible to walk south along the sea wall past some chalets to the mouth of Willoughby High Drain, and then alongside the drain, turning left into a road, which runs south parallel to the coast. The gardens along the road and the scrub at the channel mouth attract migrants.

FURTHER INFORMATION

Grid refs: Sandilands Pit TF 530 802; Huttoft Bank Pits TF 536 792; Huttoft Car Terrace TF 540 786; Anderby Creek TF 551 761; Wolla Bank/Wolla Bank Pit TF 556 749; Chapel Six Marshes/Chapel Pit TF 559 742; Chapel Point TF 561 732
Lincolnshire Wildlife Trust: tel: 01507 526667; web: www.lincstrust.org.uk

226 NORTH LINCOLNSHIRE COAST: Mablethorpe to Grainthorpe Haven (Lincolnshire)

OS Landranger 113 and 122

This stretch of coastline is well positioned to receive migrants which, despite relatively limited coverage, have included some first-class rarities. It also attracts an excellent range of wintering wildfowl and waders.

Habitat

The dunes and beaches of the Humber gradually narrow as the estuary extends south to Mablethorpe. Tidal sand and mud are backed in turn by saltmarsh and still-evolving dune system. Dunes of variable ages support extensive thickets of Sea-buckthorn and Elder, separated by damp slacks which often have freshwater marshes dominated by rushes, sedges and reeds, or areas of willows and Sallows. Active management has created some freshwater pools, and dotted along the coast are isolated stands of mature deciduous trees. In the north of the area (from Howden's Pullover northwards) there are also several borrow pits inland of the sea wall.

Most of the area is protected. Between Mablethorpe North End and Saltfleet Haven 5 miles of dunes, foreshore and beach make up the Saltfleetby–Theddlethorpe Dunes NNR, with 952 ha managed by Natural England and an additional 43 ha by the LWT, while the LWT Donna Nook–Saltfleet Reserve covers another 1,150 ha along 6 miles of coast between Grainthorpe Haven and Saltfleet, abutting the NNR.

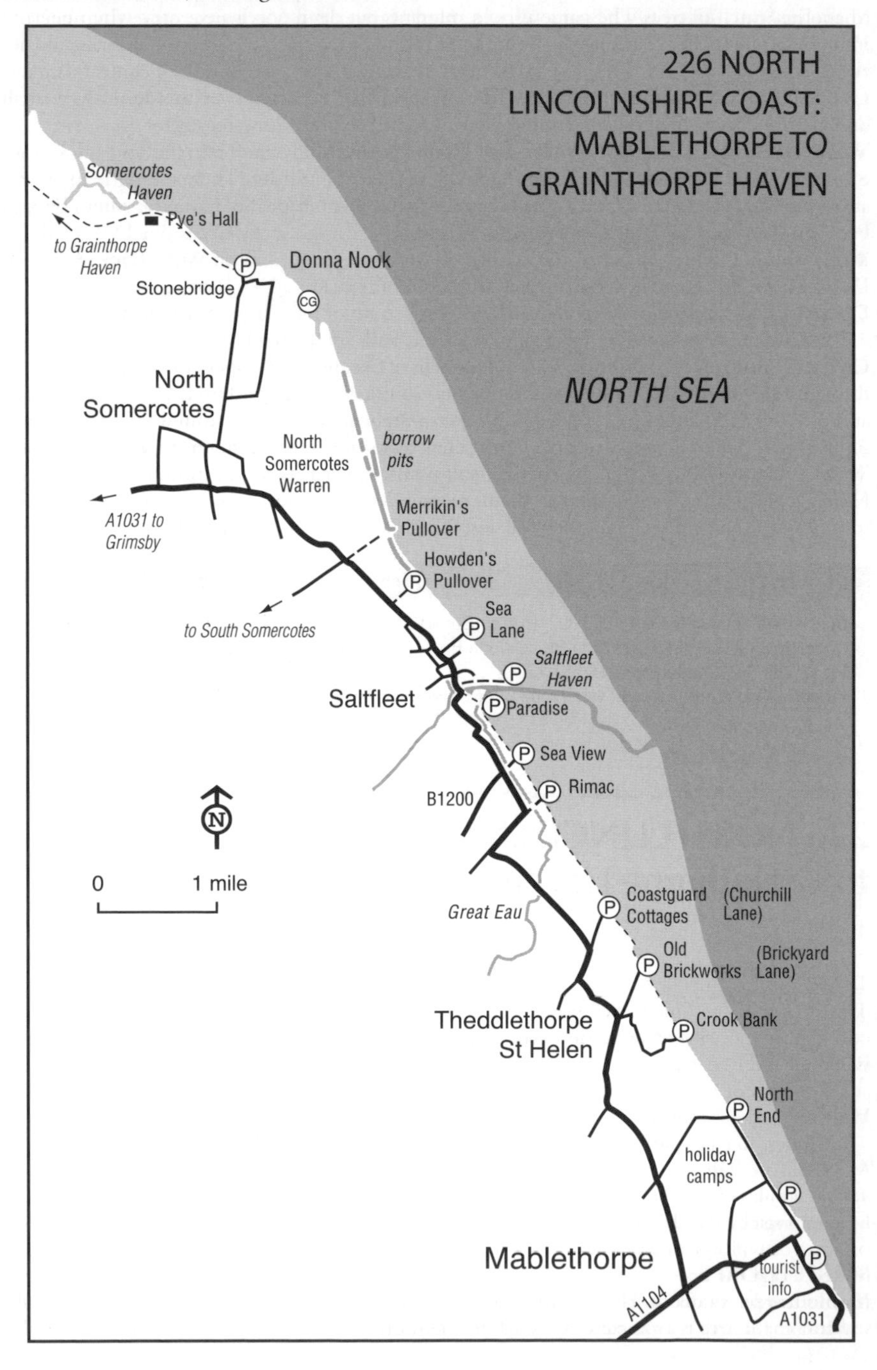

Species

In winter there are Red-throated Divers and Great Crested Grebes offshore, as well as Common Scoters and Common Guillemots. Other divers, grebes and seaducks are erratic in their appearances. Brent Goose is regular in some numbers, as are Wigeon, Pintail and Shoveler. Short-eared Owl, Sparrowhawk, Hen Harrier and Merlin are quite frequent and Peregrine reasonably regular. Waders include Jack Snipe and Woodcock in the dunes (and the former also on the foreshore at Rimac) as well as the usual coastal species. Glaucous Gull, especially from January, and Kittiwake may join the commoner gulls. On the saltmarsh and tideline Twite and Snow Bunting are reasonably common (the latter favouring the saltings at Rimac, between Pye's Hall and Donna Nook and in the Theddlethorpe area), but Shore Lark and Lapland Bunting are much less regular, the former has declined greatly but is still possible at Rimac or Donna Nook, and the latter favours areas of damp grassland around Pye's Hall and Rimac. Rock Pipit is common on the foreshore and occasionally Water Pipit may be found in 'fresh' habitats (notably on pools near Rimac car park), together with Water Rail. The Sea-buckthorn attracts large numbers of Fieldfares, Redwings and Starlings, Stonechats and sometimes a few Blackcaps.

Landbird migrants can be excellent in spring and autumn, with a long list of vagrants recorded in the area. In addition to the usual chats, warblers and flycatchers, scarcer migrants to search for in early spring include Short-eared Owl, Ring Ouzel, Black Redstart, Firecrest and the Scandinavian subspecies of Rock Pipit (on the saltmarshes; the somewhat similar Water Pipit favours 'fresh' habitats). In May–June these may be joined by Marsh Harrier, Hobby, Wood Warbler and Pied Flycatcher. Rarer but near-annual late-spring migrants include Bluethroat, Marsh Warbler, Red-backed Shrike, Golden Oriole and Common Rosefinch. Numbers and variety of landbird migrants are usually greater in autumn. In August–September search for Wryneck, Bluethroat, Icterine and Barred Warblers and Red-backed Shrike, all of which are scarce but annual, and may be joined from late September by Ring Ouzel, Black Redstart, Siberian Stonechat, Yellow-browed and Pallas's Warblers, Red-breasted Flycatcher, Great Grey Shrike and Lapland Bunting. Richard's Pipit is something of a speciality, with the dunes north of Rimac car park and the region between Howden's Pullover and Pye's Hall being the best areas. Late autumn can witness arrivals of Long-eared and Short-eared Owls, Woodcock and a few Bearded Tits. A variety of waders occurs on passage and may include Whimbrel, Spotted Redshank (regular in Grainthorpe Haven), Curlew Sandpiper, Little Stint and, sometimes, Wood Sandpiper, with the best range in autumn, but in spring 'trips' of Dotterels are fairly regular, notably on the fields at Donna Nook, and large flocks of Sanderlings may pass through.

At sea, onshore winds in autumn may prompt a passage of Gannets, Fulmars and Manx Shearwaters, and sometimes Leach's Petrel and Sooty Shearwater, but in general the area is unproductive for seawatching, being too exposed with no obvious headlands, though the beach shelters at Mablethorpe offer some height and protection. Arctic Skua is regular, attending the flocks of gulls and terns, but Great Skua is quite scarce and Pomarine and Long-tailed Skuas uncommon. In spring and autumn Little Gull and Common, Arctic, Sandwich, Little and Black Terns are often present, with Kittiwake in autumn, and a notable late-summer concentration of terns.

Breeding species include Little Egret, Shelduck, occasionally Teal, Water Rail, Oystercatcher, Ringed Plover and, notably, Little Terns in some numbers (best seen when feeding and loafing at the mouth of Saltfleet Haven). Marshy areas have Grasshopper (at Rimac), Reed and Sedge Warblers, and occasionally Short-eared Owl; Lesser Redpoll and Nightingale are both scarce.

Access

These are listed south–north, from Mablethorpe to Donna Nook, and divided between Mablethorpe, Saltfleetby–Theddlethorpe Dunes NNR and Donna Nook–Saltfleet LWT reserve.

MABLETHORPE

Mablethorpe seafront The concrete beach shelter by the tourist information is useful for seawatching. There is a car park 200 yards to the south.

Mablethorpe North End Follow the minor road north from Mablethorpe seafront alongside the dunes to the signed car park at North End (or turn off the A1031 halfway between Theddlethorpe and Mablethorpe on a minor road signed North End). From here there is access to the dunes and beach.

SALTFLEETBY–THEDDLETHORPE DUNES NNR

Access is at six points off the A1031, connected by a footpath. Limited parking is available at all six, which are listed south–north. There is open access except to the danger zone – visitors must comply with safety regulations displayed at the entrances.

Crook Bank In Theddlethorpe St Helen, south of the church, turn east off the A1031 at the right-hand bend onto a metalled track and fork right, following Sea Lane for 1 mile, to Journey's End.

Old Brickworks Follow directions as for Crook Bank, but fork left to Brickyard Lane. Check the copses of trees in Theddlethorpe village.

Coastguard Cottages Leave Theddlethorpe St Helen north on the A1031 and after 0.5 miles, at the sharp left-hand bend, turn right into Churchill Road and follow this past Sea Bank Farm to the car park.

Rimac Continue north on the A1031 and, after 1.5 miles, the road bends sharp right, and after a further 0.5 miles sharp left. At this corner take the rough track that leads straight on, over the bridge across the Great Eau, to the car park. A saltmarsh viewpoint lies ahead and though a 1.5-mile walk to the sea is necessary in order to seawatch, it can be worthwhile. There are stands of willows and Sallows in the freshwater marsh, some interesting freshwater pools, and a belt of willows borders the landward side of the reserve.

Sea View Farm Continue north on the A1031 and, opposite the junction with the B1200, turn east on the metalled track to Sea View Farm. There is a copse of mature trees by the car park, attractive to migrants.

Saltfleet Haven (Paradise) In Saltfleet village the A1031 crosses the Haven and immediately to the south the road bends sharply west while a track leads straight on past Gowts Farm (and crosses the main dyke) to Paradise, where there are some mature trees (the copses in Saltfleet village are also worthy of scrutiny).

DONNA NOOK–SALTFLEET LWT RESERVE

Access is at five points, all off the A1031, with limited parking facilities; these are listed south–north. There is open access to the beach and foreshore but do not enter the danger area, which stretches from Pye's Hall south to Howden's Pullover; when the ranges are in use (usually on weekdays) marker boards are sited at all access points and red flags fly on the shore that is in use; the marker boards should not be passed and, although access to the sea wall and dunes is usually still possible, there is a great deal of disturbance.

Saltfleet Haven (north bank) In Saltfleet village turn east off the A1031 immediately north of the bridge over Saltfleet Haven on a rough track along the bank of the Haven to a small car park. From here footpaths lead north on either side of the dune ridge and it is possible to walk to the mouth of the Haven (but caution is required as the tidal creeks here are very dangerous), which is a good area for loafing and feeding gulls and terns.

Sea Lane, Saltfleet At the north end of Saltfleet village take Sea Lane for 0.25 miles to the car park (café, open in summer, and toilets). The sea wall here gives good views over the saltmarsh. It is possible to walk south to Saltfleet Haven or north to Howden's Pullover.

Howden's Pullover Midway between Saltfleet and North Somercotes, Howden's Pullover is signed from the A1031. A rough track leads to the car park on the sea wall. Walk south to Saltfleet village or north on the sea wall or foreshore, or through the dunes to Donna Nook reserve.

Merrikin's Pullover Just south of North Somercotes turn east off the A1031 on unmetalled roads opposite the turn to South Somercotes (NB there is no car park here).

Donna Nook/Stonebridge Follow signs from North Somercotes to Donna Nook, taking the minor road north for 2 miles to the car park at Stonebridge. A public footpath runs north-west along the sea wall and it is possible to walk the dunes or along the beach, with patches of cover for migrants, especially the Sycamores and Elders around the site of the old Pye's Hall, 1 mile from the car park (the obvious stand of mature trees south of Stonebridge car park is on the RAF base, with no access). Further on, waders roost at Grainthorpe Haven. The Donna Nook reserve proper can be reached by walking south from Stonebridge for 1.5 miles along the beach before entering the dunes south of the coastguard lookout.

FURTHER INFORMATION

Grid refs: Mablethorpe seafront TF 510 847; Mablethorpe North End TF 497 869; Crook Bank TF 489 883; Old Brickworks TF 484 892; Coastguard Cottages TF 478 902; Rimac TF 467 917; Sea View Farm TF 465 924; Saltfleet Haven (Paradise) TF 457 933; Saltfleet Haven (north bank) TF 464 935; Sea Lane, Saltfleet TF 456 943; Howden's Pullover TF 448 952; Merrikin's Pullover TF 443 957; Donna Nook/Stonebridge TF 421 998
Lincolnshire Wildlife Trust: tel: 01507 526667; web: www.lincstrust.org.uk
Natural England, Lincolnshire Office: tel: 01522 561470; email: eastmidlands @naturalengland.org.uk

227 TETNEY MARSHES (Lincolnshire)

OS Landranger 113

This area near the mouth of the Humber Estuary holds a selection of migrant and wintering wildfowl and waders, and has an important colony of breeding Little Terns.

Habitat

This RSPB reserve covers 1,500 ha of saltmarsh and dunes between Humberston Fitties and Northcoates Point, with some areas of cover, notably stands of Hawthorn, at Northcoates Point, and two large brackish pools (the MOD Pools) behind the sea wall that are especially attractive in spring. Just inland of the sea walls are extensive areas of arable farmland, including a large, relatively recently reclaimed area at Tetney.

Species

The area is excellent for raptors in winter, including Hen Harrier, Sparrowhawk, Merlin, Peregrine and, in some years, Short-eared Owl. At high tide there are good views of wildfowl from the sea walls, with up to 700 Dark-bellied Brent Geese (check also for Pale-bellied Brents), Wigeon and, sometimes, Common Scoter offshore. A variety of the usual open-shore waders are present, including Grey Plover, Bar-tailed Godwit and Sanderling, and the wader roost can be seen on higher tides from the end of the Haven track, but otherwise only distant views are possible from the sea wall. Jack Snipe can, with perseverance, be found in the marshy areas and a few Ruffs winter on the fields. Among flocks of finches and buntings check for Twite and Lapland and Snow Buntings, and Rock Pipit is frequent in the saltmarsh gutters; check also for Water Pipits around the canal at Tetney Lock.

At passage times the MOD Pools and adjacent borrow pits (and those nearer the Haven) attract waders, typically including Greenshank and Green Sandpiper, and sometimes also Wood Sandpiper. Other migrant waders favour the saltmarsh and open shore, and may include Golden Plover, Whimbrel, Black-tailed Godwit, Turnstone, Knot, Sanderling, Curlew Sandpiper and Little Stint. Dotterel is annual on passage in the Reclaimed Fields, the fields by the road to Horseshoe Point, or those near Low Farm. Other regular migrants include Marsh Harrier, Common and Sandwich Terns (with a notable late-summer concentration of loafing terns, especially on evening spring tides) and small numbers of passerines such as Wheatear and Whinchat.

Breeding birds include Yellow Wagtail, Oystercatcher, Shelduck and up to 80 pairs of Common Redshanks and 100 pairs of Little Terns, which nest on the shore north of the MOD Pools. Little Egrets are now resident along this coast.

Access

RSPB Tetney Marshes Leave the A1031 at Tetney or North Cotes, taking the minor roads to Tetney Lock, and then the no through road just north of the bridge over Louth Canal, following this east for 400 yards to the sharp left-hand bend, where there is limited roadside parking. Follow the footpath for 1.25 miles along the canal bank to the embankments separating the saltmarsh from the borrow pits and fields. A circular walk of 3 miles on the embankments encircling the reclaimed fields covers most of the habitats, and at the east end of this circuit, near Northcoates Point, a gate accesses the old sea wall and a track which parallels the airfield fence for 1.5 miles to Horseshoe Point. This passes the MOD Pools and a track east across the marsh between the two larger pools leads to the shore. The MOD Pools attract passage waders in spring but subsequently tend to dry out, although they are temporarily flooded by spring tides in late summer. Breeding Little Terns can be viewed where the track passes through the dunes, and to protect the terns the foreshore between the Haven and MOD Pools is out of bounds in summer (as is the Haven track beyond the warden's caravan). Visitors should avoid tern nesting areas and heed warning notices.

Horseshoe Point Access is also possible from the A1031 between North Cotes and Marshchapel, taking the signed turn to Horseshoe Point (Sheepmarsh Lane). After 2.5 miles there is a small car park by the sea wall. This is an excellent spot for open-shore waders. Walking

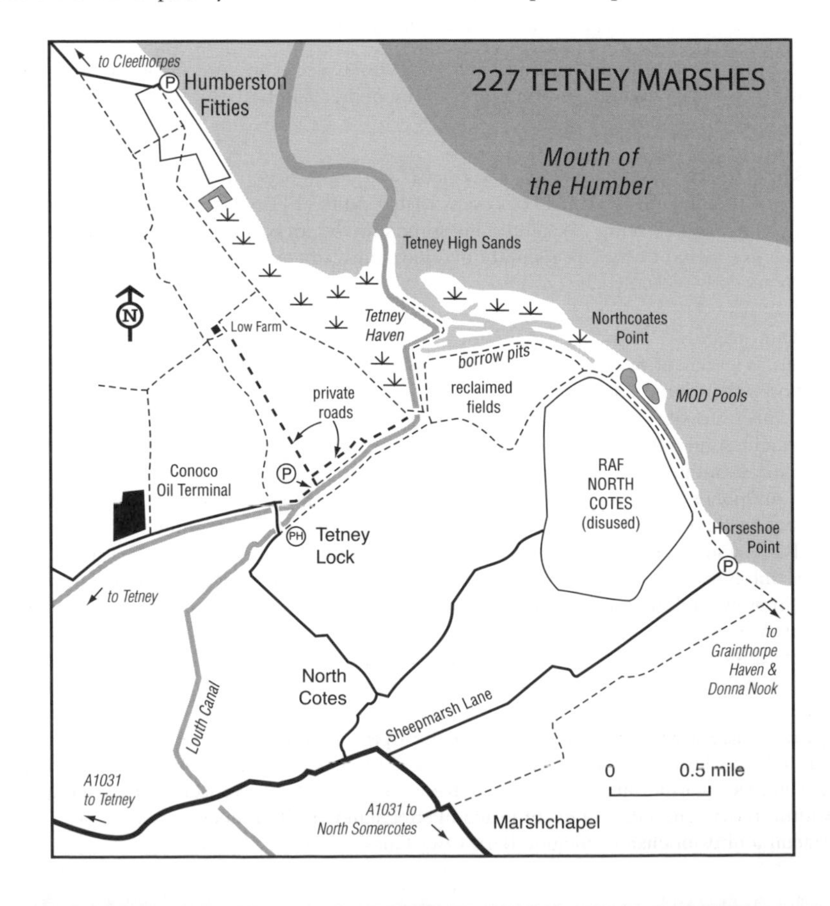

north along the bank, the MOD Pools are reached after 1 mile and, by continuing, there is access to Northcoates Point and Tetney Marshes RSPB reserve.

Humberston Fitties From the A1031 south of Cleethorpes take the minor road at the roundabout east to Humberston and, after 1 mile, turn south at the next roundabout to Humberston Fitties, where there is a car park by the sea wall. Follow the dunes south through the pines on the sea bank by Fitties holiday camp, to the yacht club, checking the gardens and pines for migrants. It is also possible to follow the sea wall for 2 miles from Humberston Fitties to RSPB Tetney Marshes. There are some more areas of bushes and scrub en route which may hold migrants, and the arable fields inland of the bank may have roosting waders and, in spring, occasionally Dotterel.

FURTHER INFORMATION

Grid refs: RSPB Tetney Marshes TA 344 024; Horseshoe Point TA 381 018; Humberston Fitties TA 331 061
RSPB Tetney Warden, RSPB Denby Dale Office: tel: 01484 861148

228 COVENHAM RESERVOIR (Lincolnshire)

OS Landranger 113

Lying 5 miles north of Louth, this relatively small reservoir lies in a flat and featureless area of arable farmland. Its proximity to the coast has resulted in a reputation for attracting seaducks, grebes and divers in winter, and passage periods can also be interesting.

Habitat

Completed in 1969, open water now covers 81 ha and the banks are steep and concrete-clad, with the outer slopes having grazed turf and some small stands of deciduous trees. The reservoir is used for a variety of water sports, and only the south-east corner is free from disturbance.

Species

Wintering wildfowl include Pochard, up to 100 Goldeneyes, small numbers of Pintails, Gadwalls, Wigeons, Shovelers and Teals, and sometimes Smew, Goosander, Long-tailed Duck, Common and Velvet Scoters or Scaup. Other waterfowl include Little and Great Crested Grebes and reasonably regular single Red-necked or Slavonian Grebes and Great Northern, Black-throated and Red-throated Divers. Rarer grebes and seaducks may make prolonged visits. Cormorant is regular and may be joined by a few Shags following gales. The reservoir is used by numbers of loafing and bathing gulls, and the roost is of interest; mostly comprising Black-headed and Common Gulls, it also holds small numbers of Herring and Great Black-backed Gulls. Mediterranean, Iceland and Glaucous Gulls are occasional, most regularly in late winter and early spring. Other wintering species include Golden Plover in adjacent arable fields, Sparrowhawk and, sometimes, Green Sandpiper, Merlin, Peregrine, Short-eared Owl, Rock Pipit, Grey Wagtail and Stonechat.

Spring migrants may include Common Sandpiper, Turnstone, Sanderling, transient flocks of Common, Arctic and Black Terns and Little Gulls, Yellow Wagtail, Wheatear, occasionally Marsh Harrier, Osprey, Black Redstart or Ring Ouzel; scarcer visitors include Temminck's Stint. On autumn passage a greater diversity of waders may be found, despite the rather unattractive concrete banks, and Little Gull and Common and Black Terns. Scarcer visitors include Black-necked Grebe, Marsh Harrier, Osprey, Red-necked and Grey Phalaropes, and coastal gales in late autumn may produce records of Leach's Petrel, skuas or Little Auk

Breeding birds in the area include Barn Owl, Cuckoo and Tree Sparrow.

Access

Leave Louth north on the A16 and, 1 mile north of Utterby, turn east on a minor road to Covenham and Grainthorpe. Continue straight on at the crossroads and the reservoir is clearly visible on the right. Park in the car park at the north-west corner. Take the steps to the top of the embankment and a footpath circumnavigates the reservoir, with the south-eastern corner the least disturbed.

FURTHER INFORMATION

Grid ref: TF 340 962

229 KILLINGHOLME HAVEN (Lincolnshire)

OS Landranger 113

On the south bank of the River Humber 3 miles north of Immingham, this rather small and compact site is notable for furnishing excellent views of passage waders.

Habitat

Comprising three flooded clay pits, the largest of which has a mosaic of shallow water, reed and sedge beds and islands with varying amounts of exposed mud, there are some mature hedgerows and a patch of Hawthorn scrub immediately to the south. The surroundings are a mix of rough grassland and extensive industrial areas. The LWT manages 32 ha around the pits.

Species

On passage Whimbrel, Black-tailed and Bar-tailed Godwits, Ruff, Greenshank, Spotted Redshank, Green, Wood, Common and Curlew Sandpipers and Little Stint are regular, especially in autumn, and there may be a notable build-up of Common Redshanks, Black-tailed Godwits and Ruffs at the same season; Avocets are increasingly regular in early spring. A variety of scarce and rare waders has also been recorded. The largest numbers of waders roost on the pits during the fortnightly spring tides. Shoveler and Wigeon also occur in autumn, and sometimes Garganey. Marsh Harrier is regular in spring, and in May a Long-eared Owl often hunts the adjacent rough grass fields in early evening. During autumn gales, especially from the north-east, seabirds may be forced into the Humber and a variety of auks, gulls, terns and skuas can be seen from the sea wall.

In winter the pits hold good numbers of Teals and Shovelers, and may attract Smew or more rarely grebes, divers or seaducks, but if water levels remain high wader variety is low. Nevertheless, Lapwing, Common Redshank and Dunlin are regular, and significant numbers of Common Snipes are present, along with a few Jack Snipes and Water Rails. Bittern and Bearded Tit are occasional, favouring the reed-filled small west pit. Long-eared Owl roosts in the Hawthorn scrub, and Short-eared Owl also winters in the area and may linger into spring.

Breeding species include Water Rail, Kingfisher, Lesser Whitethroat and Grasshopper, Reed and Sedge Warblers, with Oystercatcher and Ringed Plover nearby.

Access

Leave the A180 Scunthorpe–Grimsby road north on the A160. This road passes South Killingholme and a complex of oil refineries before reaching a roundabout after 3 miles. Turn left here, pass under the railway bridge and, after a further 100 yards, turn left on Rosper Road. After 2 miles, where this road bears to the left, turn right, signed North Killingholme Haven. The two larger pits lie to the right after 1 mile, either side of the railway crossing. The road has limited parking and is very busy in working hours, and the pit to the west of the railway is hard to view due to the tall vegetation; the best views are from the Humber sea wall on the north-eastern side of the main pit, or from the hide adjacent to the road.

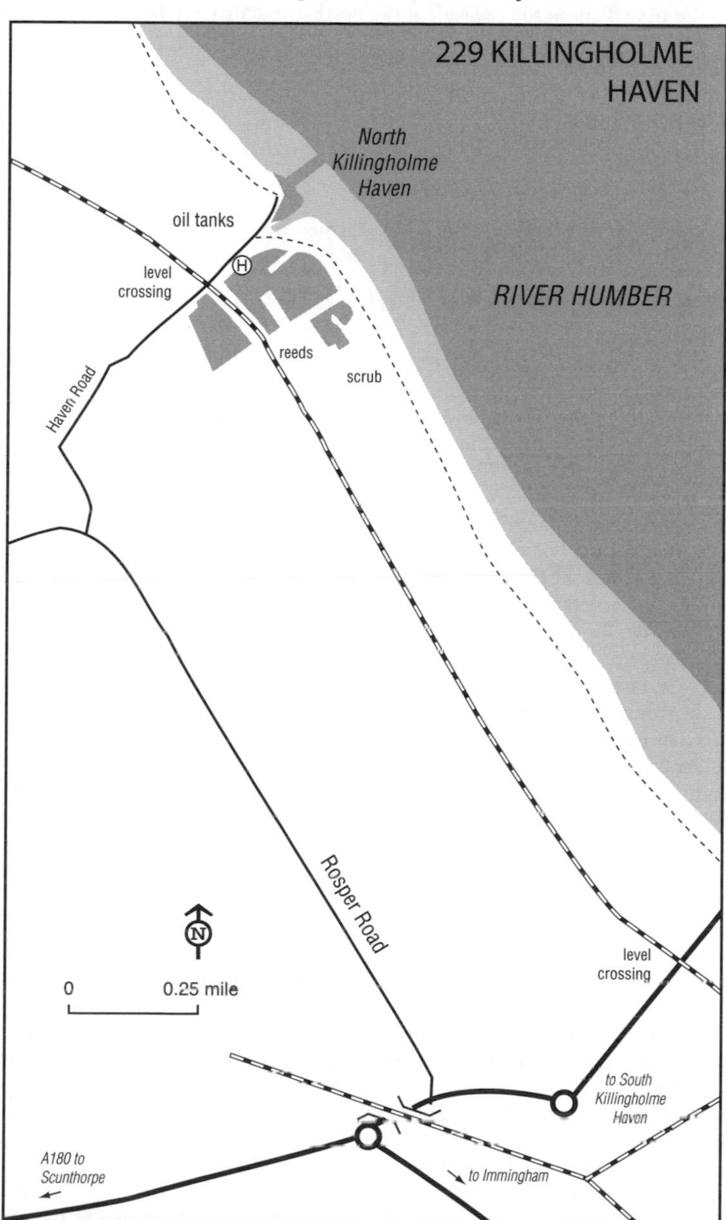

FURTHER INFORMATION

Grid ref: TA 164 199
Lincolnshire Wildlife Trust: tel: 01507 526667; web: www.lincstrust.org.uk

230 THE INNER HUMBER (Lincolnshire)

OS Landranger 112 and 113

The Humber is one of the major rivers in Britain and forms an important landmark for migrating birds. A complex of estuary habitats and borrow pits borders its southern shore, offering great potential to the more adventurous birdwatcher.

Habitat

The inner parts of the Humber Estuary are formed by rich muddy silts, especially at the confluence of the Rivers Trent and Ouse, with some limited areas of saltmarsh and rough grassland, all bounded by many sea walls.

Species and Access

The pace of birdwatching in this vast area is largely determined by the tides, and ideally, a visit should be timed to coincide with high water in order to view roosting waders, seaducks and, in season, seabirds. During spring and autumn the fortnightly spring tides are ideal, as these relatively high tides usually flood the saltmarshes and force waders onto drier ground to roost. High water on spring tides usually occurs at late evening and early morning. The intervening neap tides leave the saltmarshes uncovered; fewer waders move within view of the sea walls, and seaducks remain distant. In winter months spring tides peak during darkness and a late-morning high tide is the best compromise.

East Halton Skitter The interest here lies in the areas of rough tidal grassland, which attract wintering Short-eared Owls, Jack Snipes, sometimes Snow and Lapland Buntings, and occasionally Twite. From East Halton village take the minor road (Towles Corner) east and then immediately north for 400 m and then bear east on Townside to the road's end at East Halton Skitter. From here walk north along the sea bank, with rough grassland to the east and some borrow pits adjacent to the bank.

Dawson City Claypits This reserve at Skitter Ness consists of flooded pits by the sea bank which attract Jack Snipe and a few passage waders, with large numbers of Curlews and Golden Plovers in the fields. In Goxhill follow signs for North End and continue north on Ferry Road to Goxhill Haven. Park on the verge and walk east along the sea wall for 1 mile; the entrance to the reserve is via a stile in the north-west corner and the hide lies along the trail between the two pits. It is possible to continue for 2 miles, with areas of rough grasslands east of the bank, to East Halton Skitter.

New Holland Large numbers of diving ducks and seaducks are attracted to grain and animal feed drifting downstream from spills at the grain terminal. Numbers have peaked at 2,000 Pochards, 1,000 Tufted Ducks, 584 Goldeneyes and 260 Scaups, and in November they may be joined by up to 250 Common Scoters and a handful of Velvet Scoters, Common Eiders and Long-tailed Ducks, while the resident Mute Swans are often joined by a few Whoopers. The 2–3 hours around high water, on calm days, is best. Inland, the fields hold large numbers of Golden Plovers and other roosting waders. Snow Bunting may be found along the sea wall and, just west of the pier, Fairfield's Pit, which is a LWT reserve, attracts passage waders and breeding Grasshopper Warblers.

Follow the B1206 bypass around New Holland towards the docks and, immediately over the level crossing, turn left between the railway and a warehouse and follow the track around the sharp right-hand bend to park on the Humber Bank. Fairfield's Pit lies to the west of this track and can also be viewed from the sea wall. For best views of diving ducks walk back to the railway crossing and turn immediately left. Turn right through the green gate by the red barrier and follow signs to the coast path for 200 yards to the sea wall.

Barton-upon-Humber A complex of flooded pits stretches almost the entire Humber shore at the foot of the Humber Bridge between Barrow Haven and Chowder Ness, forming a

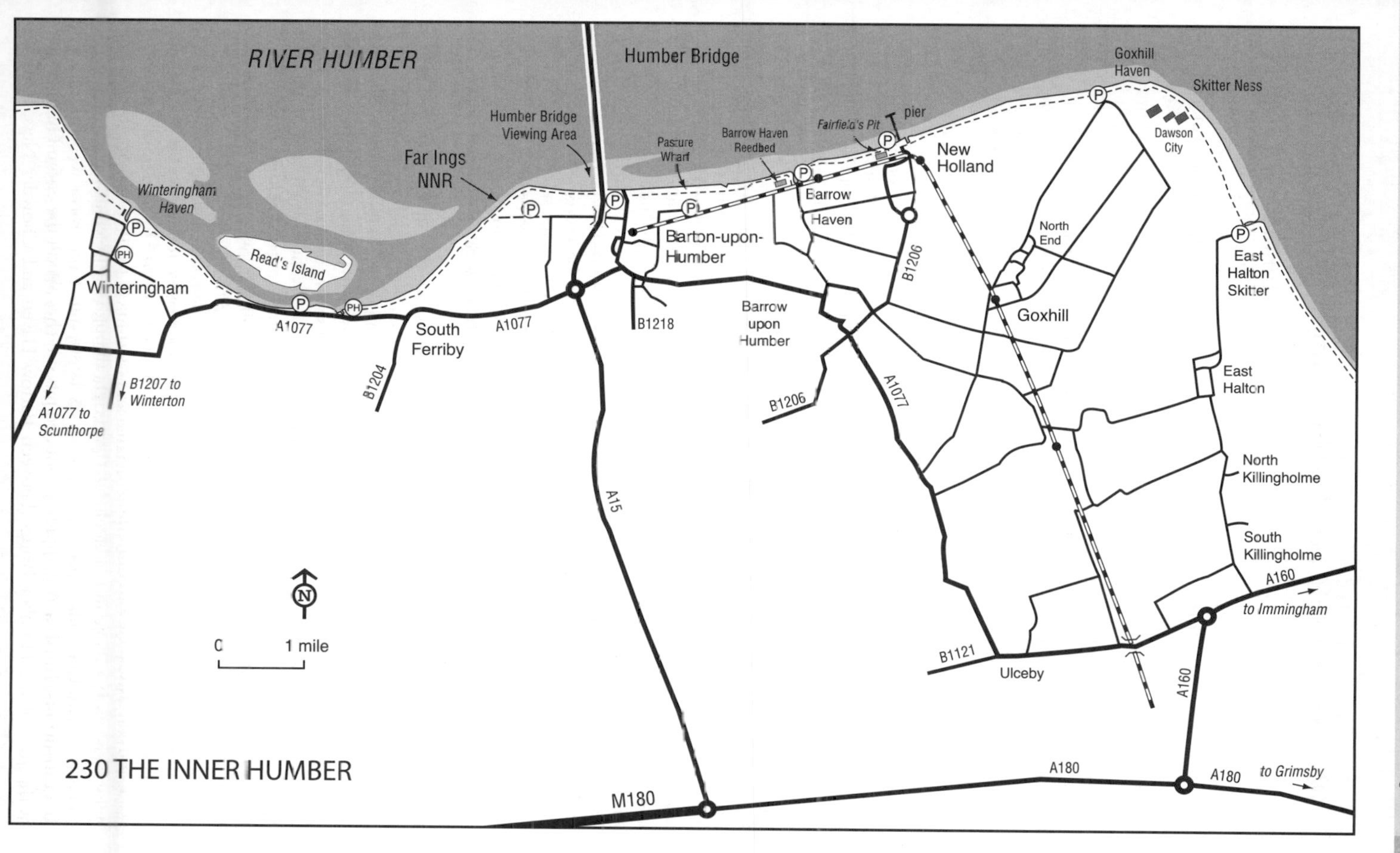

230 THE INNER HUMBER

mosaic of open water, reedbeds and scrub. Many are used for fishing or water sports but several are LWT reserves. They attract wildfowl, including up to 50 Scaups and sometimes divers, rarer grebes, Smew and Goosander. Up to three Bitterns winter but are typically elusive. Common Tern and a few Bearded Tits breed, and Garganey and Grasshopper Warbler appear on passage and sometimes nest. Several species of terns occur on passage. Beyond the sea bank, numbers of open-shore waders and seabirds may be seen on the river during autumn storms.

Barrow Haven From the A1077 in Barrow-upon-Humber take the minor road signed to Barrow Haven and follow this for 1.5 miles to the car park on the right just past the railway crossing (by the Old Ferry Wharf). Take the bridge westwards across the Haven and turn north through the gate to the sea wall. Immediately on the left a stepped path leads to a hide overlooking LWT Barrow Haven Reedbed reserve, and a footpath follows the bank west for 2 miles to Barton Waterside, giving views over the foreshore and a series of pits.

Barton Clay Pits/Pasture Wharf LWT reserve Leave Barton-upon-Humber east on the A1077 and turn north along Falkland Way. After 1 mile the road crosses a railway and after a few hundred yards turns sharply east and passes a series of pits (part of the LWT reserve, with limited parking at the southern end). After a further 0.5 miles there is a car park just before the end of the metalled road, and from which the Humber Bank footpath gives access to more pits and the saltmarsh.

The Humber Bridge Viewing Area Follow signs from the A15 or Barton-upon-Humber town centre to the Viewing Area; from the car park there is access to the Humber bank. Walk west to view the pits west of the bridge, and just west of the Haven Mouth the shelter at the Old Boathouse can be used to watch the river in rough weather for passing seabirds, which are most likely in north-easterly to south-easterly winds on a rising tide.

Far Ings NNR Follow signs as for the Viewing Area and, just north of Barton-upon-Humber railway station, turn west and follow Far Ings road for 1 mile to the visitor centre. Alternatively, turn north off the A1077 0.3 miles west of the roundabout at the junction with the A15 onto a minor road signed to the Clay Pits and, after 1 mile, turn left at the T-junction into Far Ings Road. The LWT visitor centre is open Saturdays, Sundays and Wednesday afternoons, and offers commanding views of the reserve. There are eight hides overlooking the pits, scrapes and foreshore, and the reserve is open in daylight hours. A variety of wildfowl, including Goosander and sometimes Smew winter on the pits, which also attract passage waders. Marsh Harrier, Shoveler, Water Rail and Bearded Tit are resident and Bittern and Grasshopper Warbler sometimes breed.

South Ferriby Bank Leave South Ferriby west on the A1077 and, after 0.5 miles, park just south of the road opposite the Hope and Anchor pub. Cross the main road and the drain to the right and bear left to the sea wall. Follow this east to a hide overlooking the foreshore. The period 2–3 hours before high tide is best.

Read's Island Formerly covering 220 ha, this island has dwindled to 60 ha, but it still attracts roosting waders and Pink-footed Geese, while the lagoons are managed for breeding waders, notably Avocets. Leave South Ferriby west on the A1077 and, after 1 mile, park in either of the two lay-bys to view the mudflats. This is one of the best areas on the estuary for waders, with the period 2–3 hours before high tide best, but on the highest spring tides most waders are forced onto the roost out of view on Read's Island up to 2 hours before high water.

Winteringham Foreshore LWT reserve Turn off the A1077 on minor roads to Winteringham and turn north on Low Burbage Road; after 0.5 miles, at the sharp left-hand bend over a bridge, park on the roadside verge. Either take the footpath to the right along the sea bank to a hide overlooking the estuary or follow the road over the bridge and turn immediately right on a track to the sea wall, which can then be followed west. Pink-footed Geese were abundant in the past and have recently returned to the Humber area, with counts of over 5,000. However, they are unpredictable and most likely to be seen flying to and from their roost on Read's Island at dawn and dusk. Thousands of Black-headed, Common and Great Black-backed Gulls, along with up to 1,500 Herring Gulls, roost well offshore, and occasionally Glaucous or Iceland Gulls loaf on the sand bars off Winteringham before going to roost.

FURTHER INFORMATION

Grid refs/postcode: East Halton Skitter TA 144 228; Dawson City Claypits TA 120 253; New Holland TA 080 243; Barrow Haven TA 062 235; Barton Clay Pits / Pasture Wharf LWT reserve TA 042 229; The Humber Bridge Viewing Area TA 027 233; Far Ings National Nature Reserve TA 011 229 / DN18 5RG; South Ferriby Bank SE 976 210; Read's Island SE 965 211; Winteringham Foreshore LWT reserve SE 934 227
Lincolnshire Wildlife Trust: tel: 01507 526667; web: www.lincstrust.org.uk
Far Ings Visitor Centre: tel: 01652 637055

231 MESSINGHAM SAND QUARRY (Lincolnshire)

OS Landranger 112

This complex of pits lies 4 miles south of Scunthorpe and harbours a selection of winter wildfowl and passage waders; it has also gained a reputation for attracting scarce migrants and rarities.

Habitat

The extraction of sand and gravel around Messingham has formed a number of flooded pits that demonstrate typical succession from open water to small reedbeds and willow and birch scrub; 40 ha are a LWT reserve and to the west lies an area of grassland (the site of an old rubbish tip) that is prone to flooding and attracts passage waders.

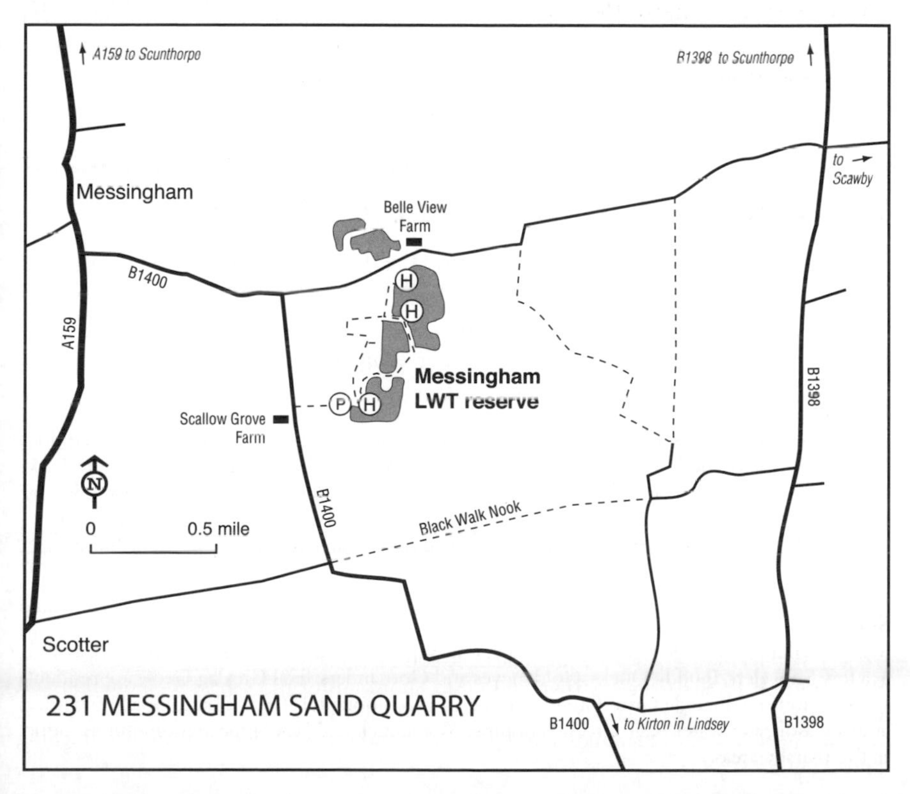

231 MESSINGHAM SAND QUARRY

Species

Wintering wildfowl include numbers of Wigeon, Gadwall and Teal, and sometimes Shoveler and Goldeneye, while Goosander and Scaup are occasional, as is Whooper Swan. Sometimes a Bittern, Jack Snipe or Chiffchaff may winter, with Siskin and redpolls regular in the waterside alders. The area is quite good for raptors, with Peregrine, Merlin, Common Buzzard and Hen Harrier seen most winters, and in some seasons the Long-eared Owl roost may be easily visible, sometimes even from the hides.

On passage the usual waders are regular, parties of Common, Arctic and Black Terns pass through and Little Gull is also possible, especially in spring. Marsh Harrier and Osprey are recorded most years in spring, while in summer there is a build-up of Lesser Black-backed Gulls with occasional single Yellow-legged Gulls.

Breeding species include Great Crested and Little Grebes, Shelduck, Shoveler, Tufted Duck, Pochard, Ruddy Duck, Lapwing, Oystercatcher, Ringed and Little Ringed Plovers, Common Snipe, Common Redshank, Turtle Dove, Kingfisher, Yellow Wagtail and Reed and Sedge Warblers. There is a notable colony of Black-headed Gulls. Sand Martins breed nearby, and concentrations of Swifts and hirundines attract Hobbies. The nearby conifer plantations and birch woods hold Sparrowhawk, Woodcock, Long-eared Owl and Willow Tit.

Access

Leave Messingham east on the B1400 towards Kirton in Lindsey and, after 1 mile, the road turns south at a sharp right-angled bend. Follow the road, and the track to the reserve car park lies to the left after 0.5 miles, opposite Scallow Grove Farm (look for the electricity pylon). A waymarked trail leads to three hides, and some of the pits can also be viewed from the minor road between the B1400 and Scawby, opposite Belle View Farm.

FURTHER INFORMATION

Grid ref: SE 908 032
Lincolnshire Wildlife Trust: tel: 01507 526667; web: www.lincstrust.org.uk

232 WHISBY NATURE PARK AND NORTH HYKEHAM PITS (Lincolnshire) OS Landranger 121

These gravel pits south-west of Lincoln attract a variety of wintering wildfowl, hold an important gull roost and have a good variety of breeding birds.

Habitat

The gravel pit complex holds a range of habitats, from bare, open areas through grassland to dense willow carr and birch scrub. To the west of the A46 lies the 150-ha Whisby Nature Park, managed by the LWT, a complex of small, medium and large flooded gravel pits surrounded by mature vegetation, while to the east of the A46 lie North Hykeham Pits, more recent workings which include the Apex Pit, a well-established deep-water pit important for wildfowl and a gull roost.

Species

Winter wildfowl include significant numbers of Wigeons, Teals, Gadwalls and Pochards, as well as a few Shovelers, Ruddy Ducks, Goldeneyes and Goosanders; feral Greylag Geese are resident. Other waterfowl include Great Crested Grebe and Cormorant. Bittern is occasionally recorded at this season, and Water Rail, Green Sandpiper, Woodcock and Jack Snipe are sometimes found in the marshy areas.

Gulls roost on the Apex Pit throughout the year but numbers peak in winter when there may be up to 20,000 Black-headed Gulls, smaller numbers of Common Gulls and up to 500 each of Herring and Great Black-backed Gulls. Careful searching in winter may also produce Glaucous, Iceland, Caspian or Mediterranean Gulls, while in summer and autumn large numbers of Lesser Black-backed Gulls roost, and sometimes one or two Yellow-legged Gulls.

On spring and autumn passage Arctic and Black Terns and Little Gull may pass through, as well as a few waders, typically Common and Green Sandpipers and Greenshank. Scarcer migrants have included Slavonian and Black-necked Grebes, Little Egret, Osprey and Garganey. Breeding species include Great Crested and Little Grebes, Little Ringed Plover, Oystercatcher, Black-headed Gull, Common Tern, Turtle Dove, Kingfisher, Cuckoo, Sand Martin, Grasshopper, Reed, Sedge and Garden Warblers, Lesser Whitethroat, Willow Tit and, notably, several pairs of Nightingales (up to 12 singing males in recent years; Coot Lake is one of the best areas). Hobby is quite frequent in summer.

Access

Whisby Nature Park is well-signed from the A46 Lincoln bypass. There is a visitor centre (Whisby Natural World Centre) but more detailed information on birds can be had from the adjacent LWT warden's office. Several marked trails through the reserve, with seven hides, four of which overlook the main pit, Grebe Lake. The reserve is open during daylight hours.

North Hykeham Pits The west end of the Apex Pit is best approached from Crow Park (off Newark Road via Thorpe Lane), with the east end accessed on foot along a lane by the Fox and Hounds pub (this is the best end from which to view the gull roost). Note that Apex Pit is

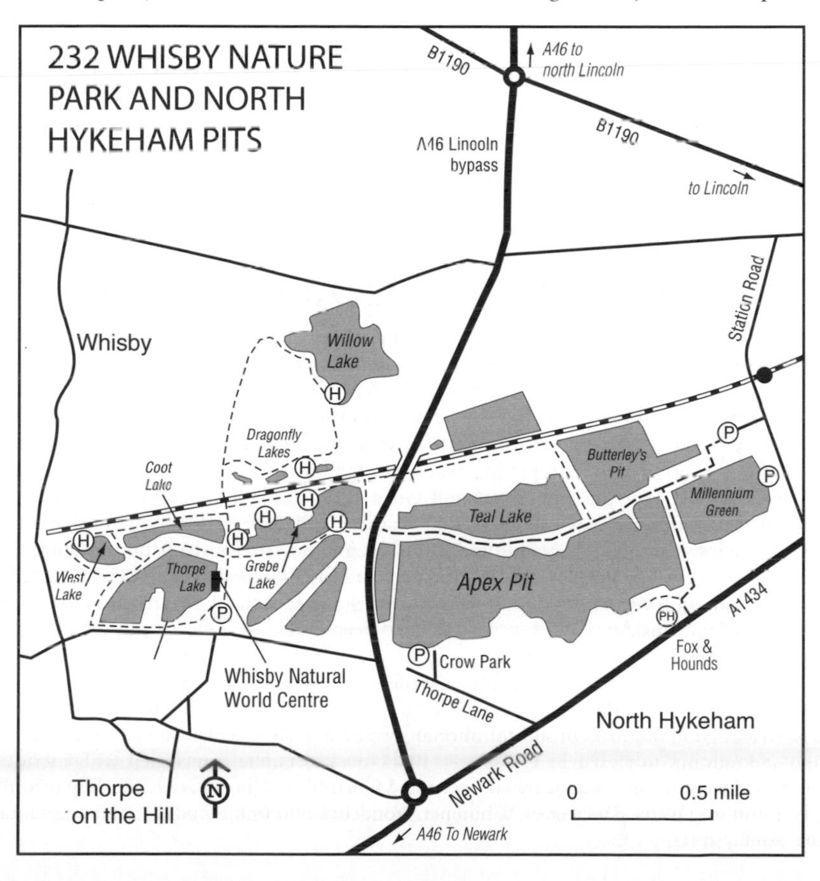

used for both sailing and fishing, and thus heavily disturbed at times. Millennium Green is best accessed from Station Road, while Teal Lake (part of the Nature Park) can be accessed from Station Road or from the Whisby Natural World Centre.

FURTHER INFORMATION

Grid refs/postcode: Whisby Nature Park SK 910 662 / LN6 9BW; North Hykeham Pits (east end) SK 933 662
Lincolnshire Wildlife Trust: tel: 01507 526667; web: www.lincstrust.org.uk
The Natural World Centre, Whisby Nature Park: tel: 01522 688 868; web: www.naturalworldcentre.com

233 LOUND (Nottinghamshire)

OS Landranger 120

This extensive complex of gravel pits lies immediately to the west of the River Idle between Retford and Mattersey in north-east Nottinghamshire. It contains a rich variety of habitats and attracts a good variety of wildfowl and waders.

Habitat

The working and disused sand and gravel pits present a typical succession of habitat from bare mud and shallow pools to willow scrub. Some older pits are now used for fishing and water sports, whilst others have been embanked and filled with pulverised fly ash, forming interesting but transient wetland habitats; when filled, they are grassed and turned over to pasture or willow plantations for biofuel. The western parts of the Idle's floodplain are very sandy and there are areas of relict heathland as well as old hedgerows, pastures and plantations, while to the east of the river there are areas of arable fields.

Species

Wintering wildfowl include several hundred each of Pochard, Tufted Duck, Gadwall, Teal and Wigeon, and notably there may be around 30 each of Goosander and Goldeneye, with occasional Smew, Scaup and Common Scoter, but the highest counts of Pintails and Shovelers are recorded in the early autumn. Both Whooper and Bewick's Swans are regularly found on the surrounding farmland. There are good numbers of Great Crested and Little Grebes, up to 2,000 Coots, and Cormorants are regular visitors throughout the year. There is a notable gull roost, mainly comprising Black-headed but at times holding several hundred each of Common, Herring and Lesser and Great Black-backed Gulls. Mediterranean and Yellow-legged Gulls are fairly regular and Glaucous and Iceland Gulls occasional. Other interesting winterers include Short-eared Owl, Peregrine, Merlin (favouring the arable areas east of the river), Water Rail, Green Sandpiper, Jack Snipe, occasionally Dunlin or Ruff, up to 6,000 Golden Plovers and 4,000 Lapwings on the adjacent fields, Short-eared Owl, Grey Wagtail and Stonechat.

On passage Marsh Harrier, Osprey, Garganey, Arctic and Black Terns, Little Gull and Kittiwake are regular, and a wide variety of waders is possible. Indeed, at both seasons passage waders are a speciality, and Lound has recorded quite exceptional totals of normally coastal species as well as good numbers of the commoner species (e.g. 100 Dunlins) and a variety in both spring and autumn to rival East Coast sites. Black-necked Grebe is recorded with increasing frequency (in common with many sites in the Midlands). Migrant passerines should not be ignored, and may include Wheatear, Whinchat, Stonechat, Common Redstart, and occasionally Rock Pipit and Ring Ouzel.

Breeding birds include Sparrowhawk, Common Buzzard, Gadwall, Pochard, Long-eared Owl, Little Ringed and Ringed Plovers, Oystercatcher, Common Redshank, Black-headed Gull (three large colonies), about 20 pairs of Common Terns, Turtle Dove, all three woodpeckers,

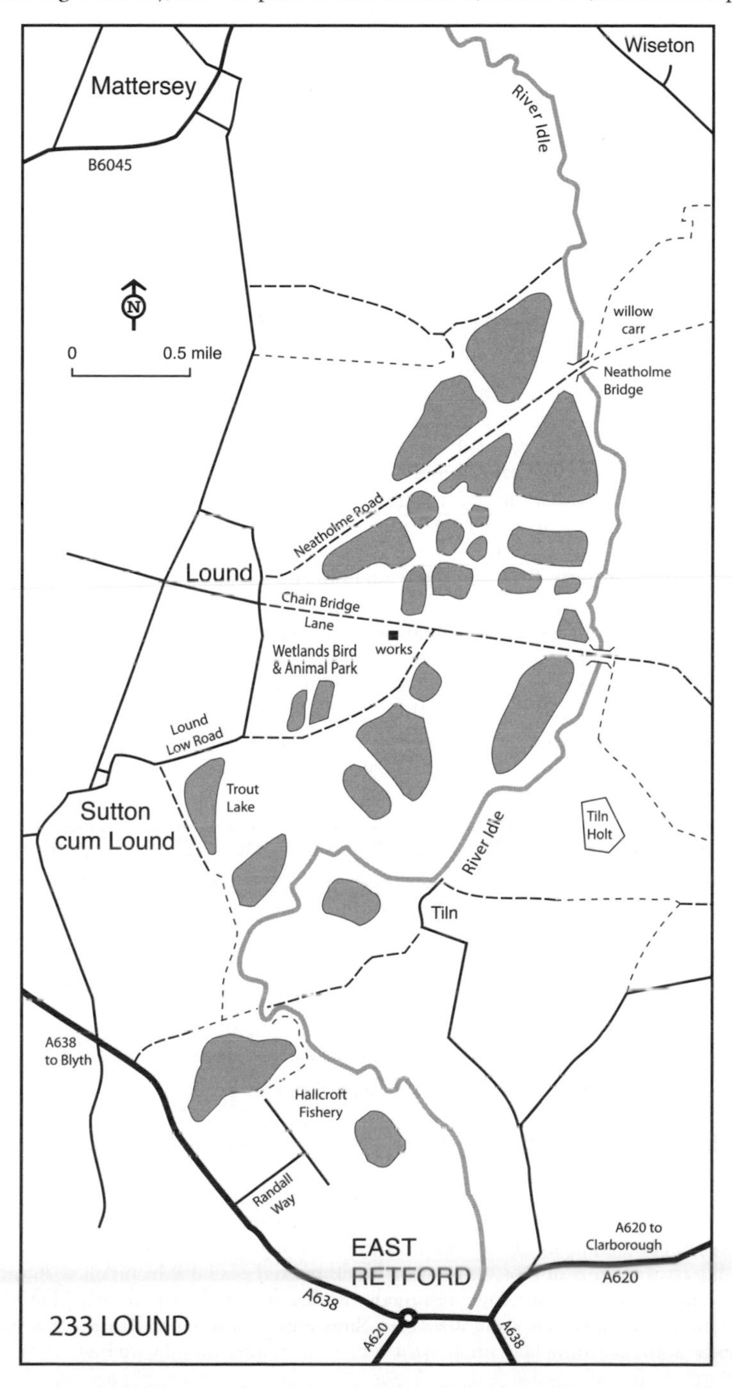

Kingfisher, Sand Martin, Yellow Wagtail, Grasshopper, Reed, Sedge and Garden Warblers and Willow Tit, occasionally also Shelduck, Pochard, Curlew and Lesser Black-backed Gull, and rarely Garganey, while a pair of Little Gulls once attempted to nest and Black-necked Grebe has bred. Visiting Hobbies are regularly seen in summer and early autumn and Marsh Harrier may summer.

Access

As is usual with working sand and gravel pits, the favourite localities, both for birds and bird-watchers, are constantly changing; good spots for birding one year quickly become featureless reclaimed land the next. Most areas of interest can be viewed from public footpaths and bridleways.

Leave Retford north-west on the A638 and after 1 mile turn right on the minor road to Sutton-cum-Lound and then in Sutton turn right signed to Lound.

Chainbridge Lane This rough but drivable gravel track runs from Lound village eastwards, eventually crossing the River Idle and over Hayton Common to Hayton village, and gives access to the main birdwatching area. Initially passing paddocks and overgrown pits, then the Tarmac and concrete works, it reaches willow plantations on the right and a wader scrape on the left (a raised viewing area). Nearer the Idle Chainbridge Wood lies to the left and the large Chainbridge Pit lagoon on the right. To view the pits to the south park carefully by the bridge and take the footpath on the west bank of the River Idle. Alternatively, walk north on the east bank to access the Neatholme Lane complex of pits, crossing the footbridge back over the river at Neatholme Bridge with a variety of pits on either side, some of which have only recently been restored. It is possible to walk back to Lound village from here to complete a loop but it is a fair distance (note that there is no vehicle access along Neatholme Lane).

Hallcroft and Bellmoor Pits Leave the A638 north-eastwards along Randall Way on the outskirts of Retford and turn left at the T-junction into Hallcroft Road. Park carefully next to the council waste site and walk along the track through the red gate to the dead end overlooking the pit. From here a footpath skirts the south-eastern shore of the pit and follows the river bank north and westwards, past the weir, to the Bellmoor Gravel Pits area.

The Sutton Trout Lake (Doughty's Pit) May hold good numbers of wintering wildfowl. View from Lound Low Road between Sutton-cum-Lound and Lound.

Wetlands Bird and Animal Park The place for a family day out, but also good birding habitat and especially good for passerines.

FURTHER INFORMATION

Grid refs/postcode: Chainbridge Lane SK 704 858; Hallcroft and Bellmoor Pits SK 692 829; The Sutton Trout Lake (Doughty's Pit) SK 690 851; Wetlands Bird and Animal Park SK 699 853 / DN22 8SB

Lound Bird Club: web: http://loundbirdclub.piczo.com

234 THE DUKERIES AT WELBECK AND CLUMBER PARKS (Nottinghamshire) OS Landranger 120

Welbeck Park is a well-known site for Honey-buzzards which can be seen from the road with no risk of disturbance, although views are usually distant and much patience may be required. Honey-buzzards are present most summers and, although reported as being much harder to see in recent years, may be undergoing a resurgence at this site; late May and early June is the best time to visit. At Clumber Park (National Trust) there is general access and a variety of woodland birds can be seen, including Hawfinches, for which the winter months are best.

Habitat

The area of ancient heathland and woodland that formerly comprised Sherwood Forest has been reduced to small fragments but successive generations of landed gentry have created estates with ornamental lakes set in mixed woodland, parkland, plantations, heathland and farmland. Welbeck Park is privately owned, but the 1,538 ha of Clumber Park are owned by the National Trust.

Species and Access

Welbeck Raptor Watchpoint From the A616 bear north onto the B6034 towards Worksop, and after 1.5 miles turn west at the first crossroads (at a sign saying Carburton). Continue along this minor road for about 1.5 miles, passing Carburton Lake, until you reach Carburton Forge Lake (the second section of the Great Lake) and the Bentinck monument on the right, with limited parking (this is a narrow road and common sense should be exercised when parking). Looking north across the lake there is an isolated deciduous wood (Cat Hills plantation). Honey-buzzards fly over this and/or the lake up to two or three times a day. Alternatively, walk another 300 yards to the gate just to the west of the monument to view from the north side of the road (the 'Welbeck Raptor Watchpoint'). Stay on the road at all times. The whole of the area north of the road is private and well-keepered, with no public rights of way. In any event, the road is the best place to see the birds.

Honey-buzzards have been summering here since at least the early 1960s and they have bred in some years. Although present late May–September, they become especially difficult to see in late August and early September. Normally there are one to three individuals present, but up to five have been seen. The optimum time is 10am–3pm on warm sunny days, with scattered cloud and only light breezes, when they may be seen soaring or in their wing-clapping display. They spend a great deal of time sitting below the canopy, however, and a long wait may be necessary for a relatively brief and distant view. They may show well, but often are not seen at all.

The area is also good for other species of raptors. Common Buzzards are common and are much more likely to be seen than Honey-buzzards. Sparrowhawks are frequent and Ospreys

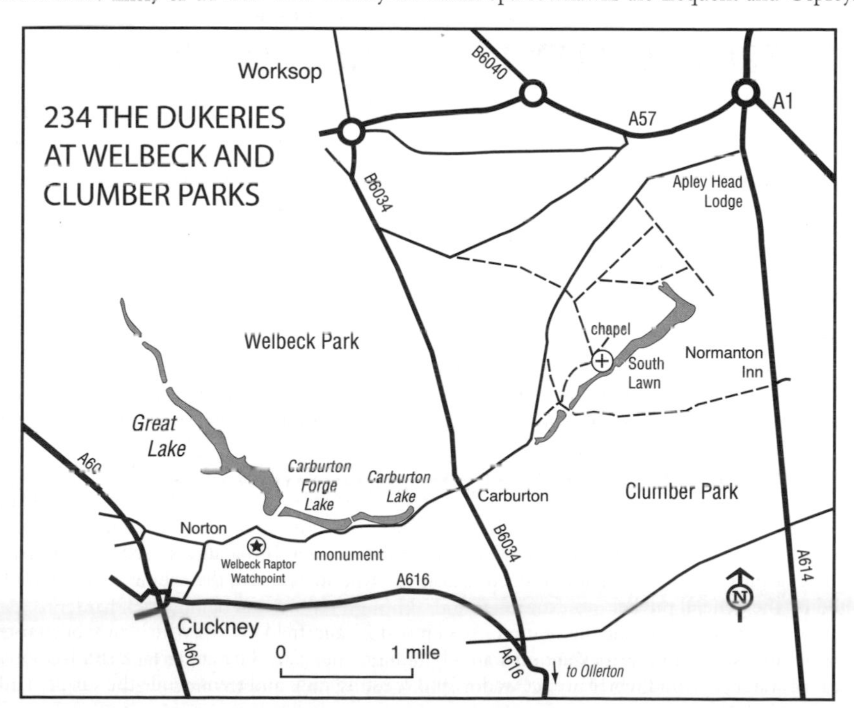

are regular in spring and summer and may linger for days or even weeks. Hobbies are also likely in late spring and summer, and Red Kite, Peregrine and Raven may visit. The woods around hold Green Woodpecker, Nuthatch, Marsh Tit and Hawfinch, while the lake attracts Gadwall.
Clumber Park This is a mosaic of parkland, broad-leaved and coniferous woodlands, scrub, wet meadows, grasslands and heathland. At the crossroads where you turn left for Welbeck Great Lake, turn right instead to Carburton and Clumber Park. Turning right after the first bridge, follow signs to the chapel and park there (fee). The Park is also accessible from the A1 via the A614 at Apley Head Lodge and Normanton Inn. Immediately behind the chapel is an area of Rhododendron, Beech, Yew and other conifers. Hawfinches are present in greatest numbers in winter but are typically elusive: check the yews and hornbeams around the chapel. Early mornings and weekdays when there is least disturbance definitely offer the best chance of finding Hawfinches (and the nearby Rufford Country Park also holds this species – try around the car park). The lake may hold wintering Gadwalls, Wigeons, Teals, Goldeneyes, Goosanders, Water Rails and occasionally Smews. Residents in the park include Little and Great Crested Grebes, Egyptian Goose, Mandarin, Woodcock, Tawny and Little Owls, all three woodpeckers (although Lesser Spotted is becoming scarcer), Marsh Tit, Nuthatch, Treecreeper and Lesser Redpoll and, while winter visitors include Stonechat, Brambling, Siskin and occasionally Common Crossbill in the conifers, Firecrest (which favours the Rhododendrons by the chapel) or a Great Grey Shrike. In summer there are Hobbies, Turtle Doves, Nightjars, Yellow Wagtails, Tree Pipits, Woodlarks (try the area of the South Lawns), a few Common Redstarts, Grasshopper and Reed Warblers and Spotted Flycatchers.

FURTHER INFORMATION

Grid refs: Welbeck Raptor Watchpoint SK 583 720; Clumber Park SK 622 742
NT Clumber Park Estate Office: tel: 01909 476592

235 TRENT VALLEY PITS (Nottinghamshire)

OS Landranger 129

This complex of pits lies in the Trent Valley on the eastern outskirts of the city of Nottingham, and comprises Colwick Country Park, the slurry pits at Netherfield, Holme Pierrepont Country Park and the adjacent A52 Pit. The Trent Valley is a major 'flyway' for migrants moving across Britain, and thus not surprisingly the area has a reputation for attracting interesting birds throughout the year.

Habitat

This vast conglomeration of gravel working shows every stage of succession, from the freshly scraped soil of new workings to mature pits with good areas of riparian reeds and willows set amidst meadows and scrub. A notable feature of the complex is that several pits have been heavily landscaped for recreational use.

To the north of the River Trent lies Colwick Country Park, where the gravel workings have been landscaped and planted with a variety of trees and shrubs, interspersed with areas of damp grassland, and bounded to the north by Colwick Woods. Colwick Lake covers 25 ha and is used for trout fishing. The other large pit, West Lake, extends over 10 ha and is used by fishermen and windsurfers year-round and, indeed, being so close to the city, the whole area is heavily used by the general public.

To the east of Colwick lie the disused gravel pits at Netherfield, two of which have been used as a dump for coal slurry; they are separated by a high causeway. The larger tank still has some areas of water and mud which attract waders and roosting gulls and terns, while the smaller tank is largely deep water, attractive to wildfowl.

To the south of the River Trent lies Holme Pierrepont Country Park, dominated by the National Watersports Centre's 1,000 m rowing course (which has attracted a selection of interesting birds), and the A52 Pit, covering 57 ha and surrounded by pasture. To the east of the A52 Pit lie more recent workings which may attract waders, gulls and terns, while to the north is the Nottinghamshire Wildlife Trust's Skylark Reserve, comprising willow carr and a pool with a specially constructed nesting platform for terns. To the east of the rowing course lie areas of grassland and the Finger Ponds.

Species

Wintering wildfowl include Wigeon, Teal, Gadwall, Shoveler, Pintail, Pochard and Goldeneye. Goosander is only an irregular visitor but Smew is annual, as are Red-necked and Slavonian Grebes, the odd diver, and Red-breasted Merganser (the last mainly on passage). Waterfowl favour Colwick and West Lakes, the Deep Water Pit at Netherfield, and the rowing course and A52 Pit, while Cormorants roost on the electricity pylons immediately east of Netherfield. In late autumn and winter there are large flocks of Golden Plovers and Lapwings (with up to 4,000 Golden Plovers roosting at Netherfield), often together with a few Ruffs, while Jack Snipe favour the more waterlogged ground. Occasionally Long-eared Owls may be found roosting in some of the denser areas of scrub, such as the railway embankment at Netherfield, and Short-eared Owls sometimes hunt over the rough grassland. Numbers of gulls roost in the area, and Mediterranean Gull is sometimes recorded, less often Caspian, Iceland and Glaucous Gulls. The area attracts good numbers of thrushes, pipits and larks, and other interesting wintering species may include Stonechat (favouring the scrub at Netherfield), Blackcap, Chiffchaff, Siskin (up to 100 in the alders at Colwick), Brambling and sometimes Firecrest.

On spring passage Black-necked Grebe and Garganey are annual, and Black and Arctic Terns and Little Gulls are something of a speciality, particularly after east winds. Wader passage can be good, with Ringed Plover, Ruff, Dunlin, Common Redshank, Greenshank and Green and Common Sandpipers regular, while Turnstone and Sanderling are also possible, especially in May. The same mix of species occurs in autumn, with a greater chance of Spotted Redshank, Wood Sandpiper, Little Stint and Curlew Sandpiper. Passage waders favour the A52 Pit, Holme Pierrepont works pits, and the slurry pits at Netherfield. In both spring and autumn, migrant passerines may include Wheatear and Whinchat.

Breeding birds include Great Crested and Little Grebes, Gadwall, Sparrowhawk, Little Ringed Plover, Common Tern (up to ten pairs on rafts at Colwick, with a late summer post-breeding gathering there), Turtle Dove, Kingfisher, Lesser Spotted Woodpecker (Colwick Hall area), Yellow Wagtail, Grasshopper, Reed and Sedge Warblers and Willow Tit. Little Egret, Hobby and Yellow-legged Gull are regular visitors in summer.

Access

Being so close to Nottingham, and largely dedicated to recreation, the whole area is subject to considerable disturbance, especially at weekends, making early morning the best time to visit. Beware also 'car crime', and avoid the quieter, more secluded car parks (especially the Starting Gate car park; the safest car parks are at the fishing lodge and outside Colwick Hall).

Colwick Country Park From Nottingham city centre take the A612 towards Colwick and once past Nottingham Racecourse follow signs for 'water user entrance only', and turn off Mile End Road onto River Road. Alternatively, take the Colwick Hall access road (from the Nottingham Racecourse entrance) and park well before the hall. Parking is also available within the park for a small fee. Access is possible using public transport: Netherfield railway station is about 1 mile away, and there are buses from Nottingham city centre. The Country Park is open from 7am to around dusk.

Holme Pierrepont Country Park Leave Nottingham city centre on the A6011 and turn north on the minor road signed to the National Watersports Centre. There are several car parks. There is open access to the rowing course, Finger Ponds banks of the River Trent and the 11-ha Skylarks Reserve of the Nottinghamshire Wildlife Trust (which is laid out specifically

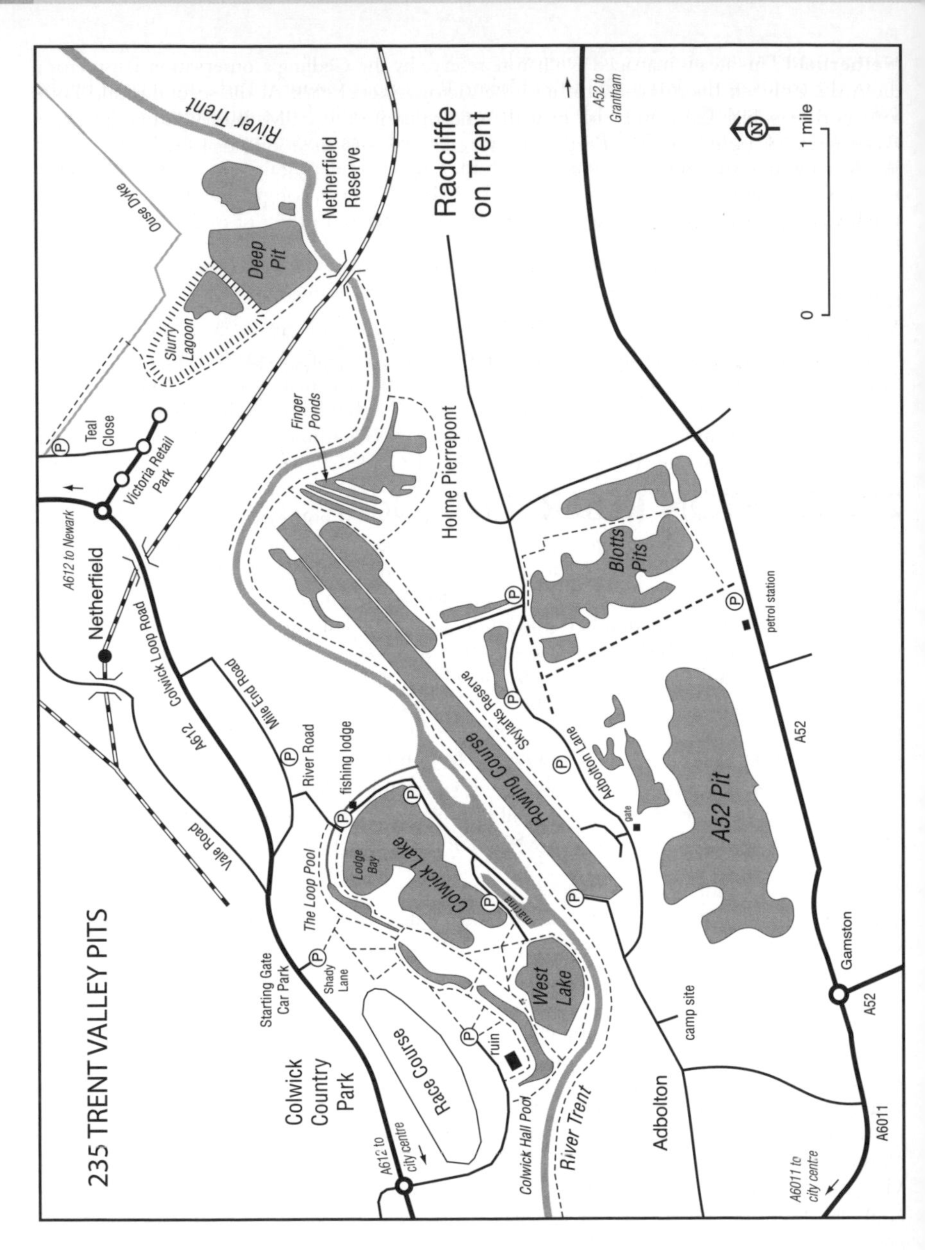

for the benefit of wheelchair users, with the entrance on Adbolton Lane). Buses run to the Watersports Centre from Nottingham city centre and, of course, the area is severely disturbed during national water sports events.

The A52 Pit This is surrounded by private farmland with restricted access, but the south-west corner can be viewed from the A52 (parking difficult), while the eastern portion can be seen from the metalled track (which also gives views of the works pits), accessed from the eastbound carriageway of the A52 from the parking area 150 yards east of the petrol station.

Blotts Pit View from the public footpath which runs from the A52 just east of the Little Chef to Adbolton Lane.

Netherfield This area is managed as a nature reserve by the Gedling Conservation Trust. From the A612 Colwick Loop Road turn into the Victoria Retail Park. At the second roundabout turn north on Teal Close and park near the end opposite the CEM factory. Cross the Ouse Dyke and walk right along the north bank for 0.25 miles to the footbridge; cross back across the dyke to enter the reserve.

FURTHER INFORMATION

Grid refs: Colwick Country Park (River Road) SK 611 399; Holme Pierrepont Country Park (Adbolton Lane) SK 615 388; the A52 Pit/Blotts Pit SK 623 380; Netherfield SK 629 410
Colwick Country Park: web: www.nottinghamcity.gov.uk
Netherfield Wildlife Group: web: www.netherfieldwildlife.org.uk
Nottinghamshire Wildlife Trust: tel: 0115 9588242; web: www.nottinghamshirewildlife.org.uk

236 OGSTON RESERVOIR (Derbyshire)

OS Landranger 119

Lying 6 miles south of Chesterfield, this reservoir has a significant winter gull roost, which regularly attracts 'white-winged gulls', and is also notable for winter wildfowl and sometimes passage waders.

Habitat

Flooded in 1958, open water now covers 83 ha, with natural banks that shelve gently on the north and west shores. The surrounding area is mainly pasture with some small stands of woodland, notably at the Derbyshire Wildlife Trust's reserve of Ogston Carr Wood, for which a permit is required (and no access January–June). The water is disturbed by sailing and also, in April–October, by trout fishing.

Species

Wintering wildfowl may include several hundred each of Pochard, Wigeon and Teal, as well as small numbers of Goosanders and sometimes Goldeneye. Shoveler, Pintail, Gadwall and Scaup are more erratic in appearance, but Little and Great Crested Grebes and Cormorant are usually present, the latter roosting here. Rarer grebes and divers may occur, as may Whooper or Bewick's Swans, and divers may make prolonged visits. The main attraction in winter is the gull roost, best viewed from the road along the west perimeter and Woolley car park. Commonest is Black-headed Gull, of which counts may top 8,000, while Herring and Lesser Black-backed Gulls may both peak at over 1,000 and Great Black-backed at over 600, but Common Gull is scarce, with just a few tens of birds. Kittiwake, Iceland, Glaucous and Mediterranean and Yellow-legged Gulls may occur in very small numbers, and Ring-billed Gull has also been recorded. Great patience is usually required, however, to pick out these rarities, and a telescope is essential.

On spring passage Arctic Tern may pass through, sometimes with Black Tern or Little Gull, and scarcer migrants have included Common Scoter and Osprey, and Hobby is regular, but wader passage is generally poor. If water levels fall in autumn, Ringed and Little Ringed Plovers, Common Sandpiper, Greenshank and Dunlin are all regular, with a smattering of scarcer species possible, such as Ruff, Oystercatcher, Whimbrel and Spotted Redshank. Other migrants may include Garganey, Osprey, Hobby, Merlin and Common Buzzard and, in late autumn, Rock Pipit.

Breeding species include Little and Great Crested Grebes, Grey Heron (15–20 pairs in Carr Wood, viewable from the road), Cuckoo, Grasshopper and Sedge Warblers, sometimes also

Little Ringed Plover, and Common Terns sometimes nest on the rafts in Woolley Bay and at the south-west end of the reservoir. Sparrowhawk, Little Owl, Kingfisher and Grey Wagtail breed in the vicinity.

Access

One mile south of Clay Cross leave the A61 at Stretton west on the B6014 towards Matlock. Follow this road for 1.25 miles to the car park at the north-east corner of the reservoir (the minor road to the hamlet of Ogston village parallels the east shore, but views are largely obscured). The B6014 then crosses the extreme north tip, separating Milltown Inlet from the main body of water. Immediately beyond this take the minor road south towards Woolley and Brackenfield, which offers good views of much of the reservoir, with a car park at Woolley, with a public hide. The Ogston Bird Club has three hides and members receive a key for these (see Further Information). Otherwise, there is no access to the water's edge.

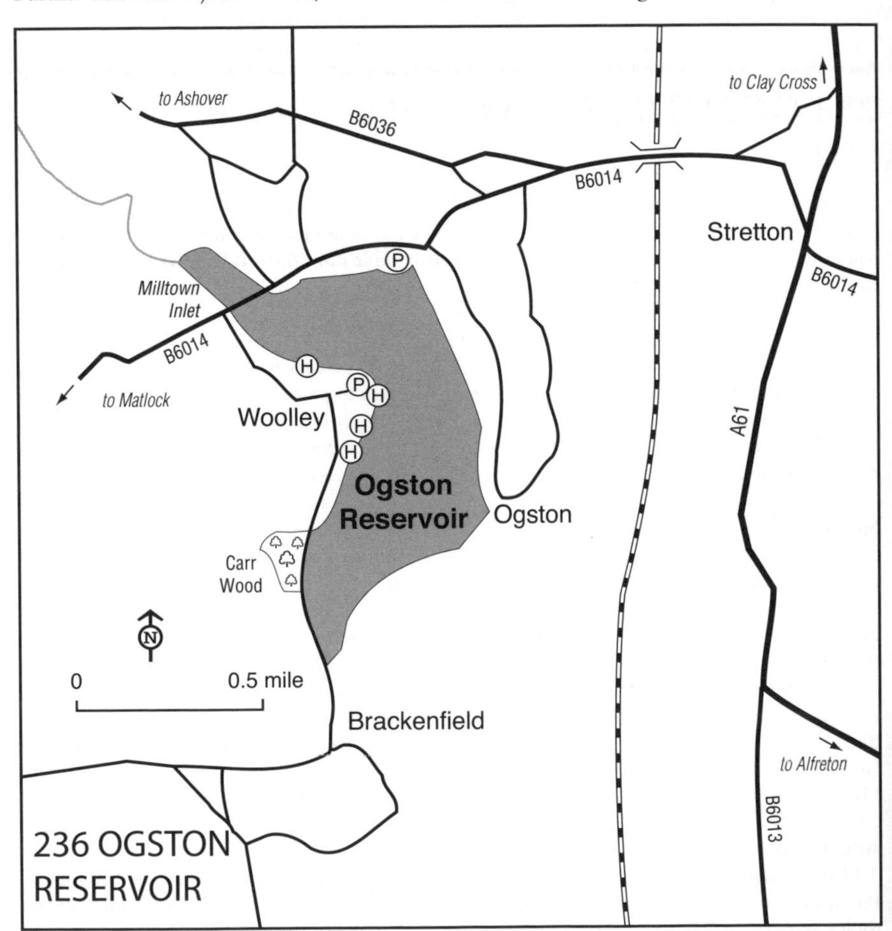

FURTHER INFORMATION

Grid ref: SK 375 609
Ogston Bird Club: web: www.ogstonbirdclub.co.uk
Derbyshire Wildlife Trust: web: www.derbyshirewildlifetrust.org.uk

237 THE UPPER DERWENT VALLEY (Derbyshire)

OS Landranger 110

Lying 12 miles west of Sheffield, this section of the Peak District is notable for its population of Goshawks which, with luck, can be viewed from the roads and public footpaths with no risk of disturbance.

Habitat

Nestled in the High Peak region, Howden and Derwent Reservoirs were constructed at the turn of the century while the biggest, Ladybower, was not completed until the 1940s. As is typical of reservoirs in upland areas, they are deep with steeply shelving banks and are relatively birdless. The surrounding valleys have been extensively planted with conifers, while the higher slopes, rising to 2,077 feet, are a mosaic of rough grassland and moorland, with some relict oak woodlands.

Species

The prime attraction is Goshawk, a species which despite much persecution has been increasing in numbers in Britain during the last three decades. Most originate, however, from escaped or released falconers' birds, and are very large and pale (being from high latitudes) making them easier to separate from Sparrowhawk. The best chance of a sighting is in February–early April in fine weather with light winds, when they display over the plantations and ridges. Sparrowhawk also indulges in similar displays. Early mornings are best and in the right weather multiple sightings of Goshawk, Peregrine, Common Buzzard, Sparrowhawk, Kestrel and Raven are possible, while Osprey and Red Kite are recorded with increasing frequency. Goshawk can be seen at other times of the year but much more luck is required.

Breeding species include resident Grey Wagtails and Dippers on the streams (and occasionally around the reservoirs, with Dipper favouring the River Derwent below Ladybower), Red-breasted Merganser and Common Sandpiper around the reservoirs, Tree Pipit on the slopes, a few Common Redstarts and Wood Warblers in relict stands of deciduous woodland, and Spotted and a few Pied Flycatchers. Resident woodland species include Nuthatch and Siskin. The moors hold Red Grouse (on tracts of Heather), Merlin, Curlew, Lapwing, Common Snipe, Golden Plover, Short-eared Owl, Ring Ouzel, Wheatear, Stonechat and Whinchat.

The area is also worth investigating in winter. Raptors may include Peregrine, Merlin, Hen Harrier and Common Buzzard. Rough-legged Buzzard sometimes occurs but is very erratic. (For wintering raptors, follow the footpath to Slippery Stones and into the Derwent Valley to view the surrounding crags.) The woodland may hold Siskin (notably in larches and alders near water), Lesser Redpolls, Bramblings and, following invasions, numbers of Common Crossbills may also be present, also favouring the larches; some may stay to breed, but they are very early nesters. The reservoirs tend to hold little of interest in winter apart from a few Goosanders.

Access

The A57 Sheffield–Glossop road passes across the valley and the south part of Ladybower Reservoir. Immediately to the west, a minor road (signed Derwentdale) runs north along the west edge of all three reservoirs, passing several small car parks from which footpaths lead to the moors. This road terminates near the north end of Howden Reservoir at Kings Tree, where there is a small car park. On Sundays and, in April–October, on Saturdays and Bank Holidays, the road is closed at Fairholmes (SK 172 893), where there is a car park and information centre. From here there is a shuttle service to the two northern reservoirs, saving the walk to Windy Corner (SK 168 927, probably the best spot for Goshawk); bicycles can also be hired.

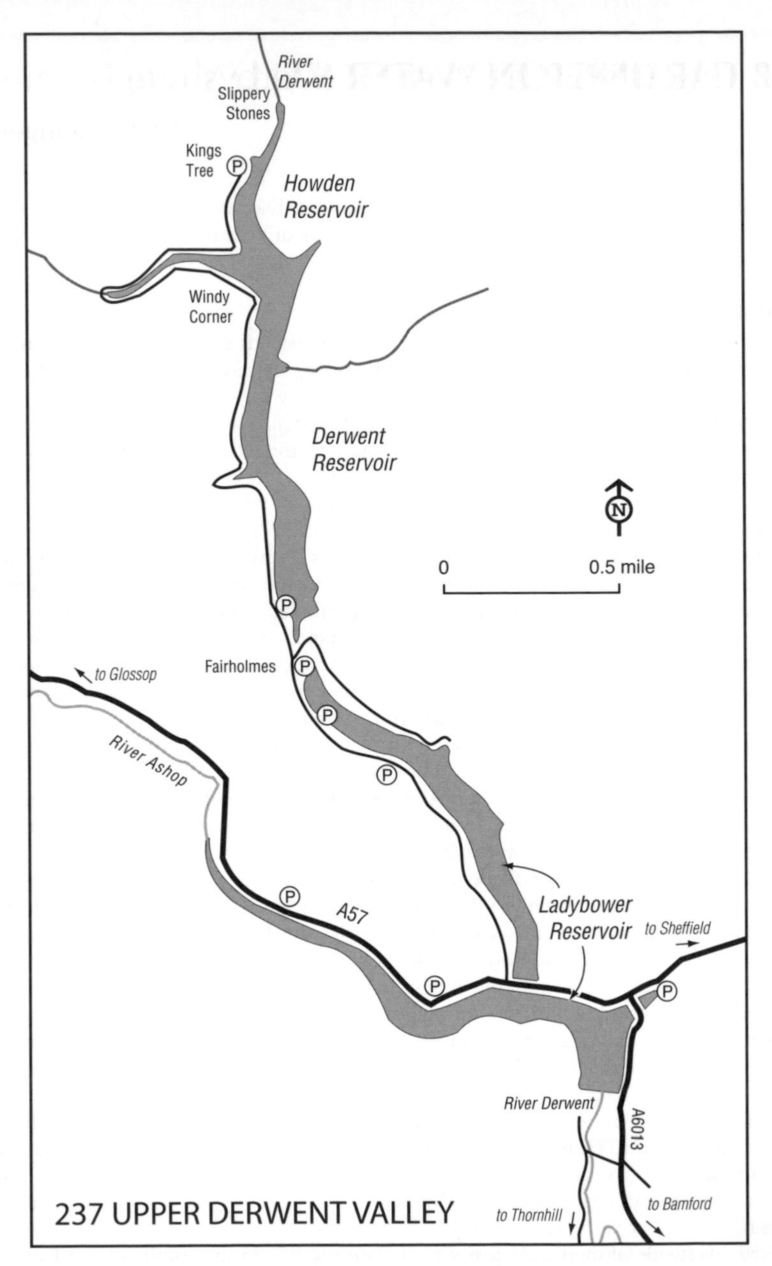

237 UPPER DERWENT VALLEY

The entire area is very popular with a range of outdoors enthusiasts and to minimise disturbance (and ensure parking space) an early-morning arrival is recommended. Note too that bad weather is possible year-round and you should be adequately prepared if you wander far from the roads, with appropriate clothing, footwear, a compass and the relevant OS map.

FURTHER INFORMATION

Grid ref: Kings Tree SK 167 938

238 CARSINGTON WATER (Derbyshire)

OS Landranger 119

Lying on the south-eastern edge of the Peak District, this large reservoir covers 300 ha and was only completed in 1991. Attracting good numbers of wildfowl, it is one of the premier birdwatching sites in the county. The reservoir is owned and operated by Severn Trent Water.

Habitat

The dam rises to an impressive 128 feet, but with gently sloping grassy banks. The surroundings are a mixture of farmland and mature woodland.

Species

Wintering wildfowl include more than 1,000 Coots, as well as several hundred Wigeons, Tufted Ducks and Canada Geese (all three have peaked at over 1,000). Teal and Pochard are also numerous and there are usually small numbers of feral Barnacle Geese, Goosanders, Goldeneyes, Shovelers and Gadwalls and sometimes a few Shelducks, Red-crested Pochards and Pintails. Smew and Scaup are occasional visitors and parties of Whooper Swans visit, especially November–December. Divers (especially Great Northern), the rarer grebes and seaducks turn up from time to time. Other waterbirds include Great Crested and Little Grebes and Cormorants. Up to 1,000 Lapwings winter, together with Golden Plovers, and there may be a few Dunlins. There is a large gull roost: over 10,000 gulls may gather in the late afternoon, mostly Black-headed, during the winter there is a chance of Mediterranean, Iceland or Glaucous Gulls, while in the autumn and early winter the large numbers of Lesser Black-backed Gulls may be joined by one or two Yellow-legged Gulls. The gull roost is usually best viewed from the Sheepwash or Lane End Hides. Feeding stations at the Visitor Centre and at the Sheepwash and Millfields car park may pull in Tree Sparrows, and other notable passerines in winter include Stonechat. Common Buzzard and Sparrowhawk are resident and Peregrine a regular visitor.

Spring migrants may include Black-necked Grebe, Common Scoter (April), Garganey and Common, Arctic and Black Terns. Osprey is often seen from spring to early autumn. Spring wader passage is variable; up to 20 species may occur, with Dunlin, Ruff, Black-tailed Godwit and Ringed Plover among the more regular species, but numbers are typically small. A similar range of species can occur in autumn, with greater numbers and variety if water levels are low. Breeding species include Oystercatcher, Lapwing, Common Redshank, Little Ringed Plover and Grey Wagtail, with Little Owl, Common Redstart and Spotted Flycatcher in the surrounding woodland and farmland.

Access

Access is from the B5035 about 4 miles north-east of Ashbourne. The visitor centre is accessed along a minor road south-east from Carsington village. The reservoir is open daily, 7am to sunset, and the visitor centre opens daily at 10am (parking fee payable). There is a heated hide just north of the car park here, overlooking Shiningford Creek. The Sheepwash car park on the north-west side of the reservoir is signed from the B5035 just south-west of Carsington village and gives access to the other three hides (Paul Stanley, Sheepwash and Lane End). The Millfields car park (fee payable) is accessed via the minor road that runs along the dam wall south-eastwards from the visitor centre. An 8-mile circuit of the reservoir is possible, but the southern part of the reservoir is well-used for recreation, making the northern half more profitable for birding.

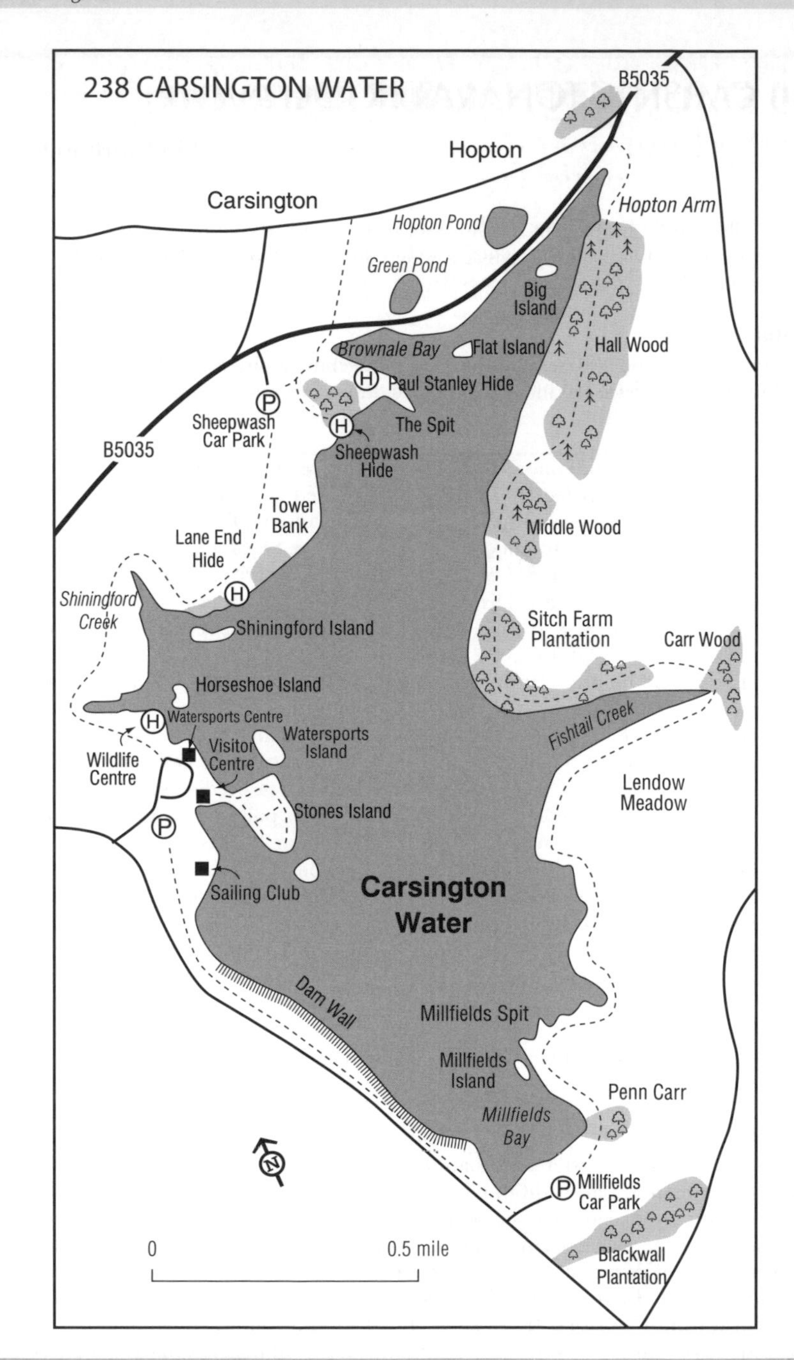

FURTHER INFORMATION

Grid ref: SK241516;
Visitor Centre: tel: 01629 540696
Carsington Bird Club: web:www.carsingtonbirdclub.co.uk

239 STAUNTON HAROLD RESERVOIR

(Derbyshire) OS Landranger 128

Only 5 miles south of Derby and slightly east of Foremark Reservoir, this water has shallow shelving banks and is attractive to passage waders and terns, while in winter scarcer grebes or divers are possible.

Habitat

Completed in 1964, open water covers 87 ha, with gently sloping natural banks. The reservoir is surrounded by farmland and areas of deciduous woodland, including Spring Wood immediately to the south-east, a Derbyshire Wildlife Trust reserve, while Calke Park to

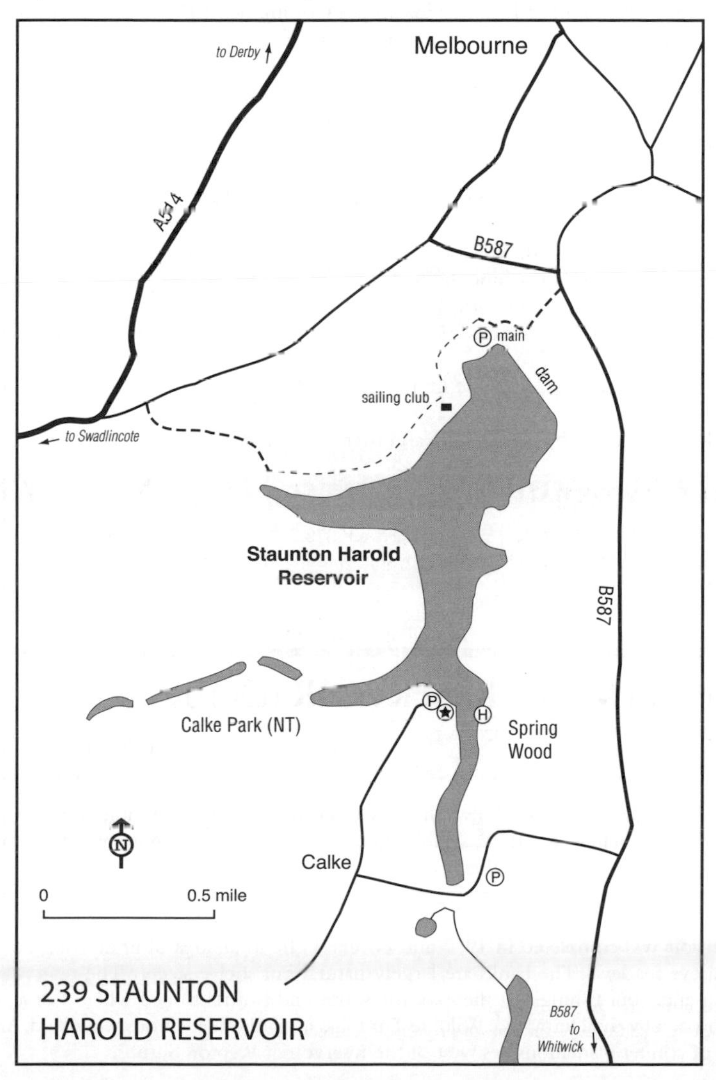

the south-west comprises an area of oak-dotted parkland owned by the NT. Sailing and fishing disturb the central and northern parts of the reservoir.

Species

Wintering waterfowl include numbers of Great Crested Grebes, Goosanders, Goldeneyes, Pochards, Wigeons and Teals, as well as Cormorant and, sometimes, small numbers of Ruddy Ducks, Shovelers and Pintails. Divers and rarer grebes are recorded relatively frequently. The gull roost is rather small and overshadowed by that at nearby Foremark Reservoir (but is best viewed from just south of the sailing club). Other notable species in winter include Grey Partridge, Little Owl, Water Rail (at the south end), Siskin and Common and Lesser Redpolls.

On spring passage Ringed and Little Ringed Plovers, Dunlin, Ruff, Common Sandpiper, Greenshank, Common, Arctic and Black Terns, White and Yellow Wagtails, and Wheatear are regular, while scarcer migrants have included Black-necked Grebe, Common Scoter, Osprey, Hobby and Little Gull. In autumn greater numbers and a wider variety of waders occur, sometimes including Little Stint and Curlew Sandpiper. Common and Black Terns may pass through, sometimes in large flocks and, again, Black-necked Grebe and Little Gull are among the scarcer visitors. Migrant passerines may include Common Redstart and Whinchat.

Breeding birds include Turtle Dove, Cuckoo, Kingfisher, a few Grasshopper and Sedge Warblers, and Willow Tit. In some years, Quails may call from adjacent fields.

Access

The reservoir is viewable at three points, all accessed off the B587 south of Melbourne.

Dam End Leave Melbourne south on the B587 and, after 0.5 miles, turn west at the signs to the main car park (fee; note the gates are locked at dusk). A footpath runs along the north shore of the reservoir, eventually reaching the A514.

Southern Arm Continue south on the B587 for 1.5 miles and turn west onto a minor road towards Calke. After 1 mile the road passes the south tip of the reservoir. This area is favoured by dabbling ducks.

Central Area Continuing along the minor road for 0.5 miles, turn north on a minor road to the viewpoint car park.

Spring Wood Access is by permit only, and there is a hide giving views over the reservoir.

FURTHER INFORMATION

Grid refs: Dam End SK 376 244; Southern Arm SK 378 219; Central Area SK 375 226
Derbyshire Wildlife Trust: web: www.derbyshirewildlifetrust.org.uk

240 FOREMARK RESERVOIR (Derbyshire)

OS Landranger 128

Lying just 5 miles south of Derby, this reservoir holds significant numbers of wintering Goosanders and, being very deep, seldom freezes, making it an important refuge for diving ducks and gulls during cold weather.

Habitat

The reservoir was completed in 1977 and covers 93 ha in an area of undulating farmland at 110 m above sea level. The banks are largely natural but shelve steeply. The reservoir is used for sailing and trout fishing, but the extreme south end is undisturbed where Carver's Rocks reserve, owned by the Derbyshire Wildlife Trust, has a stand of deciduous woodland. An extensive area of conifer plantations lies west of the reservoir at Repton Shrubs.

Species

Winter waterfowl include significant numbers of Great Crested Grebes and Cormorants, also Pochard, Goldeneye and, notably, up to 200 Goosanders. Small numbers of Wigeons, Teals and Shovelers may also occur, and all three divers have been recorded. The gull roost peaks at 10,000–15,000 birds, largely Black-headed Gull but with numbers of Herring, Lesser and Great Black-backed and Common Gulls. Glaucous and Iceland Gulls are recorded most winters, occasional Kittiwakes are most likely in late winter to early spring, and records of Yellow-legged and Mediterranean Gulls are increasing. The gull roost is best viewed from the shelter of the toilet block in the main dam car park. Other notable species in the area include Grey Partridge, Little and Barn Owls, and Tree Sparrow, while Carver's Rocks holds Siskin and Lesser Redpoll.

On spring passage Common and Arctic Terns, Common Sandpiper and Yellow Wagtail are regular, and Osprey is something of a local specialty, with immatures sometimes making prolonged visits. On autumn migration Little Gull, Common Tern and small numbers of waders occur, though water levels have to drop markedly to expose much mud.

Breeding birds include Woodcock, Turtle Dove, Cuckoo, Lesser Spotted Woodpecker, Tree Pipit, Grasshopper Warbler (at the entrance to Carver's Rocks) and Marsh Tit.

Access

Turn west off the A514 at Ticknall on the minor road to Milton and, after 1 mile, turn south through the gates (locked at dusk) to the reservoir. There are two car parks overlooking the north end of the reservoir and a footpath leads south along the east shore to Carver's Rocks reserve. Carver's Rocks is also accessible from the A514 1.25 miles south of Ticknall (again, the gates are locked at dusk).

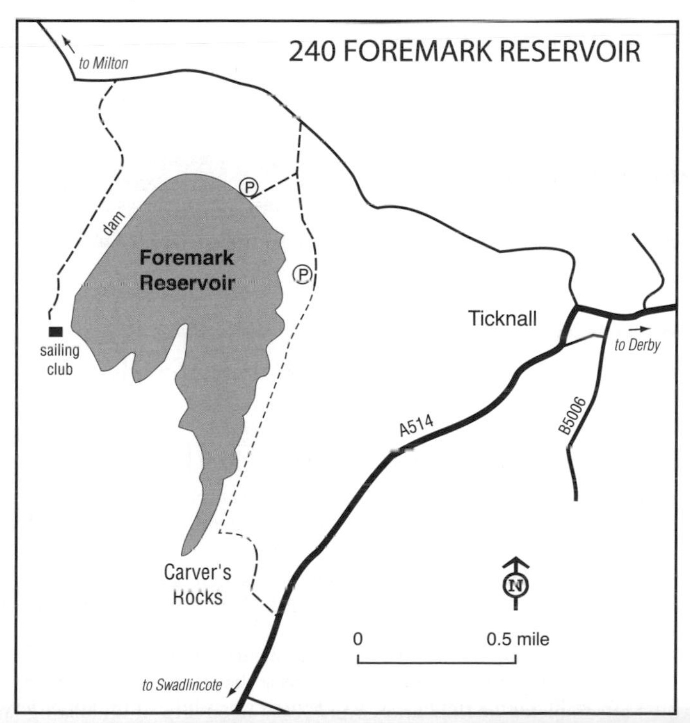

FURTHER INFORMATION

Grid ref: SK 334 243

241 COOMBES AND CHURNET (Staffordshire)

OS Landranger

The scenic RSPB reserve of Coombes Valley on the edge of the Peak District holds a good range of woodland birds, including good populations of Pied Flycatchers and Common Redstarts, best seen May–mid June.

Habitat

Nestled in the valley of the Coombes Brook, a tributary of the River Churnet, the reserve is dominated by deciduous woodland, with Sessile Oak the main species of tree, with hay meadows and pastures at the higher levels.

Species

Residents include Common Buzzard, Sparrowhawk, Woodcock, Great Spotted Woodpecker, Grey Wagtail and Dipper. In summer look for Tree Pipit, Pied and Spotted Flycatchers, Common Redstart, Common and Lesser Whitethroats and the fast-declining Wood Warbler may still cling on here; look for roding Woodcocks at Clough Meadow in the centre of the reserve.

Access

Turn south off the A523 Leek–Ashbourne road, about 3 miles east of Leek and just east of Bradnop, on the minor road to Apesford (signed RSPB Coombes Valley). The reserve lies to the left after 1 mile and is open daily 9am–9pm or dusk if earlier: last entry 2 hours before closing. The visitor centre is open during busy periods. The main trail loop, which is steep in places, is 1.5 miles long, a shorter loop of 1.3 miles cuts through Clough Meadow, and there is a short trail, just 0.5 miles long.

FURTHER INFORMATION

Grid ref: SK 009 534
RSPB Coombes Valley: tel: 01538 384017; email: coombes.valley@rspb.org.uk

242 BLITHFIELD RESERVOIR (Staffordshire)

OS Landranger 128

Blithfield is the largest reservoir in the West Midlands and is important for wildfowl and passage waders. It provides interesting birdwatching year-round, especially in autumn and winter. Part is managed as a reserve by the West Midland Bird Club.

Habitat

Blithfield covers 324 ha, has largely natural banks (extending for 9 miles) and is surrounded by agricultural land. The east shore has been extensively planted with conifers and there is some deciduous woodland. Areas of marsh fringe the inflows at the north end, with willow scrub and alders, the latter especially on the upper reaches of the River Blithe. If the water level drops in autumn extensive mudflats are exposed.

Species

In winter Red-necked Grebe and divers (especially Great Northern) are occasional. Large numbers of ducks occur: up to 1,600 Wigeons and 600 Teals have been noted, as well as Ruddy Duck and a few Gadwalls, Shovelers and, sometimes, Pintails. There may be up to 80 Goosanders and 80 Goldeneyes and Smew is a regular visitor, while seaducks sometimes appear, especially during hard weather (though Common Scoter and Scaup may visit at any time of year). Several hundred Greylag and Canada Geese are resident, while Cormorant, though regular, is generally present only in early morning. The large gull roost often has one or two Glaucous and/or Iceland Gulls, Yellow-legged, and Mediterranean Gulls. The best time is Christmas–March, especially for Iceland Gull, and a telescope is essential. Siskin and redpolls are regular visitors in surrounding trees, notably the alders by the River Blithe, and Peregrine and Raven may visit.

Passage periods can be very good. Black-necked Grebe may be recorded (especially in spring) and Osprey and Hobby have appeared in most recent autumns and sometimes stay long periods. Wader numbers are generally low in spring, but in autumn large numbers may be present, mostly Ringed Plover and Dunlin. In addition, Little Stint and Curlew Sandpiper occur regularly, especially in September. Ruff, Whimbrel, Black-tailed Godwit, Jack Snipe, Spotted Redshank and Green Sandpiper are also regular, with a chance of Sanderling and Turnstone. Common, Arctic and Black Terns and Little Gull are quite frequently present. Pintail and Garganey often occur in autumn, with Rock Pipit regular late in the season. Breeding birds are unexceptional but include Great Crested Grebe, Sparrowhawk, Little Owl, Kingfisher, Lesser Spotted Woodpecker and Willow Tit.

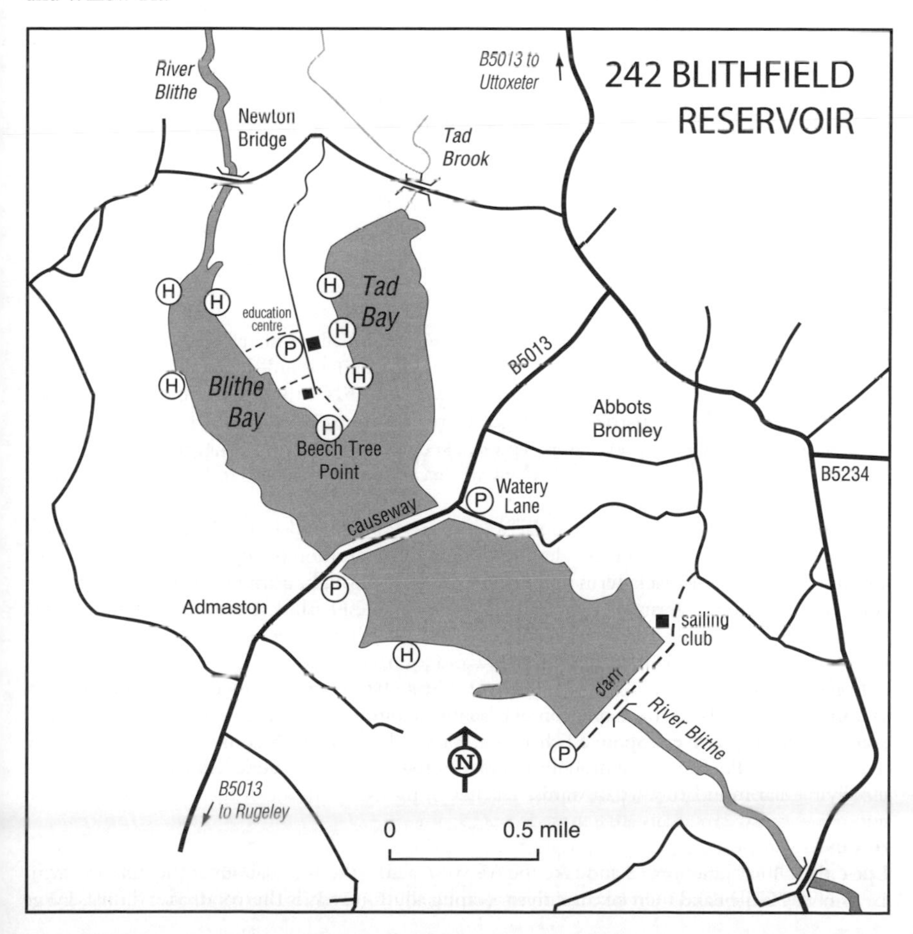

Access

The B5013 Rugeley–Uttoxeter road crosses the reservoir via a causeway (parking at either end), affording views of much of the open water, including that used by the gull roost. Some of the reservoir is also viewable from Watery Lane, but access is otherwise by permit only, issued on behalf of the South Staffordshire Waterworks Company by the West Midland Bird Club. There are eight hides and permit holders can walk the entire shoreline, though the north half is less disturbed and generally has the best variety of birds. On a half-day visit park near to the Education Centre and walk to either Tad or Blithe Bays where many of the ducks and waders congregate, and from there around Beech Tree Point.

FURTHER INFORMATION

Grid ref: causeway SK 054 235
West Midland Bird Club: web: www.westmidlandbirdclub.com

243 BELVIDE RESERVOIR (Staffordshire)

OS Landranger 127

Approximately 7 miles north-west of Wolverhampton and just south of the A5, this relatively small reservoir is important for breeding, moulting and wintering wildfowl and for passage waders. It is a reserve of the West Midland Bird Club.

Habitat

Built as a canal-feeder, the reservoir covers 73 ha and has largely natural banks. It is situated in mixed farmland, with a small wood on the south-east shore; otherwise the immediate surroundings are largely unimproved pasture.

Species

Wintering wildfowl include numbers of Shovelers (occasionally as many as 500), Teals, Wigeons, Pochards, Goldeneyes, a few Ruddy Ducks and Goosanders and, especially in late winter, Gadwall and Shelduck. Divers occur most years, with Great Northern Diver the most frequent. Bewick's Swans and small parties of grey geese are also annual visitors. The gull roost holds up to 20,000 Black-headed, 1,000 Lesser Black-backed and 500 Herring Gulls, and Yellow-legged, Glaucous, Iceland and Mediterranean Gulls are occasional. Peregrine, Goshawk, Red Kite and Raven may visit at any time of year.

On passage some of the more regular waders are Ringed and Little Ringed Plovers, Ruff, Whimbrel, Little Stint, Dunlin, Sanderling, Green and Common Sandpipers, and Greenshank. Common, Arctic and Black Terns and Little Gull pass through, usually in small numbers, but large flocks of Arctic Terns may occur and as many as 200 Black Terns have been counted. Other migrants include Black-necked Grebe and Marsh Harrier (especially in spring) and Garganey (especially in late summer), Osprey, and a variety of passerines.

Breeding species include Little and Great Crested Grebes, Shoveler, Tufted and Ruddy Ducks, Common Buzzard, Common Redshank, Common Snipe, Little Owl, Grasshopper, Reed and Sedge Warblers, Spotted Flycatcher, Willow Tit and Tree Sparrow; Shelduck, Gadwall, Teal, Garganey, Pochard, Common Tern and Oystercatcher have nested. A pair of Hobbies is usually regular from late April onwards.

Access

Leave the M6 at junction 12 and take the A5 west, across the roundabout at the junction with the A449 at Gailey, and then take the third turning south (towards Brewood) after 2 miles. After

1 mile turn right, over the canal and though the hamlet of Shutt Green, parking at the car park on the right adjacent to the wood. Access is by permit only, issued by the West Midland Bird Club (see Further Information). There are three hides available to permit holders. Otherwise, a small section of the west shore can be viewed from the footpath south of the A5 (but note that there is no parking along this private road).

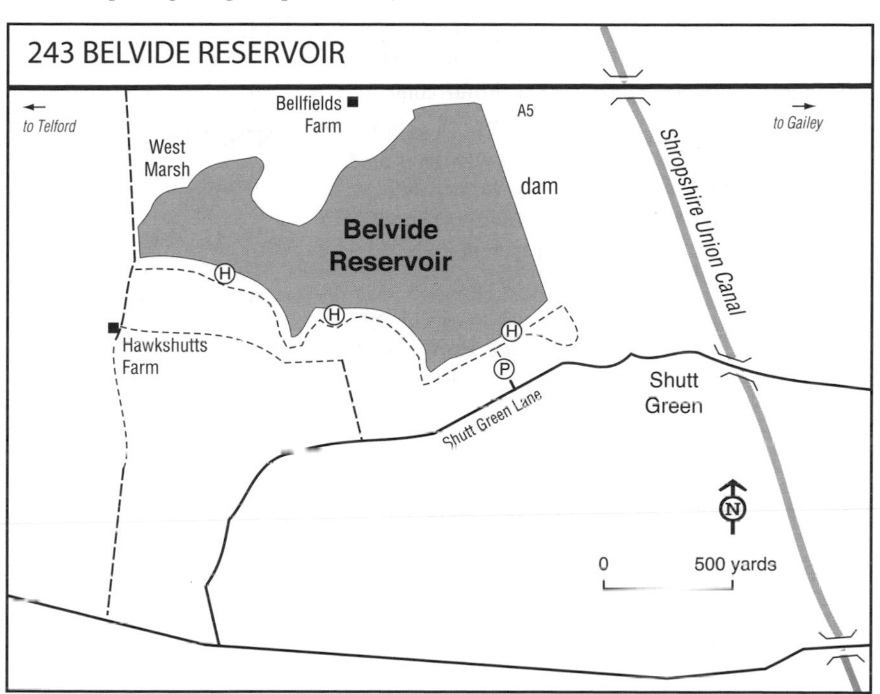

FURTHER INFORMATION

Grid ref: SJ 868 097
West Midland Bird Club: web: www.westmidlandbirdclub.com

NORTHERN ENGLAND

CHESHIRE & WIRRAL

244 The Inner Dee Estuary
245 Hilbre and Red Rocks
246 The Wirral Coast
247 Frodsham Lagoons and the Weaver Bend

GREATER MANCHESTER

248 Pennington Flash Country Park

LANCASHIRE & NORTH MERSEYSIDE

249 Seaforth and Crosby Marina
250 Martin Mere
251 South Ribble Marshes
252 Pendle Hill
253 Forest of Bowland
254 Stocks Reservoir
255 Heysham
256 Morecambe Bay
257 Leighton Moss

CUMBRIA

258 South Walney
259 Hodbarrow
260 St Bees Head
261 Bassenthwaite Lake Osprey Viewpoint
262 Haweswater
263 Geltsdale
264 Bowness-on-Solway and Campfield Marsh

YORKSHIRE

265 Pugneys Country Park
266 Wintersett Reservoir and Anglers Country Park
267 Swillington Ings and St Aidan's Wetland
268 Fairburn Ings
269 Derwent Valley
270 Dearne Valley
271 Potteric Carr
272 Blacktoft Sands
273 Whitton Sand and Faxfleet Ponds
274 Cherry Cobb Sands and Stone Creek
275 Spurn Point
276 Hornsea Mere
277 Tophill Low
278 Bridlington Harbour
279 Flamborough Head
280 Bempton Cliffs
281 Filey
282 Scarborough and Scalby Mills
283 Wykeham Forest
284 Arkengarthdale and Stonesdale Moor

CLEVELAND

285 Scaling Dam Reservoir
286 Coatham Marsh and Locke Park
287 South Gare
288 North Teesside
289 Seaton Carew, North Gare and Seaton Snook
290 Hartlepool

DURHAM

291 Hurworth Burn Reservoir
292 Upper Teesdale
293 Washington WWT and Barmston Pond
294 Whitburn Coast

NORTHUMBERLAND

295 Seaton Sluice to St Mary's Island
296 Druridge Bay area
297 The Farne Islands
298 Lindisfarne

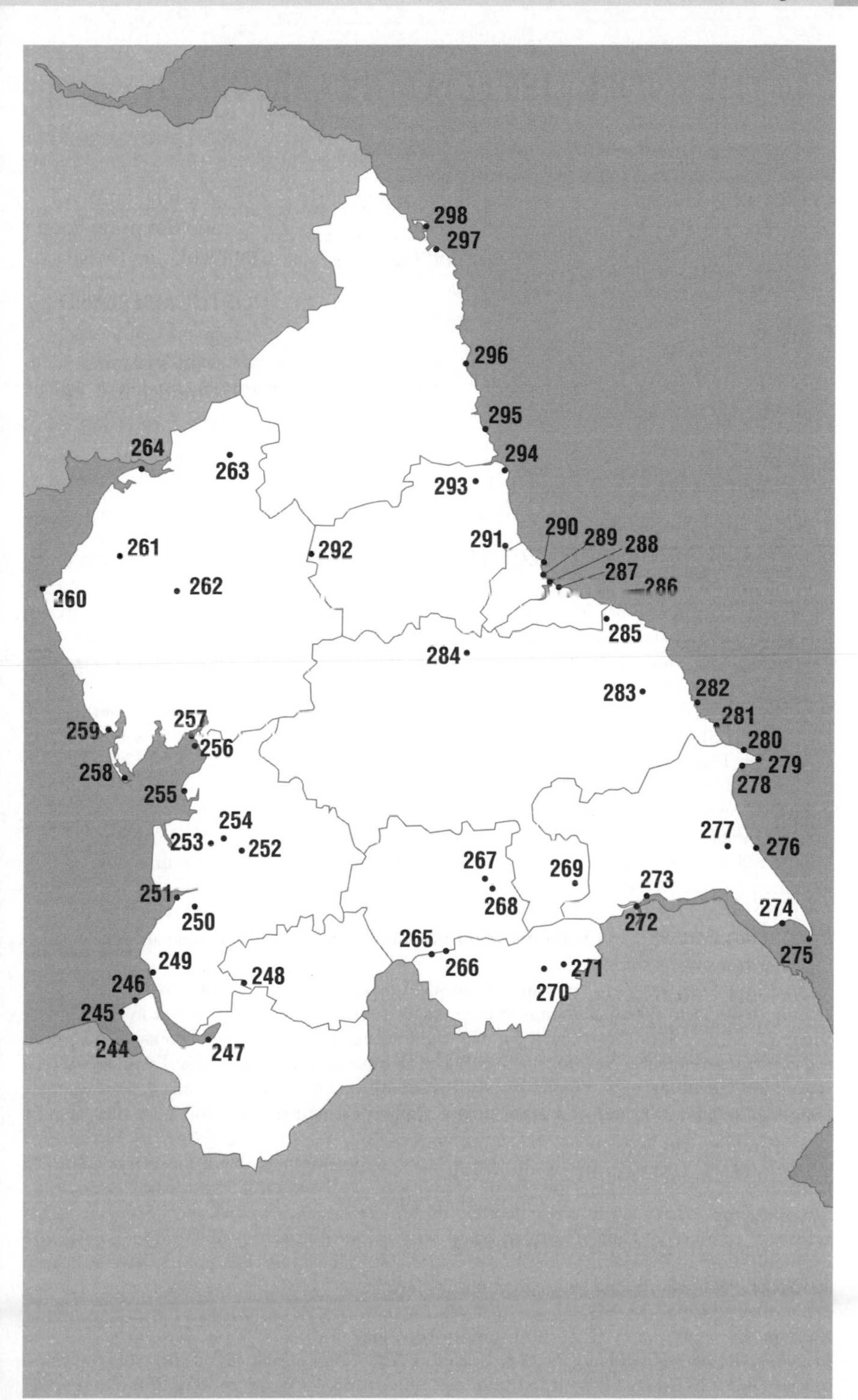
298
297
296
295
294
264
263
293
290
289
288
291
261
292
287
286
260
262
285
284
283
282
281
257
259
280
256
279
258
278
255
254
277
276
253
252
267
269
273
268
251
272
250
274
265
266
271
275
249
248
270
246
245
244
247

244 THE INNER DEE ESTUARY (Cheshire & Wirral)

OS Landranger 117

The Dee Estuary supports internationally important populations of ten species of wader and three of wildfowl. Along the Cheshire shore of the inner estuary there are several excellent sites, holding year-round interest (though quiet in midsummer). The RSPB has extensive holdings on the estuary, from Burton Point to Gayton.

Habitat

The east shore of the estuary (east of the canalised Dee) has extensive areas of inaccessible mudflats backed by large saltmarshes; the upper, drier parts of the marsh are grazed. The hinterland is a mixture of farmland and some heavy industry.

Species

Wintering wildfowl include large numbers of Pintails, as well as Shelduck, Teal and Wigeon. Diving ducks such as Red-breasted Merganser and Goldeneye occur in the river channel, and may be distantly visible across the saltmarshes. Once famous for its geese, protection has resulted in the return of Pink-footed Goose, and a flock of up to 500 over-flying Pink-footed Geese is now a regular feature of the estuary. Small numbers of Bewick's and Whooper Swans are frequent, and roost at Inner Marsh Farm. The Dee holds large numbers of waders; dispersed at low water, the estuarine species such as Oystercatcher, Bar-tailed Godwit, Curlew, Common Redshank, Knot and Dunlin are best seen roosting on the upper saltmarshes at high tide, especially at Burton. Scarcer species include occasional wintering Greenshanks and Spotted Redshanks (e.g. on Decca Pools). These concentrations attract wintering raptors; Merlin, Hen Harrier and Short-eared Owl, as well as the resident Peregrines, Sparrowhawks and Barn Owls. Hen Harriers roost at Parkgate, with up to six birds in recent winters. The saltmarsh gutters hold wintering Rock Pipits, and there are much smaller numbers of Water Pipits (which favour 'fresher' habitats, such as flooded fields or pools on the upper saltmarshes). The grazed marshes and adjacent fields hold large flocks of Golden Plovers and Lapwings. Large numbers of finches and buntings feed along the strandline, and the commoner species may be joined by Tree Sparrow or Twite and variable numbers of Bramblings. Corn Bunting may still hang on and there are sometimes Lapland Buntings (usually only seen when the marshes flood). The saltmarsh cover harbours Water Rails and Jack Snipes, but these are seldom seen unless forced out by the very highest tides.

Passage also brings large numbers of waders, with a greater variety than in winter. Whimbrel is more frequent in spring, but autumn is generally better; Ruff, Common, Green and Curlew Sandpipers, Little Stint, Greenshank and Spotted Redshank may all occur. Black-tailed Godwit is present in variable numbers for much of the year, especially at Inner Marsh Farm. Black Tern is frequent, especially in the upper estuary. Passerines may include White and Yellow Wagtails, Wheatear and Ring Ouzel, and other notable migrants include Spoonbill, Garganey, Osprey and Spotted Crake.

Breeding species include Shelduck, Oystercatcher, Common Redshank and a large colony of Black-headed Gulls at Inner Marsh Farm, as well as Avocets and Little Egrets (the roost of Little Egrets at Inner Marsh Farm has topped 250 birds). There are sometimes a few Stonechats, and Grasshopper, Reed and Sedge Warblers breed, while Common Tern may visit from colonies on the Welsh shore of the estuary. Common Buzzard, Peregrine and Raven are resident in the area and Hobby and Marsh Harrier are frequent in summer.

Access

Heswall Beach (OS Landranger 118) Leave the A580 in Heswall west at the traffic lights on minor roads into Lower Heswall and head downhill to Banks Road, parking in the car park at its west end overlooking the estuary. From here walk north for 200–500 yards along the shore to view. This may now be the best area for waders on the Cheshire shore of the Dee, with a rising tide from 2.5 hours prior to high water onwards being best.

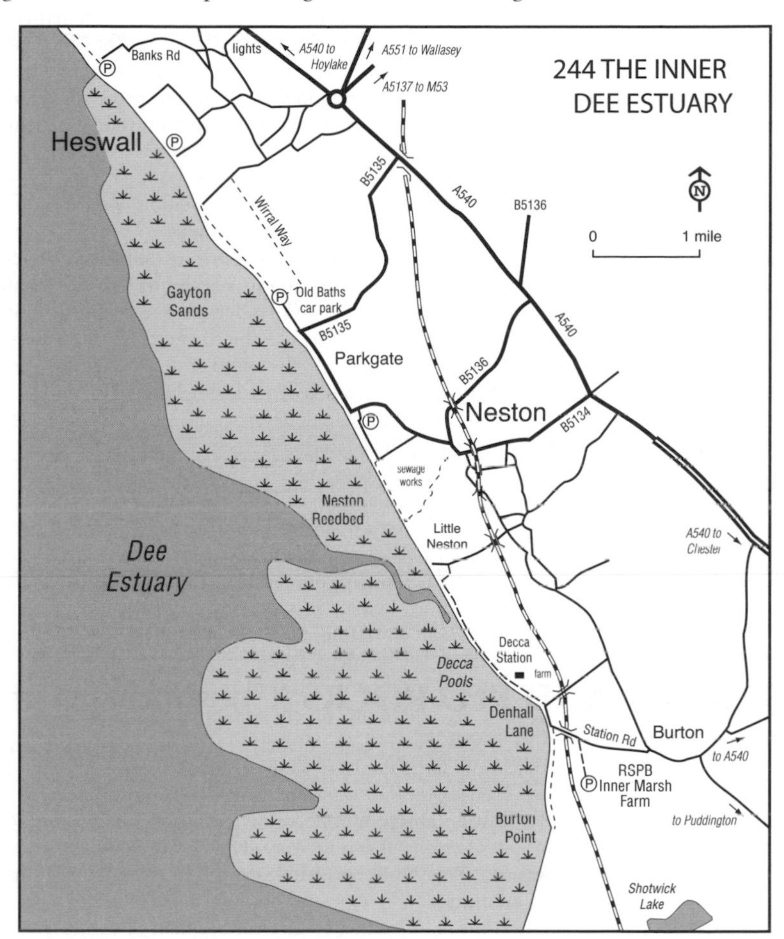

Gayton Sands Leave the A540 on the B5136 or B5134 west to Neston. After passing under the railway in the town, take the B5135 to Parkgate and follow the road sharply right along the edge of the saltmarsh. At the right-angle right turn (where the B5135 turns away from the coast), continue straight on to the Old Baths car park. There are good views from here and, for the more active, a public footpath runs north-west along the shore. The area is best in the period 1–2 hours either side of high water. Parkgate is especially good for wintering wildfowl, notably up to 10,000 Pintails (i.e. the entire Dee population). On high 'spring' tides, roosting waders include up to 200–300 Black-tailed Godwits, and on the very highest tides (33 feet plus) Water Rail and Jack Snipe are forced to leave the marshes, only to be preyed upon by Grey Herons. It is also a good area for raptors (though the Hen Harrier roost has sadly declined). Other attractions are flocks of finches and buntings, and a few Water Pipits.

The Wirral Way This long-distance footpath runs parallel to the coast north-west of Neston along the line of a dismantled railway; a footpath leads inland from the Old Baths car park to connect with the Wirral Way. The track-side scrub and adjacent arable land attracts a variety of warblers in summer, and finches and buntings in winter.

Neston Reedbed This small reedbed near the sewage works holds Reed and Sedge Warblers in summer, and small numbers of Grasshopper Warblers can be found in the scrubbier margins. In Neston, turn south off the B5135 where it meets the coast to a car park, and from here follow the footpath south along the edge of the saltmarsh to the reedbed.
Burton Marshes Leave the A540 on minor roads into Burton (follow signs for Ness Gardens) and follow the minor road through the village towards Neston. As you leave the village turn left into Station Road, crossing the railway after about 0.5 miles and continuing for a few hundred yards to view the saltmarshes at Denhall Lane. Waders may be present on the pools here, and the area is good for wildfowl.
Decca Pools From Denhall Lane, Burton, it is possible to follow the public footpath north to Little Neston. Just beyond Denhall House Farm (and below Decca Station) a series of pools is particularly attractive to waders and egrets. View from the embankment.
RSPB Inner Marsh Farm Lying in an area of former saltmarsh at Shotwick Fields, and separated from the estuary by the Wallasey–Wrexham railway, a series of artificial pools and scrapes now attracts an excellent variety of wildfowl and waders. Leave Burton on Station Road (as for Burton Marshes) and after 0.5 miles, just before the road crosses the railway, turn left along a narrow track signed Burton Point Farm. At the end is the small RSPB car park. The reserve is open 9am–9pm (or dusk if earlier), closed Tuesdays. There is one hide.

FURTHER INFORMATION

Grid refs: Heswall Beach SJ 254 814; Gayton Sands/The Wirral Way SJ 273 791; Neston Reedbed SJ 282 777; Burton Marshes/Decca Pools SJ 302 736; RSPB Inner Marsh Farm SJ 305 741
RSPB Dee Estuary: tel: 0151 3367681; email: deeestuary@rspb.org.uk

245 HILBRE AND RED ROCKS (Cheshire & Wirral)

OS Landranger 108

Lying at the mouth of the Dee Estuary, Hilbre is well known as a site for roosting waders (immortalised by legions of bird photographers) and as an excellent seawatching station, with Leach's Petrel a speciality. The adjacent Red Rocks peninsula on the mainland is also attractive to waders and can offer exciting autumn seawatching, and both areas attract small numbers of migrant passerines.

Habitat

Lying about 1.5 miles off Hoylake are three sandstone islands: Little Eye covers only 0.2 ha, Little Hilbre is larger and Hilbre, the largest island, is 4.5 ha in extent with cliffs rising to 55 feet on the west shore. The islands are sparsely vegetated and separated from the mainland for 3 hours either side of high water. The islands form an LNR, managed by Wirral Council. At Hoylake, sand from the mouth of the River Dee has formed two low dune ridges which sandwich a dune-slack marsh with stands of reeds and alder and willow scrub. Inland is the Royal Liverpool Golf Course and, at the north end of the dunes, areas of sandstone are exposed at Red Rocks, forming a short promontory, with the furthest rocks (Bird Rock) cut off at high tide. The 4-ha Red Rocks Marsh is a reserve of the Cheshire Wildlife Trust.

Species

The wader roost on Little Eye and/or Little Hilbre has declined in recent years but remains one of the attractions of the area. Wintering waders include large numbers of Oystercatchers, Curlews, Black-tailed Godwits, Dunlins and Knots, smaller numbers of Grey and Ringed

Plovers, Bar-tailed Godwits, Sanderlings and Turnstones, and up to 30 Purple Sandpipers. A similar range of species may roost at Bird Rock, Red Rocks, and at low water these waders (except Purple Sandpiper) are scattered over the estuary. In winter there are also small numbers of divers off the estuary mouth, mostly Red-throated, with Great Northern scarce and Black-throated only occasional. Small numbers of Red-breasted Mergansers and Goldeneyes frequent the river channel and mouth, together with a few Common Scoters, Common Guillemots and Razorbills. Other wildfowl include Shelduck and Wigeon, and there is a flock of up to 100 Pale-bellied Brent Geese around Hilbre (small numbers of Dark-bellied also occur). Gulls regularly include Mediterranean and there is a passage of Little Gulls in March and April. Snow Bunting is sometimes found in the dunes, with Stonechat in the scrubby areas. The concentrations of waders attract Peregrine, and other raptors may include Sparrowhawk and Merlin.

In spring and autumn small falls of migrants occur, such as White and Yellow Wagtails, Whinchat and Wheatear. The best conditions are light to moderate south-east to south-west winds, coupled with poor visibility and perhaps rain or drizzle. However, by 2 hours after dawn, most have left Hilbre, though some falls occur later in the day. The scrub at Red Rocks tends to hold birds for longer; being on the west coast, however, scarce migrants are just that, scarce! Visible passerine migration can also be interesting and usually occurs in the first 4 hours after dawn.

Autumn is best for seawatching. From September to early November, following 2–3 days of north-west gales, Leach's Petrel is virtually guaranteed, along with small numbers of Manx Shearwaters, and occasionally Sooty Shearwater and Sabine's Gull. A deep low centred on the Faeroes and moving slowly east will produce the necessary gales. Fulmar, Gannet, Kittiwake and Arctic Skua are regular offshore in autumn, but Great and Pomarine Skuas are scarce and Long-tailed is rare. The occasional shearwater or skua also occurs in early spring. Late-summer congregations of terns occur off the estuary mouth, including Little Terns and occasionally also Black or Roseate.

Breeding species include Shelduck at Hilbre, with Grasshopper, Reed and Sedge Warblers at Red Rocks.

Access

Hilbre Reached on foot from West Kirby, but cut off for at least 2 hours either side of high tide. The walk to the island takes about 1 hour and you should start out, at the very latest, 3 hours before high tide when going to, or returning from, Hilbre. The safest route is to start from the slipway on the promenade opposite Dee Lane (just north of the Marine Lake, see below for access), walk towards Little Eye, the smallest of the three islands, keeping it on your right. As soon as you pass Little Eye turn right and continue on the sand passing Middle Eye on your left. Between Middle Eye and Hilbre take the rough track over the rocks towards the small tidal pool. Once off the rocks turn left towards the gate at the south end of Hilbre. (Do not walk directly to Hilbre Island, do not cross from Hoylake, and do not attempt the crossing in foggy conditions; permits are required by groups of more than six.) The paddocks and bungalows are private. Hilbre Island Bird Observatory operates a ringing station and seawatching hide. No accommodation is available. The period around high tide is most productive for seawatching, and the hide can be used by arrangement with the observatory, otherwise the north end of Hilbre, above the Lifeboat House, is a good spot to watch from, providing conditions are not too bad.

Red Rocks From West Kirby, walk north along the foreshore, past Red Rocks Marsh to Red Rocks Point. Alternatively, from the A540, turn north-west towards the coast at the roundabout by Hoylake Station and after 200 yards take the first left into Stanley Road and park. The Point lies at the far end of the road. The scrub around the dune slacks attracts migrant passerines (especially the small poplar plantation at the north end), as do the trees and gardens along Stanley Road. Access to Red Rocks Marsh is restricted to the foreshore or the boardwalk on the inland side of the marsh. Bird Rock off Red Rocks Point is an island at high tide, and when undisturbed holds roosting waders. The Point is also a worthwhile seawatching station, offering a similar range of species to Hilbre, but from a more accessible spot.

West Kirby Marine Lake This may hold one or two seaducks, grebes or divers, especially in the early morning. Leave the A540 at West Kirby Station and follow signs for the Marine Lake west to the beach.

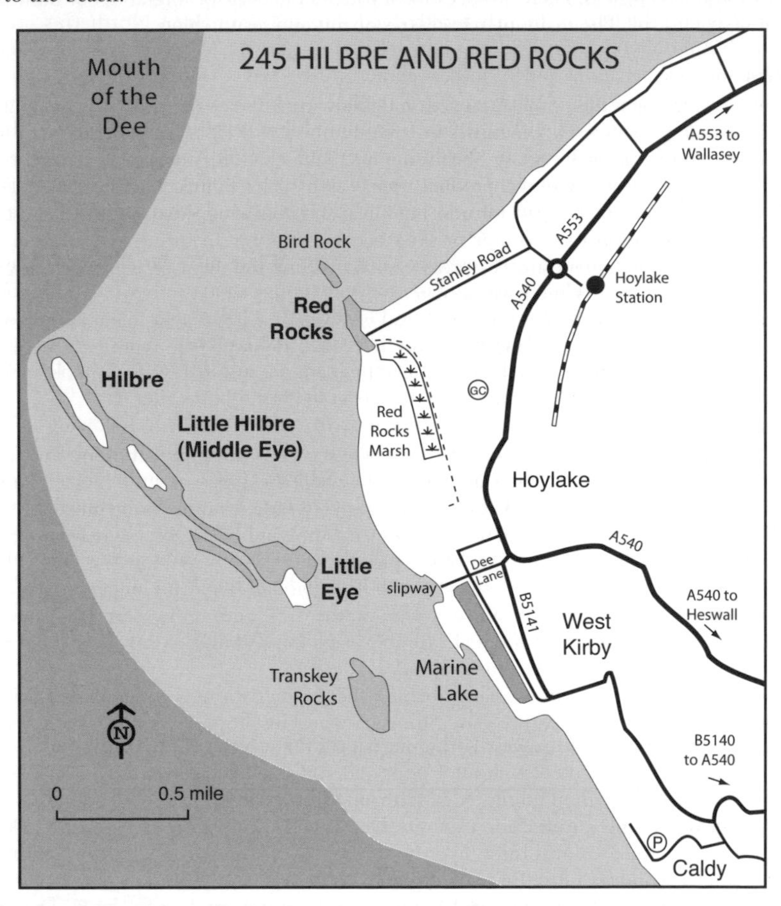

FURTHER INFORMATION

Grid refs: Hilbre/West Kirby Marine Lake SJ 210 867; Red Rocks (Stanley Road) SJ 211 887
Hilbre LNR: tel: 0151 648 4371/648 3884 (advice on tide times); web: www.wirral.gov.uk
Hilbre Island Bird Observatory: web: www.hilbrebirdobs.blogspot.com
Cheshire Wildlife Trust: tel: 01948 820728; web: www.cheshirewildlifetrust.co.uk

246 THE WIRRAL COAST (Cheshire & Wirral)

OS Landranger 108

In autumn, north-west gales push large numbers of seabirds inshore, notably Leach's Petrel, and the Wirral is one of the best sites in the country for this species, with a variety of other seabirds also on offer. In spring and autumn there is also the possibility of passerine migrants.

Habitat

The north coast of the Wirral Peninsula is guarded by a system of fixed dunes, backed by market gardens and built-up areas. There are occasional patches of cover for migrants, but numbers of birds are usually small. The main interest centres on autumn seawatching.

Species

Autumn seawatching is the prime attraction, and following 2–3 days of north-west gales Leach's Petrel is virtually guaranteed, sometimes in large numbers, as well as small numbers of Manx Shearwaters, and occasionally Sooty Shearwater and Sabine's Gull. A deep low centred on the Faeroes and moving slowly east will produce the necessary gales. Fulmar, Gannet, Kittiwake and Arctic Skua are regular offshore in autumn, but Great and Pomarine Skuas are scarce and Long-tailed rare. The occasional shearwater or skua occurs in spring. In spring and autumn small numbers of passerine migrants may be present, such as White and Yellow Wagtails, Wheatear and Ring Ouzel, and several rarities have occurred.

In autumn and winter small numbers of waders occur along the coast, with Turnstone and sometimes Purple Sandpiper around the fort at New Brighton. Throughout the year, gulls should be carefully checked – Mediterranean and Little are frequent.

Access

Dove Point, Meols Offers seawatching from the shelter of a car and the potential of the closest views of seabirds, but only at high tide (when the sands of East Hoyle Bank are completely covered). On lower tides and in calmer weather, numbers of gulls and terns, sometimes including Mediterranean Gull, roost on the sands here. Access is from A553 at Hoylake, driving to the north-east tip of the promenade.

Leasowe Lighthouse A good place to seawatch from, though it may no longer be possible to take a car onto the sea wall. Leave Wallasey west on the A551 and, at the sharp left-hand bend a few hundred yards beyond the hospital, take the rough track straight ahead along the sea wall; the lighthouse is clearly visible 0.5 miles ahead.

Leasowe Common Lying immediately behind the sea wall, this is a good area for migrants in spring and autumn, and has breeding Stonechat and Grasshopper Warbler. From Leasowe Lighthouse follow Lingham Lane to the south, checking the hedges for migrants.

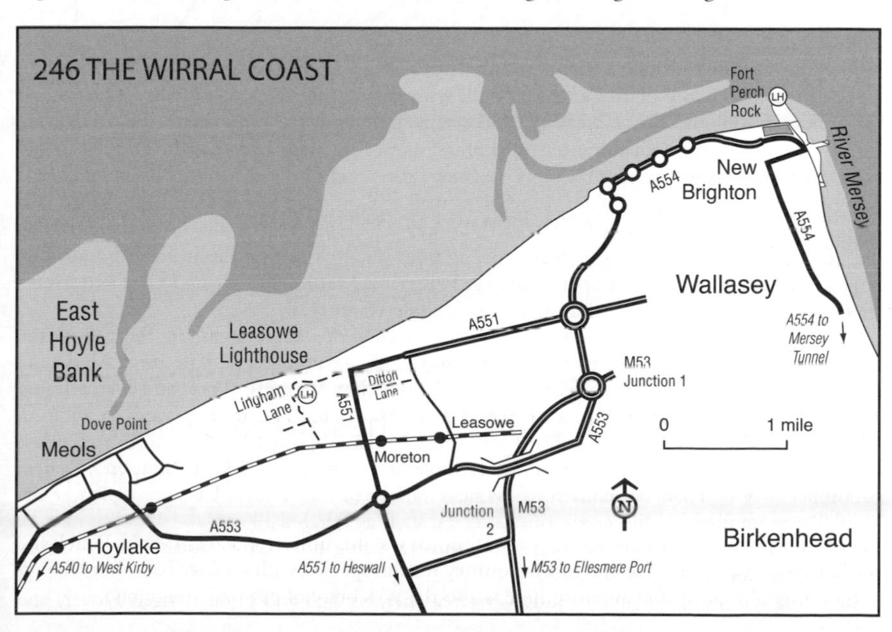

Ditton Lane This runs between the A551, 0.3 miles north of Moreton Station, and Wallasey. The willow and alder scrub here may hold good numbers of migrants.
New Brighton Good for seawatching, with the period around high tide preferable (but not essential as at Meols); Leach's Petrel may be seen trying to exit from the mouth of the Mersey (and are sometimes observed upriver as far as Seacombe Ferry). The A554 follows the coast for 1 mile to the marine lake and lighthouse, and the shelters near the Fort Perch Rock are ideally placed for seawatching, as is a parked car along the seafront.

FURTHER INFORMATION

Grid refs: Dove Point, Meols SJ 234 907; Leasowe Lighthouse/Leasowe Common SJ 262 918; Ditton Lane SJ 262 912; New Brighton SJ 309 945

247 FRODSHAM LAGOONS AND THE WEAVER BEND (Cheshire & Wirral) OS Landranger 117

This area, sandwiched between the M56 and the Manchester Ship Canal, attracts a variety of waders and wildfowl in winter and on passage, and is especially favoured by Curlew Sandpiper and Little Stint in autumn.

Habitat

Dredgings from the Manchester Ship Canal are pumped into embanked lagoons at Frodsham. When full, these slowly dry out, resulting in a succession of habitats from bare mud to pasture. Together with the surrounding farmland they form a complex of freshwater and brackish habitats. Beyond the ship canal, Frodsham Score is an area of grazed saltmarsh and to the north, where the River Weaver flows into the Manchester Ship Canal, the resultant estuarine habitats add further variety.

Species

Wintering wildfowl include variable numbers of Shelducks, Wigeons and Teals, which often favour Frodsham Score or the Weaver Bend, with smaller numbers of Pintails. Pochard and Goldeneye favour the River Weaver. Wild swans and geese are only erratic visitors, as are Great Crested Grebe, Cormorant, Scaup and Smew, but these may also be found on the river, especially in cold weather. Waders on the pastures and on Frodsham Score include large numbers of Lapwings and Golden Plovers, as well as Common Snipe, Common Redshank, Dunlin and sometimes Ruff or Black-tailed Godwit. The estuary attracts more of these species, and sometimes a wintering Little Stint; on big tides waders roost on the wetter lagoons. The concentrations of prey attract Sparrowhawk, Peregrine, Merlin, Short-eared Owl and sometimes Hen Harrier. Other notable winterers include Water Rail.

Garganey may occur on spring passage, as can a variety of waders, including Ringed Plover, Common Sandpiper, Greenshank, Sanderling and Ruff, and Little Gull, Common, Arctic and Black Terns. Passerine migrants include White and Yellow Wagtails, Rock and Water Pipits, Wheatear and Whinchat. Autumn passage is more protracted and more varied; in addition to the waders listed above, Turnstone, Black-tailed Godwit, Spotted Redshank, Knot, Curlew and Green Sandpipers, and Little Stint are possible, and in some years Little Stint and Curlew Sandpiper may be relatively numerous, while a variety of rare waders has appeared over the years. Large numbers of the commoner species, such as Ringed Plover and Dunlin, may be forced off the Mersey to roost on the lagoons during 'spring' tides. Other notable migrants may include Spoonbill and Spotted Crake.

Breeding species in the area include Oystercatcher, Ringed and Little Ringed Plovers and

Grasshopper, Reed and Sedge Warblers. Hobbies are regular at the Swallow roost in late summer and Little Egrets and Peregrines are seen year-round.

Access

The Weaver Bend Leave the M56 at junction 12 and take the A56 south into Frodsham. Just before a set of traffic lights turn right into Ship Street and follow the road to the right and then turn left to a bridge over the motorway. Cross this and follow the public footpath to the gate (some birdwatchers drive to the gate, but the road is private beyond the motorway). Continue along the footpath to the Weaver Bend. The patch of mud around the small island by the ICI Tank is especially favoured by waders.

Frodsham Lagoons In Frodsham continue along the A56, past the main shopping area and then turn right into Marsh Lane. Cross the motorway bridge onto the public footpath that follows the private track. The track soon forks, to the right is Tank No. 5 and the Weaver Bend, while the left track passes along the south edge of No. 5 Tank and skirts No. 6 Tank.

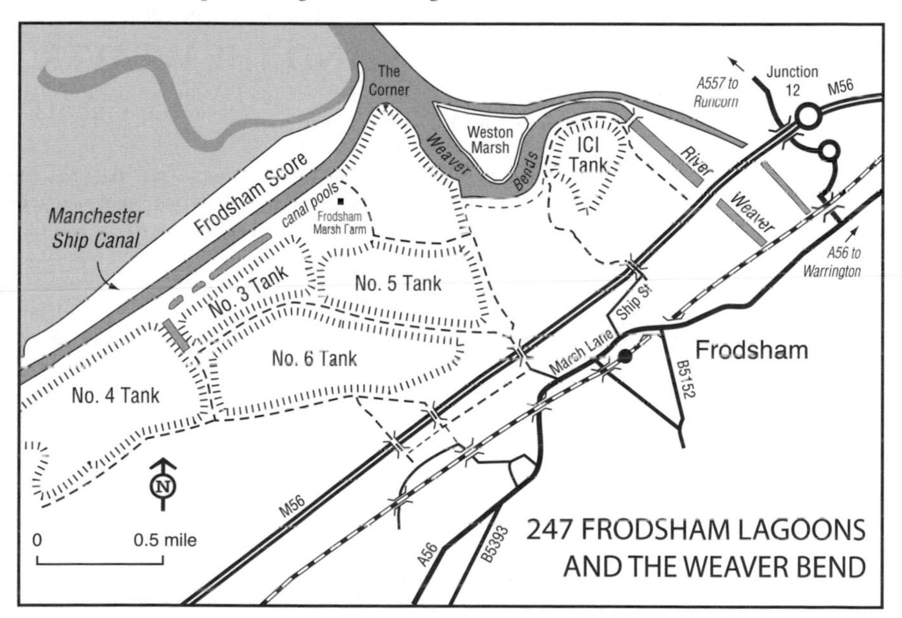

FURTHER INFORMATION

Grid refs: The Weaver Bend (Ship Street) SJ 520 785; Frodsham Lagoon (Marsh Lane) SJ 512 779

248 PENNINGTON FLASH COUNTRY PARK

(Greater Manchester) OS Landranger 109

This 200-ha country park on the outskirts of Leigh shot to fame in 1994 when Britain's first Black-faced Bunting spent several weeks there. On a more prosaic level, it regularly attracts a variety of wildfowl and waders, often including a few Smews, and Long-eared Owl also sometimes winters. The site is owned and managed by Wigan Council.

Habitat

Subsidence due to coal mining has resulted in a large 'flash' – 70 ha of open water. The north-west quadrant was filled with colliery waste (Ramsdale's Ruck, now grassland and birch scrub), while the south part was filled with domestic refuse. The remainder forms a large, shallow lake, with another 8 ha of smaller pools and scrapes in the north-east reserve area and some stands of reeds. The surrounding land comprises expanses of rough grassland, meadows and scrub, while areas of derelict land and landfill have been landscaped and planted with a variety of trees and shrubs, and developed as a golf course.

Species

Wintering wildfowl include several hundred each of Mallard, Teal, Pochard and Tufted Duck, with smaller numbers of Shovelers, Wigeons, Goldeneyes and often a few Ruddy Ducks. Shelduck, Gadwall, Scaup, Red-breasted Merganser, seaducks and Whooper and Bewick's Swans are scarce, but occasional Smew and Goosander may take up winter residence. Other notable waterfowl include Water Rail and large numbers of Cormorants, but divers, the rarer grebes and Bittern are only occasional. Up to 100 Common Snipes may be present, and large numbers of Golden Plovers and Lapwings winter in the agricultural areas to the south and may visit at times. Occasionally Water Pipit, Common Redshank and Dunlin also use the area. A few Long-eared Owls may winter between mid-November and mid-March; traditionally they are best looked for roosting in dense thickets near New Hide. A large gull roost forms in winter; mostly Black-headed Gull, there are also good numbers of Lesser Black-backed and Herring Gulls, occasionally Mediterranean, Iceland and Glaucous Gulls, while Yellow-legged and Ring-billed Gulls have been recorded. Great Spotted Woodpecker does not breed in the Country Park and is only a visitor, as are Tree Sparrow and Siskin, while Brambling is regular at the feeding station in most winters.

On spring migration small numbers of waders may appear, including Oystercatcher, Ringed and Little Ringed Plovers, Curlew, Black-tailed Godwit, Dunlin and Common Sandpiper, with Sanderling and Turnstone possible in May. Common, Arctic and Black Terns may also pass through. Black-necked Grebe has been regular in spring, and Little Egret turns up with increasing frequency. Other migrants include White Wagtail and sometimes Water Pipit and Wheatear (try Ramsdale's Ruck), while Common Scoter, Garganey, Marsh Harrier, Osprey and Yellow Wagtail are scarce visitors at this season. Autumn passage is more protracted, but with many of the same species possible, though Greenshank, Spotted Redshank, Green Sandpiper, Little Stint and Ruff are more likely. Migrant passerines, such as Common Redstart, Spotted Flycatcher and Tree Pipit, are also possible.

Residents include Great Crested and Little Grebes, Tufted Duck, Kingfisher, Willow Tit (2–3 pairs breed and the species is regular at the feeding station in winter), and Lesser Redpoll, while Grey Heron is almost always present. Other breeding species (mostly present in small numbers) include Shelduck, Gadwall, Shoveler, Pochard, Oystercatcher, Ringed and Little Ringed Plovers, Common Redshank, Common Tern (on rafts in the main flash), Grasshopper, Reed, Sedge and Garden Warblers, and Lesser Whitethroat.

Access

Leave the A580 (signed for the Country Park) north onto the Leigh by-pass. After about 1 mile turn south-west (left) onto the A572 and the entrance to the Country Park is on the right after 400 yards. There is a car park at the main entrance or you can follow the track around the east perimeter of The Flash to park at the information centre. A network of well-made paths and tracks permits exploration of the entire area, especially the reserve area on the north-east margin of The Flash, while tracks along the south and west shores are productive in autumn and winter. There are seven hides, the Horrocks Hide being one of the best, and the feeding station at Bunting Hide provides excellent views of Willow Tit, up to 30 Bullfinches and occasionally Water Rail as well as many common woodland species. The Country Park is popular in summer, and much of the area away from the sanctuary area is disturbed by water sports and anglers, especially at weekends.

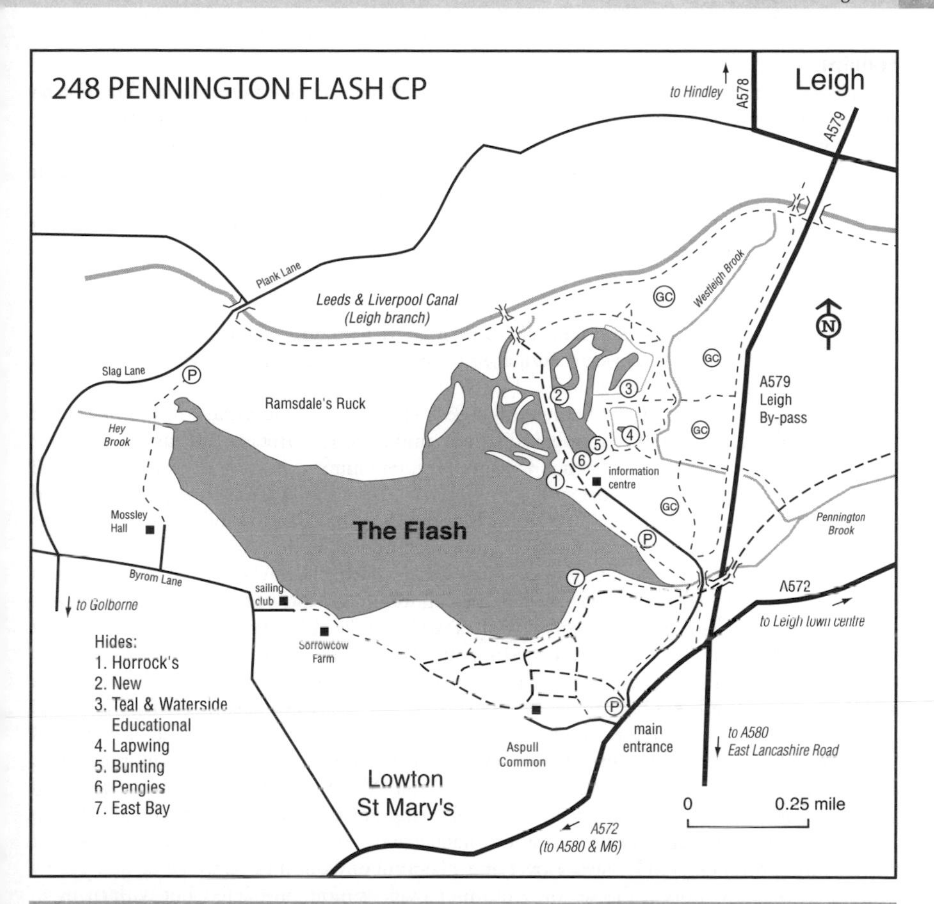

FURTHER INFORMATION

Grid ref/postcode: SJ 643 984 / WN7 3PA
Pennington Flash Country Park Visitors Centre: tel: 01942 605253; email: pfcp@wlct.org

249 SEAFORTH AND CROSBY MARINA
(Lancashire & North Merseyside) OS Landranger 108

Lying within Liverpool Freeport, Seaforth NR has an enviable reputation as a centre for the close study of gulls, the venue for the largest spring concentrations of Little Gulls in Britain, and a first landfall for ship-hopping American passerines, while adjacent Crosby Marina attracts occasional seaducks sheltering from harsh weather, and small numbers of waders.

Habitat

Seaforth consists of 30 ha in the heart of the Liverpool Docks, comprising two lagoons surrounded by tipped infill, and a small reedbed. The reserve is owned by the Mersey Docks and Harbour Company and managed by the Lancashire Wildlife Trust. Crosby Marina lies immediately to the north and is heavily disturbed by water sports.

Species

Gulls are the major focus at Seaforth. All of the commoner species are regularly recorded in large numbers, but scarce visitors are the great fascination. Small numbers of Kittiwakes are present year-round. Little Gull is present offshore most of the year, but is seldom seen in winter unless winter storms bring birds in from the Irish Sea to shelter. They are regular, however, on spring passage from the last week of March to the first week of May, when up to 500 gather. Small numbers may summer and flocks are also recorded in autumn, especially following onshore winds. Mediterranean Gull is similarly recorded in very small numbers almost daily throughout the year, most regularly in September–early April (especially mid–late March). Yellow-legged Gull is most frequent in mid-June to July (immatures) and mid-October to November (adults). Glaucous and Iceland Gulls, on the other hand, are most regular in winter–early spring. Ring-billed Gull is also annual, with spring (immatures) and early winter (adults) being the most likely periods.

Wintering wildfowl include Shelduck, Teal, Pochard, Scaup, Goldeneye and Red-breasted Merganser, and other notable waterbirds include numbers of Cormorants (huge numbers – up to 900 – may shelter here from storms), Common Snipe and a few Jack Snipes. Good numbers of waders roost on the pools at Seaforth, mostly Oystercatcher and Common Redshank, with smaller numbers of Ringed Plovers, Turnstones, Bar-tailed Godwits, Dunlins and Knots. Occasionally Snow Bunting is found around the marina.

On passage large numbers of Common Terns are recorded, peaking at over 2,000 in August, and small numbers of Arctic, Little and Sandwich Terns are also regular in small numbers, but Roseate and Black Terns are scarce. Waders may include Whimbrel, Black-tailed Godwit and Common Sandpiper, with Curlew Sandpiper, Sanderling and Little Stint most likely in autumn. Migrant passerines include numbers of White Wagtails in spring, with Yellow Wagtail and Wheatear at both seasons. Several vagrant American sparrows have been recorded, and although these almost certainly arrived aboard ships, they are officially deemed 'wild' unless known to have been fed en route! Autumn gales may prompt an excellent passage of seabirds, notably Leach's Petrel, which is regular after north-west gales. Manx and Sooty Shearwaters and all four species of skuas are also possible (though Long-tailed is, of course, scarce), and Storm Petrel is occasionally seen after late-summer gales.

Breeding species include Ringed Plover and a handful of Common Terns.

Access

Seaforth NR Access to this port area is strictly controlled and restricted to permit holders, who must be members of the Lancashire Wildlife Trust – contact the trust for the latest information. Exit the M57/M58 at their northern/western terminus (junction 7) and take the A5036 south-westwards. At the junction with the A565 exit the dual carriageway following signs for the Freeport (dock gates 99–103). The Freeport entrance is on Crosby Road. Three hides overlook the pools and the bund used by roosting and loafing birds. For waders the period 2–3 hours before high tide is best, but gulls are less governed by the tides, with late morning and afternoon being the best periods.

Crosby Marina Access immediately north of the Freeport entrance, parking at the marina. Though heavily disturbed, some parts of the shoreline are quieter and the surrounding short turf attracts a few passage waders. Seawatching is best along the estuary sea defences near the coastguard lookout.

FURTHER INFORMATION

Grid refs/postcode: Seaforth NR SJ 318 971 / L21 1JD; Crosby Marina SJ 317 977
Wildlife Trust for Lancashire, Manchester and North Merseyside: tel: 01772 324129; web: www.lancswt.org.uk

250 MARTIN MERE (Lancashire)

OS Landranger 108

Martin Mere is a WWT Refuge and intensive management has produced an excellent wetland area that attracts large numbers of ducks, Pink-footed Geese and swans in winter, as well as a variety of waders on passage.

Habitat

In the centre of the south Lancashire mosses, Martin Mere was once a large lake. But, like the rest of the mosses, the Mere was drained to leave an area of winter flood water. This relic of past glories was purchased by the WWT in 1972, the area of flood increased and many permanent pools and scrapes established, surrounded by damp pastures.

Species

In early winter the Lancashire Pink-foot population converges on Martin Mere and can peak at 34,000 prior to dispersal. Many leave Lancashire (for Norfolk) but some feed on Altcar Moss and others move to Marshside (see page 384), and large numbers continue to spend the day at Martin Mere, being augmented by others coming to roost. The Pink-feet are often joined by family parties of other geese, and sometimes by lone Snow Geese or Canada Geese of one of the small subspecies, both of which are likely to be genuine vagrants from North America. There are also several hundred feral Greylag Geese and smaller numbers of feral Barnacle Geese in the area (as well as captive birds from the collection, just to confuse the issue). A herd of up to 1,900 Whooper Swans has built up, together with up to 100 Bewick's Swans, some of which arrive from the Ribble in the evening. Ducks are numerous. Wigeon has topped 25,000 and Teal can peak at 8,300, while up to 700 Pintails can be present in early winter. Smaller numbers of Shelducks, Gadwalls, Shovelers and Pochards occur, and often a few Ruddy Ducks, Goldeneyes, and sometimes Scaups and Goosanders. Green-winged Teal (another American vagrant and now considered a species) has been almost regular in recent winters. Short-eared Owl, Hen Harrier, Common Buzzard and Merlin are regular in winter and on passage and the area is especially good for Peregrine. Corn Bunting and Tree Sparrow may be seen in winter around the refuge.

Many of the same species can be seen in the surrounding mosses (now merely fields, though still liable to occasional flooding), by touring lanes in the area, especially those across Altcar, Plex and Halsall Mosses. The area to work is the rectangle between the A5147 in the east and A565 in the west, and the A570 in the north and B5195 in the south.

On passage a variety of waders occurs, with Ringed and Little Ringed Plovers, Curlew, Dunlin, Common Sandpiper, Spotted Redshank, Greenshank and Black-tailed Godwit being likely. Martin Mere also boasts an impressive list of rare and scarce birds, of which Temminck's Stint and Pectoral Sandpiper are seen most years. Golden Plovers and Lapwings congregate in autumn and winter and may be joined by over 100 Ruffs. Garganey and Common, Arctic and Black Terns may also be recorded on passage, with very small numbers of Yellow Wagtails and, in spring, White Wagtail. Also in spring, the persistent searcher may sometimes find trips of Dotterels in the surrounding mosses, especially on Altcar Moss.

Breeding species include Shelduck, Gadwall, Shoveler, Pochard and large numbers of Mallards. Small numbers of other species, as well as 'pricked' or otherwise injured geese and swans, may also summer. A recent addition is Avocet, which bred for the first time in 2004, and other breeding species are Lapwing, Common Snipe, Common Redshank, Barn Owl and Sedge Warbler. Oystercatcher, Little Ringed Plover and Ruff have also bred, while numbers of Black-tailed Godwits may summer, and Marsh Harrier and Hobby are now a regular sight over the reserve in summer.

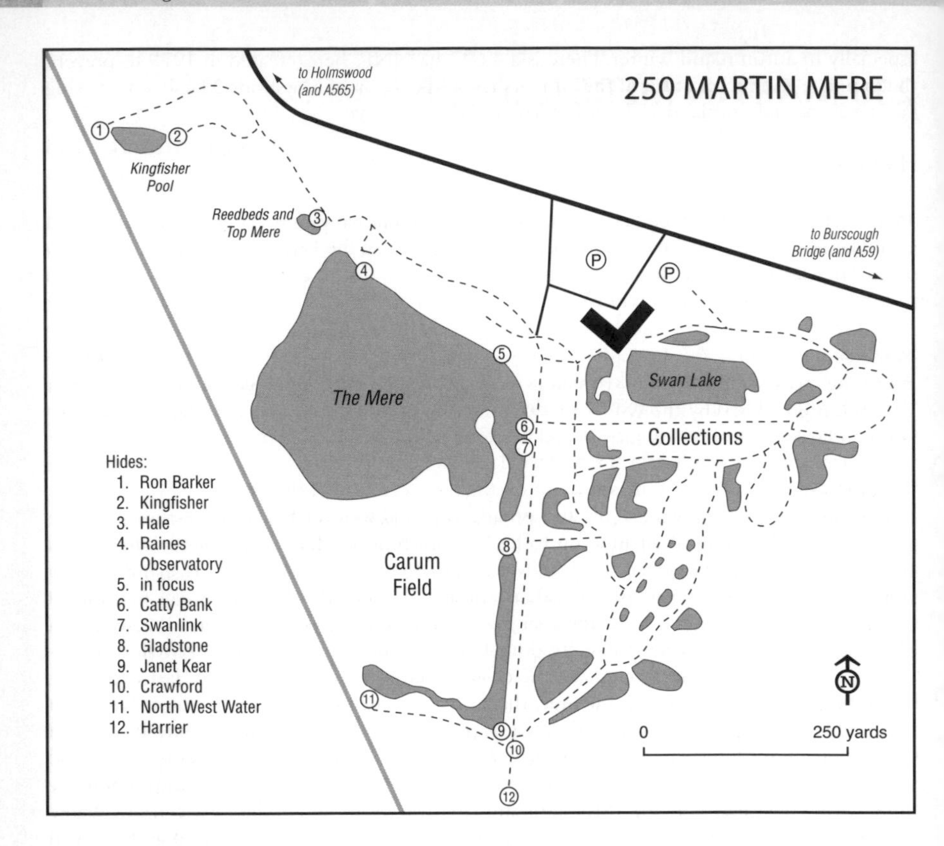

Access

The refuge is on a minor road between the B5246 at Holmeswood (south of Mere Brow on the A565) and the A59 at Burscough Bridge, and is signposted from Burscough Bridge and Mere Brow. Open daily from 9.30am until 5pm (November–February; closed Christmas Day) or to 5.30pm (rest of the year), there is an entrance fee for non-members of the WWT. There is an education centre, exhibition hall and a collection of captive wildfowl and flamingos, as well as 12 hides.

FURTHER INFORMATION

Grid ref/postcode: SD 428 143 / L40 0TA
WWT Martin Mere: tel: 01704 895181; web: www.wwt.org.uk

251 SOUTH RIBBLE MARSHES (Lancashire)

OS Landranger 102 and 108

Lying between Southport and Blackpool, the Ribble Estuary is the most important wildfowl site in the UK, supporting over a quarter of a million waders and wildfowl each winter, and is an internationally important site for 20 species of birds, including Pintail and nine species of wader, particularly Knot, Sanderling and Bar-tailed Godwit. This provides excellent birdwatching, especially in autumn and winter. There is a 4,697-ha NNR here, created in 1979 to prevent the drainage and reclamation of the saltmarshes, while 94 ha of Marshside Marsh, comprising saltmarsh, coastal grassland and pools, form an RSPB reserve.

Habitat

The outer estuary is sandy, attracting tourists to Southport, Lytham St Anne's and Blackpool. The inner estuary is also mostly sandy, but fortunately attracts fewer people. The Ribble has one of the largest areas of tidal flats and saltmarsh in the country, the latter mainly on the southern shore. The saltmarshes are grazed in summer by cattle and inland of the sea walls lie reclaimed pastures, which may flood in winter.

Species

Wintering species include Pink-footed Goose, which favours Marshside and may peak at 10,000 birds, and is often joined by singles or family parties of other species, notably a Snow Goose that has been returning to the area for several years and is generally considered to be wild. There are also occasionally a few Bewick's or Whoopers Swans, but those that do still visit the area have taken to using the saltmarsh on the northern side of the estuary at Warton Bank. Most of the geese and swans that use the estuary fly inland to roost at WWT Martin Mere (see page 381). Other wildfowl include up to 81,000 Wigeons, large numbers of Pintails, as well as Shelduck and a few Gadwalls and Shovelers. The grazing marshes hold large numbers of Golden Plovers and Common Redshanks, as well as a few Ruffs. At high tide small numbers of seaducks may appear offshore, including Common Eider, Scaup, Goldeneye and Red-breasted Merganser. Waders on the shoreline are dominated by Knot (up to 42,000) as well as hordes of other common species. Up to 1,000 Black-tailed Godwits may be present, and occasional Little Stints, Spotted Redshanks and Greenshanks winter. The wader roost can be watched from Marine Drive, by parking at the sand depot. Hen Harrier, Peregrine, Merlin and Short-eared Owl can all be seen. Wintering passerines include Rock Pipit, Stonechat and often Twite, and very occasionally Lapland Buntings on the saltmarshes (these are very elusive and only likely to be detected in flight, giving their distinctive *tricky-tick* call), while the rather erratic Snow Bunting prefers sandier areas. On the highest tides, Water Rail and Jack Snipe may be forced from the saltmarshes to seek shelter on drier ground.

In spring a trickle of passerines follow the west coast, notable species including White Wagtails bound for Iceland, Wheatear (the larger and brighter 'Greenland' Wheatear passing in May), and Ring Ouzel. Other regular migrants include Marsh Harrier and Garganey. Spring movements of waders are rapid, but autumn passage is more leisurely, with large influxes from late July, and huge flocks of moulting waders on the estuary. Wintering species are joined by Curlew Sandpiper, Little Stint and up to 4,000 Black-tailed Godwits. In autumn, prolonged westerly gales can push seabirds inshore, occasionally including Leach's Petrel and Manx Shearwater, but the area is not particularly productive for seawatching, with sites such as Heysham and Rossall Point to the north and Seaforth to the south being better (see pages 379, 388 and 390).

Breeding birds include several thousand pairs of Black-headed Gulls, a few hundred pairs of Herring and Lesser Black-backed Gulls and Common Terns on the marshes, as well as a few Arctic Terns, large numbers of waders, particularly Common Redshank, as well as Lapwing and Avocet at Marshside Marsh, Shelduck, Shoveler and Corn Bunting. Black-tailed Godwits may oversummer.

Access

Southport Marine Lake Lying at the base of Southport pier, this is rather disturbed and only Cormorant, Goldeneye and Red-breasted Merganser are regular. In winter gales divers, Shag and seaducks sometimes seek shelter here, while Snow Bunting may be found on the nearby beach. In autumn prolonged westerly gales, blowing for two days or more, can push seabirds onshore, and at high tide a few Leach's Petrels and Manx Shearwaters may be seen from the pier, but it is not as good a vantage point as other classic north-west seawatching sites such as

Heysham, Rossall Point, the Ainsdale-Formby coast or the Mersey mouth.

Southport Marine Drive This follows the coast north-east from Southport Marine Lake to the Banks roundabout on the A565, with a car park by the Marine Lake; stopping on the road is otherwise prohibited, though there are several pull-offs. The golf course at Marshside attracts small numbers of migrants during passage.

RSPB Marshside Accessible from the small car park by the sand works on Marine Drive, the reserve is open daily 8.30am–7pm and there are two hides, 200 and 600 yards from the car park (both of which are wheelchair-accessible and often afford some close views of waders), a viewing platform and three viewing screens. The trails plus existing footpaths allow a circular walk of up to 4.3 miles. The fields inland of the road 0.5 miles north of the car park are excellent. Large flocks of Pink-feet spend the day feeding here, usually roosting on Southport Sands. The area attracts raptors, including Hen Harrier, Peregrine and Merlin. The fields often hold Golden Plover and large numbers of Black-tailed Godwits. On the seaward side of the road, Crossens Marsh is not attractive to feeding waders, but has a huge roost on the highest tides in spring and autumn (30 feet or more). Twite can sometimes be found on the saltings, and Lapland Bunting is a very occasional visitor. The reserve holds some of the highest densities of breeding Lapwings and Common Redshanks in Britain and now has a strong colony of Avocets, which bred here for the first time in 2002.

Ribble Estuary National Nature Reserve In the NNR, Crossens, Banks, and Hesketh Out Marshes are visible from a public footpath along the sea wall. Access is from Crossens Pumping Station, just off the A565, and at Hundred End (on a minor road north from the

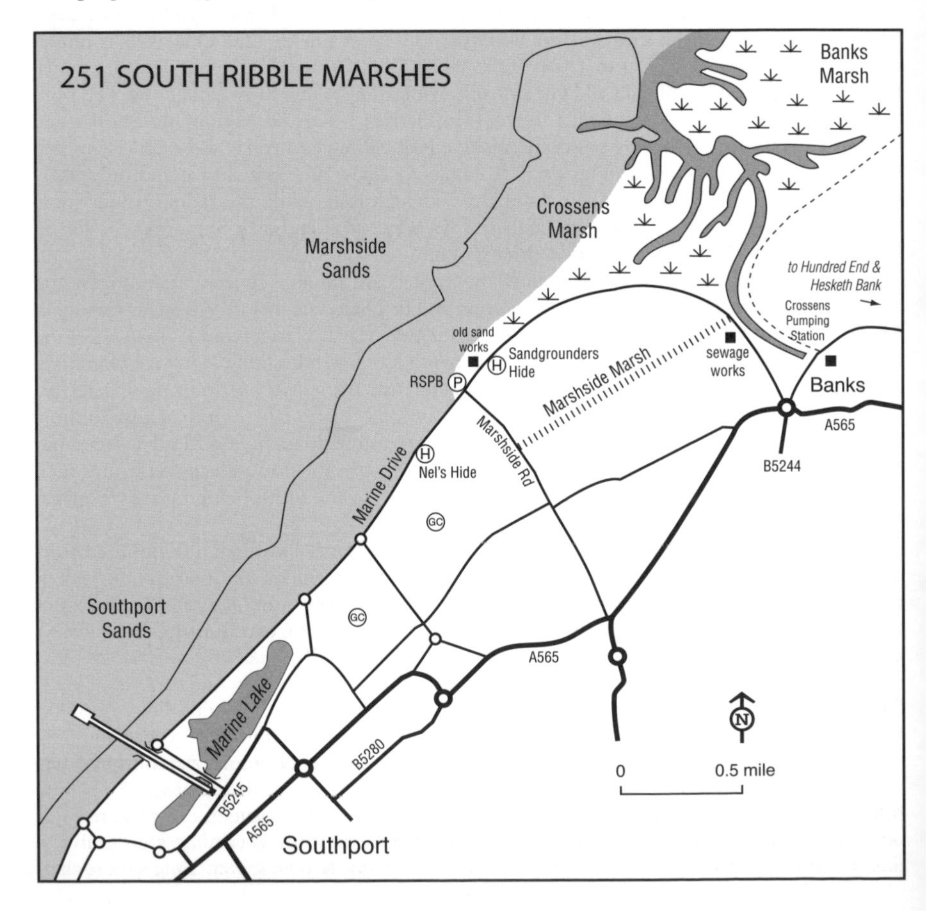

A565). It is 4 miles each way, and you are unlikely to see more than from the more convenient Marine Drive. Hesketh Out Marsh is, however, now an RSPB reserve (it opened to the public in 2009). A viewing shelter overlooks a large tidal scrape and the reserve car park (open 8am to 6pm) can be reached along Dib Road, a rough private track, signed from the minor road between Hundred End and Hesketh Bank, saving the long walk from Crossens.

FURTHER INFORMATION

Grid refs: Southport Marine Lake/Marine Drive SD 336 184; RSPB Marshside SD 352 204; National Nature Reserve (RSPB Hesketh Out Marsh) SD 430 225
Natural England, Ribble Estuary Site Manager: tel: 01704 225624; web: www.naturalengland.org.uk
RSPB Marshside: tel: 01704 226190

252 PENDLE HILL (Lancashire)

OS Landranger 103

Lying on the south-east flanks of the Ribble Valley, this is a traditional stopover for migrant Dotterels in spring, and is best visited in mid-April to mid-May.

Habitat

The lower slopes of the hill are occupied by pastures bounded by dry-stone walls or hedges. These grade into rocky, bracken-covered slopes on the upper hill, with sparse grass and heather at the summit. Predictably, numerous streams run off the hill, providing variety.

Species

Trips of Dotterels may be present from mid-April, but the largest numbers are likely to be seen in the first two weeks of May, prior to departure for the breeding grounds (recent counts have been as high as 32, but trips are usually rather smaller). Note that, although regular, the species is far from guaranteed. Other migrants may include the larger and brighter Greenland subspecies of Wheatear, Whinchat and Ring Ouzel, along with Golden Plover and Curlew. Visiting raptors may include Hen Harrier, Merlin and Short-eared Owl (but all are thin on the ground here).

Breeding species include small numbers of Peregrines, Red Grouse and Golden Plovers, with Wheatear on the rocky slopes and around dry-stone walls, Common Redstart in areas of scattered bushes and trees, and Common Snipe and Common Redshank in wetter pastures (sadly, both Ring Ouzel and Twite have been lost as breeding birds). Look for Grey Wagtail and Dipper along streams.

Access

Leave the M65 at junction 13 north on the A682 at Barrowford and, after 0.75 miles, turn north-west on minor roads to Barley. Continue through the village on the minor road towards Downham and, after a further 1 mile, there is a popular pull-in and a public footpath follows a track west to Pendleside and Pendle House and then a steep, stepped path ascends to the summit. A less strenuous alternative, which takes about the same length of time, is to take the 'landslide trail', which veers south, or left as you look at the hill from Pendleside, up the hill. It is considerably less steep than the dreaded steps. This path takes you past an area of large fallen rocks, which are frequented by Wheatears in spring and autumn and have also produced several Black Redstarts. The fields below the landslide can be particularly good for Ring Ouzel in spring. When you reach the plateau of the hill, turn right and walk towards the summit area. You can either return via the steps or back down the 'landslide trail'. There are also many footpaths around the lower slopes. Dotterel often occurs on spring passage and occasionally in autumn as well on the summit plateau, particularly around the area by the trig point but also anywhere on the short grass from here to the north-west slopes of the hill. The area is popular with walkers, especially at weekends, so an early start is advisable but the sought-after birds, Dotterel and Snow Bunting, can often be confiding despite the hordes of brightly-coloured hikers.

FURTHER INFORMATION

Grid ref: SD 814 416

253 FOREST OF BOWLAND (Lancashire)

OS Landranger 41

A vast area of uplands between Lancaster and Clitheroe, this area is important for moorland birds, and is the stronghold for England's beleaguered Hen Harriers.

Habitat

The area consists of a mosaic of unimproved pastures, heather moorland, blanket bog and forestry.

Species

The extensive heather moorland of Bowland holds resident Red Grouse, Hen Harrier, Peregrine, Merlin, Short-eared Owl and Raven, with Dipper and Grey Wagtail along the streams. These are joined in summer by Common Sandpiper, Curlew, Oystercatcher, Cuckoo, Tree Pipit, Stonechat, Wheatear, and there are still a few Ring Ouzels in the craggier areas, while the similarly declining Whinchat remains here in good numbers. Breeding species in the small patches of deciduous woodland include Common Redstart, Spotted Flycatcher and occasionally also Wood Warbler. The conifer plantations hold Siskin and Lesser Redpoll, resident Long-eared Owls, and good numbers of Common Crossbills during invasion years. The delightful old stone barns of Bowland provide nest sites for Little, Tawny and Barn Owls, the latter enjoying a recent increase in numbers in this area. A speciality is Eagle Owl. Presumed to be of captive origin, a pair has taken up residence in the Dunsop Valley, breeding successfully since 2007.

Dotterel is occasionally reported on spring passage at Langden Head (SD5851), which is best approached from Langden Valley but it is a 7-mile hike, the last mile or so being over boggy moorland. In winter the area is much quieter, but Hen Harriers can still be found, and the remaining breeding birds are joined by birds from further afield, together with a few Merlins, Peregrines and Short-eared Owls.

Access

Note that bad weather is possible year-round and you should be adequately prepared if you wander far from the roads, with appropriate clothing, footwear, a compass and the relevant OS map.

Dunsop Valley Running at first through forestry, the upper reaches of the valley are open, with some rocky areas. From the car park in Dunsop Bridge follow the footpath from just east of the bridge, northwards up the valley; this crosses the river after 0.5 miles to join the main road (note that there is strictly no public vehicular access up the Dunsop Valley). Follow the road up the valley through forestry, which is easy going on the flat, until it forks to Whitendale and Brennand Farms.

Good spots to scan for raptors are the 'Witcher Well' (spring), to the west of the road at SD 651 521 and looking east from the road at the start of Whitendale (SD 655 535). To look for

Eagle Owl scan the rocky areas on the heather-clad hillsides on the east side of Whitendale below the 'stone man' (a 3-mile walk each way with a short climb at the end). A walk along the public footpath to the head of the valley (another 2.5 miles) should be rewarded in spring and summer by sightings of Whinchat and Ring Ouzel.

Langden Valley A steep-sided valley with deciduous woodland, heather moorland and rocky outcrops. Park in the parking area by the entrance to Langden Intake, on the minor road between Dunsop Bridge and Abbeystead (SD 631 512) and take the footpath westwards past the water company buildings and up the valley; initially passing through larches (check for Common Crossbill), the valley soon opens out. Follow the footpath along the valley (there is a fork where the paths split into a higher or lower option after a mile but they join again in less than another mile). A good vantage point is the hillside just north-west of Langden Castle, which is actually more of a sheep shelter than a castle (SD 604 504), where keen local raptor watchers gather in spring to watch the sky-dancing display of Hen Harriers from a safe distance.

FURTHER INFORMATION

Grid refs: Dunsop Valley SD 660 501; Langden Valley SD 631 512

254 STOCKS RESERVOIR (Lancashire)

OS Landranger 41

Lying on the eastern flanks of the Forest of Bowland, Stocks covers 192 ha and was completed in 1932. It is around 2 miles long and up to 0.75 miles wide, and is owned by United Utilities. This large isolated body of water, lying at 590 feet above sea level, is the premier birding site in east Lancashire.

Habitat

The reservoir has natural banks and the catchment for Stocks Reservoir, also owned by United Utilities, is divided between three farms with upland pastures and moorland, and the plantations of Gisburn Forest, 1,245 ha, which is leased to the Forestry Commission and consists largely of plantations with some enclaves of native woodland.

Species

On passage Osprey is regular, especially in the spring (the first three weeks in April are by far the best for this species), as are Common Scoter, Common, Arctic and Black Terns and Little Gull, all of which pause at Stocks while taking a short cut across the Pennines to and from the Irish Sea. A variety of waders occurs in small numbers.

Breeding birds include Red-breasted Merganser, over 1,000 pairs of Black-headed Gulls, which have been joined by a few pairs of Mediterranean Gulls, and Little Ringed Plover. The woodland and rough pasture around the reservoir can also be productive with Green Woodpecker, Siskin and Lesser Redpoll being joined by Cuckoo (Bowland is one of Lancashire's last strongholds of this declining bird), Tree Pipit, Common Redstart, Grasshopper, Willow and Garden Warblers and Pied Flycatcher in spring and occasionally Great Grey Shrike in winter. Common Crossbills are present in invasion years and occasionally stay to breed.

In winter there is a large gull roost, mainly Black-headed and Common Gulls, as well as an occasionally very impressive roost of thousands of Starlings, which can be seen swirling around the reservoir at dusk, often attracting Peregrines and Sparrowhawks. The reservoir can also hold good numbers of waterfowl in winter with up to 600 Wigeons, 300 Pintails and 500 Teals (the latter flock is now responsible for several records of Green-winged Teal).

Access

Leave the B6478 at the Stephen Moor crossroads, between Tosside and Slaidburn, on a minor road to the northern end of the reservoir and park after 1.5 miles at the School Lane car park. Take the trail from the gate at the north-west end of the car park and fork left to the first hide; the main footpath continues to a second hide after about 0.5 miles. There is an 8-mile circular walk around the reservoir, connecting with the extensive trail system in Gisburn Forest.

FURTHER INFORMATION

Grid ref: SD 732 564

255 HEYSHAM (Lancashire)

OS Landranger 96, 97 and 102

Heysham Harbour is one of the few sites in Britain producing Leach's Petrel following south-west to west gales (west to north-west elsewhere). However, the main emphasis of seawatching, as discovered relatively recently, is the spring passage of seabirds, especially Arctic Tern and skuas, although numbers of the latter are fewer than on the Solway.

Habitat

The two nuclear power stations are surrounded by stands of cover, and the north-east sector of the compound has been made into a reserve with a mosaic of habitats, including wetland. Offshore, the warm-water outflows can be particularly attractive to gulls and terns.

Species

On spring passage in mid-April to mid-May numbers of seabirds, particularly Arctic Tern, can be seen moving north; early mornings are best, preferably with a rising tide and an east wind. Red-throated Diver, Gannet, Kittiwake and skuas are also seen in variable numbers in this period. April–August gales can produce Fulmar and more terns.

In September–October south-west to west-north-west gales (especially after strong west winds) should produce Leach's Petrel (on the second day of the storm and thereafter), as well as Arctic and some Great Skuas. Other seabirds are unlikely; shearwaters are rare, as is Gannet (Leach's Petrel is often commoner than Gannet in autumn). The north harbour wall is the best place to watch from, although petrels should be visible, albeit more distantly, from the outfall hide. The warm-water outflows attract gulls and terns, including Mediterranean and Little Gulls, and Black and Arctic Terns.

The power station lights act like a huge lighthouse. Sizeable falls of passerines occur in spring and autumn, and have included scarce migrants; indeed, Heysham boasts a very impressive list for a west coast site, including remarkable numbers of Yellow-browed Warblers. The best conditions are either south-east winds around the west flank of an anticyclone, or south-east winds ahead of a warm front with the edge of the cloud cover coinciding with dawn.

In winter small numbers of Purple Sandpipers are occasionally found (though the large and once-regular flocks have now abandoned the area), and a variety of other common waders occurs; in recent winters a spectacular high-tide roost of Knots has appeared on the helipad at a viewing range of c. 50 m. Up to 30,000 have occurred in November–February. Winter gales may produce Little Gull and Kittiwake, and oddities such as Great and Pomarine Skuas or Little Auk may be caught up in such movements.

Access

Heysham Harbour Straightforward from the A589. On entering Heysham turn right at the

Moneyclose Inn traffic lights, and after about 0.25 miles turn left at the T-junction (by the helipad). Almost immediately turn right and then follow the road, which eventually narrows to a 'private' road, to the tip of the breakwater. This is the best seawatching station. Waders roost on the helipad.

Heysham Nature Reserve Enter Heysham on the A589 and turn left at the Moneyclose Inn traffic lights into Moneyclose Lane. After 250 yards, turn right to the reserve car park. Any bushes around the power station can hold birds, especially those by the old observation tower, along Moneyclose Lane, and bordering the golf course and caravan site (please ask permission at reception before entering the caravan site). The reserve is open at all times, with vehicle access to the car park possible at least 10am–6pm (can be longer in summer and shorter in winter). The reserve is managed by the Wildlife Trust for Lancashire, Manchester and North Merseyside, and is also the base for Heysham Bird Observatory, which shares an office at the car park with the Wildlife Trust and can provide details of the latest sightings.

Red Nab Follow directions as for the nature reserve but park carefully at the Ocean Edge caravan site car park. Proceed to the shore on foot, turning right and following the sea wall to the outfalls. Waders roost on Red Nab (although not on the highest tides). The outfalls are best 2–3 hours before high water, especially during onshore winds, but may be birdless around high tide.

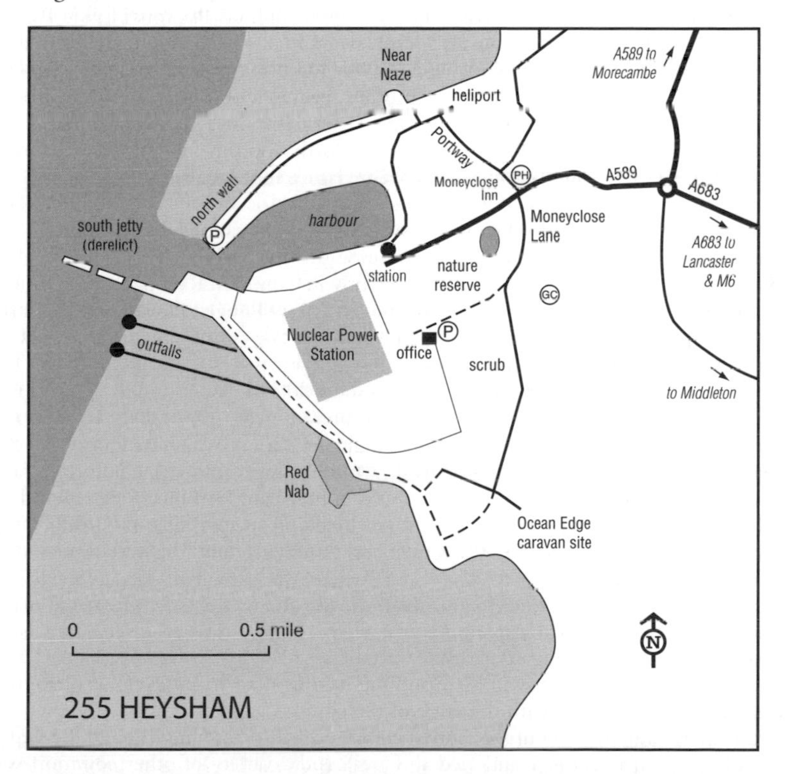

FURTHER INFORMATION

Grid refs: Heysham Harbour SD 396 600; Heysham Nature Reserve SD 406 598; Red Nab SD 407 590

Wildlife Trust for Lancashire, Manchester and North Merseyside, Heysham Nature Reserve Manager: tel: 07979 652138; email: rneville@lancswt.org.uk; web: www.lancswt.org.uk

Heysham Bird Observatory: web: www.heyshamobservatory.blogspot.com

256 MORECAMBE BAY (Lancashire and Cumbria)

OS Landranger 96, 97 and 102

Morecambe Bay is the largest and, for birds, the most important intertidal area in Britain, attracting up to 200,000 waders in winter. It is of international importance for Shelduck, Wigeon, Pintail, and nine species of waders, particularly Knot. The best time to visit the bay is in winter and during passage periods. The RSPB manages a large reserve between Silverdale and Hest Bank.

Habitat

Morecambe Bay covers about 120 square miles of tidal mud and sand, and is a complex of five estuaries, those of the Rivers Wyre, Lune, Keer, Kent and Leven. There are large mussel beds and the flats are fringed by heavily grazed saltmarsh, especially in the upper reaches of the rivers.

Species

In winter small numbers of Red-throated Divers appear offshore. Raptors include Peregrine, Merlin, Short-eared Owl and occasionally Hen Harrier. Pink-footed Goose is regular, feeding inland and roosting in the Bay. As many as 1,000–4,000 use sandbanks in the Lune, and they are joined by small numbers of geese of other species, with European White-front and Barnacle perhaps the most regular. The Pilling/Cockerham area is also favoured. Greylag Geese concentrate on the RSPB reserve at Carnforth, roosting on the Keer Estuary or at Leighton Moss; a separate population of Greylags frequents the Lune Estuary. Other wildfowl include Pintail and a few Shovelers. Seaducks occur in small numbers, Scaup, Goldeneye and Red-breasted Merganser sometimes being joined by Common Scoter, Common Eider or Long-tailed Duck. The commonest wader is Knot, which can total more than 50,000. Oystercatcher, Curlew, Bar-tailed Godwit, Common Redshank and Dunlin are also numerous, with small numbers of Turnstones and a few Ringed Plovers and Sanderlings. Purple Sandpiper is regular at Heysham and Fleetwood. The saltmarshes attract Rock Pipit, but Snow Bunting and Twite are uncommon.

Passage periods produce more waders but fewer wildfowl. Most waders are commoner in autumn, from July, than in spring. Exceptions are Knot (up to 70,000), Ringed Plover and Sanderling. Autumn passage also produces a greater variety, which may include Curlew Sandpiper, Little Stint and Spotted Redshank. Other passage visitors include terns, and in autumn the right conditions can produce seabirds at Heysham Harbour or Fleetwood.

Small numbers of waders oversummer, when breeding species include Shelduck, Red-breasted Merganser and Shoveler. Oystercatcher, Redshank and Lapwing breed on the saltmarshes, together with Ringed Plover and Avocet; the latter two species are local but Avocets can be seen in spring and summer from the Allen and Eric Morecambe Hides. Wheatear breeds at Carnforth slag tips, but the commonest birds around the estuary are Meadow Pipit and Skylark.

Access

It is important to note the state of the tide, as the sea can be up to 7 miles distant at low water. Fortnightly 'spring' tides are generally best and predictions for Liverpool are more or less correct. Access is at the following points.

WYRE ESTUARY AREA (OS Landranger 102)

Attractions include seawatching at Rossall Point, Fleetwood, and concentrations of Icelandic Black-tailed Godwits on passage.

Rossall Point, Fleetwood Lying at the west end of the promenade near the coastguard station, the Point occasionally attracts a few Purple Sandpipers as well as Turnstone, which may number 500 by spring. There are often Common Scoters and Red-breasted Mergansers offshore, sometimes Red-throated Diver, and occasionally Snow Bunting on the foreshore. In September–October westerly or south-westerly gales push seabirds inshore, notably Leach's Petrel (see Heysham, page 388, for details). The coastguard tower at Rossall Point is open to Fylde Bird Club members for scawatching.

Knott End-on-Sea Leave the A588 on the B5377/B5270 to Knott End-on-Sea from where you can walk east along the shore for 2.5 miles to Fluke Hall; the saltmarsh holds wintering Twites, and there is a high-tide roost of Sandwich Terns in the autumn.

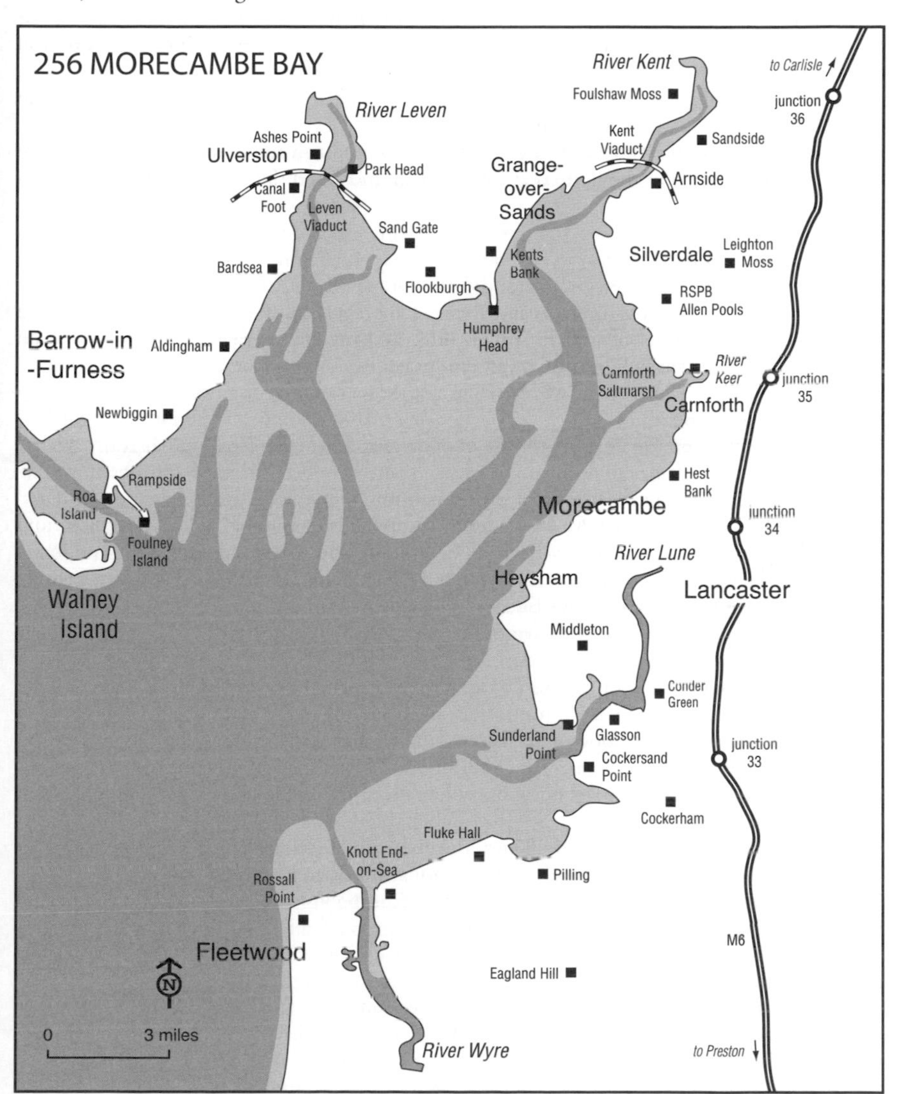

LUNE ESTUARY AREA (OS Landranger 102)

As well as the usual wildfowl and waders, this area is excellent for Pink-footed Goose, which feeds on farmland at Cockerham Moss and roosts on Pilling Sands.

Pilling The wader roost is best seen from Lane Ends car park, on the sea wall just off the A588 north of the village. It is also worth visiting Fluke Hall on a rising tide, by turning off to Pilling from the A588 and following signs in the village to the car park on the shore.

Cockerham–Pilling–Eagland Hill The low fields in this area are good for Pink-footed Goose and Whooper Swan, especially in mid-February to mid-March. Numbers vary between 500 and 11,000, and the flocks often include family parties of other geese. It is necessary to drive the maze of lanes to find the geese.

Cockersand Point Half a mile north of Cockerham on the A588, a lane leads west to Bank End Farm, from where a path follows the coast north-west for about 2 miles to Cockersand Point. Alternatively, leave the A588 at Thurnham on the minor road to Cockersand Abbey, just a short distance from the Point. There is a wader roost on lower tides, seaducks offshore, and late-summer congregations of terns.

Conder Green Turn west off the A588 in Conder Green at the Stork Inn and park by the disused railway embankment in the shore car park.

Glasson Dock Take the B5290 off the A588 at Conder Green to Glasson Dock. Waders are visible from the road over the disused railway embankment and the area is productive during passage periods.

Middleton Salt Marsh Follow signs from Heysham south on minor roads to Middleton, turning right at the first junction in the village, signed Middleton Sands. Park at Potts Corner from where there are good views of the sands and a footpath runs south-east through the marsh. There is a large wader roost (this site and Pilling Lane Ends are the best places to guarantee large numbers of a variety of waders at high tide), and late-summer concentrations of terns. Sunderland Point is worth checking for migrant passerines in the right conditions.

KEER ESTUARY AREA (OS Landranger 97)

Morecambe Promenade Good for waders at low water, on neap tides waders roost around the jetty, while on higher tides numbers of smaller species roost on the new groynes along the seafront north to Teal Bay. For seaducks watch from the Stone Jetty (just behind the Midland Hotel); Red-breasted Merganser, Goldeneye, Scaup and Common Eider is regular, and Long-tailed Duck and Common Scoter also occur, along with Great Crested Grebe and Red-throated Diver; all are best seen on calm days 2–3 hours before high water. Gulls can include Mediterranean. A similar range of seabirds as at Heysham Harbour may occur following gales, though the area is not worthwhile if winds are west-north-west or from further north (rather than from the west or west-south-west).

RSPB Hest Bank At the south tip of Morecambe Bay and once the best place to see large numbers of waders, but saltmarsh erosion has taken its toll and now only small numbers of Curlews and Oystercatchers roost, with most of the smaller waders now using the groynes on Morecambe seafront. Wildfowl and raptors are now the area's principal attractions. Cross the railway at the level crossing at Hest Bank signal box on the A5105 (plenty of parking space) and walk north for good views.

Carnforth Saltmarsh Reached by taking the footpath along the north bank of the River Keer near Cote Stones Farm. Leave the A6 into Carnforth and take the minor road past the station towards Silverdale. After about 1 mile turn left at the junction and at a sharp right-hand bend take the road straight ahead to Cote Stones, forking left after 100 yards to the riverbank. The old slag heaps offer good vantage points. The *Juncus*-dominated inner saltmarsh is good for Jack Snipe, there is a wader roost, and the slag banks may attract Twite and Snow Bunting, especially in cold weather.

RSPB Eric Morecambe and Allen Pools Two lagoons have been constructed on the saltings of the RSPB reserve, overlooked by hides on the sea wall. This is a prime spot to watch waders and a good area for Greylag Goose. Several hundred pairs of Black-headed Gulls breed (and Mediterranean Gull bred in 1997), and migrants have included Garganey, Mediterranean and Little Gulls, and Black Tern, with Marsh Harrier wandering from nearby Leighton Moss (see page 394). Leave the Carnforth–Silverdale road at Crag Foot (just south of Leighton Moss), passing under the railway bridge to the car park. Hides are open 9am–9pm (or dusk when earlier).

KENT ESTUARY AREA (OS Landranger 96 and 97)

Notable are flocks of Greylag Geese that frequent the upper estuary, especially the mosses at Foulshaw and Brogden, and Meathop Marsh, between the Kent Viaduct and Holme Island, which are sometimes joined by small numbers of White-fronted, Pink-footed or Barnacle Geese. Diving ducks favour the deeper channel around Kent Viaduct, and the limestone whaleback of Humphrey Head provides shelter in west and north-west storms, and in such conditions seabirds may occasionally be seen from Kents Bank Railway Station.

Upper Kent Estuary, east shore The south side of the upper Kent Estuary can be viewed either from the east end of Arnside promenade, looking over the railway wall, or from Sandside promenade, 1.5 miles north of Arnside. Waders roost but not on the highest tides. This area is good during passage and there is a large gull roost, which is the most north-westerly site for multiple records of Mediterranean Gull, especially in July–August, and sometimes attracts scarcer species such as Glaucous Gull.

Upper Kent Estuary, west shore Leave the A590 south at the Derby Arms, Witherslack, on the minor road to Ulpha. After 2 miles take the footpath towards Sampool Bridge and view Brogden and Foulshaw Mosses from the sluice at Crag Wood.

Lower Kent Estuary The lower Kent Estuary can be viewed from the footpath to New Barns and Blackstone Point from the west end of Arnside promenade. There is a wader roost, except on higher tides.

Kents Bank Railway Station A good vantage point, with large numbers of Pintails and Shelducks, and the possibility of seaducks, grebes and divers in rough weather.

Humphrey Head Follow signs from Flookburgh, off the B5277. The Head is a reserve of the Cumbria Wildlife Trust and is a good site to observe visible migration, while a number of scarce migrants have occurred in recent years.

LEVEN ESTUARY AREA (OS Landranger 96 and 97)

The extensive saltmarshes at Flookburgh are good for waders, raptors, gulls, and sometimes grey geese. The Leven Estuary has the usual mixture of wildfowl and waders, occasionally including Whooper Swan and grey geese.

Flookburgh Marshes One of the most important areas for roosting waders, which favour West Plain, but may also use East Plain (note that the wader roost at Out Marsh has declined in importance during recent years). West Plain can be viewed by walking west from West Plain Farm (south on a minor road from Flookburgh to Cark airfield) on the Cumbrian Coastal Path to Cowpren Point. High 'spring' tides are best.

Sand Gate Marsh At Flookburgh, off the B5277, proceed west along the main street of the village to a vantage point on the shore beyond Sand Gate Farm to view Sand Gate Marsh. The area has a wader roost on higher tides and is productive for wildfowl, especially on a rising tide, with a large gull roost.

Leven Estuary, east shore The east side of the Leven can be reached from the B5278 1.5 miles north of Flookburgh, along the minor road to Old Park and the Park Head car park on the shore. Walking north from here along the road, the viaduct just south of Low Firth offers good views over the estuary.

Leven Estuary, west shore The Leven Estuary is best viewed from the west shore. From the A590 in Ulverston take the minor road east, past the entrance to the Glaxo chemical works, to the Bay Horse Inn at Canal Foot to view. More extensive areas can be accessed by taking a minor road from the A590 at Newland (just north of Ulverston) to Plumpton Hall. Walk north along the shore from here, under the railway and on towards Ashes Point. There is a wader roost and it is a productive area during passage.

Bardsea to Roa Island The A5087 parallels the coast from Bardsea to near Roa Island. Waders roost at Bardsea (a track runs north-east off the main road to Wadhead Scar, just north of the turn-off to the village; distant views can also be had of Chapel Island, which holds roosting wildfowl and waders), Aldingham (take the minor road into the village to view the shore) and Newbiggin (visible from the main road). Seaducks occur off Bardsea, including Red-breasted Merganser, Goldeneye and Scaup, as well as Red-throated Diver.

Foulney Island From the A5087 at Rampside take the minor road to Roa Island and, about halfway along the causeway to the island, park and follow the track south-east over the granite-block causeway to Foulney Island, which is a Cumbria Wildlife Trust reserve. The island is a good watchpoint for wildfowl (numbers of Common Eiders and sometimes other seaducks) and waders, and, in bad weather, also seabirds, while passerine migrants also occur. In winter Twite and occasionally Snow Bunting may be found along the shore, with raptors over the rough ground. In summer, numbers of Common, Arctic, Sandwich and Little Terns and Black-headed Gulls breed at the Slitch Ridge at the south tip.
Walney Island See page 396.

FURTHER INFORMATION

Grid refs: Rossall Point, Fleetwood SD 311 476; Knott End-on-Sea SD 351 485; Cockersand Point (Bank End Farm) SD 441 528; Conder Green SD 457 562; Glasson Dock SD 446 560; Middleton Salt Marsh SD 413 571; Morecambe Promenade SD 430 644; RSPB Hest Bank SD 468 666; Carnforth Saltmarsh SD 489 716; RSPB Eric Morecambe and Allen Pools SD 475 737; Upper Kent Estuary, east shore (Arnside promenade) SD 459 789; Upper Kent Estuary, west shore SD 449 811; Lower Kent Estuary SD 453 786; Kents Bank Railway Station SD 396 755; Flookburgh Marshes SD 368 742; Sand Gate Marsh SD 355 757; Leven Estuary, east shore SD 336 787; Leven Estuary, west shore SD 313 777; Bardsea (Wadhead Scar) SD 308 745; Foulney Island SD 233 655
RSPB Morecambe Bay: tel: 01524 701601; email: leighton.moss@rspb.org.uk
Fylde Bird Club: www.fyldebirdclub.org

257 LEIGHTON MOSS (Lancashire)

OS Landranger 97

Lying close to the north-east corner of Morecambe Bay, this is one of the RSPB's premier reserves. Its attractions include breeding Marsh Harrier and Bearded Tit, and Bittern still clings on here as a breeding bird (just one booming male in 2009), although the population is supplemented by birds from elsewhere during the winter, when a good variety of wildfowl is also present. A visit is worthwhile at any time, other than when the Moss is completely frozen in cold weather, but to see Bittern, May–June or winter, when hard frosts may force them to feed in the open, are probably best.

Habitat

Once an arm of the sea, Leighton Moss was embanked, drained and ploughed, but then allowed to re flood in 1917. It is now a freshwater marsh, extensively overgrown with reeds, willow and alder carr, with large areas of open water and specially constructed scrapes and islands.

Species

Winter brings large numbers of ducks, including Teal, Wigeon, Pintail, Shoveler, Pochard, increasing numbers of Gadwalls and a few Goldeneyes and Goosanders. Feral Greylag Goose is resident, and other (probably wild) birds also frequent the adjoining saltmarsh; Whooper Swan and other goose species are often seen on passage in late autumn and early winter, regularly including Barnacle Geese. Bittern is present, though difficult to see, as is Water Rail, but both may become bolder if the pools freeze over. Water Pipit is occasional around the muddy edges, with Siskin and Lesser and Common Redpolls in the carr, and Hen Harrier may visit, joining resident Sparrowhawk, Common Buzzard and occasional Barn Owl. Hawfinch is resident in Silverdale, and with patience and luck can be seen around Woodwell. Other woodland birds include Woodcock, all three woodpeckers, Nuthatch and Marsh Tit; the last two are regular at the feeders near the reserve centre. There was an influx of Cetti's Warblers in 2009 but it remains to be seen whether or not the species will take up permanent residence.

Breeding species include Bittern, Marsh Harrier, Common Buzzard, Teal, Shoveler, Gadwall, Pochard, Oystercatcher, Black-headed Gull and one of Lancashire's very few pairs of Great Black-backed Gulls, and occasionally also Garganey, while Spotted Crake has bred twice. Little Egret is now common, especially in autumn when they can be seen by day feeding on the pools of the Eric Morecambe complex and the dykes of the adjacent grazing marsh or at dusk coming to roost in the trees of the Island Mere, viewable from the public hide on the causeway or the road to Yealand Redmayne (although parking spots along this road are very limited). This roost now regularly exceeds 50 birds and has occasionally attracted both Cattle and Great White Egrets. The reedbeds hold Bearded Tit, which may be seen coming to special 'grit trays' on the causeway, especially in September–October. The reeds also hold Reed Warbler in summer, with Sedge and occasionally Grasshopper Warblers in scrubbier areas.

A fortunate visitor may find a Jack Snipe in early spring or late autumn, feeding quietly along the edge of the reeds, or perhaps a Garganey. Osprey and Spoonbill are also regular on passage. Little Gull and Black Tern are more frequent in spring, but late summer and autumn is the best period for passage waders, which can include Common and Green Sandpipers, Spotted Redshank and more than 1,000 Icelandic Black-tailed Godwits. There is also a regular autumn gathering of up to 40 Greenshanks. The autumn roost of up to 100,000 Starlings attracts hunting Sparrowhawks, and sometimes a Merlin or Peregrine. Hobby is now a regular summer visitor taking advantage of the large supply of dragonflies.

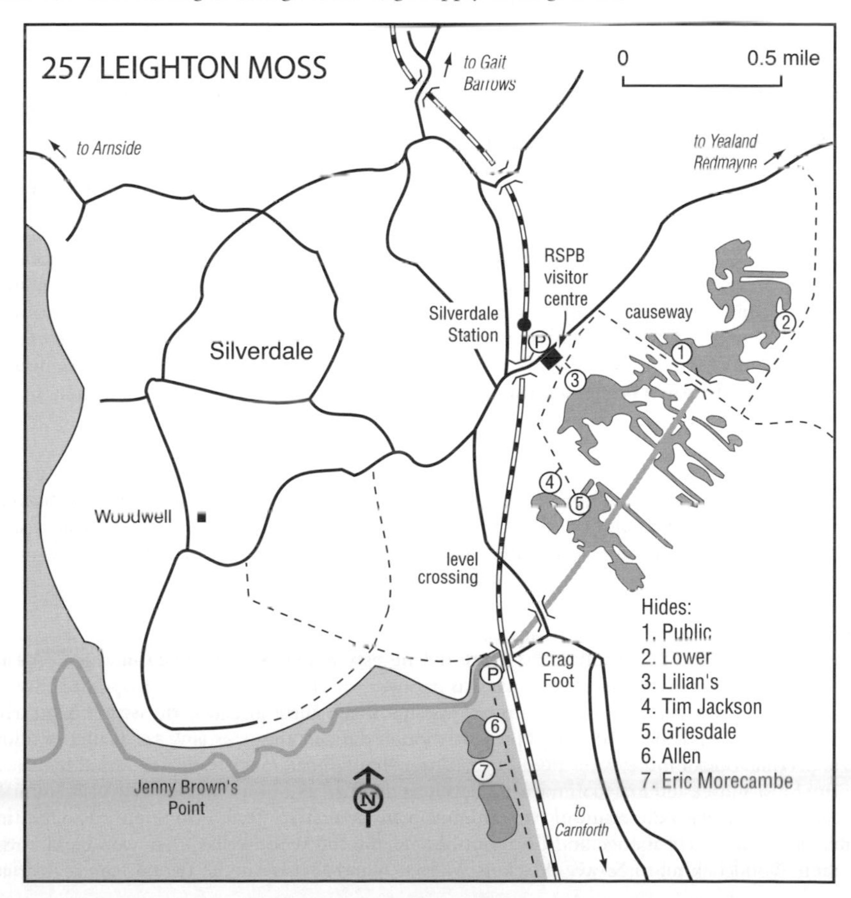

Access

Leave the M6 at junction 35 (signposted Carnforth), then follow the A6 north (signposted to Milnthorpe). Brown tourist signs direct you to Leighton Moss from the A6 through the villages of Yealand Redmayne and Yealand Storrs. The car park and visitor centre are well signed at Myers Farm, near Silverdale Station. Alternatively, coming from Carnforth, follow signs to Silverdale, passing through the village of Millhead. Shortly after entering Warton, turn left and follow the road until you come to a T-junction beyond the level crossing, where you turn right, then right again just before Silverdale Station. From the car park a causeway runs across the reserve, with a public hide that is always open. Otherwise the reserve, with four additional hides and three trails, is open daily (except Christmas Day) 9am–9pm (or dusk if earlier), and the visitor centre daily 9.30am–5pm (9.30am–4.30pm November–January). Four of the hides are wheelchair-accessible, as are some of the trails.

The north end of the RSPB's Morecambe Bay Reserve is adjacent to Leighton Moss; the Eric Morecambe and Allen Pools are especially convenient (see page 394). Nearby, Warton Crag Quarry (SD 491 724) offers a superb opportunity to watch breeding Peregrines and Ravens as well as Little Owl from the public car park, accessed off Crag Lane, the minor road between Warton and Crag Foot.

FURTHER INFORMATION

Grid ref: SD 478 750
RSPB Leighton Moss: Tel: 01524 701601; email: leighton.moss@rspb.org.uk

258 SOUTH WALNEY (Cumbria)

OS Landranger 96

South Walney is the south part of Walney Island, which forms the west flank of Morecambe Bay and thus naturally shares its huge population of waders. This position also concentrates migrants and has led to the establishment of a Bird Observatory. A few semi-rarities and rarities are recorded annually. The best times for migrants are April–early June and August–early November, but for wildfowl and waders winter is better. An area of 130 ha at South Walney is managed as a reserve by the Cumbria Wildlife Trust.

Habitat

There is a variety of habitats: low dunes and sandy beaches are interspersed with areas of gravel, Bracken, marsh, and brackish and fresh water, with one substantial patch of elder. The limited cover is concentrated around the five Heligoland traps and coastguard cottages. Saltmarsh and mudflats fringe the east side of South Walney, and the Spit protects the sheltered tidal basin of Lighthouse Bay and is used by large numbers of roosting waders.

Species

Large movements of diurnal migrants occur during passage periods, and in spring a few Marsh Harriers are usually recorded. Calm, overcast or hazy weather is most likely to produce a fall of night migrants. Among the commoner species, Black Redstart, Red-breasted Flycatcher, Yellow-browed and Melodious Warblers and Firecrest are annual. Seawatching during or after strong west winds in August–September can produce Fulmar, Gannet, Manx and sometimes Sooty Shearwaters and Leach's Petrel, while Great and Cory's Shearwaters are virtually annual. Another possibility is Pomarine Skua, though you are more likely to see Great and Arctic Skuas, and large numbers of Kittiwakes, Common, Arctic and Sandwich Terns, Common Guillemot, Razorbill and Common Scoter. Migrant waders usually include some of the scarcer species such as Curlew Sandpiper or Little Stint, especially in autumn.

Wintering species include Red-throated Diver on the sea, with up to 5,000 Common Eiders and 2,000 Common Scoters, as well as a handful of Velvet Scoters, Red-breasted Merganser, Scaup and Goldeneyes. There are also large numbers of Wigeons, Teals and Shelducks, as well as Whooper Swan, Greylag and Brent Geese, Pintail and Shoveler, and occasional visitors include Black-throated and Great Northern Divers, Red-necked Grebe and Long-tailed Duck. Merlin and Peregrine are daily visitors, and there are sometimes Barn and Short-eared Owls or Hen Harrier too. The whole of Walney Island attracts raptors, so keep a sharp look-out from the car. Little Auk is almost annual following north-west gales, and there have been spectacular hard-weather movements of wildfowl and passerines. Waders include up to 12,000 Oystercatchers, 12,000 Knots and 3,000 Dunlins, and significant numbers of other common species. The Pier Hide affords good views of the Spit, which is used by roosting waders.

Breeding birds include 1,200 pairs of Common Eiders, about 200 pairs of Sandwich Terns and a few each of Common, Arctic and Little Terns, 20,000 pairs of Herring Gulls, 30,000 pairs of Lesser Black-backed Gulls and about 60 pairs of Great Black-backs. There are small numbers of Shelducks, Oystercatchers and Ringed Plovers. Peregrine breeds in the area and is often seen.

Access

Cross the road bridge from Barrow-in-Furness to Walney Island, turn left at the traffic lights, and after 400 yards fork right by the King Alfred Hotel into Ocean Road. Turn left after about 0.5 miles into Carr Lane (signed for the caravan site) and continue for about 1.25 miles to Biggar

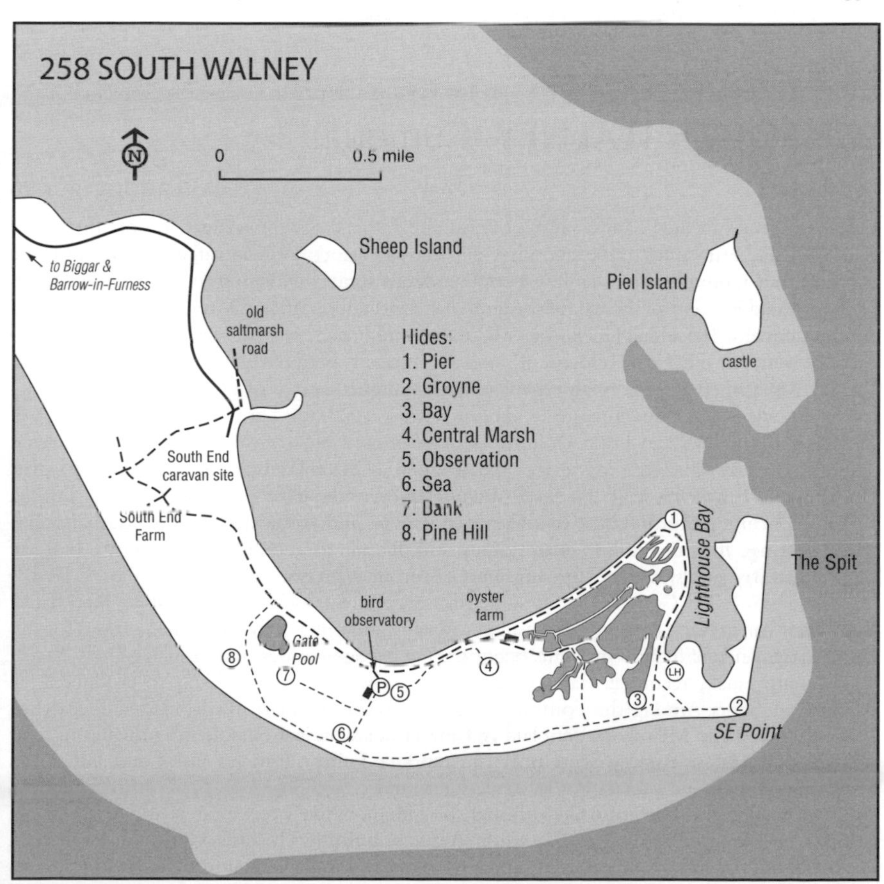

village and on past the rubbish tip. Just before the South End caravan site fork right, signed to the nature reserve, onto a rough track which cuts across the island past South End Farm to the reserve. South Walney Nature Reserve is open all year, 10am–5pm (or 4pm September–April). There is a permanent warden. There are two trails (3 and 2 miles long) and 8 hides – some of which face the Irish Sea and Morecambe Bay, excellent for seawatching in wet and windy weather. The bird observatory offers some accommodation – contact the secretary for details.

FURTHER INFORMATION

Grid ref: SD 215 620
Cumbria Wildlife Trust: web: www.cumbriawildlifetrust.org.uk
Walney Bird Observatory: web: http://walneybo.blogspot.com

259 HODBARROW (Cumbria)

OS Landranger 96

Lying north of the Duddon Estuary and Barrow-in-Furness, this RSPB reserve holds a ternery and numbers of wildfowl and waders, especially on passage.

Habitat

Formerly an industrial site, part of the area has been developed as a holiday complex, but the major portion is now a reserve, with a complex of lagoons and smaller pools, marshy areas, overgrown lime tips, slag banks and scrub.

Species

Breeding species include Sandwich, Common and Little Terns; the ternery lies on a specially created island in the south of the lagoon, which also has breeding Oystercatcher, Ringed Plover, Lapwing and Common Redshank. Other breeding species include Great Crested and Little Grebes, Shelduck, Red-breasted Merganser, Sparrowhawk, Stock Dove, Barn Owl, Lesser Whitethroat and Sedge and Grasshopper Warblers, and Peregrine may visit.

In late summer there is a build-up of moulting Red-breasted Mergansers, and numbers may peak at 200. Other possible migrants include Garganey, a variety of waders including Black-tailed Godwit, Ruff, Knot, Turnstone and, in some years, Little Stint and Curlew Sandpiper, as well as Black Tern and Little Gull. Areas of scrub and other cover can harbour passerine migrants, occasionally including scarcer species such as Black Redstart, Ring Ouzel and Pied Flycatcher. As at other west coast sites, White Wagtail is a feature of spring migration.

Winter brings small numbers of divers and grebes, and wildfowl may include Whooper Swan, Wigeon, Teal, Gadwall, Pochard, Scaup, Goldeneye and Long-tailed Duck, with divers and seaducks especially likely following winter storms. Waders include Golden Plover, Black-tailed and Bar-tailed Godwits, Sanderling, Turnstone and notably a few Spotted Redshanks and Greenshanks. Such concentrations attract raptors, and gull flocks should be perused for the occasional Glaucous, Iceland or Mediterranean.

Access

Leave the A5093 by Millom Station, taking Devonshire Road into the centre of Millom. Turn right into Mainsgate Road (signed to the reserve), and after about 0.5 miles turn left at the T-junction and the car park is just beyond the entrance to the rubbish tip. From here follow the track around the lagoon for 1.4 miles to the hide (a byway open to all traffic, it is possible to drive, but the track has many pot-holes and is very bumpy). The hide overlooks the ternery on the island, which is also used by roosting waders and wildfowl. In autumn and winter, visits timed to coincide with high tide are likely to be most productive, when waders arrive to roost. Alternatively the reserve may be accessed by car from Haverigg village. From the roundabout by the Harbour Hotel, take the road over the small river bridge and follow it towards the caravan park. Just before the caravan park gate, turn right to the sea wall. The reserve is open at all times.

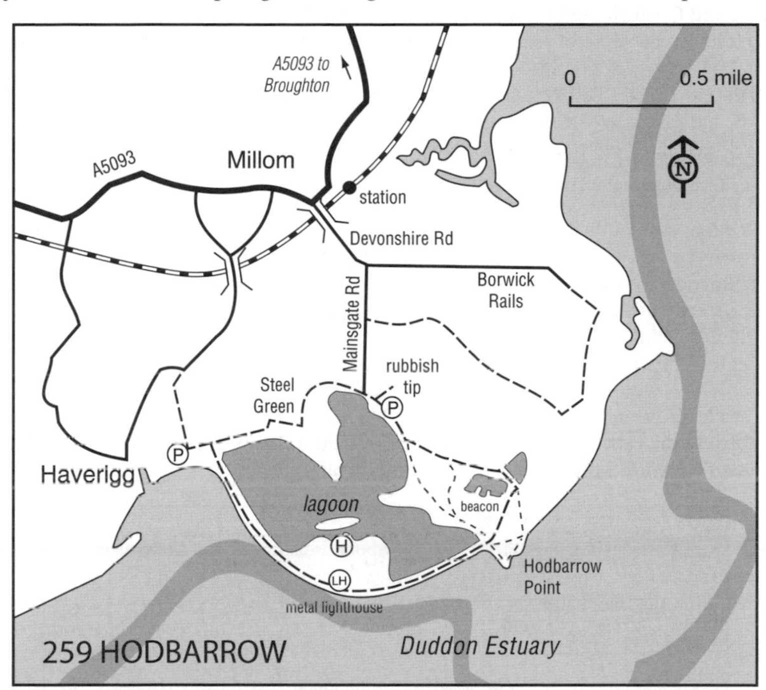

259 HODBARROW

FURTHER INFORMATION

Grid ref: SD 174 790
RSPB Hodbarrow: tel: 01697 351330; email: campfield.marsh@rspb.org.uk

260 ST BEES HEAD (Cumbria)

OS Landranger 89

To the west of the Lakeland fells, St Bees Head holds a colony of seabirds, including England's only breeding Black Guillemots. The Head is an RSPB reserve, and is best visited April–mid-July for breeding birds and in autumn for seawatching.

Habitat

Red sandstone cliffs rise to 300 feet and are topped by areas of gorse and bramble (especially at Fleswick Bay) and grassland, with large fields ringed by dry-stone walls in the hinterland.

Species

The 5,000 pairs of breeding seabirds include large numbers of Common Guillemots, but fewer Razorbills and only a few Black Guillemots and hard-to-see Puffins; Black Guillemot should be looked for around the base of the cliffs at Fleswick Bay. There are also Fulmar and Kittiwake, about 600 pairs of Herring Gulls, a single pair of Great Black-backed Gulls, and, in recent years, Cormorants (look back from North Head across Fleswick Bay to South Head to view the 100-strong colony). Other breeding species include Raven, Peregrine, Little Owl, Rock Pipit, Stonechat and Corn Bunting.

The Head can be good for seawatching, especially in strong south-west winds. Red-throated Diver and Manx and Sooty Shearwaters are possible, especially in late summer and autumn, as are Storm and Leach's Petrels, together with Gannet, Great and Arctic Skuas and Sandwich, Common and Arctic Terns. Small numbers of passerine migrants may occur along the cliffs and in the well-vegetated gully at Fleswick Bay.

Access

Leave Whitehaven south on the B5345. At the staggered crossroads in St Bees, where the B-road turns left, continue straight ahead on Abbey Road and then turn right after 400 yards at the T-junction into Beach Road. The car park is a further 0.5 miles. From there take the cliff path (steep and uneven in places) north over the metal footbridge at the north end of the promenade and continue for about 2.5 miles to the lighthouse at North Head (the path continues to Whitehaven). There are three observation points on North Head overlooking the seabird colonies, with the best views from Fleswick Bay north to the lighthouse. Alternative access is to leave Whitehaven south on the B5345, turning west after about 2 miles on minor roads to Sandwith. Park near the post office and walk 2 miles west, past Tarnflat Hall, along the private road to North Head. The reserve is open at all times.

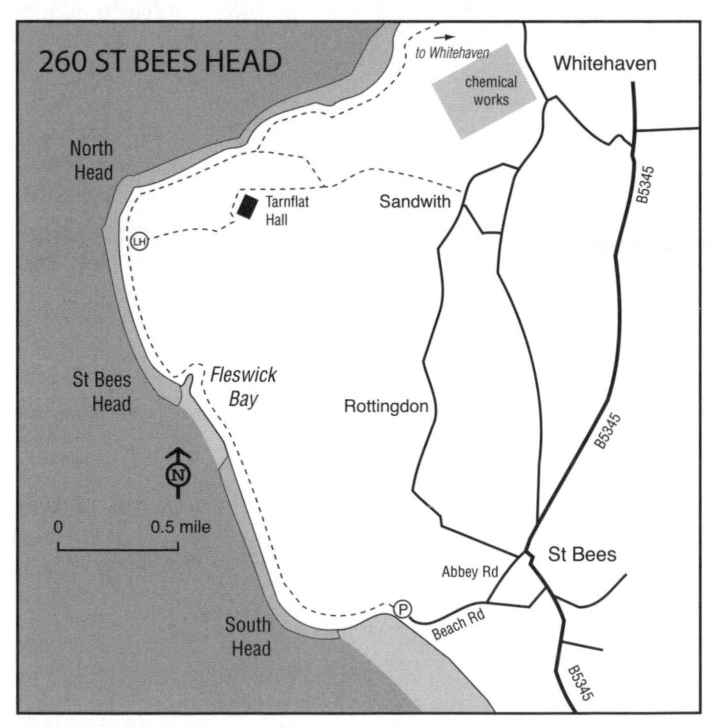

FURTHER INFORMATION

Grid ref: NX 960 118
RSPB St Bees Head: tel: 01697 351330; email: stbees.head@rspb.org.uk

261 BASSENTHWAITE LAKE OSPREY VIEWPOINT (Cumbria)

OS Landranger 89

Ospreys returned to England as a breeding bird in 2001, when a pair took up residence at Dodd Wood, adjacent to Bassenthwaite Lake, just north-west of Keswick. The birds were encouraged to nest with the help of a purpose-built nest provided by the Forestry Commission and the Lake District National Park Authority and public viewing facilities were laid on.

Habitat

This is the fourth largest lake in the Lake District but one of the shallowest. At the southern end of the lake there is an area of marsh, reeds and scrub known as Bassenthwaite Bog.

Species

As well as Ospreys, the lake attracts a range of wintering wildfowl. Bassenthwaite Bog holds Water Rail and Grasshopper Warbler, and Marsh Harrier occasionally visits.

Access

The Dodd Wood viewpoint is open April–mid August and situated about 3 miles north of Keswick off the A591, and the car park lies opposite the entrance to Mirehouse. From the car park the viewpoint is a 15–20 minute uphill walk. The lower viewpoint is open daily 10am–5pm, giving excellent views of Ospreys fishing over the lake. The upper viewpoint is 0.5 miles from the lower viewpoint, around 30 minutes walk via forest road and has spectacular views of the nest site 400m away. It is open 10.30am–4.30pm. Note that the Osprey Bus links Keswick with a round Bassenthwaite Lake public transport route taking in the Dodd Wood viewpoint.

There is a hide at the southern end of the lake, accessed from a lay-by on the A66.

FURTHER INFORMATION

Grid ref: NY 235 281
Lake District Osprey Project: www.ospreywatch.co.uk

262 HAWESWATER (Cumbria)

OS Landranger 90

This area of Lakeland fells and valleys around the scenically attractive Haweswater reservoir in the valley of Mardale is famous as the home of England's only breeding Golden Eagles, and also holds a range of other upland birds. A large part of the area is an RSPB reserve.

Habitat

The fells are bisected by rocky streams and crowned by rocky crags, while in the valley near the dam, Naddle Forest comprises steep woods of oak and mixed conifers. Haweswater itself is artificial, and is 4 miles long by 0.5 miles wide.

Species

The Golden Eagles occupy Riggindale Valley near the head of the reservoir and have bred here since 1969, though they have been unsuccessful in recent years and since 2004 just a single male is in residence; perhaps the presence of walkers on High Street, just above the nest site, causes too much disturbance. This lone male still displays through spring in the hope of attracting a new mate. As with any raptor, sightings are most likely in fine weather (but note that the late afternoon light can be poor). Peregrine, Raven and Ring Ouzel also breed among the high crags and rocky screes, with Wheatear around the boulders on the valley floors. Above the road, mixed oak, ash and birch woodland holds good populations of Sparrowhawk, Common Buzzard, Woodcock, Pied Flycatcher, Garden and Wood Warblers, Tree Pipit and Common Redstart. The reservoir edge and feeder streams are good for Dipper, Grey Wagtail, Common Sandpiper and Goosander, with Teal and Greylag Goose around the reservoir. The adjoining pine and larch woods hold Sparrowhawk, Siskin and, in some years, Common Crossbill is also seen. On the island at the south end of the reservoir is one of the few inland breeding colonies of Herring and Lesser Black-backed Gulls and there are also ground-nesting Cormorants (of the British subspecies *carbo*).

In winter there is a large gull roost (up to 12,000 birds, mainly Common and Black-headed Gulls), and Teal, Wigeon and Goldeneye are regular with occasional sawbills and divers.

Access

From junction 39 on the M6, take the A6 north and in Shap follow signs for the minor road to Haweswater and Bampton Grange. Bear left in Bampton Grange, past the church and over the bridge, for Burnbanks and Mardale. The route passes the dam and then a narrow road follows the south-east flank of Haweswater (with views of the lake and the adjoining fells en route), with a small car park at the road terminus at the south end of the reservoir. Access to the reserve is possible at all times along rights of way, and the eagle observation post, 1.25 miles from the car park, is open at all times but is only manned 11am–4pm, April–August, on Saturdays, Sundays and Bank Holidays. Naddle Forest is best accessed via the public footpath from the road above the dam. The area is very popular with tourists, thus weekends and especially Bank Holidays are busy.

FURTHER INFORMATION

Grid ref: NY 469 107
RSPB Haweswater: tel: 01931 713376; email: haweswater@rspb.org.uk

263 GELTSDALE (Cumbria)

OS Landranger 86

Lying in the north Pennines to the east of Cumbria, this RSPB reserve is a good area for Black Grouse, birds of prey and breeding waders, and is best visited April–early July.

Habitat

The RSPB reserve comprises a mixture of moorland, blanket bog and hill pastures, and includes Tindale Tarn and the landscape-scale Bruthwaite Pasture Woodland, a plantation of 100,000 native trees on the hillside above Stagsike.

Species

Red and Black Grouse are resident, but the small population of Black Grouse is usually rather elusive; late winter and early spring is probably the best time to see them. Other residents include Common Buzzard, Barn Owl and Raven, and in some seasons also some of the handful of Hen Harriers breeding in England (Geltsdale is one of only two areas that regularly support this much-persecuted species). Breeding summer visitors include Merlin, Golden Plover, Curlew, Common Snipe, Lapwing, Common Redshank, Short-eared Owl,

Cuckoo, Tree Pipit, Whinchat and Stonechat, with Ring Ouzel around the craggier areas, Common Redstart, Wood Warbler and Pied Flycatcher in the woodlands and Common Sandpiper, Grey Wagtail and Dipper on the streams.

In autumn and winter the moorlands are very quiet, but a variety of wildfowl may be seen on the tarn, including Teal, Goldeneye and Goosander, and sometimes also Whooper Swan or Smew.

Access

From the A69 near Brampton, take the A689 towards Hallbankgate and Alston. In Hallbankgate, fork right (south-east) on the minor road in front of the Belted Will pub and follow this to the reserve car park at Clesketts. The reserve is open at all times, with the information point at Stagsike cottages (around 40 minutes' walk from the car park) open 9am–5pm. There are four trails: the Stagsike Trail at 2.8 miles, the Bruthwaite Trail, 1.8 miles and the Moorland Trails, 2.5 miles and 5 miles. Wheelchair access can be prearranged to the information point, and will soon be possible to the Tarn Viewpoint.

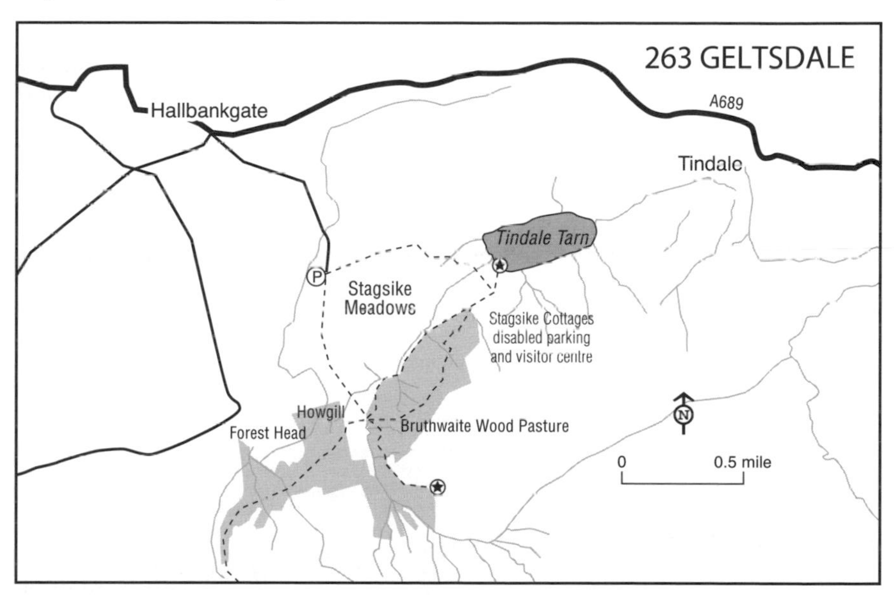

FURTHER INFORMATION

Grid ref: NY 588 584
RSPB Geltsdale: tel: 01484 861148; email: northernengland@rspb.org.uk

264 BOWNESS-ON-SOLWAY AND CAMPFIELD MARSH (Cumbria)

OS Landranger 85

The Solway is one of Britain's largest estuaries and is of international importance for both wildfowl and waders. The northern, Scottish, shore is perhaps the more famed ornithologically, but the Cumbrian shore is notable too, both for wildfowl and for the regular spring passage of Pomarine Skuas moving east along the Firth between mid-April and mid-May.

Habitat

The south shore of the estuary is bordered by extensive areas of saltmarsh both east and west of Bowness-on-Solway. Here, where the saltmarsh narrows (and terminates) at the narrowest point of the Firth, there was formerly a railway bridge, of which the remnants of its embankment, Herdhill Scar, now form a useful seawatching station. To the south of the coast there are areas of wet grassland and raised bog at Bowness Common.

Species

Wintering wildfowl include Whooper Swan, up to 10,000 Barnacle Geese and 15,000 Pink-footed Geese. North Plain Farm and Rogersceugh Farm hold good numbers of Barnacle Geese in December and January, with Pink-footed Goose peaking in February and March. The fields around Cardurnock, Anthorn or Newton Marsh are other favoured feeding grounds if geese cannot be found on the Campfield Marsh reserve. Other wildfowl include Greylag Geese and large numbers of Wigeons and Pintails. On the sea occasional Red-necked Grebes, Common Guillemots and Scaups are seen (high water being best). Waders include Lapwing and Golden Plover on the grazing marshes and Grey Plover, Dunlin, Knot, Bar-tailed Godwit, Turnstone and Sanderling on the shore. Large numbers of the commoner gulls roost on the estuary. The huge numbers of birds attract raptors, including Hen Harrier, Peregrine, Merlin and Short-eared Owl. Look for Twite and Snow Bunting along the tideline and Rock Pipit in the saltmarsh gutters, which may also hold Jack Snipe.

On spring passage Pomarine, Great and Arctic Skuas may be observed moving east over the estuary, and in autumn Arctic and Great Skuas and Kittiwake may seek shelter in the Firth, especially in stormy conditions. A variety of waders may appear, including Whimbrel, Greenshank, Green and Common Sandpipers, Little Ringed Plover and, most frequently in autumn, Wood Sandpiper, Spotted Redshank, Curlew Sandpiper and Little Stint. There is a notable build-up of Black-tailed Godwits at Bowness in August–September. Other notable migrants have included Garganey, Little Gull and Black Tern, and oddities such as Little Egret and Spoonbill have occurred.

Residents include Teal, Shoveler, Oystercatcher, Lapwing, Common Redshank, Curlew, Common Snipe, Barn Owl, Grey Wagtail and Stonechat. Ruff may oversummer and areas of scrub hold breeding Grasshopper Warbler and Lesser Whitethroat.

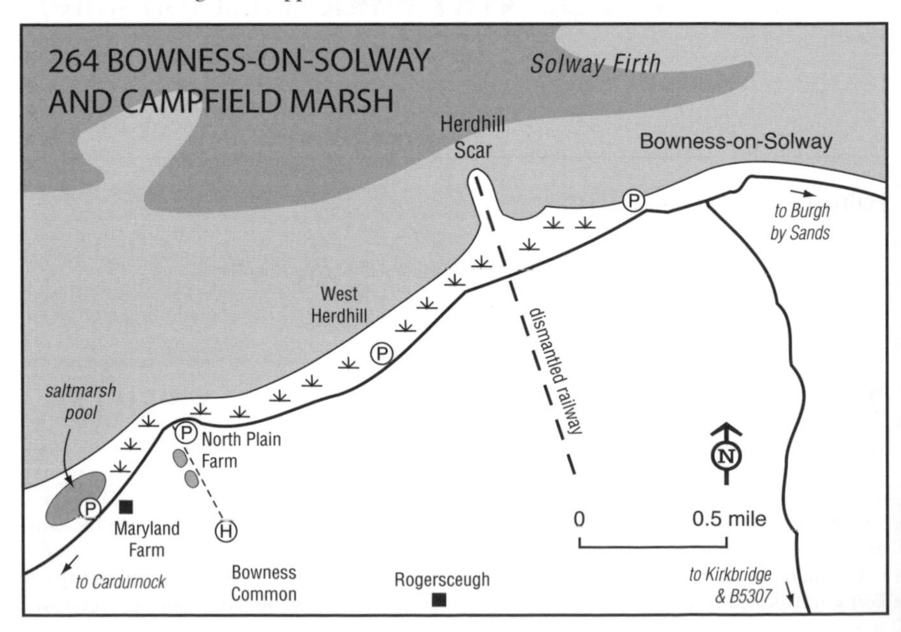

Access

Bowness-on-Solway Leave Carlisle west on the B5307 and, after about 13 miles, turn north near Kirkbride on the minor road to Bowness-on-Solway. In Bowness turn left at the T-junction where the road meets the coast and park after 0.25 miles at the RSPB car park to view the estuary.

Herdhill Scar From the RSPB car park walk west along the road for 0.5 miles to the point where the dismantled railway crosses the road. From here it is a scramble through gorse and over sandstone blocks for 600 yards to the tip of Herdhill Scar (it can be hazardous when wet or otherwise slippery; the Scar is not part of the RSPB reserve). This is the best spot to observe spring skua passage. Pomarine Skua is possible during the second half of April and the first two weeks of May. The best weather conditions are west winds, coupled with cloud and showers. The skuas fly east over the estuary and presumably continue overland to the North Sea.

RSPB Campfield Marsh: saltmarsh This holds the largest wader roost on the Solway. Follow directions to Bowness-on-Solway as above, turning left in Bowness along the coast road. Park in the lay-bys at West Herdhill or Maryland Farm to view north over the marsh and estuary, including the wader roosts and saltmarsh pool at Maryland Farm.

RSPB Campfield Marsh: mire and wet grassland To view the mire and wet grassland park at North Plain Farm and take the 0.5-mile trail south to the hide for breeding waders and wintering wildfowl. This trail continues for 1.5 miles across fields and bog to Rogersceugh Farm, and using public footpaths it is possible to complete a 7-mile circular walk back to North Plain Farm via Bowness-on-Solway.

FURTHER INFORMATION

Grid refs: Bowness-on-Solway/Herdhill NY 220 626; RSPB Campfield Marsh (Maryland Farm): saltmarsh NY 194 612; RSPB Campfield Marsh: mire and wet grassland NY 197 615
RSPB Campfield Marsh: tel: 01697 351330; email: campfield.marsh@rspb.org.uk

265 PUGNEYS COUNTRY PARK (West Yorkshire)

OS Landranger 110 and 111

Lying on the southern marches of Wakefield, this area attracts a range of waterbirds throughout the year. The 101-ha country park is owned and managed by Wakefield Council.

Habitat

The Country Park is an area of restored and landscaped gravel pits containing three lakes surrounded by farmland with, to the south-west, a small wood. The larger of the lakes covers 40 ha and is used for water sports, but nevertheless attracts wildfowl in winter. To the south a smaller pool (10 ha) is managed as a reserve, with well-vegetated banks, including a reedbed. To the north lies another, smaller pool.

Species

Wintering wildfowl include up to 100 Wigeons and Pochards, as well as small numbers of Goldeneyes, Gadwalls, Teals and Shovelers. Other notable winter visitors include Water Rail, and large numbers of Golden Plovers use the surrounding fields. Occasional visitors in winter include divers, rarer grebes, Bittern, Whooper Swan, Pintail, Goosander and Smew. Cormorants are often present and there is a respectable gull roost, which sometimes holds Glaucous, Iceland or Mediterranean Gulls.

On migration a variety of commoner waders occurs, as well as Common, Arctic and Black Terns and Lesser Black-backed Gull, and Shelduck is sometimes noted in spring. Reed and

Sedge Warblers are breeding summer visitors to the south lake, and Little Ringed Plover and Common Tern have nested. Residents include Great Crested and Little Grebes and Sparrowhawk.

Access

Leave the M1 north at junction 39 on the A636 towards Wakefield. After 0.75 miles turn right at the roundabout (signed for the Country Park) on the A6186 and turn into the car park after 100 yards. From here footpaths encircle the larger lake. To reach the reserve, turn right from the car park and follow the path around the west end of the lake, branching right through the wood to a hide overlooking the south lake, or continuing along the main path to the hide on the north shore. The small north pool (Cawoods) can be viewed by walking left from the car park. The main lake is heavily disturbed in summer, especially at weekends, making early-morning visits preferable. The Country Park is open from 9am, typically to 4.30pm in winter and 9.30pm in summer, but times vary.

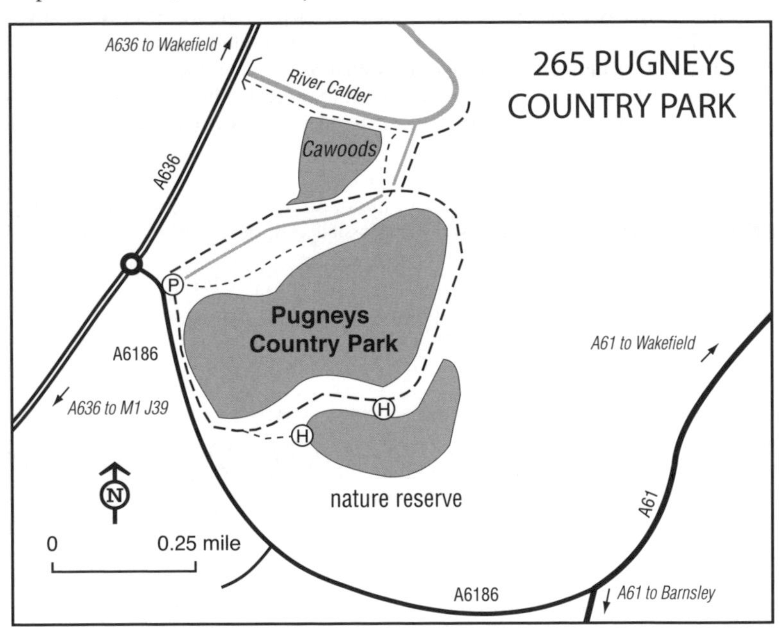

FURTHER INFORMATION

Grid ref/postcode: GR 328 179 / WF2 7EQ
Pugneys Country Park: tel: 01924 302360; web: www.wakefield.gov.uk

266 WINTERSETT RESERVOIR AND ANGLERS COUNTRY PARK (West Yorkshire) OS Landranger 110 and 111

Lying south-east of Wakefield this area contains a broad variety of habitats and attracts an equally wide range of birds, especially waterbirds. It is worth visiting throughout the year, but especially in spring and autumn.

Habitat

The largest water in the complex, Wintersett Reservoir (the 'Top Reservoir') is bounded by farmland with some stands of deciduous woodland and large areas of willow scrub, as well as stretches of marshy and rocky shore; it is used for water sports. To the west lies Cold Hiendley Reservoir (the 'Low Lake'), which has mature deciduous woodland at the west end and much riparian vegetation, and borders Haw Park, an area of conifer woodland that is now an LNR owned by Wakefield Council. Anglers Country Park Lake is a former opencast coal mine and following restoration is surrounded by areas of open meadow and recently planted trees, interspersed with rough ground. The 'Pol' is a small scrape on the south-west flank of the Lake.

Species

Winter wildfowl in the complex include a few Bewick's and Whooper Swans and occasional White-fronted and feral Greylag Geese, and sometimes also a few Pink-footed Geese. Ducks include numbers of Wigeons, Pochards and Tufted Ducks, with smaller numbers of Gadwalls, Teals, Pintails, Shovelers, Goldeneyes and Goosanders (up to 20 roost on Anglers Country Park Lake). Bittern is a regular winter visitor to the rather limited areas of reeds at Wintersett Reservoir, and Water Rail, Jack Snipe and Stonechat also winter. Occasional visitors include divers and rarer grebes, Merlin, Short-eared Owl and Smew. A substantial gull roost, on either Wintersett Reservoir or Anglers Country Park Lake, may hold occasional Glaucous, Iceland or Mediterranean Gulls.

On passage in both spring and autumn Black-necked Grebe may turn up, as well as Cormorant, Shelduck, Hobby, Little Gull, Common, Arctic and Black Terns and, especially in spring, Garganey, Marsh Harrier and Osprey are possible. A variety of waders occurs on passage, sometimes including scarcer inland species such as Oystercatcher, Whimbrel or godwits. In late summer, parties of Common Crossbill sometimes appear in Haw Park.

Residents include Great Crested and Little Grebes, Sparrowhawk, all three woodpeckers (though Lesser Spotted is scarce), Kingfisher, Little and Long-eared Owls and Willow Tit. Tree Pipit, Common Redstart, Whinchat, Grasshopper, Sedge and Reed Warblers and Lesser Whitethroat are breeding summer visitors.

Access

Wintersett Reservoir Leave Wakefield south on the A61 and turn left on the B6378 towards Walton and then, after 0.25 miles, right on the B6132 towards Royston. After 2 miles turn left on a minor road to Ryhill. Pass under the railway bridge and then alongside Haw Park, turning left at the sign for Cold Hiendley and Ryhill. After 1 mile turn left again, at the sign to Anglers Country Park. This road passes the south-east flank of Wintersett Reservoir and then, 0.7 miles from the turning, the road cuts across the north-east arm of the reservoir, separating the main water from Botany Bay (park by the causeway to view). The water may be disturbed by sailing or windsurfing, and the wildfowl often fly to Anglers Country Park Lake. (Alternatively, the area can be accessed off the A638 south-east of Wakefield, following signs to Crofton and turning left in the village towards Ryhill, bearing left in Wintersett, just past the Anglers Retreat pub, to the reservoir.)

The north shore of the reservoir can be viewed by leaving the Anglers Country Park car park (see below), crossing the road and walking to West Riding Sailing Club. Bear right for 50 yards to view the water, and it is possible to follow this path along the north and west shores, eventually reaching a causeway from which Cold Hiendley Reservoir can also be seen (this is a good spot for Water Rail).

Anglers Country Park Continue from Wintersett Reservoir for 0.5 miles and turn left at the sign for the Country Park. The car park lies to the right after 0.5 miles. (The Country Park is also signed off the A638 Wakefield–Doncaster road at Crofton.) From the car park a path encircles the lake, passing a hide overlooking the 'Pol' scrape on the west flank of the lake and also accessing a hide on the west shore, which is open at the same times as the park centre (open daily except Mondays (open Bank Holiday Mondays), 1 April–31 October, 11am–4pm, winter 10.30am–3.30pm).

Haw Park Walk west from Anglers Country Park along the road, taking the track to the right through a gate to the park, where there is a network of paths.

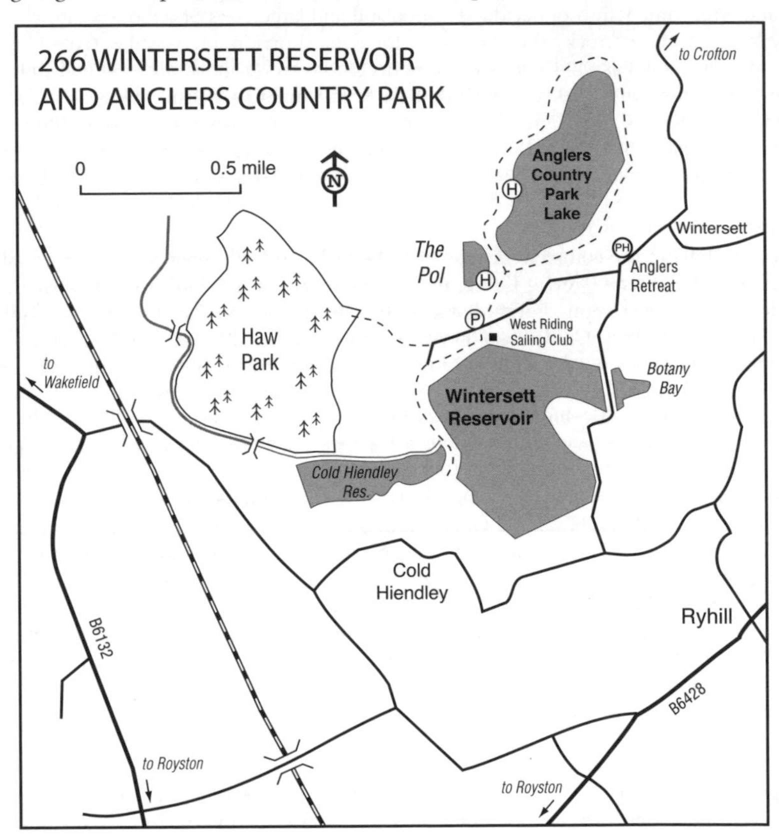

FURTHER INFORMATION

Grid refs: Wintersett Reservoir (Botany Bay) SE 380 150; Anglers Country Park SE 374 153
Anglers Country Park: tel: 01924 303980; web: www.wakefield.gov.uk

267 SWILLINGTON INGS AND ST AIDAN'S WETLAND (West Yorkshire) OS Landranger 104

Lying south-east of Leeds, this area of mining subsidence in the Aire Valley is very attractive to wintering wildfowl, roosting gulls and, in spring and autumn, passage waders.

Habitat

Over the years mining subsidence in the area bounded to the south by the River Aire has produced a number of shallow flashes, and though most are now filled in, Astley Lake was restored as a reserve in 1987. Shallow with gently shelving banks, it contains 14 small islands. Immediately to the west lie two areas of water: Park Lake and Oxbow Lake. To the east, within a loop of the river, St Aidan's Lake is a larger and more extensive sheet of water. Further east still a huge area of former open-cast coal workings is being reshaped into an exciting wetland reserve.

Species

Winter wildfowl include significant numbers of Wigeons, Gadwalls, Teals, Pochards, Goldeneyes and Ruddy Ducks, and smaller numbers of Goosanders and Pintails, while scarcer visitors include Whooper and Bewick's Swans, Greylag Goose, Smew and Peregrine. Merlin is fairly regular in the area at this season, with Golden Plover in the fields and Siskin in Fleet Plantation. Bittern is a regular winter visitor to Astley Lake, if secretive, and Water Rail is usually present. Large numbers of gulls roost on St Aidan's, and Kittiwake and Glaucous and Iceland Gulls are sometimes present in the roost.

On passage Black-necked Grebe is occasional in spring (several pairs bred here in the 1940s), as is Common Scoter, and Water Pipit can be found in the early season (try the filter beds on the right before the railway). Other possible migrants include Little Gull, Sandwich, Arctic and Black Terns, and less regularly Garganey, Marsh Harrier or Osprey. A variety of waders is recorded on passage, sometimes including Grey Plover, Sanderling, Little Stint, Wood or Curlew Sandpipers, godwits and Turnstone.

Residents include Little and Great Crested Grebes, Cormorant, Canada Goose, Shelduck, Sparrowhawk, Kingfisher and Green Woodpecker, while Common Tern and a few Sedge and Reed Warblers are breeding summer visitors. Hobby is regular in summer and in 2009 Avocets bred on Astley Lake, continuing their spread to inland sites in Britain.

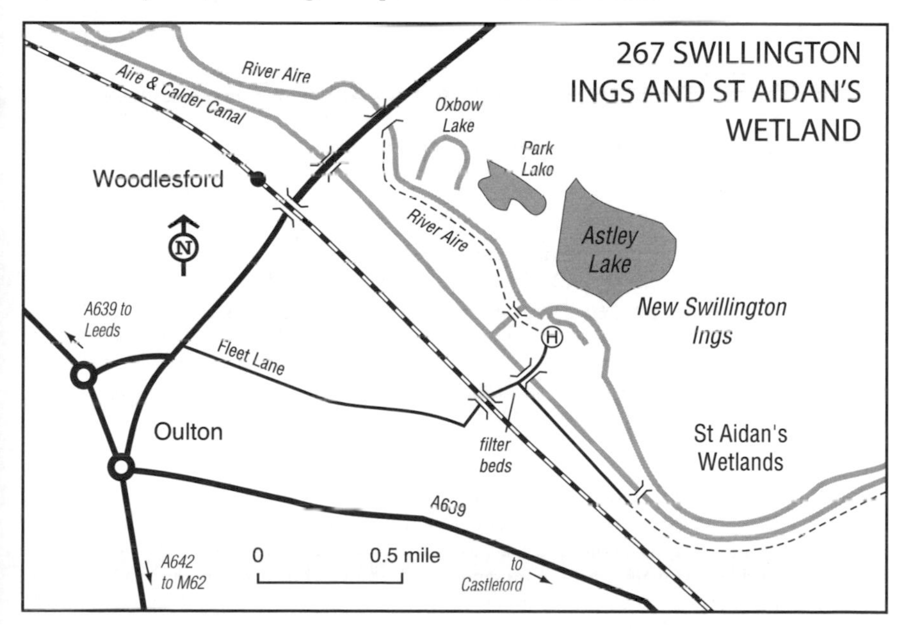

Access

Astley Lake From the A642 in Oulton turn east (just south of the Kwik-Save supermarket and opposite the Old Masons Arms pub) into Fleet Lane. Follow this for 1.25 miles, under a railway bridge and over the Aire and Calder Canal, to the car park adjacent to Bayford oil terminal. An elevated hide looks north from here over the river and Astley Lake; managed by the New Swillington Ings Bird Group, the hide is open to the public at weekends and on Bank Holidays, with access at other times by arrangement. From the car park it is possible to walk west along the south bank of the river to Fleet Plantation (a stand of birch and alder) and on to view more wet meadows and pools.

St Aidan's Wetland This 400-ha site is currently being transformed by its owners, UK Coal, into a large reserve that will include a 50-ha reedbed, lakes with islands, scrapes, 50 ha of wet grassland, scrub, and plantations of native trees. The complex will include much of Swillington Ings. There are no access arrangements at the time of writing, but upon completion of the restoration the RSPB will take over management of the site and develop visitor facilities. In the meantime, the area, which has already attracted many interesting birds, can be viewed from the bank of the River Aire. From the car park at the oil terminal (see above) walk east to view Fleet Lane Pond and then to Lemonroyd Lock, just beyond which is a viewpoint for St Aidan's Wetland. Alternatively, turn south-east off Fleet Lane just before the bridge over the canal and drive for 0.5 miles to a footbridge, which gives views of St Aidan's Wetland.

FURTHER INFORMATION

Grid ref: Astley Lake SE 381 284
Swillington Ings Bird Group: web: sibg1.wordpress.com

268 FAIRBURN INGS (West/North Yorkshire)

OS Landranger 105

These shallow lakes in the Aire Valley were formed by mining subsidence and the variety of habitats created has proved very attractive to birds within an otherwise industrial area. Fairburn lies alongside the A1, 4 miles north of the A1/M62 interchange. It is best during migration seasons or in winter. The 286-ha reserve is managed by the RSPB on behalf of Leeds City Council.

Habitat

Once an area of extensive flood meadows, subsidence has produced a mosaic of low spoil heaps and permanent lakes (over one-third of the reserve is open water). The pools at the east end of the reserve have little fringing vegetation and subsidence has continued to reduce the reedbeds, though floating islands have been constructed to help compensate for the loss. Abandoned spoil heaps have been colonised by a variety of plants and planted with trees. At the west end of the reserve, low-lying pasture is still subject to flooding and subsidence, and the resulting shallow flashes are attractive to wildfowl and waders.

Species

Many wildfowl winter including a herd of Whooper Swans, which usually feed in the fields during the day, arriving to roost in the evening. Bewick's Swan may also occur. There are also large numbers of Coots, Wigeons and Teals, as well as Gadwall, Pintail, Shoveler, Pochard and Goldeneye, and occasionally Scaup, Goosander and Smew. Bitterns winter with increasing regularity, and divers or rarer grebes are sometimes recorded. A large gull roost frequently attracts Glaucous or Iceland Gulls, sometimes Mediterranean Gull and, in early spring or late autumn, Kittiwake. Lesser Redpolls feed on the alders and birches and are sometimes joined by a few Siskins. Up to 7,000 Golden Plovers roost in the area, and Merlin and Peregrine may visit.

Small numbers of waders occur on passage, sometimes including Whimbrel, Ruff, Black-tailed Godwit, Greenshank, Spotted Redshank and Green Sandpiper. Tern passage can be notable, particularly in spring following east winds. The main species are Common, Arctic and Black Terns (this being one of the best sites in Yorkshire for the latter). A few Little Gulls are regular at these times. In July–August Common Scoter is fairly frequent. In autumn large numbers of hirundines can be seen, many of them roosting on the reserve. Hobby, Red Kite

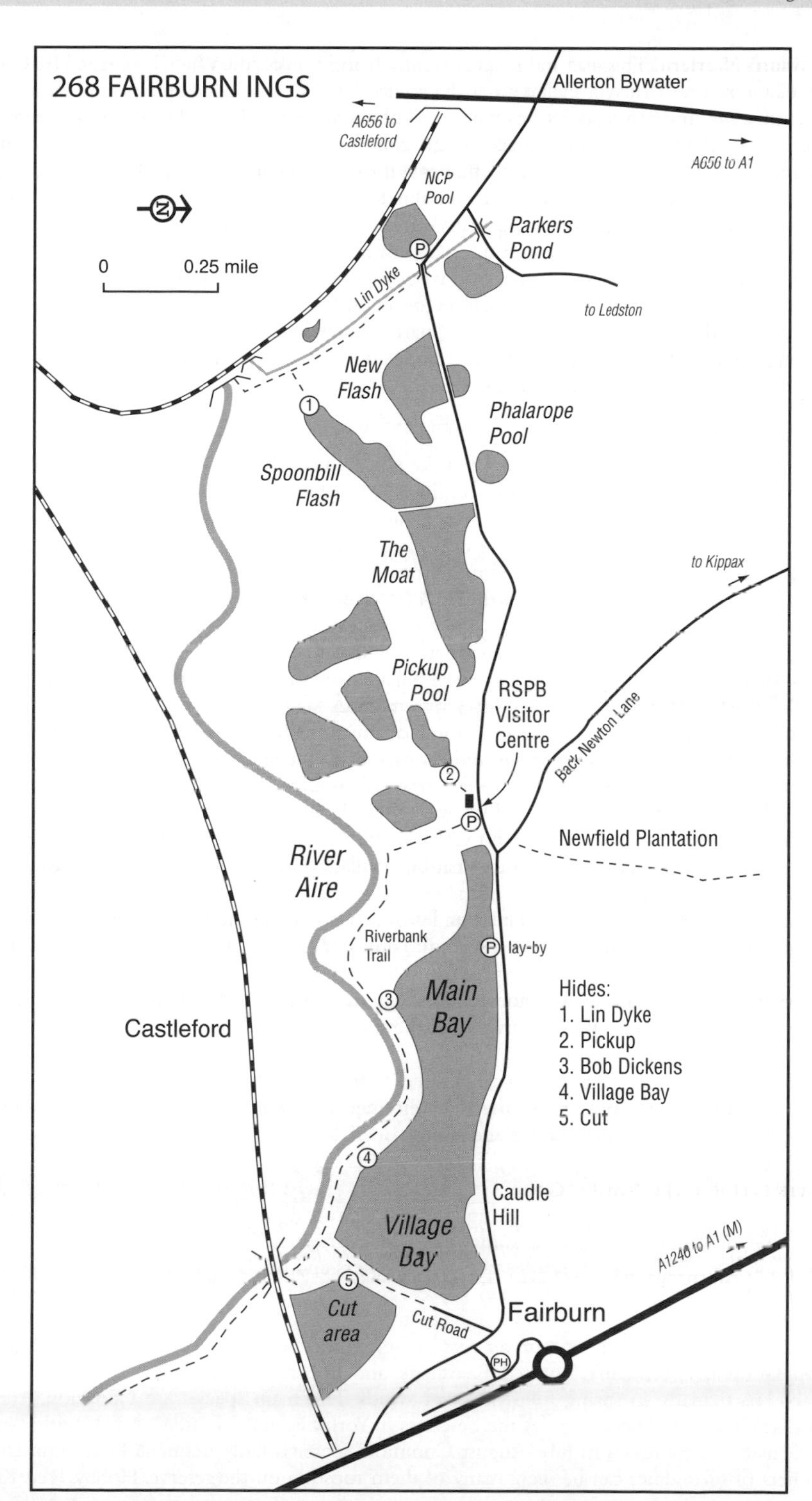

268 FAIRBURN INGS
Allerton Bywater
A656 to Castleford
A656 to A1
NCP Pool
Parkers Pond
0 0.25 mile
Lin Dyke
to Ledston
New Flash
Phalarope Pool
Spoonbill Flash
The Moat
to Kippax
Pickup Pool
RSPB Visitor Centre
Back Newton Lane
Newfield Plantation
River Aire
Riverbank Trail
lay-by
Main Bay
Hides:
1. Lin Dyke
2. Pickup
3. Bob Dickens
4. Village Bay
5. Cut
Castleford
Caudle Hill
Village Bay
A1246 to A1 (M)
Cut area
Cut Road
Fairburn
PH

and Marsh Harrier are regular and Osprey is occasionally recorded. Water Rail is resident and Spotted Crake has occurred several times in autumn.

Summer is comparatively quiet. Several species of wildfowl breed including Gadwall, Shoveler, Garganey and Pochard, and occasionally Garganey and Shelduck. Lapwing, Common Snipe and Common Redshank favour the shallow flashes at the west end of the reserve, Kingfisher breeds, and Little Ringed Plover and Reed Warbler almost reach the northerly limits of their breeding ranges here. A colony of Black-headed Gull nests on Priestholme and Common Tern also breeds; Mediterranean Gulls may visit the Black-headed Gull colony. The scrub and deciduous woodland attract Green and Great Spotted Woodpeckers (and occasionally also Lesser Spotted Woodpecker), a variety of common warblers (occasionally including Grasshopper Warbler) and Whinchat. Little Egret, Common Buzzard, Sparrowhawk and Little Owl are resident. Willow Tit and Tree Sparrow both maintain healthy populations in this area, making the reserve a good place to catch up with these declining species.

Access

Access to the reserve is from junction 42 of the A1(M). Take the A63 westwards for 0.25 miles and then turn south at the roundabout on the A1246 (the old A1) to Fairburn village. At the village turn right at the roundabout, then right again at the Wagon and Horses pub into Gauk Street, and then right at the T-junction into Caudle Hill.

Cut Road From Caudle Hill turn left into Cut Road and park. From here follow the road to the Cut, a lane bordered by large hawthorns. This gives views of Village Bay and then accesses the Village Bay and Cut Hides on the causeway among the spoil heaps on the south flanks of Village Bay and Main Bay. The hides are open at all times.

Main Bay Continuing west along Caudle Hill, a lay-by on the left after about 1 mile gives views over Main Bay.

Visitor Centre Continuing west along the road for a further 0.5 miles (and bearing left at the fork in the road), the visitor centre and car park lie to the left. This is open daily 9am–4pm; maps and guides are available here, as well as up-to-date information on species present on the reserve. From here a boardwalk (the Discovery Trail) leads past the feeding stations (good for Tree Sparrow) to Pickup Hide, which overlooks the Pickup Pool, a shallow scrape, while the Riverbank Trail follows the River Aire to Fairburn village. The visitor centre and boardwalk are wheelchair-accessible.

Newfield Plantation Rather than bearing left, fork right along the road towards Ledston and Kippax and take the footpath along the south flank of Newfield Plantation, which is a good area for woodland species.

Phalarope Pool Continue west for 1 mile from the visitor centre along the road towards Castleford, and Phalarope Pool, a series of shallow, reed-fringed pools at Ledston Ings, lies to the north of the road.

Lin Dyke After a further 0.5 miles, a car park lies to the south of the road and a trail leads along Lin Dyke through the west end of the reserve to the Lin Dyke Hide overlooking Spoonbill Flash (this is the best area for waders and swans).

FURTHER INFORMATION

Grid refs/postcode: Cut Road SE 470 278; Main Bay SE 458 279; Visitor Centre SE 452 278 / WF10 2BH; Phalarope Pool SE 437 277; Lin Dyke SE 432 275
RSPB Fairburn Ings: tel: 01977 628191; email: fairburnings@rspb.org.uk

269 DERWENT VALLEY (East/North Yorkshire)

OS Landranger 105

This area is prone to seasonal flooding and can hold significant numbers of wildfowl in winter, while at Wheldrake Ings, owned by the Yorkshire Wildlife Trust (YWT), careful management of the habitat provides year-round interest, with exceptional numbers of breeding ducks and waders, and a variety of passerines. The area forms the 467-ha Lower Derwent Valley NNR.

Habitat

The Wheldrake Ings Reserve lies in the flood plain of the River Derwent, and comprises seasonally flooded meadows with stands of willow along the riverbanks and an area of permanent open water. Around the car park lie mature hedges, scrub and rough ground, which are attractive to passerines. To the south, towards Aughton and Bubwith Ings, water levels in the meadows are more closely controlled and the river valley floods much more rarely.

Species

The numbers and variety of wintering wildfowl depend upon the extent of the flood waters. In good seasons, both Bewick's and Whooper Swans occur, as well as up to 5,000 Wigeons and several thousand Teals and Pochards, with smaller numbers of Gadwalls, Pintails, Shovelers, Goldeneyes and Goosanders. Occasionally, rarer grebes or divers may appear in winter, as can parties of Bean, Pink-footed or White-fronted Geese. Sparrowhawk is resident and in winter may be joined by Short-eared Owl, Peregrine, Merlin, Hen Harrier and, sometimes, Goshawk. Large flocks of Golden Plovers and Lapwings winter, together with small numbers of Dunlins, Ruffs, Curlews and Common Redshanks.

On passage, depending on water levels, a variety of waders may occur, including Whimbrel, Common and Green Sandpipers, Dunlin, and sometimes Ruff or Black-tailed Godwit may stay late into spring and even display; there have been up to 25 lekking male Ruffs. Black-necked Grebe appears most springs and has bred, as do parties of Common Tern and sometimes Arctic and Black Terns and Little Gull. Hobby occurs in late summer, and there is a chance of Marsh Harrier or Osprey in both spring and autumn.

Breeding species include Little Grebe, Cormorant (at Wheldrake Ings), Greylag Goose, Shelduck, Shoveler, Garganey, Wigeon, Gadwall, Teal, Pintail, Pochard, Marsh Harrier, Water Rail, many Lapwings, Common Snipe, Curlew, Common Redshank, Black-headed Gull, Turtle Dove, Barn and Little Owls, Kingfisher, Yellow Wagtail, Grasshopper, Sedge and Reed Warblers, and Corn Bunting also nest. Outstanding is the population of the highly secretive Spotted Crake (30+ calling birds in 1998; they are best heard at night but extremely hard to see). Corncrake and Quail, equally secretive, have also bred in recent years, but the chances of seeing these are slim.

Access

Access in the NNR is restricted to the nine hides and marked footpaths.

Wheldrake Ings Leave the A64 York ring-road on the A19 towards Selby. After 1 mile turn left at Crockey Hill to Wheldrake village and continue through the village towards Thorganby. The road bends sharp right and, after 0.5 miles, turn left between two old stone gateposts onto an unsigned sealed track (Ings Lane), arriving at the car park after 0.25 miles. From here cross the bailey bridge over the River Derwent and turn right, over the stile, to follow the riverbank footpath south. After 0.25 miles a hide overlooks the main ings (often flooded in winter) and after a further 0.5 miles, by the wind pump, a track forks left to a second hide overlooking the main flash. The reserve is open at all times and access is limited to the marked trails leading to the four hides.

Bank Island This lies just 200 yards north of the entrance to Wheldrake Ings. A track leads to

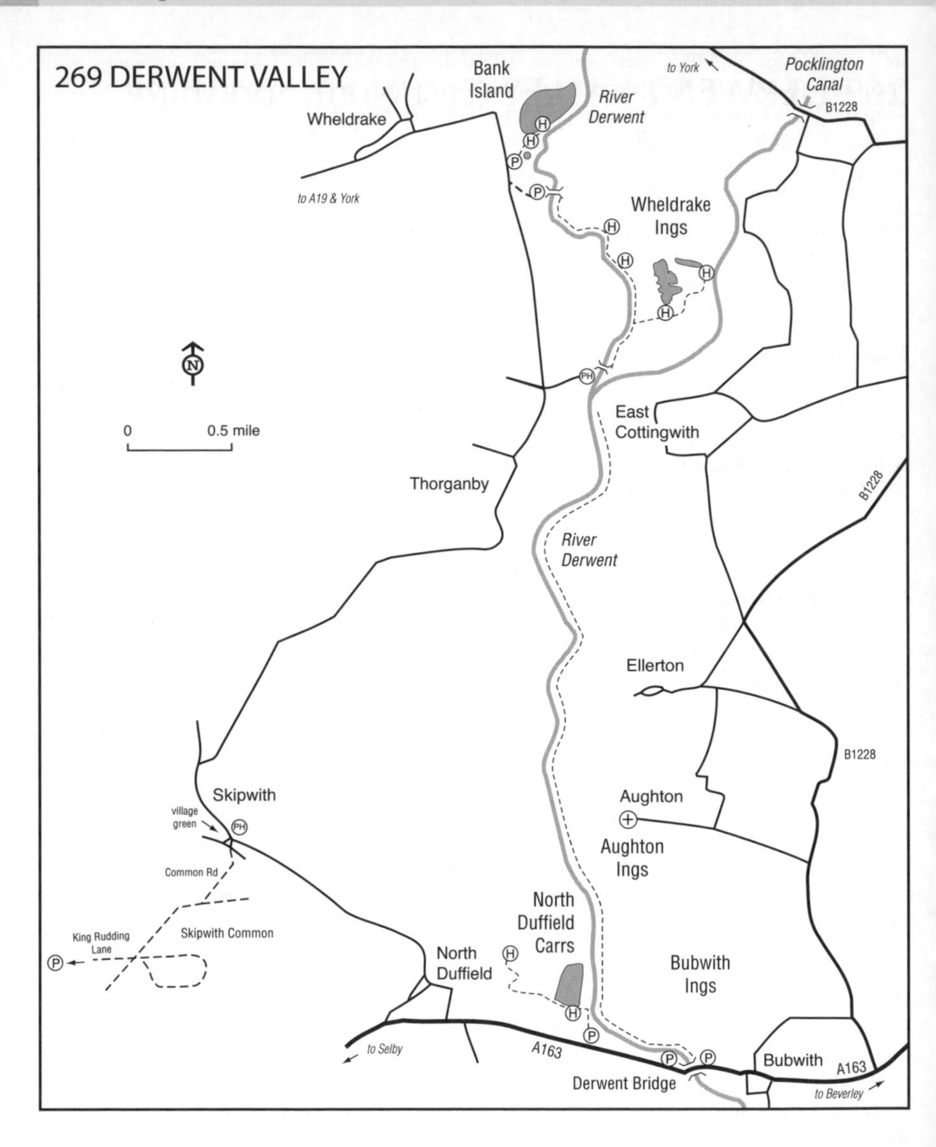

two hides overlooking a large area of flooding (sometimes dry in summer) and grazing meadows.
North Duffield Carrs Leave North Duffield east on the A163 towards Bubwith and turn north into the car park after 0.75 miles. From here walk 150 yards to the hide, which overlooks a large, well-vegetated pool on the west bank of the river and the meadows beyond.
Bubwith and Aughton Ings Continuing east towards Bubwith on the A163 the road crosses a narrow bridge over the River Derwent, with parking on both the east and west sides of the bridge. In winter it may be worth walking north along the east bank of the river to view Bubwith Ings and Aughton Ings for some 2 miles from the bridge.
Skipwith Common This has breeding Sparrowhawk, Woodcock, Turtle Dove, Long-eared Owl, Tree Pipit, Common Redstart, Lesser Whitethroat and a few Grasshopper Warblers; it is also well known for Nightjars, but the species has been scarce or absent in recent years. Access is from the triangular village green in Skipwith village, taking Common Road southwards from the end of the green. This enters woodland and comes to an information board. There are open

areas suitable for Nightjars after a further 0.25 miles and 0.75 miles (although anywhere with open vistas could hold the species); at the latter point turn right (west) at the crossroads into King Rudding Lane to reach the car park after a further 0.5 miles; continuing on this track, you come to the A19 near Riccall in 1.25 miles.

FURTHER INFORMATION

Grid refs: Wheldrake Ings SE 694 443; North Duffield Carrs SE 696 366; Bubwithy and Aughton Ings SE 707 364; Skipwith Common SE 643 374 (King Rudding Lane car park) / SE 664 383 (Skipwith Green)
Yorkshire Wildlife Trust: tel: 01904 659570; web: www.ywt.org.uk
Natural England: web: www.naturalengland.org.uk

270 DEARNE VALLEY (South Yorkshire)

OS Landranger 111

The Dearne Valley cuts through one of the most industrialised areas of England, but a series of interesting wetlands survive, hosting large numbers of birds, and two sites are now RSPB reserves. Tree Sparrow, breeding and passage waders and large numbers of wintering Golden Plovers are all specialities, and there are excellent visitor-friendly facilities.

Habitat

Along the river valley at Old Moor habitats include lakes, damp grassland and copses, while Bolton Ings boasts 43 ha of reedbeds.

Species

Residents include Great Crested and Little Grebes, Common Snipe, Common Redshank, Lapwing, Black-headed Gull, Kingfisher, Barn and Little Owls, Stonechat and Tree Sparrow – the provision of large numbers of nest boxes has resulted in a healthy population of this declining species. In summer look also for Oystercatcher, Ringed and Little Ringed Plovers, Cuckoo, Yellow Wagtail and Sedge and Reed Warblers.

On passage a variety of waders may occur, including Black-tailed Godwit, Green Sandpiper, Greenshank, Ruff and Ringed Plover; Hobby, Osprey, Little Egret, Spoonbill, Garganey, Avocet and Black Tern may occur.

In winter there are large numbers of Lapwings and up to 8,000 Golden Plovers have been recorded at Old Moor. Wildfowl include Goldeneye, Wigeon, Teal, Shoveler, Gadwall, and sometimes Goosander. The concentration of prey species attracts raptors, including Peregrines, and in recent winters Bittern has been recorded.

Access

RSPB Old Moor Leave the M1 at junction 36 and take the A61 (Barnsley). At the small roundabout, continue straight ahead on the A6195 (Doncaster) for about 4 miles. After passing the Morrison's superstore, follow the brown RSPB Old Moor signs. Alternatively, leave the A1 at junction 37 (Doncaster), follow the A635 towards Barnsley, then follow the brown RSPB Old Moor signs. From 1 November to 31 January the visitor centre and shop are open 9.30am–4pm with the gates open until 4.30pm; 1 February–31 October the visitor centre is open 9.30am–5pm, and the gates open until 5.15 pm. The reserve is closed on Christmas Day and Boxing Day. There are five hides plus another overlooking the bird-feeding garden. There are two viewing platforms in the pond and picnic area.
RSPB Bolton Ings From the car park at Old Moor walk east along the Trans Pennine Trail.

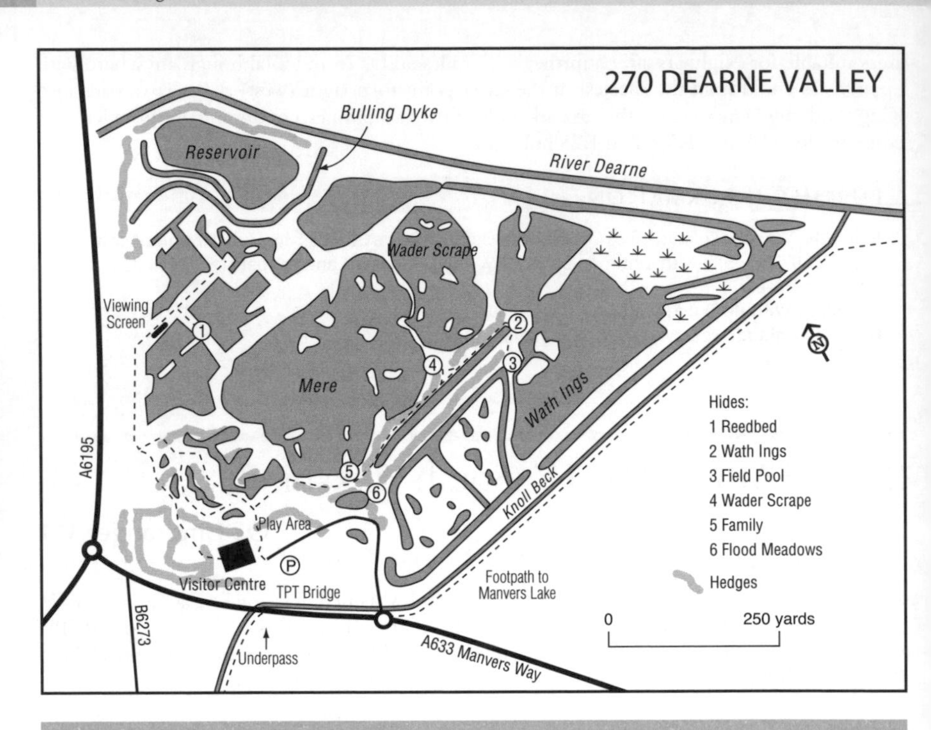

FURTHER INFORMATION

Grid ref/postcode: RSPB Old Moor SE 422 022 / S73 0YF
RSPB Old Moor: tel: 01226 751593; email: old.moor@rspb.org.uk

271 POTTERIC CARR (South Yorkshire)

OS Landranger 111

Lying just 2 miles south-east of Doncaster city centre, this 200-ha wetland is a reserve of the Yorkshire Wildlife Trust. A wide variety of birds is present throughout the year, making a visit worthwhile at any time, and it is a particularly good site for wintering Bittern.

Habitat

An area of relict fenland, the carr was partially drained before succumbing to subsidence and reverting to wetland. It is now a complex of subsidence and artificial pools, drainage dykes and areas of wetland, with extensive areas of reeds (notably at Low Ellers and Decoy Marshes), together with willow carr, birch woodland, more mature stands of oak in Black Carr Wood, and grassland. Railway embankments, both used and disused, crisscross the area.

Species

Breeding species include Great Crested and Little Grebes, Shelduck, Gadwall, Shoveler, Teal, Pochard, Sparrowhawk, Common Buzzard, Grey Partridge, Water Rail, Oystercatcher, Ringed and Little Ringed Plovers, Woodcock, Common Redshank, Lapwing, Black-headed Gull (about 250 pairs), Stock Dove, Kingfisher, Green and Great Spotted Woodpeckers, Yellow Wagtail, Reed and Sedge Warblers and Willow Tit. The site's greatest claim to fame must surely be the presence in 1984 of a pair of Little Bitterns, which raised three young (the first confirmed breeding record in Britain). Avocet has also started to breed on the new, southern sector of the reserve (Huxter Well) and Little Egret is an increasingly regular visitor. Black-necked Grebe has bred on several occasions and is often present in summer, while Garganey, Common Tern, Turtle Dove, Lesser Spotted Woodpecker, Grasshopper Warbler and Lesser Redpoll have also bred, but less regularly.

On spring migration east winds may bring parties of Arctic and Black Terns. A variety of waders passes through in spring and autumn, depending on water levels, including Greenshank, Common and Green Sandpipers and occasionally Black-tailed Godwit. Less common visitors include Garganey, Marsh Harrier, Osprey, Peregrine, Merlin, Hobby, Kittiwake and Little Gull.

Wintering wildfowl include Wigeon, Gadwall, Teal, Pintail, Shoveler, Pochard and Goldeneye, with Smew and Goosander sometimes present. There may be large numbers of Lapwings and Golden Plovers, as well as gulls, and Yellow-legged and Mediterranean Gulls are possible. Look for Jack Snipe with the resident Water Rail around the pools, and wintering passerines may include Grey Wagtail, Stonechat and Siskin. Bittern is regularly found in winter, with up to five or six birds (Piper Marsh is a good spot) and indeed, has latterly been present all year. Irregular visitors include Cormorant, Whooper and Bewick's Swans, Pink-footed and Greylag Geese, Merlin, Short-eared Owl and Bearded Tit.

Access

Leave the M18 at junction 3 north on the A6182 towards Doncaster. After 1 mile, at the first roundabout, take the third exit, signed 'no through road', and turn right after 50 yards into the reserve car park. Coming from Doncaster, take the A6182 (White Rose Way) and turn left at the roundabout near B&Q. The reserve is open daily from 9am, last entrance 5pm (fee for non-members of the Trust), the café open daily 10am–4pm, and the Field Centre open on Sundays and holidays. There are 6 miles of marked trails and 14 hides, with those nearest the field centre overlooking a passerine-feeding station and a small pool frequented by Water Rail and Kingfisher.

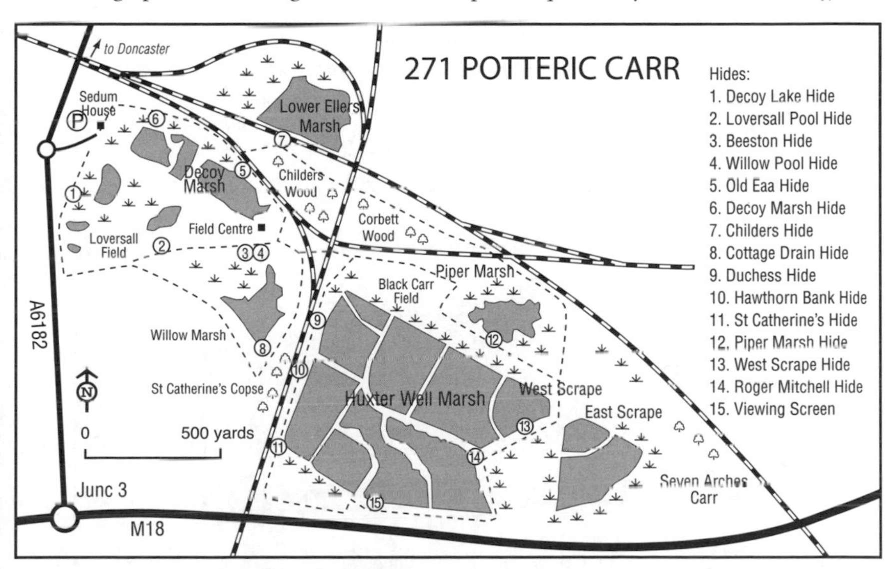

FURTHER INFORMATION

Grid ref: SE 599 003
Yorkshire Wildlife Trust: tel: 01904 659570; web: www.ywt.org.uk
Friends of Potteric Carr: web: www.potteric-carr.org.uk

272 BLACKTOFT SANDS (East Yorkshire)

OS Landranger 112

Blacktoft Sands is an RSPB reserve lying at the confluence of the Rivers Ouse and Trent on the south side of the Humber Estuary. Several uncommon species breed and the artificial lagoons attract a variety of waders.

Habitat

The east part of the reserve consists of an area of mudflats and saltmarsh at the confluence of the Trent and the Ouse, while the majority of the area, bounded to the south and west by a flood bank, is a tidal reedbed (the largest in England), within which six brackish lagoons have been excavated. Other habitats include areas of flooded grassland and willow scrub.

Species

Large numbers of commoner ducks winter on the Humber Estuary and visit the reserve, including Wigeon, Teal and Pochard, and a few Goosanders, Red-breasted Mergansers and Goldeneyes may also be present. Flocks of Pink-footed Geese and wild swans, especially Whooper, sometimes fly over. Wintering waders include up to 2,500 Golden Plovers, and Jack Snipe is sometimes present. Short-eared Owl, Merlin, Peregrine and Hen Harrier are regular, with Merlins and up to five Hen Harriers roosting in the reeds. Bitterns occasionally overwinter and Little Grebe, Sparrowhawk, Water Rail, Barn Owl, Kingfisher, Bearded Tit, Tree Sparrow and Reed Bunting are resident, with large numbers of Bearded Tits sometimes present in winter.

During migration seasons a wide variety of waders pass through and are best viewed from the hides. Many are often only visible at high tides when their feeding areas on the Humber Estuary are covered. Characteristic species include Little Stint, Curlew Sandpiper, Ruff, Black-tailed Godwit, Greenshank and Spotted Redshank, and several rare waders have been found in recent years. The best periods for waders are May and July–August with greater numbers during the latter. In spring, Temminck's Stint, Black Tern and Little Gull are occasional, and other irregular visitors include Osprey, Montagu's Harrier, Hobby, Little Egret and Spoonbill.

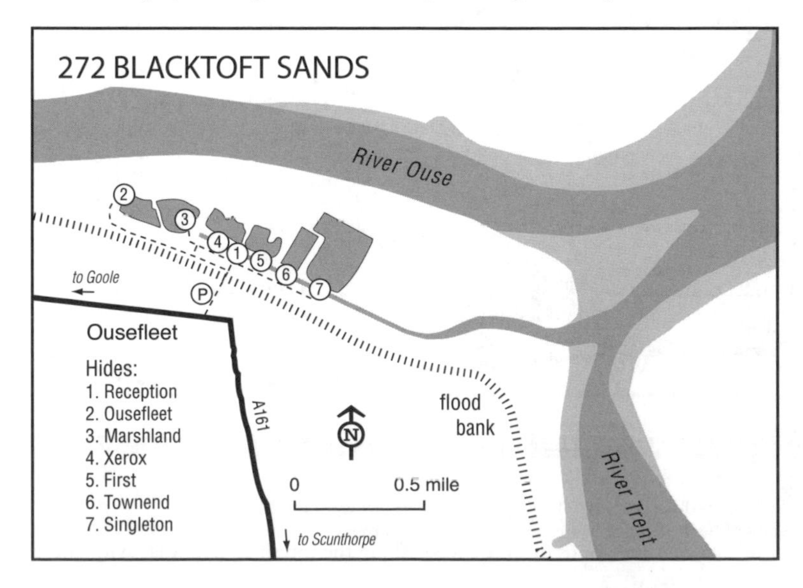

In summer, Common Redshank, Lapwing and Common Snipe nest on the lagoons alongside Pochard, Teal, Shoveler, Gadwall, Shelduck and feral Canada and Greylag Geese, and Ruddy Duck has also bred. Some 30-40 pairs of Avocets and a few pairs of Little Ringed Plovers breed on the lagoons and one or two pairs of Short-eared Owls and Grasshopper Warblers nest in grassy areas. Large numbers of Reed Warblers breed in the reedbeds, together with Sedge Warbler, Bittern and a few pairs of Marsh Harriers. Blacktoft is an important breeding site for Bearded Tit (usually about 40 pairs but up to 100 pairs in some years).

Access

Leave Goole on the A161 to Swinefleet and Crowle (signed RSPB). Turn left at the mini-roundabout in Swinefleet and then right at the next T-junction on a minor road to Reedness. Continue for about 5 miles, through Reedness, Whitgift and Ousefleet and, approximately 0.4 miles east of Ousefleet, turn left into the signed reserve car park, immediately before the road bends sharply right. The reserve is open daily 9am–9pm (or dusk if earlier), with seven hides giving good views of the lagoons and reedbeds. All trails and hides are wheelchair-accessible. The Reception Centre/hide is open 9am–5pm April–October, weekends only 9am–4pm November–March. High 'spring' tides push the greatest numbers and variety of waders from the Humber Estuary onto the lagoons.

FURTHER INFORMATION

Grid ref: SE 843 232
RSPB Blacktoft Sands: tel: 01405 704665; email: blacktoft.sands@rspb.org.uk

273 WHITTON SAND AND FAXFLEET PONDS

(East Yorkshire) OS Landranger 106

Lying at the head of the Humber Estuary, just below the confluence of the Ouse and the Trent (and almost opposite the RSPB's Blacktoft Sands reserve), this area attracts very large numbers of wildfowl and waders, and forms part of the Humber Wildfowl Refuge.

Habitat

Whitton Sand is a large sandbank, about 2 miles long. Formerly only covered on the highest tides (over 16 feet), it is now a permanent island. In consequence it is an important roost site, being the only exposed bank in the upper Humber on low and medium tides. On the Humber shore to the west lie Faxfleet Ponds, two lagoons fringed by sallows and reeds, and there are also large stands of reeds along the shore.

Species

Waders in winter should include Golden and Grey Plovers, Bar-tailed Godwit, Dunlin, and sometimes Knot, Sanderling and Turnstone. Wintering wildfowl include up to 3,000 Pink-footed Geese, which fly to feed south of the Humber in Lincolnshire and sometimes roost overnight on Whitton Sand. Up to 200 feral Greylag Geese are resident, and a few Dark-bellied Brent Geese visit. Wigeon, Teal and Pintail (September–October) occur in large flocks, and small numbers of Shovelers, Pochards, Scaups, Goldeneyes and Goosanders can also be expected. Occasionally, parties of Whooper or Bewick's Swans may be present. In winter, raptors may include Peregrine, Merlin, Sparrowhawk and sometimes Hen Harrier or Short-eared Owl. Large numbers of gulls are present, especially Black-headed and Common Gulls, roosting on the Sand. Rock Pipits may be found along the tideline, a few Bearded Tits can be seen around the ponds or in the reeds along the foreshore, together with Water Rails, and small parties of Twites may pass through in early spring.

On passage in spring and autumn a variety of waders may be found, such as Little Ringed Plover, Whimbrel, Ruff, Greenshank, Spotted Redshank, Green Sandpiper and Black-tailed Godwit, with Little Stint and Wood and Curlew Sandpipers visiting some years. Common Scoter is sometimes seen flying west over the estuary in the evening (July–August). Passage periods may also bring Marsh Harrier and occasionally Osprey, and small numbers of Common, Arctic, Sandwich and Black Terns and Little Gulls, and larger numbers of Lesser Black-backed Gulls. The ponds hold a large hirundine roost, and a variety of common passerine migrants may occur.

Breeding species on the ponds include Little and Great Crested Grebes and Sedge and Reed Warblers, while Barn and Little Owls are resident.

Access

From the M62 junction 38 take the B1230 south-westwards towards Newport and Gilberdyke. Continue through Newport and turn south on the minor road to Faxfleet, which is reached in just over 3 miles. At the southernmost point of the village, where the road bends sharp right, carry straight on to a small car park by the Humber Bank.
Whitton Sand From the car park walk north-east for 0.75 miles along the Humber Bank, past Faxfleet foreshore, to view Whitton Sand from Bowes Landing. Check the foreshore for waders en route and, for roosting waders on a high tide, the hour before high water is best at Bowes Landing; on lower tides flocks will be scattered and viewing more difficult.
Faxfleet Ponds Walk west from the car park for 0.25 miles along the Humber Bank.

FURTHER INFORMATION

Grid ref: SE 863 240

274 CHERRY COBB SANDS AND STONE CREEK

(East Yorkshire) OS Landranger 113

This area along the north bank of the Humber harbours one of the largest wader roosts on the estuary and, over the years, has also attracted a number of major rarities.

Habitat

Cherry Cobb Sands comprise the largest expanse of raised saltmarsh on the north shore of the Humber, bordered by flat, featureless farmland.

Species

Wintering wildfowl include Brent Goose, Wigeon, Teal and Pintail. Good numbers of waders are present, especially Grey Plover, Curlew, Knot, Dunlin, Common Redshank and Bar-tailed Godwit. Occasionally small numbers of Bewick's Swans may be found. On higher tides roosting waders, notably Curlew, Golden Plover and Lapwing, use the fields east of Stone Creek. Wintering raptors in the area may include Peregrine, Merlin, Sparrowhawk and, sometimes, Hen Harrier.

On passage the typical estuarine waders are usually present, including Whimbrel, Ruff, Greenshank and sometimes Black-tailed Godwit, Little Stint or Curlew Sandpiper.

Access

Leave Hull east on the A1033 and, after passing through Thorngumbald and adjacent Camerton, turn sharp right on a minor road signed 'Paull'. After 200 yards turn left signed to Cherry Cobb

Sands and Stone Creek. Continue on this minor road for 2.25 miles, passing the hamlet of Thorney Crofts and follow the road left, now almost parallel to the Humber Bank, for about 3 miles until you reach Stone Creek. (Alternatively, leave the A1033 just west of Keyingham on the minor road to Cherry Cobb Sands.)

Cherry Cobb Sands Park by the Humber Bank and follow it on foot north-west to view Cherry Cobb Sands for waders (a rising tide is best).

Stone Creek/Sunk Island From the parking area walk left along the track to the bridge over the creek and then cut right, back to the Humber Bank. Follow this for about 1 mile to the small wood, checking for migrants en route.

To check the fields to the east, drive left along the track from the parking area for a few yards to the bridge over Stone Creek and follow the made-up road left for about 400 yards to park where the road leaves the creek, scanning the surrounding fields. Continuing, the road runs through farmland, which may hold plovers and raptors, to Sunk Island Farm and it is then possible to cut north back to the main A1033 at either Ottringham or Patrington.

FURTHER INFORMATION

Grid ref: TA 235 188

275 SPURN POINT (East Yorkshire)

OS Landranger 113

Lying 32 miles south-east of Hull (on a road to nowhere), in 1946 Spurn was the site of the first Bird Observatory to be established on mainland Britain and it remains an excellent place to observe migrants and rarities. Spurn NR is owned by the Yorkshire Wildlife Trust.

Habitat

Spurn is a narrow sand and shingle spit, about 3.5 miles long, extending into the mouth of the Humber. The base of the peninsula is farmed, and includes Beacon Ponds and Easington Lagoons, but the spit really begins at the observatory, and is only about 30 yards wide at the Narrow Neck. Indeed, it has been breached by the sea several times in recent years, although the road is always maintained to give access to the tip. Sea-buckthorn forms extensive and largely impenetrable thickets, with occasional elder bushes and areas of Marram Grass. To the seaward side there is a narrow beach, while on the Humber shore are the extensive mudflats of Spurn Bight.

Species

Visible migration commences in March with Lapwing, Starling, Rook, Jackdaw and Chaffinch on the move. A peculiarity of Spurn is that birds appear to be moving the 'wrong' way in spring, that is south or south-south-east. In March–April Rock Pipits of the Scandinavian subspecies may occur on the saltmarsh and lagoons. East winds in late March–April will bring Black Redstart, and later on flycatchers, chats and warblers; Wryneck, Bluethroat, Icterine and Marsh Warblers, Red-backed Shrike and Common Rosefinch occur most years. Easterlies also bring movements of Black, Common and Arctic Terns in the first few hours of daylight.

Late summer sees passages of waders and terns offshore, especially in strong west or south-west winds, as well as large movements of Swift. From late August to early November, north-west to north-east winds may prompt a passage of seabirds. Regular are Sooty Shearwater, skuas (sometimes including Pomarine and occasionally Long-tailed), and, from October, Little Auk, while Little Gulls can peak at 100 on a very good day. Large numbers of waders frequent the estuary, with a good variety on Beacon Ponds and Easington Lagoons. Spotted Redshank is regular, with Little Stint and Curlew Sandpiper in some years.

Autumn passerine migration commences in late July with movements of Sand Martins, Swallows and Pied and Yellow Wagtails. August produces a trickle of warblers, mainly Willow, but towards the end of the month and into September the pace increases with Common Redstart, Whinchat, Wheatear and Pied Flycatcher joining a wide variety of warblers. East winds are most likely to produce interesting birds and there is a chance of Wryneck, Red-backed Shrike, Bluethroat, Icterine or perhaps even a Greenish Warbler. Spurn is *the* locality on the British mainland for Barred Warbler. In October diurnal migrants (pipits, Linnet, Greenfinch and sometimes Twite) increasingly dominate the scene, but there can be large falls of Robins, thrushes or Goldcrests. Flocks of Meadow Pipits may occasionally be joined by a Richard's, and in late October Spurn has consistently produced records of Pallas's Warbler, as well as Yellow-browed Warbler and Red-breasted Flycatcher. Long-eared Owl is regular and may be flushed from the Buckthorn; Short-eared Owl tends to pass straight through. There can be large arrivals of thrushes in bad weather, especially Blackbird.

Winter is comparatively quiet. Large numbers of Red-throated Divers can occur offshore and wildfowl on the Humber include Brent Goose, often around the Narrow Neck. Hen Harrier, Merlin and Short-eared Owl may, with luck, be seen and in early winter there are often large numbers of thrushes feeding on the Buckthorn berries. Snow Bunting is regular, numbers varying from a few dozen to several hundred. Hard-weather movements can occur: fleeing birds can be seen moving south in early morning, especially if it is clear, and some may pause to feed.

Breeding birds include a few pairs of Little Terns around Easington, and Marsh Warblers have bred in a couple of recent seasons.

Access

Spurn Head Leave Hull on the A1033 through Keyingham and on to Patrington, and then take the B1445 to Easington, continuing on a minor road to Kilnsea. The road bends sharp left and then, at the junction (with the caravan site to the left), turn sharp right and continue for 1 mile to the entrance of the reserve. The Canal Scrape and hide lie to the right just before the gate. The Observatory is based at Warren Cottage at the base of the peninsula and offers simple self-catering accommodation, and there is a resident warden.

Access to the peninsula is unrestricted, apart from the Bird Observatory garden, the various RNLI, pilots' and coastguards' buildings, and the Point Camp area. There is a charge for cars using the road along the peninsula, usually collected near the Observatory. An Information Centre is situated outside the Bird Observatory, and there is a car park there and at the tip of the peninsula.

Visible migration is best watched from the Narrow Neck, 0.75 miles south of the gate. Movements usually peak in the first few hours of daylight and west winds are most productive. This can also be a good place to seawatch from. Otherwise migrants can occur anywhere but the Buckthorn is difficult to work. It is better to concentrate on:

- The trees and hedges around Kilnsea, especially at the churchyard and near the Crown and Anchor pub.
- Beacon Lane, which is bordered by hedges.
- The Canal Zone, an area of farmland with hedges and ditches.
- The Big Hedge, a mature belt of hawthorns which runs east–west across the base of the peninsula, about 100 yards north of the entrance gate.
- The trees and bushes around Warren Cottage.
- Chalk Bank, which is especially good for larks, pipits, wheatears and Snow Bunting.
- Near Chalk Bank, a hide on the Humber Shore overlooks a raised sand and shingle beach, used by roosting waders.
- The Point Camp area, especially the sycamores and elders around the old parade ground, and the stands of Sea-buckthorn by the lighthouse.

Easington Lagoons Just outside Easington, heading towards Spurn, there is a sharp right-hand bend. At the bend carry straight on along the track to park at the clifftop. Walk south from here to the lagoons and hides. The lagoons are good for waders if water levels are low, especially at

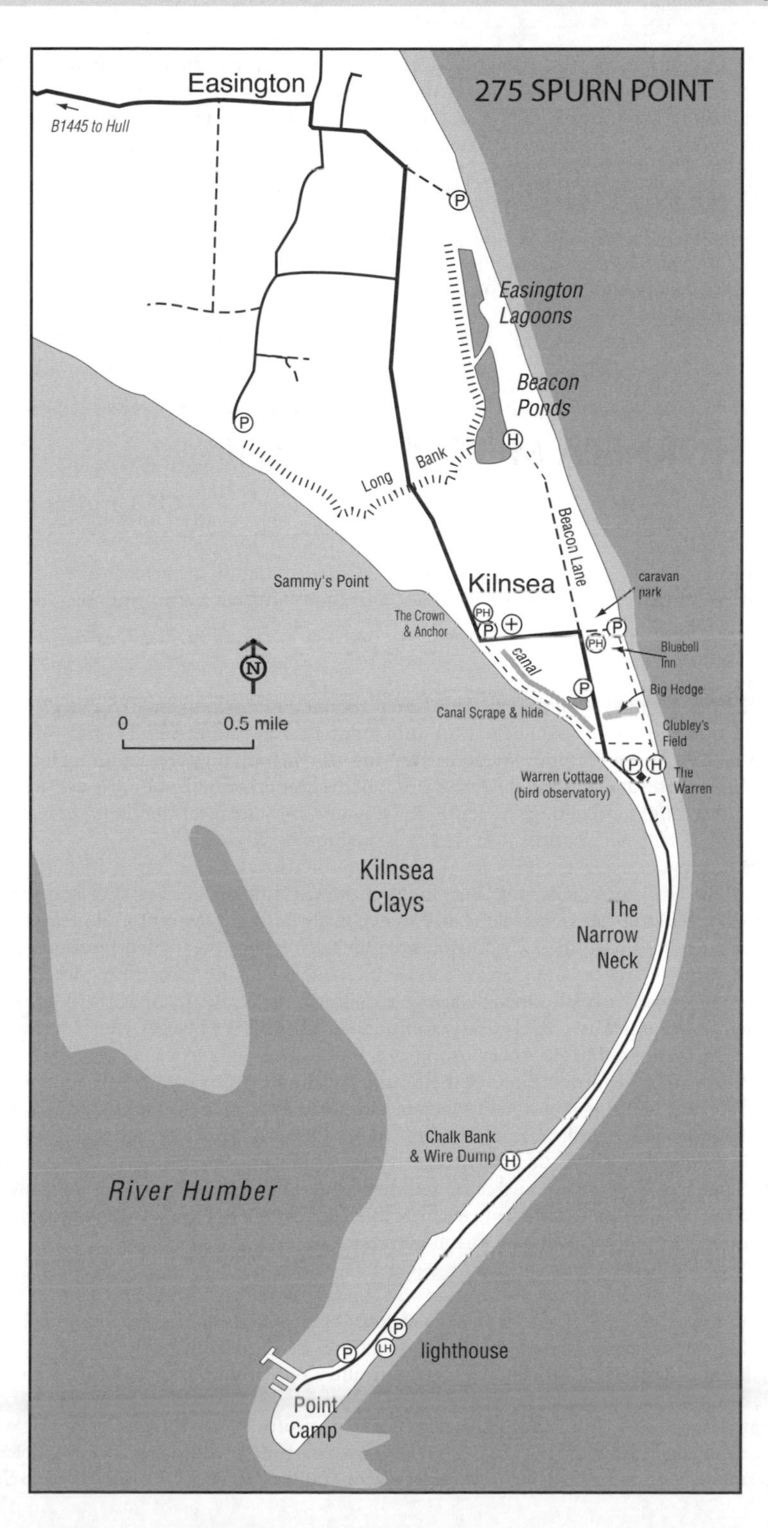
275 SPURN POINT
Easington
B1445 to Hull
Easington Lagoons
Beacon Ponds
Long Bank
Beacon Lane
Kilnsea
Sammy's Point
The Crown & Anchor
caravan park
Bluebell Inn
canal
Big Hedge
Canal Scrape & hide
Clubley's Field
Warren Cottage (bird observatory)
The Warren
0
0.5 mile
Kilnsea Clays
The Narrow Neck
Chalk Bank & Wire Dump
River Humber
lighthouse
Point Camp

high tide. In winter they attract the occasional duck or grebe and there are often many Snow Buntings, occasionally Lapland Buntings and raptors.
Beacon Ponds Lying north of Kilnsea, access is by walking north from Beacon Lane or by continuing south from Easington Lagoons. Birds are similar to Easington Lagoons.

FURTHER INFORMATION

Grid refs/postcode: Spurn Head (bird observatory) TA 420 148 / HU12 0UG; Easington Lagoons TA 407 187; Beacon Ponds TA 416 158
Spurn Bird Observatory: tel: 01964 650479; web: www.spurnbirdobservatory.co.uk
Yorkshire Wildlife Trust: tel: 01904 659570; web: www.ywt.org.uk

276 HORNSEA MERE (East Yorkshire)

OS Landranger 107

This large natural lake lies very close to the sea and is highly attractive to wildfowl and migrants, also holding a notable concentration of Little Gulls in early autumn. It is worth a visit at any season.

Habitat

Lying just under 1 mile from the sea, the Mere is about 1.5 miles long and 0.5 miles wide, with the 120 ha of open water being used for sailing and fishing (the east end is most disturbed). Bordering the town of Hornsea at the seaward end, the surrounding vegetation includes grassy fields, with large reedbeds at the west end, stands of deciduous woodland on the north and west sides, and reeds and scrub along the south shore.

Species

In winter there are large numbers of Wigeons, Pochards and Tufted Ducks, notable concentrations of 100–200 Gadwalls and Shovelers, and occasionally as many as 400 Goldeneyes. A few Goosanders are also regular, and occasional seaducks, notably Long-tailed Duck, may occur. The resident feral Greylag and Canada Geese and Mute Swans are sometimes joined by small numbers of Bewick's or Whooper Swans or feral Barnacle Geese. Around the Mere's fringe a few Jack Snipe and Water Rails find cover, and occasionally a Hen Harrier is seen. Bittern and Bearded Tit are rather irregular.

In early spring a few Rock Pipits of the Scandinavian subspecies occur. Subsequent passage brings Garganey and possibly Marsh Harrier, Osprey and terns, including Black and Arctic.

Breeding species include Great Crested Grebe, Pochard, Shoveler and Gadwall. In the waterside vegetation there are Reed and Sedge Warblers. Sparrowhawk, Great Spotted Woodpecker and Treecreeper nest in the stands of trees. There are late-summer concentrations of Cormorants, gulls and terns, most notably Little Gull, which is generally present early May–October and can peak at over 100. In midsummer, good numbers of Manx Shearwaters may be seen offshore at Hornsea.

Red-necked, Black-necked and Slavonian Grebes may occur in ones and twos in autumn and winter, but are irregular, as is Red-throated Diver. In autumn both Ferruginous Duck and Red-crested Pochard have appeared several times and a careful check through the ducks is worthwhile. Little Gull continues to occur, often with migrant terns and, as in spring, Black Tern is regular. If water levels fall, exposing some mud, a variety of waders may occur. There is a large roost of hirundines in the reedbeds, and these are hunted by Sparrowhawk and occasionally Hobby. There may be migrants in the surrounding woods, fields and hedges, and sometimes these include scarcer species such as Wryneck or Red-backed Shrike.

Access

Access is only possible at two points (the B1244 parallels the north shore but there are no viewpoints along this road).

Kirkholme Point From the B1242 in Hornsea town centre follow signs to 'The Mere and Car Park' via the track to Kirkholme Point. An information centre is situated in The Bungalow left of the entrance to the boating complex and is open at weekends, May–August. Parking is available in business hours, and there is a café. The Point is a good spot to look for gulls and terns over the Mere.

South Shore Travelling south out of Hornsea on the B1242 towards Aldbrough, fork right into Hull Road (opposite the garage). Park after about 800 yards in Mere View Avenue on the left and take the public footpath via the gate on the other side of Hull Road along the south shore of the Mere. The first 0.5 miles give the best views (and the stretch is known as 'First Field'), after this reeds and trees obscure the water, although the footpath continues to the west end at Wassand.

There is no other access to the reserve.

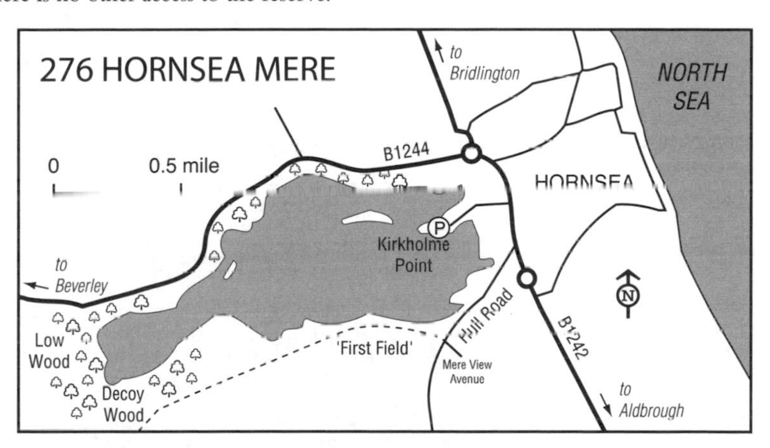

FURTHER INFORMATION

Grid refs: Kirkholme Point TA 198 472; South Shore TA 198 465

277 TOPHILL LOW (East Yorkshire)

OS Landranger 107

This complex of reservoirs lies north of Hull, between Beverley and Great Driffield. Sympathetic management in recent years has produced a range of habitats and the site now holds year-round interest (although noted for winter wildfowl and passage waders) and a reputation for attracting the rare and unusual.

Habitat

Lying in the valley of the River Hull and just 9 miles from the east coast, the varied habitats include two large reservoirs, as well as lagoons, marshes, scrub and some small stands of woodland. The reserve is owned by Yorkshire Water and managed in collaboration with Tophill Low Wildlife Group.

Species

Winter wildfowl usually include large numbers of Teals, Wigeons and Tufted Ducks, as well as Gadwall, Pintail, Shoveler, Pochard and Goldeneye (numbers of some of these peak in autumn). Parties of Whooper and Bewick's Swans may be present, and other waterfowl include Cormorant (numbers peak in autumn) and occasionally the rarer grebes or divers, Smew, Goosander and seaducks, notably Long-tailed Duck, Common Scoter and Scaup. The reserve has an impressive list of rare diving ducks to its credit. Large numbers of Lapwings and Golden Plovers winter in the area. Short-eared Owl is an occasional winter visitor. There is a very large gull roost, mainly Black-headed and Common Gulls. Jack Snipe join resident Water Rail in dense cover at the water's edge, while Bearded Tit sometimes appears in late autumn, and notable winter passerines include Rock Pipit, Siskin, redpolls and, sometimes, Brambling.

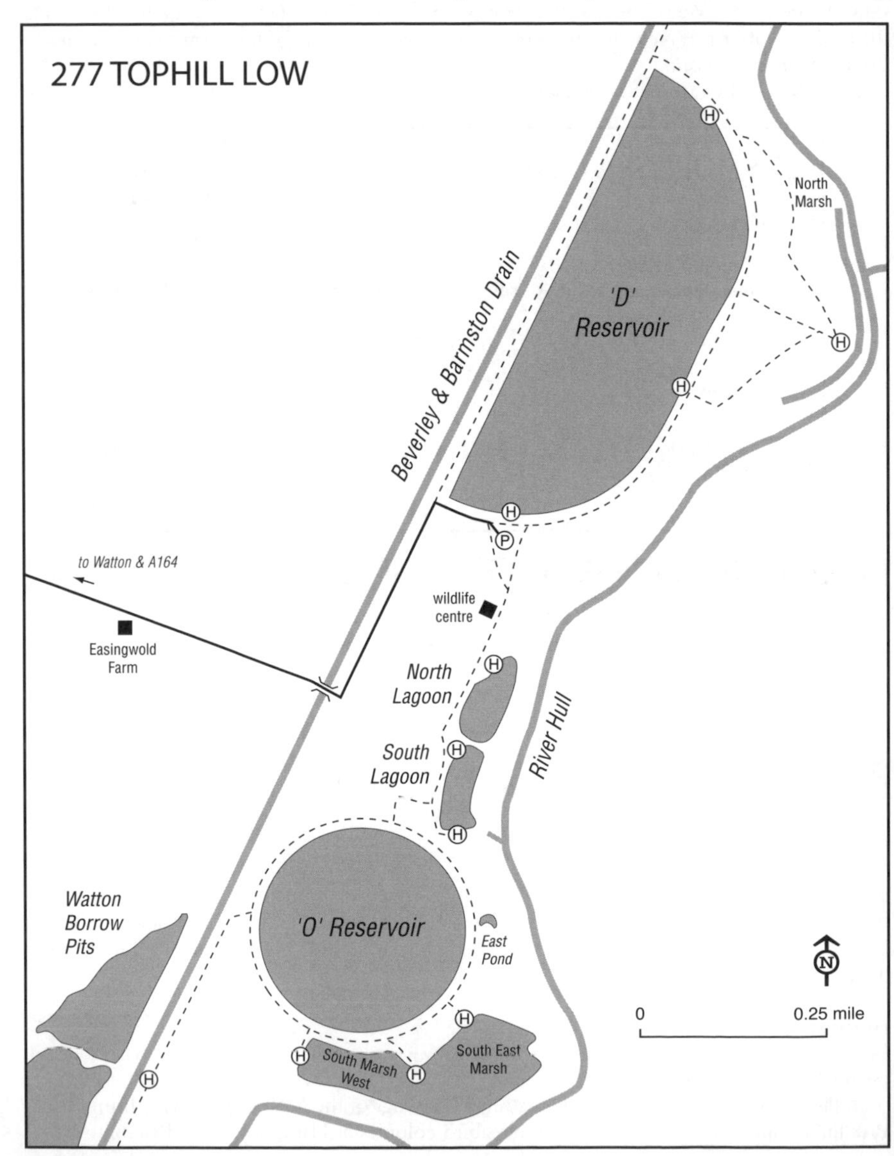

On passage Garganey and Osprey are occasionally seen, and Spotted Crake has been recorded several times. A variety of waders occurs, including Dunlin, Ruff, Greenshank, Spotted Redshank, Green and Common Sandpipers and Ringed Plover, with less frequent species including both godwits, Whimbrel, Little Stint and Wood and Curlew Sandpipers. Common, Arctic and Black Terns are regular, and Sandwich Tern and Little Gull are sometimes recorded. Small numbers of passerine migrants such as Tree Pipit, Yellow and White Wagtails, Wheatear and Whinchat occur.

Residents included Little and Great Crested Grebes, up to 500 Greylag Geese, Shelduck, Sparrowhawk, Common Buzzard, Water Rail, Little Ringed Plover, Barn and Little Owls, Kingfisher, and Grasshopper, Sedge and Reed Warblers.

Access

Turn east off the A164 about 6 miles north of Beverley at the south end of Watton (signposted Tophill Low Pumping Station) and follow the brown tourist signs along an unclassified road for 4 miles. Turn right at the main gates and follow the road round to the car park. The reserve is open daily except Mondays and Tuesdays (but is open on Bank Holiday Mondays), 9am–6pm in summer, 9am–4pm in winter. Day permits are available from the car park. A well-marked system of paths accesses 12 hides.

FURTHER INFORMATION

Grid ref/postcode: TA 072 485 / YO25 9RH
Yorkshire Water, Tophill Low Nature Reserve: tel: 01377 270690; web: www.yorkshirewater.com/leisure-time/tophill-low.aspx

278 BRIDLINGTON HARBOUR (East Yorkshire)

OS Landranger 101

Bridlington Harbour and beach are worth a look, especially in winter for gulls, waders and, in rough weather, seabirds sheltering in the harbour.

Habitat

The harbour is sheltered by piers and to the south is South Beach, part of a series of strands that extends from Bridlington to Spurn Head.

Species

In winter, divers, grebes, Cormorant, Shag, seaducks and auks may seek shelter in the harbour, especially during rough weather, when they can give excellent views. At low tide the exposed mud attracts numbers of commoner waders, with Turnstone, Purple Sandpiper and Rock Pipit on the harbour walls and rocks on the seaward side of the South Pier. When fishing boats are unloading their catches the large assemblage of gulls regularly includes Kittiwake and sometimes Glaucous, Iceland or Mediterranean. Otherwise, the gulls may be found loafing on South Beach (especially on falling or low tides), together with a variety of waders, notably Grey Plover, Sanderling and Turnstone.

Access

In Bridlington follow signs for South Beach, with parking available near the harbour and, especially in winter, in nearby side streets.

Mid-May to mid-September there are regular sailings of the MV *Yorkshire Belle* from North Pier in Bridlington Harbour to view the seabird colonies at Flamborough and Bempton. The boat carries 200 passengers and full facilities are available on board. Special extended cruises for birdwatchers are also operated, in conjunction with the RSPB, some of which are 'pelagics' in August–October to search for skuas and shearwaters. See also page 430 for details of cruises from North Landing on Flamborough Head.

FURTHER INFORMATION

Grid ref: TA 185 665
MV *Yorkshire Belle*: tel: 07774 193404 or 07950 648838; web: www.yorkshire-belle.co.uk
RSPB Seabird Cruises: tel: 01262 850959; email: bempton.cruises@rspb.co.uk

279 FLAMBOROUGH HEAD (East Yorkshire)

OS Landranger 101

Flamborough Head provides the best seawatching on the east coast and is rivalled as a seawatching site in Britain only by west Cornwall. Landbird migrants also occur in significant numbers and have included many rarities. In recognition if its importance in migration studies, a bird observatory has recently been formally constituted.

Habitat

The chalk cliffs of Flamborough Head rise to 250 feet and the peninsula projects over 6 miles into the North Sea. Largely comprising farmland, the network of hedges and some small plantations provide cover for migrants. Variety is provided by a couple of golf courses and a large pond near the Head. Danes Dyke, an ancient fortification, runs north–south at the base of the headland and holds much more extensive stands of cover, as does South Landing. Along the clifftop there are also areas of scrub and stunted hedges that attract migrants, notably at Selwicks Bay.

Species

There are always Gannets offshore at Flamborough and Fulmar is common all year, with larger numbers in spring and early autumn. Occasional 'blue' Fulmars, dark birds from northerly populations, are seen. Manx Shearwater is also quite common March–November and, of the rarer shearwaters, Sooty is regular July–October and can total several hundred on good days, and small numbers of Balearic Shearwaters occur in late summer. Great (August–October) and Cory's (scattered records April–October) Shearwaters are both much rarer. Little Shearwater, the rarest of all, has been seen on a handful of occasions in June–October. Arctic is the commonest skua; peak passage is in the last ten days of August and the first half of September. Great Skua can also appear in large numbers, up to 100 in a good day, especially in September and sometimes October. Though Pomarine Skua occurs in spring, autumn is better. It usually occurs in small numbers, only a handful even on a good day. Singles are coming through by late August and are increasingly frequent in September. Occasionally, it appears in large numbers and in these 'invasions' most are seen October–early November. Long-tailed Skua is very uncommon and tends to be associated with the main Arctic Skua passage in the last few days of August and September. Sabine's Gull is also uncommon, with a mean of about 15 a year, but sometimes as many as 100 Little Gulls can be seen in a day, especially in late autumn. Other species that appear on spring and autumn seawatches are divers (mainly Red-throated), ducks, waders, gulls and terns, especially Common, Arctic and Sandwich, but also including Black. In late autumn Little Auk may occur, occasionally in large numbers. For the breeding seabirds, which are present offshore in large numbers for much of the year, see Bempton Cliffs (page 431).

During passage periods there are occasionally large falls of migrants on the Head, with good numbers of thrushes, warblers, flycatchers and finches. These follow the same pattern as at other east coast sites. Wryneck, Common Redstart, Black Redstart, Ring Ouzel, Whinchat, Wheatear and Pied Flycatcher occur regularly, together with a variety of warblers. Icterine and Yellow-browed Warblers, Firecrest, Red-breasted Flycatcher and Ortolan Bunting are annual in very small numbers, and Flamborough is *the* locality on the east coast for Yellow-browed Warbler. Late spring and late autumn are generally best for scarce migrants and rarities.

In winter, fishing boats going to and from Bridlington pass close to the Head, and their attendant gulls should be checked for Glaucous and Iceland. All three divers are regular in winter, though Great Northern is scarce. A regular group of Black-throated Divers winters in Bridlington Bay and can sometimes be seen from the Head. Shag, auks, Scaup, scoters and especially Common Eider are frequent offshore in small numbers. The occasional Arctic or Great Skua may appear, and much less often a Pomarine. There may be large movements of seabirds in severe weather at this season. Peregrine can appear anywhere in the area. Other notable winter visitors include a few Snow Buntings and Rock Pipits of the Scandinavian subspecies, which can be recognised in early April as they begin to assume summer plumage.

Residents include Feral Pigeon (many appearing to be wild-type Rock Doves).

Access

Access to the area is via the B1255 from Bridlington. Flamborough Head is private farmland, with access only along roads and public footpaths. In particular, the gardens around the Head are private and residents' privacy should be respected at all times.

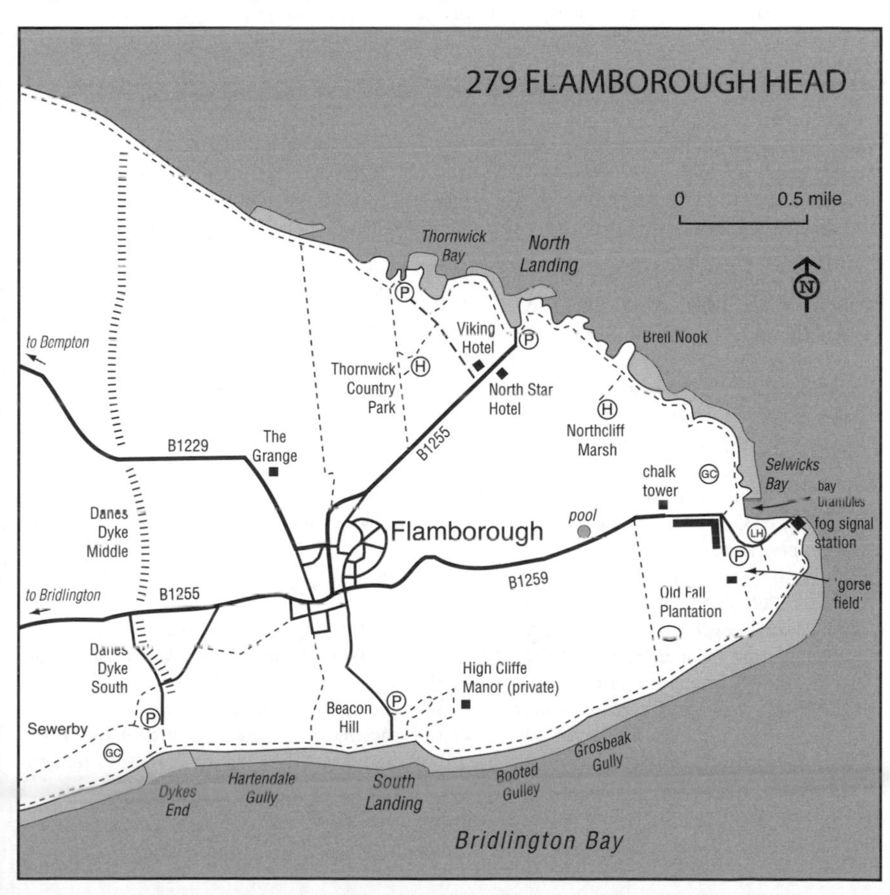

THE OUTER HEAD

Fog Signal Station Park in the public car park and walk the metalled road to the foghorn. This is the traditional seawatching site, in the shelter of the walls of the building or on the grassy ledges below (do not cross the fence and take great care on the potentially treacherous cliff edge). Seabirds can be seen at all times, but the best numbers and variety are likely when there is a strong north-west to east wind coupled with poor visibility. The Cliff End Café, on the right just before the car park, is a regular meeting point for birdwatchers.

Gorse Field An area of rough grazing with a dense stand of gorse lies immediately east of the car park, viewable from the permissive path running south from the car park to Head Farm and on to the South Cliff Path.

South Cliff Path This runs from the fog station towards Old Fall. There is little cover but in autumn the stubble fields and areas of rank grass may hold larks, pipits and buntings, notably Lapland Bunting and occasionally Richard's Pipit.

Old Fall Hedge Perhaps the best area for large numbers of migrants. Park in the car park by the new lighthouse and walk back along the road to the entrance to the footpath (signed 'New Fall'). Old Fall Plantation lies immediately to the east and is also very attractive, but is private and any temptation to enter must be resisted; with patience most of the birds in the wood can be seen from the public footpath (a mounted telescope is useful).

Roadside Pool This is easily viewed from the road and the surrounding sallows may hold migrants (it is a potentially good area for Bluethroat). Park in the car park by the new light and walk back to the area.

Selwicks Bay A scenic bay just north of the car park, the Bramble at the head of the Bay offers a first landfall for tired migrants (keep to the well-marked path in this sometimes precipitous area). The areas of low scrub along the north cliff path to North Landing may also be worth a look.

Golf Course A good area for larks, pipits, wagtails and Wheatear, especially in early morning. The course is private but can be seen from the north cliff path or the road.

Breil Nook One of the best areas on the Head for breeding seabirds, with spectacular numbers present in April–July. Access is from Selwicks Bay or North Landing along the clifftop footpath.

North Cliff Marsh A short path running south-west from Breil Nook gives access to a hide overlooking two pools.

NORTH LANDING AND THORNWICK BAY

North Landing From Flamborough follow signs to North Landing. Park in the car park and follow any of the many footpaths along the clifftop or around the hedges. Between Easter and late August there are regular boat trips, weather permitting, from North Landing to view the seabird colonies on the north side of the Head (see also page 427 for details of cruises from Bridlington).

Thornwick Bay Approaching North Landing on the road from Flamborough, turn left just before the Viking Hotel on the track to Thornwick Bay (car park and café). En route you pass a small area of reeds on the left which may hold migrants.

Thornwick Country Park The pools are viewable from a hide, accessible from the Thornwick Bay road.

SOUTH LANDING AREA

South Landing One of the best areas on the Head for migrants, but the extensive areas of cover can make it hard (and time-consuming) to work thoroughly. Follow signs from Flamborough village along South Sea Road to the pay-and-display car park. Public footpaths surround the wood and it is also worth checking the boulder beach; particularly productive areas are the ravine running south to the sea and the bridge at the east end of the wood. Follow the footpath east along the clifftop to Booted and Grosbeak Gullies, with more areas of cover attractive to migrants.

Beacon Hill The highest point on the Head and well positioned for observing visible migration and also any raptors in the area. On entering Flamborough on the B1255 from Bridlington turn first right and drive south through Hartendale housing estate. There is limited

parking near Beacon Farm and from here follow the footpath south to the hill.
Danes Dyke South Another large area of cover which, like South Landing, may be hard to work but can hold numbers of migrants. Signed from the B1255, park in the car park at the south end, with free access to all of the woodland south of the B1255; the ravine nearest the sea is perhaps the most productive area. A footpath leads west across Sewerby golf course, which may hold larks, pipits, wagtails, Wheatear etc. especially in early morning.

FURTHER INFORMATION

Grid refs: New lighthouse car park (for Fog Signal Station, Gorse Field, South Cliff Path, Old Fall Hedge, Selwicks Bay, Breil Nook) TA 254 706; North Landing TA 238 719; Thornwick Bay TA 231 722; South Landing TA 230 695; Beacon Hill TA 225 699; Danes Dyke South TA 215 696
Flamborough Bird Observatory: web: www.flamboroughbirdobs.org

280 BEMPTON CLIFFS (East Yorkshire)

OS Landranger 101

Lying on the north flank of the great chalk outcrop of Flamborough Head, Bempton Cliffs support one of the largest seabird colonies in Britain, with over 200,000 birds, and are best visited late April to mid-July.

Habitat

Bempton Cliffs rise to 400 feet and extend for 4 miles to Flamborough Head, and are topped by mixed farmland.

Species

Between April and August there are around 38,000 pairs of Kittiwakes between Bempton and Flamborough, making it the largest colony in Britain, despite recent declines, as well as 60,000 Common Guillemots, 15,000 Razorbills, several hundred Herring Gulls and Fulmars, and a few Shags. Although numbers have declined sharply in recent years, there are around 1,000 Puffins, but note that they are present until July only. By contrast, there are over 6,000 pairs of Gannets and the colony is still growing – in 1948 the first three nests were found and just a few continued to breed for many years until the 1960s when the colony slowly but steadily began to increase; by 1970, there were 24 nests and in 1980 there was a spectacular increase to 280 nests. Rock Dove also breeds on the cliffs, as do Peregrine, Jackdaw and Rock Pipit, with Sedge and Grasshopper Warblers and Tree Sparrow around the clifftop fields.

In the winter Short-eared Owl may be seen over the fields and Tree Sparrow is a regular visitor to the feeding stations.

Access

Leave Bridlington north on the minor road to Buckton and turn east on the B1229 to Bempton, from where you take Cliff Lane (by the White Horse Inn) to the reserve car park and visitor centre. Alternatively, coming from the north, leave the A165 just south of Reighton (opposite The Dotterel Inn) on the B1229 to Bempton. The reserve is open at all times and access to the cliffs is unrestricted but it can be very dangerous, especially when wet. A footpath runs along the clifftop for the length of the reserve with five clifftop observation points (the most distant 0.5 miles from the visitor centre). The visitor centre is open 10am–5pm March–October, 10am–4pm November–February; phone to check Christmas opening times. See page 427 and page 430 for details of cruises to view the seabird cliffs from North Landing at Flamborough and from Bridlington.

Migrants can be found at Bempton in season, and the most productive area has traditionally been the area of scrub and boggy ground near the seaward end of Hoddy Cows Lane, which runs north from near the pond in Buckton (0.5 miles west of Bempton village). This area can also be accessed by walking west on the clifftop path from the reserve car park.

FURTHER INFORMATION

Grid ref: TA 197 738
RSPB Bempton Cliffs: tel: 01262 851179; email: bempton.cliffs@rspb.org.uk

281 FILEY (North Yorkshire)

OS Landranger 101

Filey has gained a reputation in recent years as a premier site for both seawatching and scarce and rare passerines (indeed, it now rivals the more traditional sites to the south, Flamborough and Spurn). It is worth a visit in spring and autumn, and also has a good range of seabirds, notably often including Little Auk in late autumn. Nearby, the small Filey Dams reserve lies just inland of the coast behind Filey town, and is owned by Scarborough Council and managed by the Yorkshire Wildlife Trust.

Habitat

This is a typical seaside town. The beach at Filey is sandy, and north of the town lies Carr Naze, a cliff promontory that terminates in Filey Brigg, a long, narrow finger of low rocks that is almost covered by the sea at high tide. The Brigg is backed to the north and south by eroded cliffs of boulder clay that are penetrated by two well-wooded gullies, Arndale and Church Ravines. Above the cliffs are a large car park and caravan site, bordered by trees and shrubs, and to the north-west arable land. Protected to the north by Carr Naze and the Brigg, Filey Bay may shelter seabirds in rough weather.

Species

Well placed for seawatching, Filey has an impressive reputation and, unlike Flamborough Head, where one is forced to watch from the top of high cliffs, the low vantage offers a better chance of seeing petrels in favourable conditions. In autumn Manx and Sooty Shearwaters are regular, with occasional Balearic Shearwaters; as at all North Sea seawatching stations, the two large shearwaters, Great and Cory's, are rare. Strong north-west winds may bring small numbers of Leach's Petrels and all four skuas are regularly recorded, though Long-tailed is scarce. Large numbers of terns use the Brigg at high tide and can include Black, together with Little Gull in late September/early October. Seawatching may also produce divers, grebes, seaducks, waders (sometimes including Grey Phalarope) and auks, including Little Auk. Both Grey Phalarope and especially Little Auk are most likely in late autumn following strong northerlies and occasionally take shelter in the bay.

As well as seabirds, Filey attracts many passerine migrants in season. Wryneck, Richard's Pipit, Icterine, Barred and Yellow-browed Warblers, Red-backed Shrike and Red-breasted Flycatcher are annual in autumn. Easterlies in spring also produce records of shrikes, together with Bluethroat. During passage periods, Filey Dams attracts passerine migrants, together with 'fresh' waders and Garganey.

Winter brings all three divers (best seen in the bay around high tide), Red-necked and sometimes Slavonian Grebes, Common and Velvet Scoters, Common Eider, Goldeneye, Long-tailed Duck and Red-breasted Merganser to Filey Bay. Gulls, including Glaucous and occasionally Iceland, follow the fishing boats to Coble Landing, and the Brigg supports a large flock of

Purple Sandpipers, as well as often large numbers of Dunlins around the grassy Country Park car park. Small numbers of Snow and Lapland Buntings are possible around the car park and coastal fields, with Rock Pipit along the shoreline.

Breeding species include Fulmar, Cormorant, Herring Gull, Kittiwake, Razorbill, Common Guillemot, Puffin, Rock Dove and Rock Pipit.

Access

Filey Brigg From the A1039 in Filey follow signs to Filey Country Park and park in the metalled area in the far right-hand corner. From the car park footpaths lead over Carr Naze to the seawatching hide at the base of the Brigg; operated by the Filey Brigg Ornithological Group, this may be locked but visiting birdwatchers are welcome when it is manned, and regular or holiday visitors can hire a key for a small charge. It is possible to scramble onto the Brigg from near the seawatching hide, but it is a treacherous descent. Access to the beach is otherwise via steps on the south side, part way along the promontory. The scrub around the Country Park car park often holds interesting migrants.

Cleveland Way, Totem Pole Field and North Cliff From the Country Park car park it is possible to follow this footpath north-west along the clifftop, checking the fields and hedgerows for migrants and wintering buntings.

Arndale Ravine The small copse is excellent for migrant passerines. Access is from the Country Park Stores in the car park, via a gate signed 'Filey Sailing Club'.

Church Ravine The large trees here also hold migrants. From the Country Park car park follow the footpath around the seaward side of the caravan site to the bottom of the ravine.

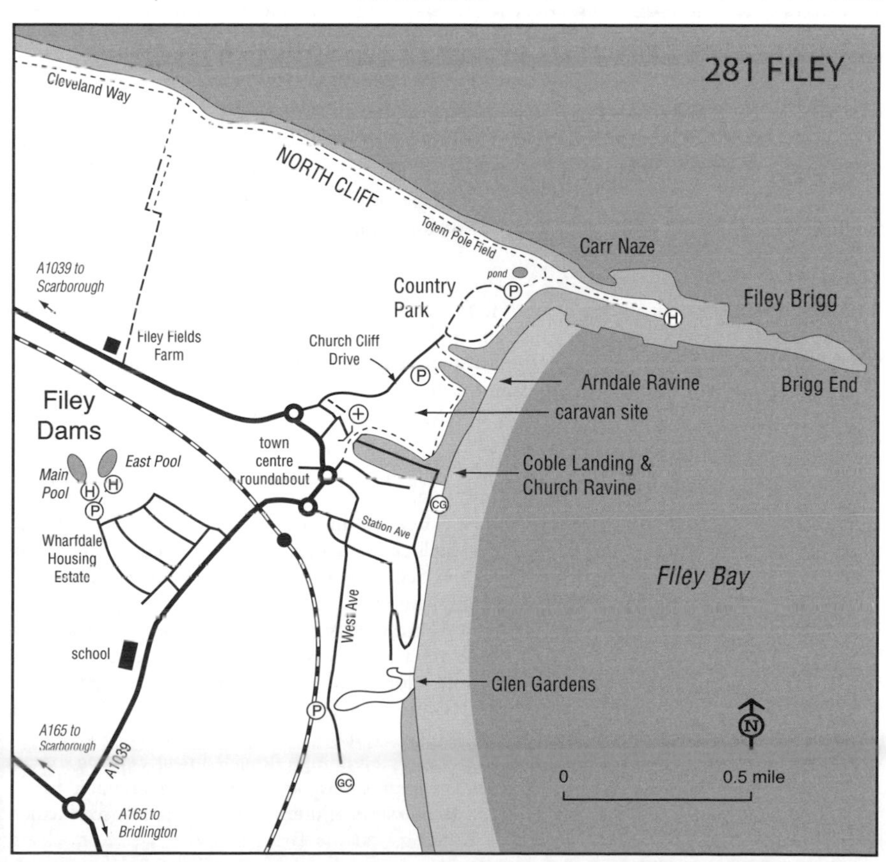

Walking up the ravine, under the footbridge, and bearing right brings you back to the Country Park entrance. Alternatively, leave the Country Park car park and turn left into Ravine Road towards Coble Landing, parking along the road.

Glen Gardens Another migrant trap. Access is from the south end of the seafront via the steps near the White Lodge Hotel, or from the town centre roundabout, heading east along Station Avenue and turning second right into West Avenue, with the entrance to Glen Gardens on the left opposite the small car park. The gardens along the entire seafront may hold migrants.

Filey Dams reserve On the west side of Filey and comprising a marshy area and artificial scrapes surrounded by farmland, this site is attractive to wildfowl, Little Egret and passage waders, with Tree Sparrows at the feeder. Leave the A1039 at the southern entrance to the Wharfdale Estate, turn left at the T-junction and follow the road round the right-hand bend for 650 yards (past several side roads) to its end at a rough car park. A hide adjacent to the car park overlooks the Main Pool, while the East Pool Hide lies 200 yards away by a signed path. There is no other access.

FURTHER INFORMATION

Grid refs: Filey Country Park TA 123 815; Glen Gardens TA 116 799; Filey Dams reserve TA 107 806
Filey Brigg Ornithological Group: www.fbog.co.uk

282 SCARBOROUGH AND SCALBY MILLS (North Yorkshire)

OS Landranger 101

Scarborough, a popular holiday resort, is also strategically situated on an east-facing coast, with the castle promontory forming a natural migrant trap. In autumn and winter a number of interesting gulls and other seabirds can also be found.

Habitat

The coast is largely built-up but Scarborough Peasholme Park and the area around the Mere provide stands of cover attractive to migrants, as do the well-wooded slopes of Castle Hill. Immediately south of Castle Hill lies the harbour, which provides shelter to seabirds, especially in rough weather. To both north and south, sandy beaches are interspersed by areas of rocky coastline, and north of Scarborough at Scalby Mills, Scalby Beck meanders across the beach, providing a bathing and loafing area for gulls.

Species

In winter look for divers offshore; the majority are Red-throated but Black-throated or Great Northern are sometimes seen. Grebes may include Red-necked and Slavonian, and other waterbirds include Cormorant, Shag, Common Scoter, Red-breasted Merganser, Goldeneye, Common Guillemot and Razorbill. In rough weather all may be seen in the harbour and, especially in late autumn, there may be Little Auk too. In such conditions look also for Grey Phalarope, which favours Scalby Mills. Gulls are a feature of the area, with regular records of Glaucous, Iceland, Mediterranean and Little. Commoner waders occur, with particularly impressive numbers of Purple Sandpipers. Residents include Kingfisher, Grey Wagtail and Dipper, and the local Rock Pipits may be joined by birds of the Scandinavian subspecies, although these cannot be identified until they assume breeding plumage in early spring.

On passage an excellent variety of passerine migrants may occur, including a notable list of rarities. Seawatching can also be interesting, with numbers of shearwaters and skuas given winds from the north.

Access

Scalby Lodge Pond This small area of fresh water is attractive to wildfowl and waders. Park on the verge of the A165 just north of the Scalby Lodge Hotel, cross the stile and follow the footpath to view the pond (telescope useful).

Scalby Mills Leave Scarborough north on the A165 towards Whitby. After 2 miles turn right into Scalby Mills Road (signed Sea Life Centre), continuing for 0.5 miles to the seafront car park. View the beach and stream from here, with good numbers of gulls and, in season, terns, especially around low water; evenings are particularly good. Look for waders on the beach and Common Sandpiper, Kingfisher, Grey Wagtail and Dipper on Scalby Beck, where it passes through a steep ravine before spilling onto the strand. In March–April, Rock Pipit of the Scandinavian subspecies is regularly seen on the beach by the stream. The slopes of the beck have areas of cover that should be checked for migrants by crossing the bridge over the beck and following the stepped path up. A good walk in winter is south from here along the seafront; if you have left your car at Scalby Mills, there are regular buses back from Scarborough.

Peasholme Park This lies near North Bay adjacent to the A165. From just north of the roundabout take Northstead Manor Drive and park in the car park adjacent to the nearby swimming pool (200 yards on the right), walking across the road to the park. The park attracts passerine migrants, and its lake may hold seaducks in rough weather.

Castle Hill Access is best from the north end of Marine Drive, parking near the small café and gift shop, and taking the path to the castle walls, bearing right to pass under an obvious

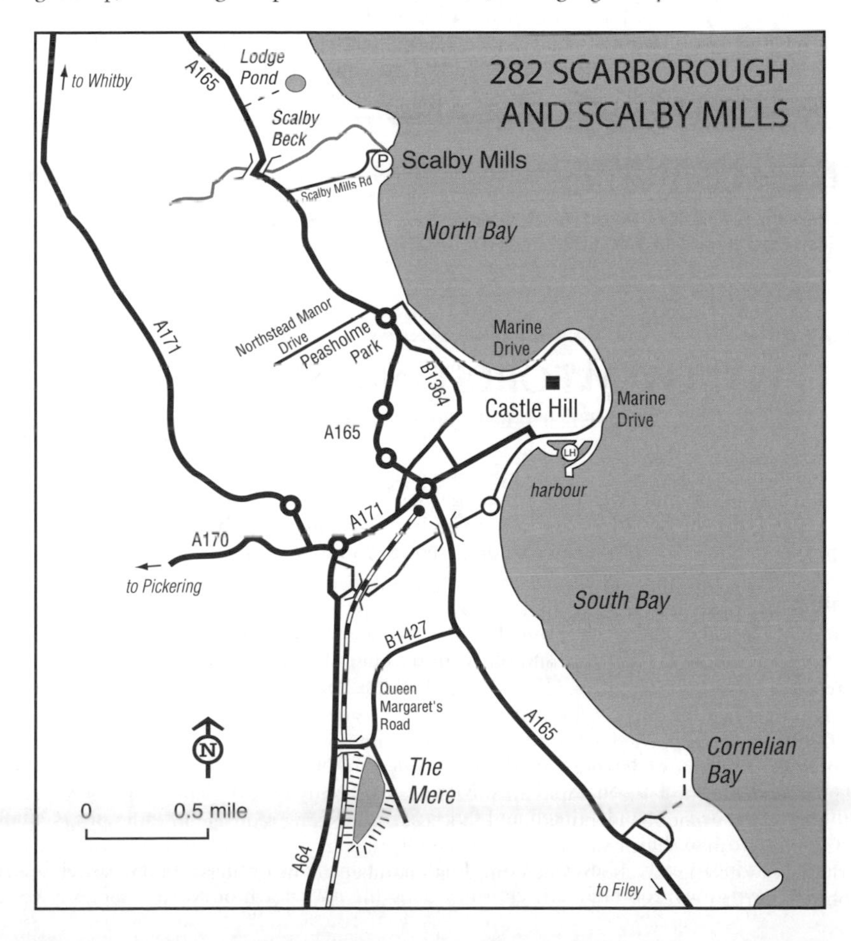

archway. The south-facing slopes have the most productive areas of cover. More than 2,000 pairs of Kittiwakes as well as a few Fulmars and Herring Gulls nest on Castle Hill cliffs, and can be watched from Marine Drive as they cruise overhead.

Marine Drive The main promenade in Scarborough, this can be a good seawatching vantage, with the advantage of being able to use a car for shelter. Marine Drive is, however, closed in really rough weather.

Scarborough Harbour Easily accessed from South Bay seafront, pedestrian access is possible to the three breakwaters (the outer, eastern pier from near the fun fair at the entrance to Marine Drive). In winter the east pier is linked to the central pier (Vincent Pier) by a footbridge, which commands a good view of the entire harbour. Gulls may be found in the harbour or loafing on nearby buildings, especially those along the west pier. In rough weather, look for divers, grebes, Shag, seaducks and auks in the harbour. Rock Pipit and Turnstone can also be found, with large numbers of Purple Sandpipers on the seaward flanks of the outer pier.

The Mere Leave Scarborough south on the A64 and, after about 1 mile, turn left on the B1427 (signed A165 Filey and Bridlington) into Queen Margaret's Road. Cross the railway bridge and the entrance to the Mere lies on the right after 300 yards. The lake is encircled by a driveable track. Though busy in summer, Reed Warbler and sometimes Kingfisher may be present, and stands of cover around the lake are attractive to migrants.

Cornelian Bay Leave Scarborough south on the A165 coast road and, after about 2 miles, turn east at the roundabout into Osgodby Hill. After 400 yards turn left into Cornelian Drive, taking the rough track from here east towards the sea, parking just beyond the large farm building on the right. The Cleveland Way runs along the clifftop and you can view the sea from here or take the path to the left, down the cliffs. The slopes have plenty of cover to attract migrants, the beach and rocky sills hold good numbers of waders, especially Purple Sandpiper, and the bay may have seaducks.

FURTHER INFORMATION

Grid refs: Scalby Mills TA 035 908; Peasholme Park TA 034 895; Castle Hill TA 051 894; Scarborough Harbour TA 048 887; The Mere TA 035 863; Cornelian Bay TA 056 860

283 WYKEHAM FOREST (North Yorkshire)

OS Landranger 101

Lying a little over 6 miles west of Scarborough, this area has become well known in recent years as a venue for watching Honey-buzzard. A visit late May–July in suitable weather offers the best hope of a sighting. The area forms part of the North Yorkshire Moors NP.

Habitat

The area has been extensively planted with exotic conifers, bordered to the north-west by Troutsdale, with areas of pasture and stands of deciduous woodland, and to the north by the River Derwent.

Species

Several pairs of Honey-buzzards are thought to nest in the region, and other raptors include Sparrowhawk and occasionally Hobby, Merlin and Goshawk (the latter especially in late summer). Residents include Green and Great Spotted Woodpeckers, Siskin, Lesser Redpoll and Common Crossbill.

Summer visitors include several pairs of Nightjars, which favour the areas of clear-fell before the trees have grown too high (any clearings along the road north of North Moor are worth checking at dusk), as well as Tree Pipit, Common Redstart and Whinchat.

In winter the forest is quiet, but raptors may be seen, including Goshawk in early spring and, in more open areas, Peregrine, Merlin or Hen Harrier. Common Crossbill may be more obvious, especially in years with high populations.

Access

To reach High Wood Brow Viewpoint turn north off the A170, 6 miles west of Scarborough, at Downe Arms Hotel in Wykeham, and drive north on a minor road signed 'North Moor' through the forest for about 4 miles to Highwood Brow. Turn left at the T-junction and park in the signed car park on the right after 300 yards. From the car park walk north for 300 yards to the observation platform, which has spectacular views of Troutsdale and the Upper Derwent Valley. Honey-buzzard is present late May–late August but is difficult to see after midsummer. The optimum time is 9am–1pm, on warm sunny days with scattered cloud and only light breezes, when they may be seen soaring or in wing-clapping display. They spend much time sitting below the canopy, and a long wait may be necessary, but views are often exceptionally good. Visiting birdwatchers are asked to keep to the viewpoint, and indeed, the chances of seeing Honey-buzzard away from there are slim. The viewpoint is managed by the Forestry Commission.

FURTHER INFORMATION

Grid ref: SE 935 887

284 ARKENGARTHDALE AND STONESDALE MOOR (North Yorkshire)

OS Landranger 91 and 92

This section of the Yorkshire Dales holds a typical selection of upland birds and is one of the better areas for Black Grouse. There is little to see, however, outside spring and early summer.

Habitat

The area is covered with *Juncus* moor interrupted by gullies and streams on Stonesdale Moor, with hill pasture in Arkengarthdale.

Species

Black Grouse are declining in Britain, for unknown reasons, and the remaining groups are best looked for when lekking in early mornings in April–May. To avoid disturbance, it is essential to stay in your car and use it as a hide. The birds often favour wet, rushy fields. Other species to look for include Red Grouse on the heather moorland between Tan Hill and Arkengarthdale, Peregrine, Golden Plover, Oystercatcher, Curlew, Common Snipe, Common Redshank, Dunlin, Short-eared Owl and Wheatear, with Common Sandpiper, Dipper and Grey Wagtail on the streams and rivers, and Ring Ouzel elusive around rocky outcrops. In the Swale Valley along the B6270 look for Goosander along the river.

Access

Arkengarthdale (OS Landranger 92) Leave the A6108 about 5 miles west of Richmond on the B6270 towards Kirkby Stephen. After c. 5 miles, turn north in Reeth on a minor road towards Langthwaite and into Arkengarthdale. After 3 miles turn right (north) on the minor road towards Scargill and the A66. The best area for Black Grouse is the moorland west of the road around Shaw Farm. The area can also be accessed by turning south off the A66 south of Barnard Castle on a minor road and following this for about 7 miles to Shaw Farm.

Stonesdale Moor (OS Landranger 91) Leave the A685 in Kirkby Stephen south on the B6259 towards Hawes and, after 1 mile, turn east on the B6270 towards Gunnerside and Richmond. After about 9 miles, just west of Keld, turn north on the minor road signed 'West Stonesdale and Tan Hill' to Stonesdale Moor. (Alternatively, approach the area from the A6108 near Richmond, via the B6270 past Arkengarthdale and on for about 13 miles to Keld.) After 2 miles there is a grassy lay-by on the left near a small bridge (NY 884 044), with another lay-by on the left after a further 1 mile; both spots may produce Black Grouse. Continuing north to Tan Hill, a right turn at the T-junction takes you to Arkengarthdale and passes through good Red Grouse habitat.

FURTHER INFORMATION

Grid refs: Arkengarthdale NZ 004 053; Stonesdale Moor NY 884 044

285 SCALING DAM RESERVOIR (Cleveland and North Yorkshire)

OS Landranger 94

Lying on the edge of the North Yorkshire Moors, this upland reservoir is surprisingly productive for wintering wildfowl and passage waders, and the surrounding moorland has resident Red Grouse and is good for raptors, especially in winter.

Habitat

The reservoir was constructed in the 1950s and covers about 50 ha, with large areas of mud exposed as water levels drop. The water is used for sailing Easter–October, but 8 ha in the south-west corner have been set aside as a sanctuary, including stands of willow carr. The reservoir lies at an altitude of 625 feet and is bordered to the south by heather moorland and a block of conifers.

Species

Wintering wildfowl include feral Greylag Goose (often with odd feral Pink-footed Geese – wild Pink-feet are rare), Wigeon, Teal and small numbers of Pochards, Goldeneyes and Goosanders (the latter usually spend the day to the south on the River Esk and roost on the reservoir); Whooper Swan, Gadwall, Pintail, Shoveler and Scaup are also fairly frequent but divers, Red-necked and Slavonian Grebes, Bean Goose, seaducks and Smew are only occasionally recorded. Other notable waterbirds include Little Grebe and Cormorant, and the gull roost may hold Glaucous Gull in winter or (especially in October and March) Mediterranean Gulls, as well as large numbers of Common and Black-headed Gulls, and there can be good numbers of Lapwings and Golden Plovers in the reservoir's hinterland. Interesting passerines may include Willow Tit (along Bog House Lane) and Brambling, and it is worth looking for Water Pipit and Stonechat. A major focus in winter is the concentration on raptors on the moorland, especially Merlin, Peregrine, Sparrowhawk and Short-eared Owl, with the possibility of Hen Harrier and Barn Owl. Afternoons and evenings are best. Long-eared Owl is sometimes found in the conifers.

Breeding species include Great Crested and Little Grebes, Teal, Woodcock, Whinchat, Lesser Whitethroat and occasionally Ringed and Little Ringed Plovers, and Black-headed Gull, and Red-necked and Black-necked Grebes may summer. Red Grouse is resident on the adjacent moors, and may be joined in summer by Merlin and Short-eared Owl, as well as Curlew, Common Snipe and Common Redshank.

On passage, Marsh Harrier and Osprey may occur in spring, together with Black-necked Grebe, Little Gull, Black and Arctic Terns and a handful of waders. If water levels fall a much greater variety of waders is possible in autumn, with Ruff, Black-tailed Godwit, Green, Common and Curlew Sandpipers, Greenshank, Spotted Redshank and Little Stint all regular. Marsh Harrier is again possible, as is Osprey.

Access

The reservoir lies immediately south of the A171, 12 miles west of Whitby. At the west end the yacht club car park gives access to a public hide overlooking the sanctuary area, while raptors are best watched for from upper parts of this car park. Walking west along the A178 for 50 yards gives access to Bog House Lane. This in turn accesses several footpaths across the moors, and in summer (22 March–31 October) access is also permitted to a path along the entire south side of the water. There is also a public car park at the east end of the reservoir.

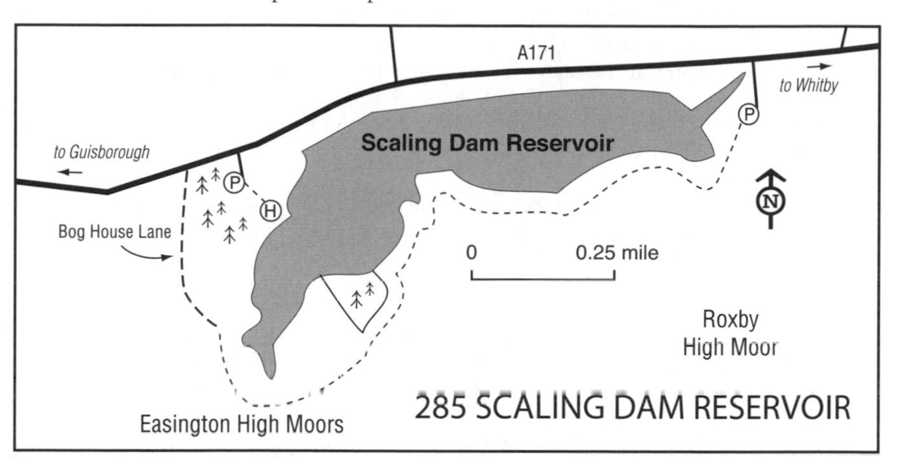

FURTHER INFORMATION

Grid ref: NZ 740 125

286 COATHAM MARSH AND LOCKE PARK

(Cleveland) OS Landranger 93

Lying immediately west of Redcar, these two adjacent sites attract a variety of wildfowl and waders, as well as passerine migrants in spring and autumn. Coatham Marsh is a 54-ha reserve of the Tees Valley Wildlife Trust.

Habitat

Coatham Marsh is bisected by the Saltburn–Darlington railway and the Fleet, an old saltmarsh channel. South of the railway a rubbish tip has been landscaped to form low hillocks and two artificial lakes (Long Lake and Round Lake), while north of the railway the relict saltmarsh is still prone to flooding, and a scrape has been excavated at the western end. Locke Park is a 10-ha town park centred on a small lake and planted with a variety of trees and shrubs. Backing the Saltburn–Middlesbrough railway, areas of cover along the line and the gardens to the north are also attractive to migrants.

Species

Early-spring migrants may include Black Redstart, Ring Ouzel and Firecrest, as well as wandering Hawfinch at Locke Park, while later in the season falls of Scandinavia-bound birds such as Chiffchaff and Pied Flycatcher may also include a Bluethroat or Red-backed Shrike. A visit to both Locke Park and Coatham Marsh can be rewarding during such falls. Passage waders can include Green and Common Sandpipers, Greenshank, Ruff and Little Ringed Plover, with Garganey, Osprey and Black Tern annual at Coatham Marsh.

In autumn the usual east coast migrants are regular and Yellow-browed Warbler is something of a speciality of Locke Park in late September and October. In addition to the waders listed above, Coatham Marsh may attract Black-tailed Godwit (which may number 50 on Middle Marsh), Whimbrel, Spotted Redshank, Curlew Sandpiper and Little Stint, and there have been several records of Spotted Crake.

Wintering wildfowl include Teal, Wigeon, Pochard, small numbers of Shovelers and sometimes Goldeneye, Scaup, Pintail or Gadwall; seaducks, divers and the rarer grebes are occasional. Large numbers of Oystercatchers may visit on Coatham Marsh, and a Jack Snipe may be found among the Common Snipes; similarly, small numbers of Ruffs or Golden Plovers occasionally join the large flocks of wintering Lapwings. Other species to watch for include Sparrowhawk, Merlin, Water Rail, Short-eared Owl and Kingfisher, and Peregrine, Stonechat, and Snow and Lapland Buntings are possible.

Residents at Coatham Marsh include Little Grebe and Common Snipe, and Shelduck, Sedge and Reed Warblers and occasionally also Little Ringed Plover and Grasshopper Warbler are breeding summer visitors. Common Tern may visit at this season.

Access

Coatham Marsh On the west outskirts of Redcar turn north off the A1085 at the traffic lights, opposite the junction with the A1042, into Kirkleatham Lane. After 0.25 miles turn west at the roundabout (towards South Gare) into York Road / Tod Point Road (aka Warrenby Road), and park on the left after 0.25 miles (just after the road rises over a disused railway) in the reserve car park. There is free access to the reserve, and a large hide overlooks the old saltmarsh at Middle Marsh (kept locked, but good views are possible from outside the hide), while a raised viewpoint overlooks the scrape. The footpath from here continues to the Oxbow, Long Lake and Round Lake, with a viewing platform overlooking Middle Marsh and then to Kirkleatham Lane.

Locke Park, Redcar Just east of Coatham Marsh, access is from Locke Road on the east perimeter and from Corporation Road (the A1085) along the south side (ignore the main entrance with the large Locke Park sign and park in the unfenced car park 100 yards to the east). Best in spring and autumn, there is usually little of note in midsummer and winter.

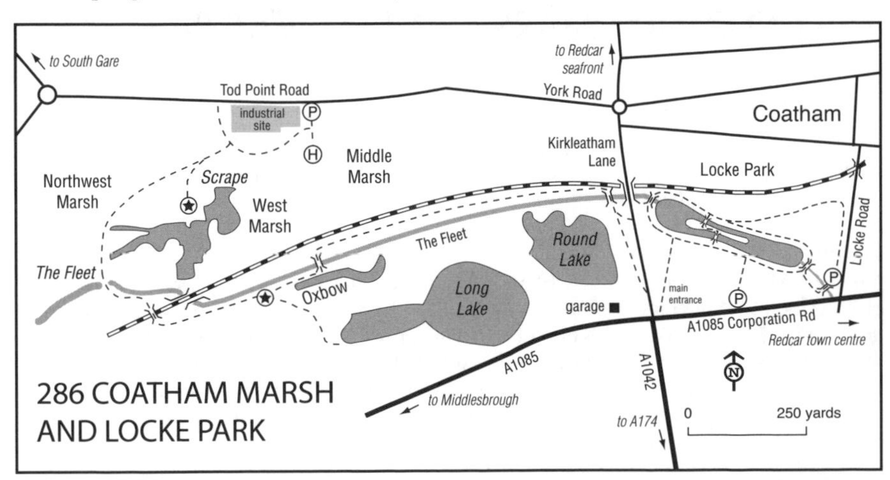

FURTHER INFORMATION

Grid refs: Coatham Marsh NZ 586 250; Locke Park, Redcar NZ 594 246
Tees Valley Wildlife Trust: tel: 01287 636382; web: www.teeswildlife.org

287 SOUTH GARE (Cleveland)

OS Landranger 93

Lying south of the mouth of the River Tees, this artificial breakwater and its immediate environs is attractive to passerine migrants, especially in spring, and also attracts concentrations of terns and skuas.

Habitat

A relict of the dunes and saltmarshes that once protected the mouth of the Tees, the area comprises a mosaic of dunes, rough grassland, mudflats, and fresh- and saltwater pools, while the Gare itself was built in the 19th century from iron-ore slag. On the south-west flanks of the Gare lie Bran Sands, which hold good numbers of the commoner estuarine waders.

Species

The Gare is at its best during spring and autumn, and a large variety of migrants can occur, including Black Redstart and Ring Ouzel in early spring, with Marsh Harrier, Garganey and Black Tern later in the season and Wryneck, Bluethroat, Red-backed Shrike and Common Rosefinch scarce migrants worth looking for in May–early June. The variety of birds is greater in autumn, with Icterine and Barred Warbler possible in August–September, in addition to Wryneck, Bluethroat and Red-backed Shrike, and a chance of Richard's Pipit, Yellow-browed Warbler, Red-breasted Flycatcher, Great Grey Shrike and Lapland Bunting later in the autumn, as well as commoner east coast migrants such as Woodcock, Common Redstart, Willow Warbler, Goldcrest and Pied Flycatcher, and oddities such as Long-eared and Short-eared Owls. Passage waders in autumn may include Greenshank, Little Stint and Curlew Sandpiper (on the estuary or the pools), as well as Green and Common Sandpipers.

The Gare can be productive for seawatching, but is somewhat overshadowed by Hartlepool to the north; except in the severest gales, most birds cross Tees Bay well out to sea. Nevertheless, a variety of shearwaters and skuas is possible, and in late-autumn or winter gales species such as Grey Phalarope and Little Auk may seek the shelter of the Gare.

Breeding species include Little Grebe, Water Rail, Ringed Plover, Common Snipe, Little Tern and Sedge Warbler, while Common Tern breeds nearby and small numbers of Common Eiders and Sandwich Terns summer. By August the large numbers of terns at the estuary mouth include small numbers of Arctic and sometimes one or two Roseates. Notably, these concentrations attract numbers of Arctic Skuas, and in some years a Long-tailed Skua (including adults with full tails) may linger.

In winter Red-throated Diver is regular on the sea, as are Cormorant, Common Eider, Red-breasted Merganser and Common Guillemot, and these are often joined by Shag, Common and Velvet Scoters and Long-tailed Duck, and occasionally a Great Northern Diver or Red-necked Grebe. The various fresh pools attract dabbling ducks, mostly Wigeon and Teal. Large numbers of gulls often include Mediterranean Gulls and occasionally a Glaucous or Iceland. Waders include the usual species on Bran Sands, including Sanderling and Grey Plover, with Purple Sandpiper on the rocks and breakwaters, together with Rock Pipit. Such large concentrations of birds regularly attract Peregrine, with Merlin and Short-eared Owl over rougher ground. Large flocks of Snow Buntings are found on the beach, especially around the mouth of the Lagoon.

Access

Just west of Redcar, leave the A1085 (opposite the junction with the A1042) north on a minor road (Kirkleatham Lane) towards Coatham. After 0.25 miles turn west at the roundabout into York Road / Tod Point Road (aka Warrenby Road). Follow this road, forking right at the roundabout over the old level crossing and then following the road round to the left and onto the Gare access road. This is a narrow private road and despite the 'no entry' signs, is open to vehicles (with the exception of one Sunday each year in September when it is closed to traffic for legal reasons). The road extends for 3 miles, almost to the tip of the breakwater, and when stopping it is best to pull right off the road to avoid obstructing traffic.

Coming from Redcar, to the east of the access road lies a golf course, and scattered along the road are bushes, which sometimes hold migrants. Next is a large freshwater pool (with other, smaller, pools further from the road) worth checking for migrant waders. West of the road are the Shrike Bushes, an area of reeds and scrub, which very often attracts migrants; just opposite the Quarries is a similar but rougher area. Further north, to the east of the road the Lagoon is an area of mud and sand that was formerly flooded from time to time, backed by saltmarsh, rough grassland and dunes; at its south end is a stand of *Suaeda* (Shrubby Sea-blite), traditional migrant habitat but difficult to work. Immediately north of the Lagoon the expanse of slag known as

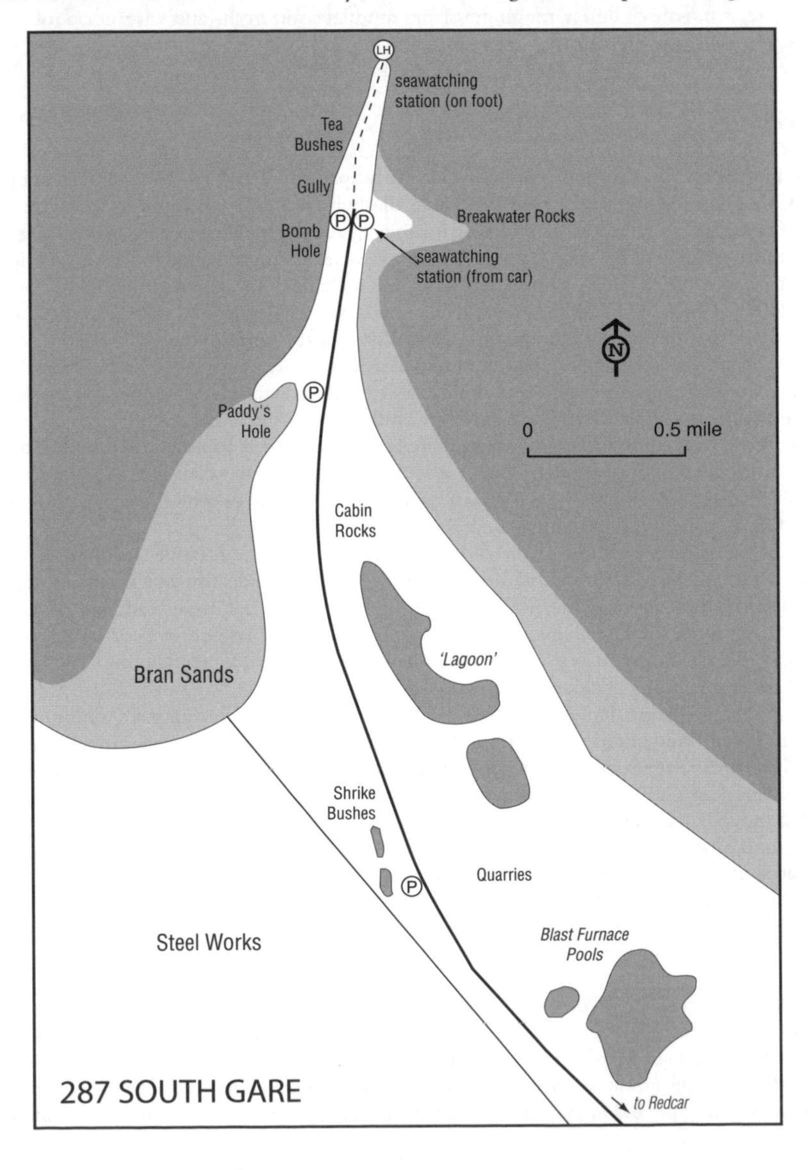

Cabin Rocks can be good for migrants. The Gare end is normally the most productive area for passerine migrants, notably the Tea Bushes, Gully, Bomb Hole and Paddy's Hole, although the limited cover means that birds may filter inland fairly rapidly. At the very end of the road it is possible to seawatch from the car around high water (the sea is too distant at low tide), otherwise seek shelter at a point almost at the tip of the Gare. Off the very tip, the Breakwater Rocks hold loafing ducks, waders and gulls.

FURTHER INFORMATION

Grid ref: NZ 556 279

288 NORTH TEESSIDE (Cleveland)

OS Landranger 93

This area holds a series of pools and tidal creeks around the Tees Estuary that, although surrounded by a grim industrial landscape, regularly hold interesting birds and have attracted an astonishing number of rare waders. Seal Sands is an NNR covering 335 ha and Saltholme an RSPB reserve.

Habitat

The Tees Estuary has been extensively reclaimed for industry, leaving just 100 ha of intertidal mud at Seal Sands. Despite this, the remaining Seal Sands retain their attraction for wildfowl and waders, holding internationally important populations of Shelduck, Knot and Sanderling. The mudflats are bounded to the north by Greatham Creek, a steep-sided channel, and west of the A178 this is bordered by areas of relict saltmarsh. Separated from the mudflats by slag reclamation banks are a number of areas of marsh, rough grazing and several pools.

Species

On the estuary and marshes wintering wildfowl include large numbers of Shelducks, Wigeons and Teals as well as Gadwall, Pintail and Shoveler. Greylag Goose is resident, and occasional wild White-fronted Geese may join them; Brent Goose and Whooper and Bewick's Swans are also sometimes recorded. Diving ducks congregate off the mouth of Greatham Creek by Seal Sands Hide and, although erratic, may include hundreds of Goldeneye and Pochard, and often Scaup; in rough weather Common Eider, Common Scoter, Red-breasted Merganser and sometimes Velvet Scoter, Goosander and Long-tailed Duck may join the commoner species. Large numbers of waders feed on Seal Sands, mostly Knots, Dunlins and Common Redshanks, but also Oystercatchers, Ringed and Grey Plovers, Curlews and Bar-tailed Godwits, along with a few Sanderlings. Golden Plover favours rough pastures and may be joined by a Ruff, Black-tailed Godwit or Spotted Redshank, and even Avocet may winter on the estuary. Such numbers of birds attract raptors, with Sparrowhawk, Peregrine and Merlin regular, Short-eared Owl favouring rough pastures, Hen and Marsh Harriers scarce but increasing, and sometimes Long-eared Owl may roost in areas of dense thorn bushes, notably at Haverton Hole and Hargreaves Quarries. Around marshy pools there may be Water Rail, Jack Snipe and, at Saltholme Pools, Bearded Tit, and gulls are abundant, with Mediterranean, Iceland and Glaucous irregularly recorded. Interesting passerines can include Rock Pipit, Stonechat and often Twite along the Long Drag. Lapland Bunting winters in very small numbers in rough pastures, but is very elusive.

On spring passage Marsh Harrier is regular, as are Garganey and Little Gull, but the main interest, as in autumn, is the waders. Regular species include Whimbrel, Ruff, Greenshank and Common Sandpiper, and almost anything is possible, with Temminck's Stint virtually annual.

Autumn passage is more varied, with waders again taking centre stage. Black-tailed Godwit, Green and Wood Sandpipers, Greenshank, Spotted Redshank, Little Stint and Curlew Sandpiper are all regular, and Pectoral Sandpiper is nearly annual. Regular rarities have included Wilson's Phalarope and White-rumped Sandpiper. Large numbers of terns, principally Sandwich and Common but occasionally including one or two Roseate and Arctic, loaf on Seal Sands and attract a few Arctic Skuas. Interesting gulls and terns may also be found on the Reclamation Pond. As in spring, Osprey, Marsh Harrier and Garganey are regular, but Black Tern is surprisingly scarce, though there have been several records of White-winged Black Tern. Wild Barnacle Geese may pass through in late autumn.

Little Egret is seen year-round and breeding species include Little and Great Crested Grebes, Gadwall, Pochard, Water Rail, Oystercatcher, Common Redshank, Common Snipe, Lapwing, Ringed Plover, some of Britain's most northerly Little Ringed Plovers and a large colony of Common Terns (at Saltholme), as well as Cuckoo, Yellow Wagtail, Whinchat, Stonechat, Reed and Sedge Warblers, and occasionally Grasshopper Warbler. In 2008 Avocets bred on the Tees Estuary, and seem likely to become well-established in the area.

Access

RSPB Saltholme Wildlife Reserve and Discovery Park The new RSPB reserve has pools (including the Haverton Hole pools), scrapes, reedbeds and grassland and is one of the best areas for migrant waders and Marsh Harrier, Garganey, Black Tern and occasionally Black-necked Grebe. The area is interesting in summer for breeding Water Rail, Little Ringed Plover, Common Tern, Whinchat and Reed and Sedge Warblers, and in winter for irregular Scaup and sometimes a roosting Long-eared Owl. Leave the A19 just north of Stockton on the A689 towards Hartlepool and, after 0.5 miles, at the roundabout, take the A1185 eastwards towards Billingham. After 4 miles, at the junction with the A178, turn right (southwards) at the roundabout and the reserve entrance lies on the right after 300 yards. (From the centre of Middlesbrough centre, the reserve can be accessed via the Transporter Bridge, every 15 minutes weekdays 7am–8pm, Saturday 11am–5.50pm, Sunday 2pm–5.30pm). Saltholme is 1.5 miles from the bridge along the A178 north from Port Clarence. The area can also be accessed on foot or by bicycle from the A1046 along the cycleway that starts at Holly Terrace. The reserve is open every day except Christmas Day, 10am–5pm (1 April to 30 September), 10am–4pm (1 October to 31 March). There is a state-of-the-art visitor centre with shop and café and three hides.

Dorman's Pool and Reclamation Pond Dorman's Pool is a shallow, slightly saline flash, probably the best wader pool on North Teesside, especially in wetter autumns when water levels do not fall far. Turn east off the A178, 75 yards south of Saltholme Pools, on the private road (Huntsman Drive) and cross the level crossing. There is a small car park on the left giving access to the Jeff Youngs Memorial Hide; this is kept locked, but a key is available to members of Teesmouth Bird Club. Further views can be obtained by turning left after a further 100 yards along a track past the south end of Dorman's Pool (it is worth stopping and scanning). Turn left at the T-junction beyond the pool and fork left again to a raised parking area for more views. The Reclamation Pond is an embanked lagoon which sometimes attracts interesting waders and is often used by roosting and loafing gulls and terns, and attracts ducks, notably Shoveler, in winter. Access is as Dorman's Pool, either viewing from the raised parking area (final fork right instead of left) or by following the track along the south shore. Note that although access to this area has always required a permit, this has seldom been enforced. Since 2007, however, access is only possible for card-carrying members of Teesmouth Bird Club.

Long Drag Pools The Long Drag marks the west perimeter of the reclaimed area of Seal Sands. Turn east off the A178 at the Seal Sands roundabout towards Seal Sands Industrial Estate and park at the private sign after 0.5 miles. From here walk north along the Long Drag to view the pools, scrub and reeds. Access is also possible from Seal Sands Hide (see below).

Cowpen Marsh Lying west of the A178 and south of Greatham Creek, this has fresh- and saltmarsh, and though once a reserve is now in industrial ownership. The area can be viewed from the main A178 (pulling completely off this busy road) but it is probably best to turn west off the road, 0.5 miles north of Seal Sands roundabout, onto a very short track, viewing Holme

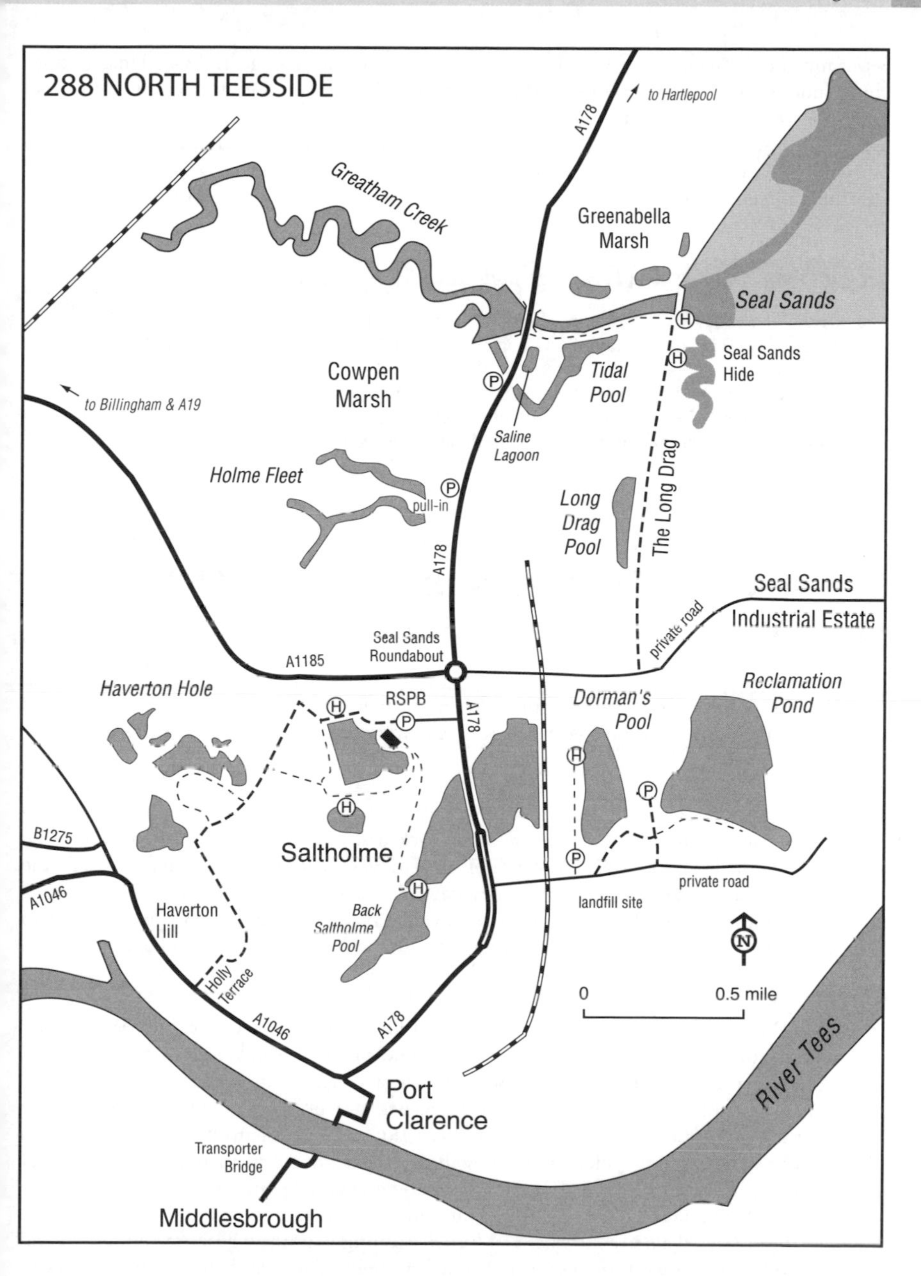

Fleet from the gate. The Teesmouth NNR car park lies west of the road after a further 0.5 miles (by the small stands of trees). From here walk north along the A178, to view Greatham Creek and the Saline Lagoon from the bridge. A public footpath runs west from here along the north bank of the river, with views from higher ground after 500 yards.

Seal Sands These hold large numbers of waders, best seen on a rising tide; at high water they are absent, but numbers of diving ducks can then be seen on Greatham Creek from Seal Sands Hide. Access is from Cowpen Marsh NNR car park, walking north along the A178 to Greatham Creek and taking the track east along the south bank of the creek for 0.5 miles to

Seal Sands public hide. From here it is possible to walk south along the Long Drag (with a public hide overlooking some pools 300 yards south of the Seal Sands Hide). Note that the road along the southern perimeter of Seal Sands is private.

Tidal Pool South of Greatham Creek and visible as you walk to Seal Sands Hide.

Greenabella Marsh An area of rough grassland, managed as a private nature reserve, north of Greatham Creek. It may attract raptors and is viewable from the track to Seal Sands Hide or the A178.

FURTHER INFORMATION

Grid refs/postcode: RSPB Saltholme Wildlife Reserve and Discovery Park NZ 505 232 / TS2 1TU; Dorman's Pool and Reclamation Pond NZ 516 229; Long Drag Pools NZ 515 236; Cowpen Marsh, Seal Sands, Tidal Pool, Greenabella Marsh NZ 508 251
RSPB Saltholme: tel: 01642 546625; email: saltholme@rspb.org.uk
Teesmouth Bird Club: web: www.teesmouthbc.com
Natural England, site manager: tel: 01429 853325; web: www.naturalengland.org.uk

289 SEATON CAREW, NORTH GARE AND SEATON SNOOK (Cleveland) OS Landranger 93

This area lies on the north side of the Tees Estuary and is particularly interesting in late summer for the concentrations of terns and attendant skuas, and in autumn/early winter for migrant passerines.

Habitat

Seaton Snook is a promontory of sand, mud and rock within the Tees Estuary, used by roosting waders, while the river mouth is protected by an artificial breakwater, the North Gare. To the north lie extensive areas of beach and dunes, backed by a golf course and the reclaimed salt-marsh known as Seaton Common. Stands of cover, attractive to migrants, lie between North Gare and the golf course, and in the cemetery at the seaside village of Seaton Carew.

Species

In winter Red-throated Diver and Great Crested Grebe may be seen on the sea, together with a few Common Scoters, Common Eiders, Red-breasted Mergansers and Goldeneyes, and there is a scattering of Common Guillemots and Cormorants. Other seaducks, Black-throated Diver and the rarer grebes may sometimes also be found. Raptors may include Peregrine, Merlin and erratically Short-eared Owl, hunting over the rough ground of Seaton Common or harassing the waders on the beach. The beach and estuary hold the commoner waders, notably Golden Plover on Seaton Common, Sanderling on the beaches and Purple Sandpiper at North Gare. Gulls are abundant and often include Kittiwake and sometimes Mediterranean Gull. Very mobile flocks of Snow Bunting haunt the beaches and dunes, and there are sometimes small numbers of Lapland Bunting and Twite in the rough pastures of Seaton Common, together with Stonechat.

Passage periods are the most exciting. Gannet, Fulmar and Manx Shearwater are regular offshore in late summer and autumn (especially if the wind has a northerly component), together with Kittiwake and sometimes Sooty Shearwater and, in late autumn, Little Auk. There may be up to 100 Little Terns on the beach in May and from late June there are gatherings of up to 2,500 terns off the estuary mouth, mostly Sandwich and Common but sometimes also Arctic and, in July–August, Roseate and Black Terns. These congregations attract up to 100 Arctic Skuas and, in some years, a Long-tailed Skua may linger too. Great and Pomarine Skuas can be seen offshore, but usually only after periods of northerly winds. Among landbird migrants, the usual chats, warblers and flycatchers are sometimes joined in spring by Marsh Harrier, Ring Ouzel, Black Redstart or Red-backed Shrike, and in autumn also by Woodcock, Wryneck, Barred, Icterine and Yellow-browed Warblers, Red-breasted Flycatcher and Great Grey Shrike. Breeding species include Ringed Plover.

Access

The area is accessed from the A178 north of Middlesbrough (and can be combined with a visit to Seal Sands and the wader pools of North Teesside, see page 443).

Seaton Snook Turn sharply south-east off the A178 300 yards north of the nuclear power station roundabout on the Zinc Works Road. After 0.75 miles park at the end of the road by the chemical works. Walk onto the sands and bear right, reaching the Snook after 650 yards. Large numbers of waders may be roosting here over the high tide, but can be anywhere along the beach. In late summer, the large tern roost attracts skuas. You can walk west along the sea wall, past the nuclear power station, for 0.75 miles to view Greatham Creek, or north to North Gare.

North Gare The most convenient route is along the minor road that leaves the A178 1 mile south of Seaton Carew (and 300 yards north of the Zinc Works Road). Follow the road for 0.5 miles to the large car park, and then walk 0.25 miles onto the breakwater, checking areas of

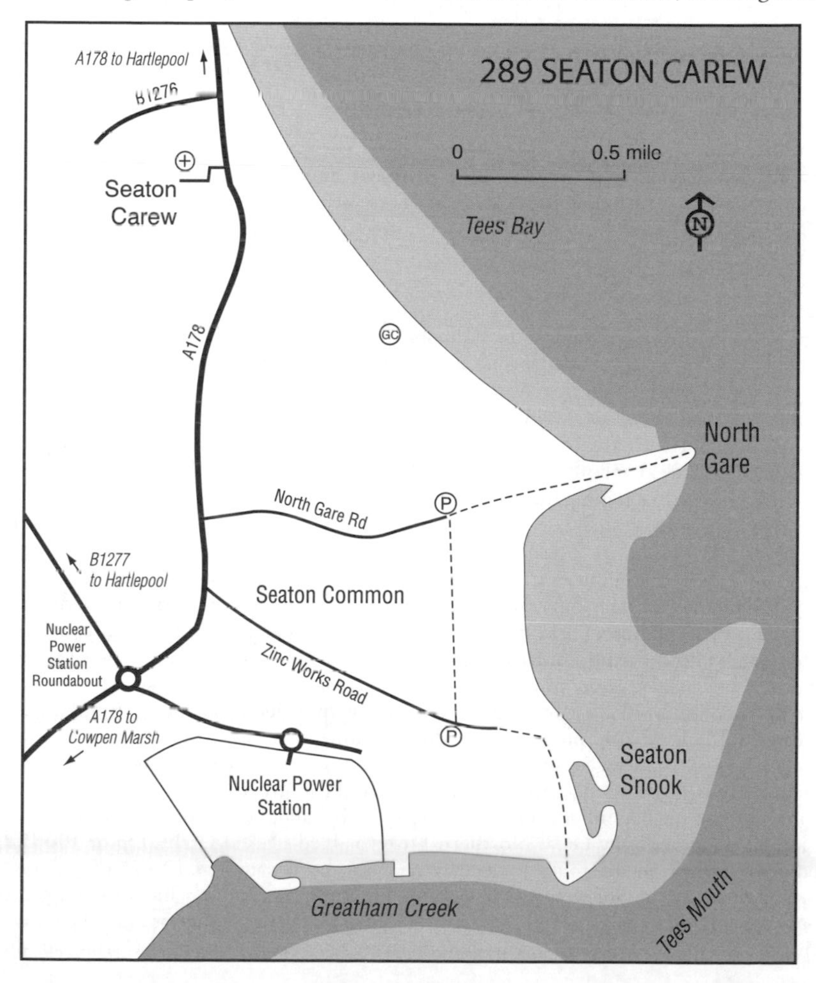

cover to the north for migrants. North Gare is good for seaducks, Purple Sandpiper and Snow Bunting in winter, and for seabirds in autumn. There is little shelter here for seawatching, and Hartlepool to the north or South Gare to the south are better in the right conditions. It is possible to walk south from the Gare along the beach to Seaton Snook.

Seaton Carew cemetery Entering Seaton Carew from the south on the A178, turn west just past the Seaton Hotel into Church Street, with the cemetery at the end of the road after just 50 yards. The churchyard holds passerine migrants, including annual Yellow-browed Warbler.

FURTHER INFORMATION

Grid refs: Seaton Snook NZ 535 273; North Gare NZ 532 281; Seaton Carew cemetery NZ 524 295

290 HARTLEPOOL (Cleveland)

OS Landranger 93

Hartlepool's geographical location on a promontory almost surrounded by the sea makes it a first landfall for many migrants, and also places it in pole position to witness movements of seabirds along the east coast.

Habitat

As with any migrant trap, arriving birds frequent any cover that is available, which in Hartlepool means gardens and the taller trees of the various cemeteries and parks. The decline of the fishing industry has reduced the town's attractiveness to gulls, but the Fish Quay Dock remains a magnet for these scavengers.

Species

Migrant landbirds can be exciting in both spring and autumn, although the latter season is invariably best. In spring, scarcer species include Ring Ouzel and Black Redstart early in the season with Bluethroat, Icterine Warbler and Red-necked Shrike possible in May–June. In autumn, east winds combined with poor weather may produce falls, and as well as the classic east coast migrants such as Common Redstart and Pied Flycatcher, scarcer species may appear, with Wryneck, Barred, Yellow-browed and Pallas's Warblers, Firecrest, Red-breasted Flycatcher and Great Grey Shrike joining the list of spring species.

Seawatching interest increases in midsummer, when numbers of Manx Shearwaters, Razorbills and Puffins are possible, but it is mid–late August that the season really gets underway. The best conditions are strong winds with a northerly component. Sooty and occasionally Balearic Shearwaters may join the regular Manx Shearwaters, and Arctic Skua is normally present offshore. These may be joined by a few Long-tailed Skuas in late August–September, with Great and Pomarine Skuas becoming more likely as the season progresses. Other seabirds include large numbers of Fulmars, Gannets, auks, Kittiwakes, and sometimes Little or even Sabine's Gulls, and numbers of terns, often including Arctic and sometimes also Roseate or Black. As autumn progresses ducks feature more heavily, including Goldeneye and a variety of seaducks. Seawatching in late autumn–winter can also produce Little Auk, sometimes in large numbers in the right conditions (i.e. northerly gales).

In winter the small numbers of Great Crested Grebes and Red-throated Divers on the sea are joined occasionally by Great Northern Diver or Red-necked Grebe. Up to 100 Purple Sandpipers winter on the rocks (especially near the breakwater), and Sanderling joins the commoner waders on Steetley Beach. Variable numbers of Common Eiders and Common Scoters are found off the Headland, sometimes joined by a Velvet Scoter or Long-tailed Duck, with Red-breasted Merganser in Hartlepool Bay. Other seabirds include Cormorant and a few Common Guillemots. Large numbers of gulls occur in the area, with Mediterranean, Iceland and Glaucous regularly recorded.

Breeding species include small numbers of Kittiwakes on the docks and old piers as well as Herring and Lesser Black-backed Gulls.

Access

The following largely follows *Where to Watch Birds in Northeast England* (Britton & Day, 2004).

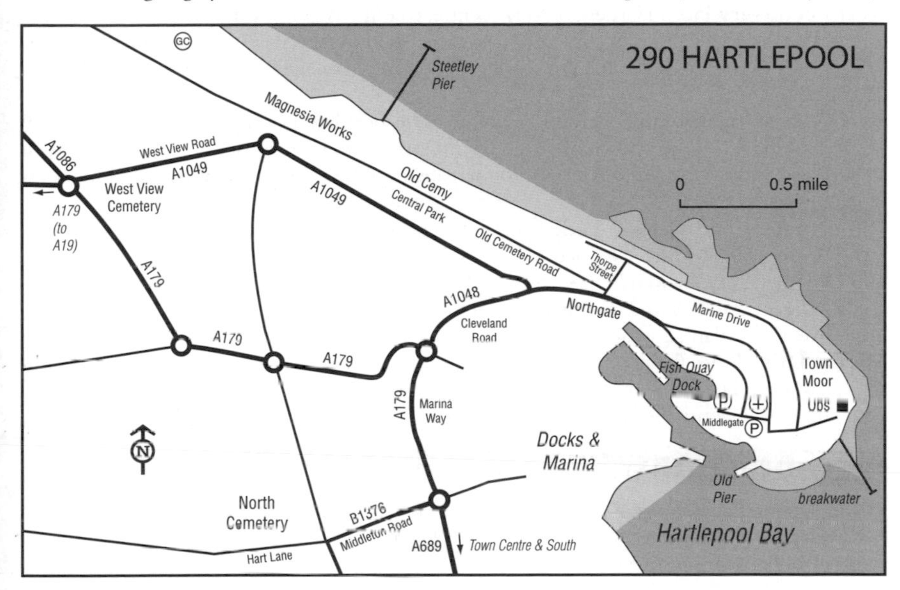

HARTLEPOOL HEADLAND

The most productive area for migrants, this lies 1.5 miles north-east of Hartlepool town centre, largely separated from it by the docks and marina complex. Leave the town centre north on Marina Way (A689 then A179) and follow this over the roundabouts to the A1048 Cleveland Road, following this round eastwards onto West View Road and then Northgate to the headland, following signs to the Maritime Museum and Fish Quay. Turn left (just before the bus depot) into Middlegate and park in the car park on the right after 100 yards. Searching for migrants here involves checking any patch of cover; the side streets from the seafront south-west of the seawatching point being especially productive. Care and consideration are needed when looking into private gardens; a good relationship has been established between local birders and residents that should not be compromised by visitors. In particular, follow any advice or instructions from local birdwatchers. Recognised hotspots include:

St Mary's churchyard Opposite the car park, access is possible if the wooden gate is open.
Fish shop trees Bordering the east side of the small park, next to the car park, near Verill's fish-and-chip shop.
The Croft The sloping park opposite the chip shop.
Olive Street trees Head along Middlegate, then turn left into Durham Street, right into Friar Terrace and left into Olive Street.
Doctor's garden At the junction of Durham Street and Friar Terrace.
Bowling green Across Marine Crescent from Olive Street, and the shrubs around the perimeter extend to the adjacent tennis courts. One of the classic spots.
Town Moor An extensive grassy area north of the bowling green, between the road and sea. Good for wheatears, pipits, etc.
Memorial garden On the seafront, beside Cliff Terrace, just south-west of the Observatory.

Hartlepool Headland 'Observatory' The best seawatching station lies near the lighthouse, just north of the breakwater at the tip of the Headland. Access is on foot from the bowling green (walking towards the sea and then turning right) or by parking by the sea at the end of Church Close (at the seaward end of Middlegate turn right and then left).
Fish Docks Quay Approaching the Headland, turn right off Northgate into Abbey Street (just before the bus depot) and park in the quayside car park. Gulls occur on the adjacent roofs as well as on the water, but have to be viewed through a fence.
North cemetery From Hartlepool town centre, take the A689 Marina Way northwards and turn left at the roundabout just past the historic quay onto the B1376 Middleton Road. Go straight over two sets of lights and the cemetery is on the right after 600 yards, with good stands of mature trees.
West View cemetery From Hartlepool town centre, bear left (not right, as for the Headland) and follow the A179 over two more roundabouts. At the third, turn right on the A1049 West View Road and the cemetery is almost immediately on the right. Again, it has good numbers of mature trees.
Old cemetery and Central Park Follow directions as for the Headland but on Northgate turn left into Thorpe Street (signed to CJC Chemicals and Magnesia) and immediately left into Old Cemetery Road. The cemetery lies on the right after 0.5 miles, and though sparsely vegetated, is near the sea and thus first landfall for some birds. Central Park is to the left, and has been extensively planted with trees.
Marine Drive Leave the town as for Old Cemetery, but continue along Thorpe Street towards the sea. Marine Drive follows the coast for 0.5 miles, and is good for seaducks; it is possible to walk around the entire Headland from the end of Marine Drive to the Old Pier.

FURTHER INFORMATION

Grid refs: Hartlepool Headland NZ 525 337; North cemetery NZ 502 330; West View cemetery NZ 495 350; Old cemetery and Central Park NZ 510 349
Teesmouth Bird Club: www.teesmouthbc.com

291 HURWORTH BURN RESERVOIR (Durham)

OS Landranger 93

Lying slightly to the north of the Teesside conurbation, this small reservoir is notable for winter wildfowl and a good wader passage, especially in autumn.

Habitat

Covering just 13 ha, the southern, dam end of the reservoir is deepest, while the northern half and the two cut-offs are shallower, with areas of mud exposed when water levels are low. The surroundings are largely farmland.

Species

Wintering wildfowl include Teal, Wigeon and Pochard, with small numbers of Goldeneyes, often a few Goosanders and sometimes Scaup or Pintail. Other notable waterbirds include Cormorant, Curlew, Common Snipe, Common Redshank and Water Rail. Occasionally Peregrine or Merlin is seen, as well as the regular resident Sparrowhawks. There is a small gull roost, largely Common and Black-headed Gulls, but occasionally including a white-winged gull.

Wader passage can be exciting, especially in autumn, but depends at both seasons upon water levels. Possibilities include Ringed Plover, Whimbrel, Black-tailed Godwit, Ruff, Greenshank,

Spotted Redshank, Green and Common Sandpipers, with Curlew Sandpiper and Little Stint in autumn. Shoveler, Common Tern and Grey Wagtail may also appear, and other possible migrants are Black-necked Grebe, Garganey, Black Tern and Little Gull.

Breeding species include Little and Great Crested Grebes, Shelduck, Oystercatcher, Little Ringed Plover, and sometimes Common Sandpiper and Common Redshank. Other summer visitors include Turtle Dove and Lesser Whitethroat, and Little Owl, Willow Tit and Tree Sparrow are resident.

Access

Leave the A19 west on the B1280 towards Wingate and, after c. 0.5 miles, turn left (south-west) on a minor road towards Trimdon, following the road right and then right again to the reservoir. Park on the right at the Castle Eden Walkway and follow the old railway track north to view. It is possible to make a circuit of the reservoir by taking the footpath westwards, c. 100 yards north of the reservoir, and then back south to the road, although high hedges obscure views of the water on the western side.

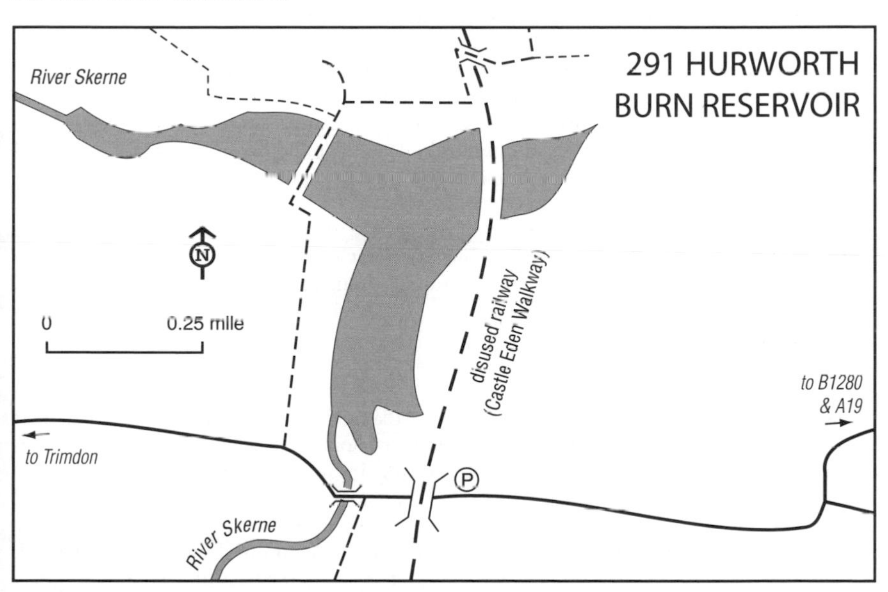

FURTHER INFORMATION

Grid ref: NZ 410 332

292 UPPER TEESDALE (Durham)

OS Landranger 92

Perhaps most famous botanically (and for the failure in the 1960s to prevent Cow Green Reservoir from damaging one of the great natural wonders of Britain), Upper Teesdale does possess ornithological attractions, notably Black Grouse (best looked for in March–April), a variety of other moorland birds and high concentrations of breeding waders (best June to early July).

Habitat

Upper Teesdale comprises a mosaic of habitats, with moorland and hill pastures on higher ground and hay meadows and areas of deciduous woodland in the valleys. There are many small tarns as well as Cow Green Reservoir straddling the border of Cumbria and Co. Durham in the Tees Valley above Cauldron Snout. Part of the area lies in the 7,387-ha Moor House-Upper Teesdale NNR.

Species

In this region Black Grouse favour areas of hill pasture, grazed by sheep, to lek. British (and especially English) populations have declined in recent years, making this a threatened species. The Langdon Beck lek is one of the best in the country with 20–30 males. Other breeding species include a variety of waders, with Oystercatcher, Golden Plover, Lapwing, Common Snipe, Curlew, Common Redshank and a handful of Dunlins and Ringed Plovers. Indeed, Upper Teesdale west of Forest-in-Teesdale holds one of the densest concentrations of breeding waders in Britain. In addition, Common Sandpiper joins resident Goosander, Grey Wagtail and Dipper on the River Tees (and Red-breasted Merganser and Goldeneye may also summer). Wheatear and Ring Ouzel are relatively common summer visitors, the latter favouring areas of crags and rocky slopes. Red Grouse frequent areas of heather, and small numbers of raptors breed, including Merlin, Peregrine, Sparrowhawk and Common Buzzard. Stands of woodland hold a variety of typical residents, including Woodcock, Great Spotted Woodpecker, Marsh Tit and Nuthatch, which are joined in summer by Tree Pipit, Common Redstart, Garden and Wood Warblers, and Pied Flycatcher. Occasionally Common Crossbill and Siskin may be found in the conifer plantations.

Winter is bleak but Black Grouse is easiest to see in late winter and early spring, Red Grouse remain on the heather and the occasional Hen Harrier, Common Buzzard, Peregrine, Merlin or Short-eared Owl may be seen hunting, occasionally joined by a Goshawk or Rough-legged Buzzard (the prime area for raptors being around the Eggleston–Stanhope road), with Raven also possible, especially in the west.

Access

Langdon Common Holding one of the largest Black Grouse leks in Britain, for the best experience the area is best visited in the first couple of hours after dawn, although some birds may be present throughout the day. Leave Middleton-in-Teesdale north-west on the B6277 and, after 7.5 miles, the road crosses Langdon Beck. After a further 0.5 miles take the minor road northwards, signed Weardale and St John's Chapel. After 0.25 miles along this minor road there is a cattle grid and after a further 0.5 miles the road crosses Langdon Beck and Black Grouse lek in the valley to the east of the road between the cattle grid and the beck. It is important to stay in your car at all times, even if the birds are distant; it is possible to pull off the narrow road onto the verge at various points. If the birds are absent, check the area north of the B6277 between Langdon Beck and the turn-off and also the fields to the east of the minor road between the turn-off and the cattle grid. Look also for Common Sandpiper and Dipper on the beck, and Red Grouse on the slopes of Three Pikes (west of the minor road).

Widdybank Fell and Cauldron Snout An excellent area for breeding moorland birds. From the B6277 at Langdon Beck turn left by the hotel on the minor road to Cow Green Reservoir. After 750 yards park by the gated track on the left (Natural England NNR sign), go through the gate and walk south on the farm track to Widdybank Farm. Through the farmyard take the footpath to the left down to the River Tees and follow the Pennine Way west along the north bank of the Tees towards Cauldron Snout (about 2 miles). The river runs through a deep valley past Cronkley Scar and Falcon Clints, with Peregrine, Merlin, Red Grouse, Ring Ouzel, Raven and Twite all possible, as well as the usual river birds. By following the track north from Cauldron Snout to Cow Green Reservoir car park, and then the road back towards Langdon Beck, a circular route of 7 miles around Widdybank Fell (which rises to 1,715 feet), good numbers of up to ten species of breeding waders may be seen. (Note that the track past Cauldron Snout is tricky, and you should be equipped for adverse weather if attempting this circuit.)

Cow Green Reservoir Take the signed turn from the B6277 at Langdon Beck to the car park overlooking this bleak upland reservoir. The nature trail running south from here is easy walking and a small number of breeding waders may be seen as well as small numbers of waterfowl.
High Force The highest waterfall in England. Leave Middleton-on-Tees north-west on the B6277 and, after 4.5 miles, park on the right by the hotel. Cross the road and take the track for 600 yards to the falls. The surrounding mixed woodland holds Woodcock, Wood Warbler, Pied Flycatcher and, in invasion years, Common Crossbill, with Common Sandpiper, Grey Wagtail and Dipper on the river.
Eggleston Common Comprising excellent heather moorland, this is a good area for raptors in winter. Leave Eggleston village north on the B6278 towards Stanhope. After 2.5 miles the Common lies to the east of the road.
Barnard Castle Woods These hold the typical deciduous woodland species, with Goosander, Dipper and Grey Wagtail on the river. Park at Barnard Castle and take the footpath from the village green (west of the castle) north-westwards along the north bank of the River Tees for about 2.5 miles.

FURTHER INFORMATION

Grid refs: Langdon Common NY 849 326; Widdybank Fell and Cauldron Snout NY 847 309; Cow Green Reservoir NY 810 309; High Force NY 885 286; Eggleston Common NY 990 280; Barnard Castle Woods NZ 048 164
Natural England: www.naturalengland.org.uk
North Pennines Black Grouse Recovery Project: www.blackgrouse.info

293 WASHINGTON WWT AND BARMSTON POND (Durham)

OS Landranger 88

Lying between the urban sprawls of Sunderland and Washington, this small area holds a variety of wetland and woodland birds, with waders on passage, and the waterfowl collection provides additional year-round interest.

Habitat

Washington is owned by the WWT and about one-fifth of the 42-ha grounds are occupied by a collection of over 100 species of swans, geese and ducks, the remainder comprising a mosaic of brackish and freshwater pools, and small stands of reeds and woodland on the north bank of the River Wear (still tidal at this point). To the north of the A1231, Barmston Pond was formed in 1970 by subsidence and is an LNR owned by Sunderland City Council. A freshwater pool, there are shallow scrapes at both ends and the pond is managed to provide ideal feeding conditions for passage waders in autumn.

Species

Wintering wildfowl at Washington include many Pochards and Teals and a few Wigeons and Shovelers, as well as up to 60 Goldeneyes, often with a handful of Goosanders. Other notable waterfowl include Great Crested and Little Grebes and Cormorant. Long-eared Owl breeds nearby and may sometimes be found roosting in areas of dense cover. Waders include up to 200 Common Redshanks, a few Common Snipes and occasionally Jack Snipe, and small numbers of Water Rails occur throughout. Wintering Siskin and redpolls may be found anywhere in the grounds. and the feeding station attracts woodland birds such as Great Spotted Woodpecker, Willow Tit and Bullfinch, with Brambling possible in late winter, and these in turn attract hunting Sparrowhawk. Barmston Pond has a similar range of waterfowl and there is some interchange with Washington. The surrounding fields may hold Merlin, Short-eared Owl and Golden Plover, and there is a possibility of a hunting Long-eared Owl at dusk.

On passage a variety of waders may occur at Washington and Barmston Pond (the latter being particularly productive in autumn), including Common and Green Sandpipers, Greenshank, Dunlin, Ruff and Black-tailed Godwit. In spring Osprey, Marsh Harrier, Common Buzzard, Hobby and Garganey may pass through, and in autumn there are concentrations of Grey Heron and Common Tern on Wader Lake and Barmston Pond, with other terns, Little Gull and Garganey scarce visitors. Waders at this season may include Spotted Redshank, Little Stint and Curlew Sandpiper, in addition to those species listed for spring.

Breeding species include a small number of Grey Heron (viewed at eye level from Heron Hide or via a video link), Sparrowhawk, feral Shelduck and Gadwall, Lapwing, Little Ringed Plover, Oystercatcher, Common Redshank, a handful of pairs of Common Terns (on Wader Pool), Stock Dove, Reed, Sedge and Garden Warblers and Willow Tit.

Access

Washington WWT From the south, turn east off the A1(M) at junction 64 just south of Gateshead on the A195. Continue on the A195 past its junction with the A182 and then, after

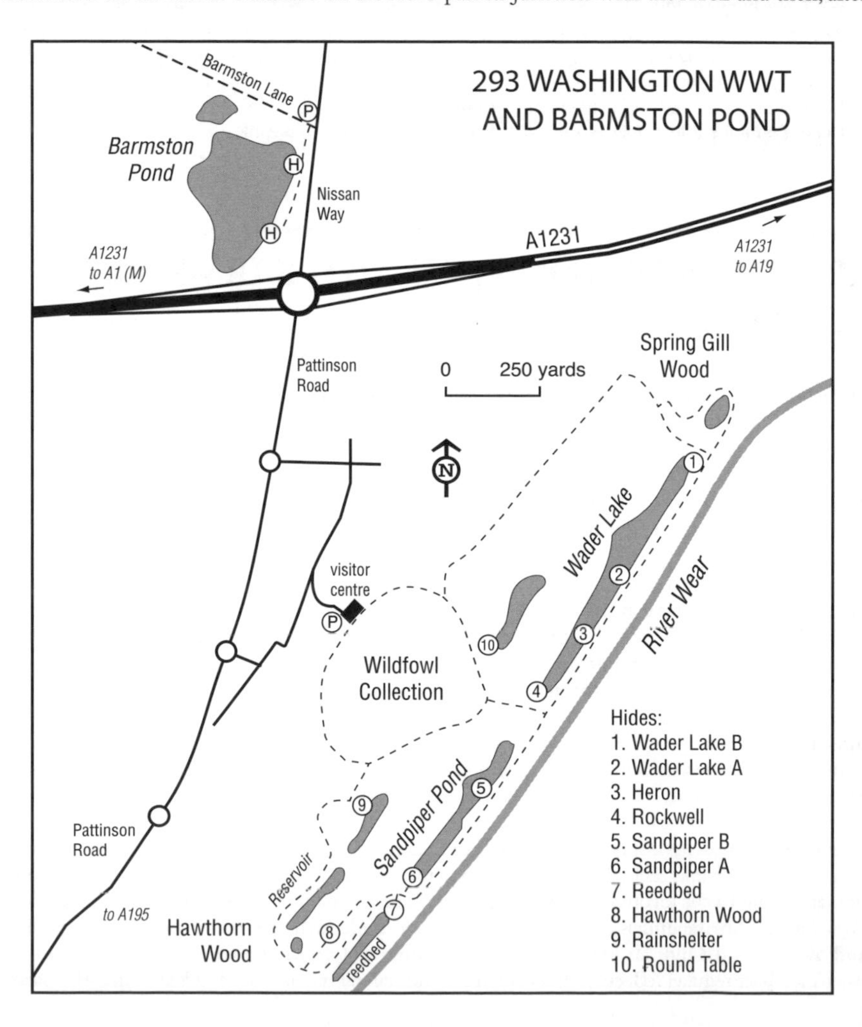

1 mile, turn right at the second roundabout on to Pattinson Road and follow signs to the WWT. Alternatively, leave the A19 just west of Sunderland on the A1231 towards Washington. After 1 mile turn left at the roundabout on to Pattinson Road, and then left again after 300 yards at the next roundabout, following signs to the WWT car park. The grounds, shop, visitor centre etc. are open daily (except Christmas Day) 9.30am–5.30pm April–October and 9.30am–4.30pm in winter. There are ten hides overlooking the pools, notably Wader Lake, which is attractive to dabbling ducks, waders, gulls and terns, and also has a heronry, and The Reservoir (from Rainshelter Hide) which attracts diving ducks and Wigeon, while the Hawthorn Wood Feeding Station is operative mid-September to Easter. The River Wear may hold grebes and Goldeneye, and can be viewed from the track along Wader Lake.
Barmston Pond Turn north off the A1231 at the roundabout signed for the Nissan works, turn first left into Barmston Lane and park on the right. From here a track follows the east shore of the pond, and there are two hides.

FURTHER INFORMATION

Grid refs/postcode: Washington WWT NZ 330 562 / NE38 8LE; Barmston Pond NZ 328 573
WWT Washington: tel: 0191 416 5454; web: wwt.org.uk/Washington

294 WHITBURN COAST (Durham)

OS Landranger 88

Due largely to the activities of a group of keen local birders, this section of coast has gained an enviable reputation for the number and variety of migrants that occur, including many rarities, and for the sometimes outstanding seawatching. The cliffs also hold a variety of breeding seabirds.

Habitat

The mouth of the River Tyne is flanked to the south by dunes at South Shields, and further south the coast is bounded by limestone cliffs (topped with areas of short turf known as The Leas), which reach 90 feet near Whitburn. At the foot of the cliffs sandy beaches and extensive rock shelves ('Steels') are exposed at low tide. As usual, any area of cover, from parks, cemeteries and gardens to the scrappiest bush, can and do attract migrant passerines.

Species

Spring migration brings a scatter of commoner species such as Wheatear, Whinchat and Common Redstart as well as scarcer species such as Ring Ouzel and Black Redstart, while Bluethroat and Marsh and Icterine Warblers are annual in May. Seawatching in spring is relatively quiet, but small numbers of Manx Shearwaters, skuas and terns are seen.

In summer the cliffs around Marsden Rock hold up to 150 pairs of Fulmars, 200 pairs of Cormorants, 3,000 pairs of Kittiwakes, 150 pairs of Herring Gulls and 25 pairs of Razorbills. Ringed Plover and Rock Pipit also nest. Offshore, Puffin, Common Guillemot and Razorbill and Common, Arctic and Sandwich Terns are regular. Notable breeding landbirds include Sparrowhawk and Little Owl (around Marsden Quarry).

Inevitably, autumn holds the greatest promise for the migrant hunter. All of the regular east coast species are recorded, especially during the classic conditions of south-east or east winds coupled with overcast skies or precipitation. Wryneck, Icterine and Barred Warblers, Red-backed Shrike and Red-breasted Flycatcher are annual, and look too for Long-eared and Short-eared Owls and Woodcock. In addition, the area is notable for rare *Phylloscopus* warblers, with regular Yellow-browed Warbler and several records of Pallas's Warbler. Richard's

Pipit is another near-annual species in late autumn. Wader passage is limited by the available habitat, but a wide variety may be recorded over the season on the beaches and 'Steels'. In autumn seawatching comes into its own. Manx Shearwater is present all summer, as are terns, with a few Roseate and Black Terns passing each year. Gannet and Fulmar are present offshore year-round. The largest numbers and variety of seabirds are recorded during strong winds from the north, and in gales resultant movements may be spectacular. From August Sooty Shearwater may occur, with up to 100 per day in the right conditions. A few Balearic Shearwaters are also seen annually, but large shearwaters are rare. Leach's Petrel is sometimes seen from shore late in the season, but Storm Petrel is rare (although they are present offshore, as demonstrated by their routine capture using tape lures at night). Perhaps the most exciting seabirds are the skuas, and all four species are possible, with Arctic and Long-tailed more likely in early autumn and Pomarine and Great in the latter part of the season (until November or even December). Large numbers of Kittiwakes are regular, as is Little Gull, with Sabine's Gull putting in an occasional appearance. Auks are regular, with Little Auk occurring in significant numbers in late autumn and early winter during strong northerly winds. Other species seen during seawatches include numbers of ducks, and a feature of October is a short passage of Barnacle and Pink-footed Geese.

Offshore in winter small numbers of Red-throated Divers are regular, and Great Northern and Black-throated Divers and Red-necked Grebe are occasionally found. Fulmar is present year-round (except sometimes for a short period midwinter), as is Cormorant (roosting on Marsden Rock) and there are often also Shags. Seaducks are uncommon, however, with just a handful of Common Eiders and Common Scoters being regular. Wintering waders include large numbers of Lapwings and Golden Plovers (the latter favouring the fields around the Observatory and Whitburn Lodge, as well as the 'Steels)', and the 'Steels' and beaches attract Oystercatcher, Turnstone, Ringed Plover, Sanderling, Dunlin, Common Redshank, Curlew and small numbers of Grey Plovers, Bar-tailed Godwits and Knots. Purple Sandpiper is something of a speciality, with up to 100 between South Shields Pier, Lizard Point, the 'Steels' and the rocks near the Observatory. Large numbers of gulls are present and Mediterranean, Iceland and Glaucous Gulls are often found. Gulls often follow fishing boats back to North Shields Fish Quay and it is always worth checking carefully for rarities such as Ivory and Ross's Gulls. Other notable wintering species include Peregrine, Short-eared Owl, Rock Pipit and occasionally Lapland Bunting, while Shore Lark, Twite and Snow Bunting are rare.

Access

South Shields Congregations of gulls around the mouth of the Tyne, especially birds following fishing boats back into harbour in the late afternoon, can be viewed by turning north off the A183 onto the B1344 (signed Riverside). There are several pay-and-display car parks overlooking the sea. In the evening gulls roost on the sea south of the pier and can be viewed from near the Gypsies Green Stadium.

South Shields Leas An area of close-cropped turf that attracts buntings and sometimes also waders. It lies immediately east of the A183 between South Shields and Marsden. Park at Trow Quarry or along the A183.

Marsden Quarry This abandoned limestone quarry has limited areas of cover and is an LNR with free access. It is good for migrants such as Ring Ouzel and Black Redstart in spring but is better during autumn falls and has attracted several rarities. Turn west off the A183 in Marsden on the A1300 and, after about 175 yards, turn south at the roundabout into Lizard Lane. The entrance to the quarry lies on the right after 500 yards, and there is limited parking. The well-vegetated grounds of adjacent Marsden Hall are private, but some of the area can be viewed from the quarry.

Marsden Bay and Rock Marsden Rock is a 90-foot-high stack lying just offshore and holding a large colony of Cormorants. Numbers of other seabirds also breed on the two smaller stacks of Jack Rock and Pompey's Pillar, and on the magnesian limestone cliffs of the mainland south to Lizard Point (together these comprise the most important seabird colonies between the Farnes and Bempton Cliffs). The rock and cliffs can be viewed from the Marsden Bay car

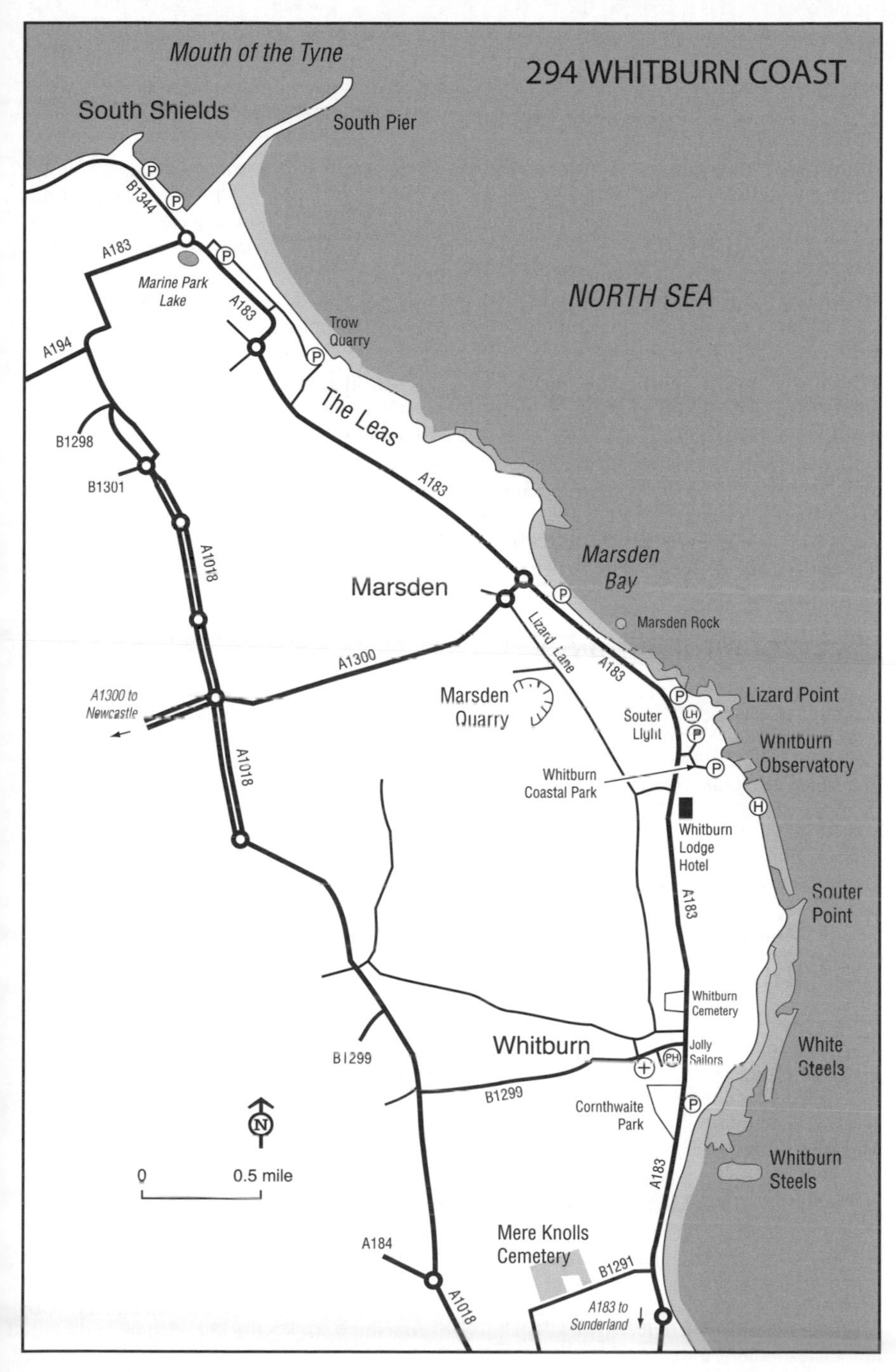

park, which lies east of the A183, about 550 yards south of the junction with the A1300, and from the footpath which follows the clifftop south from there.

Lizard Point Together with Whitburn Observatory and Souter Point, this is the best seawatching station, and has the advantage of easy access. Park off the A183 just north of the Souter Lighthouse (it is possible to seawatch from the car here in bad weather).

Whitburn Coastal Park Accessed off the A183 just south of the lighthouse. An area of rough grassland with some cover (especially along the edge of the housing estate to the south), attractive to migrants.

Whitburn Observatory The large seawatching hide here provides shelter in rough weather. Normally locked, keys can be obtained from the Durham Bird Club. The best access is from the southern car park of the Coastal Park, walking a short distance south and east to the clifftop.

Whitburn cemetery This small cemetery may hold migrants. It lies west of the A183 in Whitburn, about 350 yards north of the Jolly Sailors pub.

Whitburn churchyard Another site to search for migrants. Turn west off the A183 at the Jolly Sailors pub in Whitburn onto the B1299. Turn first left into Church Lane and the churchyard lies on the right after 100 yards.

White Steel and Cornthwaite Park Whitburn Rock shelves (the 'Steels') flank the cliffs for over 1 mile at Whitburn, and attract numbers of waders. Park to the east of the A183, on the southern outskirts of Whitburn (just on the edge of the City of Sunderland), to scan the 'Steels' with a telescope. Cornthwaite Park lies west of the road and can be attractive to migrants, with the best areas of cover at the northern end.

Mere Knolls cemetery This holds large numbers of mature trees attractive to migrants. On the northern outskirts of Sunderland turn west off the A183 at the traffic lights into the B1291 (Dykelands Road). The cemetery lies north of the road after 300 yards and it is possible to park on the roadside.

FURTHER INFORMATION

Grid refs: South Shields NZ 368 679; South Shields Leas NZ 380 667; Marsden Quarry NZ 397 646; Marsden Bay and Rock NZ 397 650; Lizard Point NZ 407 644; Whitburn Coastal Park NZ 407 640; Whitburn Observatory NZ 411 634; Whitburn cemetery NZ 407 621; Whitburn churchyard NZ 405 616; White Steel and Cornthwaite Park NZ 407 613; Mere Knolls cemetery NZ 398 599
Durham Bird Club: www.durhambirdclub.org

295 SEATON SLUICE TO ST MARY'S ISLAND (Northumberland)

OS Landranger 88

This section of coast has three principal attractions, wintering waders, spring and autumn passerine migrants, and seawatching, with both Seaton Sluice and St Mary's Island being well-established watchpoints.

Habitat

Much of this coast is bounded by areas of rocky sills that, together with the adjacent sandy and muddy beaches, hold numbers of waders. To the north, Rocky Island at Seaton Sluice is a traditional seawatching venue, while to the south, St Mary's Island (marked by an old lighthouse) is also a good seawatching station, though inaccessible at high tide. It also has some areas of cover, which may hold migrants, as may the rough ground along the low clifftops of the mainland and the fields immediately inland of the coast, but Whitley Bay Cemetery, with stands of trees and shrubs, is perhaps the best area for migrants.

Species

Autumn is perhaps the peak season. Fulmar and Gannet are almost constantly present offshore, and auks and terns from colonies on the Farnes are regular in summer and autumn, whatever

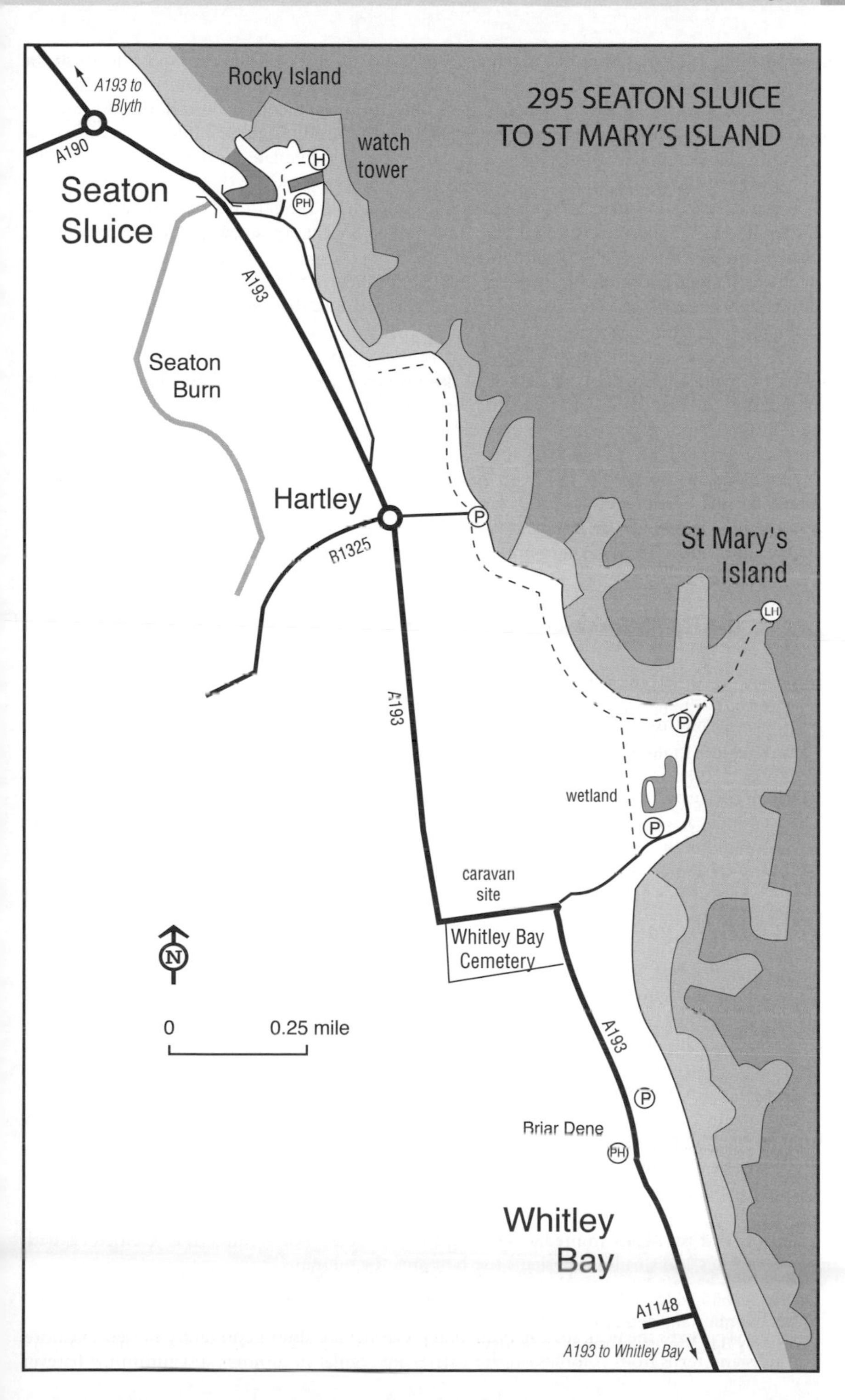
295 SEATON SLUICE
TO ST MARY'S ISLAND
A193 to Blyth
Rocky Island
A190
watch tower
Seaton Sluice
A193
Seaton Burn
Hartley
B1325
St Mary's Island
A193
wetland
caravan site
Whitley Bay Cemetery
N
0
0.25 mile
A193
Briar Dene
Whitley Bay
A1148
A193 to Whitley Bay

the weather, attracting marauding Arctic Skuas, and from midsummer Manx Shearwater is often present offshore. From late August strong to gale-force winds from the north may prompt movements of seabirds. Sooty and Balearic Shearwaters may join the Manx, and Sabine's Gull and Long-tailed Skua are possible. As autumn progresses, Pomarine and Great Skuas become more likely, and geese and ducks appear, while late-autumn and winter gales may produce Little Auk.

Passerine migrants can be interesting, and the typical scarce migrants of the east coast, Icterine, Barred and Yellow-browed Warblers and Red-backed Shrike, are recorded most years, and the area has produced a good list of rarities.

Spring passage is quieter, but seabirds include numbers of terns, and passerines may include Ring Ouzel and Black Redstart, with Bluethroat and Red-backed Shrike possible in May–early June.

In winter good numbers of waders are present along the coast, especially around St Mary's Wetland, with 1,000+ Golden Plovers, Lapwings and Oystercatchers, and smaller numbers of Knots, Common Redshanks, Turnstones, Sanderlings, Curlews, Grey and Ringed Plovers and Purple Sandpipers. The clifftop fields hold wintering thrushes, finches and buntings (sometimes including Snow or Lapland Buntings), with Stonechat around bushier areas, which attract raptors such as Peregrine, Merlin and Short-eared Owl. Large numbers of Black-headed Gulls roost on the sea, and both Glaucous and Iceland Gulls are sometimes recorded. Common Eider is resident, and wintering seaducks may also include Common Scoter, Red-breasted Merganser and Goldeneye, and Red-throated Diver is also regular.

Access

Seaton Sluice At the north end of Seaton turn east off the A193 immediately south of the bridge over the Seaton Burn and park just south of the harbour by the Kings Arms pub. Cross the wooden footbridge opposite onto Rocky Island, walk straight to the sea and then around the island to the white wooden tower. This is the Northumberland and Tyneside Bird Club's seawatching hide and is kept locked, with access only to members, but it is possible to seawatch from the shelter of the wall here. Alternatively, find a sheltered position on the mainland near the Kings Arms.

Hartley Turn east off the A193 at the roundabout in Hartley, taking the minor road to the radio masts by the sea, where there is a small car park. From here, footpaths run north along the coast to Seaton Sluice and south to St Mary's Island, giving views of the coastal fields and beaches.

St Mary's Island (Bait Island) Leaving Whitley Bay north on the A193 turn sharply east immediately as the dual carriageway ends (and the main road turns sharp right), onto a road leading to the point overlooking the island. There are large car parks at the point and midway along this road. The island is only accessible on foot, and is cut off from the mainland around high water; it is best to depart as the tide falls and return to the mainland as the tide rises (details of access and tide times are available from the Tourist Office). There is a seawatching hide on the island, operated by North Tyneside Council, and keys and information on opening times etc. are available from the information centre on the island (where there is also a café).

St Mary's Wetland Viewable from the fist car park on the St Mary's Island road, and from various screen viewpoints along the footpath on the western side.

Whitley Bay Cemetery Leaving Whitley Bay north on the A193, park at the cemetery entrance on the left just before the dual carriageway ends.

Briar Dene A scrub-filled gully attractive to migrants. Access from the car park on the A193, crossing the road and entering immediately north of the Briar Dene pub.

FURTHER INFORMATION

Grid refs: Seaton Sluice NZ 337 767; Hartley NZ 343 757; St Mary's Island (Bait Island), St Mary's Wetland NZ 349 747; Whitley Bay Cemetery NZ 346 744; Briar Dene NZ 349 739
Northumberland and Tyneside Bird Club: www.ntbc.org.uk
Whitley Bay Tourist Information Centre: tel: 0191 200 8535; web: www.visitnorthtyneside.com

296 DRURIDGE BAY AREA (Northumberland)

OS Landranger 81

This 5-mile stretch of the Northumberland coast forms a single conservation area and contains several excellent birdwatching sites, holding interest throughout the year. Seabirds, wildfowl, waders and migrant landbirds provide an exceptionally rich 'mixed bag' of interesting species.

Habitat

Druridge Bay extends for c. 6 miles along the Northumberland coast and is fringed for most of its length by sand dunes and a narrow sandy beach, with the mouth of Chevington Burn (Chibburn Mouth) attracting loafing and bathing gulls and terns, and areas of low rocks at both the northern and southern extremities. Formerly a mixture of farmland and collieries, open-cast mining in the 1970s devastated much of the hinterland, but with its demise these areas of industrial dereliction have been 'landscaped' and returned to a variety of uses, including conservation. Notably, immediately inland of the coast lie a series of relatively small areas of open water. Much of the former farmland cover was stripped during mining operations, but has now been replaced by shelter belts and plantations of conifers and mixed woodland trees, as well as native scrub species planted by various conservation bodies.

Species

Wintering waders on the foreshore include Ringed and Grey Plovers, Turnstone, Curlew, Bar-tailed Godwit, Common Redshank, Sanderling, Dunlin, Knot and Purple Sandpiper, with good numbers of Lapwings and Golden Plovers around the fields; the latter are sometimes joined by a handful of Ruffs (especially at Druridge Pools). Common Eider and Common Scoter are present offshore year-round, and are joined in autumn and winter by Red-breasted Merganser, Goldeneye, a few Long-tailed Ducks and Velvet Scoters, and sometimes Scaup. The sea also has small numbers of Red-throated Divers and Slavonian Grebes, and occasionally other divers or rarer grebes. The freshwater pools are favoured by Shelduck, Teal, Wigeon, Gadwall, Shoveler, Pochard, and sometimes small numbers of Pintails. A handful of Smews is often present, commuting between the various freshwater pools. The pools may also attract seaducks, especially during periods of bad winter weather. Bittern is an increasingly regular visitor, especially to the Chevington Lagoons. Flocks of Whooper Swan are regular, but wide ranging (although they favour the area around Warkworthlane Ponds), and there is occasionally also a few Bewick's Swans. Occasional parties of Greylag Geese may be found in the coastal fields, but Pink-footed, Barnacle and Pale-bellied Brent Geese are irregular. The dunes and coastal fields may also hold flocks of Twite and Snow Bunting (often favouring the fields where cattle are fed), and less frequently the elusive Lapland Bunting. Such concentrations of birds attract Peregrine, Merlin, Short-eared Owl and sometimes Hen Harrier. In winter Mediterranean, Iceland and Glaucous Gulls are occasionally recorded together with the commoner gulls.

On spring passage both Black Tern and Little Gull are possible, as are Marsh Harrier, Osprey, Hobby and Garganey. In autumn, and to a lesser extent in spring, a variety of waders may occur, both on the foreshore and around the various pools. Little Ringed Plover, Ruff, Whimbrel, Greenshank and Common Sandpiper are likely at either season (as is the rarer Wood Sandpiper or Black-tailed Godwit), but autumn is more likely to produce Spotted Redshank, Green and Curlew Sandpipers and Little Stint. Cormorant, Gannet and Fulmar are almost constantly present offshore. Autumn seawatching can be productive, with divers, Red-necked Grebe, Manx and sometimes Sooty Shearwaters, Arctic and sometimes Great, Pomarine and even Long-tailed Skuas, Kittiwake and auks all possible in the right conditions (usually strong winds with a northerly component). In late autumn and winter, storms may promote movements of Little Auk along the coast, and sometimes produce a Grey Phalarope. In both spring and autumn a superb array of passerine migrants has been recorded, with many scarce and rare species. Migrants may be found in any cover on the coast, with no area particularly favoured (and apparent concentrations of rarities being due to numbers of birdwatchers rather than birds). Scarcer species such as Ring Ouzel and Black Redstart are regular, as are the typical east coast autumn specialities: Wryneck, Icterine and Barred Warblers and Red-backed Shrike, and anything is possible.

Breeding species include Little Grebe, Shelduck, Sparrowhawk, Ringed Plover, Common Sandpiper, Barn and Little Owls, Sand Martin (in some years), Rock Pipit, Yellow Wagtail, Whinchat, Stonechat (in the dunes), Wheatear, Grasshopper, Sedge and Garden Warblers, Spotted Flycatcher, Tree Sparrow and a few Corn Buntings, and sometimes Pochard, Gadwall and Shoveler. Marsh Harriers bred at Chevinton in 2009 – the first breeding in the county for over 100 years. Breeding birds on Coquet include up to 13,000 pairs of Puffins, about 4,000 pairs of Black-headed Gulls and 2,000 pairs of Sandwich Terns, as well as several hundred pairs of Common Eiders, Fulmars and Common and Arctic Terns, and, most notably, around 30 pairs of Roseate Terns. Of course, all these can be seen along the entire stretch of coast throughout the spring and summer (gulls and terns bathing and loafing on the pool at Hauxley), together with resident Herring and Great Black-backed Gulls and, in spring and summer, Lesser Black-backed Gull.

Access

RSPB Coquet Island Lying 1 mile off the coast at Amble-by-the-Sea, this small island holds important colonies of seabirds, notably Common Eider, Fulmar, Kittiwake, 18,000 pairs of Puffins, Black-headed Gull and Common, Arctic, Sandwich and a few pairs of Roseate Terns (although Roseate Terns are most likely to be seen at Hauxley NR). No landing is permitted, but for the 'virtual' birdwatcher, during the breeding season, live images of nesting seabirds, including Puffins and four species of terns, are viewable at the Northumberland Seabird Centre on Amble quay (open March–August, 9am–5pm, September–November, Thursday–Sunday, 10am–4.30pm). To see the breeding birds in the flesh take one of the boat trips around the island operated by the Seabird Centre. Otherwise, the breeding species can be seen all along this coast as they commute to and fro, with the coastguard lookout at Hauxley providing one of the closest vantage points.

Hauxley NR A reserve of the Northumberland Wildlife Trust (NWT) centred upon an artificial freshwater pool, the reserve also has areas of reedbed and scrub. The area is at its best from spring to early autumn, and is probably the best site on the east coast to find Roseate Tern, which sometimes visit to loaf or bathe (the Tern Hide being the best viewpoint), while waders use the pool as a roost site. Leave the A1068 east on the minor road to High Hauxley, about 1 mile south of Amble. As the road passes through the village it turns sharply right and then, after 300 yards, at the right-angle left bend, turn south at the sign to the reserve car park. The reception hide is open daily 10am–5pm (summer) and 10am–3pm (winter). Four hides overlook the main pool, and the surrounding fields are also worth checking. To the south, the dunes and foreshore form part of Druridge Bay CP, while to the north the dunes near the coastguard lookout are an LNR. The entire stretch of coast can be interesting, with the dunes east of the Tern Hide best for autumn seawatching, while the area of the lookout affords views in summer of seabirds going to and from Coquet Island. (Note that the wood at Low Hauxley has no public access.)

Hadston Carrs Turn east off the A1068 1 mile south of the turning for Hauxley (and about halfway to the turn for Druridge Bay CP) on the minor road to Hadston Carrs, turning south at the seafront. This gives easy access by car for seawatching.

Druridge Bay Country Park (Hadston Country Park) Leave the A1068 around Broomhill at the signed turning to the Country Park. There are several lay-bys overlooking Ladyburn Lake, with the main car park, visitor centre (including a blackboard detailing the latest sightings) and toilets at the seaward end of the lake. In winter, the car park at the watersports slipway also gives good views. A footpath circles the lake, and from the main car park access through the dunes to the beach is also possible (with the option of walking north to Hauxley or south to

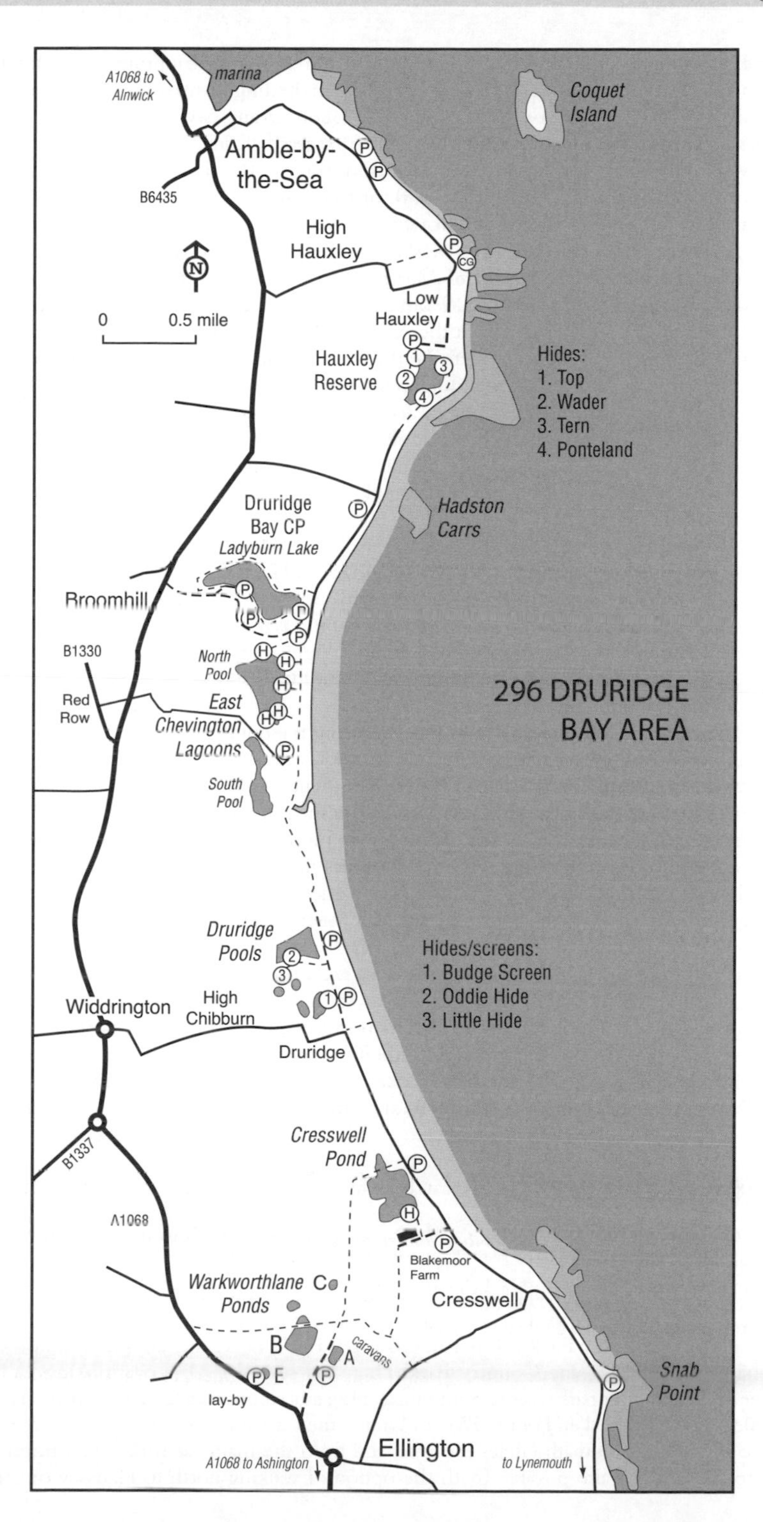

A1068 to Alnwick
marina
Coquet Island
Amble-by-the-Sea
B6435
High Hauxley
CG
Low Hauxley
0 0.5 mile
Hauxley Reserve
Hides:
1. Top
2. Wader
3. Tern
4. Ponteland
Druridge Bay CP
Ladyburn Lake
Hadston Carrs
Broomhill
B1330
North Pool
Red Row
East Chevington Lagoons
South Pool
296 DRURIDGE BAY AREA
Druridge Pools
Hides/screens:
1. Budge Screen
2. Oddie Hide
3. Little Hide
Widdrington
High Chibburn
Druridge
B1337
Cresswell Pond
A1068
Blakemoor Farm
Warkworthlane Ponds
Cresswell
caravans
lay-by
Snab Point
Ellington
A1068 to Ashington
to Lynemouth

the Chibburn Mouth). The area is at its best in winter, and is somewhat prone to disturbance in summer, as all but the western end of Ladyburn Lake is used for water sports April–September.
East Chevington Lagoon A reserve of the NWT, comprising two large lakes with fringing reedbeds, grassland and newly planted woodland. Access is from the A1068, turning into a minor road to the coast opposite the minor road into Red Row (access is also possible along the footpath from the south-eastern car park of the Country Park to the north). The large area of shallow open water attracts wildfowl and passage waders, and there are several hides.
Druridge Pools NR A reserve of the NWT, this area is best from late autumn to late spring for wildfowl and waders. The main pool is a product of reclamation following the cessation of open-cast mining, but the Trust created the two south scrapes; these pools are separated by fields that are prone to flooding. Leave the A1068 at Widdrington east on the minor road to Druridge and, after 1.5 miles, turn north (at the second sharp right-hand bend) signed NWT and NT. After paying the car park fee (non-members), drive north parallel to the coast. Within a few hundred yards, park on the right to access Budge Screen (through a gap in the hedge on the other side of the road), which overlooks the wader scrapes and wet fields, and after 0.5 miles another parking area accesses the raised walkway to the Oddie Hide overlooking the main, deep-water pool, and the Little Hide south of the track views a small pool and the wet fields to the south.
Cresswell Pond A product of subsidence, this shallow brackish lagoon is fringed by saltmarsh and reeds, separated from the sea by a narrow sandbar and surrounded by pasture. Managed by the NWT, it is of interest year-round, but is best in spring and autumn. Leave the A1068 east at Ellington roundabout and turn left in the village on the minor road to Cresswell. Continue north, parallel with the coast, towards Widdrington, and park at the bottom of the track to Blakemoor Farm. Walking up the track the gate to the British Alcan Hide lies on the right by the first building. Continuing north along the road for a further 0.25 miles, the small car park (opposite the track to Warkworth Lane) gives views of the pond and it is also possible to walk east through the dunes to seawatch.
Warkworthlane Ponds These small subsidence pools lie 1 mile south-west of Cresswell Pond (but 9 miles south of the village of Warkworth!). The surrounding fields are good for Whooper Swan (sometimes as many as 100), but the ponds are most interesting in spring and autumn for passage waders. Access is from the A1068, 0.25 miles north of Ellington roundabout, turning off north-west towards Warkworth Lane Cottage and then parking carefully by the caravan site reception area. Follow the footpath past the touring caravans, chalets and Pond A, and then bear left before the main area of residential caravans to meet the main bridleway between Ellington and Highthorn. Walk left along this for 0.25 miles to view Pond B (the largest) or go straight on at the footpath through the trees to view Pond C. The small Pond E can be viewed from a lay-by on the A1068 0.75 miles north of Ellington roundabout. (Alternatively, from the car park at Cresswell Pond, follow the footpath around the north edge of the pond for 0.7 miles to Ponds C and D, and then continue on the main bridleway to Ponds A and B.)
Snab Point Accessed along the minor road south from Cresswell, the car park here is a convenient point from which to seawatch.

FURTHER INFORMATION

Grid refs/postcode: RSPB Coquet Island (Amble quay) NU 269 048 / NE65 0AA; Hauxley NR NU 282 024; Hadston Carrs NU 277 008; Druridge Bay Country Park (Hadston Country Park) NZ 271 997; East Chevington Lagoon NZ 271 984; Druridge Pools NR NZ 275 963; Cresswell Pond NZ 283 945; Warkworthlane Ponds NZ 274 928; Snab Point NZ 300 928
Northumberland Seabird Centre: tel: 01665 710835; web: www.northumberlandseabirdcentre.co.uk
Northumberland Wildlife Trust: tel: 0191 284 6884; web: www.nwt.org.uk; Hauxley Nature Reserve and Visitor Centre, Low Hauxley, Amble, Morpeth, Northumberland: tel: 01665 711578
RSPB Coquet Island: tel: 0191 233 4300

297 THE FARNE ISLANDS (Northumberland)

OS Landranger 75

Lying off the Northumberland coast between Bamburgh and Seahouses, the Farne Islands hold superb, easily accessible seabird colonies, and the short boat trip is guaranteed to add to the excitement. The best time to visit is May–early July. The islands belong to the NT and are an NNR.

Habitat

These small islands, comprising ten larger outcrops and numerous smaller rocks, totalling 28 in all, lie 1.5 miles or more offshore. Most are rocky stacks, but Inner Farne and Staple Island are flatter, providing habitat for nesting terns.

Species

Over 65,000 pairs of 14 species of seabirds breed. Puffin is the most numerous, with around 55,600 pairs, followed by Common Guillemot (32,500 pairs), Kittiwake (4,700 pairs), Arctic Tern (2,250 pairs), Sandwich Tern (1,400 pairs) and Shag (about 1,000 pairs). There are several

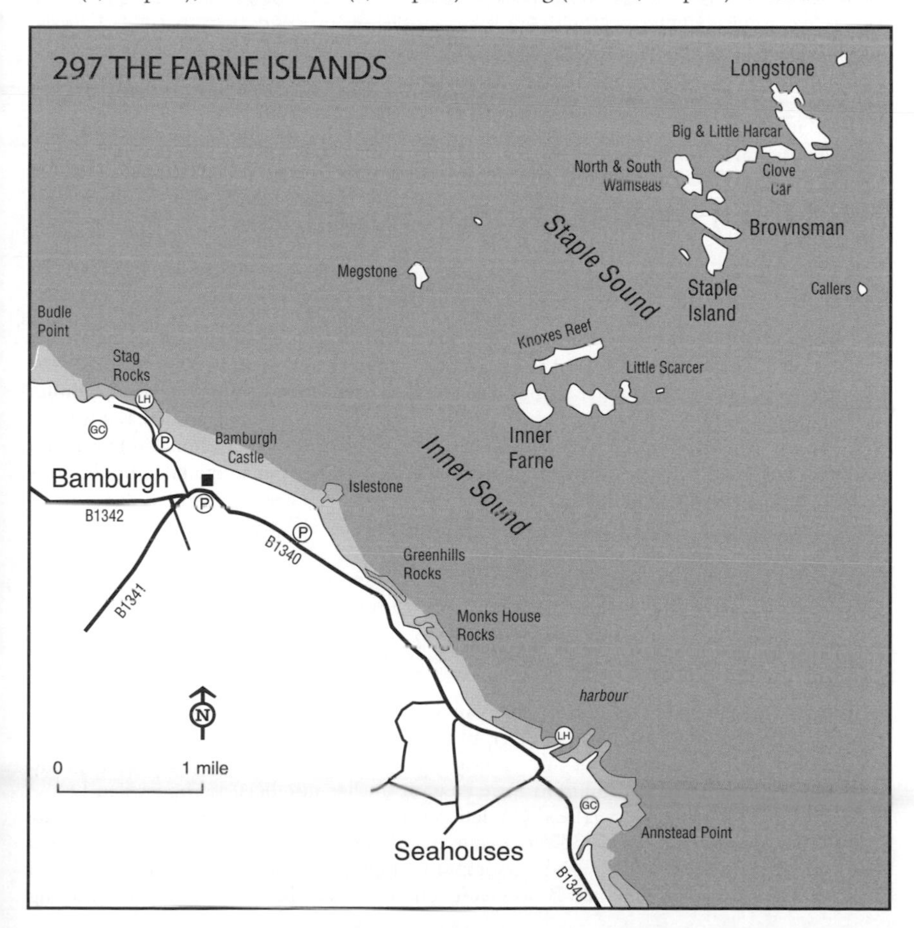

hundred pairs of Cormorants, Fulmars, Lesser Black-backed, Herring and Black-headed Gulls, Razorbills and Common Terns, and there are sometimes one or two pairs of Roseate Terns. The terns breed on Inner Farne, Staple Island and Brownsman. Other breeding birds include 650 pairs of Common Eiders, as well as Mallard, Ringed Plover, Oystercatcher, Rock Pipit, Pied Wagtail, Swallow and sometimes Shelduck.

Passage brings divers, shearwaters, waders and skuas to this coast, as detailed under Lindisfarne–Seahouses, and passerine migrants to the islands, although for obvious reasons, these are usually seen only by the islands' wardens.

Access

Weather permitting, boats run daily from Seahouses Harbour, from April to late September. Tickets are available from the boat kiosks at the quay, but it is best to book in advance, especially during busy holiday periods. Seahouses Tourist Office can advise; one of the boatmen most regularly used by birders is Billy Shiels (of the MV *Glad Tidings*). Sailings commence at 10am, and several itineraries are available, either landing on Inner Farne and/or Staple Island, or non-landing (check on booking). Half-day trips allow only 1 hour ashore, however, and as there is so much to see, a full-day trip may be preferable. An entrance fee is payable on landing (NT members free) and leaflets are available for the nature trails on the islands.

FURTHER INFORMATION

NT Farne Islands: tel: 01665 720651
Seahouses Tourist Office: tel: 01665 720884
Billy Shiels/MV *Glad Tidings*: tel: 01665 720308; web: www.farne-islands.com

298 LINDISFARNE (Northumberland)

OS Landranger 75

Lindisfarne, on the Northumberland coast, is an excellent area for birdwatching during winter and passage periods, especially for divers, grebes, wildfowl and waders. The area supports internationally important populations of Pale-bellied Brent Goose, Wigeon, Knot, Dunlin and Bar-tailed Godwit. On passage a wide variety of passerines has been recorded. Lindisfarne NNR covers 3,541 ha and includes Budle Bay. The adjacent coast between Bamburgh and Seahouses has essentially the same population of divers, grebes and seaducks as Lindisfarne, and also attracts passage migrants.

Habitat

The intertidal flats are sheltered from the sea by the dune systems of Goswick Sands, Holy Island and Ross Links. Budle Bay is another sheltered tidal basin, while the 4-mile stretch of coast from Bamburgh to Seahouses has several rocky sections.

Species

In winter Red-throated Diver and Slavonian Grebe are quite common offshore; Black-throated is the second commonest diver but is distinctly scarce, as is Great Northern. Red-necked Grebe is regular but uncommon, and the sea also has Cormorant, Shag, and many Common Eiders and Common Scoters, with smaller numbers of Long-tailed Ducks and Red-breasted Mergansers, a few Goldeneyes and Velvet Scoters, and more erratically Scaup. Good numbers of Greylag Geese occur, mainly in the fields between Budle Bay and Ross, or on the flats of Budle Bay, and are sometimes joined by small numbers of Pink-footed Geese or a few White-fronted and Bean Geese. The flats hold the largest flock of wintering

Pale-bellied Brent Goose in Britain. They belong to the declining Svalbard and Franz Josef Land populations, which number only c. 5,000 birds. These winter in Denmark and at Lindisfarne, which holds half of the population, numbers peaking at c. 2,500 in December. Dark-bellied Brent Goose may also be present, and numbers of Barnacle Geese pass through in autumn and spring. A flock of 20–30 Whooper Swans uses the fields and flats and is sometimes joined by a few Bewick's. Wigeon has peaked at 40,000 and typically 10,000 are present, while Shelduck and Teal are also common, with a few Pintails and Pochards. Peregrine, Merlin and Short-eared Owl are regular, and there is sometimes a Hen Harrier. Of waders, Dunlin and Knot are the commonest and there are the usual other estuarine species, but Sanderling and Purple Sandpiper occur in relatively small numbers (the latter favours Nessend on Holy Island and Stag Rocks at Bamburgh). There is occasionally a wintering Spotted Redshank or Greenshank. Occasionally Glaucous Gull puts in an appearance and Iceland is even more rarely seen. Snow Bunting can be quite common, especially on Ross Links. Shore Lark, Twite and Lapland Bunting are only occasional. Grey Wagtail and sometimes Dipper can be found at Warren Burn outlet.

Seawatching can be good during passage periods. Gannets and Manx and sometimes Sooty Shearwaters are noted offshore in autumn, together with six species of tern, though Little is scarce and Roseate quite rare. Arctic Skua is almost constantly present in autumn, and sometimes Great too, but Pomarine and Long-tailed are unusual. Waders frequent the flats, but it is worth checking any area of fresh water as well. Black-tailed Godwit, Greenshank, Spotted Redshank and Wood Sandpiper occur at both seasons, but Little Stint and Curlew Sandpiper almost exclusively in autumn. Large falls of migrants are occasional in autumn, including huge arrivals of thrushes, and sometimes Woodcock, Jack Snipe and Long-eared Owl in late autumn. Wryneck, Shore Lark, Black Redstart, Bluethroat, Red-backed Shrike, and Ortolan and Lapland Buntings may occur at either season, but all are scarce and a visitor would be lucky to see any of these. The equally elusive Barred and Yellow-browed Warblers, Red-breasted Flycatcher and Great Grey Shrike almost exclusively occur in autumn, as does Waxwing, albeit sporadically.

Breeding species include small numbers of Fulmars (on Lindisfarne Castle and at Coves Bay), Common Eider, Shelduck, Ringed Plover and Oystercatcher, as well as Common, Arctic and Little Terns. Seabirds from the Farnes can be seen moving along the coast, as well as Gannet and Manx Shearwater from further afield.

Access

Within the NNR, permits are not required for access to the shore or causeway, but remember that farmland is private.

Holy Island The island is c. 3 miles long by 1.5 miles wide at the widest point. From the Snook, areas of sand dunes extend along the north shore, via a rather narrow strip at the Links, to Emmanuel Head. South of the Links lies farmland, and the village, priory and castle. The road to the island is well signed off the A1 at West Mains Inn. It is impossible to use the causeway for at least 2 hours either side of high water, and the strength of the wind and the height of the tide must also be taken into account. Tide tables are displayed on the causeway and Berwick Tourist Information or the Northumberland County Council's website also give details of the tides. The causeway itself is a well-made tarmac road. The following areas are worthwhile.

1. In the dunes of the Snook there is a scattering of bushes and clumps of willows and hawthorns, notably in the garden of Snook House and the bushes and pools to the west, which may harbour migrants, while the dunes attract pipits. It is possible to park along the access track.
2. Off Sandon Bay and North Shore look for Red-breasted Merganser and Shag, and Sanderling on the beach. It is possible to park along the road (tides permitting) and walk through the dunes to the North Shore. The Links and dunes area also attract migrants, and may harbour Long-eared or Short-eared Owls or Woodcock in late autumn.
3. The fields at Chare Ends attract larks, finches and buntings, sometimes including Twite or Lapland Bunting.

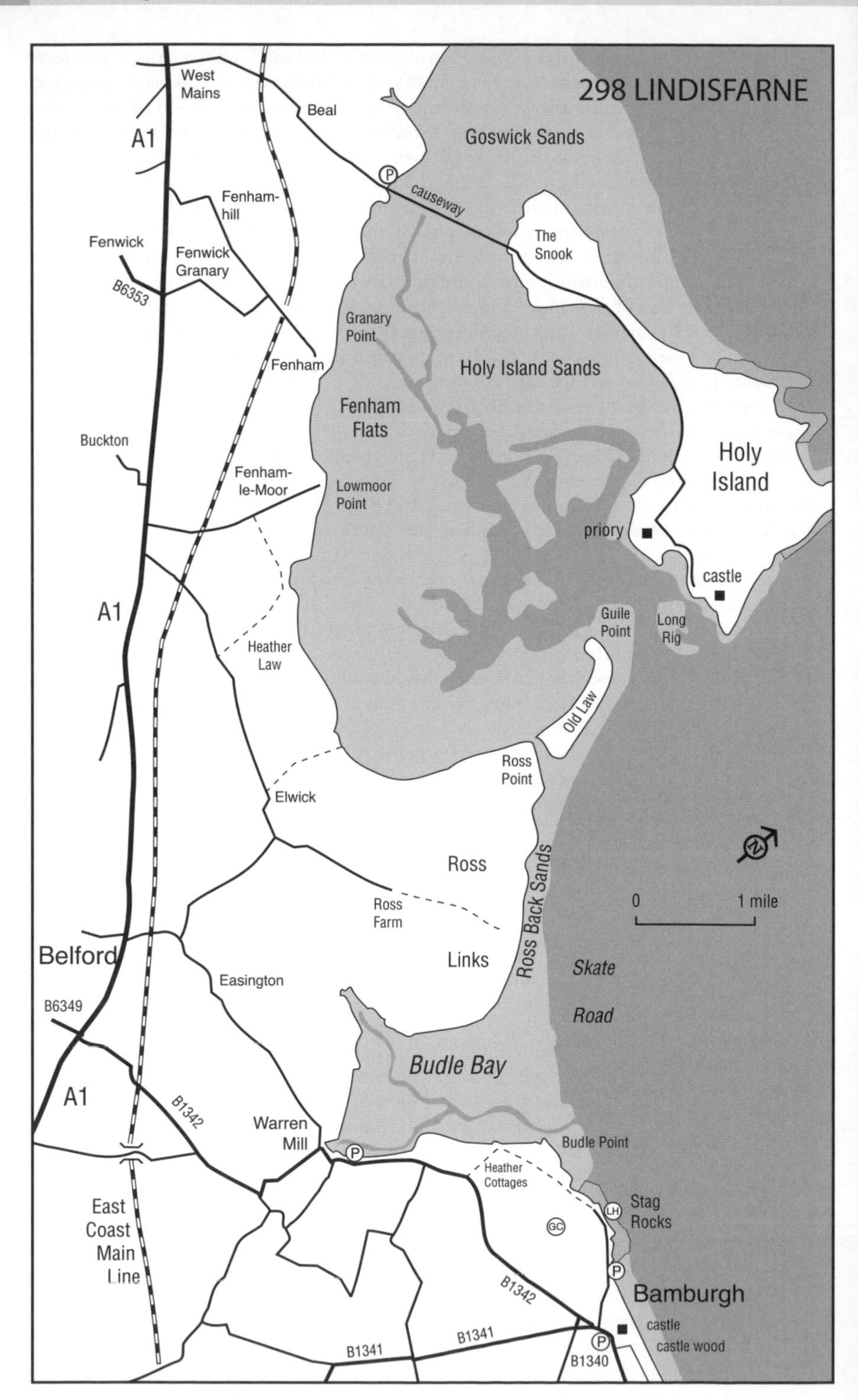
298 LINDISFARNE
West Mains
Beal
A1
Goswick Sands
causeway
Fenham-hill
Fenwick
Fenwick Granary
B6353
The Snook
Granary Point
Fenham
Holy Island Sands
Fenham Flats
Buckton
Holy Island
Fenham-le-Moor
Lowmoor Point
priory
castle
A1
Guile Point
Long Rig
Heather Law
Old Law
Ross Point
Elwick
Ross
Ross Back Sands
0
1 mile
Ross Farm
Belford
Links
Skate Road
Easington
B6349
Budle Bay
A1
B1342
Warren Mill
Budle Point
Heather Cottages
East Coast Main Line
Stag Rocks
B1342
Bamburgh
castle
castle wood
B1341
B1341
B1340

4. The Heugh (an igneous dyke by the coastguard tower south of the priory, complete with benches), affords views over Fenham Flats and south towards Ross Back Sands, and notably of Long Rig and Black Law for roosting waders and gulls. The adjacent harbour also attracts gulls and is worth checking for divers and grebes, especially on a rising tide, while the freshwater pools in the nearby fields attract passage waders (over the wall behind the car park kiosk).
5. Any area of cover around the village is worth checking in spring and autumn for migrants, but especially the gardens east of the road between Chare Ends and the Lindisfarne Hotel, the trees opposite the Island Oasis Café and Open Garden Retreat, the trees around the main car park and St Mary's churchyard, and the gardens on the eastern periphery of the village. Migrants tend to leave the island during the day, thus in suitable weather an early-morning visit is best.
6. The walled garden just north of the castle attracts migrants and the close-cropped turf by the lime kiln larks, wheatears and buntings.
7. The stunted hedges along the Straight and Crooked Lonnens, and the garden of the house north of the Crooked Lonnen, also attract migrants, with waders, finches, buntings and sometimes wildfowl in adjacent fields. Access is past the 'visitor farm', 300 yards north of the main car park.
8. The Lough attracts 'fresh' waders and duck, and holds breeding Little Grebe, Shoveler and 200–300 pairs of Black-headed Gull. The Paul Greenwood Memorial Hide overlooks the eastern end.
9. You can seawatch from the high dunes at Snipe Point on the north coast or the cliffs at Coves Bay or Nessend, from the shelter of the white stone beacon at Emmanuel Head (the best point) or Castle Point, but these are probably no better than Bamburgh or Seahouses.

Holy Island is simply too large an area to cover in its entirety in a single day, and thus a long weekend or even a one-week visit could be contemplated. A half-day visit, preferably in the morning, provides sufficient time to cover a variety of habitats, with a circular walk from the main car park, along either the Straight or Crooked Lonnen to the coast, to Emmanuel Head, and then completing the square along the other lonnen. Another circular walk would be to check the gardens around the village, then the Heugh and Castle, returning along the Crooked Lonnen.

Holy Island Causeway From the A1 at West Mains, take the minor road to Beal. Continue to the shore and causeway (from the base of which a public footpath runs along the shore). The causeway to Holy Island, with lay-bys either side of the white refuge tower, is the best viewpoint to see large numbers of waders, which favour Holy Island Sands and Fenham Flats. Seaducks may sometimes be seen on the channel by the refuge, with Whooper Swan on the flats, especially in autumn, while this is also an excellent vantage point to observe the hordes of wintering wildfowl, notably Wigeon and Pale-bellied Brent Goose.

Fenham Flats There is open access to the foreshore here, although it may be disturbed by wildfowling 1 September–20 February. This is a good area for Pale-bellied Brent Goose and Wigeon, and is best an hour or so either side of high water. Access is from the A1 at three points, although parking is limited at all of these. Note that it is unwise to venture onto the mudflats.

1. The minor road to Fenwick Granary and Fenham (opposite the B6353 turning to Fenwick), continuing to the shore.
2. The minor road to Fenham-le-Moor and on to Lowmoor Point, where there is a public hide. Good for waders and wildfowl on the rising tide.
3. A signed wildfowlers footpath runs north from the hamlet of Elwick for c. 1 mile to the shore. From here two circular routes are possible: north to Lowmoor Point, returning via Fenham-le-Moor and the footpath to Heather Law, or rather longer, east to Guile Point, returning via Ross Back Sands and Ross Farm.

Ross Back Sands A good area to view divers, grebes, seaducks and other wildfowl on Skate Road, with Sanderling and Grey Plover on the beaches, Snow Bunting around the dunes and a variety of raptors around the coastal pastures. Terns breed around Guile Point. A footpath runs east from Ross Farm (park on the verge) along a tarmac road and then over Ross Links to Ross Back Sands, with access north towards Guile Point along the shore (note that Old Law and Guile Point are separated from Ross Point at high tide).

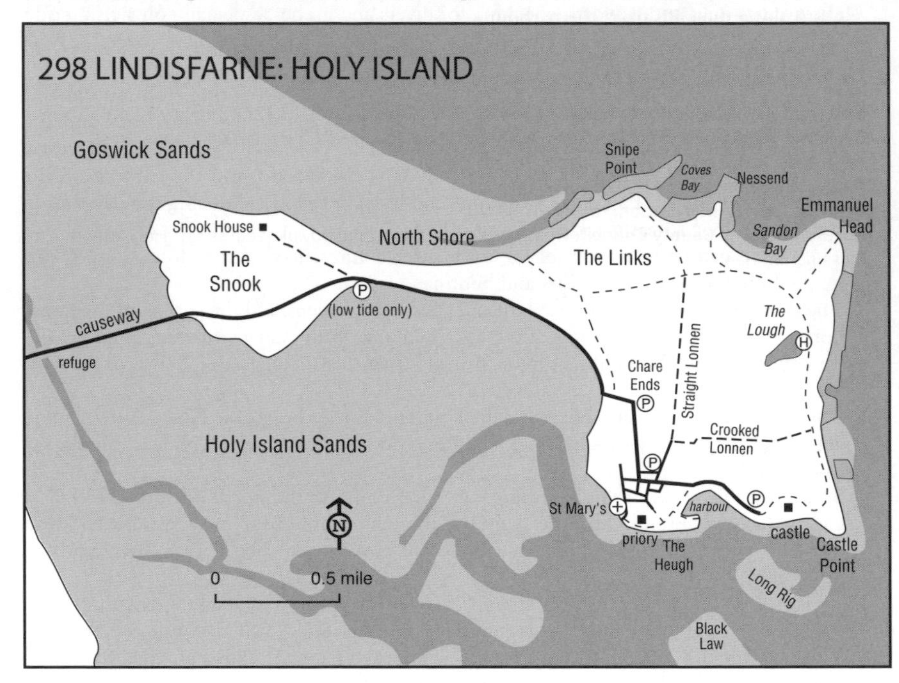

Budle Bay Very good for wildfowl and waders, especially if Fenham Flats are disturbed by shooting.

1. The B1342 between Warren Mill and Budle offers views over the south-east shore of the bay, and parking is possible along the verge. There is also roadside parking at Warren Mill.
2. The minor road west from Warren Mill towards Easington, with the 'Belford' lay-by on the north side of the road, gives distant panoramic views of the bay and fields between Easington and Ross, which may be used by Greylag Goose when not in the bay.
3. From Budle a footpath runs to Heather Cottages and the golf course, past Budle Point, which is a good spot to view the bay and sea. The path then skirts the golf course to Bamburgh lighthouse.

Bamburgh to Seahouses This section of the coast is bounded by large areas of rocks and extensive sandy beaches, notable for divers, grebes and seaducks. On the landward side, areas of cover attract passerine migrants.

1. The B1340 parallels the coast from Bamburgh to Seahouses, with open access to the foreshore.
2. From the centre of Bamburgh you can drive along a narrow, concealed minor road (the Wynding) to the lighthouse. Beyond the houses, it is possible to park by the foreshore on the cliff at Stag Rocks (marked as 'Harkess Rocks' on the OS map), using the wall of the lighthouse as shelter to view the sea; this is the best spot for divers, grebes and seaducks.
3. Bamburgh Castle Wood and the numerous stands of bushes in the dunes to the south should be checked for passage migrants. There are car parks opposite the castle and in the dunes to the south.
4. Seahouses Harbour attracts small numbers of gulls, including Glaucous and occasionally Iceland, and holds large numbers of Common Eiders in winter.
5. The public park/bowling green opposite the north pier in Seahouses should be checked

for passerine migrants in season (see The Farne Islands,page 465).

6. It may be worth seawatching from near the old lookout at Annstead Point (known as Snook or North Sunderland Point on the OS map), accessible by following the footpath south along the cliffs from the harbour.

FURTHER INFORMATION

Grid refs: Holy Island, The Snook NU 107 433; Holy Island village NU 127 421; Holy Island Causeway NU 078 427; Fenham Flats, Lowmoor Point NU 096 397; Ross Back Sands, Ross Farm NU 135 372; Budle Bay, Warren Mill NU 149 345
Natural England: www.naturalengland.org.uk
Berwick Tourist Information: tel: 01289 330733
Northumberland County Council: web: www.northumberland.gov.uk (search for 'Holy Island crossing timetable')

WALES

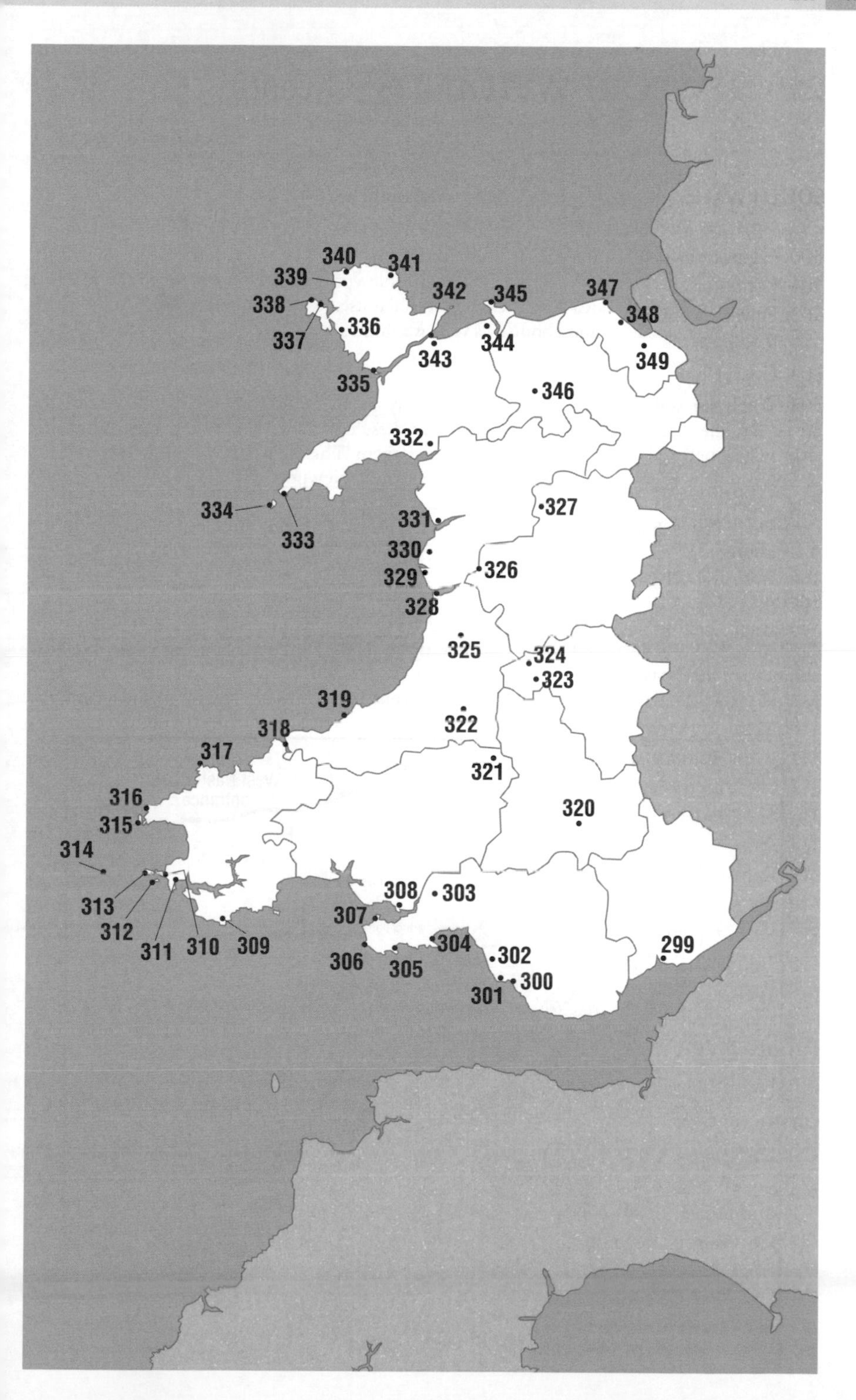
340
341
339
342
345
347
338
348
336
337
343
344
349
335
346
332
334
333
327
331
330
326
329
328
325
324
323
319
322
318
317
321
316
320
315
314
308
303
313
307
312
304
311
310
309
306
305
302
299
301
300

299 NEWPORT WETLANDS (Gwent)

OS Landranger 171

The Newport Wetlands Reserve, created as mitigation for the loss of mudflats of Cardiff Bay, lies on the northern shore of the Severn Estuary and covers over 865 ha between Uskmouth in the west to Goldcliff in the east. It is owned and managed by the Countryside Council for Wales through a partnership with the RSPB and Newport City Council. Since the reserve's construction in 1999, the area has quickly established itself as a top birdwatching locality with vagrant American waders almost annual as well as a good supporting cast of other scarce species.

Habitat

Behind the sea walls that keep the Severn at bay lie the Caldicot Levels, and west of the River Usk the levels of Peterstone, and together these comprise one of the largest remaining areas of unimproved wet grassland in lowland Britain. The Newport Wetlands Reserve lies on the Caldicot levels to the south-east of the city of Newport. The reserve is diverse, with deep water pools in the west (remnants of ash settling pools associated with the now defunct Uskmouth Power Station), reedbeds, shallow scrapes and grazing meadow, all alongside the Severn foreshore.

Species

Wintering wildfowl include Wigeon (up to 2,000), Shoveler, Gadwall, Teal, Pintail, Pochard, Goldeneye and Shelduck, and waders include up to 8,000 Dunlins, 2,900 Lapwings and small numbers of Black-tailed Godwits. The resident Water Rails may be most obvious in cold weather and a single Bittern has wintered in recent years. Possible raptors include Hen and

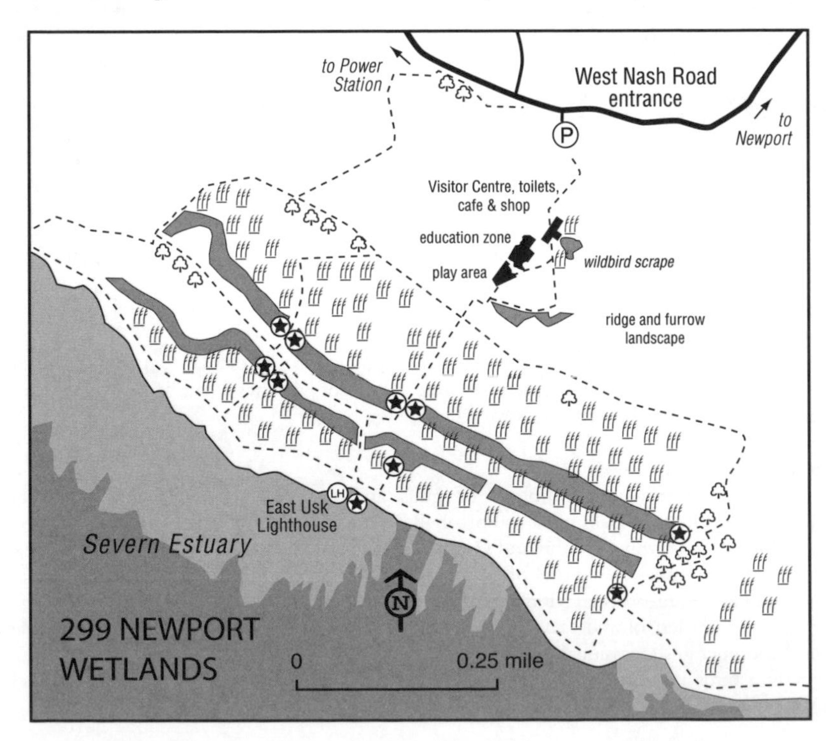

Marsh Harriers, Common Buzzard, Sparrowhawk, Merlin, Peregrine and Short-eared Owl. Up to 50,000 Starlings use the reedbeds to roost in winter, and their pre-roost gatherings make an impressive spectacle.

Breeding birds include Great Crested and Little Grebes, Yellow Wagtail, Spotted Flycatcher, Common and Lesser Whitethroats and Reed and Sedge Warblers, which join the resident Cetti's Warblers (at least 30 pairs) and Bearded Tits. The site hosts Wales's first Avocet colony and Lapwing, Redshank, Oystercatcher and Little Ringed and Ringed Plovers breed on the wet grassland and saline lagoons. Little Egret and Little Owl are resident and Hobby and Spoonbill may visit.

On passage a variety of waders may be seen, including Green and Curlew Sandpipers, Little Stint, Greenshank, Spotted Redshank and Whimbrel, as well as Garganey. Several Aquatic Warblers (a globally threatened species) have been found in the reedbeds and no doubt many more of this ultra-skulker have passed through unseen.

Access

From the M4 junction 24 take the A48 to Newport Retail Park then turn towards the steel-works and follow the brown 'duck' signs to the reserve car park on West Nash Road, between Nash Village and Uskmouth Power Station. The reserve is open daily, 9am–5pm, and there is an RSPB visitor centre, café, shop and children's play area. From the car park there are way marked trails varying in length from 1.7 miles to 2.7 miles.

FURTHER INFORMATION

Grid ref: ST 334 834
Countryside Council for Wales, Southern Team, enquiries helpline: tel: 0845 1306229; web: www.ccw.gov.uk
RSPB Newport Wetlands: tel: 01633 636363; web: www.rspb.org.uk/newportwetlands

300 OGMORE ESTUARY (Glamorgan)

OS Landranger 170

This relatively small estuary, tucked away to the east of Porthcawl, holds large numbers of wintering waders and gulls, which regularly include scarcer species such as Mediterranean and Iceland Gulls.

Habitat

To the south of the river mouth the coast is flanked by low rocky cliffs at Ogmore-by-Sea, but to the north it is guarded by the extensive dune system of Merthyr Mawr Burrows. Upstream, there are stands of woodland towards Merthyr Mawr.

Species

The estuary supports huge numbers of gulls and careful checking regularly produces Mediterranean Gull; the gulls roost at Portobello Island. Wintering waders include Turnstone, Ringed Plover, Sanderling, Dunlin and Common Redshank, with Oystercatcher occurring in especially large numbers. Wildfowl often include Goldeneye on the river channel, and sometimes Scaup and Red-breasted Merganser.

On passage a variety of waders occurs, including Whimbrel, Black-tailed Godwit, Greenshank, Green Sandpiper and occasionally also Wood Sandpiper. Wheatear also occurs.

Breeding birds in the area include Shelduck, Common Redshank, Rock Pipit, Stonechat and Wheatear.

Access

Leave the A48 c. 3 miles south-east of Bridgend west on the B4524 to Ewenny (or take the B4265 south to Ewenny c. 0.5 miles east of Bridgend) and then follow the B4524 south-west to Ogmore-by-Sea.

Portobello Island Two miles beyond Ewenny (and c. 1 mile before reaching the coast) pull off the road onto the grassy area on the right (by the bye-laws sign, near Portobello House). From here walk north-east along the river to view Portobello Island, a good area for roosting gulls and waders, and also for wintering Water Pipit.

Ogmore River Continue towards Ogmore-by-Sea on the B4524 and park in the pull-in on the right after c. 0.5 miles. This offers views of the river channel, with Goldeneye and other ducks possible in winter.

Ogmore-by-Sea Continue towards the sea and park in the car park just north of the village. This offers good views of the river mouth and the area of saltmarsh on the north shore. It is also possible to follow the coastal footpath south from here, with resident Rock Pipit, Wheatear on passage, a chance of Peregrine and Chough (Choughs sometimes also forage on the short turf at the north end of the car park).

Newton, Porthcawl Alternative access to the estuary is from the north, along the coast from the car park in Beach Road, Newton.

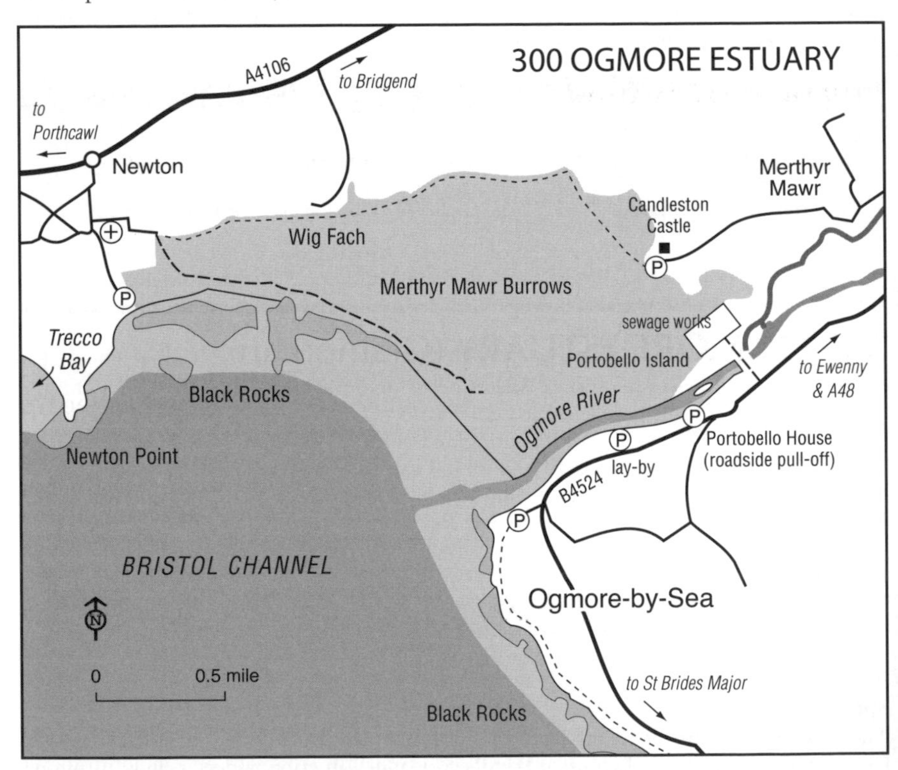

FURTHER INFORMATION

Grid refs: Portobello Island SS 873 762; Ogmore River SS 868 760; Ogmore-by-Sea SS 861 755; Newton, Porthcawl SS 836 769

301 KENFIG (Glamorgan)

OS Landranger 170

Kenfig, north of Porthcawl, has a variety of habitats attracting a range of wildfowl, raptors and waders in winter and on passage, and it is the best site in Wales for wintering Bittern. The area is an NNR.

Habitat

Kenfig Pool is a natural 28-ha dune-slack lake, surrounded by areas of reeds and sallows and set within c. 485 ha of dunes (both mobile and fixed), with areas of wet dune-slack between them. The reserve is flanked on the seaward side by Kenfig Sands. At Sker Point there is a small area of rocky shore and the tiny Sker Pool attracts waders, especially after bad weather.

Species

Winter wildfowl include Teal and Pochard, usually smaller numbers of Wigeon, Shoveler, Gadwall (up to 80) and Goldeneye, and occasionally Smew, Scaup and Long-tailed Duck. Numbers, especially those of diving ducks, increase when nearby Eglwys Nunydd Reservoir is disturbed by water sports (most likely at weekends). Divers, rarer grebes and Whooper and Bewick's Swans are irregular. Around the pool there are often Water Rail, Jack Snipe, a few Chiffchaffs, Cetti's Warbler and Firecrest. Bitterns are regular in winter (although often typically elusive), and two or three birds may be present. Merlin and Peregrine hunt over the area, and occasionally Hen Harrier and Short-eared Owl are recorded. In the fields at Sker Farm there are usually several hundred Golden Plovers. On the beach and rocks at Sker Point are Grey Plover, Sanderling, Purple Sandpiper and Turnstone.

Breeding species include Great Crested Grebe, Oystercatcher, Ringed Plover, Common Snipe and Common Redshank, with Stonechat and Grasshopper Warbler in rough vegetation in the dunes, and Cetti's, Reed and Sedge Warblers around the pool. Teal, Shoveler, Garganey and Ruddy Duck have also nested.

Passage occasionally brings Garganey and Scaup to the pool and Purple Heron has occurred several times in the spring. Waders occur on both Kenfig and Sker Pools and the beach; the range of habitats means that a good variety is often present. Notable are relatively large numbers of Whimbrels in both spring and autumn. Common, Sandwich and Black Terns are regular migrants, but Arctic and Little Terns and Little Gull are less frequent. Small numbers of passerine migrants occur, and Aquatic Warbler has been recorded several times in August–September. Seawatching from Sker Point can be worthwhile, although there is no shelter. In May–August on a rising tide in early morning, Manx Shearwater, Fulmar, Gannet and Common Scoter can be seen, the latter two mainly in late summer, when there are also occasionally Great or Arctic Skuas. A west wind is likely to push birds inshore and produce the best views. In September, gales can produce occasional Storm or even Leach's Petrels.

Access

Leave the M4 at junction 37 (Pyle) south on to the A4229 towards Porthcawl. After 0.5 miles turn north (right) on the B4283, under the motorway and through Cornelly past the Greenacre pub. After c. 1 mile turn west (left, signed to the reserve) at the crossroads, on a minor road over the motorway, past the Angel pub, and straight on for 0.5 miles to the car park. The reserve centre is open Monday–Friday 2pm–4.30pm, weekends and Bank Holidays 10pm–4.30pm. The reserve is managed by Bridgend County Borough Council and there are no restrictions on access, except to the reedbeds on the west shore during the breeding season (March–July). Otherwise it is possible to walk around the pool and there are two hides (always open), one at the south-west corner of the pool (best for Bitterns) and the other on the north shore, approached via a 100-yard boardwalk through the reedbed from the north inlet (reached via a track that starts at the bus shelter on the road near the Prince of Wales Inn, and convenient for those arriving by bus from Porthcawl/Bridgend). Sker Point is reached by walking though the dunes from the car park.

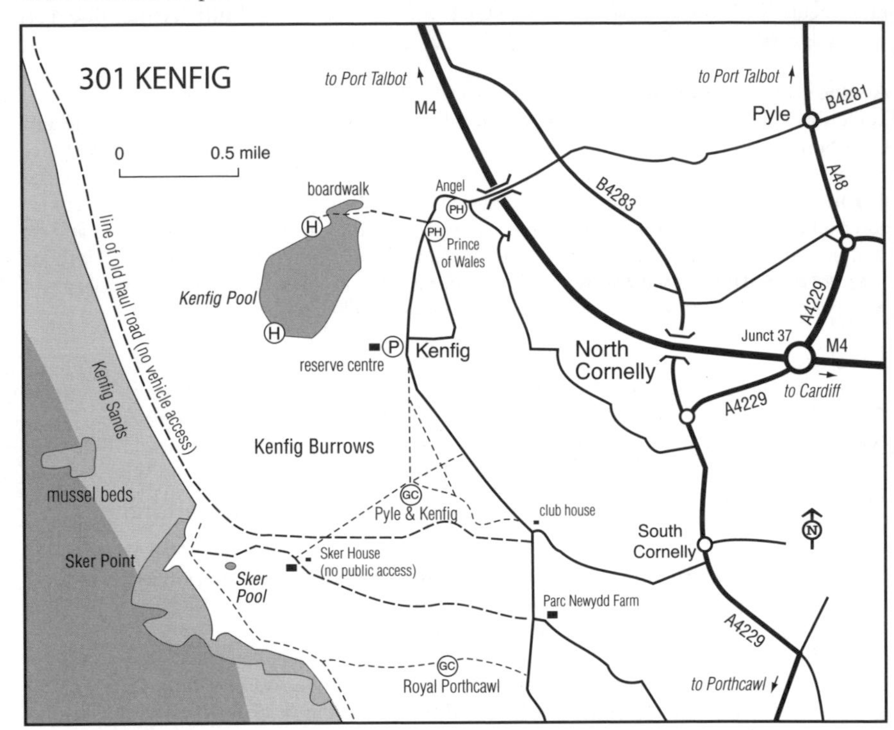

FURTHER INFORMATION

Grid ref/postcode: SS 801 810 / CF33 4PT
Kenfig NNR warden: tel: 01656 743386; web: www.meadowgarden.co.uk/hometable.htm

302 EGLWYS NUNYDD RESERVOIR (Glamorgan)

OS Landranger 170

This 80-ha reservoir attracts a range of waterfowl in winter as well as holding a substantial winter gull roost.

Habitat

The concrete-banked water is surrounded by scrub.

Species

In winter Tufted Duck and Pochard are frequently joined by divers, grebes, Scaup, Smew, Goosander and seaducks. There are few waders on passage, but Common and Arctic Terns and Little Gulls are quite often seen. The scrub surrounding the reservoir is worth checking, especially in passage periods, when it may hold good numbers of the commoner warblers and Goldcrests and perhaps also Firecrest. Autumn storms annually blow migrants onto the reservoir; in the past this has included Leach's Petrel, skuas and Grey Phalarope. The winter gull roost regularly holds several Mediterranean Gulls and occasionally also Little, Iceland, Glaucous, Yellow-legged or Ring-billed Gulls.

Access

Leave the M4 at junction 38 and, from the roundabout, take the lane past the British Oxygen Company works to the reservoir. Parking is, strictly speaking, not permitted inside the gates (your car may be locked in) but most birders do so and rely on anglers or sailors to unlock the gate in the event that it is closed (bona fide birdwatchers have not been challenged); otherwise enter on foot through the same gate.

FURTHER INFORMATION

Grid ref: SS 792 857

303 CWM CLYDACH (Glamorgan)

OS Landranger 159

Lying just 7 miles north of the centre of Swansea, this woodland RSPB reserve holds a typical range of Welsh species, and is best visited in May–June.

Habitat

Oak woodland with some stands of birch and beech, and with ash and alder on damper ground along the Afon Clydach. The slopes above the woodland have areas of heather and Bracken.

Species

Breeding species include Sparrowhawk, Common Buzzard, Tawny Owl, Common Redstart, Wood and Garden Warblers, Pied and Spotted Flycatchers, Marsh Tit, Nuthatch, Treecreeper and Raven, with Tree Pipit, Whinchat and Wheatear at the woodland edges and on open ground above it, and Dipper and Grey Wagtail along streams. Red Kites may visit. Winter is typically quiet, but visitors may include Woodcock, Kingfisher, Siskin and both redpolls.

Access

Leave the M4 north at junction 45 on the A4067 and, after c. 1.6 miles, turn left (west) at the crossroads in the centre of Clydach on a minor road. The reserve car park is adjacent to the New Inn pub in Craig-cefn-Parc after 2 miles. The reserve is open at all times, with access along the signed path by the river that is easy to start with and navigable by wheelchair users with helpers, but links up with a more rugged trail. Outwith the reserve, by following the road for another 3 miles past the upper car park, higher ground can easily be accessed to scan for raptors, with a picnic site overlooking Lliw Reservoirs.

FURTHER INFORMATION

Grid ref: SN 684 026
RSPB Cwm Clydach: tel: 01654 700222; web: www.rspb.org.uk

304 BLACKPILL (Glamorgan)

OS Landranger 159

Blackpill lies immediately west of Swansea. The beach is similar to hundreds of others but, since Britain's first Ring-billed Gull was found here in 1973, the species has occurred regularly, together with up to 50 Mediterranean Gulls, which peak late July–September, numbers slowly dropping to around 20 over the winter.

Habitat

Blackpill is the last part of the shore of Swansea Bay to be covered at high tide. Clyne Stream runs out across the beach and provides birds with bathing facilities, while roosting gulls use the sand bar, sea and beach.

Species

The high-tide gull roost, often of several thousand birds, consists mainly of Black-headed and Common Gulls, but each year one or two Ring-billed Gulls are recorded, most frequently in February–April, and often stay for long periods. Mediterranean Gull is commoner and may be found at any time, though early spring to July is best. Iceland and Glaucous Gulls are sometimes recorded in winter, and Kittiwake also occurs. The usual waders are present at Blackpill, which is used as a roost, including Grey Plover and Sanderlings. On the sea there are occasionally divers or seaducks (Common Eider, Common and Velvet Scoters, Red-breasted Merganser and Goldeneye are possible), and a Peregrine is often present.

Little Gull is regular in March–June and small numbers of Curlew Sandpipers are frequent in autumn; other possibilities include Little Stint and Black-tailed Godwit, and there is a notable concentration of up to 500 Sanderlings. Offshore in late summer and autumn there are sometimes a few skuas and small numbers of terns, which may include Black Tern.

Access

Leave Swansea on the A4067. The road runs beside the beach and the best place to watch from is the boating pool, between the Clyne and the B4436 turning. The period around high tide is usually the most productive on neap tides; an earlier arrival is desirable on 'spring' tides.

FURTHER INFORMATION

Grid ref: SS 619 906

305 OXWICH (Glamorgan)

OS Landranger 159

Oxwich Bay, on the south shore of the Gower Peninsula, is an NNR covering 266 ha. An excellent range of habitats attracts an appropriate diversity of birds, and a visit is worthwhile at any time, especially in spring and summer.

Habitat

Two miles of sandy beach are backed by dunes and bordered to the north and south by limestone cliffs. Nicholaston Pill flows across the beach attracting roosting gulls and waders. Behind the dunes are freshwater marshes with areas of open water, c. 40 ha of reedbeds, and Alder and willow carr. The small saltmarsh is rather dry and consequently unattractive to birds. To the north and south, Nicholaston and Oxwich Woods are both largely deciduous.

Species

In winter there are sometimes Great Northern Divers or Common Scoters offshore, and on the marsh occasionally Gadwall or Shoveler. Green Sandpiper may winter, and there is occasionally a Bittern, especially in early spring. Turnstone and Purple Sandpiper frequent the rocks and Sanderling the beach. Mediterranean, Glaucous and Iceland Gulls are occasionally found in the gull roost. Kingfisher and Grey Wagtail are regular. The woodland and carr sometimes have Siskin and Lesser Redpoll, and often Woodcock, Blackcap, Chiffchaff and Firecrest.

Garganey and Marsh Harrier are occasional spring visitors, and Purple Heron has been recorded several times, while more regular migrants include Common and Sandwich Terns. Offshore, Fulmar, Manx Shearwater, Gannet, Shag and Common Scoter may appear in spring and summer. Osprey has been recorded in autumn, and Hobby may appear late on summer evenings, hunting roosting hirundines.

Breeding birds on the marsh include Little Grebe, Shelduck, Teal, Pochard and Water Rail. The resident Cetti's Warblers are joined in summer by several hundred pairs of Reed Warblers as well as Sedge and Grasshopper Warblers. The woods have Common Buzzard, a heronry (Penrice Woods), Sparrowhawk, all three woodpeckers, Nuthatch, Treecreeper and Marsh and Willow Tits.

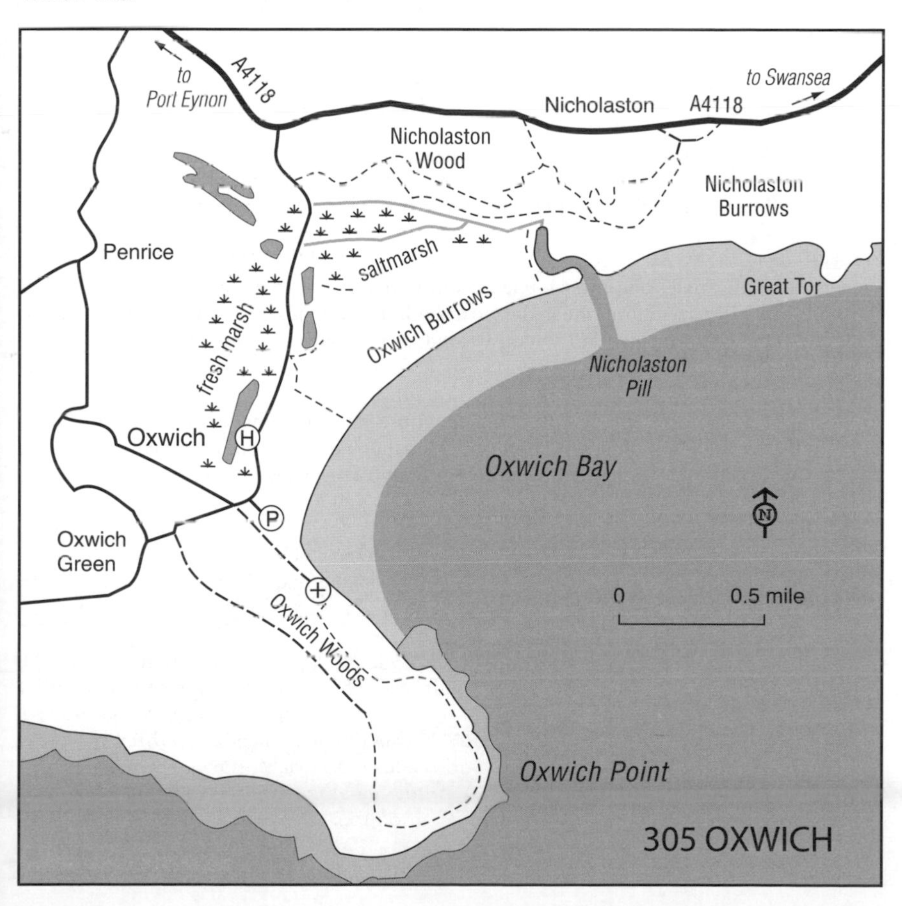

Access

Oxwich village is on a minor road south off the A4118; good views of the marsh are possible from this road. The reserve is managed by the Countryside Council for Wales and there is open access to the dunes and beach, and marked paths in Nicholaston and Oxwich Woods (the lower path in the former has good views of the marsh). The freshwater marsh and carr are closed to casual visitors, though there is a hide and marsh tower affording good views over two sections of the area. A suggested route is to park at the car park, walk along the beach, over Nicholaston Pill, through Nicholaston Wood to the road, and back along it through the marsh to the car park. The latter is a good vantage point from which to scan the bay, especially the sheltered area in the lee of the headland. Oxwich Point also has views of the bay and can be reached by following the footpath south-east from the village past the church and through Oxwich Wood.

FURTHER INFORMATION

Grid ref: SS 501 864
Countryside Council for Wales, Western Team: enquiries helpline: 0845 1306229; web: www.ccw.gov.uk

306 GOWER COAST AND WORMS HEAD (Glamorgan)

OS Landranger 159

Between Worms Head and Port Eynon Point c. 7 miles of coast are managed by the Countryside Council for Wales (CCW), the Wildlife Trust of South and West Wales and the NT. The entire area provides fine clifftop walking, but the real attractions are Worms Head in late April–July for breeding seabirds, and Port Eynon Point for summer and autumn seawatching.

Habitat

Worms Head is a mile-long grass-topped promontory at the south-west tip of the Gower Peninsula that is separated from the mainland at high tide. The adjacent coast has areas of scrub along the cliffs, terminating at Port Eynon Point, Gower's southernmost point.

Species

At Worms Head there are several hundred pairs of breeding Razorbills, Common Guillemots and Kittiwakes and a few Fulmars, Shags and Herring and Great Black-backed Gulls. Small numbers of Puffins may be present in summer, though they are often elusive. Non-breeding

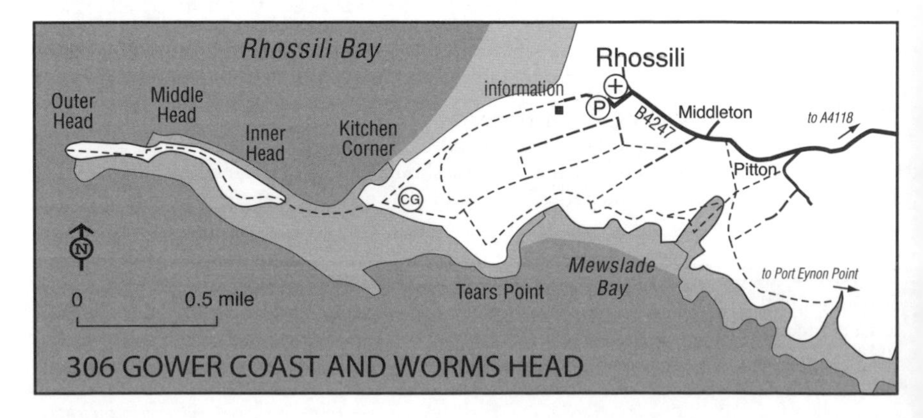

Cormorant, Gannet and Manx Shearwater can often be seen offshore in spring and summer, shearwaters most commonly in early morning. Erratically, at any time except midsummer, large flocks of Common Scoters, sometimes with a few Velvets, are off the Worm. With luck, Peregrine and Chough may be seen, and other breeding birds include Rock Pipit, Dartford Warbler, Stonechat, Wheatear, Jackdaw and Raven on the clifftop.

A few Arctic and sometimes Great Skuas, as well as Gannet, shearwaters, auk, and terns occur on passage. Particularly at Port Eynon, south-west gales may bring very small numbers of Storm Petrel (June–August), Sooty Shearwater (late July–early August) and Pomarine Skua (most likely in September). Passerine migrants such as Wheatear, warblers and, sometimes (in late autumn and winter), Black Redstart occur along the clifftop.

In winter there are Red-throated Divers offshore and occasionally Great Northerns, as well as Common Eider and Red-breasted Merganser. The rocks at the Worm usually attract a flock of Purple Sandpipers and Turnstones (some of which may oversummer).

Access

Worms Head Park in the large car park in Rhossili (large NT visitor centre and shop) and walk c. 1 mile along the clifftop path to the disused coastguard's lookout overlooking the Head. The rocky causeway is only passable for 2.5 hours either side of low water. It is essential to check the times of the tides before you cross or risk being stranded for 7 hours (Coastwatch, based at the old coastguard's lookout, can advise). The best route across the causeway is on the north side. The seabirds are on the north of the Head, visible from certain points on Middle Head (many nesting ledges are only visible from a boat). Visitors must keep to the footpath.

Port Eynon Point At the terminus of the A4118. Park just south of the village and follow the footpath south for 0.5 miles past the youth hostel to the Point. A footpath runs east along the coast from the Worm to Port Eynon Point, and is accessible in several places from the B4247.

FURTHER INFORMATION

Grid refs: Worms Head (Rhossili) SS 414 880; Port Eynon Point SS 467 851
Countryside Council for Wales, Western Team: enquiries helpline: 0845 1306229; web: www.ccw.gov.uk
National Trust, Rhossili Visitor Centre: tel: 01792 390707; email: rhossili.shop@nationaltrust.org.uk
Wildlife Trust of South and West Wales: tel: 01656 724100; web: www.welshwildlife.org
National Coastwatch, Worm's Head: tel: 01792 390167; web: www.nci.org.uk

307 BURRY INLET, SOUTH SHORE (Glamorgan)

OS Landranger 159

The south shore of this inlet is largely owned by the NT, while Whiteford Burrows NNR covers 806 ha of the dunes and foreshore. The inlet's populations of Oystercatcher, Knot and Pintail are internationally important. Ideally a visit should be in winter, timed to coincide with the fortnightly 'spring' tides, which usually occur in early morning and evening.

Habitat

The River Loughor forms a broad estuary, the Burry Inlet, into which flow four smaller rivers. Two-thirds of the Inlet is tidal flats and the remainder saltmarsh (one of the largest in Britain), concentrated on the south shore. Whiteford Burrows, a large dune system covering more than 200 ha and extending for 2 miles north–south, guards the south entrance to the Inlet and has been planted in places with Corsican Pines.

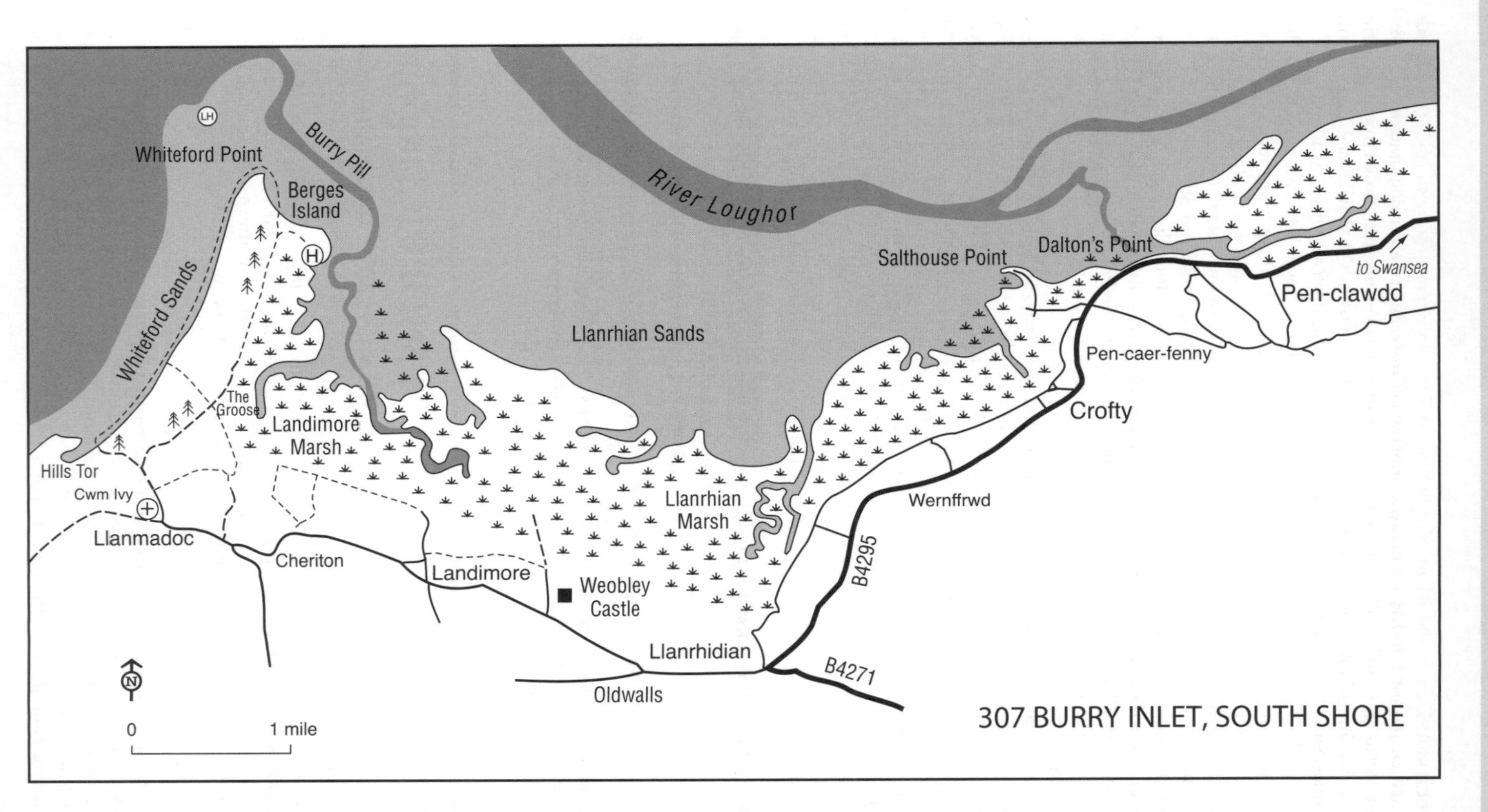

Whiteford Point
Burry Pill
River Loughor
Berges Island
Whiteford Sands
Dalton's Point
Salthouse Point
to Swansea
Pen-clawdd
Llanrhian Sands
Pen-caer-fenny
The Groose
Landimore Marsh
Crofty
Hills Tor
Cwm Ivy
Llanrhian Marsh
Wernffrwd
Llanmadoc
Cheriton
Landimore
Weobley Castle
B4295
Llanrhidian
B4271
Oldwalls
0
1 mile
307 BURRY INLET, SOUTH SHORE

Species

In winter, Great Crested and Slavonian Grebes are regular in small numbers, but Black-necked Grebe is no longer guaranteed. Red-throated Diver may also be seen, and sometimes Great Northern. Wildfowl include up to 1,000 Brent Geese, 200 Shelducks, large numbers of Wigeons, Pintails and Teals, and a few hundred Shovelers. Up to 200 Common Eiders frequent the mussel beds around Whiteford Point, along with small numbers of resident Red-breasted Mergansers. Common Scoter can be seen west of the Point, and irregularly very small numbers of Scaups, Long-tailed Ducks, Velvet Scoters and Goldeneyes also occur here or in the river channel. Waders include Sanderling and maxima of 17,000 Oystercatchers, 1,500 Curlews, 3,000 Knots, 2,800 Dunlins, 600 Common Redshanks and 300 Bar-tailed Godwits. Several hundred Turnstones frequent the mussel beds off the Point, together with a few Purple Sandpipers. Small numbers of Greenshanks and Spotted Redshanks winter, mainly at Whiteford but also at Llanrhidian Marsh, together with a few Green Sandpipers and Black-tailed Godwits. Common Buzzard and Sparrowhawk are resident and regularly joined in winter by Merlin, Peregrine, Hen Harrier and Short-eared Owl.

In spring there can be large numbers of Whimbrels. Other migrants include terns, especially Common and Sandwich, but sometimes also Black.

In summer Common Eider is still present, as are numbers of non-breeding terns. Around Cwm Ivy you may find Raven or Grasshopper Warbler, but Whiteford is generally quiet at this season.

Access

Penclawdd to Crofty The B4295 gives views of the shore from Penclawdd to Crofty.

Salthouse Point This old causeway projects into the inner estuary. Leave the B4295 at the crossroads in Pen-caer-fenny on the minor road north to the Point.

Llanrhidian Marsh Accessible from the minor road between Crofty and Llanrhidian which parallels the B4295. Old earth platforms at Wernffrwd and near Crofty make useful vantages, especially at high tide. The area is good for ducks, with Greenshank and Green Sandpiper on autumn passage; both species often winter.

Weobley Castle The heavily grazed saltings below Weobley Castle are used by roosting waders on high 'spring' tides. Large numbers of Golden Plovers winter and the area also attracts raptors. Good views can be obtained from the castle.

Whiteford NNR Continue on unclassified roads south of Landimore to Llanmadoc and Cwm Ivy. Follow the signed footpath past the pine plantations to the marsh. The track winds through the dunes for c. 2 miles to Whiteford Point, giving views of Groose and Landimore Marsh from the stile before plunging back into the conifers. Just after the path re-emerges from the plantations, a hide overlooks the Inlet at Berges Island, a good spot for Brent Goose. Black-necked and Slavonian Grebes may be seen from the hide on a rising or falling tide. Further on, Whiteford Point is used by roosting waders on lower tides, and the sea has grebes and divers. Cormorants roost on the abandoned lighthouse, and Common Eiders and Brent Geese frequent the rocks and mud at its base. A suggested route would be to visit the hide on a rising tide, walk to the Point on the falling tide and finally back to Cwm Ivy along Whiteford Sands, a total of 5 miles.

FURTHER INFORMATION

Grid refs: Penclawdd to Crofty SS 532 957; Salthouse Point SS 523 958; Llanrhidian Marsh (Wernffrwd) SS 515 942; Weobley Castle SS 478 926; Whiteford NNR (Cwm Ivy) SS 438 936

Countryside Council for Wales, Western Team: enquiries helpline: 0845 1306229; web: www.ccw.gov.uk

Wildlife Trust of South and West Wales: tel: 01656 724100; web: www.welshwildlife.org

National Coastwatch, Worm's Head: tel: 01792 390167; web: www.nci.org.uk

308 NATIONAL WETLAND CENTRE WALES, LLANELLI (Carmarthenshire)

OS Landranger 159

Overlooking the north shore of the Burry Inlet and lying just south of Llanelli, this WWT refuge covers over 180 ha and has gained a reputation for attracting unusual birds, as well as regular wintering wildfowl, waders and raptors. Little Egrets are usually present and breed nearby.

Habitat

Opened in 1991, the centre's habitats include 66 ha of saltmarsh and mudflats bordering the north shore of the Burry Inlet, together with wet meadows, reedbeds and specially constructed lagoons and scrapes.

Species

Wintering wildfowl include Red-breasted Merganser, Goldeneye, Teal, Wigeon, Pintail, Brent Goose and family parties of Whooper Swans, and occasionally Bewick's Swan and Scaup. Waders include numbers of Oystercatchers, Curlews and Common Redshanks, and Black-tailed Godwit is present year-round. A handful of Greenshanks and Spotted Redshanks may winter and Golden and Grey Plovers, Knot and Bar-tailed Godwit sometimes occur in small numbers. Water Rail and Jack Snipe also winter but are typically secretive. Raptors may include Hen Harrier, Merlin and Short-eared Owl, as well as the resident species.

On passage a wider variety of waders may be present, including Whimbrel, Ruff and Green and Common Sandpipers and, in autumn, also Wood and Curlew Sandpipers and Little Stint. Common, Arctic and Sandwich Terns may visit, and Garganey is possible. Spoonbill, Mediterranean Gull, Marsh Harrier and Osprey are occasional visitors.

Residents in the area include Little Grebe, Greylag Goose (feral), Shelduck, Gadwall, Shoveler, Pochard, Common Buzzard, Sparrowhawk, Peregrine, Water Rail, Common Redshank, Barn Owl, Kingfisher and Raven. Little Egret has been almost constantly present on the reserve or

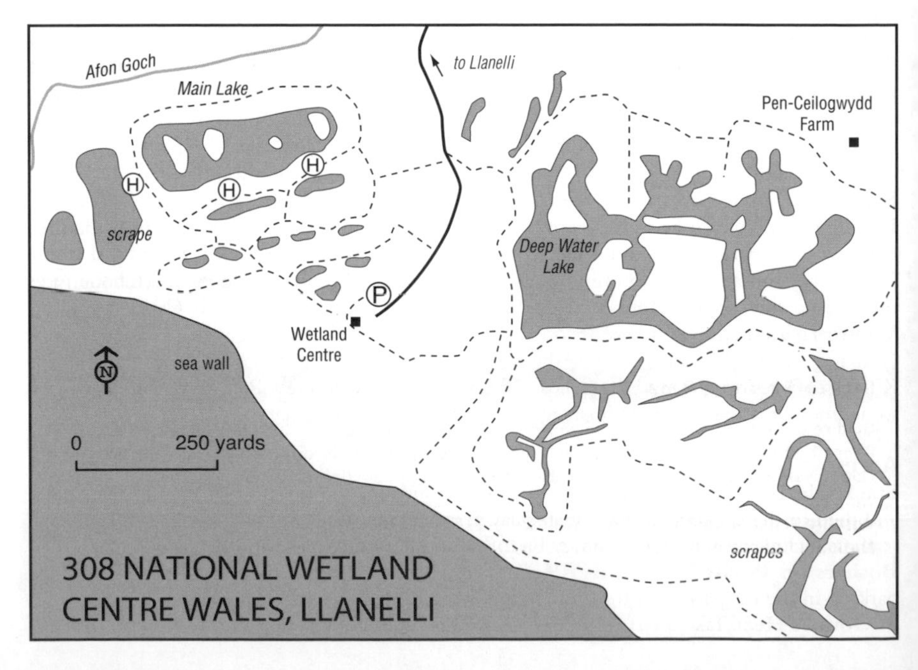

308 NATIONAL WETLAND CENTRE WALES, LLANELLI

nearby estuary in recent years with numbers peaking at around 300 in August and September. Breeding summer visitors include Grasshopper, Sedge and Reed Warblers, and Common and Lesser Whitethroats.

Access

The reserve is off the A484 Llanelli–Swansea road, via the minor road off the roundabout (2 miles east of Llanelli) to Penclacwydd (signed with a duck). Open daily 9am–5pm (grounds open till 6pm in summer; closed Christmas Eve and Christmas Day). Facilities include a visitor centre, restaurant, six hides and the heated Lagoon Observatory. All trails are level and laid with tarmac or compacted gravel. For disabled visitors there are accessible toilet facilities, mobility scooters available for hire, and low-level viewing windows in some hides. There is also a large collection of captive waterfowl.

FURTHER INFORMATION

Grid ref/postcode: SS 532 987 / SA14 9SH
WWT, Llanelli: tel: 01554 741087; web: www.wwt.org.uk/visit-us/llanelli

309 THE CASTLEMARTIN PENINSULA (Pembrokeshire)

OS Landranger 158

Largely a live-firing range used for tank training, this peninsula has a wild and unspoilt coast with some good seabird colonies, as well as breeding Chough, while Bosherston Ponds hold a variety of wintering wildfowl. Part of the area forms the 233-ha Stackpole NNR, managed in association with the NT.

Habitat

The coast is bounded almost exclusively by cliffs, only infrequently broken by small coves and bays. The area immediately inland is covered by short turf, with scrub in more sheltered places. Bosherston Ponds are artificial and were formed in the 18th and 19th centuries by the construction of dams across what had been a sea-drowned marshy valley, cut off from the sea by sand dunes. The ridges between the ponds are cloaked in deciduous woodland.

Species

Breeding seabirds include Fulmar, Shag, Common Guillemot, Razorbill, a handful of Puffins, Kittiwake, Herring Gull and a few Great and Lesser Black-backed Gulls. Manx Shearwater and Gannet can be seen offshore throughout the summer. Peregrine, Raven and Chough are resident on the cliffs. Look also for Rock Pipit and Stonechat.

Wintering waterfowl on Bosherston Lakes include Gadwall, Pochard, Scaup, Goldeneye and Goosander in small numbers, while the reeds hold Water Rail and often a wintering Bittern. Kingfisher is resident around the ponds. Offshore, there may be a few Red-throated Divers and sometimes Great Northern. Fulmars return to the colonies by January.

Access

Stackpole Head From Pembroke, follow signs south for the unclassified road to Stackpole, and then to the car park at Stackpole Quay. It is a 1-mile walk south to the Head. Breeding seabirds include Common Guillemot, Razorbill and a handful of Kittiwakes.
Bosherston Ponds Take the B4319 south from Pembroke and turn south to Bosherston, parking in the car park near the church, following footpaths to the ponds.
St Govan's Head Take the B4319 south from Pembroke and turn south to Bosherston, bearing

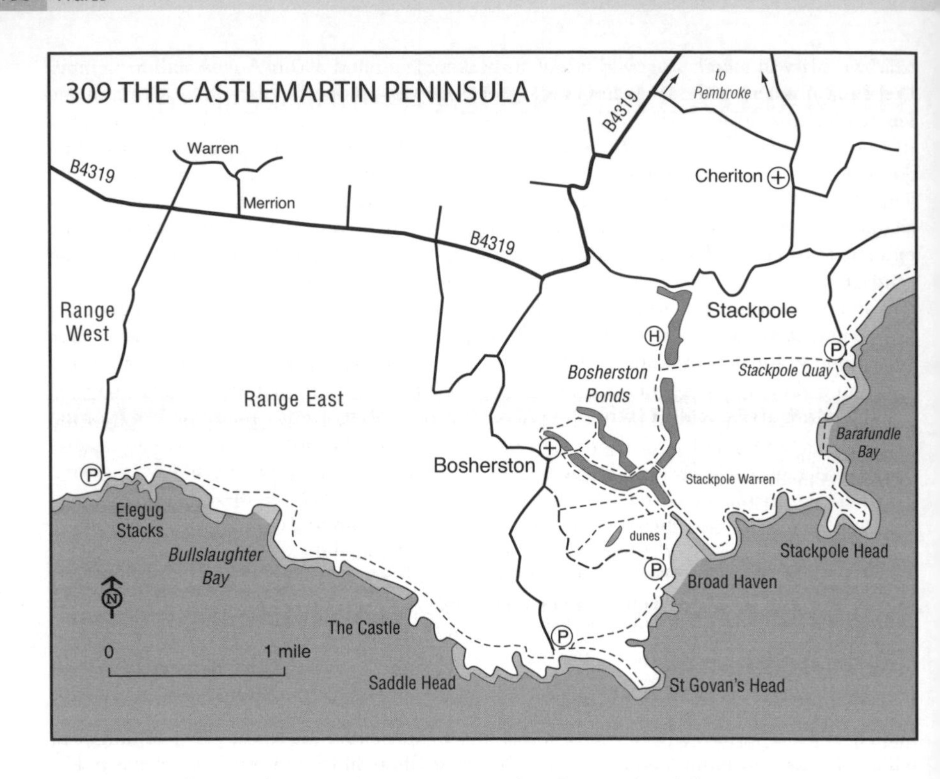

right in the village to the car park at St Govan's Chapel (note the cliff-nesting House Martins here). From the car park walk c. 0.5 miles south and east to the Head. Seawatching here may produce Manx Shearwater (especially in early mornings and evenings) and Gannet.

Elegug Stacks (Stack Rocks) Turn south off the B4319 immediately west of Merrion Barracks onto a minor road, parking at the coast after c. 1.75 miles. The road passes through a live-firing range, and there is no access when the gate is shut and red flags are flying (normally 9am–4.30pm for day firing or 7pm–midnight for night firing; check the 24-hour answering service of the Castlemartin Range Office on 01646 662367). The four stacks lie just off-shore and are excellent for watching breeding seabirds at very close range, including c. 6,000 Common Guillemots, 650 Razorbills, 200 pairs of Kittiwakes, 100 pairs of Herring Gulls, smaller numbers of Fulmars and Lesser Black-backed Gulls, and a few Shags and Great Black-backed Gulls. Stonechat and Chough are regular here.

Range East The Pembroke Coast Path runs east from Elegug Stacks to the car park at St Govan's Chapel, but is closed during firing (as above).

FURTHER INFORMATION

Grid refs: Stackpole Head SS 990 958; Bosherston Ponds SS 967 948; St Govan's Head SR 966 930; Elegug Stacks (Stack Rocks) SR 925 946

Countryside Council for Wales, Western Team: enquiries helpline: 0845 1306229; web: www.ccw.gov.uk

310 MARLOES PENINSULA (Pembrokeshire)

OS Landranger 157

This area holds some interesting breeding birds, notably Peregrine and Chough, and offers the chance of seeing seabirds without venturing offshore to the breeding islands. Much of the coast is owned by the NT.

Species and Access

Wooltack Point Large numbers of Manx Shearwaters assemble offshore each evening in the breeding season and, depending on the weather, may be seen from the mainland. Seabirds are best seen from c. 2 hours before until 2 hours after high tide from Wooltack Point (follow the footpath a few hundred yards north-west from Martin's Haven); at other times they tend to congregate in Broad Sound and St Brides Bay, and views are more distant.

Marloes Peninsula From the coastal footpath, on either the north or south sides of the Peninsula east of Marloes, Peregrine, Common Buzzard, Raven and Chough can be seen, and Stonechat and Grasshopper Warbler breed in the rough ground. Passerine migrants occur and have included rarities.

Marloes Mere Leave Marloes west on the minor road past Marloes Court and park after c. 1 mile at the junction at the Marloes Sands car park. From the corner, follow the track west past the Runwayskiln youth hostel and the mere, where there is a hide (continuing, this track connects with the coast path opposite Gateholm). Alternatively, walk north from the car park towards Treehill Farm and, after 0.25 miles, take the track to the west to a second hide. The mere largely comprises areas of rushes and cotton-grasses with some small pools. Numbers of wildfowl winter, including Wigeon, Teal and up to 50 Shovelers.

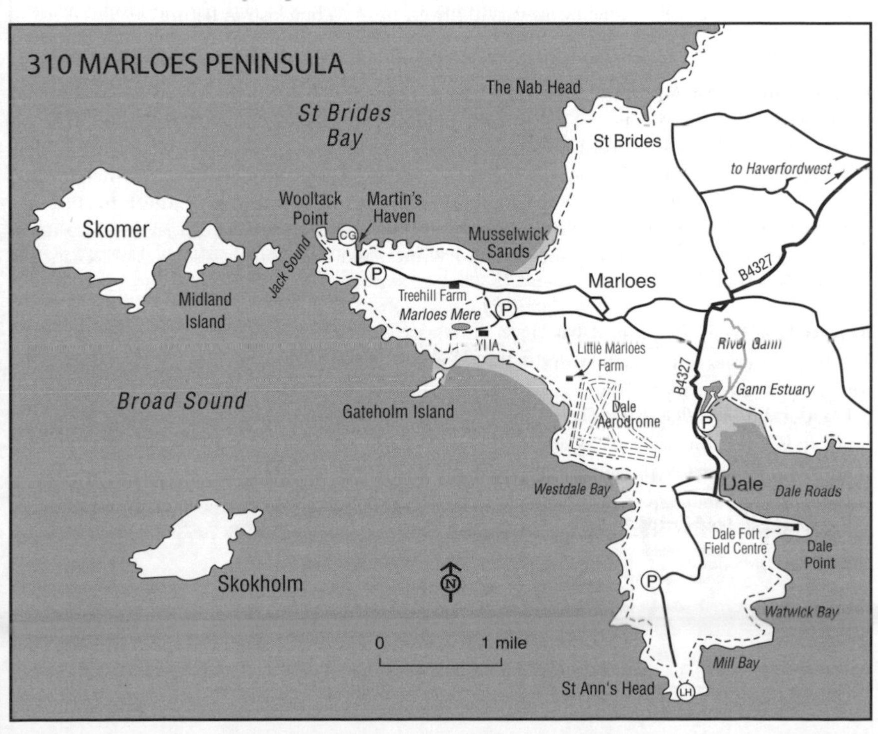

Dale Aerodrome Worth checking in autumn, when post-breeding gatherings of up to 40 Choughs have been recorded in September–October (and Buff-breasted Sandpiper has been seen several times in September). From Marloes a narrow track runs south to Little Marloes Farm and a footpath follows the runways south and east to Dale.
Westdale Bay A good spot for Chough. Follow the track west from Dale Castle for 0.5 miles to the Bay.
St Ann's Head Another chance for Chough. Take the minor road south from Dale to the lighthouse.

FURTHER INFORMATION

Grid refs: Wooltack Point SM 761 088; Marloes Mere SM 779 082; Dale Aerodrome SM 790 075; Westdale Bay SM 805 057; St Ann's Head SM 805 031

311 GANN ESTUARY (Pembrokeshire)

OS Landranger 157

This small estuary can be interesting in winter and passage periods. The sizeable gull roost may hold up to 30 Mediterranean Gulls.

Habitat

Close to the mouth of the Milford Haven Estuary and facing into the sheltered waters of Dale Roads, a shingle ridge separates the sand and shingle beach from a small area of saltmarsh, where gravel extraction has left several pools varying from fresh to brackish in nature.

Species

Winter brings a few Red-throated Divers and occasional Great Northern and Black-throated Divers and Slavonian Grebes to Dale Roads, especially in rough weather, with Little Grebe on the lagoon. Goldeneye and Red-breasted Merganser are regular on the sea. Wintering waders include Grey and Ringed Plovers and Turnstone, and regularly also Greenshank, which favours the upper part of the estuary and the high-banked channel of the river. Little Egrets can be seen year-round, with up to 20 during the winter months. The gull roost may hold up to 30 Mediterranean Gulls during the autumn and winter, and occasionally also Iceland, Glaucous or Ring-billed Gulls.

Access (see map on page 489)

The lower reaches can be seen from the B4327 north of Dale, with a small car park at the south-west end of the estuary affording good views of the flats. From here, it is possible to walk north below the shingle ridge that separates the estuary from the lagoon and then across the tidal ford over the River Gann. The area is best in the period around high water and is heavily disturbed in summer, when early-morning visits are likely to be most productive.

FURTHER INFORMATION

Grid ref: SM 808 066

312 SKOKHOLM ISLAND (Pembrokeshire)

OS Landranger 157

In 1933 Ronald Lockley established Britain's first bird observatory on Skokholm, building the country's first Heligoland traps. Ringing ceased in 1976 and the island is now managed as a reserve by the Wildlife Trust of South and West Wales. The island is now best known for breeding seabirds and migrants, and the best time to visit is May–September.

Habitat

One mile long, the island covers 107 ha, with cliffs rising to 160 feet. The exposed west and south-west sides have a short sward, extensively excavated by Rabbits, Puffins and Manx Shearwaters. The more sheltered east side is dominated by Bracken, with some rocky outcrops. The island's summit has a network of dry-stone walls and dykes which, together with the rocky scree and cliffs, shelter nesting Storm Petrels. There are two ponds, the more southerly being surrounded by a marsh. The only buildings are the lighthouse and old farm buildings.

Species

Breeding birds include c. 45,000 pairs of Manx Shearwaters (numbers are slowly increasing) and c. 1,000 pairs of Storm Petrels. Both tubenoses are strictly nocturnal on land but are very vocal and can be heard at night, and be seen at twilight or on moonlit nights. Other seabirds include 135 pairs of Fulmars, Razorbills (812 individuals), Common Guillemots (1,455 individuals), Puffins (4,900 individuals), and 320 pairs of Herring, 2,890 pairs of Lesser Black-backed and 51 pairs of Great Black-backed Gulls (all counts refer to 2007). Peregrine, Common Buzzard, Oystercatcher, Rock Pipit, Wheatear, Chough and Raven also breed. Although present, albeit intermittently, from late March, Puffins are best seen from early June, when feeding young. The breeding auks depart by mid-August (and Puffin slightly earlier, by the end of the first week of August).

Passage periods bring the usual thrushes, chats, warblers and flycatchers. Spice is provided by the possibility of Wryneck, Black Redstart, Bluethroat, Icterine, Melodious and Barred Warblers, Red-breasted Flycatcher and Lapland Bunting, and there have been numerous other rarities.

Access

The island's accommodation and other facilities are currently being upgraded and prospective visitors should contact the Wildlife Trust of South and West Wales for the latest details (phone Island Bookings, 01239 621600 or 01239 621212). The Dale Sailing Company offers 'Skomer and Skokholm Safaris', boat trips that circumnavigate the islands without landing, including evening trips to witness the dusk gatherings of Manx Shearwaters.

FURTHER INFORMATION

Wildlife Trust of South and West Wales: tel: 01656 724 100; web: www.welshwildlife.org
Dale Sailing Company: tel: 01646 603110/01646 603107; web: www.dale-sailing.co.uk

313 SKOMER ISLAND (Pembrokeshire)

OS Landranger 157

One mile offshore, Skomer has a large and easily accessible seabird colony. It is best visited in late April to mid-July, though migrants may be seen in autumn. The island is an NNR, owned by CCW and managed by the Wildlife Trust of South and West Wales.

Habitat

Skomer covers 290 ha. The central plateau of grassland, Bracken and heath has masses of Bluebells and Red Campion in spring, when the cliff slopes are a blaze of Thrift and Sea Campion. The sward is dissected by a maze of Rabbit, Puffin and shearwater burrows.

Species

Breeding seabirds include Fulmar (595 pairs), Manx Shearwater (c. 128,000 pairs, apparently 50% of the world population!), Storm Petrel (100 pairs), Puffin (c. 6,000 pairs), Common Guillemot (16,977 individuals), Razorbill (4,561 individuals), Kittiwake (2,067 pairs), Lesser Black-backed Gull (c. 10,550 pairs), Herring Gull (434 pairs) Great Black-backed Gull (114 pairs), Cormorant (10 pairs) and Shag (four or five pairs; all figures refer to 2006). Other breeding species are Teal, Shoveler, Common Buzzard, Peregrine, Curlew, Oystercatcher, Little and Short-eared Owls, Rock Pipit, Wheatear, Raven and Chough. The Wildlife Trust has a camera in a shearwater burrow, relaying pictures to a TV monitor in the Information Room, affording visitors views of the underground activities. Nevertheless, an overnight stay is essential to experience the spectacle of thousands of shearwaters returning to their burrows. By late August and early September the adults have departed, but the full-grown young are then leaving their burrows at night to exercise their flight muscles before making a final departure. Otherwise, Manx Shearwaters may come inshore in rough weather and can sometimes be seen from a boat, especially on the Seabird Spectaculars. Puffins are best seen from early June, when they are feeding young. The breeding auks have departed by mid-August (with Puffins leaving by the end of the first week of August).

Skomer attracts numbers of passerine migrants but its size and extensive cover make them harder to find than on Skokholm. Wintering birds include Short-eared Owl, Hen Harrier and Merlin.

Access

Skomer is open daily between Good Friday or 1 April (whichever is the earlier) and 31 October (except Mondays but including Bank Holidays: on other Mondays the boat does not land on Skomer but offers round-island cruises; note that Skomer is also closed for four days in late May or early June for the annual bird count). The *Dale Princess*, operated by the Dale Sailing Company, sails from Martin's Haven where there is an NT car park (fee). Boats leave at 10am, 11am and noon (with additional departures depending on demand), return boats starting at 3pm, allowing c. 5 hours on the island. Access is restricted to the marked paths and a landing fee is payable to the island's warden. A fairly steep flight of steps leads up to the island's centre, and there is covered seating and toilets at the Old Farm Complex, c. 1 mile from the landing point. No booking is required for the boat, which takes passengers on a first-come first-served basis. During particularly popular periods (May–July) an early arrival may be required to secure a place. The boat also offers hour-long 'Skomer Cruises', circumnavigations of the island in the early afternoon, without landing. During rough weather, especially north winds, the boat may be delayed or even cancelled, so always check on sailing times. The Lockley Lodge Information Centre at Martin's Haven run by the Wildlife Trust, situated close to the NT car park, is open Tuesdays–Sundays and Bank Holiday Mondays, 9am–4pm (or return of last boat from the island). On several evenings a week the Dale Sailing Company offers 'Evening Seabird Spectaculars' aboard the *Dale Princess* out of Martin's Haven to see rafting Manx Shearwaters (and permitting good views of Skomer cliffs and the colonies).

FURTHER INFORMATION

Wildlife Trust of South and West Wales: tel: 01656 724 100; web: www.welshwildlife.org
Lockley Lodge Information Centre: tel: 01646 636234
Dale Sailing Company: tel: 01646 603110/01646 603107; web: www.dale-sailing.co.uk

314 GRASSHOLM (Pembrokeshire)

OS Landranger 157

Lying 10 miles offshore, Grassholm covers 9 ha. The 39,000 pairs of Gannets form the third-largest colony in the world and make up around 10 per cent of the world population (records go back to 1860, when there were just 20 pairs).

Species

Besides the many Gannets, other seabirds breeding on Grassholm include Shag, Razorbill, Common Guillemot, Kittiwake, and Herring and Great Black-backed Gulls. The island is (frustratingly) also very good for small birds on passage, and is an NNR.

Access

Landing is not allowed on Grassholm but the RSPB, in conjunction with Thousand Island Expeditions, organises the 'RSPB Islands Adventure', a combined boat trip which circumnavigates Grassholm before moving on to land on Ramsey Island. Boats leave the lifeboat station in St Justinians, 1 mile west of St David's, at 9.30am and depart Ramsey island at 4pm. Advance booking is essential. The Dale Sailing Company also organises trips circumnavigating Grassholm, departing from Martin's Haven or Neyland Marina, and the round trip varies from 2.5 to 3 hours.

FURTHER INFORMATION

Thousand Island Expeditions: tel: 01437 721721 or 721686; web: www.thousandislands.co.uk
Dale Sailing Company: tel: 01646 603110/01646 603107; web: www.dale-sailing.co.uk

315 RAMSEY ISLAND (Pembrokeshire)

OS Landranger 157

The second largest of the Pembrokeshire islands (after Skomer), Ramsey is an RSPB reserve and NNR, and worth visiting in spring and summer for its breeding seabirds and Choughs.

Habitat

The island covers 263 ha and was, until recently, farmed (current management includes some agricultural practices). The lower, east side of the island largely comprises fields with several ponds, while the south portion is mostly heather and Bracken. There are two hills, Carn

Llundain in the south, rising to 450 feet, and the slightly lower Carn Ysgubor in the north. The coast, especially in the west, is bounded by high cliffs, with spectacular sea caves and inlets, and Ramsey has a series of satellite islets.

Species

Breeding seabirds include Fulmar, Common Guillemot, Razorbill, Kittiwake and c. 1,000 pairs of Manx Shearwaters, although the latter come ashore only at night. There are no Puffins on Ramsey. A programme to eradicate Brown Rats was completed in 2000 resulting in improved breeding success for burrow- and ground-nesting birds, and in 2008 five occupied Storm Petrel burrows were discovered. Other breeding birds include Peregrine, Common Buzzard, Lapwing, Wheatear, Raven and Chough.

Like all of the Pembrokeshire islands, Ramsey attracts migrants during passage periods and has hit the headlines with records of transatlantic vagrants (including Britain's first Indigo Bunting in 1996). Note that the early-evening gatherings of Manx Shearwaters can sometimes be seen from nearby shores.

Access

The island is accessible daily, 1 April or Easter (whichever is earlier) to 31 October, weather permitting. Boats leave the lifeboat station in St Justinians, near St David's, at 10am and noon, and return at noon and 4pm. There is a limit of 80 visitors per day and it is essential to book in advance – Thousand Island Expeditions has the sole landing rights. A variety of other trips to and around the island are available and the RSPB, in conjunction with Thousand Island Expeditions, also organises trips combining Grassholm and Ramsey Island, see the Grassholm Island account for details. The main trail is c. 3.5 miles long, but can be divided into two loops by taking the short cut (closed March–July). There is an information centre and WCs, but only light refreshments are available on Ramsey.

FURTHER INFORMATION

RSPB Ramsey Island: tel: 07836 535733; email: ramsey.island@rspb.org.uk
Thousand Island Expeditions: tel: 01437 721721 or 721686; web: www.thousandislands.co.uk

316 ST DAVID'S PENINSULA (Pembrokeshire)

OS Landranger 157

This area of relict maritime heath with a mosaic of heathland, willow scrub and patches of open water, set amid mixed farmland, is notable for its range of raptors, especially in winter, regularly including Hen Harriers, and for the resident Choughs on the coast.

Species

Residents include Peregrine, Common Buzzard and Sparrowhawk, and breeding species on the heaths and rough ground include Stonechat and Grasshopper and Sedge Warblers. Passage occasionally brings Garganey to the pools, but the lack of muddy margins means that they are unattractive to passage waders.

In winter a greater variety of raptors may be seen. Merlin and Short-eared Owl join the resident species, while Hen Harriers are at low density and possible anywhere in the day, but up to six have been recorded roosting on Dowrog Common. It is best to watch from the road and keep a sharp eye in all directions, as the harriers arrive low and fast and quickly drop to the ground (though they may fly again and move to a different roost site, making accurate counts more difficult). Small numbers of Wigeons and Teals occur on the various pools and are sometimes joined by Pintail, Shoveler or small parties of Whooper and Bewick's Swans. Water Rail is regular in winter, and Bittern has been recorded most winters on Dowrog Pool.

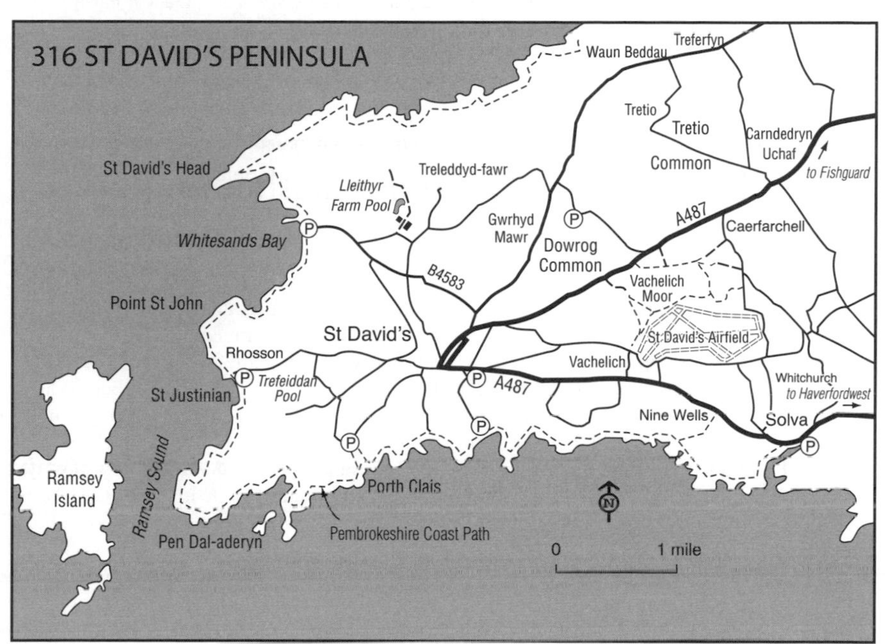

Access

Dowrog Common Comprising an extensive tract of wet and dry heath with several pools (notably Dowrog Pool) and areas of fen in the upper reaches of the River Alun, this 101-ha reserve of the Wildlife Trust of South and West Wales is crossed by a minor road which leads north off the A487 2 miles east of St David's (towards Gwrhyd Mawr), with a small car park near the north-west end.

Tretio Common Crossed by a minor road north off the A487 at Carnhedryn Uchaf, 3.25 miles east of St David's.

Vachelich Moor (St David's Airfield Common) This lies north of the now disused St David's Airfield. Turn south off the A487 on the minor road to Caerfarchell, c. 3 miles east of St David's, viewing the common to the west after 0.5 miles. The area may also attract harriers.

Pen Dal-aderyn and Point St John From the car park at St Justinian follow the coastal footpath south towards Dal-aderyn or north towards Point St John for a chance of Chough.

Lleithyr Farm Pool Leave the A487 just outside St David's north on the B4583, follow the road round the right-angle left bend, and after a further 1.25 miles, turn right (east) on the narrow minor road to Lleithyr. After 0.3 miles, where the road bends sharply right, follow the footpath along the track past the farm to view the pool on the left after 0.3 miles.

St David's Head From St David's, follow the B4583 to the car park at Whitesands Bay. From here follow the coastal footpath north towards St David's Head, looking for Chough en route.

Porth Clais and Nine Wells These two short, narrow valleys on the south side of the Peninsula can be excellent on occasion for passerine migrants, especially in autumn. They have attracted rarities and given more intensive watching could rival better-known sites in south-west Britain. Porth Clais is accessible along a minor road south-west from St David's, with public footpaths running down both sides of the valley. Nine Wells lies on the A487, 2.25 miles east of St David's, and footpaths similarly run south to the sea.

FURTHER INFORMATION

Grid refs: Dowrog Common SM 772 277; Tretio Common SM 785 284; Vachelich Moor (St David's Airfield Common) SM 796 263; Pen Dal-aderyn and Point St John SM 724 252; Lleithyr Farm Pool SM 747 271; St David's Head SM 734 271; Porth Clais SM 740 242; Nine Wells SM 787 249
Wildlife Trust of South and West Wales: tel: 01656 724 100; web: www.welshwildlife.org

317 STRUMBLE HEAD (Pembrokeshire)

OS Landranger 157

This is the top seawatching station in Wales and one of the best in Britain. Some seabirds can be seen throughout spring and summer (mostly from the nearby breeding colonies on the Pembrokeshire islands), but late August–early October is the best time to visit for the scarcer species. An added bonus is resident Chough in this area.

Habitat

Strumble Head lies at the west end of the Pencaer Peninsula and rises to a maximum of c. 100 feet, comprising grass and heather-clad slopes with rocky outcrops.

Species

Manx Shearwater is regularly seen offshore March–October, especially in early mornings and evenings. Cormorant, Shag and Gannet are also present almost year-round, and other regulars in spring and summer include Fulmar, Razorbill, Common Guillemot, Kittiwake and other gulls.

The best autumn seawatching conditions are a south-west gale veering west or north-west. Shearwaters, including Sooty and Balearic, may occur, together with Leach's Petrel in late autumn. Great, Cory's and Little Shearwaters and Sabine's Gull have been recorded. Great and Arctic Skuas are regular, Pomarine less so and Long-tailed is rare (but annual). Red-throated Diver, Common Scoter, Little Gull, Kittiwake, auks and terns (including Black Tern) are other regulars. Passerine migrants too, should not be ignored.

Breeding birds include Fulmar, Herring Gull, Chough, Raven, Stonechat and Grasshopper Warbler, and Peregrine is often present.

Access

Strumble Head Well signed along minor roads from the A40 at Fishguard Harbour, there is a car park directly opposite the lighthouse rock and, while it is possible to seawatch from the car here, the best place is the disused War Department building below the road on the right, just before the car park, which has been renovated as an observation centre by the Pembrokeshire Coast NP Authority. Dawn onwards is best. The coastal footpath east from the Head towards Carreg Gybi is good for Chough.
Fishguard Harbour This can be good for divers, grebes and seaducks, with several Mediterranean Gulls usually present from July through to the spring; Little and Iceland Gulls are sometimes recorded.
Dinas Head Leave Fishguard east on the A487 and, after c. 3 miles, turn north at Dinas Cross on the minor road to Bryn-henllan. Continue north through the village to the car park at the base of the Head. The coastal footpath circumnavigates the Head with a chance of Chough, while Razorbills and Common Guillemots nest on Dinas Island.

FURTHER INFORMATION

Grid refs: Strumble Head SM 895 412; Fishguard Harbour SM 947 379; Dinas Head SN 005 398

318 TEIFI MARSHES AND WELSH WILDLIFE CENTRE (Pembrokeshire/Ceredigion) OS Landranger 145

Lying near the mouth of the Afon Teifi immediately south of Cardigan, this site holds a variety of the commoner wildfowl and waders, and, notably, breeding Cetti's Warbler. The area includes a 107-ha Wildlife Trust of South and West Wales reserve.

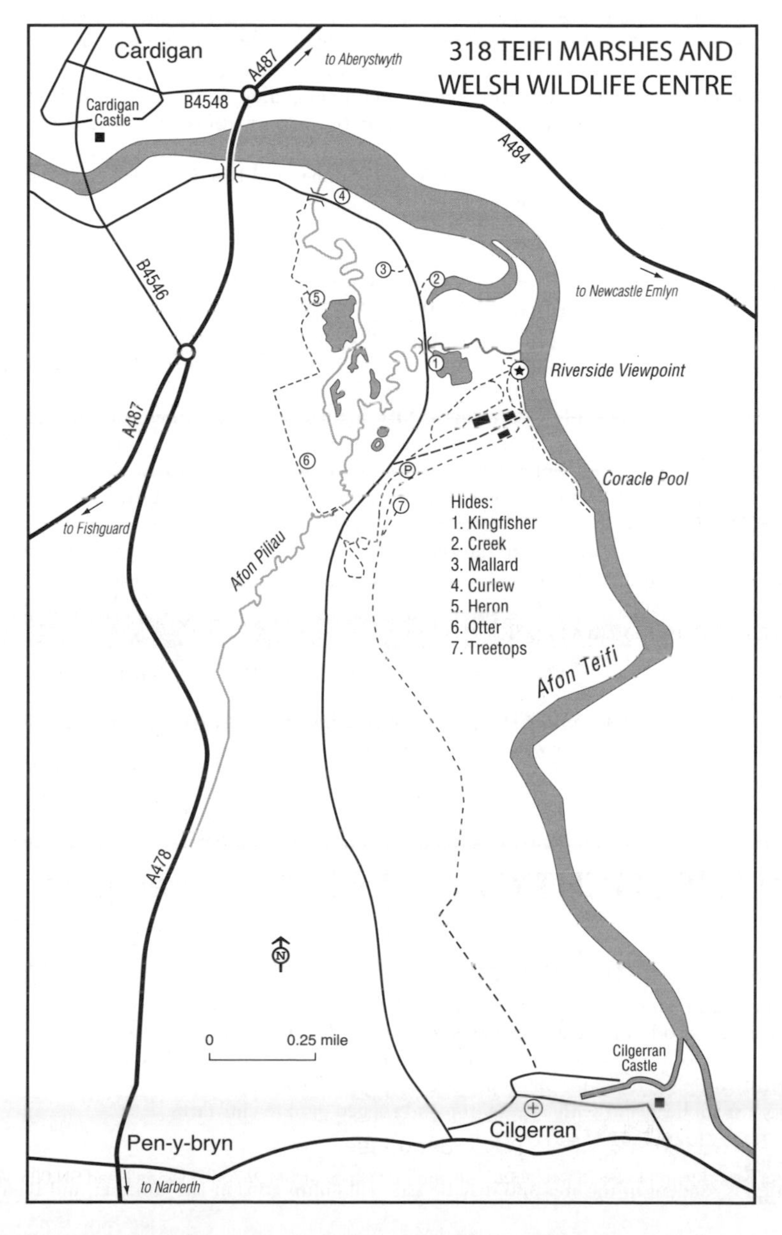

Habitat

The centre's varied habitats include the upper estuary, fresh and saltwater reedbeds, grazing marshes and pools, and the wooded gorge of the river.

Species

Wintering wildfowl include Shelduck, Wigeon and Teal, with Goldeneye, Red-breasted Merganser and Goosander on the river, and wintering waders include Oystercatcher, Ringed Plover, Dunlin, Bar-tailed Godwit, Curlew and Common Redshank. Water Rail is often present in some numbers, but is typically elusive. Little Egrets may be seen all along the Teifi, from the marshes to the river mouth, with up to four wintering. Raptors in the area include Sparrowhawk, Common Buzzard and sometimes Peregrine, and interesting passerines include Lesser and Common Redpolls, Siskin and occasionally Brambling.

On passage a greater variety of waders occurs, including Ruff, Common and Green Sandpipers, Spotted Redshank and Greenshank, and scarcer migrants have included Garganey, Osprey, Marsh Harrier and Bearded Tit.

Breeding birds include Water Rail, Cetti's, Grasshopper and Reed Warblers, and Barn Owl, Kingfisher, Grey Wagtail and Dipper.

Access

Welsh Wildlife Centre Leave the A487 Cardigan bypass at the west roundabout south on the A478 and, after c. 2 miles, turn east (left) at Pen-y-bryn on the minor road to Cilgerran, signed to the centre. After c. 1 mile (before the village) turn north on a minor road to the centre car park (a further 1.5 miles distant; free to Trust members; £5 to non-members). There is also pedestrian access along the river from Cardigan. The reserve is open all year. There is a visitor centre constructed largely of wood and glass, which gives panoramic views over the Teifi River, Cardigan town and woodland. It has a shop, information area and café, and is open 10.30am–5pm from Easter to end of September. From the end of September until December it closes at 4.30pm and is closed Monday–Tuesday. The reserve has seven hides and four trails.
Lower Teifi Estuary The lower reaches of the estuary north of Cardigan can be seen from the B4546 on the west shore, and from the B4548 on the east shore, the latter road running along the shoreline towards the mouth.

FURTHER INFORMATION

Grid refs/postcode: Welsh Wildlife Centre SN 186 449 / SA43 2TB; Lower Teifi Estuary (B4548) SN 163 493
Wildlife Trust of South and West Wales, Welsh Wildlife Centre: tel: 01239 621600; email: wwc@welshwildlife.org; web: www.welshwildlife.org

319 THE LOCHTYN PENINSULA (Ceredigion)

OS Landranger 145

This peninsula on the coast c. 6 miles south-west of New Quay is a well-known site for Chough, and also holds a few breeding seabirds as well as Red Kite.

Habitat

This is a small headland with a grassy top and rugged granite cliff-faces.

Species

Chough is resident in the area and may be especially numerous in late summer, and Red Kite is a regular visitor. Small numbers of Common Guillemots, Razorbills and Kittiwakes also breed. Although the breeding cliffs are difficult to view (being more visible from a boat), the birds can be seen on the sea and flying to and fro. Look for Peregrine along the cliffs. In winter, small numbers of Red-throated Divers are present offshore and Common Scoter may be seen throughout the year.

Access

Leave the A487 north at Brynhoffnan on the B4334 to Llangrannog. From the seafront follow the coast path north for c. 1 mile to the Peninsula (parking can be tricky in summer).

FURTHER INFORMATION

Grid ref: SN 310 541

320 LLANGORSE (Breconshire)

OS Landranger 161

This large inland lake attracts a variety of wildfowl and holds an interesting gull roost. It is best from autumn through to the spring.

Habitat

At 150 ha, the lake is the largest area of natural fresh water in South Wales with a range of wetland habitats from wet meadows and marshes to sedge and reed swamp.

Species

Wintering wildfowl include Pochard, Wigeon and Teal, together with smaller numbers of Gadwalls, Shovelers, Goldeneyes and Goosanders. Bewick's and Whooper Swans, Shelduck, Pintail, Long-tailed Duck, Smew and Red-breasted Merganser are occasional visitors. Bitterns and Cetti's Warblers visit the reedbeds most winters, and there have been occasional sightings in spring and early summer. Up to 2,000 Lesser Black-backed Gulls come here to roost in the autumn, with much smaller numbers at other times. Herring, Common and Yellow-legged Gulls may be present in small numbers.

On passage small numbers of waders may pass through, and scarce visitors in recent years have included Marsh Harrier, Osprey, Hobby, Mediterranean Gull, Kittiwake, Roseate Tern and Bearded Tit.

Breeding species include Great Crested and occasionally Little Grebes, Water Rail, Common Sandpiper, Yellow Wagtail and Reed and Sedge Warblers.

Access

The lake lies 5 miles south-east of Brecon, just south of the village of Llangors on the B4560. From the village centre take the minor road westwards (a little north of the church) and then turn south (left) to the car park near the lake shore. The southern shore may be more conveniently accessed from the lakeside church of Llangasty-Talyllyn (SO 133 262) from where a footpath skirts much of the southern and western shore. Due to the boating activities, it is best to visit either early or late in the day.

FURTHER INFORMATION

Grid ref: SO 128 272

321 GWENFFRWD–DINAS AND AREA
(Carmarthenshire) OS Landranger 146 or 147

The RSPB reserves of Dinas and Gwenffrwd, c. 10 miles north of Llandovery in the upper Tywi catchment, total around 2,800 ha. Their range of habitats and birds is typical of this part of Wales. May–June is the best time to visit; summer visitors have arrived and are singing, and there is a good chance of Red Kite. Nearby Crychan Forest holds Goshawk, best looked for in fine weather in March–April.

Habitat

RSPB Dinas covers just 52 ha and comprises a steeply wooded knoll at the confluence of the Doethie and Tywi, rising to more than 1,000 feet, with hanging oak woods and small areas of alder carr and marshland. Gwenffrwd is far larger and reaches 1,343 feet at Cefn Gwenffrwd, with areas of moorland, hanging oak woods and farmland, as well as woodland in the valleys.

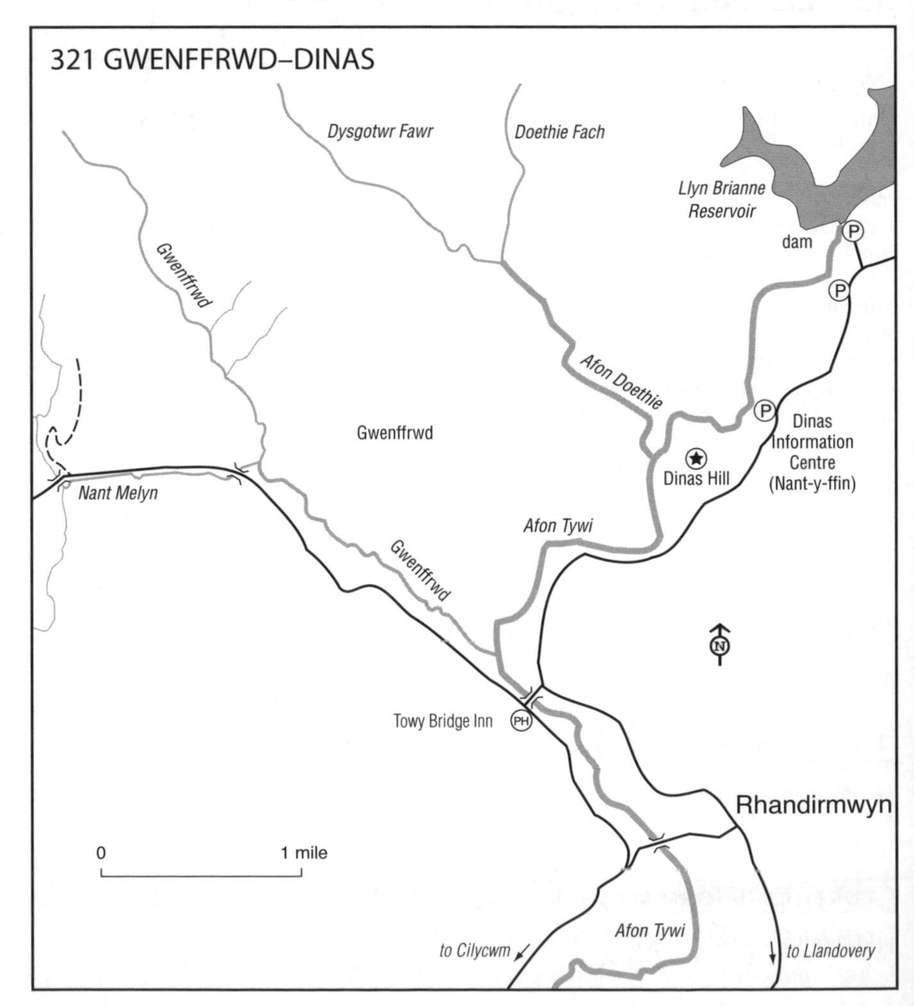

Species

Resident raptors include Red Kite, Common Buzzard and Sparrowhawk, and Merlin and Peregrine can sometimes be seen, as can Raven. A few Red Grouse persist on higher ground, but rampant over-grazing has greatly reduced their numbers and they are hard to find. In summer they are joined by Wheatear, with Tree Pipit and Whinchat in areas of scattered trees. Along streams and rivers Goosander, Common Sandpiper, Grey Wagtail, Sand Martin and Dipper breed, and Kingfisher may also visit. Summer visitors to woodlands include Common Redstart, Wood Warbler and Pied Flycatcher, which join resident Woodcock, Green, Great and Lesser Spotted Woodpeckers, Marsh and Willow Tits, Nuthatch, Treecreeper, Siskin and Lesser Redpoll.

Access

RSPB Dinas Leave Llandovery north on unclassified roads to Rhandirmwyn and continue towards Llyn Brianne dam. The Dinas Information Centre is at Nant-y-ffin, after a further 4 miles, and is open 10am–5pm Easter–late August. The 2-mile nature trail (which is rough and rocky) and reserve are open at all times.

RSPB Gwenffrwd For access to Gwenffrwd report to the Dinas Information Centre (the number of visitors is restricted by limited parking).

Nant Melyn reserve At Rhandirmwyn turn west and cross the Tywi via Rhandirmwyn bridge. Turn right at the T-junction for Cwrt-a-Cadno and continue for 2.7 miles past the Towy Bridge Inn to the bridge over the Nant Melyn. Here a footpath leads uphill to the north through an area of deciduous woodland. Nant Melyn is a reserve of the Wildlife Trust of South and West Wales.

The Irfon Valley A good area for Dipper. Leave the A483 north-west at Llanwrtyd Wells on a minor road towards Abergwesyn. This parallels the river, with several parking places from which to scan for Dipper, Grey Wagtail and Common Sandpiper.

Crychan Forest (Dyfed; OS Landranger 160) A very extensive area of conifer plantations that holds Goshawk, best looked for March–April. Five miles north-east of Llandovery, turn east off the A483 on a minor road opposite the Talgarth pub in Cynghordy. Almost immediately, fork left and follow the road for c. 2.5 miles to a picnic site at Esgair-fwyyog. From here take the main track to a large clearing, which offers a good view over the forest. Alternative access points are Sugar Loaf picnic site, east off the A483, 2.75 miles north of Cynghordy, and Banc Cefngarreg, reached by forking right (not left) after turning in Cynghordy and following the narrow road south-east for c. 1 mile.

FURTHER INFORMATION

Grid refs: RSPB Dinas SN 788 471; Nant Melyn reserve SN 729 466; The Irfon Valley SN 864 477; Crychan Forest SN 837 412, Esgair-fwyyog; Sugar Loaf SN 837 428; Banc Cefngarreg SN 819 390

RSPB Gwenffrwd–Dinas: tel: 01654 700222; email: gwenffrwd.dinas@rspb.co.uk

Wildlife Trust of South and West Wales: tel: 01656 724 100; web: www.welshwildlife.org

322 CORS CARON (TREGARON BOG) (Ceredigion)

OS Landranger 146

Tregaron, c. 15 miles south-east of Aberystwyth, is a good area for Red Kite, especially in winter. Cors Caron NNR, to the north of the village, covers 840 ha of Tregaron Bog.

Habitat

Cors Caron contains areas of raised bog covered with heather, grass, moss and patches of birch and willow carr. Peat cutting to the east of the Afon Teifi has created some small flashes, and a scrape has been created in front of the observation tower. The surrounding area is a mixture of deciduous woodland and pasture rising to more rugged hills.

Species

Red Kite, Sparrowhawk, Goshawk, Common Buzzard, Peregrine and Raven are resident in the area. Kites can be seen just about anywhere. Barn Owl, Water Rail and Dipper occur all year. A good spot for the latter is Tregaron Bridge.

Winter brings wildfowl: Wigeon, Teal and Mallard form the bulk of the wintering duck, with smaller numbers of Gadwalls; around 20 Whooper Swans commute between the bog and surrounding small lakes. The resident raptors are joined by Hen Harrier, Short-eared Owl and sometimes Merlin; up to eight Hen Harriers roost on the bog and there may be up to 60 Red Kites in the area. Other visitors include Stonechat and redpolls.

Summer visitors to the bog include Marsh Harrier, Cuckoo and Sedge Warbler, with Grasshopper Warbler and Lesser Redpoll in scrubbier areas. Teal, Curlew, Common Snipe and Common Redshank breed, and there is a Black-headed Gull colony and a heronry; Little Egrets and Hobbies are regular visitors. Small numbers of Tree Pipits, Common Redstarts,

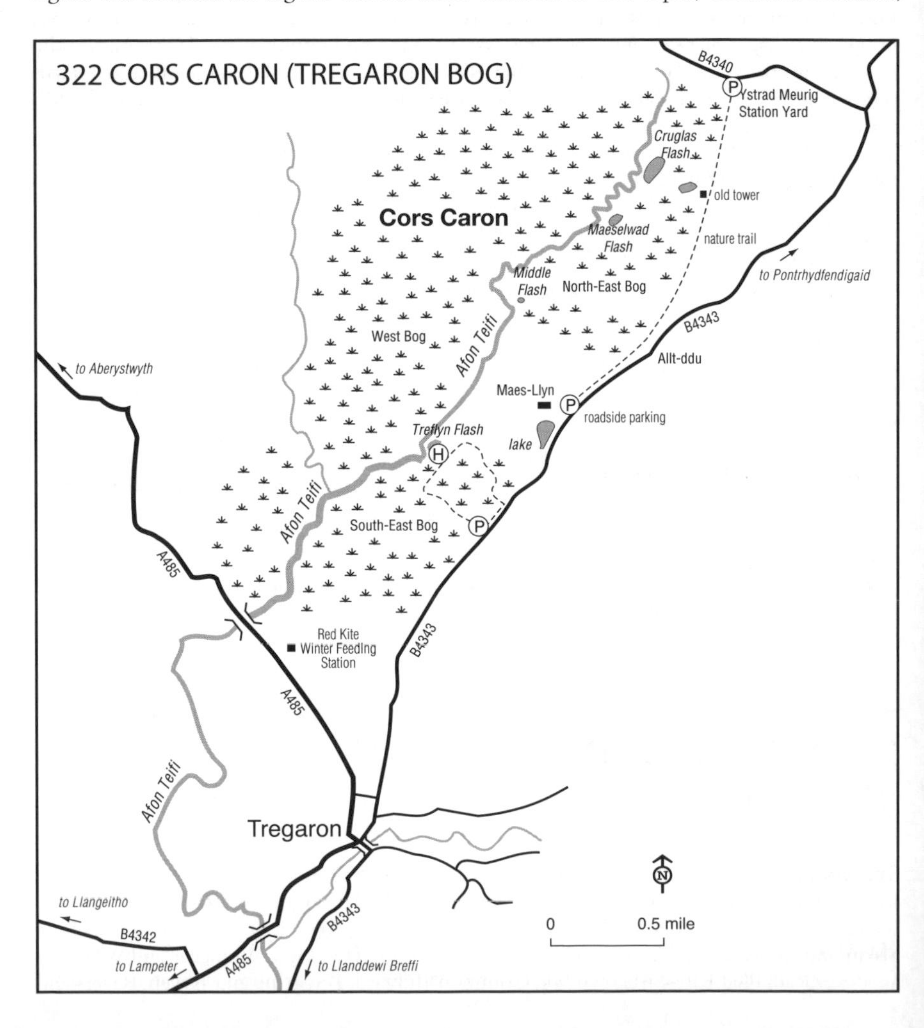

Whinchats, Wood Warblers and Pied Flycatchers breed in the remnant oak woods and scrubby hillsides, especially on the east side of the B4343, joining the resident Nuthatch, Treecreeper and Willow Tit. On passage small numbers of waders may be seen on the scrape.

Access

Cors Caron NNR Owned by the Countryside Council for Wales, access to the NNR is along a 2-mile boardwalk across the south-east bog, with a hide en route. The trail starts at the car park on the B4343. The northern and western parts of the bog can be seen from the trail which follows the disused railway line from the car park north of Maesllyn Farm on the B4343 (c. 2.5 miles north of Tregaron) for 3.75 miles to Ystrad Meurig Station Yard on the B4340. The trail forms part of the Ystwyth Trail, which goes all the way to Aberystwyth, and passes an abandoned observation tower. The more demanding Riverside Walk, accessed from near the start of the trail along the old railway, runs for 4 miles through the heart of the reserve beside the River Teifi.

Tregaron to Llanwrtyd Wells The minor road east from Tregaron to Llanwrtyd Wells crosses hill country with oak woods and plantations. Red Kite is possible as well as the usual woodland species in summer.

Tregaron to Devil's Bridge (OS Landranger 135 or 147) Continuing north from Tregaron to Devil's Bridge, the triangle formed by the B4343 between Pont-rhyd-y-groes and Devil's Bridge and the minor road that leaves the B4343 at Pont-rhyd-y-groes to the north-west and bends back after 2 miles north-east to Devil's Bridge, is a good area to look for Red Kite, especially in early spring.

FURTHER INFORMATION

Grid ref: Cors Caron NNR (Maesllyn Farm) SN 695 631
Countryside Council for Wales, Western Region Reserves Team: enquiries helpline: 0845 1306229; web: www.ccw.gov.uk
Wildlife Trust of South and West Wales: tel: 01656 724 100; web: www.welshwildlife.org

323 WYE–ELAN WOODS (Radnorshire)

OS Landranger 147

Lying in the Wye and Elan Valleys near Rhayader, these woodlands hold a typical range of breeding birds, including Common Redstart, Wood Warbler and Pied Flycatcher, and are best visited in May–June. Nearby, at Gigrin Farm, a Red Kite feeding station offers superb views of this, the most graceful of raptors, through the winter.

Habitat

The woodlands in the upper Wye Valley are some of the best in Wales for birds, and the adjacent hill farms and moorland areas have the usual range of upland birds. Gigrin Farm lies in the Wye Valley between 700 and 1,200 feet and is entirely grazed. The RSPB has three woodland reserves in the area (Dyffryn, Cwm and Cwm yr Esgob), and an upland reserve at Carngafallt, on the plateau south-west of the confluence of the Elan and the Wye.

Species

Breeding species in the woodlands include Woodcock, Lesser Spotted Woodpecker, Common Redstart, Wood and Garden Warblers, Pied Flycatcher and Willow Tit, and areas of conifers hold Siskin and Lesser Redpoll. At woodland fringes look for Tree Pipit, Whinchat and Wheatear and, overhead, Red Kite, Sparrowhawk, Common Buzzard, Peregrine and Raven. Rivers and streams hold Goosander, Common Sandpiper, Kingfisher, Grey Wagtail and Dipper.

In autumn rowans attract Redwings, Fieldfares and sometimes Ring Ouzels. In winter the uplands are largely devoid of birds and the woodlands are usually very quiet, but at this season the feeding station at Gigrin Farm provides a guaranteed Red Kite spectacle.

Access

Gigrin Farm Red Kite feeding station Leave Rhayader south on the A470 towards Builth Wells and after c. 0.3 miles turn left (east) to Gigrin Farm (signed). The farm opens at 1pm and there is an entry fee. The birds are fed daily at 2pm in winter (3pm BST from April to October) and several hides permit excellent views, but the kites are tame and will swoop low over the track even with people present. Numbers vary but from autumn onwards through the winter as many as 200 birds may be present at this site together with Common Buzzards, Ravens and other corvids. Trails are accessible for disabled visitors.

Dyffryn Wood Leave Rhayader south on the A470 and after 1 mile park in the lay-by as woodland appears on the left (east) side of the road. Access to the reserve is possible at all times. The A470 follows the River Wye and it is worth stopping where possible to scan the river for waterbirds.

RSPB Carngafallt This area of c. 260 ha is dominated by heather, with woodland and meadows on the lowers slopes. The main access point is just off the B4518 c. 3 miles south-west of Rhayader, at the eastern end of Elan village where the road enters woodland at a cattle grid (RSPB information sign). From here access is along public rights of way (some steep and rugged). In addition, from here the minor road to Llanwrthwl (off the A470 3 miles south of Rhayader) runs along the southern margin of the plateau.

FURTHER INFORMATION

Grid refs/postcode: Gigrin Farm Red Kite feeding station SN 978 676 / LD6 5BL; Dyffryn Wood SN 976 668; RSPB Carngafallt SN 936 652
Gigrin Farm: tel: 01597 810 243; web: www.gigrin.co.uk
RSPB Carngafallt: tel: 01654 700222; email: ynys-hir@rspb.org.uk

324 ELAN VALLEY (Radnorshire)

OS Landranger 147

This area, much favoured by tourists, provides a typical cross-section of the birds of Welsh uplands and waterways, including Peregrine and Red Kite. Late April–June is the best time to visit.

Habitat

The area comprises a mosaic of habitats – moorland, blanket bog, woodland, rivers and reservoirs – covering c. 70 square miles in the catchment of the Rivers Elan and Claerwen. Set aside to protect the watershed, the Elan Estate is owned by Dwr Cymru Welsh Water, and mostly vested in the Elan Valley Trust, with largely open access. Craig Goch, Penygarreg, Garreg-ddu and Caban-Coch Reservoirs lie in the valley of the River Elan and were constructed in the late 19th century to supply drinking water to Birmingham, while Claerwen Reservoir in the valley of the same name was opened in 1952 (the system now also supplies parts of south and mid-Wales). Penygarreg, Garreg-ddu and Caban-Coch Reservoirs are bordered by stands of Sessile Oak woodland, and there are also large blocks of conifers. Claerwen and Craig Goch Reservoirs lie at higher altitudes, and upland areas are largely sheep-walk, dominated by Purple Moor-grass, with rather restricted areas of heather.

Species

Breeding species include Great Crested Grebe, Teal, Goosander and Common Sandpiper around the reservoirs and Grey Wagtail and a few Dippers along streams and rivers. Red Grouse (in heather-dominated areas), Common Snipe, Golden Plover and a very few Stonechats are present in the uplands all year, while Dunlin, Tree Pipit, Whinchat and Wheatear are summer visitors; the latter is common and may be seen from the roads, but Golden Plover and Dunlin are found only on the highest moors. Common Buzzard, Peregrine and Red Kite are resident and usually easy to see, but Merlin is rare and Short-eared Owl usually present only in good vole years. (Woodcock, Curlew, Lapwing and Ring Ouzel have been lost as breeding species.) Sessile Oak woodland holds breeding Common Redstart, Wood Warbler and Pied Flycatcher, with small numbers of Lesser Spotted Woodpeckers. The conifers hold Goshawk and Siskin and, in invasion years, Common Crossbill.

In winter, wildfowl on the lakes may include Pochard, Goldeneye and Goosander, and the surrounding moorland occasionally produces a Hen Harrier or Merlin.

Access

Leave Rhayader west on the B4518. The reservoirs are well signed and much of the four waters in the Elan Valley can be seen from surrounding minor roads. A visitor centre with ample parking lies near Caban-Coch dam and is open daily 10am–5pm mid-March to late October. Offering a shop, cafe and exhibitions, there is also an information desk with full details of the 80 miles of leafleted walks and nature trails, and a Countryside Ranger service. The Elan Valley Trail runs for 6 miles from the visitor centre to Craig Goch dam beside the reservoirs, and Cnwch Wood Nature Trail climbs through excellent Sessile Oak woodland. There are also car parks at Penygarreg and Craig Goch dams, and at the dam of the massive Claerwen Reservoir. Upland species are best sought in the watershed between the Elan and Claerwen Valleys (open access). A less strenuous option is to follow the road north and west from Pont ar Elan at the head of Craig Goch reservoir to Cwmystwyth, scanning en route for raptors.

FURTHER INFORMATION

Grid ref/postcode: Visitor Centre SN 928 646 / LD6 5HP
Elan Valley Trust, Elan Valley Visitor Centre: tel: 01597 810898; web: www.elanvalley.org.uk

325 VALE OF RHEIDOL AND DEVIL'S BRIDGE (Ceredigion)

OS Landranger 135

This attractive valley holds a good variety of woodland and riverside birds and there is a kite feeding station that attracts large numbers of Red Kites.

Habitat

The lower reaches of the River Rheidol comprise a narrow flood plain with the river meandering past oxbow lakes and gravel islands; there are also several gravel pits. Upstream it is dammed to form the Gwm Rheidol Reservoir and beyond this the valley is cloaked with extensive stands of Sessile Oak woods, while at Devil's Bridge there is a spectacular waterfall. At Devil's Bridge the Coed Rheidol NNR is managed by the Countryside Council for Wales, while Coed Simdde Lwyd Reserve is owned by the Wildlife Trust of South and West Wales.

Species

Around the reservoir Goosander (has bred), Red-breasted Merganser, Common Sandpiper, Kingfisher, Sand Martin, Dipper and Grey Wagtail frequent the river, and Tree Pipit,

Common Redstart, Wood Warbler and Pied Flycatcher the woods. At Devil's Bridge the usual species occur in summer and Red Kite, Sparrowhawk, Common Buzzard and Raven may be seen overhead.

Access

Vale of Rheidol Approximately 5 miles east of Aberystwyth on the A44 take the minor road south at Capel Bangor to Dolypandy. Continue, paralleling the Afon Rheidol, for 3.5 miles to the Powergen Information Centre at the dam, from where a nature trail circuits the reservoir.
Coed Simdde Lwyd In the lower reaches of the valley, this reserve is accessible along a very narrow road past the Powergen Information Centre, with extremely limited parking at SN 703 787, from where a footpath runs upslope.
Devil's Bridge Reached by narrow-gauge steam train from Aberystwyth (Easter to mid-September) and via the A4120. At the bridge there is a 400-foot cascade. From the car park above the falls a path descends through mixed woodland.
Nant yr Arian Some of the best views of the valley can be obtained from Nant yr Arian, at the head of the Melindwr, a tributary of the Rheidol, and close to where the A44 reaches the first summit on its eastwards route across the county. There is a car park, Forestry Commission visitor centre (open in summer 10am–5pm, in winter 11am–dusk) and numerous trails, popular with walkers and mountain bikers, as well as a small pool, the side of which is a kite feeding station. The kites are fed daily around 3pm in summer, 2pm in winter, and there may be over 100 Red Kites in attendance during the winter months. RSPB volunteers man an observation hut at the edge of the car park.

FURTHER INFORMATION

Grid refs/postcode: Vale of Rheidol SN 698 795; Coed Simdde Lwyd SN 703 787; Devil's Bridge SN 739 768; Nant yr Arian SN 718 813 / SY23 3AD
Wildlife Trust of South and West Wales: tel: 01656 724 100; web: www.welshwildlife.org
Forestry Commission, Bwlch Nant yr Arian Forest Visitor Centre: tel: 01970 890453; web: www.forestry.gov.uk/bwlchnantyrarian

326 DYFI FOREST (Meirionnydd)

OS Landranger 124 and 135

This area, bounded by the Afon Dyfi (River Dovey) to the south-east and the A470 and A487 to the north and west, is worth exploring for moorland and woodland birds in spring and early summer.

Habitat

The hills rise to 2,213 feet, with areas of moorland fragmented by conifer plantations.

Species

Red Grouse, Wheatear and the occasional Golden Plover and Ring Ouzel breed on the moors, while the plantations have Lesser Redpoll, Siskin and, sometimes, Common Crossbill. Deciduous woods hold Common Redstart, Wood Warbler and Pied Flycatcher, with Common Sandpiper and Dipper on the river and streams. Common Buzzard may be seen anywhere.

Access

Access is via the roads and footpaths, especially from the A487 and a minor road that parallels the River Dulas on the west and east sides of its well-wooded valley.

FURTHER INFORMATION

Grid ref: River Dulas SH 770 099

327 LAKE VYRNWY (Montgomeryshire)

OS Landranger 125

The largest remaining area of heather moorland in Wales is in the Berwyn Mountains, and the RSPB has a reserve at Lake Vyrnwy, c. 15 miles east of Dolgellau. Mostly managed by agreement with Severn Trent Water, the Forestry Commission and the Lake Vyrnwy Hotel, the total area of the reserve is c. 10,500 ha. The best time to visit is May–June for summer visitors to the woodlands, but the reserve as a whole is worth visiting at any season.

Habitat

Lake Vyrnwy is the largest man-made lake in Wales, nearly 5 miles long and covering 450 ha. It was formed when the Afon Vyrnwy was dammed in 1891. Fringed by scrub woodland and meadows, heather and grass moorland rise to nearly 2,000 feet, with a network of streams and large areas of conifer plantations, and c. 80 ha of deciduous woodland.

Species

The moorland has a fine range of breeding species including Common Buzzard, Hen Harrier, Merlin, Red Grouse, Curlew, Short-eared Owl, Wheatear, Whinchat, Stonechat and Raven. Peregrine and Red Kite also visit the area (and Peregrine sometimes nests on the dam, the nest viewable via a closed-circuit TV relay to the RSPB visitor centre). Ring Ouzel is scarce but can sometimes be found near Eiddew Waterfalls. Areas of deciduous woodland hold Woodcock (which may be seen roding over Grwn-oer trail), Tawny Owl, Green and Great Spotted Woodpeckers, Willow Tit, Nuthatch and Treecreeper, and these are joined in summer by Common Redstart, Garden and Wood Warblers and Spotted and Pied Flycatchers. In the conifers there are Goshawk, Sparrowhawk, Grasshopper Warbler (young plantations), Common Crossbill, Lesser Redpoll and Siskin. The woodland fringes attract Tree Pipit and, sometimes, Black Grouse. Along the streams and around the reservoir Great Crested and Little Grebes, Grey Heron, Teal, Red-breasted Merganser, Goosander, Common Sandpiper, Kingfisher, Grey Wagtail and Dipper occur.

Waders occasionally appear on passage, especially at the north end of the reservoir, and Osprey may pass through. In winter, wildfowl include Wigeon, Teal, Pochard, Goldeneye and Goosander, and there is a roost of Black-headed Gulls, while Great Northern Diver may also turn up.

Access

Lake Vyrnwy Take the B4393 from Llanfyllin (north-west of Welshpool) to Llanwddyn, from where the road circumnavigates the lake. The reserve is open at all times, with access along public roads and footpaths. The RSPB information centre and shop lies on the minor road 100 yards south of the west end of the dam. It is open daily 10.30am–5.30pm from 1 April to 31 October, 10.30am–4.30pm from 1 November to 24 December and 10.30am–dusk (1 January to 31 March). In spring and summer there may be a video link to interesting nests (e.g. Peregrine). The Coed y Capel Hide (wheelchair-accessible) is close to the visitor centre and the Centenary and Lakeside Hides lie on the north-east shore of the reservoir (accessible either by car or on the trails). There are five nature trails: the Rhiwagor Trail (2 miles), Llechwedd-du Walk (5.5 miles if walked as a circular route); Ty Uchaf Walk (5 miles if walked as a circular route); Craig Garth Bwlch Trail (3 miles) and Grwn Oer Trail (1–2 miles).

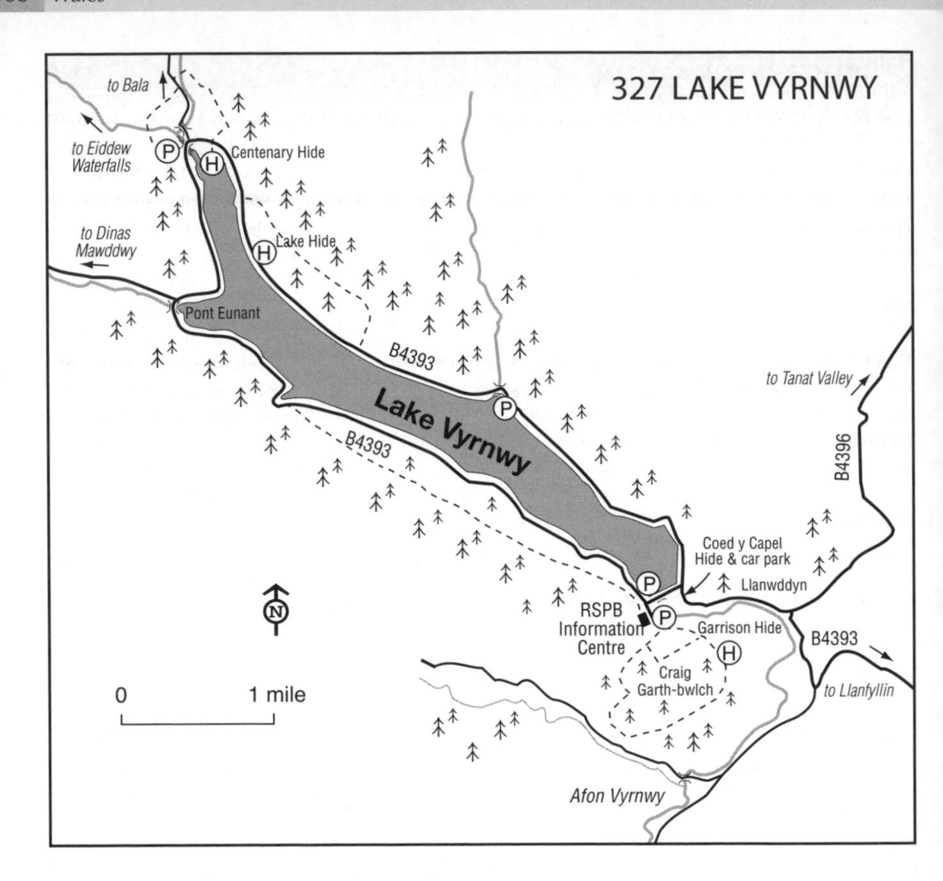

Minor roads north and west of Lake Vyrnwy The two minor roads that lead north and west from the head of the reservoir are very good for moorland species. The south road leaves the B4393 at Pont Eunant and climbs to c. 1,600 feet before dropping into the upper watershed of the River Dyfi and Dinas Mawddwy. The other heads north over the watershed and descends past the conifer-clad hills of Penllyn Forest through Cwm Hirnank to Bala.

FURTHER INFORMATION

Grid ref: Lake Vyrnwy SJ 016 191
RSPB Lake Vyrnwy: tel: 01691 870278; email: vyrnwy@rspb.org.uk

328 DYFI ESTUARY (Ceredigion and Meirionnyd)

OS Landranger 135

The Dyfi (or Dovey) is the largest estuary emptying into Cardigan Bay. A small flock of wintering Greenland White-fronted Geese is of special interest, and at all times of year the area's range of habitats should guarantee an interesting day. The estuary, Ynys-las Dunes and Cors Fochno (Borth Bog) form an NNR covering 2,290 ha. Ynys-hir, an RSPB reserve covering c. 400 ha, lies on the south shore of the estuary with an additional large area in the adjacent Llyfnant Valley.

Habitat

Approximately 5 miles long by 1.5 miles at its widest, three-quarters of the estuary comprise sand flats and the remainder saltmarsh, concentrated on the south shore and near the seaward end. The mouth of the estuary is guarded to the south by the extensive dune system of Ynys-las, and to the south-east lies Cors Fochno (Borth Bog) which, at c. 570 ha, is the largest unmodified raised bog in Britain. Further inland, the south shore is fringed by extensive water meadows, and south of the A487 lie open hillsides and conifer plantations, as well as deciduous woodland. Oak woodland borders the north shore of the estuary.

Species

Winter brings a flock of c. 60 Greenland White-fronted Geese, which feed on the saltmarsh or on Cors Fochno and adjoining arable land, and roost on the estuary sands. Present late October–early April, the birds are best seen from the RSPB reserve on a high rising tide. In recent years, there have also been c. 300 Barnacle Geese wintering in the area. Whooper Swans may visit and Bewick's Swan sometimes appears in February–March. Other wildfowl include significant numbers of Shelduck, Wigeon, Mallard and Teal, and small numbers of Pintail, Goldeneye and Goosander, while Red-breasted Merganser is always present in the channel. The Merlin, Peregrine, Sparrowhawk, Red Kite and Hen Harrier and sometimes Marsh Harrier occur in winter, and the commoner waders are joined by small numbers of Sanderlings on the beach and Black-tailed Godwits on the estuary. Offshore in winter there are often up to 100 Red-throated Divers off the estuary mouth, and large numbers of Great Crested Grebes, occasionally with a Great Northern Diver or a few Scaup, Long-tailed Duck or Velvet Scoter. Parties of Common Eider and large numbers of Common Scoter are present almost year-round, as are Cormorant, Shag and Gannet.

Spring passage brings a slightly wider variety of waders, and concentrations of up to 1,000 Curlews on the estuary. Up to seven Mediterranean Gulls have been seen in early spring at

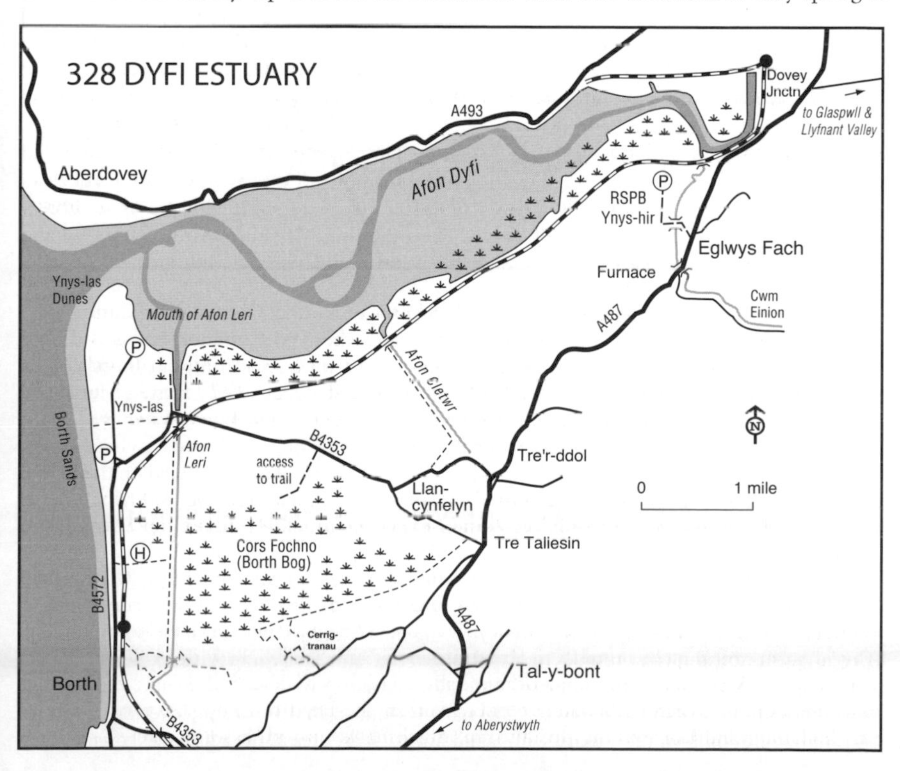

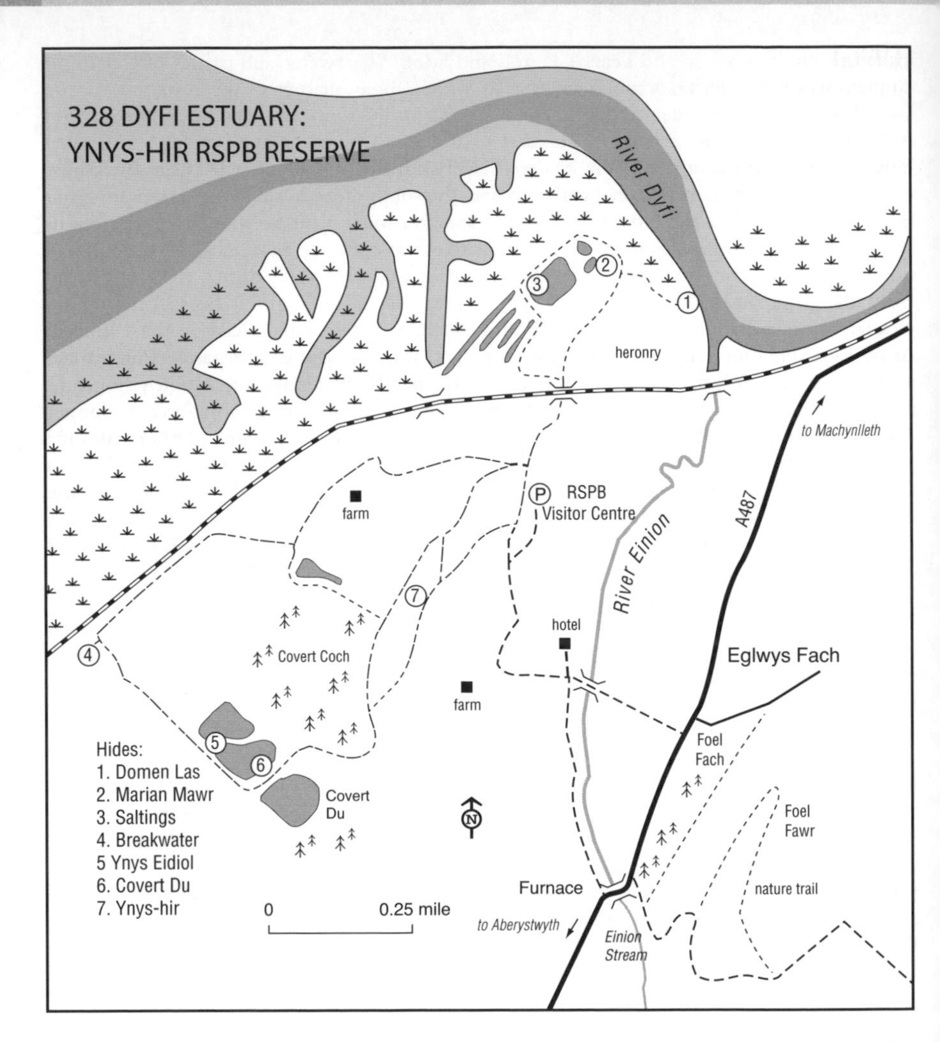

the Borth gull roost, where there have also been records of Iceland, Glaucous and Little Gulls.

Breeding species on the saltmarshes include Oystercatcher and Common Snipe, and Dyfi Marshes is now the most important breeding site for Lapwing and Common Redshank in Wales. A few pairs of Ringed Plovers nest at Ynys-las, and Shelduck, Red-breasted Merganser, Teal, Shoveler, Sparrowhawk, Common Buzzard, Grey Heron, Little Egret, Water Rail, Barn Owl, all three woodpeckers, Tree Pipit, Stonechat, Whinchat, Common Redstart, Grasshopper, Sedge and Wood Warblers, many Pied Flycatchers, Willow Tit and Raven all breed on the RSPB reserve at Ynys-hir. Goosander breeds in the area and is often seen in spring, and Kingfisher is frequently seen on the RSPB reserve at Domen Las at the mouth of the Einion Stream. Red Kite is a regular visitor.

Midsummer is rather quiet, but autumn brings passage waders, including Black-tailed Godwit, Knot, Green Sandpiper, Greenshank and Spotted Redshank; as in spring, there is a concentration of Curlews on the estuary. Ospreys are regular visitors, and large numbers of terns gather off the river mouth in late summer, mostly Sandwich, with smaller numbers of Common, Arctic and Little and occasionally Black or Roseate, while in late summer huge numbers of Manx Shearwaters can be seen in the bay from Ynys-las, particularly in early morning and late evening. In autumn, north gales may push some interesting birds inshore, including Storm and Leach's Petrels and Sooty Shearwater, and parties of Common Guillemots are regular in October and April.

Access

Ynys-las Dunes Leave the A487 on the B4353 and turn north at Ynys-las on the minor road to the dunes. There is a car park, information centre (open daily 9.30am–5pm from Easter until the end of September) and nature trail. From here you can view the river mouth and look out to sea – early morning and high tide are best for seawatching. Several shingle ridges attract concentrations of gulls and waders at high tide, especially during passage periods. Gulls also roost on the sea anywhere between Ynys-las and the cliffs at Borth, and can be viewed from the B4353 between Ynys-las and Borth.

Afon Leri The sand flats behind the dunes attract waders and a good place to watch from is the mouth of the Afon Leri. A short track north off the B4353 follows the west bank of the river, affording views of the flats, and a footpath on the east bank accesses the saltings. The tidal channels of the Afon Leri and Afon Cletwr are worth checking in late summer for passage waders.

Cors Fochno A circular route of approximately 1.2 miles provides public access to Cors Fochno. It runs south along a track from the B4353 0.5 miles west of Llancynfelyn; parking is very limited. The bog and Aberleri water meadows can also be viewed from the Afon Leri embankment where a public footpath follows the west bank from the B4353, via a railway bridge, south to Borth. After 1.3 miles it connects with another public footpath west to the B4353, with a public hide overlooking the Aberleri Marshes. Although of great importance in other respects, the bog is nowadays of limited interest for birds, although Red Kites often show well here and Hen Harrier may be found in winter.

RSPB Ynys-hir Off the A487 c. 6.5 miles west of Machynlleth, the entrance is signed in Eglwysfach. Report to reception on arrival. The reserve is open daily 9am–9pm (or sunset if earlier) and the visitor centre and shop 9am–5pm April–October and 10am–4pm November–March (closed Monday–Tuesday). There are several nature trails, with the two main circular routes of 1.5 miles and 3 miles, and seven hides giving views of the estuary, wader scrape and woods (the best place on the Dyfi to watch waders is the Saltings Hide at Ynys-hir on a rising tide). The trails are not suitable for wheelchair users.

Llyfnant Valley The minor road off the A487 through the Llyfnant Valley to Glaspwill makes an excellent walk for woodland species, with Dipper on the small river.

Einion Stream This also has Dipper and Grey Wagtail, with Red-breasted Merganser and Common Sandpiper on the higher reaches. A minor road from Furnace follows the stream (well above it in places) and there is a good tea shop at the end.

Aberystwyth The nodal point for RSPB Ynys-hir (and the Vale of Rheidol), there are often up to a dozen wintering Purple Sandpipers on the castle rocks while wintering Black Redstarts are almost annual around the harbour buildings or the scant remains of Aberystwyth Castle.

FURTHER INFORMATION

Grid refs: Ynys-las Dunes SN 609 940; Afon Leri SN 617 931; Cors Fochno (off B4353) SN 636 926; RSPB Ynys-hir SN 682 961; Llyfnant Valley SN 707 975; Einion Stream SN 700 943; Aberystwyth Castle SN 578 816

RSPB Ynys-hir: tel: 01654 700222; email: ynys-hir@rspb.org.uk

Countryside Council for Wales, Western Team, enquiries helpline: 0845 1306229

Ynys-las Information Centre: tel: 01970 872901

329 BROAD WATER (Meirionnydd)

OS Landranger 135

This small estuary holds a selection of wildlife and waders and has some interest all year.

Habitat

Broad Water is a saline lagoon formed in the estuary of the Afon Dysynni.

Species

The estuary has small numbers of wintering wildfowl and can be good for waders at low tide. Large numbers of gulls roost. The river mouth is good for gulls and terns plus seaducks.

Access

The estuary of the Afon Dysynni is accessed via the minor road north of the A493 in Tywyn, paralleling the railway, to reach the estuary near its mouth. A public footpath follows the south shore of the Dysynni (in places), east from the mouth to Bryncrug.

FURTHER INFORMATION

Grid ref: SH 566 028

330 CRAIG-YR-ADERYN (Meirionnydd)

OS Landranger 124

Inland along the Dysynni Valley a precipitous crag towers almost 700 feet above the farmland of the valley floor. This is Craig-yr-Aderyn (Bird Rock), which holds a unique colony of breeding Cormorants, as well as Choughs.

Habitat

The crag is a steep and mainly bare rocky hill, which was formerly on the coast but is now stranded 6 miles inland from the receding sea.

Species

Up to 40 pairs of Cormorants breed (of the British subspecies *carbo*), as well as Chough and Jackdaw, and Raven, Peregrine and Common Buzzard occur in the area. Outside the breeding season Cormorants roost on the rock, and may be seen in late afternoon and evening, and there may be up to 50 roosting Choughs (the rock is best viewed in the evening, as the light is against you in the morning).

Access

Leave the A493 at Bryncrug (2.5 miles north-east of Tywyn) on a minor road along the valley. The rock is on the right after 3 miles, with roadside parking at the junction, and is a reserve of the North Wales Naturalists' Trust.

FURTHER INFORMATION

Grid ref: SH 640 067

331 MAWDDACH ESTUARY (Meirionnydd)

OS Landranger 124

This extremely scenic estuary holds a variety of commoner wintering wildfowl and waders, while the surrounding woodlands hold all the usual Welsh species. A visit at any time is likely to be productive, though May–June is best for woodland birds.

Habitat

The Afon Mawddach rises within Snowdonia NP and reaches the sea at Barmouth, after passing through c. 7 miles of sand- and mudflats. The mouth of the estuary is guarded by a narrow spit, which extends north from Fairbourne on the south shore, and in its shelter a relatively small saltmarsh has developed. Arthog Bog, now an RSPB reserve, is a relict of once-extensive raised bog on the southern shore of the estuary. The upper reaches of the estuary are bounded by extensive stands of deciduous woodland and conifer plantations, including the RSPB reserve of Coed Garth Gell, which largely comprises birch and Sessile Oak.

Species

Small numbers of waders winter, including Oystercatcher, Golden and Ringed Plovers, Turnstone, Curlew, Bar-tailed Godwit, Common Redshank and Dunlin. Occasionally Greenshank may winter. Wildfowl include Wigeon, Teal, Goldeneye and Red-breasted Merganser, while interesting passerines include Dipper and, sometimes, Siskin and Lesser and Common Redpolls.

On passage a variety of waders occurs including Whimbrel and all of the wintering species, although being in the far west, birds such as Little Stint and Curlew Sandpiper are infrequent. In late summer there are congregations of Common and Sandwich Terns off the estuary mouth, and sometimes Arctic and Little Terns. Osprey has been recorded on passage.

Breeding species include Shelduck, Red-breasted Merganser, Oystercatcher, Common Snipe and Common Redshank, with Common Sandpiper, Grey Wagtail and Dipper by the river and its tributaries. Arthog Bog holds Water Rail, Sedge and Grasshopper Warblers and Lesser Redpoll. The surrounding hills and woodlands have resident Sparrowhawk, Common Buzzard, Peregrine, all three woodpeckers, Nuthatch, Treecreeper and Raven, joined in summer by Cuckoo, Tree Pipit, Whinchat, Common Redstart, Wood and Garden Warblers, and Pied and Spotted Flycatchers.

Access

Penmaenpool There is an RSPB information point at the old signal box at Penmaenpool on the A493, with views over the upper reaches of the estuary.

Penmaenpool–Morfa Mawddach Walk From Penmaenpool signal box it is possible to walk west for c. 5.5 miles to Morfa Mawddach Sation (from where you can cross to Barmouth via a 1-mile long footbridge). This walk gives excellent views over the entire lower estuary. Walking east along the old railway from Penmaenpool signal box gives views over the upper estuary, including some stands of reeds and willow scrub.

RSPB Arthog Bog Accessible at all times via the Penmaenpool–Morfa Mawddach Walk from Morfa Mawddach Station; the single trail joins with the Mawddach trail to form a flat and reasonably even circular route, 1 mile long. The bog has a selection of breeding passerines, including Grasshopper Warbler and Common Whitethroat.

Fairbourne Bar The minor road that follows the railway north from Fairbourne gives views of the mouth of the estuary, which is a good area for terns.

Barmouth/Cutiau Marsh On the north shore of the estuary, this area of saltmarsh can be viewed from lay-bys on the A496 c. 2 miles east of Barmouth.

RSPB Coed Garth Gell Comprising woodland and heathland, this reserve is accessible at all times along the footpath from the lay-by on the A496 near the Borthwnog Hotel. There are two circular routes, 1.25 miles and 1.5 miles long, both rugged and steep in places; Nightjar is possible on the higher, heathy sections. The visitor centre contains steep steps.

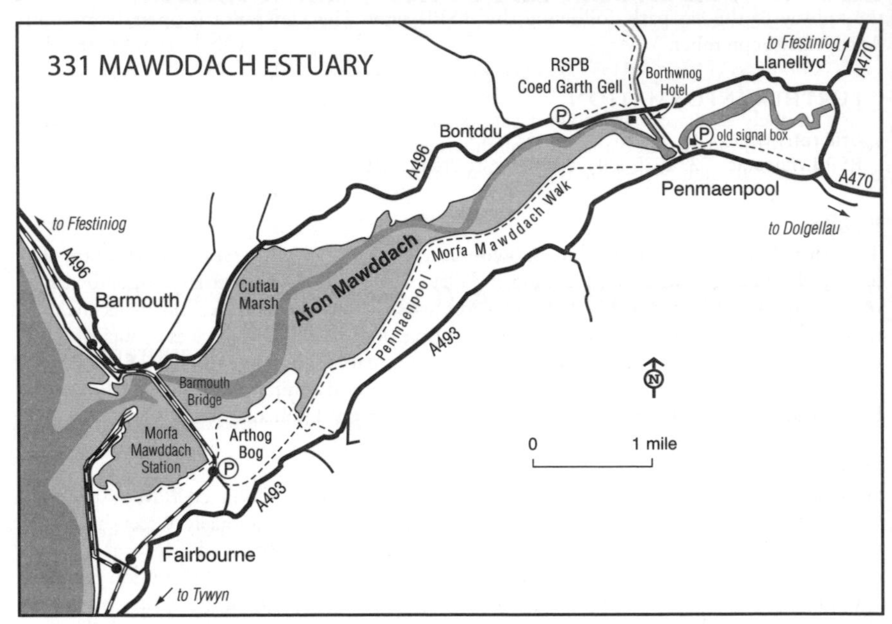

FURTHER INFORMATION

Grid refs: Penmaenpool SH 695 185; RSPB Arthog Bog (Morfa Mawddach Station) SH 629 140; Fairbourne Bar SH 614 146; Barmouth/Cutiau Marsh SH 634 171; RSPB Coed Garth Gell SH 683 190
RSPB Mawddach: tel: 01654 700222; email: mawddach@rspb.org.uk

332 GLASLYN OSPREY PROJECT (Meirionnydd)

OS Landranger

Ospreys first bred in Wales in 2004 and since then large numbers of visitors have admired them from the viewing site at Pont Croesor.

Habitat

The picturesque Glaslyn Valley lies in Snowdonia. The Ospreys nest in a tall pine and fish the three local rivers.

Species

Besides the Ospreys, this is a good area for raptors generally with Common Buzzard, Peregrine, Red Kite and Goshawk all regularly seen. Common woodland and riverine birds also frequent the area.

Access

Leave Porthmadog northwards on the A487, turn right onto the A498 at Tremadog and, after c. 2.5 miles, right again at Prenteg onto the B4410. The entrance to the project lies to the right after 500 yards at the sharp bend in the road. Turn into the site and continue over the railway to the car park, following the RSPB signs. The viewpoint is open 10am–6pm April–early September.

FURTHER INFORMATION

Grid ref: SH 593 413
RSPB Wales: tel: 029 2035 3008; email: cymru@rspb.org.uk

333 LLEYN PENINSULA (Caernarfon)

OS Landranger 123

This can be as good as Bardsey for migrants, and rarities have also been recorded, while resident Chough is an added bonus.

Habitat

The areas of interest comprise a sheltered, well-vegetated valley, and a high grass- and heath-topped headland with rugged cliffs.

Species

In summer, Fulmar, Gannet, Shag, Cormorant, Herring and Great Black-backed Gulls, Kittiwake, Razorbill and Common Guillemot may be seen around the cliffs or offshore, with Manx Shearwater possible, especially in the evenings. Peregrine, Little Owl, Rock Pipit, Stonechat, Wheatear, Raven and Chough also haunt the cliffs, with Garden and Sedge Warblers and Common and Lesser Whitethroats at Porth Meudwy.

In spring and autumn migrants may turn up in any area of cover, and the commoner species may be joined by Turtle Dove and Common and Black Redstarts, Whinchat and Ring Ouzel, while several rarities have been recorded. Offshore at Braich y Pwll, especially in autumn, Sooty Shearwater, Leach's Petrel and a variety of terns may occur, especially in strong west or north-west winds.

Access

Porth Meudwy This valley is attractive to migrants. Leave Aberdaron west on the minor road to Uwchmynydd and turn left after 0.5 miles. Park in the lay-by after a further 0.5 miles and follow the track east through the steep-sided valley to the beach, checking for migrants in the dense Bracken, gorse and honeysuckle en route. It is possible to take the steps to the clifftop path which runs south to Pen y Cil, and in summer small numbers of seabirds, as well as Peregrine and Chough, can be seen.

Braich y Pwll This is the headland opposite Bardsey. Leave Aberdaron west on the minor road to Uwchmynydd and follow the road to the coastguard buildings. There is plenty of parking along this road and the headland is crossed by several footpaths. It is good for seabirds and Chough in summer, with potential for seawatching and migrants.

FURTHER INFORMATION

Grid refs: Porth Meudwy SH 158 259; Braich y Pwll SH 142 255

334 BARDSEY ISLAND (Caernarfonshire)

OS Landranger 123

Bardsey lies 2 miles off the Lleyn Peninsula and the Bird Observatory was established in 1953. Several scarcer migrants are nearly annual, and Bardsey has a very long list of rarities to its credit. The best times to visit are late April–early June and August–early November, and ideally a visit should be timed to coincide with the first ten days after a new moon to have the best chance of witnessing a lighthouse 'attraction'. The lighthouse is an NNR.

Habitat

Bardsey covers 180 ha and is dominated by the 548-foot high Mynydd Enlli (the 'Mountain'), the upper areas of which are covered by Bracken and rocky outcrops. The steep east slopes are riddled with Manx Shearwater and Rabbit burrows. The Observatory garden, the four withy beds and a small pine plantation provide cover for migrants. The south of the island has a lighthouse, which in certain weather conditions following a new moon can attract hundreds or even thousands of nocturnal migrants. Nights with thick cloud cover or a 'smoke haze' obscuring the horizon are best, but such conditions formerly resulted in many fatalities. A 'false' light was built nearby in 1978 and has greatly reduced the death toll. Large numbers of birds are now trapped and ringed and an 'attraction' is certainly a unique experience.

Species

Spring migration commences in late March with the usual assortment of chats, warblers and flycatchers as well as diurnal migrants, but late spring is best for rarities, with Hoopoe and Woodchat Shrike among the more regular.

Ten species of seabird breed, notably c. 3,000 pairs of Manx Shearwaters, as well as small numbers of Fulmars, Shags, Razorbills, Common Guillemots, Herring, Lesser Black-backed and Great Black-backed Gulls, and Kittiwakes. Peregrine, Oystercatcher, Little Owl, Rock Pipit, Stonechat, Wheatear, Raven, and about seven pairs of Choughs add interest. Small numbers of non-breeding Storm Petrels may come ashore at night.

Return passage begins in late July with a trickle of warblers and these are joined by chats, Goldcrest and flycatchers as the autumn progresses – light east or south-east winds are best. October is the premier month for migration, with the possibility of large falls of night migrants such as Redwing and Blackbird and movements of Skylarks, Starlings and finches. Icterine, Melodious and Yellow-browed Warblers, Firecrest and Red-breasted Flycatcher are all nearly annual. Bardsey also has potential for seawatching. There may be large numbers of Manx Shearwaters, Kittiwakes, Common Guillemots and Razorbills, as well as Gannets, terns and Arctic Skuas, plus a few Sooty Shearwaters and Great Skuas, while Leach's Petrel, Pomarine Skua, Sabine's Gull and Little Auk are all scarce but regular; the largest and most varied movements occur in September–early October following strong west or north-west winds.

Access

The island boat leaves Porth Meudwy, 1 mile south-west of Aberdaron at 8.30am on Saturday or the first suitable day thereafter; delays are possible in either direction during bad weather. Day trips can be organised from Porth Meudwy. The Observatory is open late March–early November and bookings are taken on a weekly basis Saturday–Saturday. Accommodation is on a self-catering basis in the large stone-built farmhouse, with five bedrooms (one single, two double/twin and two family rooms – solo visitors may have to share the family rooms with other guests of the same sex). There are two seawatching hides and another in a small bay. In addition, the Bardsey Island Trust has several holiday cottages for hire on the island.

FURTHER INFORMATION

Bardsey Bird Observatory: web: www.bbfo.org.uk
Bardsey Island Trust: tel: 08458 112233; web: www.enlli.org

335 NEWBOROUGH WARREN (Anglesey)

OS Landranger 114

The Warren and adjacent coast harbour an excellent variety of wildfowl, waders and raptors. The treeless south-east section of the Warren and 1,556 ha of its coastal perimeter is an NNR, managed by the Countryside Council for Wales (CCW). The best time to visit is during winter and passage periods, but summer is also interesting.

Habitat

The Warren is an area of sand dunes covering over 1,200 ha, two-thirds of which (Newborough Forest) have been planted with Corsican Pine. South-west of the dunes is Ynys Llanddwyn, a rocky promontory connected to the mainland by a narrow strip of sand which is covered on the highest tides. North of the Warren is the Cefni Estuary, largely comprising the extensive Malltraeth Sands, but with some saltmarsh along its south shore. At the head of the estuary, and separated from it by an embankment known as the Cob, is Malltraeth Pool. This is an excellent spot for passage waders, as are the adjacent fields when flooded (although recently water levels have varied erratically, and when high there are few waders). In the dunes is a small pool, Llyn Rhos-ddu, while Llyn Coron, a larger lake, lies north of the Warren. Finally, on the south flank of the Warren lies the Braint Estuary.

Species

In winter, Cormorant, Shag, all three divers (particularly Great Northern and Red-throated), and Great Crested and Slavonian Grebes are joined offshore by numbers of Common Scoters, and occasionally Common Eider or Long-tailed Duck. There are up to 250 Pintails and many Wigeon (up to 1600 on Llyn Coron), plus Goldeneye and a few Shoveler and Gadwall. Flocks of Canada and feral Greylag Geese roam the area, and occasional Pink-footed and White-fronted Geese join them. Both Whooper and Bewick's Swans may winter. Peregrine, Merlin, Common Buzzard, Hen Harrier, and Barn and Short-eared Owls can be seen, and waders include Knot and Grey Plover, and sometimes Spotted Redshank and Greenshank, with Turnstone and Purple Sandpiper on Ynys Llanddwyn. Small flocks of Snow Buntings may occur along the shore.

Passage can be good, with a wide variety of waders. Little Stint, Sanderling, Curlew Sandpiper, Black-tailed Godwit and Spotted Redshank may all occur.

Breeding species include Great Crested and, sometimes, Little Grebes, Shelduck, Teal, Red-breasted Merganser, Oystercatcher and Ringed Plover, while Herring, Lesser Black-backed and Black-headed Gulls, Tree Pipit and Whinchat nest in the dunes, together with Stonechat and, occasionally, Short-eared Owl. Sparrowhawk, Coal Tit, Siskin, Lesser Redpoll and, sometimes, Common Crossbill breed in the plantations, together with the elusive Golden Pheasant. At Ynys Llanddwyn there are breeding Cormorants and a few Shags, Rock Pipits and Stonechats, and Turnstone summers. Often there are terns offshore, mostly Common, Arctic and Sandwich, but occasionally also Little or Roseate. Other breeding species in the area include Barn and Little Owls, Grasshopper, Reed and Sedge Warblers, and Common and Lesser Whitethroats. Newborough was famous in the past for its breeding Montagu's Harriers and Merlins, but sadly these have long since gone.

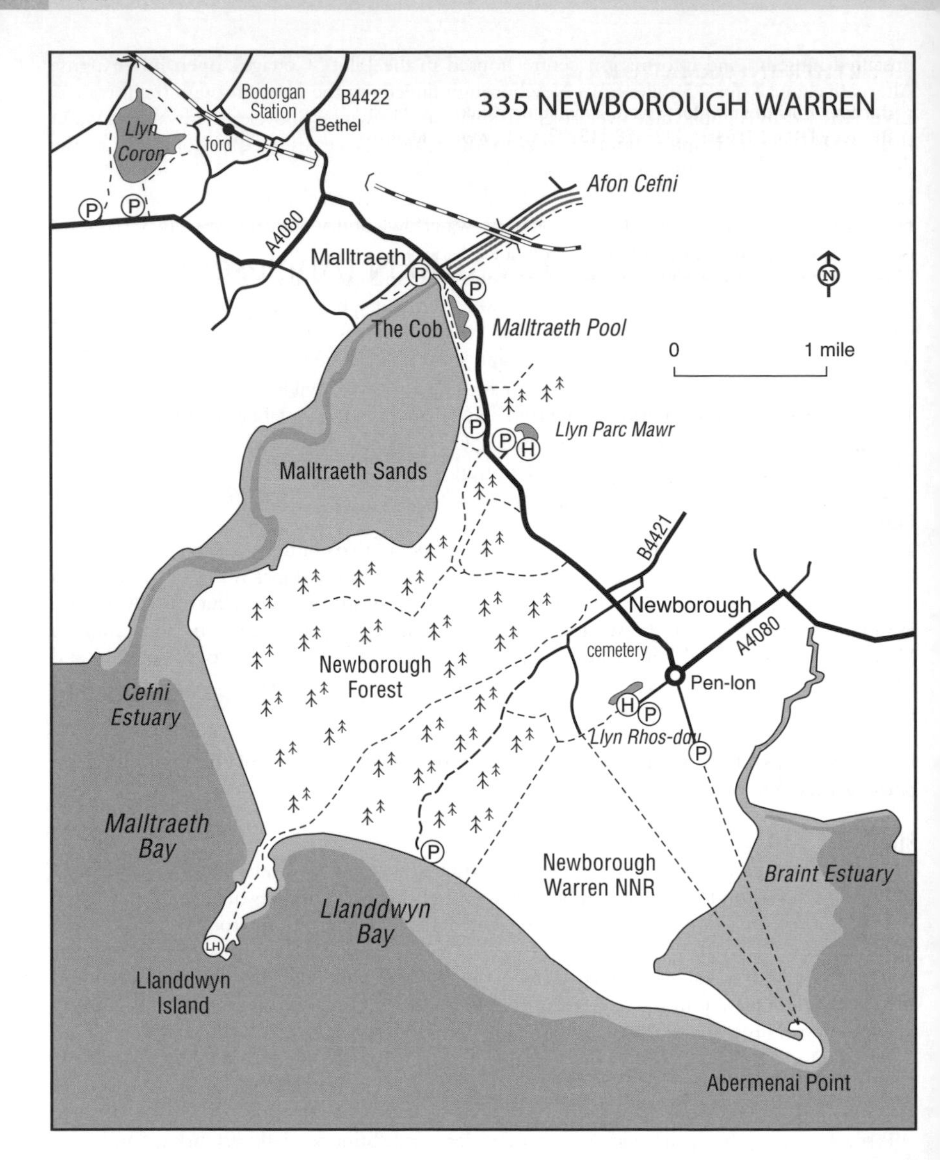

Access

Malltraeth Pool This can be viewed from the A4080 and the Cob (be careful not to disturb birds if using the Cob). The River Cefni upstream of the bridge can also be good for wildfowl.

Cefni Estuary Walk south from the Cob along the seaward side of the forest to view the saltmarsh. In winter a few Hen Harriers roost in this area, as well as Ravens; other raptors are also frequent.

Ynys Llanddwyn An excellent spot in winter. Divers are often present offshore, together with Slavonian and Great Crested Grebes and Common Scoter. Purple Sandpiper and Turnstone occur on the rocky shore. Drive south-west from Newborough on a minor road past the cemetery and then follow the track through the forest to the FC car park (information centre open in summer). Walking west along the shore, the island is reached after 1 mile; there is an unstaffed museum and information centre housed in the Pilots' Cottages, open July–August. Alternatively, walk for 2.5 miles from Newborough directly to the island through the forest.

Llyn Rhos-ddu A public hide overlooks this small lake. In winter Shoveler, Gadwall and Water Rail can be seen, and raptors hunt the surrounding dunes. Access is via the short track to the small car park from the A4080 at Pen-lon.

Llyn Parc Mawr This pool has a hide, accessible along forest tracks from the A4080 c. 1 mile north-west of Newborough. The pool attracts waterfowl, and woodland passerines come to drink, sometimes including Common Crossbill.

Llyn Coron Footpaths run along the west, south-east and north shores of the lake, accessible from car parks along the A4080 via a short walk across dry heathland. Llyn Coron supports wintering wildfowl, sometimes including wild swans, as well as breeding Great Crested Grebe, Stonechat, Grasshopper and Sedge Warblers, and summering Ruddy Duck.

FURTHER INFORMATION

Grid refs: Malltraeth Pool/The Cob SH 411 672; Ynys Llanddwyn SH 405 634; Llyn Rhos-ddu SH 426 647; Llyn Parc Mawr SH 413 670; Llyn Coron SH 379 693
Countryside Council for Wales, northern team: tel: 0845 1306229; web: ccw.gov.uk

336 VALLEY WETLANDS (Anglesey)

OS Landranger 114

Lying just north of the RAF Valley airbase, Llyn Penrhyn is just one of the many lakes on Anglesey, and supports a typical range of species. It is an RSPB reserve.

Habitat

The lake is surrounded by reedbeds, areas of sedges and willow carr. To the south lies an area of heathland with some rocky outcrops.

Species

Wintering wildfowl may include Bewick's Swan, Wigeon and Goldeneye, as well as resident breeding wildfowl. Bittern is sometimes recorded. The area attracts raptors, including Sparrowhawk, Hen and Marsh Harriers, Peregrine, Merlin and Short-eared Owl.

Breeding species include Little and Great Crested Grebes, feral Canada and Greylag Geese, Gadwall, Teal, Shoveler, Pochard, Oystercatcher, Water Rail, Lapwing, Common Redshank and Cetti's, Grasshopper, Reed and Sedge Warblers; Marsh Harrier is sometimes present in summer.

Access

Leave the A55 dual carriageway c. 2 miles west of Bryngwran signed Bodedern and Caergeiliog. Turn left (south) at the roundabout, following MOD signs for RAF Valley. Continue through the village of Llanfihangel yn Nhowyn and, following the road downhill, the lake appears on the right-hand side and immediately beyond it the reserve car park lies to the right (next to a white gate with an RSPB logo; if you go over the railway bridge you have gone too far). From the car park public footpaths (not suitable for wheelchairs) skirt the south and west flanks of the lake.

FURTHER INFORMATION

Grid ref: SH 313 765
RSPB Valley Wetlands: tel: 01407 764973; email: south.stack@rspb.org.uk

337 HOLYHEAD HARBOUR (Anglesey)

OS Landranger 114

The harbour is worth visiting in winter for divers, grebes and seaducks.

Habitat

A sheltered harbour surrounded by piers and a long breakwater.

Species

In winter divers are usually present in the harbour, especially Black-throated and less often Great Northern. Red-necked Grebe is also sometimes seen, late winter being the best period. One or two Black Guillemots are regular, and seaducks may include Common Scoter and Red-breasted Merganser.

In autumn Gannet, Manx Shearwater and a variety of terns may be seen offshore, but in good seawatching conditions, Penrhyn Mawr at RSPB South Stack or Point Lynas are better.

Access (see map opposite)

In Holyhead, park near the end of the A5 near the sailing club or near Soldiers Point. There is access on foot along the breakwater, though many of the birds can be seen from its base at Soldiers Point.

FURTHER INFORMATION

Grid ref: Soldiers Point SH 236 834

338 SOUTH STACK (Anglesey)

OS Landranger 114

Situated 3 miles west of Holyhead, on Holy Island at the west extremity of Anglesey, South Stack cliffs and their hinterland of maritime heath form an RSPB reserve. The best time to see the seabirds is May–June.

Habitat

The reserve includes both North and South Stacks, and these rise to nearly 400 feet. South Stack is connected to the mainland by a footbridge and a flight of steps that gives access to its lighthouse. Landward of the cliffs, areas of heather and gorse heath rise to 853 feet at Holyhead Mountain, and there is a separate area of heathland to the south, at Penrhos Feilw Common.

Species

Nine species of seabird breed, including Fulmar, Kittiwake, Herring, Great Black-backed and Lesser Black-backed Gulls, a few pairs of Shags, up to 3,000 Common Guillemots, and smaller numbers of Razorbills and Puffins (often only a few Puffins are on view, bobbing on the sea). Other breeding species include Peregrine, Kestrel, Stock Dove, Little Owl, Rock Pipit, Raven and four or five pairs of Choughs, which can be seen around the cliffs and RSPB car park, and in winter in fields along the approach road. The heath has breeding Stonechat and Common Whitethroat, with Lapwing and Common Redshank on Penrhos Feilw Common.

Gannet and Manx Shearwater can sometimes be seen offshore in summer and autumn, the shearwaters being commonest in July–August when up to 1,000 a day occur; they are especially numerous in early morning and evening. Common and Arctic Terns are also present offshore and, in spring, Great Northern and Red-throated Divers, Common Scoter and Arctic Skua are regular, and Pomarine Skuas probably annual. No special conditions are required for seawatching at this season, and interesting birds may be seen at any time of day. In autumn, by contrast, an early-morning watch in strong north-west winds is recommended: Sooty Shearwater, Storm and Leach's Petrels, and Pomarine and Long-tailed Skuas are possibilities, while Great and Arctic Skuas and large numbers of Kittiwakes and auks are regular.

On passage the heaths attract harriers, Short-eared Owl, Whimbrel and other waders (sometimes including Dotterel), Wheatear, Ring Ouzel and warblers, and there can be large movements of diurnal migrants in late autumn. In winter Red-throated Diver, Common Scoter and Red-breasted Merganser occur offshore, while Hen Harrier and Merlin hunt the heaths. Choughs may have formed large nomadic flocks, making them difficult to locate at times, though they are often in the grassy fields bordering the approach road to the reserve, and they may roost at Holyhead Breakwater Quarry Country Park (see map).

Access

South Stack Cliffs From the A5 in Holyhead follow signs from just before the ferries via Llaingoch and Twr to South Stack. There is a large car park and access is possible at all times along the public paths. The RSPB information centre at Ellin's Tower (a short walk from the main car park) overlooks the main auk colony, and provides a video link to the ledges; there are also good views from the steps to the lighthouse and from the lighthouse itself. Ellin's Tower is open daily Easter–September, 10am–5.30pm, and during this period a warden is present. Visitors should keep to the roads and paths.

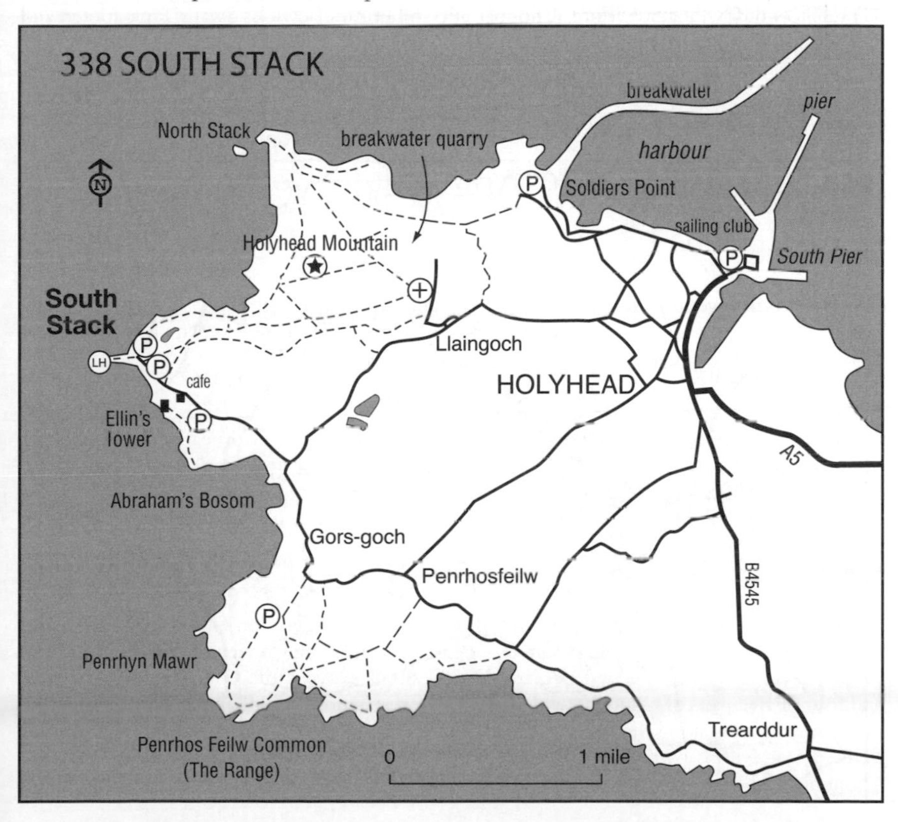

Penrhos Feilw Common (The Range) Follow minor roads south from South Stack or leave the A5 at Dyffryn on the B4545 to Trearddur and then take the minor coast road through Penrhosfeilw, parking at Gors-goch. There is open access to Penrhos Feilw Common along tracks from the road south of South Stack. Penrhyn Mawr is a good spot to seawatch from; boulders and shattered rocks provide some cover.

FURTHER INFORMATION

Grid refs/postcode: South Stack Cliffs SH 211 818 / LL65 1YH; Penrhos Feilw Common (The Range) SH 216 803
RSPB South Stack: tel: 01407 764973; email: south.stack@rspb.org.uk

339 LLYN ALAW (Anglesey)

OS Landranger 114

Habitat

The lake covers 315 ha and is surrounded by reedbeds and areas of fen vegetation, as well as some mixed woodland.

Species

Wintering wildfowl include Mute, Whooper and sometimes Bewick's Swans, Pink-footed and occasionally White-fronted Geese, Ruddy Duck, Wigeon, Gadwall, Teal, Shoveler and Pochard. Seaducks may occur, including Long-tailed Duck. Common Snipe, Curlew and Golden Plover may be found around the lake and in surrounding pastures, with Peregrine, Merlin, Hen Harrier, Sparrowhawk and Barn and Short-eared Owls all hunting the area.

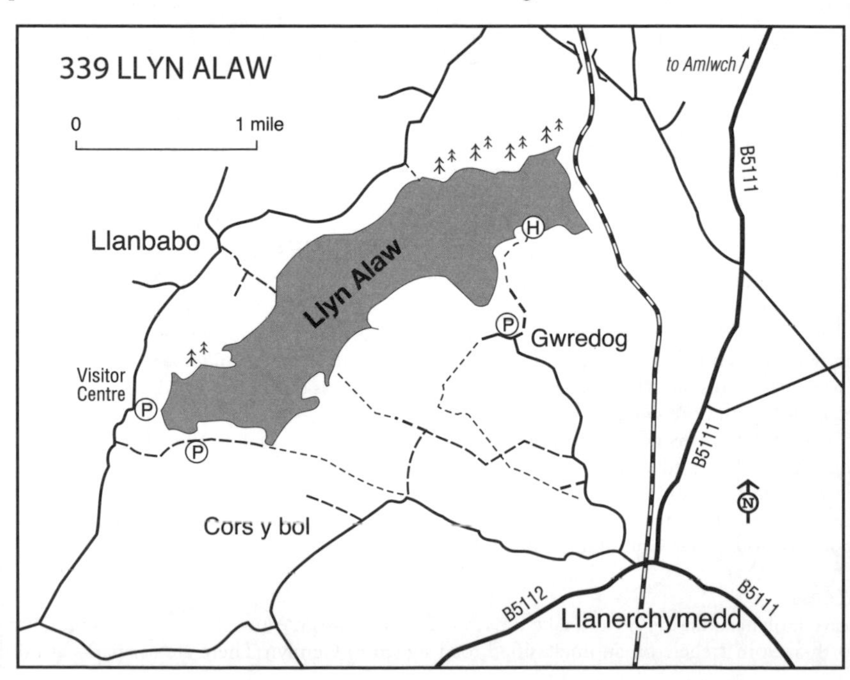

On passage small numbers of commoner waders occur, as well as some scarcer species such as Spotted Redshank, Wood Sandpiper and, in some autumns, Little Stint and Curlew Sandpiper. There may also be terns and sometimes Little Gull.

Breeding species include Great Crested Grebe, Grey Heron, feral Greylag and Canada Geese, Gadwall, Oystercatcher, Common Tern, Whinchat, Grasshopper and Sedge Warblers, and Common and Lesser Whitethroats.

Access

The lake is approached via minor roads from Llanerchymedd on the B5111 and is well sign-posted from all directions. There are three car parks, with a visitor centre at the south-west end of the lake and a hide overlooking the north-east sector. The area is managed as a reserve by Welsh Water and there is open access, except to the sanctuary area at the north-east end.

FURTHER INFORMATION

Grid ref: SH 373 855
Llyn Alaw Visitor Centre: tel: 01407 730762

340 CEMLYN LAGOON (Anglesey)

OS Landranger 114

On the north coast of Anglesey, this small lagoon holds important colonies of breeding terns, in some years including Roseate Tern. During passage and winter periods, a variety of wildfowl and waders occurs.

Habitat

A shingle storm beach has sealed off the entrance to Cemlyn Bay and formed a saline lagoon. The area is owned by the NT and managed as a reserve by the North Wales Wildlife Trust.

Species

Breeding species include variable numbers of Sandwich, Common and Arctic Terns. Sandwich Tern has peaked at over 1,000 pairs but conversely Roseate Tern has all but disappeared from Anglesey and can no longer be expected at this site (just off birds appear on passage; the formerly important Anglesey population has, apparently, moved across the Irish Sea to nest at Rockabill in Co. Dublin). Other breeding species include Shelduck, Red-breasted Merganser, Oystercatcher, Ringed Plover, Common Redshank and Black-headed Gull. Others in the area include Yellow Wagtail, Wheatear, Sedge Warbler and Common Whitethroat.

On passage Little Tern may join the locally breeding species and waders can be interesting, with Grey Plover, Sanderling, Common, Curlew and Purple Sandpipers, Whimbrel and Turnstone all possible. Manx Shearwater may occur at sea, as can Leach's Petrel and a variety of skuas during onshore winds.

In winter numbers of wildfowl use the lagoon, including up to 400 Wigeon, as well as Teal, Shoveler, Pochard and Goldeneye, and Bewick's and Whooper Swans are regular visitors. Hen Harrier, Peregrine and Barn Owl can be found hunting in the general area. Offshore, look for Red-throated and Great Northern Divers, Great Crested Grebe, Common Eider, Common Scoter, Razorbill and Common Guillemot.

Access

Leave Holyhead on the A5 and take the A5025 north towards Amlwch. After c. 11 miles, turn north-west in Tregele on an unclassified road signed to Cemlyn. There are car parks at both the east (Traeth Cemlyn) and west (near Bryn Aber) ends of the shingle ridge. The lagoon can be viewed from the shingle ridge, but it is best to avoid 'skylining' as this will inevitably alarm the birds (late April and August visitors are requested to walk on the seaward side of the ridge). During the breeding season access to some areas is restricted, and a summer warden is present.

FURTHER INFORMATION

Grid refs: Bryn Aber SH 329 936; Traeth Cemlyn SH 336 932
North Wales Wildlife Trust: tel: 01248 351541; web: www.northwaleswildlifetrust.org.uk

341 POINT LYNAS (Anglesey)

OS Landranger 114

Situated at the north-west tip of Anglesey, this is an excellent seawatching station.

Habitat
A narrow promontory projecting north for c. 0.5 miles, with a lighthouse at the tip.

Species
Manx Shearwater may be seen offshore throughout summer and autumn, and other regular species include Fulmar, Gannet and Kittiwake, with large numbers of terns in early autumn, sometimes including a few Black, and Razorbill and Common Guillemot in late autumn, with occasional Puffin, Black Guillemot or Little Auk. All four species of skua are possible in autumn, though Pomarine and Long-tailed are scarce, and Leach's Petrel and Sooty Shearwater may occur during north-west autumn gales. Numbers of visible passerine migrants also pass through.

In winter small numbers of divers and grebes may be present offshore, and seaducks can include Red-breasted Merganser, Common Eider, Scaup and Goldeneye.

Resident species around the head include Rock Pipit, Stonechat and Raven.

Access
From the A5025 south-east of Amlwch follow minor roads via Llaneilian to the Point, parking just before the gate on the lighthouse road. The period immediately after dawn is best for seabirds.

FURTHER INFORMATION

Grid ref: SH 479 934

342 ABER (Caernarfonshire)

OS Landranger 115

The Lavan Sands, north-east of Bangor, hold internationally important numbers of Oystercatcher and Curlew, and specialities are the wintering Black-necked and Slavonian Grebes and concentrations of moulting Great Crested Grebes and Red-breasted Mergansers. Traeth Lafan LNR covers over 2,430 ha of the sands.

Habitat

The mouth of the Afon Ogwen, together with Bangor Flats, forms the muddy west extremity of Lavan Sands, which can be up to 3 miles wide at low water and extend east to Llanfairfechan.

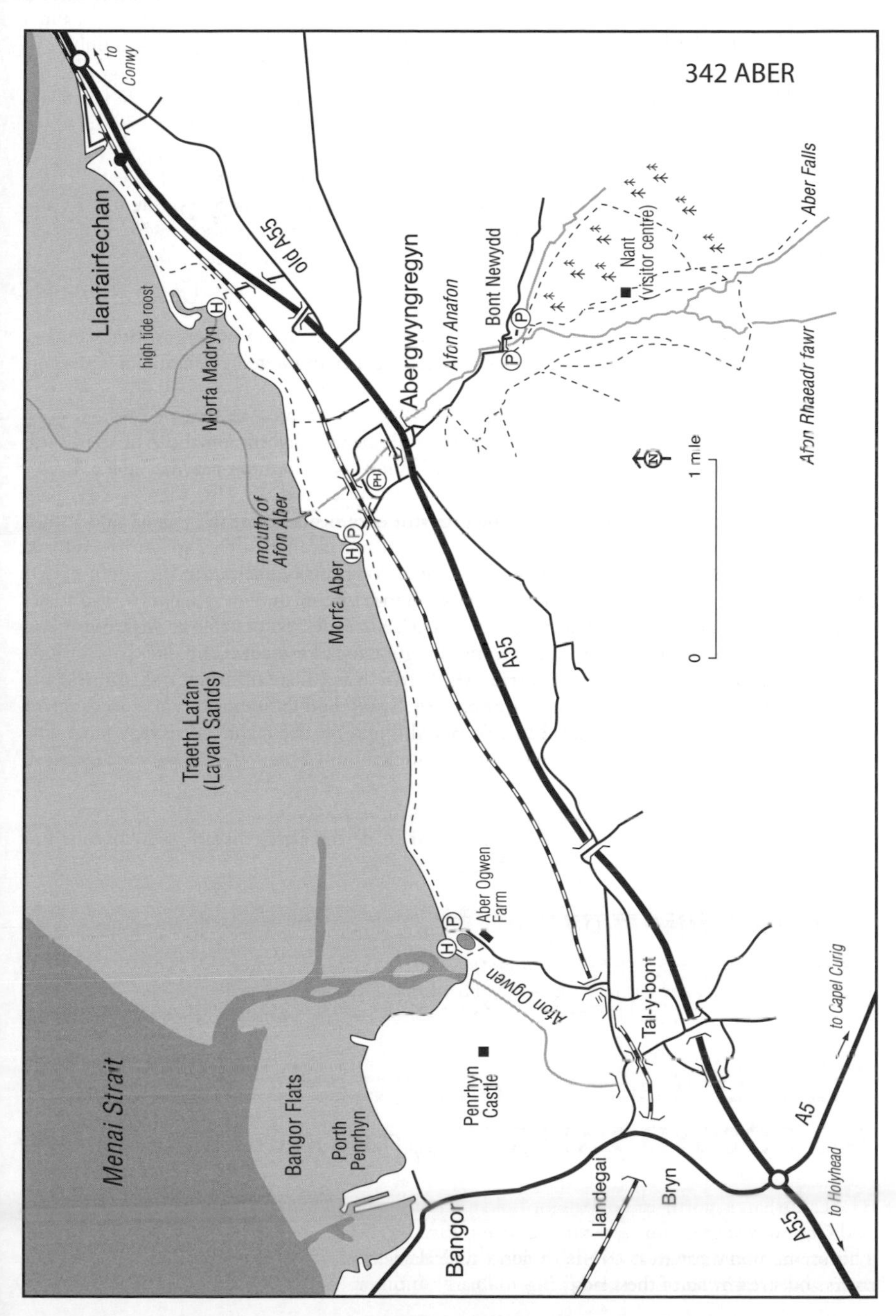

Species

In winter there are numbers of Great Crested Grebes and Red-throated Divers offshore, as well as a few Great Northern Divers. Black-throated Diver is uncommon, tending to occur only on passage. Small numbers of Slavonian and Black-necked Grebes are regular, together with Red-breasted Merganser, Goldeneye and Common Scoter, and a few Long-tailed Ducks and Velvet Scoters. Other wildfowl include large numbers of Wigeons, Teals and Shelducks, and a few Pintails and Shovelers. Commoner waders include internationally important numbers of Oystercatchers and Curlews, and these are joined by a few Greenshanks and Turnstones, and occasional Spotted Redshanks and Black-tailed Godwits. Water Rail, Kingfisher, Chiffchaff and Firecrest occur in small numbers around marshy areas and pools, and there is occasionally a Water Pipit at the sewage farm.

Passage brings a greater variety of waders and there is a concentration of Great Crested Grebes and Red-breasted Mergansers on the sea in early autumn; over 500 of the former and 400 of the latter have been recorded. Common and Sandwich Terns are also present on passage. Breeding species include Shelduck, while Peregrine is resident in the area.

Access

Divers, grebes and seaducks are best looked for around high water, and waders are best seen 1–2 hours either side of high tide. A footpath follows the shore from Aber Ogwen to Llanfairfechan, with access at the following points.

Porth Penrhyn Accessible from the A5122 in Bangor, with views over Bangor Flats.

Aber Ogwen Leave the main A55 North Wales Expressway north on the A5122 towards Bangor, turning right after 1 mile to Tal-y-bont, and then taking a minor road left after a further 0.5 miles to the coast beyond Aber Ogwen Farm. At the end of this lane there is a car park and just before this a track leads through the gate to a hide overlooking the estuary and a small pool. This is a good spot for wintering Water Rail, Black-tailed Godwit, Spotted Redshank, Greenshank and Kingfisher, and is generally productive for ducks and waders.

Morfa Aber Leave the A55 dual carriageway at Abergwyngregyn, where a minor road passes under the railway to the shore, and there is a car park and hide. It is possible to walk north-east from here to the mouth of the Afon Aber, another good area for waders and ducks.

Llanfairfechan Sewage Farm (Morfa Madryn) Leave Llanfairfechan west on the old A55. After 0.75 miles there is a farm shop on the left and directly opposite a lane leads under the A55 dual carriageway to the railway. Follow this and on the right is a sewage farm. The filter beds, which regularly attract wintering Chiffchaff and Firecrest, can be watched from the lane. Continue on foot over the railway to the shore, where there is a hide. Waders roost on the small promontory.

Llanfairfechan Offshore from Llanfairfechan promenade there are regularly small numbers of divers, Black-necked and Slavonian Grebes, and seaducks.

FURTHER INFORMATION

Grid refs: Porth Penrhyn SH 591 724; Aber Ogwen SH 615 723; Morfa Aber SH 647 731; Llanfairfechan Sewage Farm (Morfa Madryn) SH 668 741; Llanfairfechan SH 678 754

343 COEDYDD ABER (Caernarfonshire)

OS Landranger 115

This scenic valley, partly an NNR, holds a typical selection of birds in the woods and on the rivers and streams, and is best in spring and early summer.

Habitat

From Aber, the river valley climbs to Aber Falls and the Carneddau uplands, passing through extensive areas of mixed deciduous woodland in the lower part of the valley and dry acid oak woodland on the higher slopes, giving way to scrub at higher levels, within grazed grassland. There are also extensive conifer plantations.

Species

On the rivers and streams are Red-breasted Merganser, Dipper, Grey Wagtail and sometimes Kingfisher, while the woods have resident Green and Great Spotted Woodpeckers, Willow Tit, Nuthatch and Treecreeper, with Siskin and, in invasion years, Common Crossbills in the conifers. Summer visitors include Common Redstart, Garden and Wood Warblers and Pied Flycatcher, with Tree Pipit, Ring Ouzel, Whinchat and Wheatear on the moors; the scree slopes around Aber Falls are particularly good for Ring Ouzel. Look for Common Buzzard, Sparrowhawk, Peregrine and Raven overhead.

Access

Leave the A55 North Wales Expressway at Abergwyngregyn, then take the minor road south to Bont Newydd, parking after 1 mile in either the Countryside Council for Wales or adjacent FC car parks. From these a nature trail follows the valley for c. 1 mile to Aber Falls, with an information centre en route. It is possible to return via an alternative route, climbing fairly steeply at first through open country and then descending through conifer plantations.

FURTHER INFORMATION

Grid ref: SH 664 719
Countryside Council for Wales, northern team: tel: 0845 1306229; web: ccw.gov.uk

344 CONWY (Caernarfonshire)

OS Landranger 115 and 116

This RSPB reserve lies on the east shore of the Conwy Estuary and was purpose-built in 1993 from spoil. As well as holding a variety of waders and wildfowl, it has already gained a reputation for attracting rare waders.

Habitat

The reserve is sandwiched between the main A55 dual carriageway to the north and the tidal reaches of the River Conwy to the south and west, and was constructed with the huge quantities of spoil produced by the construction of the Conwy road tunnel. Two large reed-fringed lagoons have been created, with numerous small islands, and the reserve also has several ponds, areas of saltmarsh beside the river and grassland.

Species

Wintering wildfowl include Shelduck, Wigeon, Teal, Gadwall, Shoveler, Pochard, Goldeneye and Red-breasted Merganser, as well as Little and Great Crested Grebes. Numbers of commoner waders are present, including Common Snipe and Lapwing, and other high-tide estuarine species such as Oystercatcher, Curlew, Common Redshank and Dunlin roost on the lagoons, perhaps joined by a Spotted Redshank or Greenshank. Water Rails skulk in the reedbeds, perhaps together with one or two Jack Snipes, and Kingfisher and Grey Wagtail may visit, while the gull flocks may hold occasional Mediterranean Gulls. Sparrowhawk, Common Buzzard, Peregrine and Raven breed in the area and are seen all year.

On passage a wider variety of waders is present, with Knot, Sanderling, Black-tailed and Bar-tailed Godwits, Whimbrel, Greenshank and Common and Green Sandpipers all possible, while scarcer species have included Wood and Curlew Sandpipers and Little Stint. Terns, including Common and Sandwich, may bathe in the lagoons, and passerines can include White and Yellow Wagtails, Sand Martin and Wheatear. Osprey and Marsh Harrier have been recorded.

Breeding species include Little Grebe, Shelduck, Oystercatcher, Lapwing, Ringed and Little Ringed Plovers, Common Redshank, Common Sandpiper, Reed and Sedge Warblers, and Common and Lesser Whitethroats. Little Egrets and Grey Herons breed in Beanarth Woods over the estuary and often visit.

Access

RSPB Conwy Leave the A55 North Wales Expressway at junction 18 (signed Conwy and Deganwy – the first junction west of the Llandudno exit, and immediately east of the Conwy tunnel) and turn south off the roundabout to the reserve (signed). The reserve is open daily 9.30am–5pm (or dusk if earlier) and there is a visitor centre with a shop, toilets and window giving panoramic views of the reserve, two trails forming a 2-mile walk circumnavigating the reserve, three hides and three viewing screens. Trails are firm and accessible by powered wheelchairs.

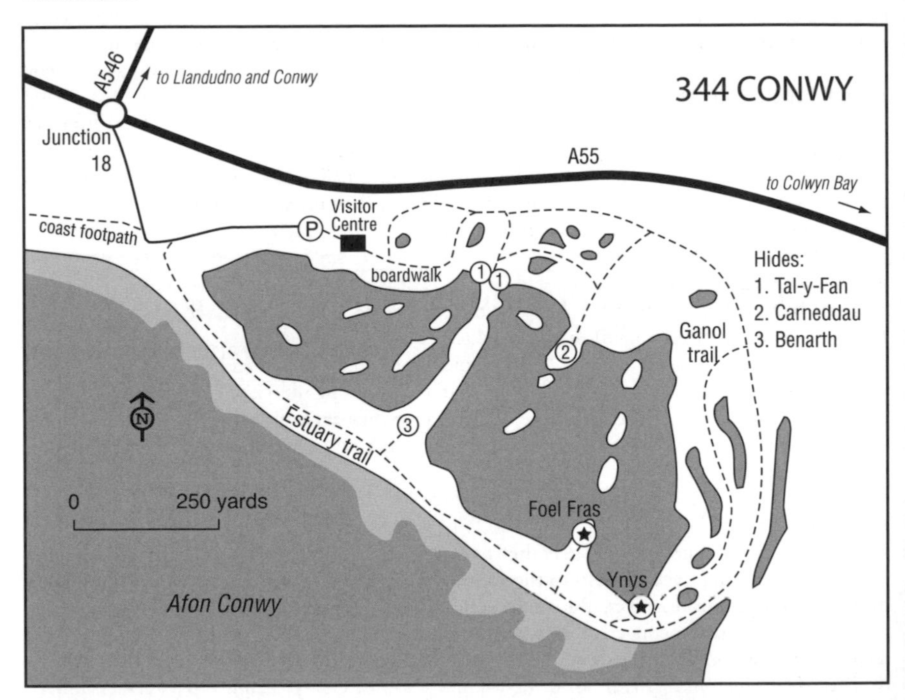

FURTHER INFORMATION

Grid ref: SH 797 773
RSPB Conwy: tel: 01492 584091; email: conwy@rspb.org.uk

345 GREAT ORME HEAD (Caernarfonshire)

OS Landranger 115

Lying immediately north of Llandudno, this massive headland forms the east boundary of Conwy Bay and is noted for its numbers of breeding seabirds.

Habitat

The Head covers c. 2 square miles and rises to 675 feet. Largely composed of limestone, it has areas of deeply fissured limestone pavement.

Species

Breeding species include c. 100 pairs of Fulmars, Cormorant, Shag, over 1,200 pairs of Kittiwakes, Herring Gull, small numbers of Great and Lesser Black-backed Gulls, 2000 pairs of Common Guillemots and 200 pairs of Razorbills. Puffin formerly nested, and occasional birds are still seen in summer, and also notable are the cliff-nesting House Martins. Other breeding species include Peregrine, Little Owl, Rock Pipit, Wheatear, Stonechat and Raven, and Chough, though not breeding here, also visits.

On passage a variety of species may occur, including Black Redstart in early spring and late autumn, and Dotterel, which has been recorded several times on the limestone pavement. Seawatching can be productive, with Leach's Petrel possible in September–October following north-west gales, together with Manx and, occasionally, Sooty Shearwaters, Gannet, Kittiwake, and Arctic, Great and Pomarine Skuas.

In winter few birds are found, but there are usually small numbers of Red-throated and Great Northern Divers and Common Scoter offshore, Turnstones and Purple Sandpipers on the Head, and Chough is most likely at this season.

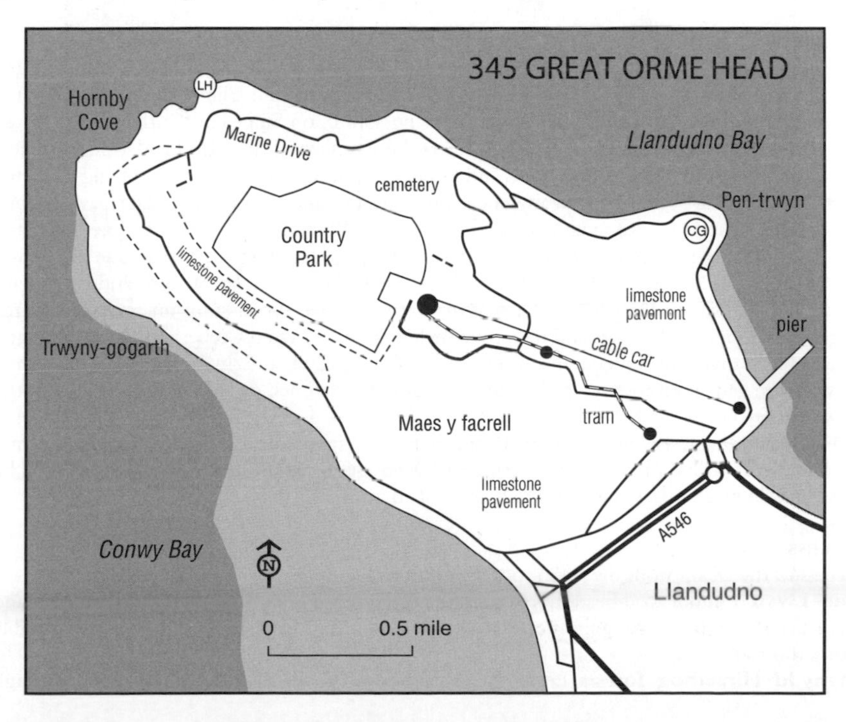

Access

Great Orme Head From Llandudno, Marine Drive encircles the Head (though parts may be closed for repair at times). A minor road also bisects the plateau and a tramway climbs to the top. From the tramway's summit, a nature trail leads west to the seabird cliffs. The whole area is a Country Park, and the best sites for passage migrants have been around the cemetery.
Rhos-on-Sea Small numbers of Purple Sandpipers, together with Turnstones, roost along the promenade here and at nearby Rhos Point.

FURTHER INFORMATION

Grid refs: Great Orme Head SH 756 843; Rhos-on-Sea (Rhos Point) SH 843 808

346 MYNYDD HIRAETHOG (Denbighshire)

OS Landranger 116

Situated in the extensive upland block between Ruthin and Betws-y-Coed, this mosaic of moorland, conifer forest and open water offers a good range of upland species. April–June is the optimum time to visit.

Habitat

Extensive areas of rolling moorland rise to 1,760 feet and are interspersed by huge blocks of conifer plantations, with Clocaenog Forest to the south-east being one of the largest areas of forestation in the principality. Mynydd Hiraethog has several areas of open water, both natural (Llyn Aled and Llyn Bran) and man-made (Aled Isaf, Alwen and Brenig Reservoirs).

Species

The moorland's specialities include breeding Golden Plovers and a handful of Dunlins (listen for the mournful cry of Golden Plover in the early morning and trilling purr of Dunlin; the latter often follows the former), as well as Lapwing, Common Snipe and Curlew. Red Grouse favour heather moors but Black Grouse prefers the interface of moorland and woodland; it is scarce and generally hard to find (look and listen for lekking males in early morning in spring and October). Raven may be seen anywhere, but Wheatear and Common Redstart favour stone walls, Ring Ouzel areas of crags and scree slopes, and Whinchat and Tree Pipit the moorland fringes. Breeding raptors may include Short-eared Owl, Hen Harrier, Merlin and Peregrine, but all are at low density; Common Buzzard is more likely to be seen. Streams and rivers have Common Sandpiper, Dipper and Grey Wagtail, the former also nesting on reservoir shores. Breeding species on the lakes and reservoirs include Great Crested Grebe, and Cormorant and Grey Heron may visit. The conifer plantations hold Goshawk and Sparrowhawk, Long-eared Owl, Siskin and, in invasion years, Common Crossbill. Hawfinch can sometimes be seen in the Alwen Valley below Pentre-llyn-cymmer.

In winter the reservoirs hold small numbers of Wigeon, Teal, Pochard, Goldeneye and Goosander, as well as itinerant family parties of Whooper Swans. Occasional raptors can still be seen, and Snow Bunting may visit the barer upland areas.

Access

Access to the area is from the A453 south-west from Denbigh.
Foel Lwyd A minor road heads north from the A453 near Pont y Clogwyn past Llyn Aled and Aled Isaf Reservoir to the high point at Foel Lwyd (1,325 feet). Upland species are possible along this route.
Mynydd Hiraethog forests Leave the A453 south on the B4501, which passes through

extensive stands of conifers. There are several parking places and exploration may be worthwhile anywhere, though sites with a panoramic view are best for scanning for raptors such as Goshawk. There is a nature trail at Pont-y-Brenig at the north-west tip of Llyn Brenig Reservoir, an archaeological trail at the north-east tip of Llyn Brenig, and visitor centres on the south shores of both Llyn Brenig and Alwen Reservoir.

Gors Maen Llwyd This 280-ha reserve of the North Wales Wildlife Trust is accessed from the B4501 immediately north of Llyn Brenig (7 miles south-west of Denbigh); park at the top car park (SH 970 580) or near the hide on the lake shore (SH 983 5754). Merlin, Hen Harrier, Red and Black Grouse, Wheatear and Whinchat occur in the area.

FURTHER INFORMATION

Grid refs: Foel Lwyd SH 913 613; Mynydd Hiraethog forests (B4501 at Pont-y-Brenig) SH 960 573; Gors Maen Llwyd SH 970 580
North Wales Wildlife Trust: tel: 01248 351541; web: www.northwaleswildlifetrust.org.uk

347 POINT OF AYR (Flintshire)

OS Landranger 116

This site, at the mouth of the Dee Estuary, is excellent for concentrations of wintering waders and wildfowl, has notable late-summer concentrations of terns, and can be good for seawatching in autumn.

Habitat

Extensive sandy flats and a small area of dunes guard the north-west mouth of the Dee Estuary, while the Point is a shingle spit surmounted by a lighthouse. In its shelter, large areas of saltmarsh have developed.

Species

In late summer and autumn there are gatherings of terns at the Point, with good numbers of Common and Sandwich Terns, often several hundred Little Terns, and careful searching may reveal Arctic or Roseate Terns too, while Black Terns also pass through in autumn. These congregations attract Arctic Skuas, which are almost constantly present. In strong north-west gales, seawatching can be productive (though, on balance, the Wirral shore may be better, see page 372). Leach's Petrel is possible after two or three days of gales, as are Manx Shearwater, Fulmar, Great, Pomarine and Long-tailed Skuas, and Sabine's Gull. Little Egret is possible year-round, and in both spring and autumn small numbers of passerine migrants, such as White and Yellow Wagtails, Wheatear and Ring Ouzel, may be present in the dunes.

In winter large numbers of wildfowl occur, including Shelduck, Wigeon, Teal, Pintail and Shoveler, with Common Scoters (sometimes in large numbers), Goldeneyes and Red-breasted Mergansers on the sea, together with Red-throated Diver, Common Guillemot and Razorbill. Scarcer wildfowl may include Pale-bellied Brent Goose. Large numbers of waders (up to 20,000) roost at the Point in the autumn and winter; mostly Oystercatcher, Knot, Dunlin and Common Redshank, others include Grey Plover, Black-tailed and Bar-tailed Godwits, Whimbrel, Greenshank, Sanderling and Turnstone. The congregations of potential prey species attract a variety of raptors, including Peregrine, Merlin, Hen Harrier and Short-eared Owl. A few Twites and Snow Buntings may occur, together with Water Pipit and, very occasionally, Shore Lark, and Ravens may visit.

Access

Leave Prestatyn east on the A548 and, after c. 2.5 miles (just beyond the end of a section of dual carriageway) turn north-west at the roundabout on Station Road to Talacre, proceed to the shore and park. An RSPB reserve with open access, the Point of Ayr is not wardened. The best place to view the high-tide wader roost is from the sea wall 0.6 miles southwards from the car park towards the colliery. Otherwise it is possible to walk west along the beach to the lighthouse, the best position for seawatching (early morning and a high tide are best) or east to the Point, but do not disturb the high-tide wader roost. Note that the whole area is badly disturbed in summer.

Up to 60 pairs of Little Terns breed at Gronant, west of the Point of Ayr. Take the road to the caravan site north of the A548 at Gronant (1.5 miles west of the Talacre turning). The breeding area on the beach is cordoned off to prevent disturbance and wardened during the breeding season.

FURTHER INFORMATION

Grid refs: Point of Ayr SJ 124 847; Gronant SJ 089 841
RSPB Point of Ayr: tel: 0151 336 7681; email: deeestuary@rspb.org.uk

348 FLINT AND OAKENHOLT MARSHES
(Flintshire) OS Landranger 117

This area of the Welsh shore of the inner Dee Estuary is freely accessible and holds much the same selection of waders and raptors as Point of Ayr to the north and Connah's Quay to the south.

Habitat

The area comprises saltmarsh and muddy estuarine foreshore.

Species

In winter large numbers of Oystercatchers, Knots, Dunlins and Common Redshanks are present, together with Grey and Golden Plovers, Bar-tailed and Black-tailed Godwits, and Turnstone. Wildfowl may include numbers of Teals and Wigeons, and Pintail is also possible. Raptors are attracted by the concentrations of birds, notably Peregrine, Merlin and, sometimes, Hen Harrier. A specialty of the area is Twite, and good numbers may be present along the tide-line or in the saltmarshes.

Access

Leave the A548 at Flint and park by the ruined castle. From here it is possible to walk both north and south along the shore.

FURTHER INFORMATION

Grid ref: SJ 247 732

349 CONNAH'S QUAY (Flintshire)

OS Landranger 117

This reserve, on the Welsh bank of the Dee, offers the chance to see large numbers of waders on passage and in winter.

Habitat

120 ha of saltmarsh, mudflats and grassland scrub, meadow and open water around Connah's Quay power station are a reserve, owned by E.ON and managed by the Deeside Naturalists' Society (DNS). A scrape and raised bank (for roosting waders) have been constructed, and large numbers occur in the 2 hours before high water, when the tide is 29.5 feet high or more.

Species

In winter, wildfowl, raptors and waders may be seen, including Pintail, Goldeneye, up to 1,000 Black-tailed Godwits and a few Spotted Redshanks, as well as the common estuarine species. Peregrine and Merlin may visit, and Twite is an irregular visitor to the saltmarsh.

On passage there is a greater selection of waders, including up to 100 Spotted Redshanks in autumn.

Access

Take the A548 north-west from Connah's Quay towards Flint; the power station entrance is on the right. There is a field studies centre, four hides overlooking the estuary and a nature trail. An advance permit is necessary and will be checked at the gate. Apply to the DNS or power station manager. Casual visitors may visit on a Public Open Day; these are held once a month on a Saturday or Sunday, generally when tides are suitable.

FURTHER INFORMATION

Deeside Naturalists' Society: web: www.deeestuary.co.uk

SCOTLAND

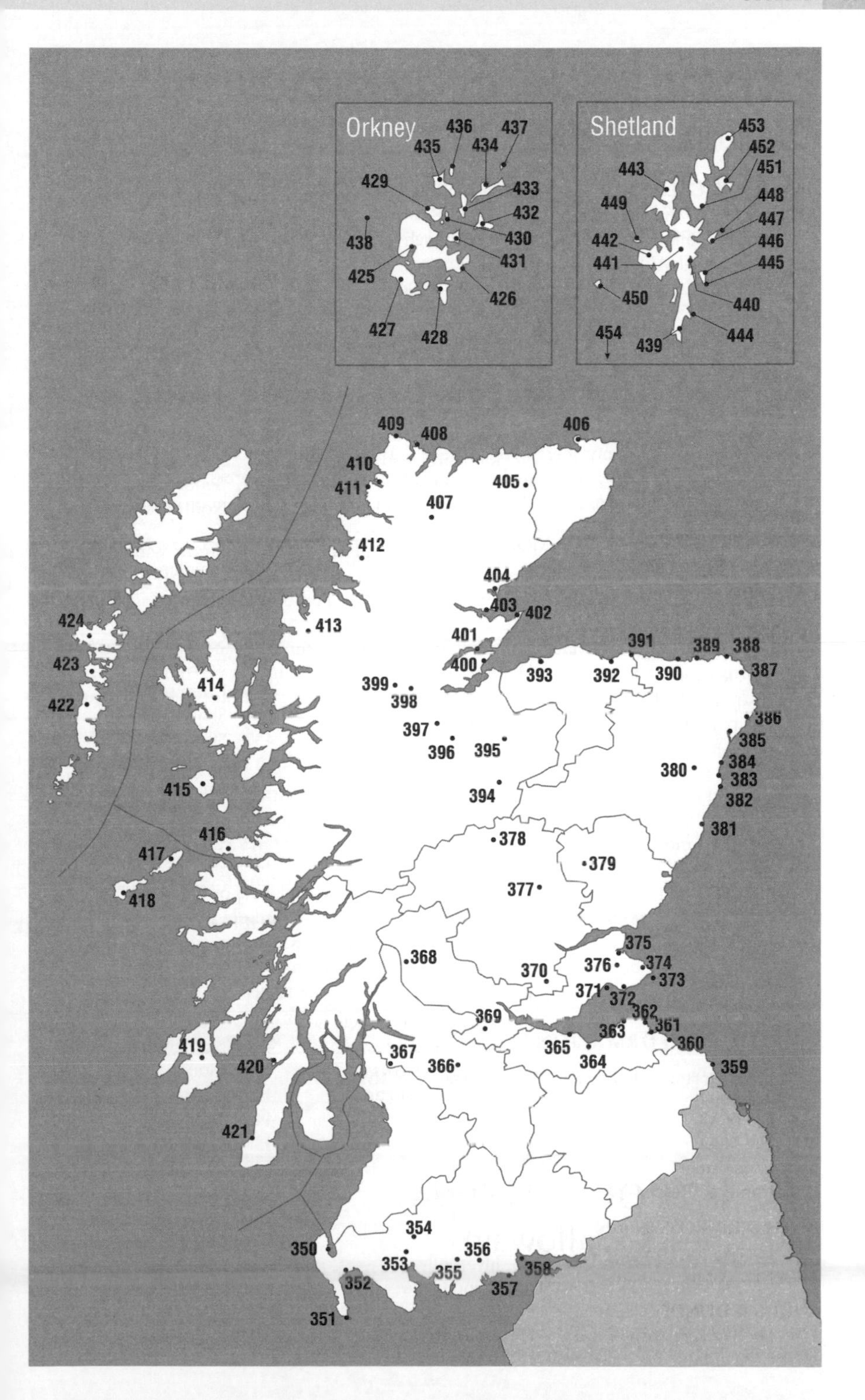
Orkney
Shetland

350 LOCH RYAN AND STRANRAER

(Dumfries & Galloway) OS Landranger 76 and 82

Loch Ryan is a large, sheltered sea loch with small numbers of divers, grebes (notably Black-necked Grebe) and seaducks. The best time to visit is winter. Nearby, Portpatrick often holds Black Guillemot and Corsewall Point is a good seawatching station, best in autumn.

Habitat

The Loch has a rather narrow shoreline, with only two significant areas of mud, around the Wig and on the south shore.

Species

Winter brings small numbers of Slavonian and Black-necked Grebes and all three divers to Loch Ryan, although Red-throated is the most numerous. Red-necked Grebe is occasional. Wildfowl include small numbers of Pale-bellied Brent Geese and seaducks include several hundred Common Eiders, with smaller numbers of Common Scoters, Scaups, Goldeneyes and Red-breasted Mergansers, and sometimes Long-tailed Ducks. There have been several King Eiders over the years (Cairnryan on the east shore has been favoured by this species). Wigeon numbers can peak at around 2,000. A handful of Black Guillemots winter, together with small numbers of waders. Iceland and Glaucous Gulls are occasional visitors, and parties of Twites may be found, especially on the Wig. Black Guillemot and white-winged gulls can also be sought at Portpatrick.

A constant stream of Gannets pass Corsewall Point in summer and autumn, as well as Cormorant, Shag and auks. Mid-August to late October brings good numbers of Manx Shearwater and Kittiwake. Especially in westerly winds, Leach's and Storm Petrels, skuas and occasionally even Sabine's Gull are also possible.

Access

Loch Ryan The A77 follows the east and south shores as far as Stranraer, while the A718 hugs the west shore north to the Wig. Most of the diving ducks and Wigeons occur on the south shore, but Common Eiders are scattered over several areas. Most birds can be seen by stopping at regular intervals and scanning the shore and water.

Portpatrick South-west of Stranraer on the A77, this is an excellent spot for Black Guillemots, which nest in the harbour walls. It is also worth checking the gulls here too.

Corsewall Point Leave Stranraer north on the A718 to Kirkcolm and follow signs for Corsewall/Barnhills, bearing right on a single-track road on reaching the lighthouse. It is possible to seawatch from a car, but views are better from the rocks.

FURTHER INFORMATION

Grid refs: Portpatrick NW 998 539; Corsewall Point NW 981 726

351 MULL OF GALLOWAY

(Dumfries & Galloway) OS Landranger 82

Lying at the southern tip of the Rhins of Galloway, this is the southernmost point in Scotland. The cliffs hold large numbers of breeding seabirds, as well as affording good views south over the Solway Firth and Irish Sea to the Isle of Man. The best time to visit is May–mid July.

Habitat

The granite cliffs rise to a maximum of 260 feet and are topped by rough grassland.

Species

Breeders on the cliffs include Fulmar, Cormorant, Shag, Kittiwake, Herring Gull, Common Guillemot, Razorbill and small numbers of Black Guillemots. A few Puffins summer, but no longer breed. Over 2,500 pairs of Gannets breed on the Scares, small rocky islets 7 miles to the east (also an RSPB reserve) and Gannet is often visible at sea; Manx Shearwaters are also frequently seen offshore, especially in September. Peregrine, Stonechat and Twite are resident on the head and Choughs sometimes visit. In spring and autumn, migrant passerines may also occur on the headland.

Access

Take the A716 south from Stranraer to Drummore, then the B7041 south through Damnaglaur and finally a minor road south to the lighthouse at the point. The reserve is open at all times. There are several shore walks along the cliffs, and good views of the colonies can be obtained from the viewing platform at the foghorn below the lighthouse (accessible down steep steps). The visitor centre is open daily, Easter to mid October, 10am–5pm (times may vary due to volunteer availability); summer warden.

FURTHER INFORMATION

Grid ref: NX 156 304
RSPB Mull of Galloway: tel: 01776 840539 (Easter–October); 01671 404975 (November–March)

352 WEST FREUGH AIRFIELD (Dumfries & Galloway)

OS Landranger 82

This disused airfield and surrounding areas attract interesting birds, especially geese, in winter.

Habitat

The short-grassed airfield is surrounded by areas of longer grass and patchy scrub.

Species and Access

Small numbers of Hen Harriers roost here from late October through the winter. Sparrowhawk, Peregrine, Merlin and Barn and Short-eared Owls can also be seen. Leave Stranraer south on the A77 and turn south-east onto the A716 and then take the B7077 towards Glenluce. Finally, after 4 miles, turn south-west onto the B7084. The harriers can be seen on the north side of the road, 1 mile south of the junction with the B7077. About 200 Greenland White-fronted Geese occur around West Freugh, but are difficult to pinpoint. Try the fields north of the B7077, east of Lochans. There are also up to 3,000 Greylags, as well as several feral Snow Geese, based on Loch Inch, Castle Kennedy.

FURTHER INFORMATION

Grid ref: NX 122 544

353 WOOD OF CREE (Dumfries & Galloway)

OS Landranger 77

This large deciduous woodland lies on the east slopes of the Cree Valley and forms part of the largest remaining stand of ancient woodland in south Scotland; it is an RSPB reserve and holds a variety of typical woodland birds.

Habitat

The wood largely comprises oak, birch and hazel, while alongside the river is an area of meadows.

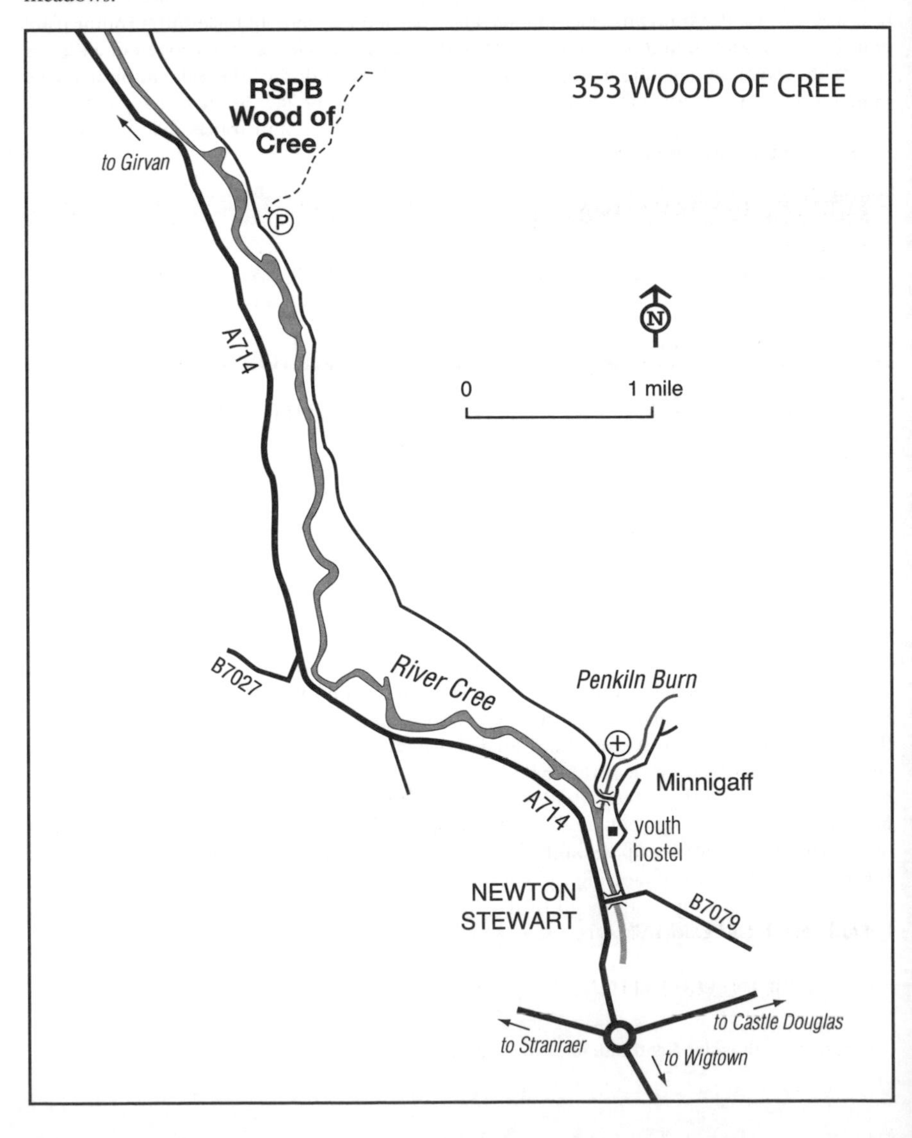

Species

Residents include Common Buzzard, Sparrowhawk, Woodcock and Great Spotted Woodpecker, with Willow Tit in the riverside scrub and Barn Owl in surrounding open areas. In May–June look also for Common Redstart, Garden and Wood Warblers, and Pied and Spotted Flycatchers, with Goosander, Common Sandpiper, Dipper and Grey Wagtail on streams and the river, Teal, Oystercatcher, Common Snipe and Water Rail in riverside meadows, Common Whitethroat and Grasshopper Warbler in the scrub and Curlew, Tree Pipit and Whinchat at the woodland/moorland interface. Ospreys are occasional visitors.

Access

The reserve lies 4 miles north-west of Newton Stewart. Just north of the town centre turn east off the A714 on the B7079. The road crosses the River Cree and, immediately beyond, turn left (north) on the minor road to Minnigaff. Follow this past the youth hostel and then bear left to cross Penkiln Burn, bearing left again beyond the church. Follow the minor road, which runs parallel to, and east of, the A714 and River Cree, for c. 4 miles to the reserve car park. The woodland trail covers 1.25 miles, to which can be added a 1.25-mile loop through a more scrubby area.

FURTHER INFORMATION

Grid ref: NX 381 708
RSPB Wood of Cree: tel: 01671 404975

354 GLENTROOL (Dumfries & Galloway)

OS Landranger 77

Part of Galloway Forest Park, this attractive valley is worthy of exploration.

Habitat

Conifer plantations with areas of relict oak and birch woodland around Loch Trool.

Species

Barn Owl, Common Buzzard, Peregrine, Raven and occasionally Hen Harrier or even Golden Eagle and Black Grouse can be seen in winter from the road, while in summer the deciduous woods around Loch Trool at Buchan, Glenhead and Caldons hold Tree Pipit, Common Redstart, Wood Warbler and Pied Flycatcher, with Common Crossbill and Siskin in the conifers and Wheatear and Dipper on the surrounding moorland.

Access

Leave Newton Stewart north-west on the A714 Girvan road and turn east after c. 9 miles on a minor road at Bargrennan, bearing right at Glen Trool village to the Glen. There are car parks at Bruce's Stone, Caldons and Stroan Bridge visitor centre, and the waymarked trails include a 4.5-mile trail circling Loch Trool.

FURTHER INFORMATION

Grid ref/postcode: Stroan Bridge NX 371 786 / DG8 6SZ
Forestry Commission Scotland: tel: 01671 402420

355 LOCH KEN (Dumfries & Galloway)

OS Landranger 77, 83 and 84

To the north-west of Castle Douglas, Loch Ken is a long, shallow loch that attracts Greenland White-fronted and Greylag Geese. To the west, the National Trust for Scotland (NTS) has a large wildfowl refuge near Threave Castle. For wintering wildfowl, October–March is the peak period, and for breeding birds May–June.

Habitat

The River Dee was dammed during a hydro-electric scheme in 1935 to form Loch Ken, a shallow lake surrounded by areas of meadows and freshwater marsh, the latter especially at the head of the loch (technically, only the north sector is known as Loch Ken, the stretch below the viaduct being the River Dee). If water levels fall in summer, expanses of mud are exposed, especially along the south-west shore. The surrounding area comprises farmland, open hillsides and woodland.

Species

Winter brings up to 120 Whooper Swans and 750–1,000 Greylags, both favouring the Threave Estate, where they are joined by Pink-footed Geese in the New Year. Around 300 Greenland White-fronts occur around Loch Ken. Ducks include Goldeneye, Goosander, Pintail (around Mains of Duchrae) and Shoveler (early winter only). Hen Harrier, Merlin, Peregrine and Barn Owl are regular, with Great Grey Shrike occasional.

In summer, breeding birds on the marshes, loch and river include Great Crested Grebe, Teal, Shoveler, Oystercatcher, Lapwing, Common Redshank, Curlew, Common Sandpiper, a handful of Common Terns, and Sedge and a few Grasshopper Warblers. Water Rails breed but are hard to see, even more so Spotted Crake, which breeds at the RSPB reserve; listen for its 'whipcrack' calls at night. Summer also brings Tree Pipit, Common Redstart, Wood Warbler and a few Pied Flycatchers.

On passage, almost any species of wader can occur, and Osprey is reasonably regular, as are terns.

Residents include Goosander, Peregrine, Common Buzzard, Red Kite (re-introduced from 2001 onwards), Sparrowhawk, Water Rail, Barn Owl, with Willow Tit in areas of damp scrub, Nuthatch, Great Spotted Woodpecker and a very few Green Woodpeckers in mature deciduous woodland, Grey Wagtail and Dipper along the streams and Siskin and Common Crossbill around conifer plantations. Black Grouse also favours the conifers, but has fallen markedly in numbers and is hard to find.

Access

Carlingwark Loch This 42-ha pool with surrounding willow and alder carr is visible from the B736 immediately south of Castle Douglas and can be circumnavigated on a 3.5-mile circular trail from the car park at Lochside Park. The loch holds diving ducks, notably Goldeneye and Goosander.

Castle Douglas to Bridge of Dee Greylag and a few Pink-footed Geese are regular, but although Bean Geese were formerly seen around Threave and south of Castle Douglas (anywhere between Gelston and Netherhall), they are now rare, erratic and unpredictable. Pink-feet can usually be seen in the pastures west of the B736, 1–2 miles south of Castle Douglas. The geese are highly mobile and you must be prepared to search the area.

Threave Wildfowl Refuge The 492-ha NTS Threave Estate lies south-west of Castle Douglas and is open daily. Access is at Kelton Mains Farm, off the A75 1 mile west of Castle Douglas (the access point for the formidable pile of Threave Castle). There is a map at the car park showing the network of trails and the location of the hides: three give views of farmland and

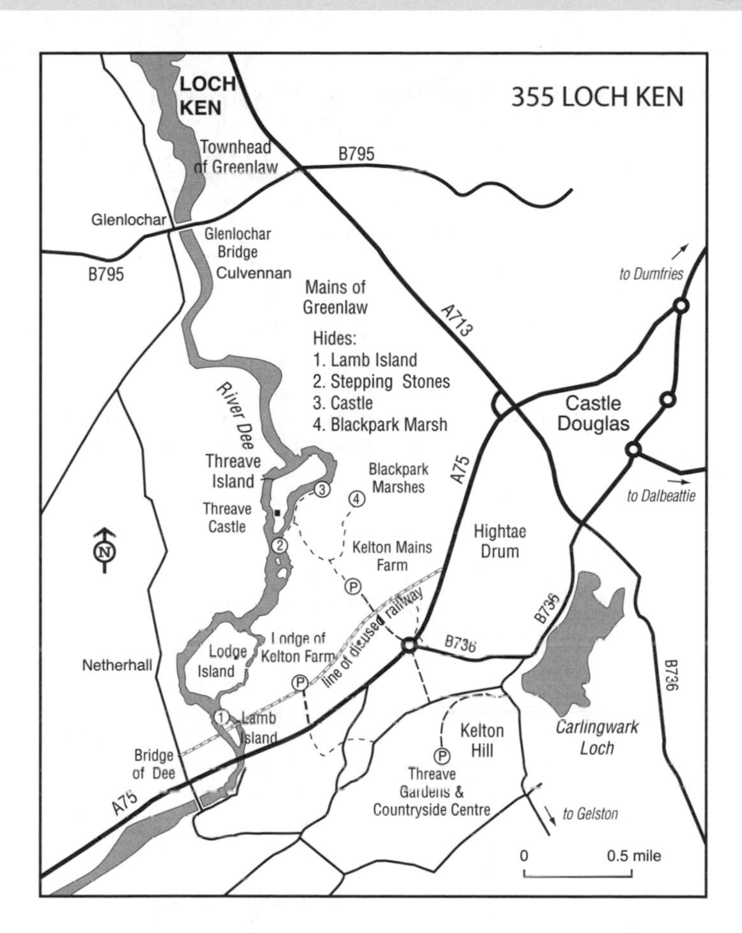

the River Dee (Lamb Island, Stepping Stones and Castle hides) and one overlooks Blackpark Marshes. Greylag and Pink-footed Geese (and occasionally a few Greenland White-fronts), and Whooper Swan favour the farmland, while ducks, including Goosander, Red-breasted Merganser and Shoveler, use the river, which is also frequented by Kingfisher; Hen Harrier is occasionally seen. There is a Countryside Centre and woodland hide at Threave Gardens on the opposite side of the A75 (entrance fee: open 1 April–30 October, 10am–5pm daily, 1 November–20 December, Friday, Saturday and Sunday only, 10am–5pm, 1 February–31 March, Friday, Saturday and Sunday only, 10am–5pm) from which information is available and from where a 1.5-mile waymarked estate walk also accesses the disused railway line and hides.

West side of Loch Ken The south section of the Loch is visible by leaving the A713 west at Townhead of Greenlaw. Cross the river on the B795 over Glenlochar Bridge, and immediately turn right. This unclassified road gives good views of the Loch as far as Mains of Duchrae, and this is the best area for geese. Greenland White-fronts favour the rough grazing immediately south of Mains of Duchrae. Be careful not to block the road with your car. Whooper Swan favours the area south of Glenlochar.

East shore of Loch Ken The A713 between New Galloway and Crossmichael gives views of the entire east shore of the loch. Greenland White-fronts favour the permanent pasture south-west of Cogarth Farm. A large herd of Whooper Swans is sometimes found west of the A713 and south of the B795 at Culvennan/Mains of Greenlaw.

Galloway Kite Trail A signposted vehicle trail, with information points and viewing locations on the A713 and A762, offers the chance to see Red Kites, but the best views of good numbers

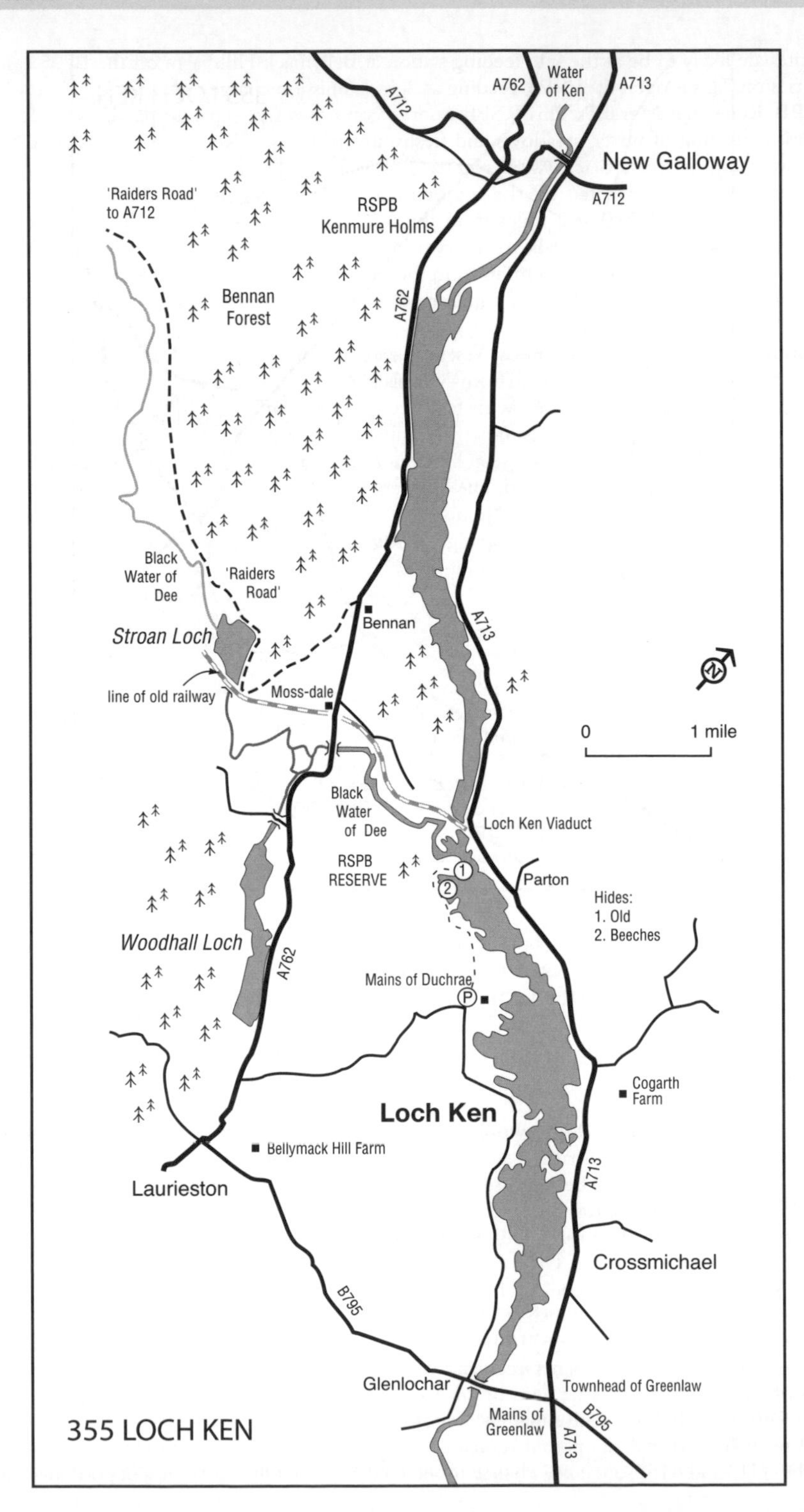

A762
Water of Ken
A713
A712
New Galloway
A712
'Raiders Road' to A712
RSPB Kenmure Holms
Bennan Forest
A762
Black Water of Dee
'Raiders Road'
Bennan
A713
Stroan Loch
line of old railway
Moss-dale
0
1 mile
Black Water of Dee
Loch Ken Viaduct
RSPB RESERVE
Parton
Hides:
1. Old
2. Beeches
Woodhall Loch
A762
Mains of Duchrae
Cogarth Farm
Loch Ken
Bellymack Hill Farm
Laurieston
A713
Crossmichael
B795
Glenlochar
Townhead of Greenlaw
Mains of Greenlaw
B795
A713

355 LOCH KEN

of kites are likely to be at the kite feeding station at Bellymack Hill Farm off the B795 east of Laurieston. This is open daily, with feeding at 2pm (admission charge).

RSPB Ken–Dee Marshes This RSPB reserve comprises five separate parcels of land, the largest consisting of water meadows and freshwater marshes at Kenmure Holms near New Galloway and between Black Water of Dee and Mains of Duchrae. There is access only to the latter. A car park is sited off the minor road at Mains of Duchrae (3.5 miles north of Glenlochar on the B795 or 2 miles from the A762, turning east just north of Laurieston), from where it is possible to walk to the two hides (1 and 1.5 miles from the car park) and the goose viewing platform (0.5 miles from the car park) through farmland, scrub, woodland and marshes. Look for Hen Harrier and Barn Owl in winter and Pied Flycatcher in summer, while Willow Tit is resident.

Bennan Forest The conifer woods west of Loch Ken are worth exploring. Following the disused railway line that runs west from Mossdale, there is a roost of Hen Harriers around Stroan Loch (best in early winter, with few after January), and it is also good for Common Crossbill, Siskin, redpolls and, in summer, Nightjar. Continuing along the railway, you reach Loch Skerrow. Black Grouse can be seen here but there is little else en route to compensate for the 4-mile walk from the main road. The 'Raiders Road' forest drive also passes Stroan Loch. This leaves the A762 at Bennan, 1 mile north of Mossdale, and passes through Cairn Edward Forest to emerge on the A712 at Clatteringshaws Dam. There are three forest walks. The road is open Easter–end October (toll). With a lot of luck the area around Clatteringshaws Loch may also produce Black Grouse.

FURTHER INFORMATION

Grid refs: Carlingwark Loch (Lochside Park) NX 765 617; Threave Wildfowl Refuge NX 746 616; Galloway Kite Trail (Bellymack Hill Farm) NX 688 651; RSPB Ken–Dee Marshes (Mains of Duchrae) NX 699 684; Bennan Forest NX 661 705
RSPB Dumfries and Galloway Office: tel: 01556 670464
National Trust for Scotland Ranger Service: tel: 01556 502575
Dumfries & Galloway Council Ranger Service: tel: 01556 502351
Forestry Commission Scotland: tel: 01671 402420
Galloway Kite Trail: web: www.gallowaykitetrail.com

356 AUCHENREOCH AND MILTON LOCHS (Dumfries & Galloway)

OS Landranger 84

Eight miles north-east of Castle Douglas, these two lochs are particularly attractive to diving ducks, and are best visited in winter.

Habitat

Both lochs are mainly deep water with diverse plant and invertebrate communities.

Species

Wintering wildfowl include Goosander, Goldeneye and Pochard. Smew is fairly regular from December, and the area has also attracted rarities such as Lesser Scaup and Ring-necked Duck.

Access

Auchenreoch Loch Lies alongside the A75 immediately west of Crocketford and is viewable from the road.

Milton Loch Leave the A75 just west of Crocketford on the minor road south to Milton. The

Loch is north of the road, and is viewable from the roadside or better, by pulling into a small turning area among the trees and then walking to the shore.

FURTHER INFORMATION

Grid refs: Auchenreoch Loch NX 815 711; Milton Loch NX 838 709

357 MERSEHEAD AND SOUTHWICK COAST
(Dumfries & Galloway) OS Landranger 84

This area lies west of Southerness Point, forming a narrow strip between the A710 and Mersehead Sands, and sharing a large population of geese and ducks with the better known Caerlaverock. Part of the area forms the RSPB Mersehead reserve and the Southwick Coast Scottish Wildlife Trust (SWT) reserve.

Habitat
The coast is bordered in succession by inter-tidal mudflats, merse (saltmarsh) or dunes, wet grassland and arable farmland, while just inland at the Southwick Coast reserve cliffs rise to c. 130 feet and support Heughwood, an intriguing stand of ancient woodland. Notable features include a rock stack, Lot's Wife, and a natural arch, the Needle's Eye, lying isolated in the marsh.

Species
The Solway Firth attracts large numbers of geese in winter, including Greylag, Pink-footed and, notably, Barnacle Geese, which may commute here from Caerlaverock; up to 9,500 of the latter may be present on the RSPB reserve. Also present are large numbers of wildfowl and waders, including up to 4,000 Teals, 2,000 Wigeons, 1,000 Pintails and a handful of Whooper Swans, raptors such as Peregrine, Merlin, Hen Harrier and Short-eared Owl, while the stubble fields may hold Tree Sparrow, Twite and a few Yellowhammers. Offshore look for Common Scoters and a few Scaups and divers. Breeding species include Shelduck, Wigeon, Teal, Shoveler, Barn Owl, Grasshopper Warbler, Spotted Flycatcher and a variety of waders, and Manx Shearwater, Gannets and terns can be seen offshore around high tide. See Caerlaverock and Southerness Point for a more detailed account (see page 545).

Access
RSPB Mersehead Turn south off the A710, c. 18 miles south of Dumfries, at the signed turn between Mainsriddle and Caulkerbush, and follow the narrow road for 1 mile to the reserve centre. There are two trails: the Wetland Trail at 1.2 miles long gives access to the Bruaich and Meida Hides, while the circular Coastal Trail covers 2.4 miles. The reserve is open at all times, the visitor centre 10am–5pm.
SWT Southwick Coast View from lay-bys alongside the A710. There is very limited parking opposite the reserve entrance at the minor road junction to Nether Clifton (c. 1.5 miles west of the junction with the B793 at Caulkerbush); from here a track leads to the Needle's Eye to view the merse. Slightly to the east (beyond Boneland Burn), the fields at Mersehead attract Barnacle Goose.

FURTHER INFORMATION

Grid refs: RSPB Mersehead NX 925 561; SWT Southwick Coast NX 913 562
RSPB Mersehead: tel: 01387 780579; email mersehead@rspb.org.uk
Scottish Wildlife Trust: tel: 0131 312 7765; web: www.swt.org.uk

358 CAERLAVEROCK TO SOUTHERNESS POINT

(Dumfries & Galloway) OS Landranger 84 and 85

On the north shore of the Solway Firth, this area holds thousands of wintering Barnacle and Pink-footed Geese. Caerlaverock NNR covers 8,184 ha, of which around 7,000 ha are mudflats and 600 ha saltmarsh, along 10 miles of coast between the mouths of the River Nith and Lochar Water, as well as 32 ha of freshwater marsh and reedbed. Caerlaverock WWT Refuge covers 607 ha, with excellent viewing facilities. The whole area provides superb birdwatching in winter, especially for those who are deprived of the spectacle of large numbers of geese nearer home.

Habitat

There are five main habitats for birds in the area – the sea, extensive mudflats, saltmarsh (known locally as 'merse', and especially prominent at Caerlaverock), rocky coastline (mainly from Carsethorn to Southerness), and finally farmland, both arable and pasture.

Species

Red-throated and sometimes Black-throated Divers winter offshore, together with Cormorant and Shag. Up to 120 Whooper Swans use the area in late September–April, and are joined by smaller numbers of Bewick's in late October; the Whoopers can be seen at close range at Caerlaverock at the 'Wild Swan feeds' which take place daily at 11am and 2pm at the Peter Scott Observatory. Barnacle Geese arrive in late September–early October and depart from mid-April to early May. They favour the merse and numbers can peak at 27,000; this is the entire Svalbard breeding population and following protection numbers have recovered from fewer than 500 birds. Note that in late winter they often move to Rockcliffe Marsh in Cumbria, and can then be seen from minor roads west of Rockcliffe Cross. Pink-footed Geese also arrive from late September and, together with several hundred Greylags, use the surrounding farmland. Numbers of Pink-feet peak in the autumn and spring, when up to 15,000 may be present, and several thousand remain through the winter. In many years there are one or more Snow Geese, though these are likely to be feral birds, perhaps those that breed near Stranraer. Confusingly, there are also often several albino Barnacle Geese. Other wildfowl include large numbers of Wigeons, Teals, Pintails (up to 4,000 in autumn – Carse Bay is especially good), and small numbers of Shovelers and Gadwalls. Up to 1,500 Scaups (exceptionally 4,500) winter in the Inner Solway between the River Nith and Blackshaw Bank, and are best seen off Carsethorn and at Southerness Point (in late winter Powfoot, south off the B724, 3 miles west of Annan, may be a better bet); they are joined by smaller numbers of Red-breasted Mergansers. The sea formerly held thousands of Common Scoters at Southerness, but only a handful are seen now. Raptors in winter include Hen Harrier, Sparrowhawk, Peregrine, Merlin and Short-eared Owl. Waders are plentiful on the mudflats, especially Oystercatcher and Bar-tailed Godwit, and there are thousands of Golden Plovers, Curlews and Lapwings in the fields. Purple Sandpiper and Turnstone occur on rocky coasts, especially at Southerness Point.

Passage brings a greater variety of waders. Large numbers of Sanderlings pass through in May, and in autumn they may be joined by Little Stint, Curlew Sandpiper, Black-tailed Godwit, Whimbrel and Spotted Redshank.

Breeding species include Shelduck, sometimes Pintail or Wigeon, Lapwing, Common Snipe, Common Redshank and Oystercatcher, with Common Tern and gulls on the saltmarsh. Grasshopper, Reed and Sedge Warblers and Lesser Whitethroat are summer visitors and the farmland holds Tree Sparrow.

Access

Kirkconnell Merse and the north section of Caerlaverock NNR The B725 from Dumfries to Caerlaverock Castle gives views, firstly of Kirkconnell Merse, a good area for

Whooper Swan, and secondly of the mudflats and merse in the north of the NNR. Pink-footed Geese are often visible from the road near the Castle.

Caerlaverock WWT Wetland Centre Take the signed turning from the B725 1 mile south of Bankend to the Centre. There are four towers, four observatories and 20 small hides looking out

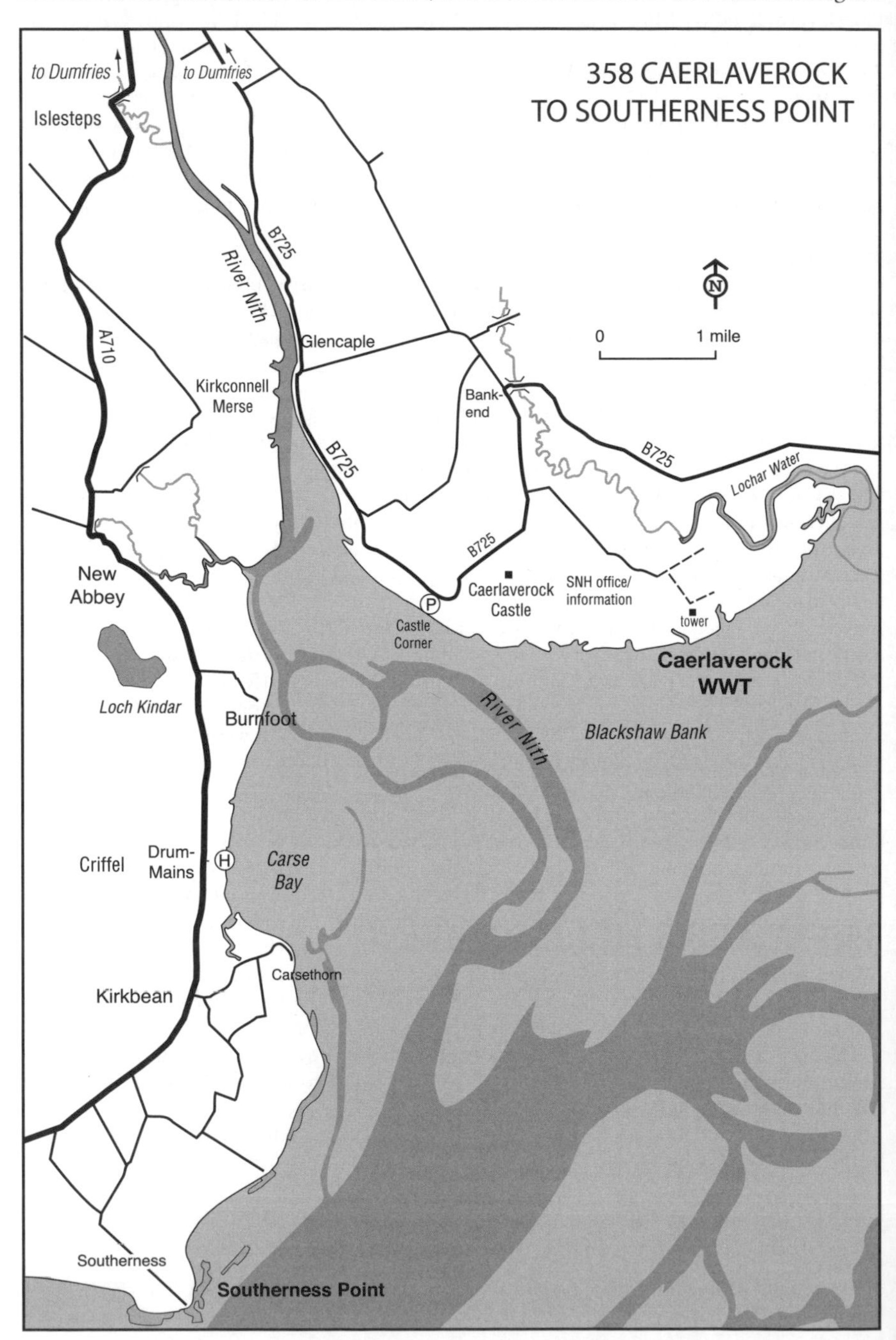

onto the reserve, and the pond in front of the main heated observatory holds up to 17 species of wildfowl in winter, while the Saltcot Merse Observatory gives views over the mudflats as well as the merse. The Centre is open daily 10am–5pm (except Christmas Day); there is an admission charge for non-members of the WWT and visitors should report to reception on arrival.
Caerlaverock NNR Most visitors to the Solway head for the WWT Wetlands Centre, but SNH also offers limited visitor facilities. There are car parks at Hollands Lane and Castle Corner, and between Castle Corner and Kenneth Bank. Public parking is also available at Powfoot, Brow Well and Carsethorn. The reserve office and information centre lies at Hollands Lane (on the same minor road as the WWT Wetland Centre), from where a network of footpaths gives access to the reserve, including the viewing platform (also accessible from Castle Corner) and a short boardwalk and hide.
Islesteps The fields around Islesteps, beside the A710 just south of Dumfries, are good for Whooper Swan.
Drummains Reedbed This small Scottish Wildlife Trust reserve overlooks the west shore of the Nith Estuary, giving views of a typical range of wildfowl and waders. Park just east of the A710 at Drum-Mains and follow the footpath east for 300 yards to the reserve.
Carsethorn On the A710 12 miles south of Dumfries, take the turning at Kirkbean to Carsethorn and drive to the waterfront. Check the sea and rocky coastline, especially for Scaup, which sometimes occurs in significant numbers.
Southerness Point Leave the A710 south on an unclassified road 1 mile south-west of Kirkbean and park at the end of the road. Divers (mainly Red-throated), Scaup, scoters, auks and the occasional skua occur off the point, with Purple Sandpiper and Turnstone on the rocks. There is a wader roost, often holding Grey Plover and Greenshank (on passage and in winter), as well as many Bar-tailed Godwits, and the area is generally very good for waders. It is worth exploring the mudflats and sandbanks east and west of the point, and the fields here sometimes hold Pink-footed and Barnacle Geese, especially in late winter.

FURTHER INFORMATION

Grid refs/postcode: Caerlaverock NNR (Castle Corner) NY 019 652; Caerlaverock WWT NY 050 655 / DG1 4RS; Islesteps NX 966 726; Drummains Reedbed NX 979 610; Carsethorn NX 993 598; Southerness Point NX 976 543
Scottish Natural Heritage, Caerlaverock NNR office: tel: 01387 770275
Caerlaverock WWT Wetland Centre: tel: 01387 770200, web: www.wwt.org.uk/gallery/117/caerlaverock.html

359 ST ABB'S HEAD (Borders)

OS Landranger 67

Twelve miles north of Berwick-upon-Tweed, St Abb's Head has 60,000 breeding seabirds, is a fine seawatching station and, during passage periods, numbers of migrants are recorded. The best time for breeding seabirds is May to mid-July, and for migrants April to early June and mid-August to October. The coast between St Abb's village and Pettico Wick is an NNR, owned by the NTS.

Habitat

The cliffs rise to 300 feet, and immediately behind the Head is a man-made freshwater pool, Mire Loch, surrounded by dense gorse bushes and birch trees that are attractive to migrants. Otherwise the hinterland is largely pasture.

Species

Breeding seabirds include small numbers of Fulmars, Shags and Herring Gulls and several thousand pairs each of Common Guillemots, Razorbills and Kittiwakes (around 7,000 Kittiwakes in 2007, a drop from a peak of 20,000 in 1989). Puffins are only present in very small numbers and consequently are hard to see. Other breeders are Rock Pipit, Wheatear, Lesser Whitethroat and Raven, while Turnstone and Purple Sandpiper haunt the rocks for much of the year. Offshore there are always Gannet and Fulmar, but seawatching is most productive in autumn, when Manx and sometimes Sooty Shearwaters, Arctic and a few Great Skuas, Little Gull and Black Tern are regular, and there are occasional Balearic Shearwaters and Pomarine or Long-tailed Skuas; strong winds between north and east are most productive. In late autumn and winter Little Auk occasionally occurs, but Red-throated Diver is regular, with the occasional Great Northern and Black-throated and small numbers of Common Scoter and Common Eider, the last especially at Starney Bay and Pettico Wick. In winter Goldeneye and Wigeon occur on the loch (and dabbling ducks also on nearby Millar's Moss Reservoir) and Sparrowhawk, Peregrine and Merlin hunt the area.

As well as the commoner chats, flycatchers and warblers, scarce migrants are recorded annually on passage. Any wind from the east is good, but south-east is best, especially if

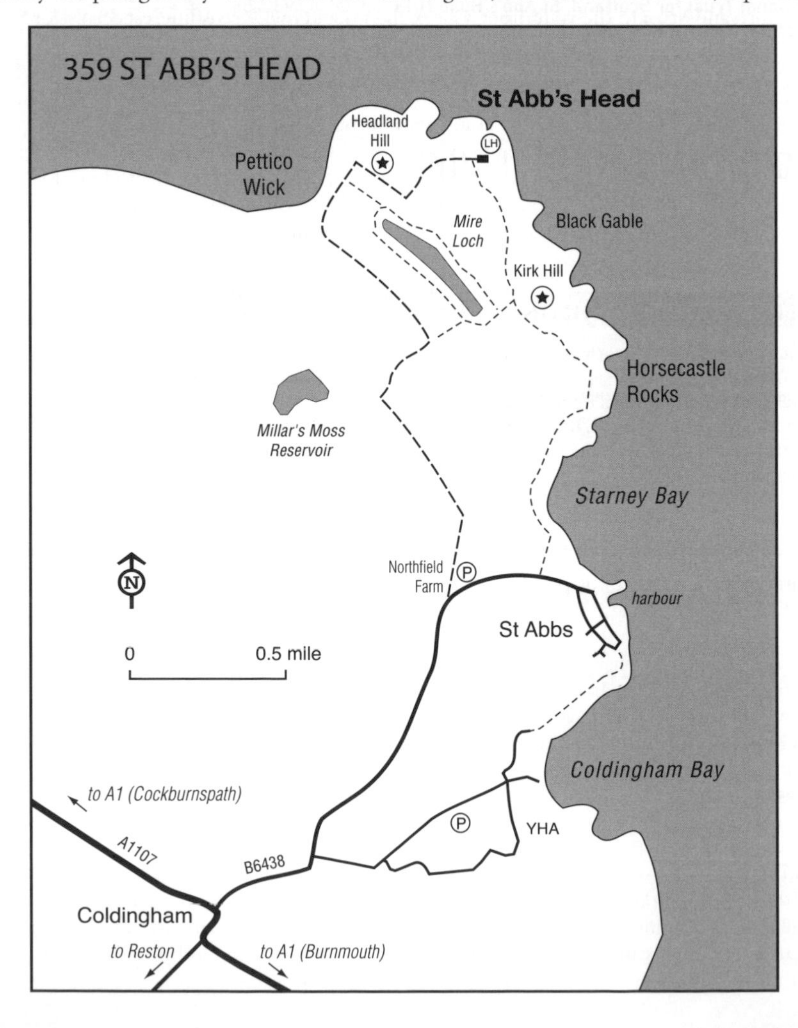

coupled with rain or drizzle. Wryneck, Black Redstart, Bluethroat and Red-backed Shrike are annual in spring, and Wryneck, Barred and Yellow-browed Warblers, Firecrest, Red-breasted Flycatcher and Red-backed Shrike equally frequent in autumn.

Access

Leave the A1 on the A1107 at Burnmouth (or, approaching from the north, at Cockburnspath), and from Coldingham take the B6438 to St Abb's Head NR. Park on the right at Northfield Farm, the NTS Visitor Centre (open daily 10am–5pm, 1 April–31 October), and from here take the footpath east for 300 yards to the track to the lighthouse and Head. This skirts the cliff and has seabirds en route, and it is possible to return via inland paths to complete a 3-mile circuit. A small car park near the Head is reserved for those with walking difficulties. Seawatch from Black Gable (the lowest point on the cliffs between the lighthouse and Kirk Hill), and during passage periods explore the cover around the loch and at Coldingham Bay.

FURTHER INFORMATION

Grid ref: NT 913 674
National Trust for Scotland, St Abb's Head NNR: tel: 0844 4932256
Scottish Natural Heritage: web: www.nnr-scotland.org.uk

360 BARNS NESS (Lothian)

OS Landranger 67

This migration watchpoint at the southern entrance to the Firth of Forth has, over the years, produced a number of interesting birds.

Habitat

This is a low peninsula with a prominent lighthouse and a rocky shore.

Species

Spring migrants can include White Wagtail, Black Redstart and, occasionally, Bluethroat, Pied Flycatcher or Red-backed Shrike.

Autumn seawatching can be good, especially in north-east winds. Manx and sometimes Sooty Shearwaters, seaducks, Arctic and Great Skuas, Black and other terns and, especially in late autumn, Pomarine Skua and Little Gull are possible, together with Gannet and Kittiwake. Passerine migrants are also weather-dependent, and as at all east coast migration watchpoints, winds from the east, combined with rain or drizzle, are the classic conditions for a fall. Among the scarcer species, Richard's Pipit, Icterine, Barred and Yellow-browed Warblers, Firecrest and Red-breasted Flycatcher have been recorded.

In winter, the sea is worth a look, especially after north or east gales, for divers, Glaucous Gull and Little Auk. Turnstone and Purple Sandpiper may be found on the rocks.

Access

Turn north off the A1 about 4 miles east of Dunbar on the minor road to Skateraw and continue on to East Barns, from where a minor road leads north to the beach car park (and from there a track leads to the lighthouse). The best spot for seawatching is just north of the lighthouse, although in the roughest conditions the higher ground by the wire dump is better. Good areas to search for passerine migrants include the lighthouse gardens, the wire dump and the campsite (access only possible October–March), with trees and scrub near its entrance and a belt of trees along the south-west perimeter.

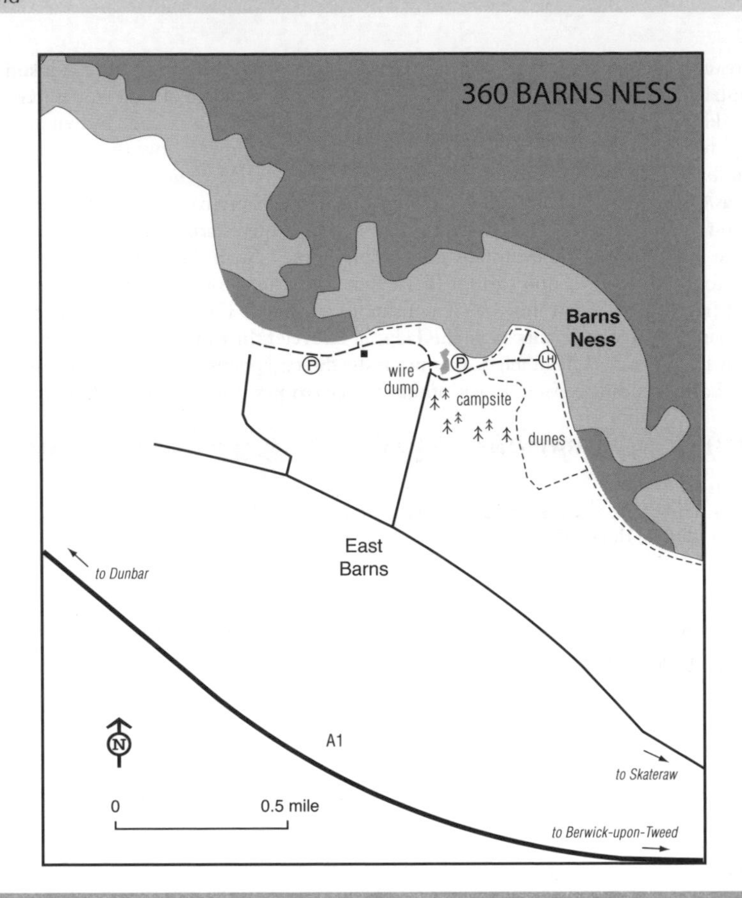

FURTHER INFORMATION

Grid ref: NT 720 770

361 JOHN MUIR COUNTRY PARK, TYNINGHAME (Lothian)

OS Landranger 67

Just west of Dunbar, the Country Park holds an excellent range of wildfowl and waders, with some notable wintering passerines, sometimes including Lapland Bunting.

Habitat

The 704-ha Country Park comprises the estuary of the River Tyne with associated mudflats and saltmarsh, bordered to the north by a rocky headland known as St Baldred's Cradle and a narrow sandy spit (Sandy Hirst), and to the south by the dunes of Spike Island. The hinterland is grassland and scrub, with a plantation of Scots Pine to the south of Spike Island.

Species

Wintering wildfowl include many Wigeon as well as Shelduck, Teal, Pintail, Goldeneye, Goosander, Common Eider and Common Scoter, and a few Velvet Scoter, Long-tailed Duck and

Red-breasted Merganser. A flock of up to 100 Whooper Swans frequents the fields south-west of the estuary, together with up to 500 Greylags and small numbers of Pink-footed Geese. The usual waders are present in winter, including Grey Plover, Knot, Bar-tailed Godwit and notably, two or three Greenshanks, with Turnstone and Purple Sandpiper below the cliffs at Belhaven and Dunbar. The concentration of potential prey attracts Peregrine, Merlin, Sparrowhawk and occasionally Hen Harrier. Look for Red-throated Diver, Shag and Cormorant offshore. Wintering passerines on the Spike Island saltmarsh can be interesting: up to 100 Twite are regular and Shore Lark and Lapland Bunting are sometimes present, with Snow Bunting possible in the dunes and on the beach. Look also for Rock Pipit and Stonechat, while Dipper can be found wintering nearby on Biel Water with Kingfisher there and on Hedderwick Burn.

On spring passage, migrants include Gadwall, Shoveler, Pintail, Sanderling, White Wagtail, Ring Ouzel, Wheatear, Whinchat, Common Redstart, and a few Spotted Redshank and Whimbrel. In autumn a few Barnacle and Brent Geese pass through, and a greater variety of waders is noted, with up to 80 Whimbrels and up to 40 Greenshanks, while Green Sandpiper, Spotted Redshank, Black-tailed Godwit, Ruff, Little Stint and Curlew Sandpiper are also worth searching for. Arctic Tern is regular and, with luck, Roseate or Black Tern may be found in late summer; loafing Kittiwakes are regular around Biel Water. Seawatching in autumn may produce Manx Shearwater, Kittiwake, and Great and Arctic Skuas.

Breeders include a few Common Eider, Shelducks and Ringed Plover, with over 200 pairs of Kittiwake nearby at Dunbar Harbour. Little Grebe and Sedge Warbler breed at Seafield Pond. The Estate woodland holds Green and Great Spotted Woodpeckers and possibly breeding Hawfinch. Look also for passage Tree Pipit, Siskin and Common Crossbill. Several hundred Common Eiders oversummer, and in late summer a moult gathering of up to 80 Goosanders is a feature, as is the number of Gannets offshore.

Access

The inner estuary and Spike Island are best in the period 2 hours either side of high tide, with waders roosting on Spike Island on 'spring' tides and on the north part of the inner estuary on neap tides.

John Muir Country Park, Linkfield car park Leave the A1 c. 1 mile west of Dunbar at the Thistly Cross Roundabout northwards on the A199/A1087 and, after 250 yards, turn north-east at the Beltonford roundabout on to the A1087 towards Dunbar. Follow this for 0.5 miles and then turn north on a minor road at the sign for the Country Park and continue on to the Linkfield car park. The saltmarsh can be viewed from the mound to the left of the toilet block, while walking out towards the bay the sandy spit at the north end of Spike Island

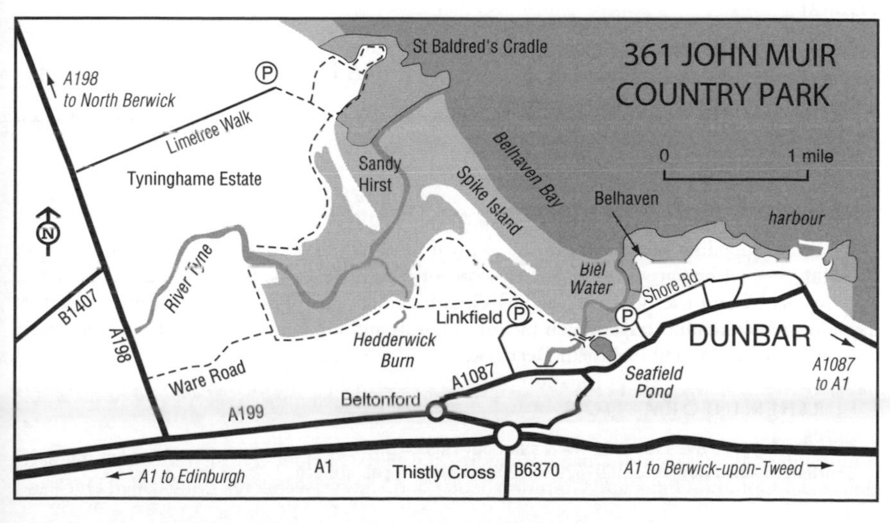

attracts roosting waders, gulls and terns, while the dunes and sandy beach attract Sanderlings and sometimes Snow Buntings.

John Muir Country Park, Ware Road Access to the west side of the bay is possible from the A198 (0.4 miles north of its junction with the A199), taking the unsigned minor road (Ware Road) to the east, but there is only very limited trackside parking. Tree Sparrows may be numerous along the lane here with passage waders on the small pools alongside; from the end of Ware Road follow the raised path northwards to view the channel of the River Tyne, a good area for waders.

Tyninghame Estate Although private, access is allowed to certain routes through the Estate. From the A198 North Berwick road turn east into Limetree Walk, parking at the end by Tyninghame Links. From here, footpaths lead to St Baldred's Cradle, the best vantage point for seawatching.

John Muir Country Park, Shore Road car park From the Beltonford roundabout continue on the A1087 past the turn to Linkfield and turn north as you enter Dunbar on to Shore Road. A footpath links this to the Linkfield car park, passing Biel Water, which is a good area for waders, and Seafield Pond, which may hold wildfowl but typically has little exposed mud.

FURTHER INFORMATION

Grid refs: John Muir Country Park, Linkfield car park NT 651 787; John Muir Country Park, Ware Road NT 626 785; Tyninghame Estate NT 627 809; John Muir Country Park, Shore Road car park NT 662 786

East Lothian District Council, countryside rangers: tel: 01620 827279; email: ranger@eastlothian.gov.uk

362 BASS ROCK (Lothian)

OS Landranger 67

Probably the most accessible large gannetry in Britain, the Bass Rock is only 3 miles off North Berwick.

Habitat

This is a fairly round 3-ha island of volcanic rock with very steep cliffs almost all the way round.

Species

An impressive 30,000–40,000 pairs of Gannets are joined by good numbers of Common Guillemots, Kittiwakes and Herring Gulls, as well as a handful of Fulmars, Shags, Common Eiders, Lesser Black-backed Gulls, Razorbills and Puffins.

Access

During the breeding season there are regular boat trips around the island from North Berwick. Several operators now run trips out of North Berwick, the best-known of which are the *Sula II* and the Scottish Seabird Centre, which also organises a limited number of day-long 'photographic' trips which include 'chumming' and landing on the Bass Rock itself; the centre is well-signed in North Berwick.

FURTHER INFORMATION

***Sula II*:** tel: 01620 892838; web: www.sulaboattrips.co.uk

Scottish Seabird Centre: tel: 01620 890202; web www.seabird.org

363 GOSFORD, ABERLADY AND GULLANE BAYS (Lothian)

OS Landranger 66

This section of the Firth of Forth attracts a wide variety of species and is worth a visit year-round. In winter seaducks, divers and grebes are prominent, and a speciality is Red-necked Grebe, which may appear as early as July. Aberlady Bay was designated as Britain's first LNR in 1952 and is managed by East Lothian Council. The reserve covers 582 ha, two-thirds of which comprises tidal sand- and mudflats and saltmarsh.

Habitat

Aberlady and Gosford Bays possess an extensive foreshore, with rocky areas at Port Seton and Gullane Point, but Gullane Bay has a relatively narrow foreshore. The beach is backed by areas of dunes and patches of Sea-buckthorn, most extensive in Gullane Bay, with a small freshwater loch (Marl Loch).

Species

Red-necked and Slavonian Grebes, as well as Red-throated and Black-throated Divers occur offshore and there can be as many as 70 Slavonian Grebes wintering along this stretch of coast (a single Black-necked Grebe has also been present in recent winters). Seaducks occur mainly in Gullane and Gosford Bays, with smaller numbers in Aberlady Bay. Up to 7,000 Common Eiders and 2,000 Common Scoters are present, with up to 800 Velvet Scoters (both Common Eiders and scoters peak in spring and autumn), and hundreds of Long-tailed Ducks, Goldeneyes and Red-breasted Mergansers. It is worth checking for Surf Scoter, which has occurred regularly, as well as Great Northern Diver. To view the grebes, divers and seaducks the period around high tide is best, ideally combined with calm weather. Up to 17,500 Pink-footed Geese winter in the area, with numbers peaking in the early winter. They feed inland, often on the fields around Fenton Barns and Drem, and roost in Aberlady Bay – wait for them to fly in at dusk. There may also be a few Barnacle Geese with them, as well as small numbers of Whooper Swans. Waders include Grey Plover, Knot, Sanderling, Turnstone and Purple Sandpiper, and sometimes Ruff in the Port Seton and Aberlady areas. Redwing, Fieldfare and Blackcap favour the Sea-buckthorn. Predators may include Merlin, Peregrine, and Long-eared and Short-eared Owls. Finch flocks on the saltmarsh and nearby stubble fields may hold Tree Sparrow, Twite, Brambling and Snow Bunting, and also occasionally Lapland Bunting or even Shore Lark.

A variety of waders occurs on passage. Whimbrel, Black-tailed Godwit, Spotted Redshank, Greenshank, and Wood and Green Sandpipers are regular, with Little Stint and Curlew Sandpiper annual, and the best variety in autumn. The area has attracted a truly outstanding list of rare waders.

Breeders include 200 pairs of Common Eiders, Shelduck, Ringed Plover, Common Snipe, Common Redshank, Common and Arctic Terns, Stonechat, Grasshopper, Sedge and Garden Warblers and Lesser Whitethroat, with Sparrowhawk in the area. Numbers of non-breeding Common Eiders and Common and Velvet Scoters are also present, as well as summering Grey Plover, Knot, Sanderling, Common Guillemot, Razorbill, and very occasionally Black Guillemot or Puffin. Fulmar, Gannet, and sometimes Manx Shearwaters can be seen offshore during the summer. Late summer and autumn is the peak period for Red-necked Grebe, with up to 40 present. The best spot is around Ferny Ness. Late-summer gatherings of terns, especially Sandwich, are a feature, while Arctic Skua regularly occurs offshore at this time, and more irregularly, Great and Pomarine Skuas.

Access

Gosford Bay Can be viewed from the road. The B1348 runs east from Port Seton, joining the A198 just before Ferny Ness, and continues along the shore of the Bay. The best spot is Ferny

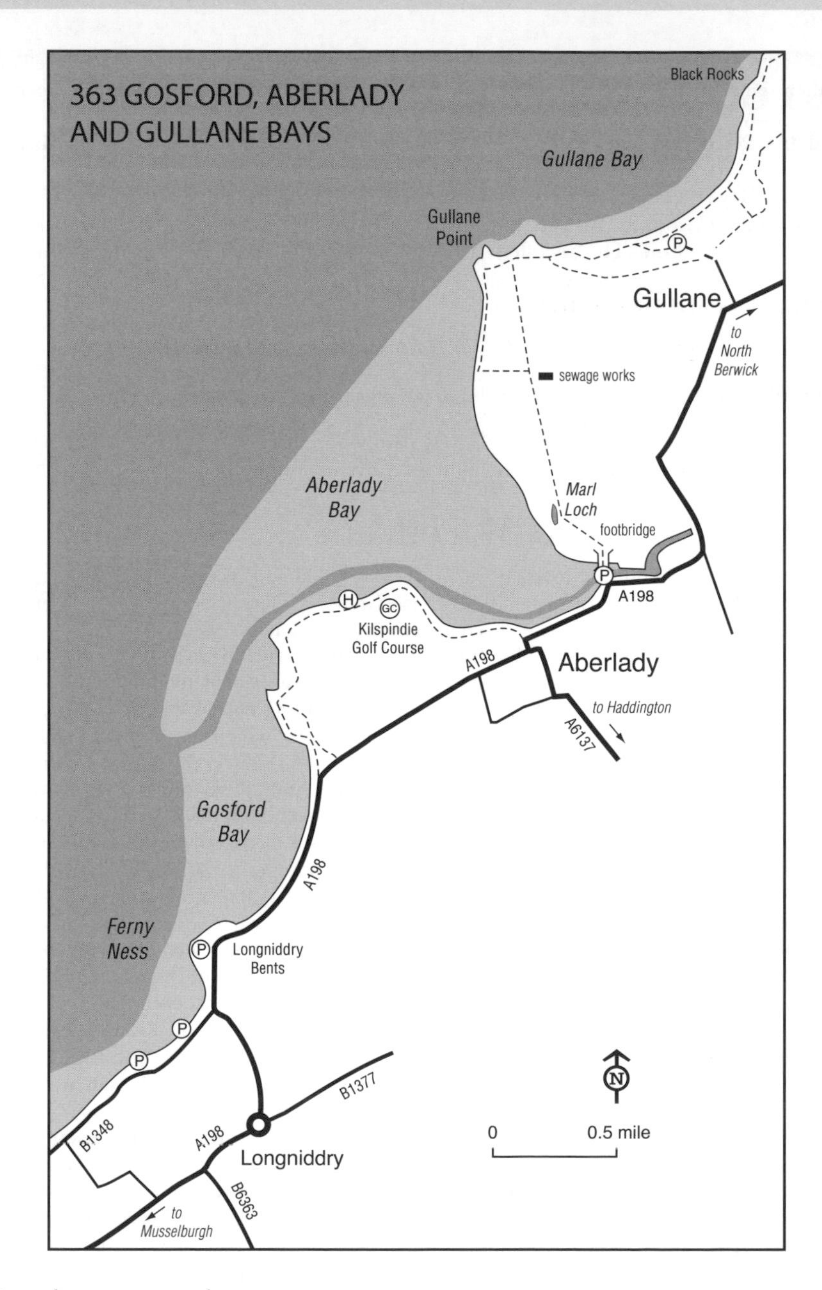

Ness, where you can park.

Aberlady Bay, south-west shore A footpath, accessible from the A198, follows the coast from Gosford Bay to Aberlady village, running along the south shore of Aberlady Bay and passing a hide midway along the shore and the Kilspindie Golf Course club house. At low tide it is best to watch Aberlady Bay from this footpath or from the A198 just east of Aberlady village. The key for the hide can be obtained from the warden by prior arrangement.

Aberlady Bay, east shore On a rising tide the east shore of Aberlady Bay is best. Leave Aberlady eastwards on the A198 and after 0.5 miles there is a small car park. From here a track

crosses the stream via a footbridge (one of the best spots for migrant waders) and Gullane Point is then signposted. To reach the shore, fork left just after the sewage farm; there is no general access away from the paths.
Gullane Point This is an excellent seawatching spot, accessed via the footpath from Aberlady Bay or via a footpath from the car park in Gullane Bay.
Gullane Bay A footpath follows the shore of the bay from Gullane Point north-east to Black Rocks, accessed from the car park near the shore (accessible north off the A198 along Sandy Loan through Gullane). The bay often has an evening roost of up to 150 Red-throated Diver.

FURTHER INFORMATION

Grid refs: Gosford Bay (Ferny Ness) NT 441 774; Aberlady Bay, east shore NT 471 804; Gullane Bay NT 476 831
Aberlady Bay Nature Reserve: tel: 01620 827847; email: jharrison@eastlothian.gov.uk

364 MUSSELBURGH (Lothian)

OS Landranger 65

Situated on the east fringe of Edinburgh, this area attracts an excellent range of seaducks, waders, gulls and terns, many of which can be watched at relatively close range, and a visit is worthwhile at any time of year, although autumn and winter are probably best.

Habitat

A series of specially constructed lagoons east of the mouth of the River Esk have been filled with ash from nearby Cockenzie Power Station, and the remaining shallow pools attract large numbers of roosting waders, gulls and terns, with passerines in the rough grassland. Offshore, the waters of the Firth of Forth attract a range of waterfowl.

Species

In winter there are Great Crested and Slavonian Grebes and Red-throated Diver offshore, and occasionally Red-necked Grebe and Black-throated and Great Northern Divers (the rarer divers being most frequent in late winter). The Firth of Forth once held huge concentrations of Scaups (peaking at 30,000–40,000 in 1968–69), but now counts seldom exceed 25. Other seaducks include small numbers of Common Eiders, Common Scoters, Goosanders and Red-breasted Mergansers, up to 500 Velvet Scoters (indeed, this species is present almost all year), 300 Long-tailed Ducks and 500 Goldeneyes. A few Common Guillemots and Razorbills are usually present, and Little Auk can occur in good numbers during some seasons. Wintering waders include large numbers of Oystercatchers, Knots and Bar-tailed Godwits, with smaller numbers of Ringed and Grey Plovers, Curlews, Common Redshanks, Dunlins and Turnstones. The concentrations of potential prey may attract a Peregrine or Short-eared Owl. The rough ground, sea wall and lagoon edges hold finches and buntings, with luck including Snow Bunting and Twite.

A wide variety of waders and terns occurs on passage, including Golden Plover, Ruff, Whimbrel, Black-tailed Godwit, Greenshank, Common Sandpiper and Sanderling, with Little Stint and Curlew Sandpiper also possible, especially in autumn. Passage waders often feed near the river mouth. Migrant passerines may include Wheatear and White Wagtail. Large numbers of the commoner gulls frequent the area, often loafing at the river mouth. Mediterranean Gull is increasingly possible and Little Gulls peak in late autumn, with odd immatures during the summer. Terns include a notable roost of up to 3,000 Sandwich and smaller numbers of Common. In late summer and autumn there are large flocks of Common Eiders, Common Scoters and Red-breasted Mergansers offshore, and at this time seawatching can be productive, especially in north or east winds, which may bring numbers of Gannet, Manx Shearwater, Kittiwake and Arctic and Great Skuas inshore.

Access

Leave the A199 Linkfield Road in Musselburgh, parking at the end of Goosegreen Crescent by the sea wall (or at the end of Balcarres Road, just beyond the old gas works). Alternatively, turn north off the B1348 Ravenhaugh Road to park at the Levenhall Links pond. Access on foot is permitted to most areas, except the active lagoons. On a rising tide the mouth of the River Esk is good for waders, ducks, gulls and terns (the light is best here in the morning). Over the period 3 hours either side of high water look for divers, grebes and seaducks on the Firth of Forth from the John Muir Way, especially from around the 'Point', and also for roosting waders, gulls and terns on Lagoon 8. The boating pond and scrapes also hold ducks and waders, but are not dependent on the state of the tide.

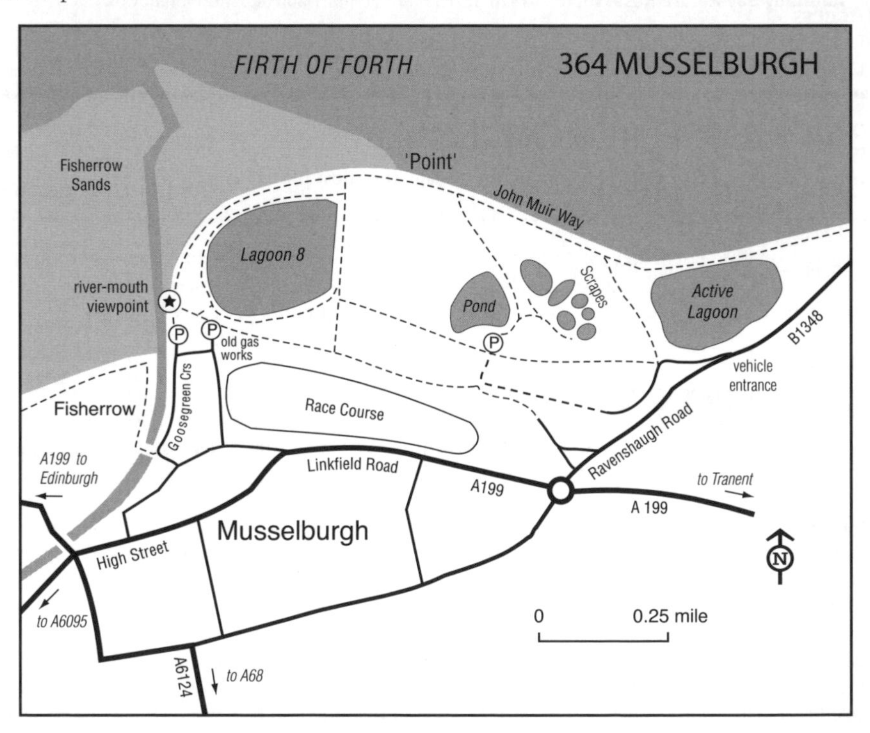

FURTHER INFORMATION

Grid ref: Goosegreen Crescent NT 346 734

365 HOUND POINT (Lothian)

OS Landranger 65

Immediately west of Edinburgh, in recent years this seawatching station has established itself among the most reliable autumn sites for Long-tailed and Pomarine Skuas.

Habitat

The Point projects into the Firth of Forth on the south shore, just east of the Forth Road Bridge.

Species

Passing seabirds generally include Fulmar, Manx Shearwater, Gannet, Kittiwake, Arctic Skua, Sandwich Tern and Common Guillemot. Large numbers of Long-tailed and Pomarine Skuas may occur in the correct conditions, and Pomarine Skua is regular in October–November; Great Skua is the rarest of the four skuas. Passing skuas often occur in flocks and move west along the Firth of Forth, gaining height to pass over the Forth bridges. They presumably make the short overland crossing to the Firth of Clyde. Other scarce visitors may include Black Tern and Little Gull, with Little Auk possible in late autumn.

Access

Leave South Queensferry east on the B924 and park just before the Forth Railway Bridge. From here it is a 2-mile walk along the coast to the Point. The period August–October is most productive (though spring can be good too), and light winds from north-west through north to south-east are best; south and west winds are usually poor. The passing of a front or clearance of sea fog may prompt a passage of skuas. Though birds can appear from any direction (and often overhead), the north and east quadrants are usually the most productive.

FURTHER INFORMATION

Grid ref: Forth Railway Bridge NT 135 783

366 BARON'S HAUGH (Clyde)

OS Landranger 64

Situated in the Clyde Valley, 1 mile south of Motherwell, this RSPB reserve attracts a variety of waders and wildfowl and is worth visiting in autumn and, more especially, in winter.

Habitat

The reserve is centred on an area of wet meadows ('haugh') alongside the River Clyde, and has some permanently flooded areas as well as parts prone to winter flooding. Other habitats include woodland, parkland and scrub. A notable feature is the upwelling of tepid water around Marsh Hide, keeping the area open even in freezing conditions in winter, resulting in concentrations of waterbirds in such conditions.

Species

Residents include Little Grebe, Grey Heron, Teal, Shoveler, Gadwall, Ruddy Duck, Water Rail, Common Redshank, Green and Great Spotted Woodpeckers, Kingfisher and Grey Wagtail. Nuthatch also occurs and, at the northernmost limits of its range here, is something of a speciality.

Breeding summer visitors include Common Sandpiper, Sand Martin, Whinchat, and Grasshopper, Sedge and Garden Warblers.

On autumn passage an excellent variety of waders can occur, with Ruff, Black-tailed Godwit, Spotted Redshank and Green Sandpiper possible. There is a late-summer concentration of up to 1,000 Lapwings.

Wintering wildfowl include up to 50 Whooper Swans, as well as numbers of Wigeon, Gadwall, Teal, Shoveler, Pochard and Goldeneye.

Access

From the centre of Motherwell (access from the M74, junction 6, or the M8, junction 6) take Adele Street, opposite the Civic Centre, south for 0.5 miles to North Lodge Avenue. Continue straight ahead along White Walk to the reserve entrance. Alternatively, turn right (west) along North Lodge Avenue and the west entrance is reached after c. 0.5 mile.

The reserve is open at all times, and there are two trails (0.5 and 2 miles) and four hides.

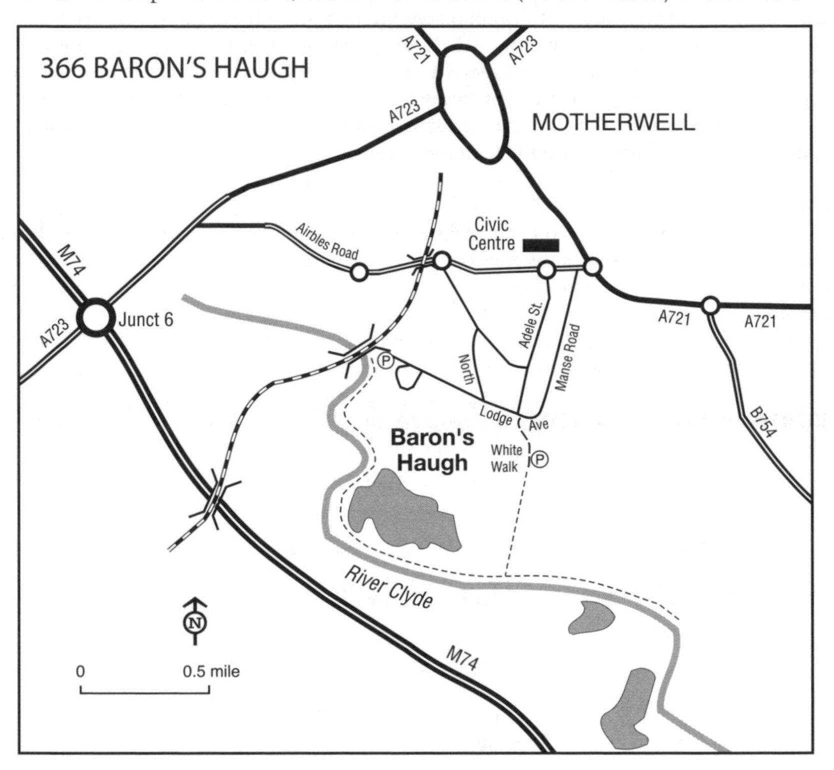

FURTHER INFORMATION

Grid ref: NS 756 552
RSPB Baron's Haugh: tel: 0141 331 0993; email: barons.haugh@rspb.org.uk

367 LOCHWINNOCH (Clyde)

OS Landranger 63

Lying 18 miles south-west of Glasgow, the best times to visit are autumn–winter for wildfowl and spring–early summer for breeding species.

Habitat

The reserve comprises two sections, Aird Meadow to the north and Barr Loch to the south, separated by the A760. Aird Meadow has some lily-covered ponds surrounded by sedge beds as well as wet meadows, areas of willow scrub and some stands of deciduous woodlands. Barr Loch was drained for a time to form farmland but the drainage system fell into disrepair in the 1950s and the area re-flooded to form firstly a marsh and now a large shallow loch.

Species

Great Crested Grebe, Teal, Pochard, Sparrowhawk, Water Rail, Common Snipe, Black-headed Gull, Dipper (on the River Calder) and Lesser Redpoll are resident, and Cormorant, Grey Heron and Raven frequently visit. Common Buzzard occasionally appears throughout the year.

Wintering wildfowl include up to 50 Whooper Swans, 700 Greylag Geese, Wigeon, Teal, c. 90 Goosanders, Goldeneye and a handful of Pintails and Smews. Raptors may include Hen Harrier and Peregrine. Numbers of gulls roost on Barr Loch, and sometimes include Glaucous or Iceland. Other winter visitors include Jack Snipe, Kingfisher and sometimes Brambling.

Breeding summer visitors include Common Sandpiper, large numbers of Sedge Warblers, as well as Grasshopper and Garden Warblers and Spotted Flycatcher.

On passage Slavonian Grebe may occur, as well as White-fronted and Pink-footed Geese, Garganey, Osprey, Marsh Harrier, Little Gull and Common Tern. From early August into autumn there is a large roost of Swallows, martins and Starlings.

Access

Turn west off the A737 towards Lochwinnoch on the A760. After 0.5 miles, having crossed the railway bridge, the reserve entrance lies to the north (right). (Lochwinnoch railway station is 200 yards from the reserve entrance.) The reserve centre is open daily (except 25–26 December and 1–2 January), 10am–5pm. There are two trails: the Dubbs Water Trail (0.5 miles) leading to a viewing area between the two lochs, and the Aird Meadow Trail (0.8 miles), leading to the far end of Aird Meadow Loch, with two hides en route; the trails and hides are open at all times.

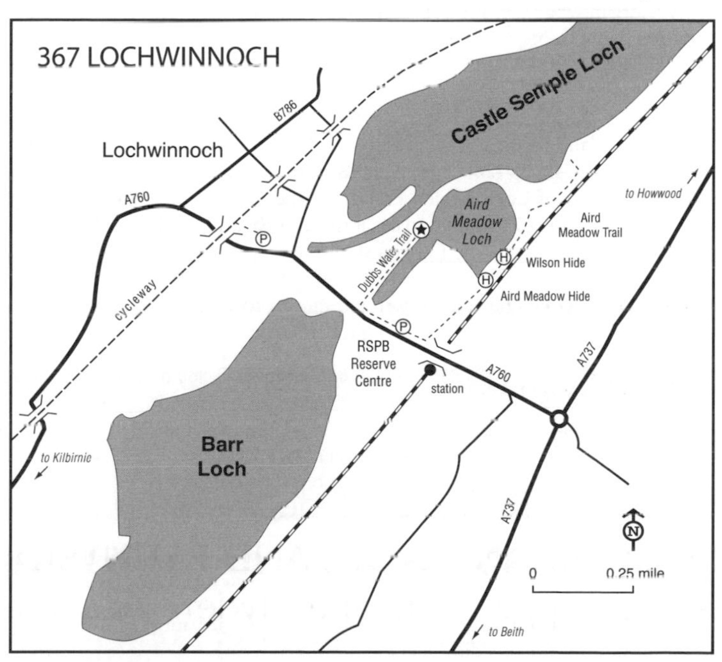

FURTHER INFORMATION

Grid ref/postcode: NS 358 581 / PA12 4JF
RSPB Lochwinnoch: tel: 01505 842 663; email: lochwinnoch@rspb.org.uk

368 INVERSNAID (Upper Forth)

OS Landranger 56

Lying on the eastern shores of Loch Lomond, this RSPB 374-ha reserve comprises Sessile Oak woodland and holds the classic trio: Common Redstart, Wood Warbler and Pied Flycatcher.

Habitat

The steep slopes along the shores of Loch Lomond are cloaked with Sessile Oak woodland that rises up to a ridge of crags, beyond which are extensive areas of moorland as well as a stand of mixed woodland.

Species

Resident species include Goosander, Woodcock, Great Spotted Woodpecker, Grey Wagtail, Dipper, Treecreeper, Siskin and Bullfinch. In spring and summer the woodland holds Common Redstart, Wood Warbler and Pied and Spotted Flycatchers, with Tree Pipit and Whinchat in the more open areas along the woodland fringes. The woodland–moorland interface is also favoured by the small population of Black Grouse, which are most likely to be seen around dawn during the lekking season in April and May. The moorland itself supports Twite, Raven, Common Buzzard and Golden Eagle (scan the ridge to the east of the reserve), while the loch should be checked for Red-throated and Black-throated Divers, Red-breasted Merganser and Common Sandpiper. Winter is very quiet, although the loch may hold wildfowl, including Goldeneye.

Access

From Aberfoyle take the B829 westwards towards Inversnaid. After 12 miles, turn left at the T-junction. The moorland section of the reserve can be accessed along the Upland Trail, which runs for 0.6 miles to the sheep fank, accessible from the car park near Garrison Farm, a little beyond the western end of Loch Arklet. For the woodland section of the reserve continue along the road for a further 0.75 miles to park at the Inversnaid Hotel. The West Highland Way runs parallel to the loch shore for the full length of the reserve (2.5 miles); following this from the north-west corner of the hotel car park for around 0.6 miles, the nature trail runs into the woodland (from just beyond the boathouse) and climbs to two viewpoints overlooking the loch before looping back down to the West Highland Way.

FURTHER INFORMATION

Grid ref: Inversnaid Hotel NN 336 088
RSPB Inversnaid: tel: 0141 3310993; email: inversnaid@rspb.org.uk

369 SLAMANNAN (AVON) AND FANNYSIDE PLATEAUS (Upper Forth and Clyde)

OS Landranger 65

A large flock of Taiga Bean Geese (*Anser fabalis fabalis*) winters in central Scotland between Glasgow and Edinburgh (about 4 miles south of Falkirk and 3.5 miles east of Cumbernauld).

Habitat

The geese roost on moorland or in fields.

Species

The Bean Geese usually arrive in late September or early October and depart by late February. Up to 300 may winter, and the flock originates in central Sweden. Other birds in the area include Pink-footed and Greylag Geese and sometimes Hen Harrier.

Access

The Taiga Bean Geese tend to be very elusive and sightings cannot be guaranteed. Do not leave the roads or enter fields and moorland ('muirs'); if you are lucky enough to locate the geese during the day it is essential to stay in the car as they will not tolerate an approach on foot. During the early part of the winter the flock stay together as one large group but by the end of November groups begin to break up to feed in different areas. The main roosting site is at the Fannyside Lochs (often on the smaller, east loch; park at the lay-by near the boating club). The regular daily routine is for the flock to flight out before dawn to a selected field, where they will spend the day; if disturbed, they move to another feeding site or onto one of the moorland areas, where they may loaf in the heather for hours at a time – a habitat in which they are almost invisible. After dusk the flock will return to their preferred roosting area and it is at this time that they are their most vocal, but they arrive to roost very late and views are limited to calling silhouettes and a splash. Note that during hard weather they may remain on their feeding areas all night.

FURTHER INFORMATION

Grid ref: Fannyside Lochs NS 804 736
Web: www.bean-geese.pwp.blueyonder.co.uk

370 LOCH LEVEN (Perth & Kinross) OS Landranger 58

East of Kinross, Loch Leven attracts large numbers of wintering wildfowl, especially Pink-footed Goose. The Loch is an NNR and the RSPB has a large reserve at Vane Farm. The best time for geese is October–March.

Habitat

The Loch covers c. 1,580 ha and has seven islands, which hold the highest concentration of breeding ducks in Britain, with over 1,100 pairs. Surrounded by farmland, the fields provide feeding areas for the thousands of geese that roost on the loch. At Vane Farm there are areas of marsh, lagoons and a scrape on the shore, while farmland, birch woods and heather moor cloak the slopes of the Vane, an 824-foot high hill.

Species

Numbers of Pink-footed Geese peak in late autumn and March, when up to 25,000 have been counted, and smaller numbers remain through the winter. They are joined by 1,000–5,000 Greylag Geese and c. 100 Whooper Swans; the geese and swans favour Vane Farm. Occasional parties of Canada, Brent, Barnacle and White-fronted Geese, or Bewick's Swans, are found in the goose flocks, and Bean Goose may occur February–April. There are often ten species of ducks wintering on the loch, including Goosander, Goldeneye, Wigeon, Shoveler, Gadwall, small numbers of Pintail and occasionally Smew. Other waterbirds include Great Crested Grebe, a few Slavonian Grebes, Cormorant, Golden Plover and Ruff. Common Buzzard, Sparrowhawk and Peregrine are resident in the area and may be joined in winter by Short-eared Owl, Hen Harrier, Goshawk and Merlin, and White-tailed Eagles have been recorded. There are often Siskins and Lesser Redpolls in areas of birch, and Long-eared Owl is occasionally found.

On passage the Loch shore at Vane Farm attracts waders, sometimes including Black-tailed

Godwit, Ruff, Spotted Redshank, Greenshank, and Green and Wood Sandpipers. Both Black-necked and Slavonian Grebes occur on migration, as does Marsh Harrier.

More than 1,000 pairs of ducks nest on St Serf's Island, using a colony of Black-headed Gulls as 'cover'. Mostly Mallards and Tufted Ducks, they also include Shelduck, Gadwall, Wigeon, Shoveler and Pochard. Ospreys are present from late March and regularly fish the loch. The breeding Black-headed Gulls may total 8,000 pairs and there are also a few pairs of Common Terns, as well as Oystercatcher, Ringed Plover, Common Sandpiper, Common Snipe and Common Redshank. Grasshopper Warbler sometimes breeds. The surrounding moorland has Red Grouse, Curlew, Whinchat and Raven, with Common Crossbill, Lesser Redpoll and Tree Pipit in the woods. An unusual sight is the handful of Fulmars that breed on nearby crags (8 miles from the sea).

Access

Loch Leven NNR Access to the shore is permitted at three places: Kirkgate Park (April–September boat trips from the jetty to Castle Island, tel: 01577 862670 for details), Burleigh Sands and Findatie. All are well signposted, with information boards at the car parks. Wildfowl, however, are best seen at Vane Farm. The Loch Leven Heritage Trail links the above car parks and the RSPB Vane Farm reserve.

RSPB Vane Farm Leave the M90 at junction 5 and take the B9097 Glenrothes road (signed RSPB), turning immediately left, then right, at the dog-leg where it crosses the B996 and continuing for 2 miles to the RSPB car park and visitor centre, which are to the south of the road. The reserve is open all year, and access is possible at all times to the car park, the Woodland Trail (which stretches for 1 mile from the car park to the top of the Vane), the Gillman Hide (overlooking a small scrape on the Loch's shore, this is reached via a tunnel

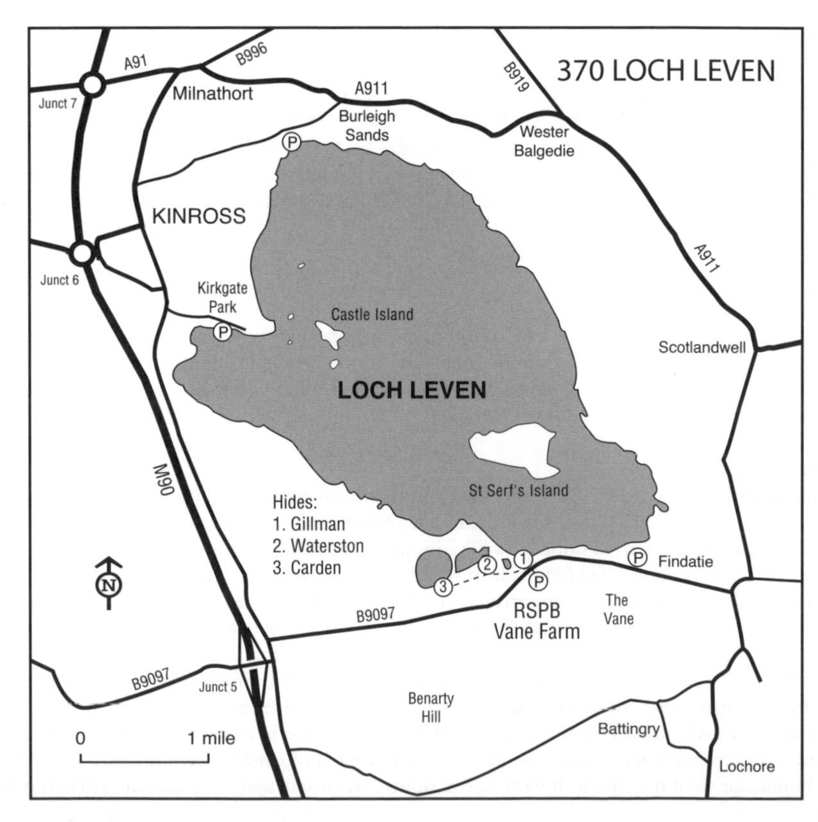

under the B9097 from the car park) and the adjacent Wetland Trail (wheelchair-accessible with care) which extends west for 0.5 miles taking in the Waterston and Carden Hides and more scrapes. The visitor centre has a coffee shop and comfortable observation room overlooking the loch and best geese fields, and is open daily 10am–5pm (closed Christmas and Boxing Days and 1–2 January). It is wheelchair-accessible.

FURTHER INFORMATION

Grid refs/postcode: Loch Leven NNR (Kirkgate Park) NO 127 018; RSPB Vane Farm NT 159 990 / KY13 9LX
Scottish Natural Heritage: tel: 01577 864439; web: www.snh.org.uk
RSPB Vane Farm: tel: 01577 862355; email: vane.farm@rspb.org.uk

371 LARGO BAY (Fife) OS Landranger 59

Largo Bay attracts large numbers of wintering seaducks, notably including regular Surf Scoter, and other interesting species include Black-throated Diver and Red-necked and Slavonian Grebes.

Habitat

The Bay extends for c. 30.5 miles and comprises sand and mud, with some rocky patches at Lower Largo. It is backed by sand dunes and urban and industrial development.

Species

On the sea, wintering Red-throated Diver are usually joined by up to ten Black-throated Divers and sometimes Great Northern, and there are usually also Great Crested, Red-necked and Slavonian Grebes. There are many Common Scoters and Common Eiders, and up to 300 Scaups, 300 Long-tailed Ducks (favouring the waters off Methil Docks) and 200 Velvet Scoters, as well as Goldeneye. Surf Scoter is annual, with up to eight present October–May, often just west of Ruddons Point. For divers, grebes and seaducks, calm conditions are best. On the beaches and rocks there are usually Grey Plover, Knot, Sanderling and Purple Sandpiper. Glaucous and Iceland Gulls are occasional, favouring Levenmouth. January–March sometimes produces records of Little Gull.

On passage there are numbers of Common, Arctic and Sandwich Terns, often a few Roseate, which can be seen on the beach almost anywhere in the bay, and sometimes Little Gull. Small numbers of Common Eiders and scoters remain all summer, with a large concentration of moulting birds in late summer, when Mediterranean Gull is also possible.

Access

Levenmouth A good area for seaducks. Park in the car park on the north-east side of the River Leven, just over the river from the former Methil Power Station, and look over the sea wall. Low tide is best for gulls. It is possible to walk along the shore for c. 4 miles from Leven to Ruddons Point, though Cocklemill Burn is difficult to cross at high tide.
Lower Largo Probably better for seaducks, with a possibility of Red-necked Grebe. View from the pier behind the Crusoe Hotel (by the harbour where the bridge crosses the Kiel Burn, reached from the A915 southwards along Harbour Wynd), or from the car park at the eastern end of Main Street.
Ruddons Point and Shell Bay Another excellent spot for seaducks, especially around high tide, and Black-throated Diver. Leave the A917 c. 1 mile north-west of Elie and drive south-west through Shell Bay caravan park for c. 2 miles to the end, from where it is a short walk to the Point (note that the gate to the caravan park may be closed, and it is then necessary to walk).

For waders, walk from the Point along the east shore of Shell Bay.

Elie In winter there is occasionally Glaucous Gull in Elie Harbour. Elie Ness, a small promontory, is the best place for seawatching around the Bay. Little Gull sometimes occurs on passage, as well as divers, Shag, auks and terns. There is direct access from the east side of Elie to the coastguards' lookout and lighthouse.

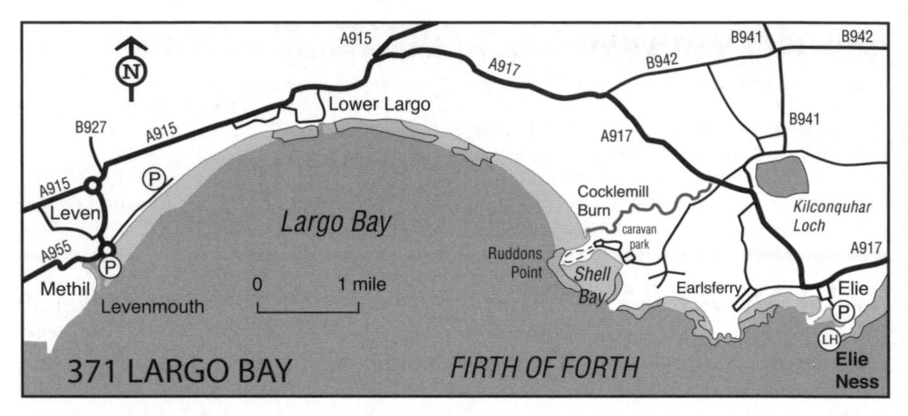

FURTHER INFORMATION

Grid refs: Levenmouth NO 381 004; Lower Largo NO 417 025; Ruddons Point and Shell Bay NO 469 004; Elie Ness NT 495 997

372 KILCONQUHAR LOCH (Fife)

OS Landranger 59

This small loch is famous for concentrations of Little Gulls in July–August, as well as holding many wintering and breeding wildfowl.

Habitat

The loch covers c. 55 ha. Shallow and fertile, it is largely surrounded by trees and, to the north, gardens.

Species

Kilconquhar Loch holds wintering ducks though, being shallow, it freezes quite quickly in hard weather. Notable are large numbers of Goldeneyes, immatures often staying into late spring and summer. Others include small numbers of Gadwalls and Shovelers. Greylag Geese roost on the loch, and there are occasionally parties of Barnacle Geese.

Numbers of Little Gulls visiting Kilconquhar Loch on passage in late July–August vary greatly. Only a few occur in some years and over 500 at once in others. The largest numbers generally occur when evening coincides with a high tide.

Breeders on the loch include Great Crested and Little Grebes and numbers of ducks, including Gadwall and Shoveler.

Access (see map above)

Leave the A917 north of Elie on the B941 to Kilconquhar. Park in the village and proceed to the churchyard, from where the loch can be viewed.

FURTHER INFORMATION

Grid ref: NO 485 020

373 ISLE OF MAY (Isle of May/Fife)

OS Landranger 59

The Isle of May lies at the entrance to the Firth of Forth. It has a large colony of seabirds, but the main attraction is migrants. It rose to fame through the work of Evelyn Baxter and Leonora Rintoul between 1907 and 1933, and the Bird Observatory was established in 1934. The island is now an NNR, and the best times to visit are late spring and autumn.

Habitat

The island covers just 57 ha and there is little vegetation; most of the cover is concentrated in the Observatory garden. The coastline varies from low on the east shore to cliffs topping 180 feet on the west.

Species

Small movements of diurnal migrants occur in spring and autumn, but potentially more interesting are the falls of night migrants in east or south-east winds during poor weather. These often include scarcer species: Wryneck, Bluethroat, Icterine, Barred and Yellow-browed Warblers, Red-breasted Flycatcher, Great Grey and Red-backed Shrikes, Common Rosefinch, and Lapland and Ortolan Buntings are almost annual, mainly in autumn. There are also large falls of birds destined to winter in Britain and Ireland, especially thrushes but also other common songbirds. These can be spectacular, for example there were 15,000 Goldcrests on one day in October 1982. The May has an excellent list of rarities.

Seawatching can be interesting. Gannet is almost always present offshore, and divers, Manx and Sooty Shearwaters, ducks, Arctic and Great Skuas, and Common, Arctic and Sandwich Terns are all regular. Few birds use the island in winter; most notable is a flock of several hundred Purple Sandpipers and Turnstones, and there are sometimes Black Guillemots or Little Auks offshore.

Breeding seabirds include Fulmar (280 pairs – all figures refer to 2007 unless specified), Shag (400–700 pairs), Puffin (68,000 pairs were counted in 2002!), Common Guillemot (c. 15,000 pairs), Razorbill (c. 2,700 pairs), Kittiwake (3,000–4,000 pairs), Herring Gull (2,800 pairs), Lesser Black-backed Gull (1,500 pairs), Great Black-backed Gulls (30 pairs), Arctic Tern (525 pairs), Common Tern (80 pairs), sporadically Sandwich Tern, as well as Common Eider (c. 900 pairs), Shelduck, Oystercatcher, Pied Wagtail and Rock Pipit

Access

Basic self-catering accommodation is available for up to six people, mid March–October (normally weekly), at the Observatory. The crossing from Anstruther on the Fife coast takes about 1 hour, and is organised when booking; delays are possible in bad weather.

Day trips are also possible: the *May Princess* sails daily from Anstruther April–September, and for seabirds it is best to visit between April and July.

FURTHER INFORMATION

The Isle of May Bird Observatory: web: www.isleofmaybirdobs.org; prospective visitors should contact the Bookings Secretary, Mike Martin, mwa.martin@btinternet.com, tel. 01738 633948
The *May Princess*: tel:01333 310 103; web: www.isleofmayferry.com
Scottish Natural Heritage: tel: 01334 654038; web: www.snh.org.uk

374 FIFE NESS (Fife)

OS Landranger 59

Fife Ness forms the east tip of the Fife peninsula, which explains its attractiveness to migrants. Both on land and sea, interesting birds can be seen. The Ness is worth visiting during spring and autumn, but only in a north-east, east, south-east or south-west wind, being otherwise usually very quiet.

Habitat

The area is a mosaic of farmland, small woods, a golf course and coastal scrub, bounded by a rocky coastline. As at any migration watchpoint, any patch of cover is worth checking. Particularly good areas are Denburn Wood, the walled garden and large trees at Balcomie Castle and Fife Ness Muir, the last-mentioned being the best. Inland of the Ness, Crail Airfield is worth a look for pipits, larks, thrushes, buntings, etc.

Species

On passage, the usual chats, thrushes, warblers and flycatchers are sometimes joined by scarce migrants. In spring, Long-eared Owl, Wryneck, Ring Ouzel, Black Redstart, Bluethroat, Firecrest and Red-backed Shrike are possible. Additional possibilities in autumn are Icterine, Barred, Pallas's and Yellow-browed Warblers, Red-breasted Flycatcher and Great Grey Shrike. Occasionally Snow or even Lapland Buntings join regular Golden Plover and Lapwing on the airfield in late autumn–winter.

Seawatching can be productive: Gannet and Shag are almost always present offshore, but in spring and autumn east-north-east winds are best, though too strong a north-east wind in spring may delay the passage of terns, which should include Common, Arctic, Sandwich and sometimes Little. Autumn brings Manx and (regularly) Sooty Shearwaters, Little Gull, terns and skuas – mostly Arctic, with some Great and small numbers of Pomarines. The small beach and pool attract passage waders.

In winter seawatching can produce all three divers, auks and seaducks. There are Purple Sandpiper and Turnstone on the rocks and large numbers of Golden Plovers on the airfield.
Summer is quieter, but large numbers of Common Eiders can still be seen, as well as some of the seabirds that breed locally – Gannet, auks, Kittiwake and terns. Sedge Warbler and Corn Bunting are fairly common breeders but Stonechat only sporadic.

Access

Please behave with the greatest consideration in this area. Park only at Kilminning or in the car park near the golf club. Around the golf course, keep to the road or coastal footpath. Always ask permission before entering private property, notably at Balcomie and Craighead Farms and when inspecting areas of cover around houses, respect peoples' privacy, especially in the early morning.

Denburn Wood This mature wood lies adjacent to the churchyard on the outskirts of Crail, and attracts migrant passerines filtering inland from the coast. There are a number of paths in the wood with free access from the road.

Kilminning coast and Crail Airfield At the sharp bend of the A917 in Crail turn north-east on the minor road (Marketgate becoming Balcomie Road) to Balcomie golf course. After c. 1 mile turn south onto the road for the go-karts; just off the main road are areas of cover attractive to migrants. Otherwise, follow the road past the entrance to the go-kart track and the abandoned airfield to park at Kilminning, from where there are paths down to the Kilminning Coast SWT reserve, an area of species-rich grassland that extends east to the 'Fish Factory' building and west back towards Crail. The old airfield may attract waders, pipits, larks, thrushes and buntings and migrants may be found anywhere along the coast – the Fife long-distance coast path can be followed from here to Fife Ness.

Balcomie Castle and Farm Members of the Fife Bird Club may access the walled garden to the south-east of the house; there is no other access to the area around Balcomie Farm.

Fife Ness Continue past Crail Airfield and Balcomie and on to Balcomie Links Clubhouse, where there is a car park. Balcomie Beach can be reached by walking east from the car park past the club house and down a small track by the practice putting green. The bushes either side of the track may hold migrants and the beach, just north of the point where the track reaches the coastal path, regularly attracts flocks of gulls and waders. To the south (and accessed from the road to the coastguard cottages), there is a seawatching hide below the coastguard station on Foreland Head; keys are available to members of the Fife Bird Club. Otherwise, it is possible to seawatch beside the hide and nearby pillbox; follow the coastal path around the front of the Coastguard Light.

Fife Ness Muir Known locally as 'the Patch', this small SWT reserve comprises the hill to the south-east of the club house, where the extensive areas of cover are attractive to migrants. There is an active ringing programme and information is available from the ringing hut. Access is only via the path that leaves the road between the golf club and the coastguard cottages, taking the footpath alongside the north side of Rose Cottage, up and over the hill, skirting the golf course and on to the south-east corner of Fife Ness Muir (note that there is no public access to Danes Dyke). It is best to park by the golf club house, as there is only very limited parking near Rose Cottage; do not park on the road, at the caravans, at the coastguard cottages or at the lighthouse.

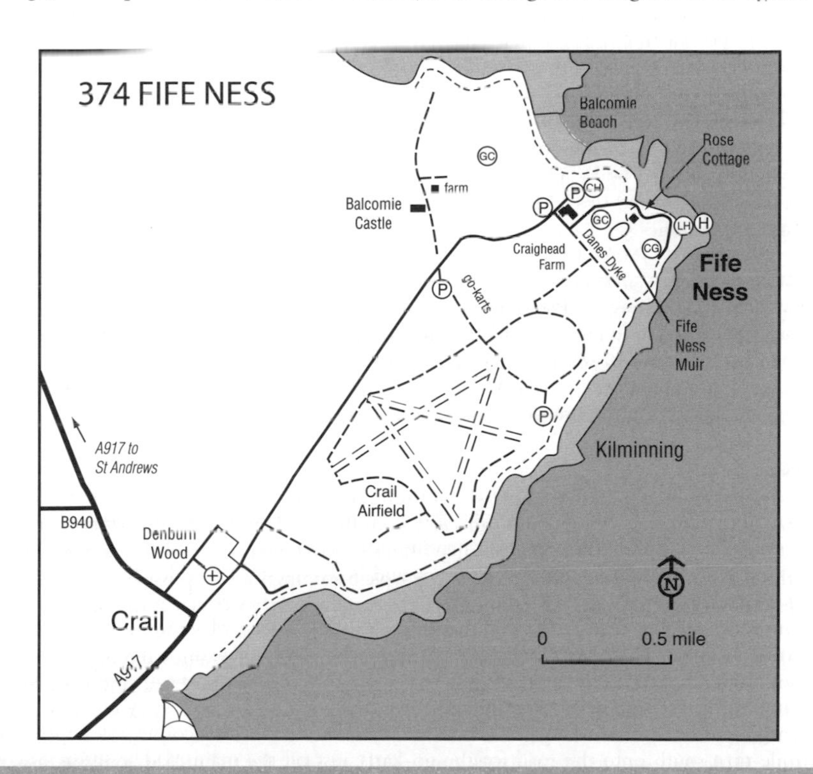

FURTHER INFORMATION

Grid refs: Denburn Wood NO 614 079; Kilminning coast NO 631 088; Fife Ness NO 631 099
Fife Bird Club: web: www.fifebirdclub.org
Scottish Wildlife Trust: tel: 0131 312 7765; web: www.swt.org.uk

375 EDEN ESTUARY (Fife)

OS Landranger 59

Good numbers of seaducks occur offshore at St Andrews, while the adjacent Eden Estuary is easy to work and holds, for Scotland, an excellent selection of waders, plus internationally important numbers of Shelducks and Red-breasted Mergansers. The entire area is good in winter and passage periods. Over 800 ha of the estuary are an LNR.

Habitat

The Eden Estuary is relatively small and is flanked at the mouth by areas of dunes. By contrast, the shore at St Andrews is very rocky.

Species

In winter, Red-throated and sometimes Black-throated and Great Northern Divers occur on the sea, together with Great Crested and occasionally Slavonian and Red-necked Grebes. There are large numbers of Common Eiders and Common and Velvet Scoters, Long-tailed Ducks and Scaups, while Surf Scoter is regular in late autumn–early winter. Other seaducks include Goldeneye, and the best area is often off the mouth of the estuary or occasionally off the Sea Life Centre at West Beach. The estuary shelters Wigeon, Teal, a few Pintails and large numbers of roosting Greylag Geese, with Goosander on the river. Waders such as Oystercatcher, Knot, Bar-tailed Godwit, Dunlin and Common Redshank are abundant, and there are up to 500 Grey Plovers (which are scarce in Scotland). Sanderlings are found around the estuary mouth, along with up to 100 Black-tailed Godwits and a few Ruffs and Greenshanks. The rocky coast at St Andrews is frequented by Purple Sandpiper and Turnstone. There are generally few raptors in the area, but Short-eared Owl, Merlin or Peregrine may be present, while Snow Bunting may be found in the dunes.

On passage, Little Ringed Plover may occur in spring in Balgove Bay, while Grey Plover, Black-tailed Godwit and Greenshank occur in larger numbers, and Little Stint, Curlew Sandpiper, Ruff and Spotted Redshank are usually seen a few times each autumn. There is a late-summer gathering of terns at Eden Mouth, and in early autumn these attract skuas, mostly Arctic. In late autumn, a few Barnacle Geese may pass through the area.

In summer, numbers of Common Eiders and scoters remain offshore, as well as a few waders, and both Peregrine and Osprey may visit.

Access

St Andrews Straightforward access to the seafront. A good place to watch from is an elevated position on East Scores Road, which runs east from the Castle in the north-east corner of the town. The rocks below the cliffs are viewable from the path between the castle and harbour, and from the harbour itself. Seaducks, grebes and Purple Sandpipers are all possible.

Out Head Access is from the A91 along Golf Place in the north-western corner of St Andrews (signed Royal & Ancient Golf Course/British Golf Museum/Sea Life Centre, and just east of the roundabout where the A91 joins the A915). Follow the road towards the beach and turn left in to West Sands Road (toll). Continue to the road's end near Out Head and park. Several paths lead to the beach and through the dunes to Out Head giving views of the mouth of the River Eden.

Balgove Hide Accessed from a car park on the A91, the key for the hide is available from the visitor centre, where up-to-date access arrangements can be checked. Balgove Bay is favoured by Pintail and Brent Geese, and it is possible to walk around the bay by keeping close to the shoreline.

Coble Shore Accessible north off the A91 c. 3.5 miles west of St Andrews, from the point it is possible to scan over the mudflats west towards Edenside and east towards Balgove Bay, while the Coble Shore Pools may attract waders.

Edenside The main concentration of birds in the estuary can be seen from Guardbridge, where the A91 crosses the River Eden. Park at the eastern end of the large lay-by just east of the bridge. A rising or high tide is best here, as the saltmarsh is used as a roost by most of the birds in the inner estuary; this is the best place to look for Black-tailed Godwit. The Fife Bird Club has a hide at Edenside Stables, giving good views of the inner estuary and the roost – access details and a key are available to club members.

Guardbridge Visitor Centre On the south side of Guardbridge Papermill there is a visitor centre that gives views of the inner estuary.

Tip Point Trail The north-west corner of the estuary can be viewed from this short path which leads east from the A919 in Guardbridge and follows the north bank of the Motray Water beside the Papermill. The birds may be quite close here, but the light is often poor.

Kincaple Den Wintering Greenshank, Black-tailed Godwit and Goosander are sometimes seen upriver of Guardbridge. Take the footpath signposted Kincaple Den to the river south off the A91.

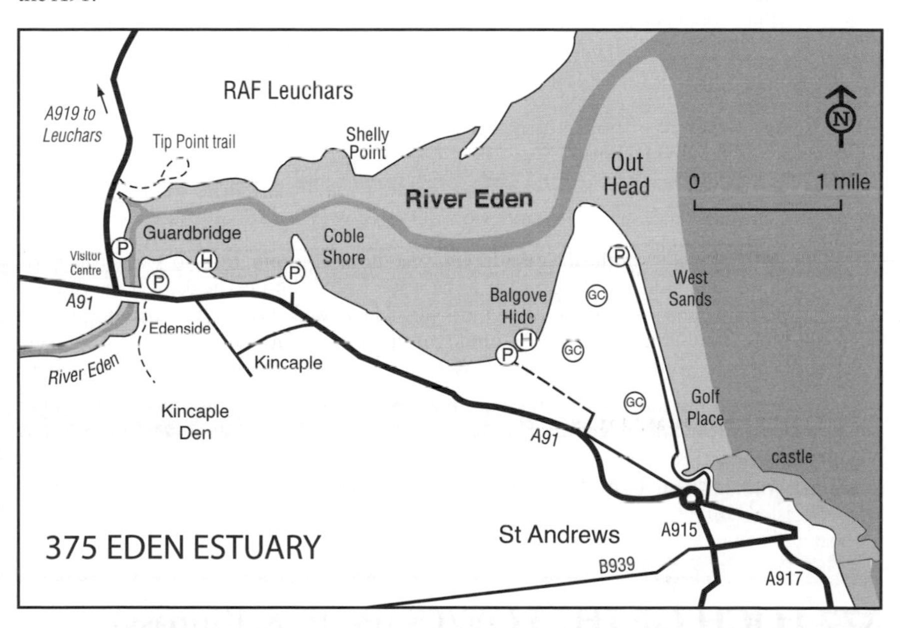

FURTHER INFORMATION

Grid refs: St Andrews (East Scores) NO 513 168; Out Head NO 498 191; Balgove Hide NO 494 173 (A91); Coble Shore NO 466 190; Edenside NO 453 188; Guardbridge Visitor Centre NO 451 191; Tip Point Trail NO 450 198
Fife Bird Club: web: www.fifebirdclub.org
Fife Coast and Countryside Trust Ranger Service: tel: 01334 473047 or 07985 707593; web: www.tentsmuir.org
Scottish Wildlife Trust: tel: 0131 312 7765; web: www.swt.org.uk

376 CAMERON RESERVOIR (Fife)

OS Landranger 59

Around 4 miles south-west of St Andrews, this water attracts wildfowl in winter, notably including both Pink-footed and Greylag Geese, while on passage a variety of waders has occurred.

Habitat

The reservoir covers c. 40 ha and is bordered in several places by stands of willows. To the south there is a narrow belt of conifers, while to the north lies grassland.

Species

Winter wildfowl include Whooper Swan, Wigeon, Pochard and sometimes Gadwall and Goldeneye. Pink-footed Geese roost at the reservoir in October–November. Numbers peaked at more than 16,000 in the 1990s but have fallen in recent years. Greylag Geese may also be present and the geese are best viewed in the hour before sunset from the hide (to reduce disturbance). Other possibilities in winter are Sparrowhawk and both Long- and Short-eared Owls.

On passage a variety of waders may occur, including Common and Green Sandpipers, Greenshank and Spotted Redshank.

Residents include Great Crested and Little Grebes, and Sedge Warbler breeds.

Access

Leave the A915 4 miles south of St Andrews west along a minor road to Cameron Kirk, turning south after 0.5 miles to the car park at the water bailiff's house. The dams at the east and west ends provide good vantage points and it is possible to walk around the reservoir, though the south shore is rather marshy. There is a hide on the north shore, but this is locked and keys are available only to Scottish Wildlife Trust (SWT) and Scottish Ornithologists' Club members.

FURTHER INFORMATION

Grid ref: NO 478 117
Scottish Wildlife Trust: tel: 0131 312 7765; web: www.swt.org.uk

377 LOCH OF THE LOWES (Perth & Kinross)

OS Landranger 53

Loch of the Lowes was the first Scottish Osprey eyrie away from Loch Garten to be widely publicised, and the birds can be seen between early April and August. The Loch is a Scottish Wildlife Trust reserve.

Habitat

The Loch of Lowes and adjacent Loch of Craiglush (connected by a canal) are shallow and very fertile, and the margins have areas of reed-grass, sedge and water lily. They are surrounded by mixed woodland.

Species

Apart from the Ospreys, raptors in the area include Sparrowhawk and Common Buzzard. Little and Great Crested Grebes, Teal, Common Sandpiper and Sedge Warbler breed around the Loch, and Slavonian Grebe has nested once. The surrounding woodland has Capercaillie and Black Grouse, though due to the restricted access these are unlikely to be seen. Resident Goshawk, Sparrowhawk, Woodcock, Green and Great Spotted Woodpeckers, Treecreeper, Lesser Redpoll, Siskin and Common Crossbill are joined in summer by Tree Pipit, Common Redstart, Garden Warbler and Spotted Flycatcher. In autumn over 1,000 Greylag Geese roost on the loch and in winter, if not frozen, the loch attracts Goosander, Goldeneye and Wigeon.

Access

Leave the A9 at Dunkeld on the A923 towards Blairgowrie and, after 1.5 miles, take the minor road on the right. The Loch is immediately north of this road, and access is restricted to the south shore adjacent to the road, where there are a number of lay-bys, and to the hide, visitor centre and car park at the west end. The visitor centre is open 10am–5pm April–September; the hide is always open.

FURTHER INFORMATION

Grid ref: NO 041 435
Scottish Wildlife Trust, Loch of the Lowes: tel: 01350 727337; email: lochofthelowes @swt.org.uk; web: www.swt.org.uk

378 PASS OF KILLIECRANKIE (Perth & Kinross)

OS Landranger 43

Lying close to the A9 between Pitlochry and Blair Atholl, on the west bank of the River Garry, this area holds an excellent selection of woodland species as well as some moorland birds. It is best in spring–early summer. Owned by the National Trust for Scotland (NTS), it was, until 2004, an RSPB reserve.

Habitat

Climbing from the dramatic narrow gorge of the River Garry, habitats include Sessile Oak woods, which rise to a plateau of pastureland, and then birch woods that climb further to expanses of moorland (up to 1,300 feet in altitude) with some crags.

Species

Residents include Common Buzzard, Sparrowhawk, Red Grouse, Black Grouse (ever-elusive, and best looked for in the mosaic of open ground and trees along wood and moorland edges), Woodcock, Green and Great Spotted Woodpeckers, Common Crossbill and Raven. In summer these are joined by Curlew, Common Snipe, Tree Pipit, Wheatear, Whinchat, Common Redstart, Wood and Garden Warblers and occasionally Pied Flycatcher. Common Sandpiper, Grey Wagtail and Dipper occur along the river. Golden Eagle and Peregrine are occasional visitors.

Access

From the A9 just north of Pitlochry, take the B8079 to Killiecrankie and follow the signs to the reserve, which is open at all times. There are several trails and the NTS visitor centre is open 10am–5.30pm from April to the end of October.

FURTHER INFORMATION

Grid ref/postcode: NN 917 627 / PH16 5LG
National Trust for Scotland, Killiecrankie Visitor Centre: tel: 01796 473233; web: www.nts.org.uk

379 LOCH OF KINNORDY (Angus & Dundee)

OS Landranger 54

This RSPB reserve is a good site for Ospreys and holds a selection of wildfowl and passage waders.

Habitat

Set in rolling farmland, this small, partially drained and very shallow, nutrient-rich loch is surrounded by marsh, willow and alder scrub, with Scots Pine on the higher ground.

Species

Breeding birds include Great Crested and Little Grebes and eight species of ducks, including Wigeon, Gadwall, Shoveler and Pochard, and Ruddy Duck has bred. Black-necked Grebes colonised the loch in 1989, reaching 11 pairs in 1994, but have now abandoned the site (and indeed, are almost lost as a Scottish breeding bird). Around 7,000 pairs of Black-headed Gulls nest on the floating mats of vegetation, while the surrounding marshes hold Water Rail, Common Snipe, Curlew, Common Redshank and Sedge Warbler. The woodland has Sparrowhawk, and Great Spotted and Green Woodpeckers. Ospreys are frequent visitors from mid-March through the summer.

On passage, waders such as Greenshank, Spotted Redshank and Ruff may pass through, and Black-tailed Godwits bound for Iceland are regular in spring. Marsh Harrier has summered and Peregrine may also visit.

From October up to 5,000 Greylag Geese use the Loch as a roost, together with small numbers of Pink-footed Geese and Whooper Swans, and occasionally also Barnacle Geese, while other wildfowl include Goosander and Goldeneye. The surrounding rough ground attracts hunting Hen Harrier and Short-eared Owl.

Access

Leave Kirriemuir west on the B951 towards Glen Isla, and the reserve lies immediately north of the road after 1.5 miles. From the car park paths lead to three hides a short distance away, and the trail (which is wheelchair-accessible) was recently extended eastwards to link the reserve to Kirriemuir. The reserve is open daily dawn to dusk (with some closures on Saturdays in September–October; phone to check).

FURTHER INFORMATION

Grid ref: NO 361 539
RSPB Loch of Kinnordy: tel: 01738 630783

380 DEESIDE (North-east Scotland)

OS Landranger 37, 43 and 44

The River Dee drains the south flank of the Cairngorm massif, and the forests and mountains of upper Deeside have a similar range of birds to Speyside with the notable exception of Crested Tit. They are, however, less disturbed and more convenient for the visitor to east Aberdeenshire. The best times to visit are March–April for residents and May–June for summer visitors.

Habitat

The Dee's wooded valley has areas of relict Scots Pine and birch woods. The surrounding slopes are bog and moorland, and as the river winds west the hills rise higher and eventually reach 3,924 feet at Beinn A Bhuird. In the valley bottom are Lochs Davan and Kinord.

Species

The high tops support Ptarmigan, which is often found around corries and moves lower in winter. It is joined in summer by Dotterel, which may be relatively numerous, and other breeding waders here and on the surrounding moorland include Golden Plover and a few Dunlins. Raven and Wheatear are widespread and Ring Ouzel quite common, but Snow Bunting is very rare – Beinn A Bhuird is probably the likeliest peak to support it. Raptors in the area include Golden Eagle, which prefers the more rugged and remote areas, Common Buzzard, Goshawk, Sparrowhawk, Hen Harrier and Peregrine, while in summer there may be Ospreys fishing the lochs or the River Dee, Merlin or even Honey-buzzard. Forest areas support Capercaillie and the woodland/moorland interface Black Grouse (but be very careful not to disturb either of these vulnerable species), and Woodcock is quite common. Among passerines, Siskin and Scottish Crossbill are resident, and Common Crossbill may breed in the conifer plantations. In summer there are also Tree Pipit, Common Redstart, Wood Warbler and Spotted Flycatcher. In the forest meadows are a few pairs of Greenshanks, and along the river and its tributaries look for Goosander, Common Sandpiper, Grey Wagtail and Dipper. These may also be seen on the lochs, together with Water Rail and Sedge Warbler.

Access

There may be restrictions on access in the deer-stalking season (July–February) and it is worth checking in advance. Exploration with the appropriate OS map is recommended. As in Speyside, visitors are asked not to search for the highly vulnerable Capercaillie and to remain on well-marked tracks and forest roads, especially during the breeding season.

Glen Tanar NNR (OS Landranger 44) Leave the B976 at Bridge o'Ess (just south-west of Aboyne) and follow the minor road for 1.5 miles to the car park at Braeloine. Continue on foot to Glen Tanar House, checking the river for Dipper, and from there follow any of the tracks into the forest. There is often Scottish Crossbill in the woods immediately beyond Glen Tanar House or on the hillock in the angle between the confluence of Water of Tanar and Water of Allachy. Capercaillie is possible, with Black Grouse in the more open woodland and scattered pines. Goshawk and Sparrowhawk can be seen over the woods, with Hen Harrier

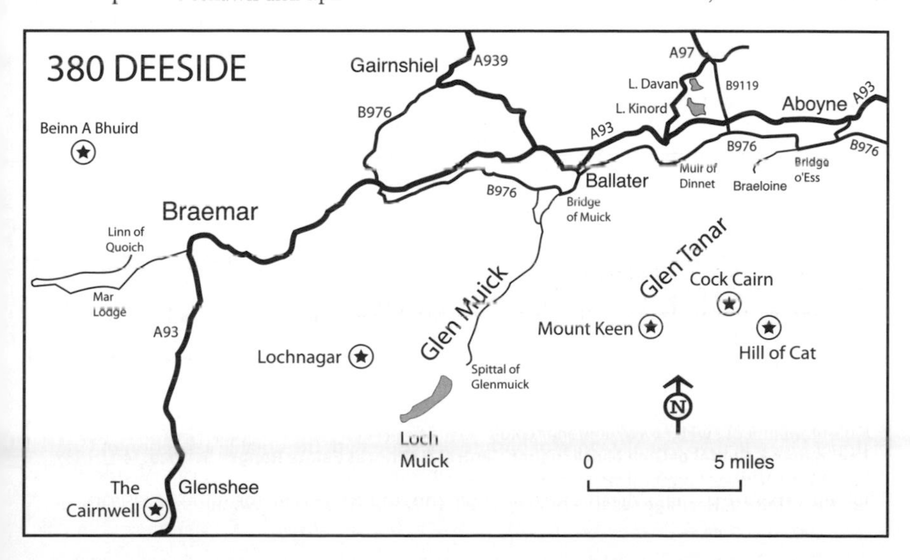

and Merlin in the open areas above the forest and Golden Eagle over surrounding ridges and summits. Mount Keen lies at the head of Glen Tanar and holds Ptarmigan. Follow the old high-level drove-road to the summit or, perhaps better, the Firmounth Road along the valley of Water of Allachy towards the high ground around Hill of Cat or Cock Cairn. Note that excursions on foot to the high tops are arduous and require proper preparation. The Braeloine Visitor Centre lies across the Water of Tanar from Braeloine and is open 10am–5pm (except Tuesdays April–September, except Tuesdays and Wednesdays October–March), providing information on the NNR and advice on access.

Muir of Dinnet NNR (OS Landranger 37 and 44) Covers c. 1,400 ha of woodland, bog and moorland surrounding Lochs Kinord and Davan, which hold Goosander and sometimes summering Goldeneye. In late autumn–winter Whooper Swan and up to 2,000 Pink-footed and 8,000 Greylag Geese use the lochs. The Burn O'Vat Visitor Centre, open at weekends, lies just off the B9119 at Milton of Logie (and is signed from the A93 Aberdeen–Ballater road). There is also a car park just east of New Kinord, with smaller parking areas on the B9119 and on the A97 to the west of Loch Kinord and Loch Davan, and an extensive network of trails.

Gairnshiel (OS Landranger 37) A good area for Black Grouse. Leave the A93 just west of Ballater north on the A939 and, after c. 5 miles, park just after the road crosses the River Gairn near Gairnshiel Lodge. The grouse may be seen lekking in the fields north-east of the river.

Glen Muick and Lochnagar (OS Landranger 44) Leave the B976 just south of Ballater at Bridge of Muick and follow the road to the car park at Spittal of Glenmuick, 8 miles from the junction. The Lochnagar & Spittal of Glen Muick Visitor Centre is open daily, 8am–6pm. From here you can walk to Lochnagar (a fairly arduous climb) where there are Ptarmigan, Red Grouse, Dotterel, Peregrine, Merlin, Hen Harrier and Golden Eagle. Glen Muick is an excellent area. Black Grouse is possible and, if you are very lucky, you may see Honey-buzzard or Wryneck in late spring. A good walk is to cross the River Muick north-west of Spittal to Allt-na-giubhsaich and follow tracks through the wood on the west bank of the river, this being a good area for Black Grouse. Another worthwhile route with fairly easy gradients is the old drove-road east of Loch Muick running south to Glen Clova.

Glenshee (OS Landranger 43) Leave Braemar south on the A93 Old Military Road and park after c. 7 miles at the car park near the foot of the ski-tows. This is an excellent site for Ptarmigan in winter, which can be seen with a telescope to the west on the slopes of The Cairnwell. In cold weather, they may descend and can then be seen close to the car park (sometimes from the car!).

Mar Lodge (OS Landranger 43) Leave Braemar west on the minor road paralleling the River Dee on the south shore. Park along the road or, after 3.5 miles, cross the river at Victoria Bridge and park at Mar Lodge. Re-crossing the river on foot, Craig an Fhithich to the west is good for Peregrine. Scottish Crossbill can be found anywhere from Victoria Bridge west to the Linn of Dee in areas of Scots Pine, and is often quite easy to find.

Beinn A Bhuird (OS Landranger 43) From Mar Lodge drive east for 1.5 miles to Linn of Quoich and park at the Punch Bowl. There is usually a barrier across the road here, but you can walk west and north along the track to the foot of the mountain. Black Grouse should be looked for on both sides of Glen Quoich (especially the north-east), and Capercaillie is possible. Continue (well prepared) to the tops for Golden Eagle, Ptarmigan and Dotterel.

FURTHER INFORMATION

Grid refs/postcode: Glen Tanar NNR (Braeloine) NO 480 965; Muir of Dinnet NNR (The Burn O' Vat) NO 429 997 / AB34 5NB; Gairnshiel NJ 295 008; Glen Muick and Lochnagar (Lochnagar & Spittal of Glen Muick Visitor Centre) NO 310 851; Glenshee NO 139 780; Mar Lodge NO 098 899; Beinn A Bhuird (Punch Bowl) NO 119 911

Glen Tanar Ranger Service, Braeloine Visitor Centre: tel: 01339 886072; email: ranger@glentanar.co.uk; web: www.glentanar.co.uk

Lochnagar & Spittal of Glen Muick Visitor Centre, Balmoral Estate Ranger Service: tel: 01339 755059; web: www.balmoralcastle.com

Scottish Natural Heritage reserve manager: tel: 01224 642863; web: www.nnr-scotland.org.uk

381 FOWLSHEUGH (North-east Scotland)

OS Landranger 45

Fowlsheugh holds one of the largest seabird colonies in the country and is readily accessible. The best time to visit is April to mid-July, and 1.5 miles of cliff are an RSPB reserve.

Habitat

The grass-topped sandstone cliffs reach over 200 feet high in places.

Species

Fulmar, Shag and Common Eider are joined by nearly 30,000 pairs of Common Guillemots, 5,000 pairs of Razorbills and small numbers of Puffins. The 30,000 pairs of Kittiwakes vastly outnumber the few Herring Gulls. Offshore there may be passing Gannets and skuas. Outside the breeding season there is little of special interest.

Access

Fowlsheugh is signposted east off the A92 at the Crawton turning, 4 miles south of Stonehaven. Park in the car park at Crawton at the end of the road and follow the stepped path down to the cliffs; from here there are good views of the colonies at several points on the 0.75-mile trail (Puffins are most likely to be seen at the furthest point on the trail).

FURTHER INFORMATION

Grid ref: NO 879 808
RSPB Fowlsheugh: tel: 01346 532017; email: strathbeg@rspb.org.uk

382 GIRDLE NESS (North-east Scotland)

OS Landranger 38

This promontory, south of the mouth of the River Dee and immediately east of the conurbation of Aberdeen, is a useful seawatching station and also attracts a wide range of passerine migrants, regularly including rarities.

Habitat

Most of the promontory is a golf course with some areas of gorse scrub, flanked allotments and gardens, all of which are bounded by rocky cliffs.

Species

Seawatching in autumn can be good. Manx and Sooty Shearwaters are frequent offshore, especially in onshore (north-east) winds. Great and Arctic Skuas are common, and there are regularly Pomarine and occasionally also Long-tailed. Passage waders may include Ruff, Black-tailed Godwit and Curlew Sandpiper, and Barnacle Goose may pass through in October. In late autumn–winter divers and numbers of seaducks, particularly Goldeneye and Common Eider, are usual (the sea just off the harbour should be checked, especially after easterly storms). Glaucous Gull is regular and Iceland Gull occasional, as well as Little Auk, with Purple Sandpiper and Turnstone on the rocks.

In spring and autumn there may be lots of Wheatears on the golf course, and interesting migrants can occur among the commoner species: Wryneck, Bluethroat and Red-backed Shrike are near annual at both seasons, with Yellow-browed, Barred and Icterine Warblers and Common Rosefinch possible in autumn.

Access

From Aberdeen city centre, take the A956 Market Street south and cross the River Dee via Victoria Bridge, continuing into Victoria Road and then turning left at the T-junction into St Fitticks Road and right into Greyhope Road, following this along the edge of the golf course to the lighthouse. There are two car parks on the Ness, both of which can be used for seawatching.

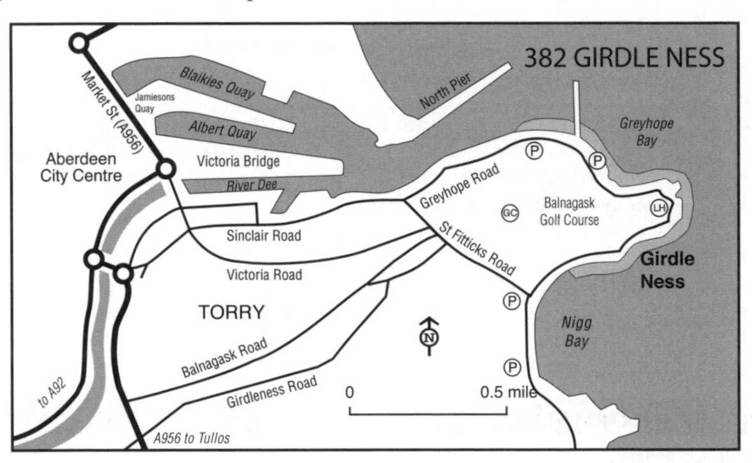

FURTHER INFORMATION

Grid ref: NJ 968 055

383 BRIDGE OF DON (North-east Scotland)

OS Landranger 38

This area, just north of Aberdeen, attracts numbers of diving and seaducks in winter and concentrations of feeding and loafing terns in summer. As with any site on this east-facing coast, it can also hold passerine migrants in spring and autumn.

Habitat

Sandy beaches and low dunes. Donmouth is an LNR.

Species

In winter off the Don Mouth many Goldeneyes and Goosanders are present, and sometimes up to 15 Scaup and a few Long-tailed Ducks. Red-throated Diver is also numerous offshore, and the river mouth is a likely spot for Glaucous Gull and sometimes also Iceland Gull in late winter. To the north, at Mains of Murcar, numbers of scoters and Common Eiders may be found (as at Blackdog, see below).

In summer there may be Goosander on the river and concentrations of Common, Arctic and Sandwich Terns, which in late summer attract Arctic and Great Skuas. Little Ringed Plover (a scarce bird in Scotland) may occur around the estuary or on the local playing fields. Offshore in autumn, winds between the north and east may prompt a passage of Manx Shearwater, Gannet, Kittiwake and skuas.

During passage periods numbers of migrants can sometimes be found in the riverside bushes. Waders too can be interesting at the river mouth.

Access

Don Mouth, north bank A minor road runs alongside the north bank of the river from the A92 immediately north of the bridge, with a car park at its east end giving views of the river mouth.
Don Mouth, south bank A minor road (Beach Esplanade) runs alongside the south bank of the river from the A92 immediately south of the bridge, with plenty of roadside parking. From here, you can walk to the river mouth or follow the foreshore south to Aberdeen beach.
Mains of Murcar Turn east off the A92, 300 yards north of the Don Bridge, into Donmouth Crescent (directly opposite its junction with the B997), and immediately fork left into Links Road towards the golf course; follow the track for 1.3 miles to its end, proceeding to the shore.

FURTHER INFORMATION

Grid refs: Don Mouth, north bank NJ 949 094; Don Mouth, south bank NJ 951 092; Mains of Murcar NJ 958 115

384 BLACKDOG (North-east Scotland)

OS Landranger 38

This area, just a few miles north of Aberdeen, holds good numbers of moulting seaducks in late summer and autumn, sometimes including a King Eider or Surf Scoter. At this time of year, it is better for these species than the Ythan Estuary to the north.

Habitat

A sandy shore backed by dunes and a golf course.

Species

This is an excellent area for concentrations of moulting seaducks late June–early September, when there are up to 10,000 Common Eider, 2,000 Common Scoter and 250 Velvet Scoter, also sometimes a King Eider or Surf Scoter. Small numbers of Scaup occur in autumn, and the concentrations of loafing terns attract Arctic and Great Skuas.

Numbers of seaducks are much lower in winter, but the commoner species are still present, together with large numbers of Red-throated Diver, Red-breasted Merganser and Long-tailed Duck. On the shore, look for Sanderling, with Snow Bunting in the dunes.

Access

Blackdog lies east of the A92, 5 miles north of Aberdeen, and a track runs from the main road (just north of Blackdog Burn) east towards the shore, with limited parking around the houses at the end. It is then a short walk to the beach.

FURTHER INFORMATION

Grid ref: NJ 958 140

385 YTHAN ESTUARY AND THE SANDS OF FORVIE (North-east Scotland) OS Landranger 30 and 38

The Ythan lies 12 miles north of Aberdeen, and together with the sea and adjacent Sands of Forvie, is an excellent area for large numbers of wildfowl, waders and terns, with the largest colony of breeding Common Eiders in Britain. Notably, at least one male King Eider may be present April–June (sometimes from November; they move to Blackdog in late summer to moult, see above). The area is worth a visit at any time of year, and much of the area is part of an NNR.

Habitat

The estuary is c. 5 miles long but only a few hundred yards wide. Flanked for half its length by the A975, it is surprisingly undisturbed and very good views are possible. The extensive mussel beds at the mouth provide food for the hordes of Common Eider, while there are small saltmarshes in the middle reaches. The Sands of Forvie, an area of dunes and moorland, lie between the river and the sea. North of Rockend the coast is flanked by cliffs, rising to 130 feet at Collieston. Four small lochs, Meikle, Little, Cotehill and Sand, lie in the north of the area.

Species

Wintering wildfowl include up to 250 Whooper Swan, which feed on the estuary and surrounding fields. Tens of thousands of Pink-footed Geese occur in autumn, together with large numbers of Greylag (the latter peak in spring); most geese roost on Meikle Loch. Passage

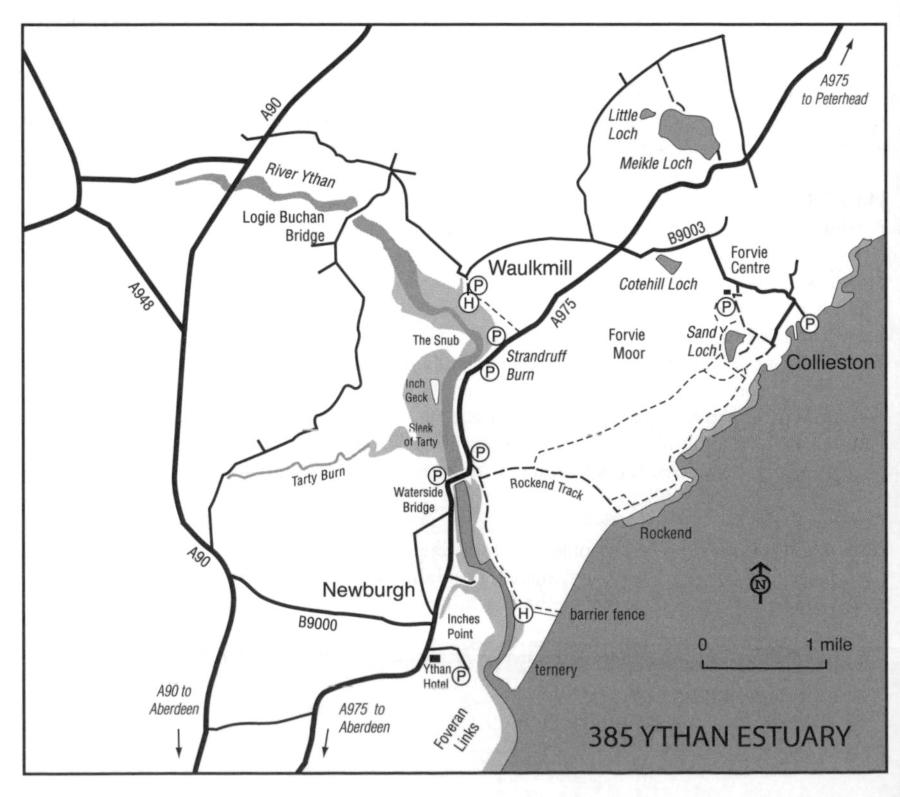

geese may include a few Brent, Barnacle, White-fronted and occasionally a Snow Goose. And, although there are far fewer geese in winter, the numbers of both Greylag and Pink-feet around the estuary and fields are still significant. Other wildfowl include Teal and Wigeon, and sometimes Gadwall, Shoveler and Ruddy Duck on the lochs. Offshore, there are small numbers of Long-tailed Duck, Common and Velvet Scoters, Goldeneye and Red-breasted Merganser, and the 1,000 Common Eider in early winter reach the full summer strength of over 5,000 by February. Red-throated Diver is also present. Waders include Sanderling and sometimes small numbers of wintering Ruff and Jack Snipe. On the beach and dunes there are often Snow Buntings, and raptors may include Peregrine and Merlin, and sometimes Hen Harrier or Short-eared Owl.

In summer there may still be several hundred Common Scoter offshore, often joined by Velvets, and occasionally a King Eider is found (usually in spring or early summer). Over 2,000 Common Eider breed on Forvie Moor, and feeding and loafing Common Eiders concentrate on the mussel beds of the lower estuary. Common, Arctic, Little and c. 1,500 pairs of Sandwich Terns breed, together with Black-headed Gull. The terns attract skuas, and both Great and Arctic are frequently present in small numbers. On the cliffs to the north there are small numbers of Fulmar, Kittiwake, Herring Gull and a few Razorbills, with Shelduck, Oystercatcher, Ringed Plover and Curlew on the dunes and moorland, and Red Grouse are resident on the moor.

A variety of waders is recorded in spring and autumn, including Greenshank and Ruff, and occasionally Little Stint, Curlew Sandpiper, Black-tailed Godwit, Spotted Redshank and Green Sandpiper, but by English standards the variety is usually quite poor. Ospreys regularly pass through in the late summer and autumn. Occasionally Garganey is recorded in spring, especially on Cotehill Loch. Little Gull and Black Tern, which favour Meikle Loch, are also possibilities. In late spring and autumn, south-east winds, especially if combined with rain or poor visibility, may bring falls of migrants, which occasionally include some of the scarcer species.

Access

There is open access to the NNR, apart from the ternery and moorland areas in April–August, and you should keep to the paths. The period 2 hours either side of high water (but not high tide itself, which is an hour later at Waterside Bridge than at the estuary mouth), is best for waders, while low tide is best for Common Eiders.

Ythan Estuary, south shore Access to the south side of the estuary is from Newburgh. Turn off the A975 just south of the village by the Ythan Hotel (signed to the beach) and continue to the car park near the golf course. Walk over the dunes to the mouth of the river, from where you can view the terneries in summer, while in winter this is an excellent spot to see divers and seaducks at sea. This is a good spot for the King Eider around high water, and is favoured by Snow Bunting in winter. Areas of bushes, which attract passage migrants, lie around the car park and golf course, in the dunes at the estuary mouth and further south on Foveran Links.

Inches Point, Newburgh The drake King Eider, if present, can often be seen in the area north of Inches Point around low water, which is reached via Inches Road off the A975 in Newburgh. Alternatively, view from the footpath on the opposite side of the estuary, running south from Waterside Bridge.

Waterside Bridge There are good views of the estuary from the A975. Heading north from Newburgh, park at the west end of Waterside Bridge to view the Sleek of Tarty and Inch Geck, which is used by roosting waders. A track follows the shore of the Sleek west from the bridge, giving the best chance of interesting waders. Continuing north there are more parking places affording views of the upper estuary.

Waulkmill Hide Turn north-west off the A975 c. 2 miles north of Waterside Bridge (opposite the B9003 turning) and after c. 1.5 miles turn left on a track along the east side of the Burn of Forvie, reaching the car park and hide after 200 yards, with good views over the inner estuary.

Logie Buchan Bridge Accessible on minor roads off the A975, this is a good spot for wintering Water Rail.

Sands of Forvie, north section Turn off the A975 on the B9003 to Collieston. Cotehill Loch is adjacent to the road soon after the turn and is good for ducks and, if the water level is low, passage waders. Nearer Collieston, the Stevenson Forvie Centre, the NNR HQ (open daily April to October), has a car park. A path leads from here past Sand Loch to the coast and Rockend Track. Sand Loch often has breeding Great Crested Grebe.
Sands of Forvie, ternery April–August the south end of the NNR is fenced off to protect breeding terns. These can be watched from a hide, reached by walking south from Waterside Bridge.
Collieston Take the B9003 to Collieston. The village gardens, churchyard, allotments and coastal scrub can attract migrants, and the height of the cliffs makes this a good spot from which to seawatch.
Meikle Loch This is the main goose roost, and is also good for ducks. Continue on the A975 for 1 mile past the B9003 turning and take a track on the left just before the crossroads. To avoid disturbing the geese, stay on this track. Small numbers of passage waders regularly include Ruff and Green Sandpiper, and there is further access via the track along the east shore.

FURTHER INFORMATION

Grid refs: Ythan Estuary, south shore NK 002 247; Inches Point, Newburgh NK 003 256; Waterside Bridge NK 001 267; Waulkmill Hide NK 003 290; Logie Buchan Bridge NJ 991 299; Sands of Forvie NK 033 289; Collieston NK 042 286; Meikle Loch NK 032 304
Scottish Natural Heritage Reserve Manager: tel: 01358 751330; web www.snh.org.uk

386 THE CRUDEN BAY AREA (North-east Scotland)

OS Landranger 30

The cliffs on this stretch of the north Aberdeen coast hold some notable colonies of breeding seabirds, while areas of cover are attractive to migrants.

Habitat

Along the coast pink granite cliffs rise to almost 200 feet, and the Bullers of Buchan is a collapsed sea cave forming an almost circular chasm (the 'pot') c. 100 feet deep, where the ocean rushes in through a natural archway.

Species

Breeding seabirds at Bullers of Buchan and Whinnyfold include good numbers of Kittiwakes and Common Guillemots, with smaller numbers of Razorbills, Puffins, Fulmars, Herring Gulls, Great Black-backed Gulls and Shags. There are usually Common Eiders offshore, as well as Gannets. Puffins are often most obvious in the late afternoon and evening, when loafing outside their burrows.

The sheltered wooded valley at Cruden Bay acts as a migrant trap in spring and autumn, and is one of the best places to see migrant passerines at both seasons. In addition to common species such as Common Redstart, Ring Ouzel and Pied Flycatcher, a number of scarce migrants may be found in the area including Wryneck, Richard's Pipit, Bluethroat, Barred, Pallas's and Yellow-browed Warblers (the latter is regular), Firecrest, and Red-backed and Great Grey Shrikes. National rarities are also seen from time to time. Tree Sparrows have been recorded breeding.

Access

Bullers of Buchan Leave the A975 c. 1 mile north of Cruden Bay on a short track east to the cliffs at Robie's Haven. From the car park a short road leads to a small group of cottages above the bay; a rough footpath leads northward in front of the cottages for 100 yards to the 'pot' and then continues northwards along the coast through the Longhaven Cliffs Scottish Wildlife Trust Reserve (also accessible down a track off the A90 at Station Farm, c. 0.5 miles north-east of the junction with the A975).

Whinnyfold Leave the A975 c. 0.75 miles south of Cruden Bay on a minor road to Whinnyfold, and walk south from the village along the cliff path to view the colonies.

Cruden Bay The small wood along the stream near the harbour attracts birds, as do the gardens in the village. Access to the wood is from the village (park at the first car park) or on a track south off the A975 just north of Cruden Bay, immediately beyond a sharp right-hand bend (difficult to find).

FURTHER INFORMATION

Grid refs: Bullers of Buchan NK 106 380; Whinnyfold NK 081 332; Cruden Bay NK 092 361

387 LOCH OF STRATHBEG (North-east Scotland)

OS Landranger 30

Midway between Fraserburgh and Peterhead, the Loch of Strathbeg is Britain's largest dune-slack lake and forms part of a large RSPB reserve. Formed in 1720 when a great storm moved millions of tons of sand to block the mouth of the bay, the loch lies just 0.5 miles from the sea. It is used by a variety of wintering wildfowl and is excellent for passage waders. Adjacent Rattray Head is a good spot in winter for all three divers, and in spring and autumn the area attracts small numbers of migrants.

Habitat

The Loch covers c. 225 ha and is very shallow – a small rise in the water level can flood the surrounding marshes and farmland. Around it are areas of freshwater marsh, large stands of reed-grass and reed and scattered patches of willow carr and fen woodland; management has including the re-routing of the feeder stream, the Savoch Burn, through 23 ha of reedbed. To the seaward, the dunes of Back Bar lead to the low-lying Rattray Head.

Species

Huge numbers of Pink-footed Goose pass through on migration, with up to 80,000 present in the area in late September and October; the majority move on southwards to Norfolk, but good numbers remain through the winter and there is a marked spring passage as they return to Iceland in late March and April. The Pinkfeet roost on the loch and in the dunes to seaward and flight out to feed at dawn; the Tower Pool Hide gives panoramic views of the area but to see this impressive spectacle an early start is necessary. Up to 10,000 Greylag Geese may also be present, and small numbers of Barnacle Geese also pass through in early October, together with a few Pale-bellied Brent and Greenland White-fronted Geese. Some Barnacle Geese also winter, while one or two Snow Geese are regularly present and several hundred Whooper Swans use the Loch in winter, peak numbers occurring in November; sometimes there are also a few Bewick's in early winter. Common Eider is resident and up to 500 Goldeneye winter, together with large numbers of Wigeon and Teal. Goosander occur for much of the year, with Red-breasted Merganser on passage, a few Gadwall are regular, and there are occasionally Long-tailed Duck and Scaup on the loch, together, especially in late winter, with Smew. Sparrowhawk, Hen Harrier, Peregrine, Merlin and Short-eared Owl are regular and White-tailed Eagle has been recorded. Bitterns are irregular winter visitors from November (try the Fen Hide, especially during periods of cold weather), and Tree Sparrows are regular at the feeders by the visitor centre. All three divers winter off Rattray Head, as well as Common Eider and Long-tailed Duck, and there are frequently Snow Bunting on the shore and dunes.

Breeding birds include Fulmar in the dunes, Common Eider, Shelduck, Shoveler, Teal and, in 1984, Ruddy Duck, while Garganey has also bred. Others are Water Rail at the reserve, Black-headed Gull, Common Tern and (irregularly) large numbers of Sandwich Terns. Grasshopper and Sedge Warblers frequent the surrounding marsh and carr with Tree Sparrows in the adjacent farmland. Both Black-tailed Godwit and Marsh Harrier are regular in summer, and Osprey annual on passage. A variety of waders occurs on passage, notably large numbers of Ruff and Black-tailed Godwit, as well as Bar-tailed Godwit, Spotted Redshank, Greenshank, Green and Curlew Sandpipers and Little Stint; Dotterels are annual in late May, and Avocet, Little Egret and Spoonbill have also been regular in recent springs. Wood Sandpiper and Temminck's Stint are also annual and Pectoral Sandpiper nearly so and other migrants can include Little Gull and Black Tern, the latter scarce in north-east Scotland. At Rattray Head a north-west or south-east wind may produce Manx and Sooty Shearwaters and skuas in August–September.

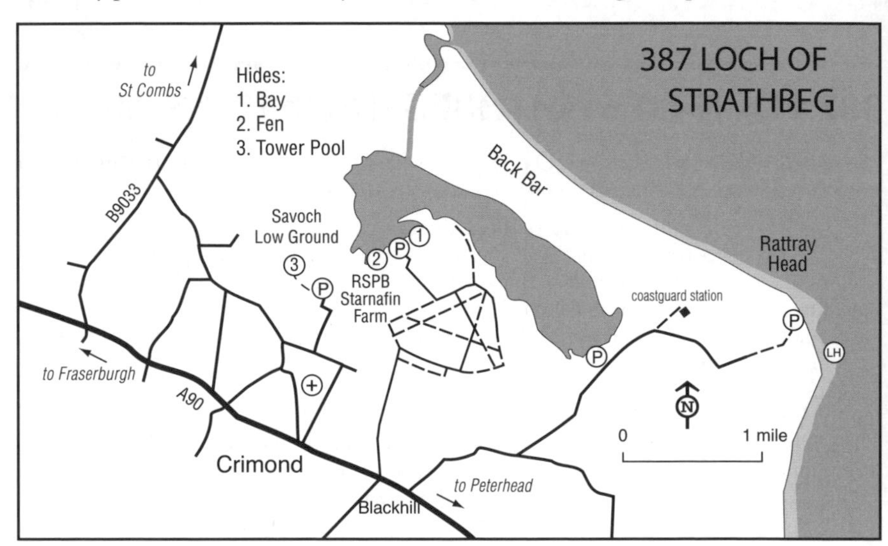

Access

RSPB Starnafin Farm Visitor Centre The Loch lies north-east of the A952 and the reserve centre is signed off the A952 in Crimond (take the minor road northwards, just east of the church, turn left at the T-junction and then, after 600 yards, turn first right to follow the track to the car park). The visitor centre is open daily 8am–6pm (or dusk if earlier) and there are views out over a selection of purpose-built wader scrapes. Tower Pool Hide, 750 yards from the visitor centre, gives a panoramic view of the main goose roosting and feeding areas and is open dawn to dusk. All trails and hides are being upgraded to improve accessibility for disabled visitors.

Loch of Strathbeg, Fen and Bay Hides For access to the loch-side Fen and Bay Hides turn north off the A952 0.5 miles south-east of Crimond church at the sign for Crimond Airfield. At the airfield cross the cattle grid to the left of the barrier and follow the RSPB signs (there are several rows of large wooden bollards along the track, but these can all be passed through safely – look for the signs or highlighted areas to pass each row); Fen Hide is 400 yards and Bay Hide 250 yards from the car park, and both hides are open dawn to dusk.

Loch of Strathbeg, south sector The south of the Loch can be viewed from the unclassified road that leaves the A952 north-east between Blackhill and Crimond; a short track leads towards the south corner.

Rattray Head coastguard station The coastguard's cottages have small areas of cover, which should be checked for migrants after south-east winds in spring and autumn, preferably in the early morning. Except during foggy weather, migrants tend to disperse quickly.
Rattray Head Continue to the lighthouse and Rattray Head, and walk either north or south along the shore, looking out to sea for divers.
St Combs To explore some of the wilder areas of the reserve park in St Combs and take the track behind the Tufted Duck Hotel into the north end of the reserve; it is possible to walk the entire length of the reserve through the fields or along the dunes and beach.

FURTHER INFORMATION

Grid refs: RSPB Starnafin Farm Visitor Centre NK 057 581; Loch of Strathbeg, Fen and Bay Hides NK 062 564; Loch of Strathbeg, south sector NK 084 574; Rattray Head NK 105 579; St Combs NK 057 629
RSPB Loch of Strathbeg: tel: 01346 532017; email: strathbeg@rspb.org.uk

388 FRASERBURGH AND KINNAIRD HEAD (North-east Scotland)

OS Landranger 30

This area of north-east Scotland is excellent for white-winged gulls, and has great potential as a seawatching station in autumn.

Habitat

Fraserburgh is a busy fishing town, bounded at its northern edge by the high Kinnaird Head.

Species

In winter Fraserburgh attracts wintering Glaucous and Iceland Gulls, sometimes in numbers. There is usually Black Guillemot on the sea and lots of Purple Sandpipers. At Kinnaird Head there are divers, often including Great Northern and Black-throated, and Black Guillemot among the masses of Common Guillemots and Razorbills.

In autumn Sooty Shearwater is seen regularly off the Head, and the large shearwaters and rarer skuas should also be possible.

Access

Fraserburgh Explore the coast from Fraserburgh Harbour west to Broadsea, where the two fish factories and sewage outfall are especially attractive to gulls. Access is along a maze of roads north off the A98 in the town.
Kinnaird Head This, the north tip of Fraserburgh, is a good spot to seawatch from in north-west and south-east winds.

FURTHER INFORMATION

Grid ref: Kinnaird Head NJ 998 675

389 TROUP HEAD (North-east Scotland)

OS Landranger 30

Lying around 8 miles west of Fraserburgh, this RSPB reserve holds Scotland's only mainland Gannet colony as well as a good variety of other species, with over 150,000 breeding seabirds present in the spring and summer.

Habitat

The cliffs at Troup Head are dramatically high and precipitous, topped with cropped grass.

Species

As well as 1,500 pairs of Gannets (there were just four in 1988), which return to the cliffs as early as mid-January, breeding birds include Fulmars, Common Guillemots, Razorbills, Kittiwakes and Herring Gulls, with smaller numbers of burrow-nesting Puffins. Shags, Common Eiders and occasionally Great Skua can be seen offshore.

Access

The reserve is signed north off the B9031 Macduff–Rosehearty road between Pennan and Gardenstown. Follow the signs to Northfield Farm, continue through the farmyard and on to a rough track to the reserve car park. Take the right-hand track through the gate and follow the fence line north, through a field and stands of gorse and on through a second gate, bearing left up and over the hill (still following the fence line) until you reach the cliffs. The track is not suitable for the mobility-impaired.

FURTHER INFORMATION

Grid ref: NJ 822 665
RSPB Troup Head: tel: 01346 532017; email: strathbeg@rspb.org.uk

390 BANFF/MACDUFF HARBOURS (North-east Scotland)

OS Landranger 29

Situated on the north coast of Aberdeenshire, this is a good site in winter for gulls and seaducks.

Habitat

The towns of Banff and Macduff sit either side of a bay, within which are harbours for both towns.

Species

In winter and on passage Red-throated Diver, Common and Velvet Scoters, and Common Eider are present offshore and may enter the harbours, and there are sometimes Great Northern and Black-throated Divers. Both Glaucous and Iceland Gulls are regular, and a Kumlien's Gull from North America wintered here for some years. Purple Sandpiper and Turnstone are common on the rocks.

On autumn passage skuas may occur offshore, and Little Auk may be present in large numbers in late autumn–early winter.

Access
The harbours and adjacent rocky coast are easily accessed north off the A98.

391 PORTGORDON TO PORTKNOCKIE (Moray & Nairn)
OS Landranger 28

The small harbours along this coast may hold white-winged gulls, as well as sheltering seaducks.

Habitat
From Portgordon east to Portknockie the coast is rocky.

Species
Large numbers of Purple Sandpiper and Turnstone winter along this coast, and offshore there may be Red-throated Diver and occasionally also Black-throated and Great Northern. Common and Velvet Scoters, Long-tailed Duck, Common Eider and Red-breasted Merganser may also be seen offshore. In rough weather, divers, seaducks, Common Guillemot, Razorbill and occasionally even Little Auk may seek shelter in the harbours. There are sometimes Glaucous and occasionally Iceland Gulls among the commoner gulls around the harbours. In autumn, especially given a strong north-east wind, Manx and Sooty Shearwaters, Gannet, and Great, Arctic and (occasionally) Pomarine Skuas may also be seen offshore, especially from Portknockie.

Access
The A990 follows the coast from Portgordon to Portknockie with straightforward access to the coast and harbours.

392 SPEY BAY AND LOSSIEMOUTH (Moray & Nairn)
OS Landranger 28

This area holds large numbers of seaducks, and up to eight Surf Scoters have been recorded. Winter and spring are the best times to visit, though seaducks are present all year.

Habitat
The Rivers Lossie and Spey discharge into a long shallow bay. There are no large tidal flats, rather a sand and shingle beach between Lossiemouth and Portgordon, backed by plantations of Scots and Corsican Pines at Lossie Forest.

Species
Red-throated Diver can be abundant offshore, especially in autumn, but wintering Black-throated and Great Northern are only present in small numbers, although with persistence both may be seen. Common Scoter and Long-tailed Duck may peak at 1,000 (occasionally as many as 5,000 scoters), and Velvet Scoter can reach 500 (and in good years 1,000). A few hundred Common Eiders are joined by small numbers of Scaup and Red-breasted Merganser, and quite often Surf Scoter. As always, the ducks can be well offshore at times. The fields between Elgin and Lossiemouth may hold Greylag Goose and sometimes Whooper Swan. Waders include large numbers of Purple Sandpiper and Turnstone at Branderburgh, otherwise the mouth of the Lossie holds numbers of the commoner species. There are sometimes Glaucous and

occasionally Iceland Gulls along the beach, especially at Lossiemouth, and large flocks of Snow Buntings winter here, especially towards the west end. In Lossie Forest, Crested Tit and Scottish Crossbill are resident, though neither is numerous.

Small numbers of Common and Velvet Scoters oversummer, together with Goosander and sometimes Surf Scoter. A few Common and Arctic Terns breed, usually on islands in the lower reaches of the Spey. In autumn there are small numbers of passage waders and Sandwich Tern may be abundant. Especially in a strong north–east wind, Manx and Sooty Shearwaters, Gannet, and Great, Arctic and (occasionally) Pomarine Skuas may also be seen offshore.

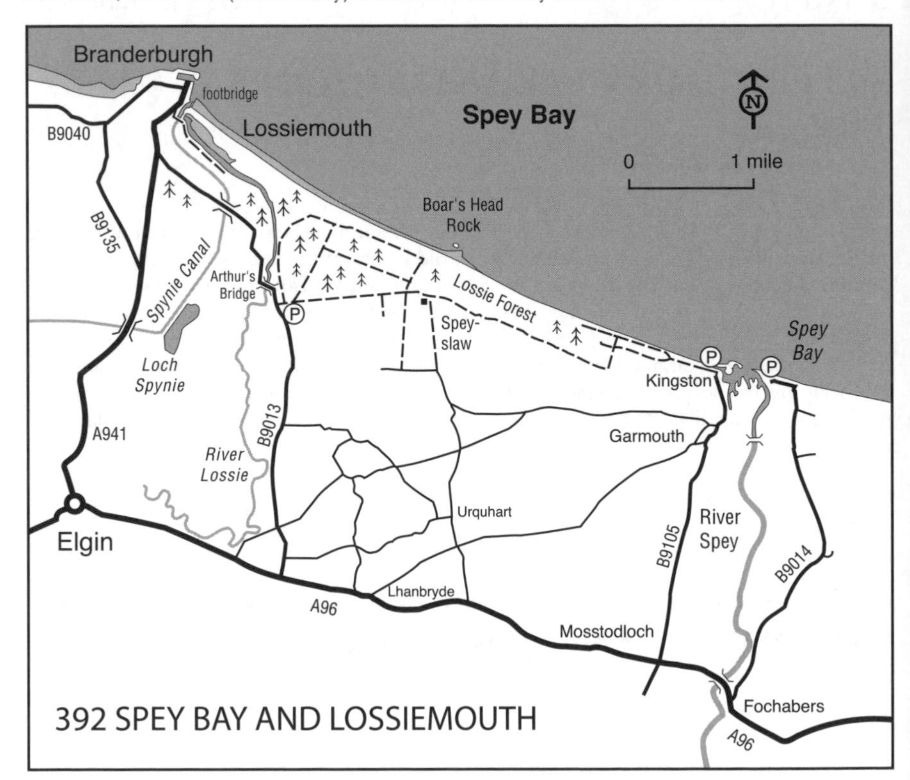

392 SPEY BAY AND LOSSIEMOUTH

Access

Lossiemouth Take the A941 from Elgin north to Lossiemouth and follow the road until it ends at Branderburgh pier and harbour. This is a good place to look for seaducks in the bay. Surf Scoter has been seen quite frequently. It is possible to cross the River Lossie on foot via a wooden bridge to work the river mouth and shore to the east. Low tide is best for waders.

Boar's Head Rock The best access to the middle of the bay; seaducks and, in autumn, divers are often numerous offshore. Turn north off the A96 on the B9013 and, after 3.5 miles, park just before Arthur's Bridge over the River Lossie. From here it is a 2-mile walk through Lossie Forest to the rock (the route can be confusing, but the OS map is of some help). Alternatively, follow minor roads north off the A96 to Speyslaw and walk north from here for 1 mile to the Rock.

Kingston Take the B9015 off the A96 north to Kingston and on to the picnic site by the river mouth. View the mouth of the river for waders at low tide or take the track west along the beach for seaducks, which may be numerous.

Speymouth Take the B9104 off the A96 at Fochabers, and drive to the coast at Spey Bay. The river mouth is good for waders at low water, and for seaducks scan the bay to locate the flocks.

Lossie Forest This is owned by the FC, and access is from the beach west of Kingston or along several tracks north off the A96. Cars are not allowed into the forest, and entry may be restricted because of activity on the firing range. The forest may be closed at times of severe fire-risk.

FURTHER INFORMATION

Grid refs: Lossiemouth NJ 238 709; Boar's Head Rock (Arthur's Bridge) NJ 254 670; Kingston NJ 339 655; Speymouth NJ 348 654

393 FINDHORN BAY AND CULBIN BAR (Moray & Nairn)

OS Landranger 27

This area is excellent in winter. Within a short distance there are concentrations of seaducks and waders, Crested Tit and Scottish Crossbill. During summer there is less to see on the coast, but in April–June the forest is at its best. The east end of Culbin Sands is an RSPB reserve of over 800 ha.

Habitat

Culbin Sands is the largest dune system in Britain and has been extensively planted with Corsican and Scots Pines, resulting in a large stand of monotonous woodland. Sandbars totalling nearly 5 miles border the coast, with some areas of saltmarsh. Findhorn Bay is an almost completely enclosed tidal basin, attractive to wildfowl and waders.

Species

On the sea, all three divers are regular in winter, though Red-throated far outnumber Black-throated and Great Northern. Large numbers of Common and Velvet Scoters and Long-tailed Duck join the local Red-breasted Merganser, and Black Guillemot is reasonably frequent. Greylag Goose and Whooper Swan favour the fields and there is a small flock of wintering Brent Geese. Waders involve the usual species, including Knot and Bar-tailed Godwit. There may be flocks of Snow Buntings along the foreshore and in the dunes.

Breeding birds include small numbers of Common Eiders, Oystercatchers and Ringed Plovers, as well as a few Common and Arctic Terns at Culbin and in Findhorn Bay (joined by non-breeding Sandwich Tern). Ospreys quite often fish in Findhorn Bay in summer. In the plantations Common Buzzard, Sparrowhawk, Long-eared Owl, Crested Tit, Scottish Crossbill and Siskin are resident. Common Crossbill may appear in late summer, especially in invasion years.

Passage brings some interesting waders, including Whimbrel and Greenshank, and less frequently Curlew Sandpiper, Black-tailed Godwit, Spotted Redshank and Little Stint. Findhorn Bay is the best place for these. Offshore there are often Gannet and Common, Arctic and Sandwich Terns. Skuas may appear, especially in north-east winds.

Access

RSPB Culbin Sands Leave Nairn on the A96, cross the river and turn first left towards the caravan park on Maggot Road. At the end of the road turn right through the caravan park to the East Beach car park. From here the 0.6-mile 'all abilities trail' gives views of the Ministers Pool; otherwise, access to the shore is possible at all times.

Loch Loy and Cran Loch Accessible on tracks north off the minor road past Kingsteps. Wildfowl include wintering Whooper Swan.

Culbin Forest Owned by the FC, this can only be entered on foot, from either Wellhill or Cloddymoss. These are accessible along a maze of lanes north off the A96. Tracks lead from the car parks through the forest to the shore and bar, and the Wellhill car park gives access to a viewpoint on Hill 99.

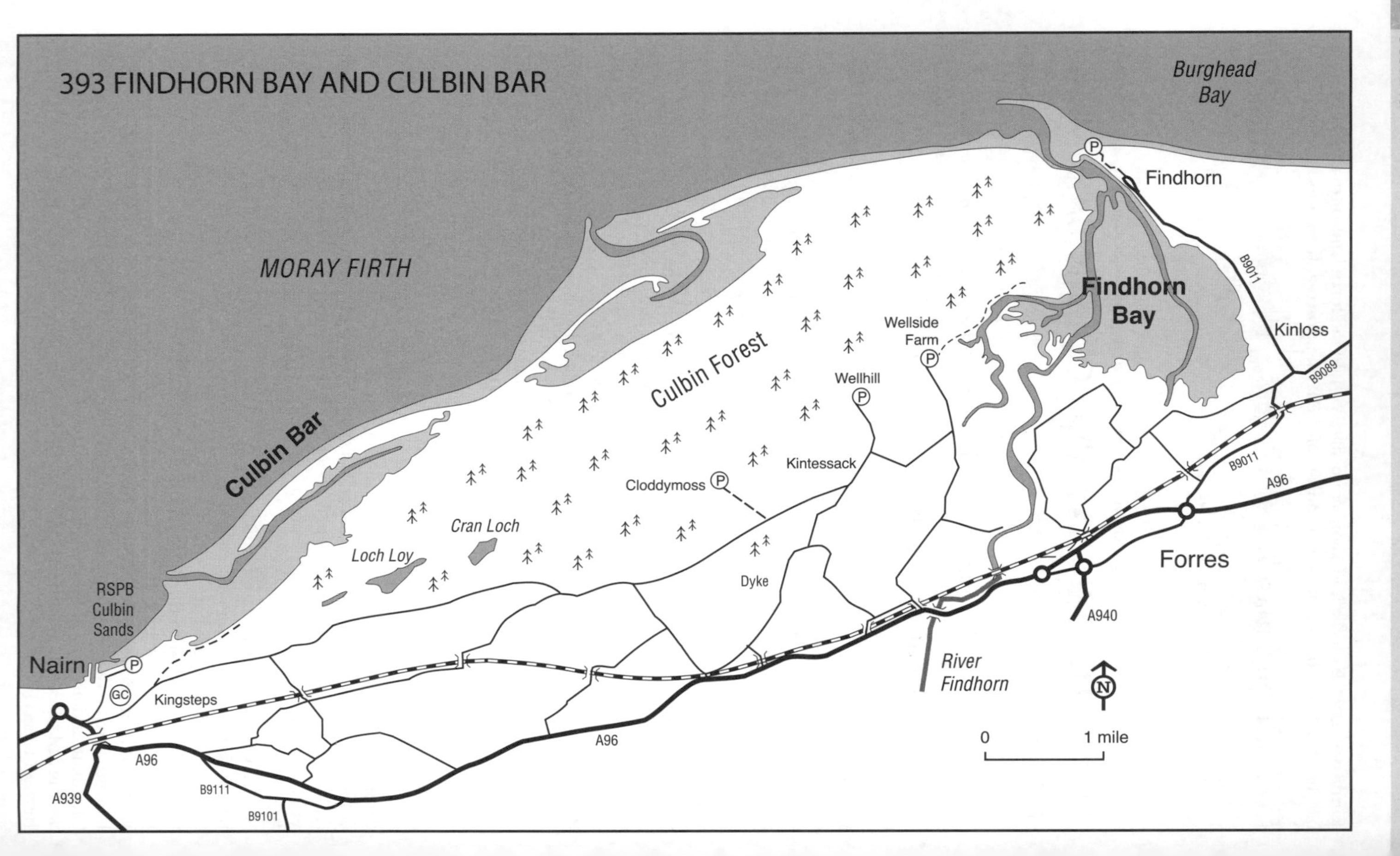
393 FINDHORN BAY AND CULBIN BAR
Burghead Bay
MORAY FIRTH
Findhorn
Findhorn Bay
B9011
Kinloss
B9089
Culbin Bar
Culbin Forest
Wellside Farm
Wellhill
Kintessack
Cloddymoss
Cran Loch
Loch Loy
Dyke
Forres
B9011
A96
A940
RSPB Culbin Sands
Nairn
GC
Kingsteps
River Findhorn
0
1 mile
A96
A96
A939
B9111
B9101

Findhorn Bay View the east shore from the B9011 Findhorn–Kinloss road. The west side can be seen by taking a minor road north off the A96, just west of the Findhorn bridge, and following this (without turning) for 2.5 miles to Kincorth House and Wellside Farm. Park carefully and follow tracks along the edge of the forest to the Bay. Access to the south side is more complicated, with several minor roads leaving Forres north to the shore.
Findhorn village A viewpoint for seaducks in winter and terns in summer offshore in Burghead Bay.
Burghead Harbour (OS Landranger 28) On the B9089 north off the A96 north-east of Kinloss, the harbour often has Common Eider, Long-tailed Duck, Black Guillemot and Purple Sandpiper, and occasionally holds Little Auk in winter. The resident gulls may be joined by a Glaucous or Iceland. Divers and seaducks may be seen at sea from the harbour or headland.
Roseisle Forest Near Burghead, Crested Tit can be found in Roseisle Forest. Leave Burghead south on the B9089 and, after c. 3 miles, turn north in Lower Hempriggs, following the track north for c. 1 mile to the picnic site near the beach. In winter the birds can be around the car park, but in summer may be more widely dispersed. Divers and seaducks may be seen offshore.

FURTHER INFORMATION

Grid refs: RSPB Culbin Sands NH 895 574; Culbin Forest NH 998 615 (Wellhill); NH 980 599 (Cloddymoss); Findhorn Bay (Wellside Farm) NJ 011 618; Findhorn village NJ 043 648; Burghead Harbour NJ 110 689; Roseisle Forest NJ 103 655
RSPB Culbin Sands: tel: 01463 715000

394 INSH MARSHES NNR (Highland)

OS Landranger 35

This RSPB reserve protects 837 ha along 5 miles of the flood plain of the River Spey between Kingussie and Loch Insh, attracting wintering wildfowl (notably Whooper Swan) and harbouring a number of interesting breeding species.

Habitat

The reserve comprises fen, rough pasture, willow carr and pools, with stands of birch and juniper on higher ground. In winter the whole area may flood, sometimes to a depth of 8 feet!

Species

The entire area regularly floods in winter and attracts Greylag Geese, up to 100 Whooper Swans and a variety of ducks. There is a Hen Harrier roost (5–15 birds), and Peregrine is regular, while wandering Red Kites from the reintroduction programme on the Black Isle may winter, but White-tailed Eagle is rare.

Breeders include feral Greylag Goose, Wigeon, Teal, Shoveler, Tufted Duck, Goldeneye (in nest boxes), Oystercatcher, Common Snipe, Woodcock, Curlew, Common Redshank, Common Sandpiper, Black-headed Gulls and Grasshopper and Sedge Warblers. Pochard, Red-breasted Merganser and Goosander nest occasionally. Water Rail breeds and Spotted Crake is sometimes present in summer, when up to five have been heard calling, but they are almost impossible to see (listen at night between mid-May and mid-June for the abrupt whip call, from the B9152 between Loch Insh and Lynchat, especially in the area opposite the Highland Wildlife Park). Wood Sandpiper also breeds nearby and sometimes feeds on the reserve. In 1968 a female Bluethroat with a nest and eggs was found, while in 1985 a pair reared five young, the first successful British breeding record. The river has Common Sandpiper and Dipper, and wooded areas Tree

Pipit, Common Redstart, Wood Warbler, Spotted Flycatcher and a few Pied Flycatchers. Osprey is a regular visitor in spring and summer and up to ten species of raptor may be present in early autumn, including Hen Harrier. On passage good numbers of Greylag and Pink-footed Geese pass through in early autumn, and Marsh Harrier is occasionally noted in spring.

Access

Stopping places on the B970 and B9152 afford views of the Marshes and Loch Insh, and this should be adequate in winter.

The RSPB information point/hide and car park lie off the B970, 0.6 miles east of the ruins of Ruthven Barracks, and 1.5 miles south-east of Kingussie (which is signed from the A9). Two nearby hides overlook the west section of the Marshes and there are three woodland trails: the Invertromie trail (a 2.8-mile circuit of the west end of the reserve through birch woods, meadows and heather), the Lynachlaggan (1.5 miles, in the north-east sector of the reserve, reached from the lay-by by the pylon just north of Insh village, giving access to a stand of birch and juniper woodland) and Loch Insh Woods trail (1.25 miles). The reserve is open at all times. The trails are not suitable for wheelchairs.

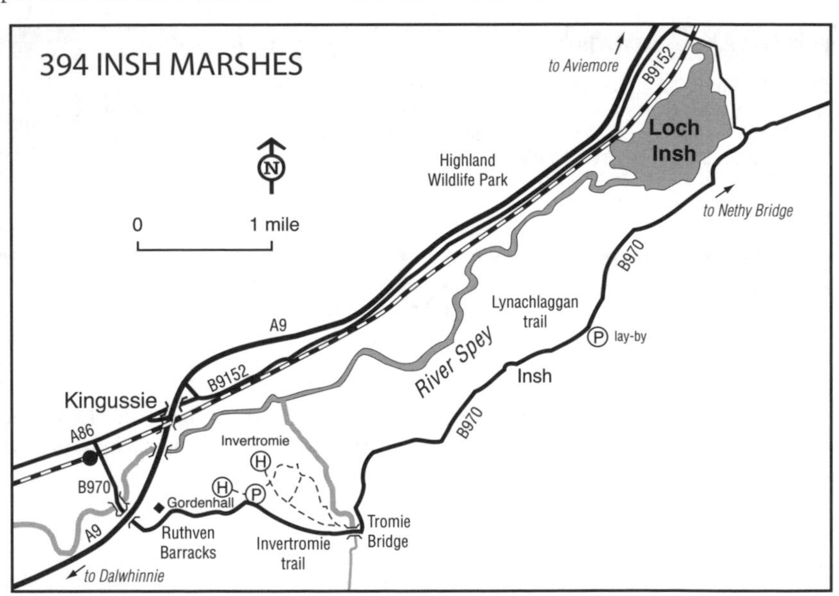

FURTHER INFORMATION

Grid ref: NN 775 998
RSPB Insh Marshes: tel: 01540 661518

395 CAIRNGORMS AND SPEYSIDE (Highland)

OS Landranger 36 and 35

The range of habitats and birds in this area is reminiscent of the high arctic and boreal forests of north Eurasia, and it supports all of the Scottish specialities. The best time to visit is April–July, although some of the more interesting species (e.g. Dotterel) do not arrive until May. There are three NNRs, including the giant Cairngorm reserve at 26,000 ha (mostly privately owned), while the RSPB owns the huge Abernethy Forest Reserve, comprising 12,500 ha.

Habitat

The region offers the exciting combination of three principal habitat types. Firstly there are the high peaks of the Cairngorms, some topping 4,000 feet (and including four of the five highest mountains in Britain), with their distinctive 'whaleback' profile. The bare, windswept summit plateau above 3,000 feet supports an arctic–alpine flora, and snow can lie in some of the high corries all year, while the lower slopes have extensive tracts of heather moorland. This wilderness is difficult to penetrate, except via the new Cairngorm Mountain Railway (which has replaced the old ski-lift). Secondly, some of the best remnants of native Caledonian pine woodland, relicts of the primeval wildwood that once covered much of Scotland, are important for breeding birds. Stands of Scots Pine, some trees over 300 years old, shade a mixture of juniper, bilberry and heather, and support Britain's only endemic bird, the Scottish Crossbill. Although not as exceptional as the pine woods, the native birch woods are equally attractive, and there are also plantations of both native and introduced conifers. The third major habitat is the valley of the River Spey and its associated lochs and marshes.

Species

Residents include Sparrowhawk, tiny numbers of Goshawks, Hen Harrier, Common Buzzard and Peregrine. There are a few Golden Eagles in the region, and these are occasionally encountered on the high tops, but by far the best area for the species is the Findhorn Valley (see page 595). Red Grouse breeds on the moors up to c. 3,000 feet while Ptarmigans are on the high tops in summer, though they may move downslope in winter, often to below the snowline (when they overlap with Red Grouse). Black Grouse favour the forest edge and areas of scattered trees, as well as birch woods. They are most easily seen when lekking in late March–early June. As they are prone to disturbance, lekking Black Grouse should be watched from a good distance or from a car. Capercaillie is a bird that every birdwatcher wants to see but the species has declined to critically low levels. It prefers the denser, more mature stands of woodland and displaying males can occasionally be found on quiet forest roads in the early morning (and again in the evening) March to mid-April but the RSPB request that birdwatchers avoid forest tracks early in the morning during April and May and attend the 'Caper Watch' at the Osprey Observation Post, where there is a high rate of success. Both Black Grouse and Capercaillie become very hard to find in summer when incubating and moulting, becoming more conspicuous again from September (and autumn can be a good time to see Black Grouse). Following successful management by the RSPB, especially the reduction of grazing and browsing, Black Grouse increased fourfold in the early 1990s to a peak of 165 lekking males in Abernethy in 1997. Numbers then fell, however, to 62 in 2009. Capercaillie numbers are stable in Abernethy, unlike the catastrophic declines reported from most other areas. Long-eared Owl is resident, but hard to find. Not so Woodcock, which can be very conspicuous in early spring when roding. Crested Tit is quite common and often located by call; it tends to move into more open areas of long heather adjacent to woods in winter. Other small woodland passerines are Siskin, Lesser Redpoll and Scottish Crossbill; the last can be scarce one year and abundant the next, but tends to be most obvious April–July. Note that Common and Parrot Crossbills also nest in these woods, so identification is critical for the purist! Snow Bunting occurs all year, though it is much commoner in winter when it can be found around the ski-tow car parks. A few pairs breed on the tops, favouring corries, boulder fields, scree slopes and the vicinity of snow fields. In open areas, both Carrion and Hooded Crows (and their hybrids) and sometimes Raven can be found.

Summer visitors include Slavonian Grebe in small numbers to several forest lochs. They are present early April–August. A few pairs of Red-throated and Black-throated Divers also breed, and occasionally visit some of the larger lochs. Breeding ducks include Wigeon, Teal, a few Shovelers, Tufted Duck, Red-breasted Merganser and Goosander.

Goldeneye is a recent colonist, and has dramatically increased, due to the provision of nest boxes, to over 100 pairs since the first breeding record in 1970. On hatching, most ducklings and their parents transfer to the River Spey. Feral Greylag Goose is a scarce but increasing breeder in the region. Osprey is present late March–August, and is most reliably seen at the Osprey Observation Post at the RSPB's Abernethy Reserve, although Inverdruie Fish Farm often attracts fishing birds, and the species may put in an appearance at any loch or river. Other raptors include a few Merlins on the heather moors, and tiny numbers of Honey-buzzards. A few Water Rails breed, for example at Loch Garten. Waders include Golden Plover and Dunlin on the moors and tops, and Dotterel, which favours short grass, moss and lichen heath on the flat summit plateaux of the mountains above 2,600 feet. Oystercatcher, Common Sandpiper, a few Ringed Plovers, Common Gull, Dipper and Grey Wagtail haunt the streams and rivers, and Greenshank and Wood Sandpiper inhabit the forest bogs in very small numbers.

Temminck's Stint, Green Sandpiper and Red-necked Phalarope have all bred in the past, and should be watched for. Great Spotted Woodpecker is joined by a few Green Woodpeckers. Tree Pipit favours woodland clearings and areas of scattered trees, with Whinchat and Wheatear on open ground and Ring Ouzel around rocky valleys and scree slopes high in the hills. Around the woods, Common Redstart and Wood Warbler are reasonably common, with the occasional Pied Flycatcher (primarily in riverine woodland along the Spey). Areas of marsh and farmland support Lapwing, Common Redshank, Common Snipe, Curlew, Black-headed Gull and Grasshopper and Sedge Warblers.

In the last 40 years several species, presumably of Scandinavian origin, have bred here, and there is potential for others to do so. On the tops Snowy Owl, Purple Sandpiper, Sanderling, Shore Lark and Lapland Bunting are all slight possibilities, with Wryneck, Fieldfare, Redwing, Icterine Warbler, Red-backed Shrike, Common Rosefinch and Brambling a little more likely in the valleys. If you see any of these, be aware that they could be breeding, and report your sightings to the RSPB and the appropriate county recorder.

Large numbers of Pink-footed and Greylag Geese occur on passage, and there are sometimes Waxwings in autumn. Smaller numbers of Greylag Geese winter, usually on farmland between Aviemore and Grantown-on-Spey, roosting at Loch Garten, as well as Whooper Swan (mainly around Boat of Garten–Nethy Bridge), and other wildfowl include Wigeon, Teal and Goldeneye. Wandering Red Kites from the reintroduction programme on the Black Isle may winter in the area, and rarely a White-tailed Eagle visits. Like many resident raptors, Hen Harrier moves to lower ground in winter.

Occasionally a Great Grey Shrike appears, although you are more likely to see the much commoner Brambling.

Access

An OS map is almost essential to navigate the many minor roads and forest tracks in the area.

Glenlivet Part of the Crown Estate, with open access to most of its roads, paths and tracks, this lies astride the B9136 and B9008 on the north-east side of the Cairngorms National Park. Of special note are the dawn 'Black Grouse Watches', organised by Glenlivet Wildlife several times a week in April and May, which offer an excellent way of getting close to lekking Black Grouse without disturbance.

Carrbridge Landmark Centre At the south end of the village of Carrbridge on the B9153, the woodland to the rear of the centre often holds Crested Tit and Scottish Crossbill. Take the trail west from the north end of the car park. (The Centre offers audio-visual shows, outdoor exhibitions, a shop and restaurant.) The north end of the village, where the River Dulnain is bridged by the B9153, is a good spot for Dipper.

Grantown-on-Spey The woodlands east of the town hold Crested Tit and Scottish Crossbill. Leave Grantown north-east on the B9102 towards Lettoch and park on the roadside by the gate after c. 0.5 miles (just past the golf course), taking the track south into the forest.

Lochindorb This loch, around 6 miles north of Grantown-on-Spey, is a regular spot for Black-throated Diver. View from the minor road south-west off the A939 at Dava.

RSPB Loch Garten and Abernethy Enclosed within the RSPB's Abernethy Forest Reserve, Loch Garten is sometimes visited by fishing Ospreys, and holds breeding Goldeneyes, a late-autumn roost of Goosanders and a winter roost of up to 1,500 Greylag Geese. The surrounding woodland has Crested Tit and Scottish Crossbill. From the B970 just north-east of Boat of Garten, turn right signed 'RSPB Ospreys', parking on the right after c. 1 mile. Crested Tit is often around the car park or along the road. There are three forest trails, including an interesting circular walk from the car park to Loch Mallachie. The road continues along the north-west shore of Loch Garten (passing the car park for the Osprey Observation Post). There is access to the reserve tracks at all times (not wheelchair-accessible), except for the large sanctuary area around the Osprey nest.

RSPB Osprey Observation Post The Ospreys breed some distance from Loch Garten within a statutory bird sanctuary and can be seen from the Observation Post, open 10am–6pm April–late August (both telescope views and CCTV available). As well as Osprey, Capercaillie is often seen from the hide in the spring and indeed, this is now one of the best sites in Britain for this declining species. The RSPB organises daily early morning 'Caper-watches' giving a good chance of seeing this magnificent bird (the centre is open for 'Caper-watch' 5.30am–8am through April to mid-May). There may be Crested Tit and Scottish Crossbill around the car park and access track. From the B970 just north-east of Boat of Garten, turn right signed 'RSPB Ospreys' and the Observation Post car park lies on the left after 1.25 miles (simply follow signs from Aviemore or Grantown-on-Spey; note that the 1.5-mile blue trail links the observation post with public transport and the Speyside Way). It is 300 yards from the car park to the hide (wheelchair-accessible).

Forest Lodge, Abernethy Forest In the south-east quadrant of Abernethy RSPB Forest Reserve, this is a good area for Crested Tit and Scottish Crossbill. The area can be accessed from Glencairn, Boat of Garten or Nethy Bridge (all on the B970) via unclassified roads, taking the road east at NH 998 167 to the car park at the Lodge. Explore the track north from the Lodge to the River Nethy, and the circular route south via Rynettin (leading to the right just before the car park), and also, a little to the west, the long forest road north to Dell Lodge. Visitors should stay on the tracks.

Tulloch Moor This area of open moorland, bog and lochans to the south-west of Loch Mallachie is a well-known and long-standing site for Black Grouse. Take the minor road east off the B970 just north of Glencairn towards Tulloch; a viewing screen has been erected alongside the road to prevent disturbance of the grouse, but great care should be taken not to block this narrow road. Birdwatchers should consider taking one of the organised trips at RSPB Corrimony (page 597) or at Glenlivet (see above).

Rothiemurchus Estate Situated on the road between Aviemore and the Cairngorm Mountain Railway, there is a visitor centre just east of Loch Morlich and marked trails through a variety of habitats. The Moormore picnic site just south of the funicular road c. 2 miles east of Coylumbridge is a favoured spot for Crested Tit and Scottish Crossbill. The quieter south-west section of the forest may produce Black Grouse. Take the minor road from Inverdruie to Blackpark and on to Whitewell, following footpaths from here to the Cairngorm Club footbridge (NH 926 078) and to the woodland between the confluence of the Am Beanaidh and Allt Druidh (a long walk).

Loch an Eilein Lies within Rothiemurchus Estate. Leave the B970 1 mile south of Inverdruie at Doune, turning south-east onto a minor road and continuing to the car park (fee) at the north end of the Loch, which holds Goldeneye, Goosander and occasionally divers. Crested Tit and Scottish Crossbill may be seen on the forest trail encircling the Loch.

Glenmore Forest Park This large area (2,645 ha) of state-owned forest adjacent to Rothiemurchus Estate includes Loch Morlich, which although disturbed by watersports holds small numbers of wildfowl and is sometimes visited by Osprey, with Dipper on the Loch and its tributary streams. Crested Tit and, with luck and perseverance, Scottish Crossbill, can be seen around the Loch. Information is available from Aviemore Tourist Office or the Glenmore Visitor Centre (on the road between Aviemore and the Cairngorm Mountain Railway).

Aviemore Peregrines nest most years on the high cliffs of Craigellachie Rock, easily viewed from the car park behind the Badenoch and Stakis Four Seasons hotels. **Craigellachie NNR** covers the birch woods on Craigellachie Rock's lower slopes. Access is from Aviemore Highland Resort (limited parking) or the waymarked trail from the Youth Hostel, taking the subway

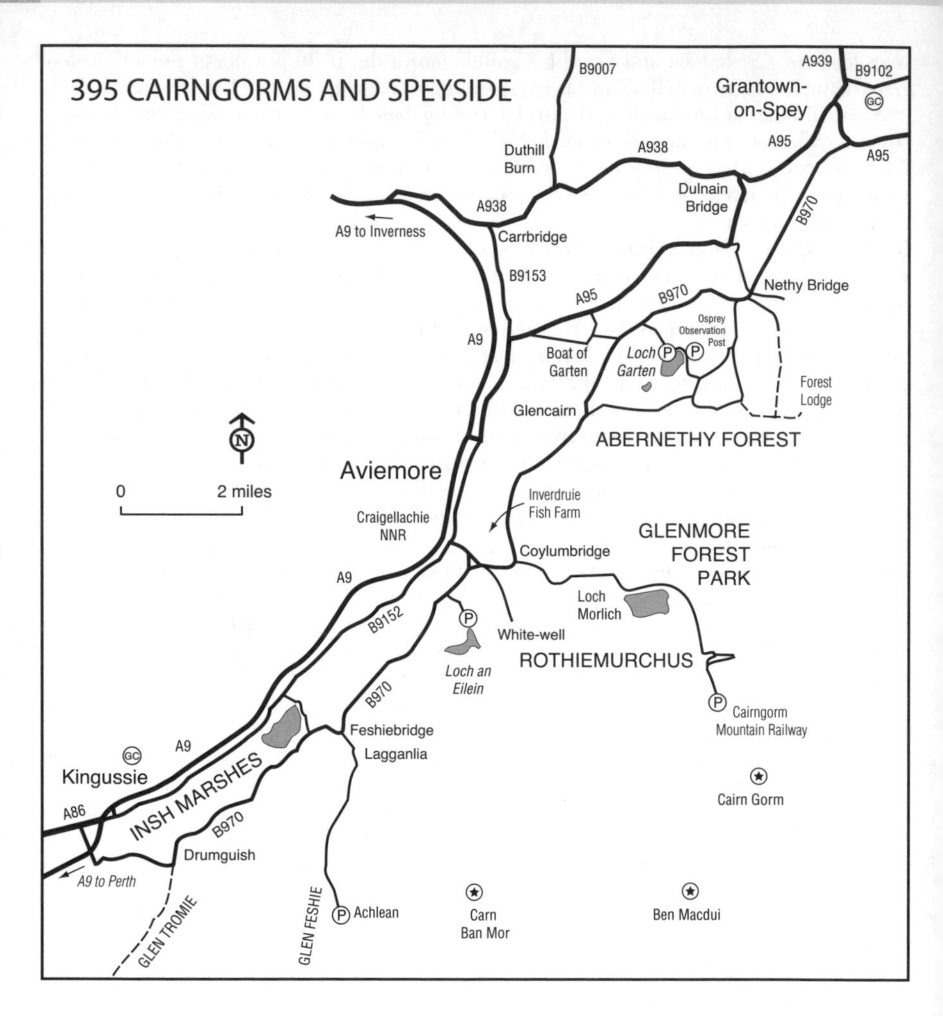

under the A9 trunk road. There is a signposted nature trail. At the south end of the town, the B970 bridge over the River Spey is a good bet for Dipper.

Inverdruie Fish Farm Offers the best chance in the region of seeing an Osprey fishing. Leave Aviemore east towards Coylumbridge on the B970 and turn left into the car park just after crossing the River Spey. There are two hides, and an entrance fee is payable. Mornings and evenings are best.

Cairn Gorm Take a minor road off the B970 at Coylumbridge and follow it for c. 8 miles, past Loch Morlich, to the car park at the foot of the ski-lift (the second, upper, car park at Coire Cas). The Cairngorm Funicular Railway operates services to the summit plateau, but during the summer season (1 May–30 Nov) visitors are not permitted to leave the top station or Ptarmigan Restaurant (this is to protect the fragile montane habitats from excessive disturbance). To access the high plateau on foot it is necessary to walk from the Coire Cas car park, a steep 1.5–2-hour walk on the well-marked path to the summit of Cairn Gorm (and walkers can take a train downhill). In the ski season the funicular gives easy access to the top for everyone. Ptarmigan, Dotterel and sometimes Snow Bunting occur on the summit plateau in summer, but due to disturbance are often difficult to find. Try walking south along the ridge path towards Ben Macdui; the further from the most accessible areas, the better the birds. In winter there are often Snow Buntings around the car park, while Ptarmigans descend to lower altitudes and, together with Red Grouse, may occur near the level of the car park; good areas to look are the valleys of Coire an t-Sneachda and Coire an Lochain south of the ski-tows.

Access to Cairngorm NNR is unrestricted, except some areas in the deer-stalking season, August–October. Note, however, that the vegetation is very fragile and it is best to stay on the well-marked trails and paths, scanning carefully with binoculars and listening for calls. The weather on the high tops is very changeable, and walkers should be properly equipped and leave notice of their intended route (even a note on the car windscreen) if they stray far from the populous areas.

Carn Ban Mor An excellent, comparatively undisturbed, yet relatively accessible area for most of the upland specialities, peaking at 3,451 feet. From Feshiebridge on the B970 follow minor roads to Achlean, where there is limited parking at the end of the road. From here it is a steep but steady 2–3-hour walk to the summit on a broad, very well-marked track. Once on the tops it is best to remain on the track and scan carefully for Ptarmigan and Dotterel, which can be very easily overlooked. Part of the Cairngorm NNR (see Cairn Gorm above).

Glen Feshie A good area for raptors, while the woods hold Crested Tit and Scottish Crossbill, with Ring Ouzel on the lower slopes. Follow directions to Achlean as for Carn Ban Mor, taking the path up the glen, which follows the river.

Glen Tromie Running south from Drumguish on the B970, this has Black Grouse and sometimes Peregrine, Merlin, Golden Eagle and Hen Harrier, with Dipper on the river. A track runs up the glen from near Tromie Bridge on the B970.

FURTHER INFORMATION

Grid refs/postcode: Carrbridge Landmark Centre NH 909 222 / PH23 3AJ; Grantown-on-Spey NJ 045 284; Lochindorb NH 975 357; RSPB Abernethy & Loch Garten NH 971 185; RSPB Osprey Observation Post NH 977 183; Forest Lodge, Abernethy Forest NJ 020 161; Rothiemurchus Estate (Moormore) NH 942 103; Loch an Eilein NH 897 085; Glenmore Forest Park (Loch Morlich) NH 963 097; Aviemore NH 893 126; Craigellachie NNR NH 894 118; Inverdruie Fish Farm NH 897 115; Cairn Gorm (Coire Cas) NH 989 061; Carn Ban Mor/Glen Feshie (Achlean) NN 852 976; Glen Tromie NN 789 994

Aviemore Tourist Office: web: www.visithighlands.com/aviemore

RSPB Loch Garten & Abernethy: tel: 01479 831476 (April–August only, 10am–5pm); email: abernethy@rspb.org.uk

Scottish Natural Heritage, East Highland Office: tel: 01479 810477; web: www.nnr-scotland.org.uk

Glenlivet Wildlife: Easter Corrie, Tomnavoulin, Glenlivet, Scotland, AB37 9JB: tel: 01807 590241; web: www.glenlivet-wildlife.co.uk

Rothiemurchus Estate Information Centre: tel: 01479 812345; web: www.rothiemurchus.net

Forest Enterprise, Glenmore Visitor Centre: tel: 01479 861220; web: www.forestry.gov.uk (search for 'Glenmore Forest Park')

396 FINDHORN VALLEY (Highland)

OS Landranger 35

This remote valley is one of the best places in the east Highlands to see Golden Eagle.

Habitat

A large area of primarily open hillsides grazed by sheep and Red Deer.

Species

Common Buzzard, Peregrine, Merlin and, in summer, Osprey also occur, with the occasional

Raven. White-tailed Eagle occasionally may visit. The river has Goosander, Common Sandpiper, Grey Wagtail and Dipper, while the moors hold Red Grouse and Golden Plover, with Ring Ouzel on the rockier sections.

Access

From the main A9, 12 miles north of Aviemore, turn west on minor roads to Tomatin Services and, just west of the old Findhorn bridge, turn south to Garbole and Coignafearn on the single-track road which parallels the north bank of the River Findhorn. Stop and scan for eagles over the surrounding ridges as the Valley becomes precipitous after c. 7 miles, continuing as necessary to the no entry sign at the road's end at Coignafearn Old Lodge, 10 miles from the junction (it is possible to continue on foot).

The road from Garbole, c. 4 miles from the Findhorn Bridge, which meets the B851 at Farr, may produce a similar range of raptors, and may be a better bet for Hen Harrier.

FURTHER INFORMATION

Grid ref: Coignafearn NH 681 152

397 LOCH RUTHVEN (Highland)

OS Landranger 35

This loch, lying just east of Loch Ness, is a well-publicised site for breeding Slavonian Grebe and is also a good area for divers and Osprey.

Habitat

Along the north shore of the loch are stands of birch woodland, otherwise the surrounding area is largely moorland, with some farmland around Tullich in the east.

Species

Up to ten pairs of Slavonian Grebes breed here (around 50% of the UK breeding population), and are present late April–early September. Other breeders include a few pairs of Common and Black-headed Gulls and Common Sandpiper, and Ospreys are regular visitors, and the loch also attracts Red-throated and sometimes Black-throated Divers. Whinchat and Sedge Warbler nest in the surrounding vegetation, with Siskin and Lesser Redpoll in the woods and Twite in more open areas. The surrounding moorland has Peregrine, Hen Harrier, Short-eared Owl and Raven. In winter Goldeneye and occasionally Smew are recorded.

Access

Leave the A9 c. 6 miles south of Inverness south on the B851. After 8 miles, turn right on the minor road towards Tullich and Dalcrombie. (Alternatively, from Inverness take the B862 south-west and, after 12 miles or just past Loch Duntelchaig, turn south-east on the minor road to Dalcrombie and Tullich.) There is a car park at the east end of the Loch near Tullich and access is restricted to the 600-yard trail (not wheelchair-accessible) along the south-east shore, leading past the boat house to the hide. The reserve is open at all times.

FURTHER INFORMATION

Grid ref: NH 638 280
RSPB Loch Ruthven: tel: 01463 715000; email nsro@rspb.org.uk

398 CORRIMONY (Highland)

OS Landranger 26

Lying south-west of Inverness and west of Loch Ness, between Glen Urquhart and Cannich, and adjacent to the better known Glen Affric, this RSPB reserve represents a long-term experiment in the restoration of semi-natural habitats from plantation woodlands.

Habitat

Covering 1,531 ha, the reserve comprises moorland and plantations bisected by the Allt Feith Riabhachain River, with areas of boggy land in the valley. There is a small patch of semi-natural Scots Pine and birch near Corrimony Falls on the slopes of Carn Bingally, but most woodland consists of plantations of Scots and Lodgepole Pines and Sitka Spruce.

Species

Breeding raptors on the moorland include Hen Harrier and Merlin, and Golden Eagle may visit, while Teal and Greenshank nest in wetter areas. Both Black and Red Grouse breed, and it is hoped to increase their numbers by positive management; Black Grouse have already increased from 16 to 35 lekking males. The woodlands hold Great Spotted Woodpecker, Crested Tit and crossbills (including Scottish Crossbill). Red-throated Diver and Osprey may visit to feed on the loch, which in winter attracts Goldeneye and Goosander.

Access

Access is from the A831, east of Cannich. At Drumnadrochit on the west side of Loch Ness, leave the A82 and head west on the A831. A minor road on the left leads to Corrimony after 8 miles. Park in the Corrimony Cairns car park, and an 8.5-mile waymarked trail (which is navigable by wheelchair with help) leads to Loch Comhnard (the trail passes through a working farm – please leave gates as you find them). In April and May guided 'minibus safaris' are operated offering a chance to see lekking Black Grouse without disturbance.

FURTHER INFORMATION

Grid ref: NH 383 302
RSPB Corrimony: tel: 01463 715000; email: nsro@rspb.org.uk

399 GLEN AFFRIC (Highland)

OS Landranger 25

This large and very scenic glen lies south-west of Inverness and is one of the richest and most diverse Scottish glens in all aspects of its fauna and flora. More than 9,000 ha are operated as a Caledonian Woodland Reserve by the FC and Glen Affric is an NNR.

Habitat

The Glen contains relicts of the once extensive Forest of Caledon, though heavy grazing pressure has prevented regeneration of native Scots Pine, leaving ancient twisted 'granny pines' with an understorey of heather and Blaeberry (the most natural pine woods, mixed with birch and rowan, are near Dog Falls and at Pollan Buide, between Loch Benevean and Affric). Restoration is under way, and away from the woodland areas there are open hillsides, a number of lochs and lochans, and the River Affric.

Species

Small numbers of Black Grouse and Capercaillies persist in the Glen, but are difficult to locate. Late winter–early spring is the best period to look, but birdwatchers should be mindful that both have undergone serious declines in recent years and are particularly prone to disturbance in the breeding season. Crested Tit is common in the pines and generally easy to find, and Scottish Crossbill, at least in some years, is also numerous. Stands of conifers also hold Siskin and Lesser Redpoll, and Redwing may be present in summer. In more open areas look for Tree Pipit, Whinchat, Stonechat and Common Redstart. Along rivers and loch-sides search for Oystercatcher, Common Sandpiper, Dipper and Grey Wagtail, with Red-throated Diver, Red-breasted Merganser, Goosander and occasionally Black-throated Diver on areas of open water. Sparrowhawk, Common Buzzard, Golden Eagle and occasionally Merlin or Osprey can also be seen (though raptors are not prominent in this area).

Access

Access is from the A831 at Cannich. From here a minor road reaches 11 miles along the Glen, with various parking points en route. From the road a number of paths lead into the surrounding area (try the trails to Coire Loch or Dog Falls). Longer routes include a circuit of Loch Affric (11 miles) or a track along the south shore of Loch Beinn a' Mheadhoin (7 miles each way).

FURTHER INFORMATION

Grid ref: Dog Falls NH 282 282
Forestry Commission: tel: 01320 366322; web: www.forestry.gov.uk/glenaffric (a detailed guide is available from the FC)
Scottish Natural Heritage, reserves manager: tel: 01479 810477; web: www.nnr-scotland.org.uk

400 FAIRY GLEN (Highland)

OS Landranger 27

On the south-eastern side of the Black Isle, just east of Fortrose, this small, scenic RSPB reserve offers a chance to see common woodland birds and flowers and is best in the spring and early summer.

Habitat

A small area of broadleaved woodland set in a steep-sided valley with a fast-flowing stream.

Species

The reserve holds a variety of common woodland birds, with Dipper and Grey Wagtail along the stream.

Access

Coming from the direction of Inverness on the A832 proceed through Rosemarkie village, passing the Plough Inn on the right and following the road round the sharp left bend. After a further 150 yards the car park lies to the right. From here a 1.25-mile trail follows the valley to two waterfalls. As the trail has some steps and slippery sections, it is not suitable for wheelchair users.

FURTHER INFORMATION

Grid ref: NH 732 580
RSPB Fairy Glen: tel: 01463 715000; email: nsro@rspb.org.uk

401 THE CROMARTY FIRTH (Highland)

OS Landranger 21

This large natural inlet on the east coast of Scotland holds good numbers of waders and wildfowl, with two RSPB reserves. It is best in autumn and winter.

Habitat

Mudflats, saltmarsh and pasture on the shores of the Cromarty Firth. At Nigg Bay the sea wall has been breached to create more extensive areas of saltmarsh.

Species

Waders include Knot, Bar-tailed Godwit and often a few Grey Plovers. Pink-footed Goose can peak at 10,000, especially on passage in late autumn and early spring. Wintering wildfowl also include Whooper Swan, Greylag Goose, Teal, Pintail (favouring the inner part of Nigg Bay) and Wigeon, and the latter can peak at 10,000 in January. The main channel may hold Red-throated Diver and Slavonian Grebe as well as a variety of seaducks, including Long-tailed Duck and internationally important numbers of Scaup. Merlin and Peregrine are occasional visitors and during the spring and summer Ospreys visit the Firth to fish.

Access

As with all tidal areas, the period around high water (and the preceeding 3 hours) are likely to produce the best views of waders.
RSPB Nigg Bay On the northern, more heavily industrialised side of the Firth. Leave Nigg northwards on the B9175 and the reserve car park lies on the west side of the road after 1 mile. It is a short walk to the hide (wheelchair-accessible). The reserve is open at all times.
Southern shore The B9163 skirts much of the southern shore of the firth and between Udale Bay and Cromarty gives views of the main channel, which should be checked for divers and seaducks, with a pull-in at the eastern end of Udale Bay (0.5 miles east of Jemimaville).
RSPB Udale Bay This lies on the south side of the Firth between Balblair and Jemimaville and a pull-in on the B9163, just west of the latter, gives access to a hide (open at all times, and wheelchair-accessible). The B9163 then turns northwards (at its junction with the B9160) and after c. 0.5 miles take the minor road northwards Newhall Point; there is another roadside viewpoint at the point giving views of the main channel.

FURTHER INFORMATION

Grid refs: RSPB Nigg Bay NH 807 730; Southern shore NH 728 652; RSPB Udale Bay NH 712 651
RSPB Nigg and Udale Bays: tel: 01463 715000; email: nsro@rspb.org.uk

402 TARBAT NESS (Highland)

OS Landranger 21

This promontory is quite good for seawatching.

Habitat

This is a scenic headland with a very striking lighthouse.

Species

In autumn, quite large numbers of Sooty Shearwaters, as well as skuas, auks and terns are seen.

Access

Take the B9165 to Portmahomack from the A9 3 miles south of Tain, and continue on a minor road to the lighthouse.

FURTHER INFORMATION

Grid ref: NH 945 872

403 DORNOCH (Highland)

OS Landranger 21

The Dornoch Firth holds large numbers of seaducks and other wildfowl, including Scaup. The populations of Loch Fleet/Embo and the Firth are closely linked and both King Eider and Black and Surf Scoters can find their way here.

Habitat

Dornoch Firth is an SPA, and has a quiet coastline with areas of saltmarsh and muddy foreshore.

Species

Wintering wildfowl include thousands of Wigeon, Teal, Shelduck, Pintail, Scaup and Common and Velvet Scoters. Waders include the usual species, notably small numbers of Turnstone and Sanderling.

On passage a variety of waders may occur, including small numbers of Grey Plovers, Knots and Bar-tailed Godwits, and a few Greenshanks. Ospreys can be seen fishing in the Firth in summer.

Access

Dornoch Take the B9168 or A949 east off the A9 to Dornoch, where several tracks lead to the shore. The best area for scoter.

Dornoch Point Reached on foot along a track south past the small airstrip.

A9 road bridge This gives good views of the central part of the Firth.

Edderton Sands and Cambuscurrie Bay Leave the A9 at the roundabout at the south end of the A9 road bridge on the minor road to Meikle Ferry. A good area for Pintail and up to 400 Scaup.

FURTHER INFORMATION

Grid refs: Dornoch NH 805 895; A9 road bridge NH 748 858; Edderton Sands and Cambuscurrie Bay NH 732 859

404 GOLSPIE, LOCH FLEET AND EMBO (Highland)

OS Landranger 21

North of the Dornoch Firth, this area holds large numbers of seaducks and often offers exceptional views of Common Eider and Long-tailed Duck. The best time to visit is probably late March–early May, when numbers peak.

Habitat

The coast is sandy towards Golspie, with dunes along the shore. About 3 miles long, the shallow saltwater basin of Loch Fleet is almost completely enclosed. An attempt to drain it resulted in the development of an extensive alder carr at the head of the Loch beyond the Mound (the embanked A9), and there are areas of both semi-natural and planted pines adjacent to the shore. 1,000 ha around the loch are an NNR managed by SNH in partnership with the Scottish Wildlife Trust.

Species

In winter all three divers occur in small numbers, though Red-throated is commonest. There are also Slavonian and sometimes Red-necked Grebes. Variable numbers of Common Eider winter, sometimes as many as 2,000, together with several hundred Long-tailed Duck, Common Scoter and Red-breasted Merganser. Many Common Eider loaf around the Loch. Velvet Scoter and Goldeneye occur in smaller numbers, and other wintering wildfowl include up to 2,000 Wigeon, Teal and very occasionally Surf Scoter.

Seaducks peak in spring and autumn, with up to 2,000 Long-tailed Ducks in April–May, but only Common Eider is numerous in summer. A King Eider was present 1973–93, with occasionally up to three in the area, but none has been seen in recent years. There are significant numbers of Greylag Geese in the area, with 400–500 in late winter, and a flock of up to 60 Whooper Swans uses the surrounding fields. Short-eared Owl is seen quite frequently, Hen Harrier and Peregrine less often. The usual waders are present in winter, with good numbers of Oystercatcher, Common Redshank, Dunlin, Knot and Bar-tailed Godwit; these roost at Skelbo Point, the beach at the mouth of the Loch and Balblair saltmarsh. Notable are small numbers of Purple Sandpipers, especially at Embo. One or two Glaucous Gulls are regular, and there are sometimes Iceland as well. Small numbers of Snow Bunting and, irregularly, Twite may be found in the dunes and along the shore.

On passage a few interesting waders occur, but are generally few and far between. In autumn, north-west or north winds, especially if it is misty or hazy, can push seabirds onto the coast, and Sooty Shearwater and skuas may occur.

Breeding birds include Fulmar, Shelduck, Common Eider and Common, Arctic and Little Terns. The surrounding woodland has resident Common Buzzard, Sparrowhawk, Siskin and Scottish Crossbill (at Balblair), and Crested Tit is present in the area but elusive. In summer look for Common Redstart, while Osprey is quite often seen fishing in Loch Fleet.

Access

Golspie The sea and beach here are of straightforward access from the A9.

Loch Fleet, Littleferry Most seaducks congregate north of the Loch's mouth. Access is along the beach, either walking south from Golspie or, better, taking Ferry Road off the A9 at the

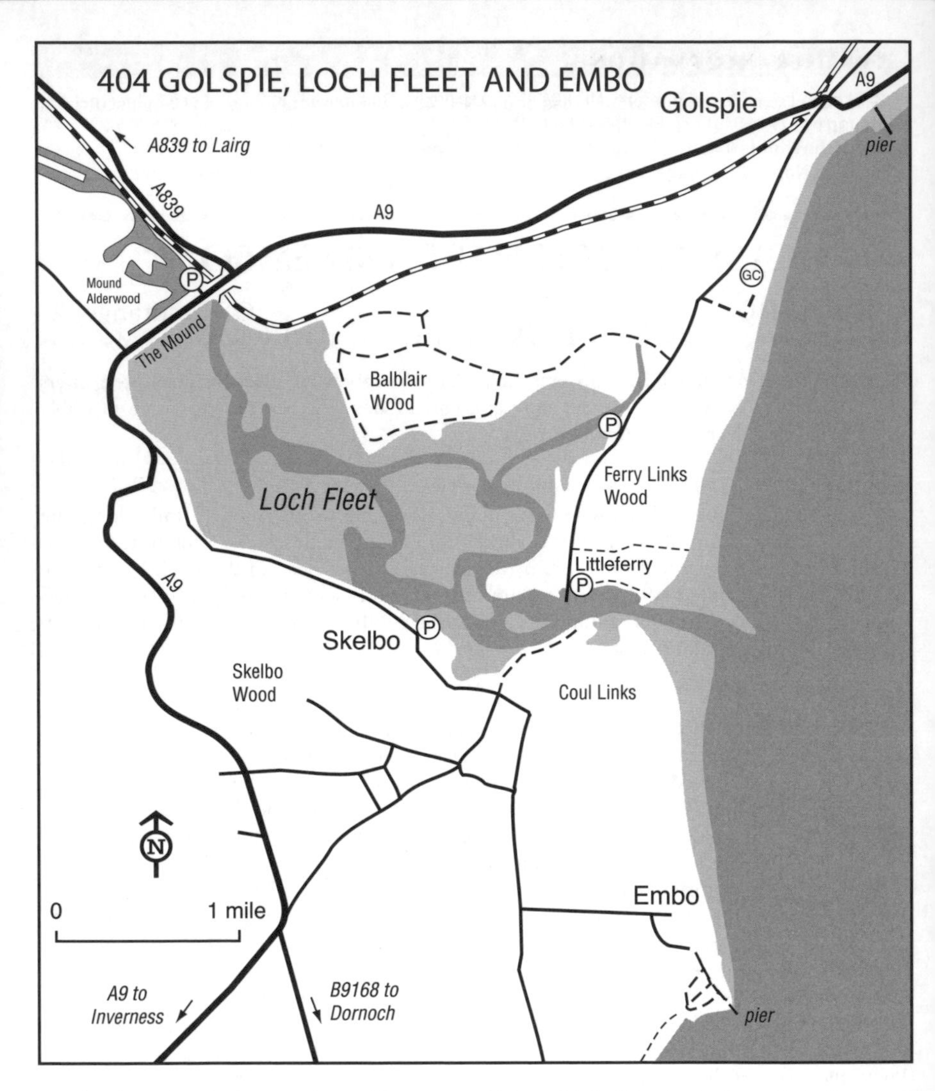

south edge of Golspie, past the golf course and Ferry Links Wood, south to Littleferry and following tracks east to the mouth.

Loch Fleet, Balblair Follow directions as for Littleferry but park where the minor road first meets the estuary at Balblair Wood to view the estuary.

Loch Fleet, The Mound The upper part of Loch Fleet, together with the freshwater pools and carr beyond the Mound are visible from the A9, with a car park at the north end of the causeway.

Loch Fleet, Skelbo The Loch can also be viewed from the minor road to Skelbo off the A9. This skirts the south shore, with a car park and information board at Skelbo.

Coul Links Carry on the minor road past Skelbo and, after 0.5 miles, park by the right-angle bend.

Embo Continue past Skelbo to the shore at Embo, a collection of caravans and chalets. There are usually seaducks off the makeshift pier (with Purple Sandpiper on the rocks) and this is a good area for Common, Velvet and sometimes Surf Scoters; if not, walk north along the beach to the mouth of Loch Fleet.

FURTHER INFORMATION

Grid refs: Loch Fleet, Littleferry NH 806 957; Loch Fleet, Balblair NH 809 969; Loch Fleet, The Mound NH 774 981; Loch Fleet, Skelbo NH 793 953; Coul Links NH 800 948; Embo NH 819 922
Scottish Wildlife Trust: tel: 0131 312 7765; web: www.swt.org.uk
Scottish Natural Heritage, reserves manager: tel: 01408 633602; web: www.nnr-scotland.org.uk

405 FORSINARD FLOWS (Highland/Caithness)

OS Landranger 10

This vast RSPB reserve covers 6,880 ha and protects an important area of deep peatland in the 'Flow Country' of Sutherland and Caithness; it is part of a larger 10,000 ha Flows NNR and is best visited in May–June.

Habitat

The 'Flow Country' consists of a vast expanse of more or less flat blanket bog interspersed by a mosaic of small pools and lochans. It is internationally important for both its flora and fauna, yet, in the 1980s, ill-conceived government subsidies and tax incentives led to large areas being damaged or destroyed by plantations of coniferous trees. Fortunately the subsidies have been removed but, as in the case of Forsinard, the only sure protection is outright ownership by a conservation body.

Species

Breeding species include Red-throated and Black-throated Divers, Greylag Goose (truly wild birds), Common Scoter, Red Grouse, Hen Harrier, Common Buzzard, Merlin, Short-eared Owl, Golden Plover, Greenshank, Common Sandpiper, Dunlin, Cuckoo, Dipper, Wheatear, Stonechat and Raven. Golden Eagle may visit.

Access

Reached along the A897, and 24 miles from Helmsdale on the A9 and 14 miles from Melvich on the north coast, the reserve is served by a railway station on the Inverness-Wick line, with trains stopping three times a day. The reserve is open at all times, and there is a visitor centre at Forsinard Station, open daily 9am–5.30pm, Easter–late October with regular guided walks. There are two circular trails, the mile-long Dubh Lochan Trail which crosses cut-over bog and then pristine blanket bog with many small lochans, and the 4-mile Forsinain Trail which passes through pastures as well as bog and forest-to-bog restoration (starting on the reserve, where it is wheelchair-accessible, it continues through Forestry Commission Scotland's Forsinain Forest and returns along the privately owned River Halladale river). The road also offers good views of the reserve.

FURTHER INFORMATION

Grid ref: NC 891 425
RSPB Forsinard Flows: tel: 01641 571225; email: forsinard@rspb.org.uk

406 DUNNET HEAD AND THURSO (Caithness)

OS Landranger 12

Dunnet Head marks the most northerly extent of the Scottish mainland and holds good seabird colonies; it is best visited in May to mid-July. Nearby Thurso Harbour regularly attracts white-winged gulls in winter–early spring, and, together with Dunnet Bay, a variety of divers and seaducks.

Habitat

Dunnet Head is topped by moorland with a few small lochs, and there are cliffs, especially on the north and east flanks. St John's Loch and Loch of Mey at the base of the Head attract waterfowl, and, to the west, Dunnet Bay is sandy, with a rocky stretch of coast between it and Thurso Harbour.

Species

Breeding seabirds on Dunnet Head include Fulmar, Common Guillemot, Razorbill, Black Guillemot, Puffin and Kittiwake. Great Skua may attempt to breed on the moorland, especially on the western flank. Rock Dove and Raven haunt the cliffs, and Twite can be found around crofts. Small numbers of Arctic Terns breed on Dunnet Links and may be seen fishing in the bay, where there are often summering Great Northern Divers.

In winter numbers of Red-throated and Great Northern Divers are present offshore at Thurso and Dunnet Bay, together with a few Black-throated Divers. Shag, Common Scoter, Long-tailed Duck, Goldeneye, Common Eider and Red-breasted Merganser are also frequent and are joined by Black Guillemot, Razorbill and Common Guillemot. In the area are small numbers of Whooper Swans, which together with a variety of wildfowl can be seen on St John's Loch. A flock of up to 100 Greenland White-fronted Geese regularly winters around Loch of Mey and Loch Heilen. Greylag Geese are also regular, and Brent Goose is occasional in Dunnet Bay. The few waders include Turnstone and Purple Sandpiper. Glaucous and Iceland Gulls are usually present, sometimes several of each. A Peregrine is often seen and, occasionally, large numbers of Snow Buntings.

Loch of Mey is attractive during passage periods and has held several rarities, and Dunnet Bay is also worth checking for migrant waders, but St John's Loch is disturbed by fishermen in summer.

Access

Dunnet Head Leave the A836 at Dunnet village north on the B855, which ends after 4.5 miles at the car park at Easter Head.

St John's Loch Lies north-east of Dunnet village and can be viewed from the A836.

Loch of Mey Approximately 3 miles east of Dunnet village, access is from the A836, taking the minor road north-west towards Harrow, and then, after 0.6 miles (just before the turn to Harrow Harbour), the track west to the James MacIntyre Memorial Hide on the loch's shore. The water level on the loch may be high and relatively uninteresting, or low, attracting passage waders.

Loch Heilen This large loch can be viewed from surrounding minor roads, notably in the east at Lochend.

Dunnet Bay Castletown pier at the south-west corner of Dunnet Bay gives views of the bay, and the A836 parallels the shore from here to Dunnet, with a car park at the east end of the bay.

River Thurso In Thurso a track runs north from the A836 along the east bank of the River Thurso towards the ruined castle, and a footpath continues along the shore.

Thurso Harbour Access to the breakwater in Thurso Harbour is straightforward from the

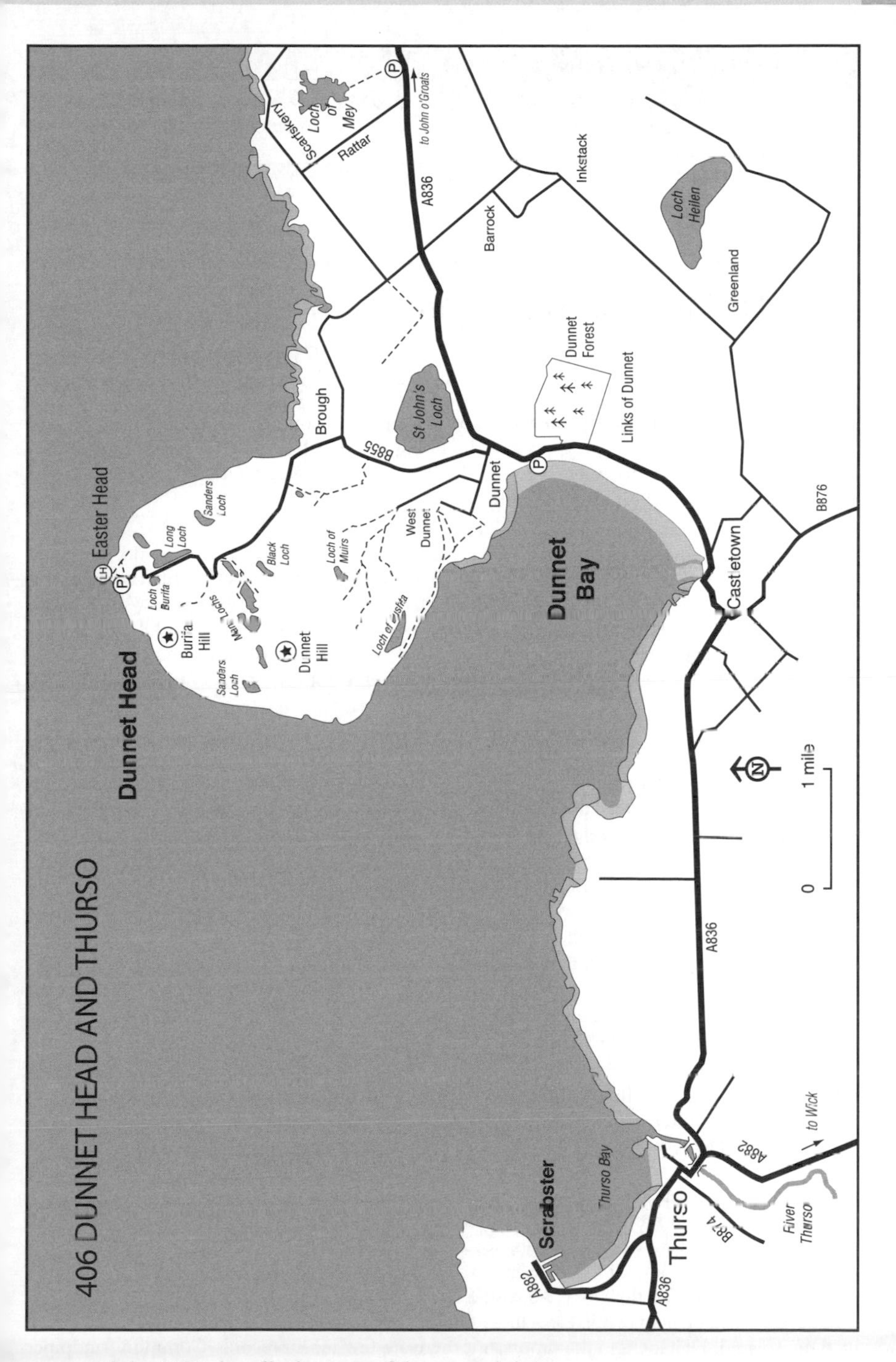

town, and the A836 also affords views of the sea and shore.

Scrabster Harbour The departure point for the Orkney ferry, this sometimes holds a few divers, seaducks and gulls.

FURTHER INFORMATION

Grid refs: Dunnet Head ND 202 766; St John's Loch ND 226 716; Loch of Mey SD 281 737; Loch Heilen ND 261 676; Dunnet Bay (Castletown) ND 201 681; River Thurso ND 121 682; Thurso Harbour ND 120 687; Scrabster Harbour ND 102 704
Thurso Tourist Centre: tel (April–October): 01847 892371; web: www.visithighlands.com

407 SUTHERLAND (Highland)

OS Landranger 9, 10 and 16

An area of sombre hills, vast expanses of moorland and flow country, this bleak but grand area is difficult to work. Distances are large and the going is rough, but the place has its own special attractions. The best time to visit is between the last week of May and early July; during the rest of the year there are few birds on the hills.

Habitat

The flow country of east Sutherland and Caithness is a mosaic of small lochans, peaty burns and blanket bogs. The stronghold of breeding Greenshank in Britain, it is threatened by forestation. One-third is owned by forestry interests and large areas have been planted with exotic species of conifer. There are small areas of native birch, mainly along watercourses, and other habitats include lochs, sea lochs and the high tops, with several 'arctic' species breeding and the potential for others to do so.

Species

Larger lochs have nesting Black-throated Diver and sometimes Red-throated, though the latter tend to favour smaller, isolated pools and lochans, flying to bigger lochs and the sea to feed. Sea lochs hold Common Eider and Arctic Tern. Especially in the east, Hen Harrier and sometimes Short-eared Owl breed, favouring young conifer plantations, usually abandoning these once the trees are more than 6–8 feet high. Young plantations also attract Black Grouse, again mainly in the east, though it is a difficult bird to find. Siskin frequents older conifer woods. Common Buzzard is uncommon, but Golden Eagle, if not common, is widespread. The way to see one is to climb a mountain, though they can be seen from the roads. Peregrine is also widespread, favouring the vicinity of cliffs and crags, especially near woodland. Merlin is scattered over moorland areas in small numbers. Ptarmigan breeds on every hill above 2,500 feet, and may sometimes be seen at lower elevations. It prefers areas with rocks, scree or boulder piles, and overlaps altitudinally with Red Grouse. Rocky screes and corries also attract the rare Snow Bunting (around long-lasting summer snowfields with nearby rocks and boulders), and you may find Dotterel on the tops.

Greenshank breeds in broad marshy valleys within the moors, and often frequents the edges of freshwater lochs once the young have fledged; Dunlin is also scattered in marshy areas. Oystercatcher, Golden Plover, Curlew and Common Snipe are widespread, but Whimbrel, Wood Sandpiper and Red-necked Phalarope are rare and erratic breeders. Other species that may have nested and are worth searching for are Temminck's Stint, Sanderling, Green Sandpiper and Turnstone. Common Gull, Teal, Wigeon and sometimes Greylag Goose breed on scattered lochs on the moors and flows. Red-breasted Merganser is commonest near the coast, but may breed at the head of glens, while Goosander favours larger rivers and lochs. Common Scoter has bred on isolated small lochs in the moors, but there are few recent records. Common Sandpiper and Dipper occur on most streams and rivers.

The commonest passerines are Meadow Pipit and Skylark. Wheatear is widespread, but Whinchat more scattered. Ring Ouzel is found around crags and rocky gullies in the hills. In birch woods look for Woodcock, Tree Pipit, Wood and Willow Warblers, Common Redstart, Spotted Flycatcher and Lesser Redpoll. These, and areas of rhododendron (which are usually found around houses), should be checked for Redwing (Britain's first breeding record was in Sutherland in 1925). Brambling bred in 1920 and may do so again. Twite favours the vicinity of old buildings, rough pasture and moorland edge, especially on the coast. It particularly likes to feed on short grazed turf around crofts, cliffs and roadsides.

Access

In this vast area it is impossible to single out one place, because birds are scattered over the available habitat. It is more a question of convenient access. The following areas are suggested as a start.
RSPB Forsinard Flows See page 603.
Loch Eriboll (OS Landranger 9) The A838 parallels most of the shore, approaching closest on the east side.
Loch Hope (OS Landranger 9) A minor road runs along the east shore from Hope, on the A838.
Ben Hope (3,040 feet, OS Landranger 9) A minor road runs south, from the A838 at Hope, along the west flank of the mountain.
Kyle of Tongue (OS Landranger 10) A tidal inlet with extensive mudflats that attract passage waders. Tongue Pier is a good place to look from, while the A838 bisects the Kyle before paralleling the shore for a short distance and connecting with an unclassified road.
Loch Loyal (OS Landranger 10) East of the A836.
Ben Loyal (2,509 feet, OS Landranger 10) Access west from the A836.
Loch Naver (OS Landranger 16) The B873 runs along the north shore.
Ben Klibreck (3,157 feet, OS Landranger 16) East of the A836.
Loch Shin (OS Landranger 16) The A838 runs along the east shore and Lairg, on the A836, is at the south end.

Exploration away from the road is necessary to see at least some of the birds, and you should be properly equipped to deal with the sometimes harsh conditions. Though access is generally not problematic, grouse shooting and deer stalking take place in late summer and autumn; notices giving details of stalking are often posted at access points. You will be unwelcome if you disturb either the hunters or the hunted, and as a general rule you should keep off the moors from early July.

FURTHER INFORMATION

Grid refs: Loch Eriboll NC 449 589; Loch Hope NC 474 573; Ben Hope (summit) NC 477 501; Kyle of Tongue NC 580 585; Loch Loyal NC 619 466; Ben Loyal (summit) NC 578 488; Loch Naver NC 608 365; Ben Klibreck (summit) NC 585 299; Loch Shin NC 467 204

408 FARAID HEAD (Highland)

OS Landranger 9

This site has smaller and less spectacular seabird colonies than Clo Mor, but is easier to reach.

Habitat

This is a high, grass-topped headland on a small peninsula.

Species

On the grassy east slopes there are Puffins, and other seabirds include Fulmar, Kittiwake,

Black Guillemot, Common Guillemot and Razorbill. During late summer and autumn Sooty Shearwater may be seen offshore.

Access

From the A838 in Durness take the minor road north-west to Balnakeil. Park and walk along Balnakeil Sands to the Head.

FURTHER INFORMATION

Grid ref: NC 391 687

409 CLO MOR (Highland)

OS Landranger 9

Mainland Britain's highest sea cliff, Clo Mor rises to 921 feet and lies c. 3 miles south-east of Cape Wrath. It is truly spectacular and has huge seabird colonies. A visit between May and mid-July is recommended.

Habitat

The cliffs are backed by a typical Sutherland landscape of moors and hills.

Species

Breeding seabirds include Fulmar, Common Guillemot, Razorbill and Black Guillemot, with one of the largest British colonies of Puffin. Large numbers of Gannets are present offshore and Peregrine and Rock Dove nest on the sea cliffs. Ptarmigan occurs on rocky areas of adjacent hills, sometimes as low as 600 feet above sea level, together with Red Grouse and a few Golden Plovers. Greenshank is present along the river valleys, and there may be summering or even breeding Great Skua. Kyle of Durness holds Red-throated Divers, and wintering Great Northern Divers sometimes remain until late spring.

Access

Leave the A838 at Kyle of Durness to Keoldale. From here a passenger ferry crosses the Kyle (crossing time 10 minutes), connecting with the road to Cape Wrath lighthouse. A minibus service runs to Cape Wrath several times daily in May–September. Ask to be dropped at the Kearvaig track, and then walk to Kearvaig, approaching Clo Mor from the west. Follow the cliffs east for c. 3 miles and then cut inland along the east flanks of Sgribhis-bheinn to meet the minibus, by arrangement, at Inshore. This is a rough cross-country walk. Access is sometimes restricted because of activity on the firing range (the minibus operator can advise).

FURTHER INFORMATION

Grid ref: NC 391 687
Durness Tourist Information Centre: tel (April–October): 01971 511259; web: www.visithighlands.com, see also www.durness.org
Kyle of Durness Ferry (John Morrison): tel: 01971 511376
Minibus service to Cape Wrath (Iris Mackay): tel: 01971 511343 or 511287; mobile: 07751 789048

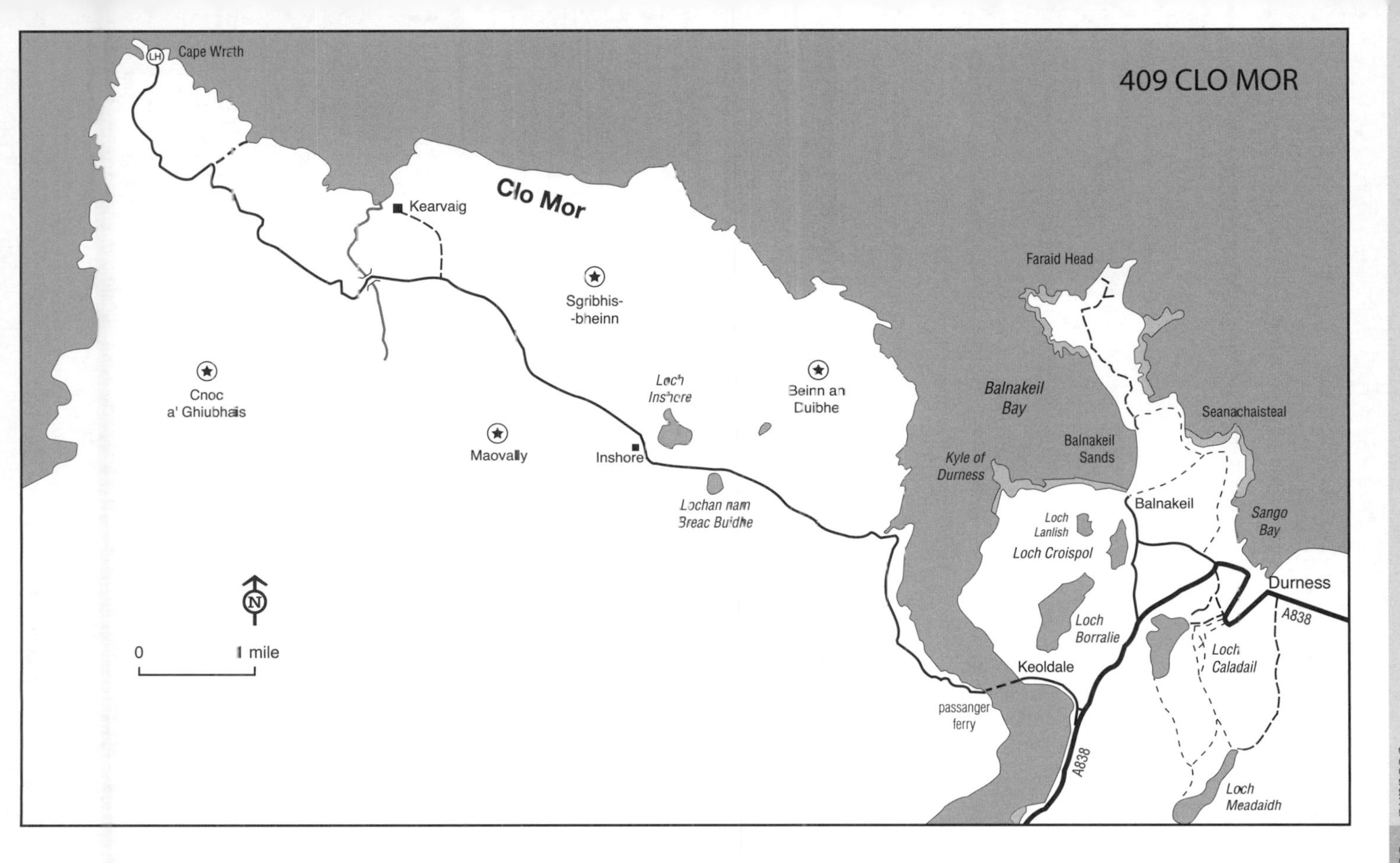
409 CLO MOR
Cape Wrath
LH
Clo Mor
Kearvaig
Sgribhis-
-bheinn
Cnoc
a' Ghiubhais
Maovally
Loch
Inshore
Inshore
Beinn an
Duibhe
Lochan nam
Breac Buidhe
Faraid Head
Balnakeil
Bay
Balnakeil
Sands
Seanachaisteal
Kyle of
Durness
Balnakeil
Sango
Bay
Loch
Lanlish
Loch Croispol
Durness
A838
Loch
Borralie
Loch
Caladail
Keoldale
passanger
ferry
A838
Loch
Meadaidh
N
0
1 mile

410 TARBET (Highland)

OS Landranger 9

Tarbet lies on the mainland, opposite Handa (see below), and is worthy of exploration.

Habitat
Tarbet is a hamlet on the coast, surrounded by moorland with lochans.

Species
The region south of Loch Laxford holds some of the typical Sutherland specialities – Black and Red-throated Divers, raptors and Greenshank.

Access
The A894 and A838 cross the area, as well as the minor roads to Tarbet. Be prepared for heavy going, and have an OS map to hand if you leave the road. Stalkers' paths and tracks are clearly marked, and there are few restrictions on access outside the deer-stalking season, which is from mid-August.

FURTHER INFORMATION

Grid ref: NC 164 488

411 HANDA (Highland)

OS Landranger 9

This superb island off north-west Sutherland holds over 200,000 breeding seabirds and is easily accessed. The island is a reserve of the Scottish Wildlife Trust and the best time to visit is May–early July.

Habitat
The island covers 310 ha and is only a few hundred yards offshore. On three coasts, cliffs rise to over 400 feet, but the south shore also has sandy beaches and dunes. Inland is rough sheep pasture and heather moor with six small lochans. There is a small plantation by the bothy.

Species
Black-throated Diver is usually present offshore in summer, and wintering Great Northern Divers sometimes stay until mid-May. Red-throated Diver, Common Eider and Shelduck breed and can be seen off the south-east coast. Approximately 100,000 pairs of Common Guillemot vastly outnumber the other seabirds, which include Razorbill (c. 9,000 pairs) and Kittiwake (c. 10,000 pairs) as well as Fulmar (c. 3,000 pairs), Shag (c. 200 pairs) and c. 800 pairs of Puffins. Black Guillemot does not regularly breed, but is often seen on the boat crossing. Both Great and Arctic Skuas breed, the former in internationally important numbers, together with Great Black-backed and Herring Gulls. Peregrine and Common Buzzard visit, Oystercatcher and Ringed Plover breed on the beaches, Rock Dove nests on the cliffs, Common and Arctic Terns on the skerries in Port an Eilein (near the landing point), and Common Snipe, Stonechat and Wheatear on the grassland and moor, occasionally joined by Red Grouse and Golden Plover.
Passage periods bring small numbers of waders to shores, and there are occasionally movements

of interesting seabirds offshore. Pomarine Skua is possible in early May, and Manx and Sooty Shearwaters in autumn.

Access

Handa is open from April to early September. Boats leave from Tarbet, which is reached from the A894 by taking a minor road to the north-west c. 3 miles north-east of Scourie. The ferry runs Monday–Saturday and the first boat usually departs at 9.30am and the last returns at 4.30pm. Crossings are on demand and weather dependent. There is a shelter and display on the quay, and a warden is present in summer.

FURTHER INFORMATION

Grid ref: Tarbet NC 164 488
Scottish Wildlife Trust: tel: 0131 312 7765; web: www.swt.org.uk

412 INVERPOLLY (Highland)

OS Landranger 15

Inverpolly is a true wilderness and magnificently scenic. It holds the usual range of highland species and is at its best from April to June.

Habitat

This remote, almost uninhabited area has a range of habitats. A rolling moorland plateau of heather and grass is interspersed with freshwater lochs (including the large Loch Sionascaig), lochans, streams and boggy hollows, and the impressive sandstone peaks of Stac Pollaidh, Cul Beag and Cul Mor, the latter reaching 2,787 feet. There are many small areas of remnant birch–hazel woodland, and the high tops form an arctic–alpine habitat. Sandy beaches and small islands flank the coast.

Species

Breeding birds include Black-throated Diver, which favours the larger lochs, including Sionascaig, and may sometimes be seen on the lochs adjacent to the A835. Red-throated Diver, Red-breasted Merganser, Goosander and Wigeon, with small numbers of Greylag Geese, also frequent the lochs, especially on the coast, where Fulmar, Shag, Common Eider and Black Guillemot breed. The high tops hold small numbers of Ptarmigan, with Red Grouse at lower elevations. Snow Bunting, which is a winter visitor to the area, may stay late in spring and perhaps even breed. Golden Eagle, Common Buzzard, Merlin and Peregrine occur, together with Raven. Though eagles may be seen from the road, exploration on foot is usually more productive. Golden Plover and Greenshank breed on the moors, with Stonechat, Wheatear and Ring Ouzel in higher, rockier gullies. Birch woodland has Woodcock, Wood Warbler, Spotted Flycatcher and sometimes Redwing. Twite occurs on the moorland edge and around crofts near the coast.

Access

The area comprises a roughly rectangular area of land between Enard Bay in the north-west and the A835 between Drumrunie and Elphin in the south-east. Access is unrestricted, but no roads penetrate the area and there are few tracks; the going can be difficult and walkers should be experienced and properly equipped. It is easier to walk the peripheral roads and, from the road on the west side of the area, just south of Inverkirkaig, the north bank of the River Kirkaig to the falls. A variety of woodland species can be seen in the glen. Note that from mid-July (i.e. after the main period of interest for birdwatchers) it is best to avoid the area or at least check with the local estate offices before accessing the hills.

FURTHER INFORMATION

Grid ref: River Kirkaig NC 085 194

413 BEINN EIGHE (Highland)

OS Landranger 19 and 25

Beinn Eighe was the first NNR to be declared in Britain and was established to protect remnant areas of Caledonian pinewood. The 4,758 ha also incorporate uplands, and is best visited May–early July.

Habitat

There are c. 180 ha of natural pinewoods with an understorey of heather, as well as c. 60 ha of birch. A further 485 ha have been replanted with native tree species. Woodland is concentrated along the south-west shore of Loch Maree. The hills rise to over 3,000 feet; the highest Ruadh-stac Mor, which peaks at 3,313 feet, is just outside the reserve. Non-wooded areas mostly consist of heather and grass moorland and bog, but on the high tops this gives way to dwarf shrubs, which mix with mosses and liverworts to form an arctic–alpine heath.

Species

Areas of relict woodland hold resident Sparrowhawk, Woodcock and Great Spotted Woodpecker, while Tree Pipit, Common Redstart and Wood and Willow Warblers are summer visitors, and join Siskin and Lesser Redpoll. Both Common and Scottish Crossbills are sometimes seen, and fortunate observers may find breeding Redwing in the birches. Crested Tit has been recorded in the area, and may eventually expand its range to incorporate the reserve's pine forests. On lochs, streams and rivers are Red-throated and Black-throated Divers, Goosander, Red-breasted Merganser, Greylag Goose, Common Sandpiper and Dipper. Golden Eagle, Common Buzzard, Merlin, Peregrine and Short-eared Owl occur, as well as Raven. Probably the best chance to see an eagle is to regularly scan the hillsides. Common moorland birds are Skylark and Meadow Pipit, joined by smaller numbers of Golden Plovers, Ring Ouzels, Wheatears, Whinchats and a few Twites. In wetter, boggier valley bottoms you may find Greenshank. Higher, Red Grouse is joined by Ptarmigan, and a thorough search of the tops may reveal summering Snow Bunting.

Access

The SNH Beinn Eighe Visitor Centre is at Kinlochewe, on the A832 1 mile north of the junction with the A896, and is open Easter–October. Three trails start from the visitor centre, the Picnic, Rhyming (400 yards) and Ridge (1400 yards) trails, and information and advice are available on the two longer trails that start at the Lochside car park at Glas Leitire (NH 000 650) on the A832 1.75 miles further north-west: the Mountain Trail extends for 4 miles and rises above the treeline to c. 1,800 feet, taking in moorland and mountain, while the Woodland Trail extends for 1 mile through typical woodland habitats (otherwise, the Pony Trail, which runs west from the Visitor Centre (leaving the road at NH 022 628), gives the best access to the highest ground).

There is unrestricted access to the hills in a large area south of Loch Maree and north of the A896, which includes the NNR; except for 1 September–21 November, which is the deer-stalking season, when permission must be sought from local estates to enter some areas.

Birdwatchers, however, are unlikely to visit during this period. The A832 follows the south-west flank of Loch Maree, with several car parks along the shore. The reserve is bordered to the south-east by the A896, but within this area there are no roads and walkers should be properly equipped.

FURTHER INFORMATION

Grid ref/postcode: Beinn Eighe Visitor Centre NH 019 630 / IV22 2PD
Scottish Natural Heritage: Visitor Centre, tel. 01445 760254; reserve manager, tel. 01445 760254; web: www.nnr-scotland.org.uk

414 SKYE (Highland)

OS Landranger 23, 32 and 33

Skye is the northernmost of the Inner Hebrides and is huge, nearly 50 miles long, with dramatic scenery.

Habitat
The island is mostly sheep-walk with some areas of forestry, no one site is outstanding.

Species
Divers, Peregrine, Golden and White-tailed Eagles, Ptarmigan, Black Grouse, Black Guillemot, Greenshank and Dipper all breed. The Cuillin Hills in the south have Ptarmigan and Golden Eagles.

Access
Exploration on foot can be worthwhile in many places. At the Aros Centre (www.aros.co.uk), just south of Portree on the A87, CCTV images from a White-tailed Eagle's nest can be viewed, and the staff can advise on seeing both White-tailed and Golden Eagles.

FURTHER INFORMATION

Grid ref/postcode: Aros Centre NG 477 424 / IV51 9EU

415 ISLE OF RHUM (Highland)

OS Landranger 39

Rhum was the centre for the successful reintroduction of White-tailed Eagle into Britain. It has many other attractions, however, including numbers of breeding seabirds. The best time to visit is April–June. The island is an NNR covering 10,700 ha, managed by SNH.

Habitat
The basic habitats are the coast (with rugged cliffs, sandy bays and maritime grassland), moorland, the higher, more exposed hilltops, relict woodland, and, importantly, reforestation. SNH is trying to re-establish woodland and has planted extensive areas with native trees, notably in the east.

Species

Breeders include Red-throated Diver, Fulmar, Shag, Common, Lesser and Great Black-backed and Herring Gulls, Kittiwake, Common Guillemot, Razorbill, Black Guillemot and Puffin, and occasionally Common or Arctic Terns. However, the Kittiwake and auk ledges are largely inaccessible, hidden at the base of the cliffs. There are c. 100,000 pairs of Manx Shearwater, which breed on the tops in the centre of the island (this is around a third of the world population). Storm Petrel occurs offshore but there is no evidence of breeding. Small numbers of Greylag Geese nest, and the species is also regular on passage. Other wildfowl include Shelduck, Teal, Common Eider and Red-breasted Merganser. Two to five pairs of Golden Eagles are known, as well as Peregrine, Merlin and Sparrowhawk. A total of 82 White-tailed Eagles was released on Rhum between 1975 and 1985, but most have dispersed and can occur anywhere in the Inner Hebrides or, indeed, as far afield as Shetland, the Outer Hebrides and Northern Ireland. They may still be seen on Rhum, but are not guaranteed, and chances are currently better on some of the other Inner Hebrides. Red Grouse and Raven breed on the hills, and waders include Oystercatcher and Ringed and Golden Plovers, with Woodcock in the wooded areas, occasionally with Long-eared or Short-eared Owls. Common Sandpiper and Dipper nest along streams. Common open-ground birds are Meadow Pipit and Skylark, together with Stonechat, Whinchat and Wheatear, and Ring Ouzel in the hills. Common Whitethroat, Willow Warbler, Chiffchaff and Siskin nest, as do a few pairs of Wood Warblers, and the number and variety of woodland birds should increase as the forest develops.

Offshore, especially in late summer, Sooty Shearwater, Gannet, and Great and Arctic Skuas are possible. On passage, Rhum holds a variety of waders and wildfowl, including Whooper Swan, Pink-footed, White-fronted and Barnacle Geese, Sanderling, Whimbrel, Greenshank and Turnstone, and numbers of common passerines. In winter there are Great Northern Diver and small numbers of seaducks offshore.

Access

Caledonian MacBrayne operate a passenger ferry from Mallaig, with several sailing a week; the journey takes 1.25–2.5 hours. The boat returns the same day and day-trips are only possible on summer Saturdays. Day trips also run from Arisaig (see Further Information) landing at Loch Scresort and allowing exploration of the area around Kinloch. However, a day trip generally permits too little time ashore to be really worthwhile (though the crossing has many birds: all three divers, Manx and sometimes Sooty Shearwaters and Storm Petrel are possibilities). Visitors who would like to stay overnight should pre-book the hostel, bothy, B & B or camp site (although camping is uncomfortable July–October because of midges). For further information, contact the SNH Reserve Office.

FURTHER INFORMATION

Scottish Natural Heritage Reserve Office: tel: 01687 462026
Caledonian MacBrayne: tel: 01687 462403; web: www.calmac.co.uk
***MV Shearwater* (day trips from Arisaig late April–late September):** tel: 01687 450224; web: www.arisaig.co.uk

416 ARDNAMURCHAN (Highland)

OS Landranger 40 and 47

The rugged Ardnamurchan Peninsula forms the westerly point of mainland Scotland. It holds a wide range of breeding species and is best visited May–October.

Habitat

The peninsula is c. 17 miles long by c. 7 miles at its widest point, and rises to 1,731 feet at Ben Hiant and 1,679 at Ben Laga. The varied habitats include heather moorland with many pools and lochans, and some very extensive conifer plantations in the inland areas, as well as stands of relict semi-natural oak woodland around the coast, bounded in many places by sandy beaches and bays.

Species

Breeders include Great Spotted Woodpecker, Wood and Willow Warblers and Spotted Flycatcher in the oak woodlands, with Tree Pipit and Common Redstart on their periphery and on the more open slopes. Dipper occurs on streams and rivers, and stands of conifers hold Siskins and Lesser Redpolls. Areas of moorland hold Red Grouse, Golden Plover, Greenshank, Ring Ouzel, Wheatear, Stonechat, Whinchat and Twite, and the entire area is a good one for raptors including Golden Eagle, Merlin and Common Buzzard; Hen Harrier and Short-eared Owl often hunt over young conifer plantations, while White-tailed Eagle and Peregrine are especially frequent along the coast. Raven should also be seen. Smaller lochans have breeding Red-throated Diver, which may visit the sea to feed, and around the coast look for Common Eider, Common Sandpiper, Common and Arctic Terns, with Gannet offshore.

On passage a wide variety of species may be present. Seawatching from the Point of Ardnamurchan may produce all three divers, Manx and sometimes Sooty Shearwaters, Storm and Leach's Petrels, Gannet, skuas, gulls and terns. Wildfowl may include Whooper Swan and Greylag, White-fronted and Barnacle Geese, and waders including Golden Plover, Bar-tailed Godwit, Greenshank and less frequently Grey Plover, Knot, Sanderling, Green Sandpiper and Whimbrel.

In winter small numbers of Great Northern Divers and Goldeneye are on the sea, and other wintering wildfowl around the coast include Wigeon, Teal and Greenland White-fronted Geese (the last in the area south of Loch Sheil). Wintering waders may include a few Greenshanks.

Access

This vast wild area could justify many days' exploration, especially for the serious walker. The following is a selection of the more interesting and accessible points.

RSPB Glenborrodale This 100-ha reserve comprises oak woodland, conifers, scrub, moorland and a section of rocky coast. The reserve is west of Glenborrodale village, immediately west of the black railings surrounding the grounds of Glenborrodale Castle, and the 2-mile trail starts west of this, where the stream crosses the road, rejoining the B8007 at the north-western tip of the reserve. Regular walks are organised (information from the RSPB's North Scotland Office or the Ardnamurchan Natural History Centre). The trail is not wheelchair-accessible.

Kentra Bay A good area for wildfowl and waders in winter and on passage. Leave the A861 in Acharacle (just north of the church) west on the B8044 towards Kentra. After c. 0.5 miles turn left at the crossroads and follow the road to the small parking area at the end. Continue on foot along the track on the south side of the bay, which reaches Camas an Lighe after 1 mile. The north side of the bay can be viewed from the B8044 beyond Kentra.

Ben Hiant A car park by the B8007 affords good views of the mountain, and persistent scanning may produce raptors, including Golden Eagle.

Loch Mudle Viewable from the B8007, this is a good spot for Red-throated Diver.

Point of Ardnamurchan From the B8007 at Achosnich a minor road leads to the Point. Seawatch from the vicinity of the lighthouse, or walk south along the cliffs to search for Peregrine and Raven.

FURTHER INFORMATION

Grid refs: RSPB Glenborrodale NM 601 608; Kentra Bay NM 649 676; Ben Hiant NM 562 616; Loch Mudle NM 540 660; Point of Ardnamurchan NM 416 672

RSPB North Scotland Office: tel. 01463 715000; email: nsro@rspb.org.uk

Ardnamurchan Natural History Centre: tel: 01972 500209;
web: www.ardnamurchannaturalhistorycentre.co.uk

417 COLL (Argyll)

OS Landranger 46

Coll is an island within the south sector of the Inner Hebrides. Lying at the west end of the island and covering 1,221 ha, the Coll RSPB reserve was established specifically to benefit Corncrake, which has declined disastrously in Britain in the 20th century, and Coll now supports around 120 pairs of Corncrakes. The island also holds numbers of breeding waders and raptors, and is at its best in spring and early summer, although winter can be good for divers and wildfowl.

Habitat

Coll is largely cloaked in Heather but the RSPB reserve includes two extensive dune systems together with *machair* (stabilised dunes of lime-rich shell sand supporting a wide variety of grasses and other wild flowers) and productive hay meadows. To the west, the reserve is bounded by the extensive sandy beaches of Feall Bay and Crossapol Bay, which are separated by the west dune system and flanked by rocky headlands.

Species

Corncrakes are present April–September, but generally cease calling in late July. They are generally easy to hear, and call most frequently at night, 10pm–5am; listen from the road between Uig and Roundhouse, but sightings are hard to come by (there is a Corncrake viewing platform next to the car park at Totronald). Early in the season is best, when the vegetation is still short. The machair has a high density of breeding waders, with Lapwing, Common Snipe, Oystercatcher and Common Redshank, and other species in this habitat include Shelduck and Wheatear. Other breeders on the reserve include Rock Dove, Raven and Twite, with Red-throated Diver, Greylag Goose, Teal and Arctic Skua in moorland areas. Hen Harrier, Peregrine, Merlin and Short-eared Owl may visit, and along the coast Fulmar, Shag, Common Eider and Common, Arctic and Little Terns nest. Gannet is often present offshore, and Manx Shearwater, Puffin, Razorbill and Common Guillemot breed on the Treshnish Isles to the south-east and are often present off Coll (look for shearwaters especially in the evening).

Wintering species include good numbers of Barnacle and Greenland White-fronted Geese, as well as 500–700 Greylag Geese. Interestingly, the RSPB reserve has a flock of c. 40 feral Snow Goose that breed nearby on the islet of Soa. Offshore, there are good numbers of divers (especially Great Northern, which may linger well into spring and can then be seen in full breeding plumage) and small numbers of seaducks, mainly Long-tailed Duck (sometimes c. 100) but also a few Common Scoter and Scaup; Feall Bay is favoured by seaducks and divers.

Access

Coll is accessible by car ferry from Oban, the boats are operated by Caledonian MacBrayne and call at Arinagour on Coll. The crossing can be excellent for seabirds, especially in spring and autumn. From Arinagour take the B8070 south-westwards: this road provides good opportunities to see moorland birds, including Red-throated Diver on the smaller lochans. At Arileod at the end of the B8070 turn either right (north) for c. 1 mile to the RSPB Totronald Stables Visitor Centre or south (forking right at the turning to the castle), to park near Crossapol Bay. Three suggested walking routes start from these two car parks. Some parts of the paths are difficult for wheelchair-users – please contact the reserve for more advice. The reserve is open all year, vehicles must not be taken onto the machair and there is no access to the hay meadows.

FURTHER INFORMATION

Grid ref: RSPB Totronald Stables NM 167 563
RSPB Coll: tel: 01879 230301
Caledonian MacBrayne: tel: 08705 650000
Tourist information: web: www.visitcoll.co.uk

418 TIREE (Argyll)

OS Landranger 46

Tiree is the westernmost of the Inner Hebrides and its open machairs and white shell-sand beaches have much in common with the Outer Hebrides. A thriving cattle-base crofting system ensures excellent management of the grasslands and machair, which hold very high densities of breeding waders, as well as almost a third of the UK's Corncrakes (390 calling males). The island also holds a wide range of breeding seabirds and wildfowl, and is at its best in spring and early summer, although autumn brings a wide selection of passage migrants including annual rarities, whilst winter is also good for divers, white-winged gulls and wildfowl.

Habitat

Tiree is largely flat and treeless, and is dominated by grassland, with around a third of the island *machair,* including extensive dune systems. The *machair* is bordered inland by fields of in-bye where grass and arable crops are grown, whilst the centre of the island is dominated by wetter acidic grasslands with some Heather, known locally as *sliabh*. There are numerous small wetlands and lochans as well as four large machair lochs that are particularly good for wildfowl

Species

Corncrakes are present April–early October, but generally cease calling at the end of July. Many call during the day in May, but they are generally easiest to hear at night, 10pm–5am; listen from the roads around Balemartine, Balephuil and Balinoe. Sightings cannot be guaranteed but are most frequent in early May when the vegetation is still short. The grasslands and wetlands have a very high density of breeding waders, with Lapwing, Common Snipe, Dunlin, Ringed Plover, Oystercatcher and Common Redshank, whilst other species in these habitats include Twite and Wheatear. The lochs and wetlands host a wide range of breeding wildfowl including Greylag Goose, Teal, Shoveler and Gadwall, and Whooper Swans regularly summer. There are scattered colonies of Arctic and Little Terns around the coast and mixed gull colonies all over the island. The cliffs at Ceann a' Mhara host Kittiwake, Shag, Fulmar, Common Guillemot and Razorbill plus a handful of Black Guillemots. Gannet, Puffin, Manx Shearwater and Red-throated Diver are often present offshore in summer, whilst seawatching in onshore winds brings Great Skua, Arctic Skua and Storm Petrel from nearby colonies, plus scarcer fare such as Leach's Petrel, Sooty Shearwater and Grey Phalarope.

Passage in both spring and autumn is often pronounced with large numbers of waders, wildfowl, seabirds and passerines passing through. Large flocks of Black-tailed Godwits, Whimbrels and Golden Plovers stage on the island and these are worth checking for scarcer species, such as Dotterel in the spring and American Golden Plover and Buff-breasted Sandpiper in the autumn. Passerine migration includes marked passage of White Wagtails, Wheatears, Meadow Pipits, Redwings and Lesser and Common Redpolls to and from Iceland/Greenland, with smaller numbers of warblers and chats including regular rarities in May/June and September/October.

Wintering species include good numbers of Whooper Swans, Barnacle and Greenland White-fronted Geese, plus over 3,000 resident Greylag Geese. Great Northern Divers are common in the bays along with a few Long-tailed Ducks and a handful of Glaucous Gulls are usually present in late winter. Wintering flocks of Twites and Skylarks attract regular Merlins and Hen Harriers, whilst Common Buzzard and Peregrine are resident. The beaches host mixed flocks of Sanderling, Dunlin, Ringed Plover and Turnstone throughout the winter, with larger numbers and additional species on passage.

Access

Tiree is accessible by car ferry from Oban, the boats are operated by Caledonian MacBrayne and call at Arinagour on Coll as well as Tiree. The crossing can be excellent for seabirds, especially in spring and autumn. The island has a good network of roads and most sites can be reached easily by car. From Scarinish, head east on the B8068 and B8069 to search Gott Bay for waders and gulls. Nearby Vaul and Salum Bays are also good, whilst Caoles provides a good vantage point for scanning Gunna Sound for divers, Harbour Porpoise and Basking Shark. West of Scarinish, Traigh Bhagh is another productive spot and the tall dunes here are a good place to scan north across the machair plain of The Reef (which has no general access) for waders and raptors. The three large lochs in the west of the island are always worth checking for waders, wildfowl and gulls, and both Loch Bhasapol and Loch a' Phuill now boast public bird hides. A walk to the headland of Ceann a' Mhara will produce large numbers of nesting seabirds in summer, while seawatching can be excellent from Balevullin in north-west winds or from Hynish in southerlies. As on other islands, vehicles must not be taken onto the machair and there is no access to the hay/silage meadows.

FURTHER INFORMATION

Caledonian MacBrayne: tel: 08705 650000
Tourist information: web: www.isleoftiree.com

419 ISLAY (Argyll)

OS Landranger 60

Adjacent to Jura and c. 12 miles off the west coast of Scotland, Islay's major draw is its wintering geese. Up to 70% of Greenland's breeding Barnacle Geese and 25% of its White-fronted Geese winter here. A visitor to Islay is thus guaranteed not only a wide diversity of species, but also a superb spectacle. The RSPB has a very extensive reserve at Loch Gruinart, covering over 1,600 ha, and also c. 1,800 ha on The Oa, the rugged southern tip of the island.

Habitat

The southernmost of the Inner Hebrides, Islay is a large island (c. 61,000 ha). The varied habitats include deciduous plantations around houses and farms, conifer plantations, extensive areas of scrub, moorland and peat bog, cliff (including the spectacular Mull of Oa), dunes, machair, freshwater and sea lochs, rivers, streams, marshes and arable land. Islay lacks high hills, but the Paps on neighbouring Jura dominate the north horizon.

Species

In winter, Great Northern Diver is quite common offshore, and there are a few Red-throated and Black-throated Divers and some Slavonian Grebes. Wildfowl include Whooper Swan (especially on Loch Gorm, but most have moved further south by midwinter) and up to 45,000 Barnacle Geese. The latter usually peak in October–November, immediately following their arrival in Britain; numbers then fall slightly as they disperse. They favour the pasture around the head of the two sea lochs, and are joined by up to 7,500 Greenland White-fronts. These tend to be more scattered, occurring in smaller, but tamer, flocks on the best grass (a recent development, as they were formerly secretive, favouring rough, unimproved grassland); a good place to see them is the RSPB's Loch Gruinart Reserve, where special management is attracting substantial numbers. Small flocks of Greylag Geese can be seen in a number of areas in winter, e.g. around Bridgend and Loch Gorm. They have bred on Islay since 1997 and are still increasing. In late summer, there can be as many as 1,000–1,500 Greylag Geese in the Gruinart-Gorm area (substantially more than the breeding population), dropping to around 300–400 during the winter. There may also be one or two wild Snow Geese, as well as vagrant Canada Geese of one of the small subspecies, and occasional Pink-footed and Brent Geese. Up to 1,200 Scaups frequent the inner part of Loch Indaal, and are almost always to be found between Bowmore and Blackrock. They are joined by Slavonian Grebe, Goldeneye, Common Eider, Red-breasted Merganser and a few Common Scoters. There are large numbers of Wigeons and Teals, and dabbling ducks concentrate on the shores of inner Loch Indaal, which also attracts small numbers of waders. Purple Sandpiper and Turnstone occur on rockier sections of coastline. Glaucous and Iceland Gulls appear in winter and, like Black Guillemot, can be found in the harbours, while Brambling and Snow Bunting may occur in reasonable numbers, depending on the year.

Red-throated Diver is resident and frequents small lochs as well as the sea and sea lochs. Breeding seabirds include Fulmar, Cormorant, Shag, Kittiwake, Common, Arctic and Little Terns, Razorbill, Common Guillemot and Black Guillemot. These are joined on the cliffs by Rock Dove, and non-breeding Arctic Skuas may be present offshore. Shelduck, Teal, Wigeon, Common Eider and Red-breasted Merganser are widespread, but Shoveler and Common Scoter are distinctly uncommon, with only a handful of pairs, and a Whooper Swan occasionally summers. Breeding waders include Oystercatcher, Ringed and Golden Plovers, Curlew, Common Snipe, Common Redshank, Common Sandpiper and a few Dunlins. A few pairs of Golden Eagle are resident, as well as Common Buzzard, Hen Harrier, Sparrowhawk, Kestrel, Peregrine, Merlin, and Barn, Tawny and Short-eared Owls, and these may be joined in winter by wandering White-tailed Eagles. Very small numbers of Red Grouse breed on the moors, and a very few Black Grouse persist in birch scrub and tussocky grass at the edge of plantations. Conifers also hold Siskin and Lesser Redpoll. In summer, Tree Pipit, Whinchat and Wood Warbler are present. Happily, Corncrakes are again widespread, especially on The Rhinns (the island total was more than 80 calling birds in 2008 and 2009). Rock Pipit, Twite and Stonechat are all widespread residents. There are up to 65 pairs of Choughs (a decline from a peak of almost 100 pairs in 1986, probably due to a succession of weather-affected poor breeding years) and Raven is widespread.

On passage Manx Shearwater and Arctic Skua are regular off Frenchman's Rocks, and Sooty Shearwater and Great Skua are reasonably frequent. Onshore gales may also produce sightings of Storm and Leach's Petrels. A variety of waders also passes through, with species such as Greenshank and Black-tailed Godwit joining the regular wintering and breeding species.

Access

Caledonian MacBrayne operates car and passenger ferries daily from Kennacraig, 7 miles south-west of Tarbert on the Kintyre peninsula, to Port Ellen and Port Askaig on Islay, and the 2-hour voyage is good for divers and other seabirds. Alternatively, Loganair (in conjunction with Flybe) operates daily flights from Glasgow International Airport. The Islay Natural History Visitor Centre (located in the same building as the youth hostel in Port Charlotte, and open April–October) acts as an information centre for birdwatchers and other naturalists. For full details of accommodation contact the tourist board. Wintering geese and raptors are widespread, and exploration with the OS map is strongly recommended, but the following spots are worth highlighting.

Gruinart Flats The south end of Loch Gruinart is an important area for geese, especially Barnacle Goose, and is part of the RSPB reserve. The B8017 crosses the flats and affords good views of the south of the reserve, while the minor road north along the Ardnave Peninsula gives views of the west. Park at Aoradh Farm on the B8017, where the RSPB visitor centre is open daily 10am–5pm. A short way to the north along the Ardnave road a path leads to a hide on a raised bank between two flooded fields, which are good for wildfowl in winter and

waders in summer (walk from the visitor centre or park in the small car park opposite the hide). Lapwing, Common Redshank, Common Snipe, Mallard, Teal and Shoveler breed in the fields, Garganey occurs in spring (and has bred, as has Gadwall) and Black-tailed Godwit also turns up on passage. In winter, the fields attract many Wigeon, Teal and Greenland White-fronts. Numbers of Barnacles roost on the Loch's saltings, and the tidal flats are also used by a variety of waders, best viewed from the road along the east shore. The whole area attracts raptors such as Hen Harrier (up to 12 roost in the area), Common Buzzard, Golden Eagle and Peregrine.
Ardnave Loch Good in winter for a variety of ducks. View from the end of the minor road along the Ardnave Peninsula. Leaving your car by the Loch, a short walk into the dunes should produce feeding Chough, as well as Snow Bunting and Twite in winter, especially where cattle are being fed.
Loch Gruinart, east shore Waders and wildfowl feeding in Loch Gruinart can be more easily seen from the east side where the minor road, which runs north from the B8017, follows the shore, than from the west side.
Sanaigmore Breeding seabirds can be seen here: drive to the end of the B8018 and walk west. Also, look for Great Northern Diver and Black Guillemot in the bay.
Loch Gorm Holds up to two pairs of Common Scoters (much-declined in recent years), and the area supports wintering Greenland White-fronted Goose and Whooper Swan. The Loch is circumnavigated at a distance by minor roads.
Saligo Bay Good for raptors, including Golden Eagle and Peregrine, over the massif to the north, and for passing seabirds. Park at Saligo Bridge.
The Rhinns The dunes at Machir Bay on the north-west coast often have feeding Chough and, together with the Ardnave Dunes, are among the best places on the island to find the species. Peregrine can be often be seen around the cliffs at the south end of Machir Bay, and the cliffs also hold small numbers of breeding seabirds. The forestry plantations on the Rhinns have very small numbers of Black Grouse and are frequented by Short-eared Owl and Hen Harrier. Greenland White-fronts feed in the fields around the peninsula.
Frenchman's Rocks This, the most westerly point on the island, is the best seawatching station on Islay. Access is from the A847 at Portnahaven, taking the minor road to Claddach and then walking to Rubha na Faing. The first hour of daylight is best (the westerly aspect can result in terrible light in the evening).
Port Charlotte Often has Black Guillemot in winter, with regular Great Northern Diver just offshore and less frequent Purple Sandpiper.
Bruichladdich On the west shore of Loch Indaal, this is a spot for Greenland White-front, and the mudflats adjacent to the A847 at Traigh an Luig to the north are good for waders. There are several lay-bys constructed specifically for birdwatchers. Bruichladdich pier is a good year-round spot for Common Scoter, and also favoured by Black-throated Diver and Purple Sandpiper.
Loch Indaal holds all three divers, up to 50 Slavonian Grebes, Common Eider, Red-breasted Merganser, 1,000+ wintering Scaups, 50+ Common Scoters (present all year) and a variety of waders. Greenland White-fronted and Barnacle Geese are very regular in the fields behind Bruichladdich (and also behind Port Charlotte).
Bridgend There are waders on the flats at Bridgend (best an hour either side of high water), Pintail and Shoveler favour the river channel, and Barnacle Geese roost in Bridgend Bay; view from the A847 and A846 north and south of Bridgend (with several lay-bys constructed specifically for birdwatchers); the best place to watch the evening flight of Barnacle Geese is a lay-by on the A847 at NR 326 629, opposite the entrance to Whin Park. The bridge over the River Sorn is a good spot to look for Dipper.
Bowmore Situated on Loch Indaal, this attracts waders, seaducks, Wigeon and Whooper Swan, with divers and grebes offshore (see Bruichladdich). The Loch can also be viewed from the B847 between Bowmore and Bridgend, with the lay-by 0.5 miles north of Bowmore by the hydro-board generating station, being a particularly good vantage point. The rubbish tip south-west of Bowmore may hold white-winged gulls, and is a good spot for Raven (10–20 may be present in the autumn), while the coast to the south-west, from Ronnachmore to Gartbreck, is good for waders.

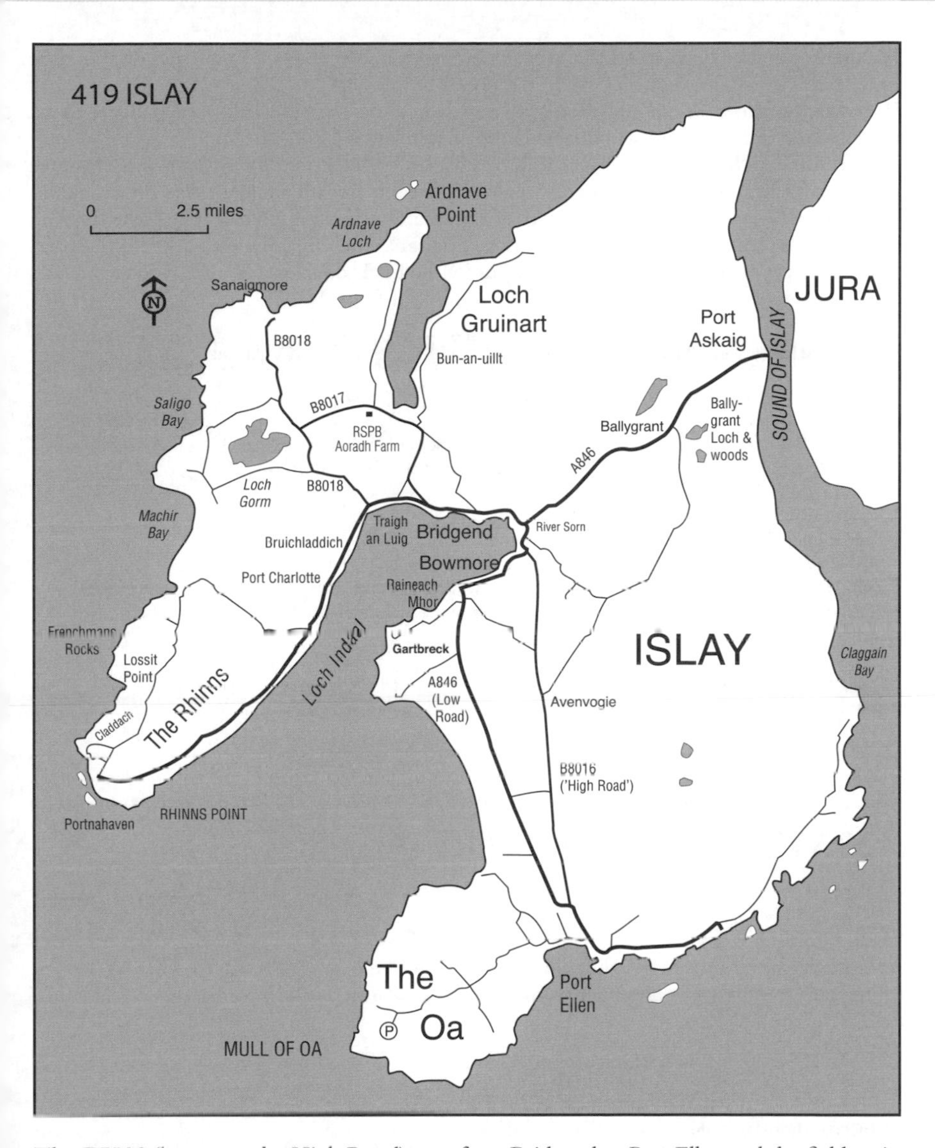

The B8016 (known as the High Road) runs from Bridgend to Port Ellen and the fields at its northern end are very good for Barnacle Geese.

The A846 (known as the Low Road) crosses an extensive area of mosses good for Greenland White-fronted and Barnacle Geese, especially the north end, while Hen Harrier is regular in summer.

Port Ellen There may be divers, grebes and seaducks around the harbour mouth.

The Oa The south cliffs are good for Golden Eagle, Peregrine, Raven and Chough, with breeding seabirds in summer and occasional Snow Bunting along the clifftops in winter. Around half of the peninsula is now an RSPB reserve, which can be reached along a minor road west from Port Ellen (turning south towards the car park near Upper Killeyan); two marked trails (1.25 miles and 2 miles) lead from the car park to the Mull of Oa.

Claggain Bay At the end of the road leading east from Port Ellen, the Bay is good for Great Northern and Red-throated Divers, with moulting Red-breasted Mergansers in late summer.

Nearby woods are good for Siskin and Lesser Redpoll, and Twite is always present on the shore in winter.

Port Askaig Holds Black Guillemot, often year-round, and there are occasionally Arctic Skuas in the Sound (from the tiny breeding colonies remaining on Jura).

Ballygrant Loch and Woods Accessible by turning south on a minor road in Ballygrant and parking by the entrance to the woods. The Loch has breeding Little Grebes, Tufted Ducks and Mute Swans, the occasional Coot in winter, and is visited by Red-throated Diver. A track leads through the woods to Port Askaig, passing Loch Allan en route, where there is a small heronry. Woodland birds include Common Redstart, Blackcap, Wood Warbler, Spotted Flycatcher, Siskin and Lesser Redpoll.

FURTHER INFORMATION

Grid refs: Gruinart Flats NR 275 672; Ardnave Loch NR 285 727; Loch Gruinart, east shore NR 297 699; Sanaigmore NR 236 706; Loch Gorm NR 235 647; Saligo Bay NR 211 663; The Rhinns (Machir Bay) NR 208 633; Frenchman's Rocks (Claddach) NR 162 535; Bruichladdich NR 264 608; Bridgend NR 326 629; Bowmore NR 320 602; Port Ellen NR 364 449; The Oa (Upper Killeyan) NR 281 422; Claggain Bay NR 462 531; Port Askaig NR 430 692; Ballygrant Loch and Woods NR 398 659

RSPB Loch Gruinart: tel: 01496 850505; email: loch.gruinart@rspb.org.uk

RSPB The Oa: tel: 01496 300118; email: the.oa@rspb.org.uk

Islay Natural History Visitor Centre: Port Charlotte, Isle of Islay PA48 7TX: tel: 01496 850288

Tourist Information Centre: Morrison Court, Bowmore, Isle of Islay: tel: 01496 810254; email: info@islay.visitscotland.com

Caledonian MacBrayne: tel: 08000 66 5000; email: reservations@calmac.co.uk

Loganair: web: www.loganair.co.uk; flights can be booked online at www.flybe.com; tel: 0871 700 2000

420 WEST LOCH TARBERT (Argyll)

OS Landranger 62

West Loch Tarbert (not to be confused with the loch of the same name in the Western Isles) is visited by a number of interesting birds in winter.

Habitat

This is a long (9.3 miles) sheltered sea loch.

Species and Access

West Loch Tarbert has all three divers in winter, as well as Slavonian Grebe, Glaucous and sometimes Iceland Gulls. These can be seen around the harbour and from the ferry en route for Islay.

421 MACHRIHANISH SEABIRD AND WILDLIFE OBSERVATORY (Argyll)

OS Landranger 68

Situated at Uisaed Point, on the west coast of the Kintyre Peninsula, and overlooking Machrihanish Bay, this site has gained prominence in recent years as the premier seawatching site on the west coast of Scotland. It is most productive in September–October.

Habitat
The rocky coastline is bordered by areas of rough grazing, giving way to upland areas.

Species
In autumn, during appropriate weather conditions, usually strong westerlies combined with poor visibility, the regular Gannets and Great and Arctic Skuas may be joined by Manx, Sooty and sometimes Balearic Shearwaters, Sabine's Gull or Long-tailed and Pomarine Skuas. Storm Petrel is regular, especially on days with poor visibility, while Leach's Petrel usually occurs after a deep depression has passed culminating in a severe north-west gale. From July to September, there is also a strong early-morning passage of waders. Golden Eagle and Twite are resident in the hinterland.

Access
Machrihanish lies about six miles west of Campbeltown, and the Observatory, accessed from the B843 just west of the village, is well signed. The seawatching hide is open daily from dawn, from Easter to the end of October.

FURTHER INFORMATION

Grid ref: NR 628 209
Machrihanish Seabird Observatory: Eddie Maguire (Warden), Lossit Park, Machrihanish, Argyll PA28 6PZ: tel: 07919 660292; web www.machrihanishbirds.org.uk

422–424 OUTER HEBRIDES (Outer Hebrides)

OS Landranger 18, 22 and 31

These islands are undoubtedly one of the most attractive parts of Britain. Supporting a range of raptors, wildfowl, and large numbers of breeding waders, the star bird is perhaps Corncrake, which maintains its British stronghold here. The best time to visit is May–July for breeding birds, but autumn could be productive for migrants. There are two large reserves, the Loch Druidibeg NNR and the RSPB's Balranald.

Habitat
The long sandy beaches of the Atlantic seaboard back onto the fertile machair, level grassland formed by a mixture of windblown shell-sand and peat, which supports the highest density of breeding waders in Britain. Often cultivated, the crops may still provide cover for nesting birds. In the machair and extending towards the east flank of the islands is a mosaic of lochs, marshes and wet grazing. The land becomes more acidic and peaty as it rises to the eastern hills, which peak at 2,033 feet on South Uist, but at only 1,139 feet on low-lying, watery North Uist. The islands are almost treeless, with only small, isolated stands.

Species
Red-throated is the commonest diver and can often be seen flying between its tiny nesting pools and fishing grounds at sea. There are about five pairs of Black-throated Diver, almost all on North Uist. They breed and fish on the larger lochs and are less often seen in the air. Great Northern is present until June, almost always on the sea. Fulmar, Cormorant and Shag can be seen around the coast at any time, but most other seabirds are summer visitors. Arctic Skua breeds mainly in the centre of North Uist, with a few on Benbecula and South Uist. Five species of gull are joined offshore by non-breeding Kittiwake, as well as the occasional summering Glaucous or even Iceland Gull. Arctic, Common and Little Terns are widespread

in that order of abundance, as is Black Guillemot. There are small numbers of Razorbill and Common Guillemot and these share the cliffs with Rock Dove.

A speciality is the several hundred pairs of Greylag Geese, deemed wild rather than feral. Their stronghold is Loch Druidibeg, which has 60–70 pairs. Outside the breeding season the flock grazes around Grogarry. Though several Whooper Swans may summer, the species has not been proved to nest since 1947. Shelduck, Teal, Shoveler, Common Eider and Red-breasted Merganser, and small numbers of Wigeon and Gadwall, are regular breeders. Raptors include Hen Harrier, with approximately 35 nesting females, Common Buzzard, Golden Eagle, Kestrel, Merlin, Peregrine and Short-eared Owl. These may be seen hunting anywhere, though eagles favour the mountains, and harriers are best seen around Loch Hallan on South Uist and Loch Mor on Benbecula. Wandering White-tailed Eagles from the re-introduction programme are a possibility, especially on the east coast. Red Grouse occur at low density on the moorland, together with small numbers of Golden Plover and Greenshank. Commoner breeding waders are Oystercatcher, Ringed Plover, Lapwing, Dunlin, Common Snipe, Common Redshank and Common Sandpiper, and a few Red-necked Phalaropes still hang on. With around 250 calling birds the Uists remain a stronghold of the Corncrake. They arrive in mid-April and are best seen early in the season before the vegetation has grown. Rush, sedge and iris beds are favoured on arrival, and thereafter hay meadows. They call most frequently midnight–4am, and only irregularly during the day. The common passerines are Skylark, Meadow and Rock Pipits, Pied Wagtail, Wren (the Hebridean subspecies), Stonechat, Wheatear, Sedge Warbler, Hooded Crow, Raven, Starling (the Shetland subspecies), Twite, House Sparrow, and Reed and Corn Buntings. The Hebridean race of Song Thrush can be quite elusive, favouring heathery hill country as well as areas of shrubs, and Blackbird is also confined to gardens and pockets of woodland on the west coast, together with Long-eared Owl in the latter.

In spring, north–north-west winds can prompt movements of skuas. Up to 271 Long-tailed and 436 Pomarine Skuas have been seen at Balranald in a day, as well as the more regular Great and Arctic. Every year since this skua passage was discovered (in 1971) and that observers have been present, Long-tailed and Pomarine have been seen, usually in the third week of May, though numbers are obviously weather-dependent. Manx Shearwater is sometimes seen in large numbers, April–August, as well as Sooty from June. Storm and sometimes Leach's Petrels occur off the west coast in strong onshore winds, but, like the shearwaters, the boat crossings are a better bet, especially for Storm Petrel. Gannet is common offshore March–November. Pink-footed, Brent and Barnacle Geese appear on passage, but rarely linger. Migrant waders may include Little Stint, Curlew Sandpiper, Black-tailed Godwit and Whimbrel, while Ruff favours the machair and crofts. Only small numbers of migrant passerines occur, notably the Greenland race of Common Redpoll and Snow Bunting, which may be abundant.

In winter, Great Northern Diver and Slavonian Grebe are widespread offshore, with just a few Red-throated Divers. Several hundred Whooper Swans winter, and Greenland White-fronted Goose is an erratic visitor in small numbers; Nunton on Benbecula, Loch Hallan and Loch Bee are the most regular areas. Other wildfowl include Scaup, Long-tailed Duck (especially between Rubha Ardvule and Grogarry on South Uist) and Goldeneye. There are internationally important populations of Ringed Plover, Sanderling and Turnstone, and smaller numbers of Knot and Purple Sandpiper are joined by a few Greenshanks and the occasional Grey Plover. Glaucous and Iceland Gulls are both regular, and after persistent north-west gales Iceland Gull may be relatively common. Small flocks of Rooks and Jackdaws winter.

Access

By Sea Caledonian MacBrayne operate car ferries to the islands from Skye (Uig) to North Uist (Lochmaddy), the voyage taking 1 hour 45 minutes, and from Oban to South Uist (Loch Boisdale), taking 5 hours 20 minutes. Cars should be booked in advance. Both crossings are good for seabirds, with a good chance of Storm Petrel and shearwaters.
By Air Flybe's franchise holder Loganair operate daily flights (except Sunday) from Glasgow to Benbecula.
Accommodation For details of accommodation, contact VisitHebrides.

Note that crofting land in the Hebrides, as elsewhere in north and west Scotland, is often unfenced. Due care should be taken, and if in doubt concerning access to any area, please ask first.

North Uist, Benbecula and South Uist are connected by two causeways, and there are large areas of essentially similar habitat bisected by the main north–south road. The following areas deserve special mention, but are only representative of the entire area.

422 SOUTH UIST OS Landranger 22 and 31

Lochboisdale Harbour This attracts Glaucous and Iceland Gulls in winter, as well as divers and seabirds.

Loch Hallan at Daliburgh At the junction of the A865 and the B888, this often holds summering Whooper Swan. Take the road west at the crossroads in the village and after 1 mile follow a track north to view from the cemetery (which avoids disturbance). Do not approach the loch over agricultural land.

The Rubha Ardvule Peninsula This is a good spot for seawatching. Access is via a road and track from the main road. Visitors must keep clear of the point when in use by the military, but this is infrequently the case.

Loch Druidibeg NNR The only restriction on access to the NNR is to the area around the south-west corner of Loch Druidibeg from April to end of August, but the loch is easily viewed the main road and B890. Species include divers, breeding Greylag Goose, waders, and sometimes summering Whooper Swan. The trees around Grogarry Lodge provide some cover for migrants.

B890 Beside the B890 to Loch Skiport (Loch Sgioport) is a grove of rhododendrons and deciduous and coniferous trees that attract migrants and have breeding Long-eared Owl, Goldcrest and Greenfinch. It is a good spot to scan for eagles over the slopes of Hecla (Thacla).

Hecla (Thacla) and Beinn Mhor These sombre peaks support Golden Eagle, Red Grouse, etc. Look for eagles from the main road or the road to Loch Skiport.

Loch Bee (Loch Bi) Excellent for wildfowl, with up to 500 Mute Swans. View from the main road crossing the loch.

Ardivachar (Aird a' Mhachair) Another good seawatching spot. Take the road west from the A865 in the north of the island to the peninsula. North Bay is good for waders.

423 BENBECULA OS Landranger 22

Balivanich (Baile a' Mhanaich) The outfall at Balivanich on the B892 attracts gulls, and is good for Glaucous and Iceland, even in midsummer. Waders frequent the adjacent foreshore and aerodrome (check with air traffic control before entering the latter).

The Benbecula–North Uist causeway This affords views of the intertidal flats for waders.

424 NORTH UIST OS Landranger 18 and 22

RSPB Balranald This has areas of beach, dunes, machair, pools and marshland, and is largely crofted. All three divers are often present in summer, and several species of raptor regularly visit. There is a dense population of breeding ducks and waders (and sometimes non-breeding Red-necked Phalaropes) as well as 10–15 pairs of Corncrake. Arctic Tern breeds and an excellent seawatching spot is the headland of Aird an Runair, the most westerly spot in the Outer Hebrides. Turn west off the A865 2 miles north-west of Bayhead, signed Houghharry, and fork left to the visitor centre at Goular Cottage. The reserve is open at all times and there is a 3-mile circular nature trail (not suitable for wheelchairs), the visitor centre is open April–August 9am–6pm; visitors should keep to the marked paths.

The 'Committee Road' This is the minor road that runs north-south across the centre of the island, and is a good area for Arctic Skua, Greenshank and Short-eared Owl.

Loch Scadavay (Loch Scadabhagh) North of the A867, this is good for divers and Arctic Skua breeds in the vicinity.

Loch Skealtar (Loch Sgealtair) North of the A867 just west of Lochmaddy, this is another good spot for divers; up to 21 Black-throated have been counted in August.

Lochmaddy Harbour Worth checking for gulls, Black Guillemot and other seabirds, and eagles, often including White-tailed, may be seen to the south over the Lees.

Newton House and Clachan The trees around Newton House and at Clachan attract migrants. Both are on the B893 in the north-east of the island. Permission to enter should be sought from Newton House and Clachan Farm respectively, but both can be seen well from the road. Other areas that have been productive in recent years are the conifer plantation on Ben Langass, south of the A867, and that on Ben Aulasary, just east of the minor road between Bayhead and Sollas.

Vallay Strand and Vallaquie Strand On the north coast (off the A865), the extensive foreshore attracts waders, especially off the B893 causeways at Oban Trumaisgearraidh.

FURTHER INFORMATION

Grid refs: Lochboisdale Harbour NF 793 193; Loch Hallan at Daliburgh NF 735 219; The Rubha Ardvule Peninsula NF 727 297; Loch Druidibeg NNR NF 789 382; B890 NF 799 387; Loch Bee NF 781 430; Ardivachar NF 740 457; Balivanich outfall NF 774 554; The Benbecula-North Uist causeway NF 839 571; RSPB Balranald NF 706 707; The 'Committee Road' NF 790 702; Loch Scadavay NF 871 670; Loch Skealtar NF 900 682; Lochmaddy Harbour NF 920 679; Newton House NF 895 783; Clachan NF 884 762; Vallaquie Strand (B893) NF 872 747
RSPB Balranald: tel: 01463 715000; email: nsro@rspb.org.uk
Scottish Natural Heritage Area Officer: tel: 01870 620238
Caledonian MacBrayne: tel: 08705 650000; web: www.calmac.co.uk
Flybe/Loganair: tel: 0871 7002000; web: www.flybe.com
Tourist Information (Visit Hebrides): Lochmaddy Tourist Information Centre, tel . 01876 500321, Lochboisdale Tourist Information Centre, tel. 01878 700286 (both open April–October); Stornoway Tourist Information Centre (open all year), tel. 01851 703088; web: www.visithebrides.com

425–438 ORKNEY (Orkney)

OS Landranger 5, 6 and 7

Orkney comprises 75 islands, 17 of them inhabited, and at the closest point lies just 6 miles from mainland Scotland. Its attractions for the visiting birdwatcher are the vast numbers of breeding seabirds, the moorland specialties and passage migrants. The best time to visit for breeding birds is May–early July, and for migrants May–early June and late August–October.

Habitat

The islands, with the exception of Hoy, are rather low and fertile. In places the coasts have cliffs, sometimes spectacular, as at St John's Head on Hoy. Other parts have beaches and sheltered bays, while Deer Sound and Scapa Flow provide a haven for large numbers of divers and seaducks. The original vegetation of Orkney was scrubby woodland but this was cleared by Neolithic and Viking settlers and replaced by moorland and bog. These form an invaluable habitat, but it would be a mistake to think that Orkney is cloaked in moorland. Large areas of moorland do remain on Mainland, Rousay and Hoy, often with scattered lochs, streams and marshes, but since the early 18th century much has been reclaimed for farming and the major land-use on Orkney is now pasturage for cattle. Some of the smaller islands are still crofted and there are tracts of maritime heath, a habitat unique in Britain to Orkney and parts of Caithness: exposure to the wind has produced a climax of Heather and Crowberry, mixed with sedges and small herbs.

Species

Fulmar, Cormorant, Shag, Kittiwake, Common Guillemot, Razorbill, Black Guillemot and Puffin are widespread on cliffs and small islands, which they share with Rock Dove and Raven. On moorland and heaths there are Arctic and Great Skuas, five species of gull, and Sandwich,

Common and Arctic Terns, these last also favouring the beaches. Gannet breeds on Sule Stack and in 2002 established colonies on Sule Skerry and Noup Cliffs (Westray) with 1,000 and 500 pairs respectively by 2009; it is widespread offshore. Manx Shearwater and Storm Petrel are rather elusive, although the latter is common.

Breeding ducks include Wigeon, Gadwall, Shoveler, Common Eider and Red-breasted Merganser, and a large proportion of Britain's Pintails. Scaup, Long-tailed Duck and Common Scoter have bred in the past. Wildfowl share the marshes and lochs with Red-throated Diver, of which there are c. 125 pairs on the islands, half on Hoy. Waders include Golden Plover, Dunlin, two or three pairs of Whimbrels and a handful of Black-tailed Godwits, but Curlew is the most conspicuous species, being abundant on the moors. Red-necked Phalarope has bred. Passerines in the marshy areas include Sedge Warbler and Reed Bunting.

The moorland supports Red Grouse and ground-nesting Kestrel, Woodpigeon, Starling and Hooded Crow. Orkney was the last refuge for the Hen Harrier when persecution had all but eliminated it from the rest of the country. Quite recently up to 100 females of this polygamous species attempted to breed in a good year but numbers have declined a little, with 60 breeding females in 2004. Sparrowhawk, Common Buzzard, Peregrine and Short-eared Owl also breed, and since the mid-1980s low of six pairs, Merlin has recovered to c. 20 pairs. Other birds on the moors, apart from the abundant Meadow Pipit and Skylark, are Stonechat and Wheatear. Twite is declining but widespread and can be found along roadside verges. Corncrake had declined to only six calling birds by 1993 and despite the instigation of the Corncrake Initiative numbers have only averaged around 20 calling birds in recent seasons (with just 17 in 2006; of late Papa Westray has been the main island for the species). Areas of trees and plantations support Rook, Willow Warbler and occasionally other woodland passerines such as Spotted Flycatcher or Garden Warbler, and Fieldfare first nested in Britain in 1967 on Orkney. Redwing has also bred.

Offshore, wintering Great Northern Diver is common and there are much smaller numbers of Black-throated Divers, which probably most frequently occurs in late winter and spring. There is a significant winter population of Slavonian Grebe and small numbers of Little Auks around the islands. Wildfowl include Whooper Swan, though many or most of these move south after early winter, unless food remains available. A flock of Greenland White-fronted Geese winters on Birsay (c. 100 birds), but there are much larger numbers of Greylags and Barnacle Geese; Greylags from Iceland have increased greatly in numbers and now over 68,000 may winter in the archipelago (including c. 10,000 local birds), and 1,600 Barnacle Geese winter on South Walls, while Pink-footed Goose has become more regular in winter in the last few years, with over 500 in East Mainland. These geese and swans occur more abundantly on passage. The breeding ducks are joined by several thousand Long-tailed Duck and a few hundred Scaup, Velvet Scoter and Goldeneye, but Common Scoter is scarce. As in summer, Curlew is abundant and Orkney supports a quarter of the British wintering population; though ubiquitous, it favours pastures and shores. Also notable are large numbers of Turnstones and Purple Sandpipers, as well as Knot, Sanderling and Grey Plover. Gulls in the harbours and along the coast should be carefully checked for Glaucous, Iceland and even Ring-billed. Other wintering species are Long-eared Owl and Snow Bunting.

In spring and autumn many of the breeding and wintering birds may be seen but there are several species which are more or less confined to these periods. A variety of waders occurs, Greenshank and Whimbrel being the most frequent, and these are joined by small numbers of Curlew Sandpiper, Black-tailed Godwit, Spotted Redshank and Green Sandpiper. Offshore there is a large, sometimes very large, autumn passage of Sooty Shearwater, especially off the northernmost isles. Pomarine and Long-tailed Skuas and Little Gull are also possible but distinctly uncommon. Other scarce migrants include Honey-buzzard and Rough-legged Buzzard. A broad variety of passerines occurs; among the commonest are Robin, Blackbird, Fieldfare, Redwing, Blackcap, Willow Warbler, Chiffchaff, Goldcrest, Brambling, Chaffinch and Siskin, and with these there are small numbers of scarcer species – most regular are Wryneck, Richard's Pipit, Black Redstart, Bluethroat, Icterine, Barred and Yellow-browed Warblers, Red-breasted Flycatcher, Great Grey and Red-backed Shrikes, Common Rosefinch, and Ortolan and Lapland Buntings. The best conditions, in both spring and autumn, as on Fair Isle and Shetland, are south-east winds. As the cream on the cake, a wide variety of rarities has been recorded and, like Shetland, Orkney has immense potential for those who wish to find their own.

Access

Full details of all transport and accommodation are available from VisitOrkney.

By Sea There is a choice of routes. NorthLink Ferries offer a service from Aberdeen to Kirkwall on Mainland Orkney (6 hours, with three or four sailings per week depending on season; their 'cruise ferries', the *Hjaltland* and *Hrossey*, are comfortable and extremely punctual; advance booking is recommended on all sailings, most especially for vehicles.) There are also three or four sailings per day, depending on season, from Scrabster to Stromness on Mainland Orkney (1.5 hours). Pentland Ferries has three or four sailings per day depending on season from Gills Bay (near John O'Groats) to St. Margaret's Hope on South Ronaldsay (1 hour), while John O'Groats Ferries provides a May –September passenger only service from John O'Groats to Burwick on South Ronaldsay (40 minutes).

By Air Kirkwall on Mainland Orkney is served by flights operated by Flybe's franchise-holder, Loganair, from Glasgow, Aberdeen, Edinburgh, Inverness and Sumburgh.

Inter-island Transport All islands are served by ferries from Mainland, with bus services from Kirkwall to the terminals at Tingwall (for Rousay) and Houton (for Hoy). The northern islands also have connections by air to Kirkwall. Full details are given under each island, but it is essential to check for the latest timetables and, if intending to take a car to the outer islands, it is essential to book in advance.

Throughout Orkney, care should be taken not to disturb breeding birds, especially divers, raptors and terns. Many of the sites described below are on private land and, if in doubt, seek permission before entering. In general you will get a very friendly reception. In addition, when searching for migrants, the nature of some of the habitats, such as crops and gardens, calls for extra discretion.

425 MAINLAND — OS Landranger 6

In winter there is an abundance of birds, with large numbers of waders, especially Curlew, throughout the whole island. Otherwise, Mainland can be conveniently divided into East and West along a line through Kirkwall, with the East mainly farmland, whilst the West still has considerable areas of moorland.

Stromness and Kirkwall Harbours Both are worth a look in winter for Glaucous and Iceland Gulls, and at Kirkwall the abattoir on Hatston Industrial Estate west of the harbour also attracts gulls.

Scapa Flow and Deer Sound Outstanding areas for divers, grebes and seaducks, in winter Scapa Flow may hold almost 800 Great Northern Divers, 120 Slavonian Grebes and 1,000 Long-tailed Ducks (with smaller numbers in Deer Sound). Waders are also numerous. The Churchill Barriers, causeways that connect Mainland to Burray and South Ronaldsay, are good vantage points for Scapa Flow, although cars are not allowed to sop on them.

Binscarth Wood Near Finstown, this woodland holds a rookery as well as a variety of small passerines and, occasionally, wintering Long-eared Owls, but there is so much cover that it is difficult to work for migrants. A track through the wood leaves the A965 just west of Finstown.

RSPB Hobbister A little over 3 miles south-west of Kirkwall, this reserve covers c. 800 ha and includes areas of moorland astride the A964 as well as low cliffs and the large, sandy Waulkmill Bay. It lies between the coast, the A964 and Loch of Kirbister. Breeding birds include Red Grouse, Hen Harrier, Merlin, Black Guillemot, Short-eared Owl, Raven and Twite (which favours the coastal path). The reserve trails are not wheelchair-accessible. Waulkmill Bay, which can be watched from the car, attracts passage waders and, from winter into the spring, Great Northern and Black-throated Divers and Slavonian Grebe are regular, as well as seaducks. Access is via a track from the car park at HY395069 or on the minor road south to Waulkmill Bay.

RSPB Birsay Moors Though much of West Mainland has been reclaimed for beef cattle farming, some large areas of moorland remain, especially around Orphir in the south, between Finstown and Kirkwall, and the area from Finstown north to the Loch of Swannay. A large

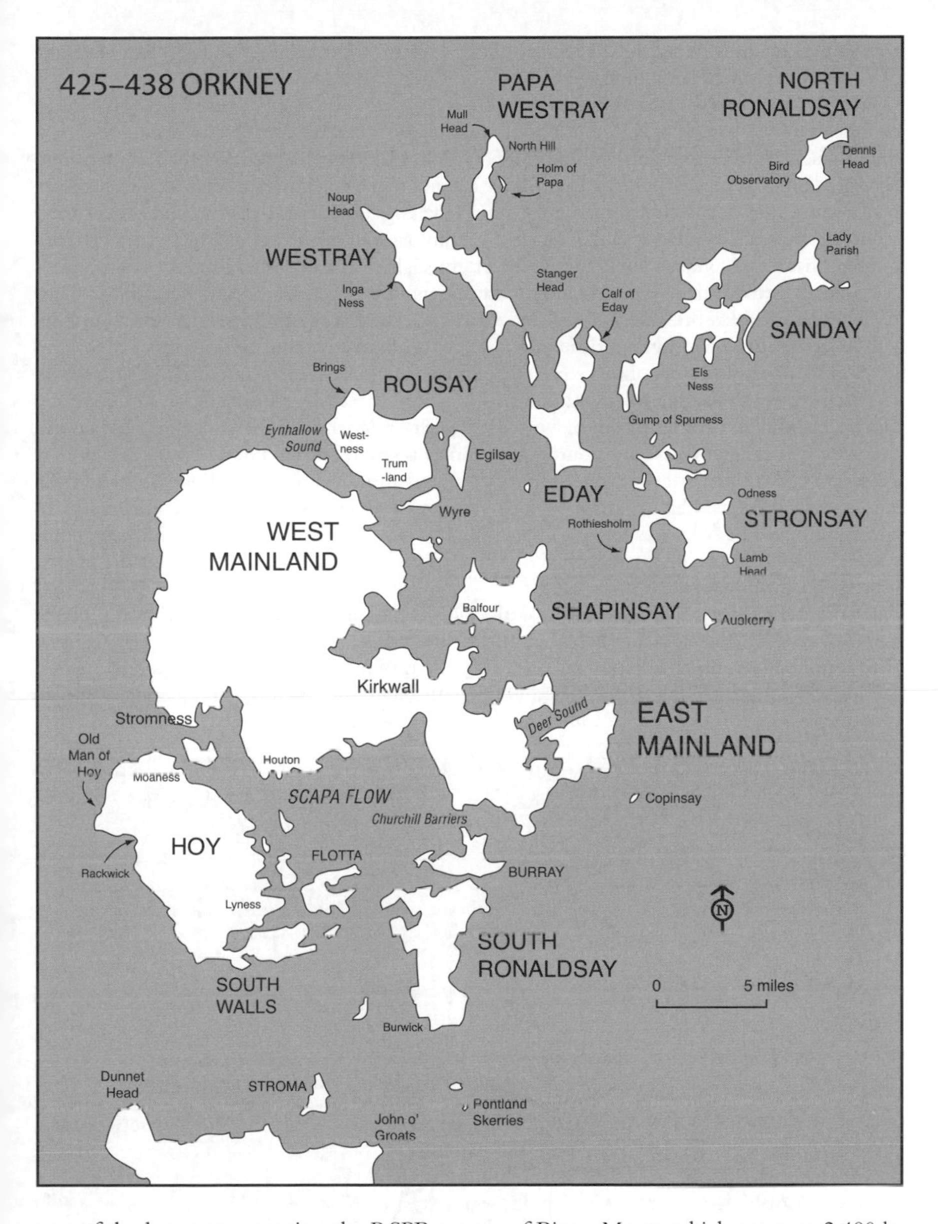

part of the last now comprises the RSPB reserve of Birsay Moors which covers c. 2,400 ha astride the B9057. Breeding species include Greylag Goose, Short-eared Owl, Hen Harrier, Great and Arctic Skuas and two or three pairs of Whimbrels. There is a winter roost of Hen Harriers at Durkadale on Birsay Moors. Access is at three main points (although the whole reserve is open access).

1. A wheelchair-accessible hide on Burgar Hill overlooks Lowrie's Water, which has breeding Red-throated Diver and other waterbirds. The entrance is signed from the A966 at Evie, 0.5 miles north-west of the B9057 turn-off, and the hide is near the wind generators.
2. View Birsay Moors from the B9057 Hillside Road between Evie and Dounby.
3. Dee of Durkadale. Turn south on a track off the minor road at the south end of Loch

Hundland and follow this to the abandoned farm of Durkadale (HY 293 252) or view from the road through the valley.

RSPB Cottascarth and Rendall Moss Set in an area of rugged moorland, this reserve comprises several parcels of land to the south and west of Rendall. Turn west off the A966 on a minor road 3.1 miles north of Finstown (just north of Norseman Garage) on the minor road to Settisgarth and then take the track north for 0.6 miles to park at Lower Cottascarth Farm; from here a track runs north-west for 0.6 miles to a hide overlooking the reserve (not wheelchair-accessible, but much of the area can be just as easily viewed from the roads). Breeding species include Short-eared Owl, Hen Harrier and Merlin, as well as a very high density of Curlews.

Loch of Banks Part of the loch and its environs is an RSPB reserve; view from the A967 or A986. Breeding ducks include Gadwall and Pintail and there is a high density of nesting waders.

RSPB Brodgar Comprising the narrow neck of land between the Loch of Harray and Loch of Stenness, and surrounding the Neolithic stone circle of the Ring of Brodgar, a World Heritage Site; the dominant habitat is damp grassland. Breeding birds include seven species of wader and eight species of ducks, with Hen Harrier, Merlin and Short-eared Owl also in the vicinity; management aims to encourage Corncrakes to return. The Loch of Harray may hold up to 10,000 wintering ducks, mostly Wigeon, Tufted Duck and Pochard, while the adjacent tidal Loch of Stenness has Long-tailed Duck and Goldeneye. Between these lochs and around Loch of Skaill there are often several hundred Greylag Geese in winter, together with Whooper Swans. The main reserve path is accessible to all. Two car parks on the B9055 provide good viewpoints.

RSPB The Loons and Loch of Banks Of the marshes and lochans, one of the best is an RSPB reserve, the Loons in Birsay, in north-west Mainland, which is a waterlogged basin with reed and sedge beds adjacent to Loch of Isbister. Breeders include Pintail, Wigeon, seven species of wader (including Black-tailed Godwit and Dunlin), Arctic Tern and Sedge Warbler. In winter up to 100 Greenland White-fronted Geese use the area, as well as waders and other

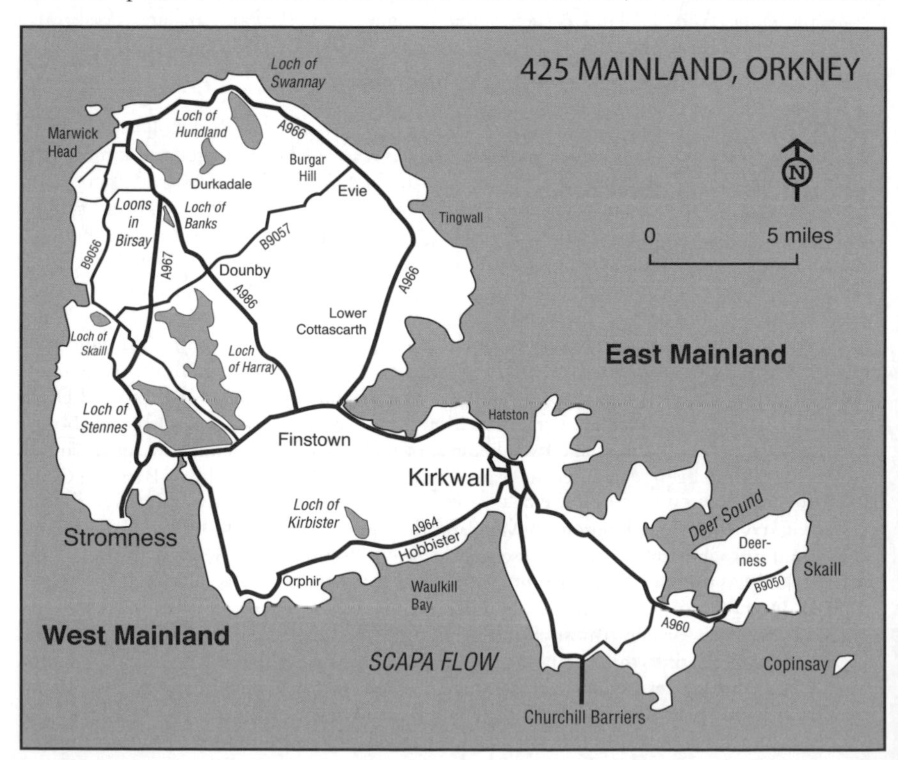

wildfowl. The only access is to the hide at HY 246 242: leave the A986, turning west at Twatt, 3 miles north of Dounby; the hide is off this minor road c. 400 yards before the junction with the B9056, and the reserve can also be viewed from this road. Disabled visitors should contact the reserve for information on accessibility.

RSPB Marwick Head On the north-west coast, this mile-long stretch of sandstone cliffs rises to 280 feet. Breeding birds include 15,000 Common Guillemots and 1,900 pairs of Kittiwakes (the latter, especially, much reduced in recent years), was well as Fulmars, Razorbills and small numbers of Puffins, while Peregrine is occasionally seen. Access is on foot from two points: either the seaward end of the minor road from the B9056 to Mar Wick bay or the car park at Cumlaquoy (similarly off the B9056 to the north of the turning to Mar Wick), from where it is a short walk (not suitable for wheelchairs) to the cliffs.

East Mainland Largely farmland and, although there is some remaining moorland harbouring typical species, the main attraction is passage migrants. Any area of bushes, crops or other vegetation is worth checking in spring and autumn.

426 COPINSAY OS Landranger 6

This 150-ha island 11 miles off East Mainland was bought by the RSPB in memory of the late James Fisher. The reserve comprises the uninhabited main island of Copinsay, three smaller islets of Corn Holm, Ward Holm, Black Holm and the Horse of Copinsay. Mainly grassland, 10 ha of arable, hay and pasture are managed for Corncrake, and this grassy interior is bounded by cliffs that rise to 250 feet. The entire south-eastern side of Copinsay is a large seabird colony, with large numbers of Common Guillemots as well as Razorbills, Black Guillemots, Puffins, Kittiwakes, Shags, Fulmars and Arctic Terns (although Common Guillemot and Kittiwake are now much reduced from their previous highs of 20,000 adults and 4,500 pairs respectively). There are more Puffins on Corn Holm and Black Holm (which, like Ward Holm, are accessible at low tide). Other species of note are Raven and Twite, and Copinsay also attracts passage migrants. From Skaill in Deerness, at the terminus of the B9050, a small boat makes day trips to the island by arrangement. Contact Sidney Foubister, tel: 01856 741252. A converted farmhouse now serves as an information and display centre.

427 HOY OS Landranger 7

This island is rather different in character to the rest of Orkney. Mainly rugged, bleak moorland, the Cuilags and Ward Hill, rising to 1,577 feet, are the highest points in the archipelago. There are few people. The scenery is spectacular, and includes the sheer 1,100-foot cliffs of St John's Head and the Old Man of Hoy, which, at 450 feet, is the highest sea stack in Britain. There is a small colony of Manx Shearwaters on Hoy, and Storm Petrel may also breed. There are c. 2,100 pairs of Great Skuas and perhaps still 50 pairs of Arctic Skuas. Many are on North Hoy but the valleys on the west side of the island are also good, especially those around Heldale Water (the continual bombardment by skuas is something to prepare for in advance; although they rarely press home an attack, a stick or tripod to hold above your head is a comfort!). As many as 1,500 pairs of Arctic Terns may nest but numbers have declined recently. Red-throated Diver, Common Buzzard, Hen Harrier, Peregrine, Merlin and Short-eared Owl all breed, though there are relatively few harriers and owls because of the lack of voles. Golden Eagle formerly bred but has seldom been seen in recent years. Winter birds include up to 1,600 Barnacle Geese on South Walls.

A ro-ro ferry runs from Houton on Mainland to Lyness (and to nearby Flotta), the voyage takes 35 minutes, while there is also a passenger service to Moaness on Hoy from Stromness, the journey taking 20 minutes; both services are operated by Orkney Ferries.

RSPB Hoy This 3,925 ha-reserve comprises most of the north-west of the island and includes Berriedale, Britain's most northerly natural woodland. It is otherwise moorland bounded by superb cliffs that are home to large numbers of seabirds (e.g. 22,500 pairs of Fulmar). There are no restrictions on access and the reserve can be explored via several long footpaths. If coming by car, leave the ferry terminal at Lyness northwards on the B9047 and towards the north of the island take the first left, signed to 'Dwarfie Stone'. From this road a short trail leads to Dwarfie Stone, but continue to the road's end at Rackwick, from where trails run westwards to the Old Man of Hoy and northwards, between the Cuilags and Ward Hill, to the road at Sandy Loch; for the more adventurous and well-prepared, a more difficult clifftop path runs north from the Old Man of Hoy to Murra. If arriving on Hoy at Moaness the best access to the reserve is at Sandy Loch. All trails are rather rough and rugged, so not suitable for the mobility-impaired.

428 SOUTH RONALDSAY — OS Landranger 7

South Ronaldsay and Burray are connected by road to East Mainland by the Churchill Barriers. The east and west coasts of South Ronaldsay have colonies of seabirds, including small numbers of Puffins, and also Raven. Otherwise the islands are mainly farmed, with small areas of heath that support nesting gulls and a few skuas, and some lochans and marshes with breeding ducks and waders. Lying on the east side of the archipelago, their major attraction is passage migrants.

429 ROUSAY — OS Landranger 6

In contrast to South Ronaldsay, Rousay is predominantly moorland and has a fine range of breeding birds, including Red-throated Diver, Hen Harrier, Merlin, Short-eared Owl, Arctic and Great Skuas and Raven. Around Westness and Trumland House there are areas of trees with the common woodland species. Rousay is served by ro-ro ferry from Tingwall in north-east Mainland, with connections to Egilsay and Wyre, operated by Orkney Ferries. The voyage takes just 25 minutes.

RSPB Trumland This reserve in the south of the island comprises c. 400 ha of dry moorland dissected by small valleys and wetter areas, with most of the moorland specialities. From the pier follow the road uphill and turn left at the T-junction at Trumland Gate Lodge, continuing past Trumland Wood to the entrance of the reserve at Taversoe Tuick Cairn. Go through the gate and follow the path to the Hass of Trumland – the two nature trails through the reserve are 1 mile and 3.1 miles long. Neither are suitable for wheelchairs.

Quandale and Brings Areas of maritime heath in the north-west have supported up to 1,000 pairs of Arctic Terns and 100 pairs of attendant Arctic Skuas, but both have declined significantly in recent years; not so Great Skua, which is slowly increasing. The west cliffs have Common Guillemot, Razorbill, Kittiwake and Fulmar; the last also nests inland on scattered rocks in the moorland.

Eynhallow Sound Offshore, the sounds between Rousay, Egilsay and Wyre support numbers of wintering Great Northern Divers and Long-tailed Ducks. View from the south coast, the Broch of Gurness on Mainland, or the inter-island ferry.

430 EGILSAY — OS Landranger 5/6

A small and relatively low-lying island, predominantly farmland, this is one of the last refuges of the Corncrake on Orkney. The RSPB reserve at Onziebust Farm is managed specifically for Corncrakes, but even so, they are not present every year. A variety of waders and Arctic Terns also breed, while in winter there are over 1,000 Greylag Geese. The island is served by ro-ro ferry from Tingwall on Mainland Orkney, operated by Orkney Ferries. To reach the reserve from the pier, walk up the road to the crossroads at the centre of the island and follow the unsurfaced track straight onwards, past the school; there is a short nature trail 0.5 miles from the pier.

431 SHAPINSAY — OS Landranger 6

This island is almost entirely agricultural land, though in the south-east there is a small area of moorland with a few pairs of Arctic Skuas. In winter up to 100 Whooper Swans may be seen around Mill Dam, an RSPB reserve which also has significant populations of breeding waders and no fewer than nine species of breeding ducks, including Pintail. The reserve lies 1 mile north-east of Balfour, turning second left from the B9059 out of the village; there is no access, but the reserve can be viewed from the roadside hide. Otherwise, the woodland in the grounds of Balfour Castle supports a variety of small passerines and are attractive to migrants. The island is connected by the ro-ro ferry to Kirkwall, operated by Orkney Ferries, and the journey takes 20 minutes.

432 STRONSAY OS Landranger 5

The island is mainly farmland, while the east cliffs between Odness and Lamb Head hold seabirds and the small lochs have breeding wildfowl, including Pintail.

Whitehall on Stronsay is connected by ro-ro ferry from Kirkwall, operated by Orkney Ferries, the voyage taking c. 95 minutes, and there are flights from Kirkwall operated by Loganair.

Rothiesholm This area of moorland on the south-west peninsula supports the best range of moorland birds in the northern group of islands, including Red-throated Diver, large numbers of breeding gulls (five species, especially Great Black-backed), Arctic Tern, small numbers of Arctic and Great Skuas and Twite. There is a small colony of cliff-nesting seabirds on the coast. Park by the old school on the B9061and follow the track east for 400 yards to the moors.

Mill Bay Large numbers of seaducks and divers.

Meikle Water Large numbers of wintering wildfowl, including up to 100 Whooper Swans and 50 Greenland White-fronted Geese.

Stronsay Bird Reserve Well placed to receive migrants, Stronsay has produced some exciting birds, and a private Bird Reserve operates on the island. Contact John and Sue Holloway (Stronsay, Orkney. KW17 2AG; tel: 01857 616363; web: www.stronsaybirdreserve.co.uk) for details of accommodation, boat trips etc.

433 EDAY OS Landranger 5

There is a large area of moorland on Eday supporting good numbers of Arctic Skuas, as well as a handful of Great Skuas and a few pairs of Whimbrels. Seabird colonies around the coast include the Calf of Eday (with large numbers of Great Black-backed Gull), Grey Head for auks, and the south cliffs, which have a large colony of Cormorants. Mill Loch in the north holds c. 8 pairs of Red-throated Divers. Near Carrick House in the north are two small woods that attract migrants.

The island is connected by ro-ro ferry to Kirkwall, operated by Orkney Ferries, and there are flights from Kirkwall operated by Loganair.

434 SANDAY OS Landranger 5

There is almost no moorland on Sanday and the low-lying coast means that there are few cliff-nesting seabirds. There may, however, be large colonies of terns, mainly Arctic, at Westayre Loch, Start Point and Els Ness, as well as breeding Arctic Skuas at Gump of Spurness in the extreme south-west. Raven regularly breeds and Short-eared Owl occasionally. Sanday is ideally placed to receive migrants, especially Lady Parish in the north-east, and the extensive beaches hold large numbers of wintering waders.

Loth on Sanday is connected by ro-ro ferry to Kirkwall, operated by Orkney Ferries, and there are flights from Kirkwall operated by Loganair.

435 WESTRAY OS Landranger 5

The island's prime attraction is the spectacular numbers of seabirds breeding along 5 miles of the west coast between Noup Head and Inga Ness, including up to 12,000 pairs of Kittiwake and 50,000 Common Guillemots. There are several lochs with breeding waterbirds, the best probably being Loch of Burness (just west of Pierowall). The island attracts migrants during passage periods, although being on the west side of the archipelago, it is not ideally placed.

Rapness on Westray is connected by ro-ro ferry to Kirkwall, operated by Orkney Ferries, and there are flights from Kirkwall operated by Loganair.

RSPB Noup Cliffs The north end of the seabird colony comprises the RSPB reserve of Noup Cliffs, straddling 1.5 miles of cliffs and forming the north-west extremity of the island. Here, Ravens haunt the cliffs and the new Gannetry already holds 500 pairs. There may be several hundred pairs of Arctic Terns on the adjacent maritime heath, together with perhaps 50 pairs of Arctic Skuas (here, as elsewhere in Orkney, the species has recently declined sharply). Follow a minor road west from Pierowall to Noup Farm, and then the track (not wheelchair-accessible) north-west for 1.5 miles to the lighthouse on Noup Head.

Stanger Head Puffins may be difficult to see at Noup Head, so it may be worth visiting Stanger Head in the south-east of the island. A track leads north-east to the Head from Clifton, which is just east of the B9066, c. 6 miles south of Pierowall. The track leads to cliffs opposite a rock stack, the Castle o' Burrian, where up to 750 Puffins have been counted.

436 PAPA WESTRAY — OS Landranger 5

Known in Orkney as Papay. Attention centres on the large tern colonies and, of course, migrants (any area of cover can produce birds). Most of the terns nest at North Hill, but there are other large colonies in the south at Sheepheight and Backaskaill, and there is a large colony of Black Guillemots on Holm of Papa, off the east coast; this island also has a small colony of Storm Petrels. The island is connected by a daily vehicle ferry to Kirkwall and by passenger ferry to Pierowall on Westray, both operated by Orkney Ferries. There are flights from Kirkwall operated by Loganair, as well as the shortest scheduled flight in the world (just 2 minutes) from Westray to Papa Westray; provided one stays at least one night on the island, fares are heavily subsidised, at just £20 return in 2009.

RSPB North Hill This RSPB reserve covers the northern uncultivated quarter of the island. It includes an area of maritime heath which until recently held one of the three largest colonies of Arctic Terns in Britain, with over 2,000 pairs, but numbers are now much reduced. There are also up c. 50 pairs of Arctic Skuas and c. 30 pairs of Great Skuas. Visitors must not walk through the tern colonies. The coast has low cliffs with Kittiwake, Common Guillemot, Razorbill and smaller numbers of Puffins and Black Guillemots (and was the last breeding site in Orkney for Great Auk); the best area is in the south-east of the reserve at Fowl Craig. The reserve is accessed from the main gate at the northern end of the island's main road (from which it is a short walk to a hide/information hut, and a 3-mile long trail (not wheelchair-friendly) gives access to the cliffs. The RSPB summer warden is based at Rose Cottage, 650 yards south-east of the reserve entrance, who may be contacted in advance to escort visitors around the colonies (tel: 01857 644240).

Mull Head at the north tip of the island offers excellent seawatching. Sooty Shearwater is regular in late summer, up to 1,300 having been seen in a day. Other species include skuas and a notable day total of 40 Long-tailed Skuas has been logged.

437 NORTH RONALDSAY — OS Landranger 5

This small island is the most north-easterly in the archipelago and is mostly crofted and very flat, with several small lochs and areas of marshland. The coastline is bounded by a dry stone wall, the 'Sheep Dyke', which restricts the kelp-eating North Ronaldsay sheep to the foreshore. This is mostly rocky, with some sandy beaches on the east and south coasts. There may be a few hundred pairs of Arctic Terns, erratically, Sandwich and Common Terns, and numbers of Black Guillemots. Other breeding birds include Raven, waders and ducks, notably Gadwall and Pintail.

The area between Orkney and Fair Isle is probably the main route for seabirds moving between the North Sea and Atlantic. North Ronaldsay, the northernmost of the Orkney archipelago, is well placed to witness these movements. Fulmars pass in vast numbers – up to 20,000 per hour. Sooty Shearwater regularly peaks at over 100 per hour, and Leach's Petrel also occurs. The best seawatching point is the hide in the north of the island, and the best conditions are following a north-westerly gale when the wind returns to south-east at first light.

Following the establishment of a Bird Observatory in 1987 and the consequent intensive coverage of the island, many rarities have been found. Indeed, in several seasons North Ronaldsay has challenged Fair Isle's hitherto unquestioned supremacy as the north's rarity hotspot. The walled garden at Holland House with its fuchsias and sycamores is attractive but many migrants frequent the crops. Large falls of the commoner species occur, and annual scarce migrants include Wryneck, Short-toed Lark, Richard's Pipit, Bluethroat, Icterine,

Marsh, Barred and Yellow-browed Warblers, Red-backed Shrike, Common Rosefinch, and Little and Rustic Buntings. A variety of passage waders and ducks includes large numbers of Whooper Swans and geese in late September–October, and among the numerous migrant Golden Plovers are regular rarities such as Pacific and American Golden Plovers, Buff-breasted and Pectoral Sandpipers.

North Ronaldsay Bird Observatory provides comfortable dormitory and guesthouse accommodation in a wind- and solar-powered low-energy building (see Further Information).

The island is connected by weekly passenger ferry to Kirkwall, operated by Orkney Ferries (sailings are weather dependent) and there are flights from Kirkwall operated by Loganair (provided one stays at least one night on the island, fares are heavily subsidised, at just £20 return in 2009).

438 OTHER ISLANDS

OS Landranger 5, 6 and 7

Four other islands are worthy of mention.

Sule Skerry This holds over 50,000 pairs of Puffins, as well as breeding Storm Petrels and Gannets, but is very difficult to reach, lying 37 miles west of Brough of Birsay (West Mainland).

Sule Stack Even more difficult to reach, it has 5,000 pairs of Gannets. It lies 41 miles west of Brough of Birsay (West Mainland).

Pentland Skerries Accessible from South Ronaldsay, these may have a large colony of Arctic Terns, along with Sandwich and Common Terns, and a small Storm Petrel colony.

Auskerry South of Stronsay, this was the site of some of Eagle Clark's pioneering studies of migration and has large colonies of Arctic Terns.

FURTHER INFORMATION

Grid refs: Binscarth Wood HY 354 140; RSPB Hobbister HY 395 069; RSPB Birsay Moors (Evie) HY 357 265; RSPB Cottascarth and Rendall Moss (Lower Cottascarth) HY 369 194; Loch of Banks (A986) HY 280 230; RSPB Brodgar HY 294 135; RSPB The Loons HY 246 241; RSPB Marwick Head (Cumlaquoy) HY 232 249; RSPB Hoy (Rackwick) ND 201 992; RSPB Trumland HY 427 275; Egilsay (Onziebust) HY 474 289; Shapinsay (RSPB Mill Dam) HY 483 178; Rothiesholm HY 622 234; Mill Bay (B9060) HY 654 279; Meikle Water (centre) HY 664 244; RSPB Noup Cliffs HY 392 499; Stanger Head (Clifton) HY 511 425; RSPB North Hill HY 495 538
RSPB Orkney: tel: 01856 850176; email: orkney@rspb.org.uk
North Ronaldsay Bird Observatory: tel: 01857 633200; web: www.nrbo.co.uk
Tourist Information (VisitOrkney): tel: 01856 872856; web: www.visitorkney.com
NorthLink Ferries, reservations and booking enquiries: tel: 0845 6000 449; web: www.northlinkferries.co.uk
Pentland Ferries: tel: 01856 831226; web: www.pentlandferries.co.uk
John O'Groats Ferries Ltd: tel: 01955 611353; web: www.jogferry.co.uk
Orkney Ferries: tel: 01856 872044; web: www.orkneyferries.co.uk
Flybe/Loganair (flights to and from Orkney): tel: 0871 7002000; web: www.flybe.com
Loganair (flights within Orkney): tel (to make reservations): 01856 872494 or 873457; web (for timetables): www.loganair.co.uk

439–454 SHETLAND (Shetland)

OS Landranger 1, 2, 3 and 4

Shetland has two major attractions for the birdwatcher: breeding seabirds and waders, and migrants, particularly rarities. There is great potential for those who wish to discover their own birds. The best times to visit are late May to mid-July for breeding birds, and mid-May to early June and mid-August to late October for migrants.

Habitat

The Shetland archipelago consists of 117 islands, of which 13 are permanently inhabited. At 60°north, or the same latitude as the south tip of Greenland, the islands have a landscape of interlocking peninsulas and voes (inlets). The rolling, dark, blanket bog-covered hills are bordered by a rocky coast. Indeed, with over 2,500 freshwater lochs as well as a labyrinthine coast, water forms a very important part of the Shetland environment. The sheltered voes often have sandy beaches but the outer coastline consists of cliffs, occasionally rising to over 1,000 feet. The sun is above the horizon for nearly 19 hours a day in summer, and it is easy to read a book in the twilight at midnight, but its rays are not strong at these latitudes. In winter there are barely 5.5 hours of daylight and Shetland can be very bleak indeed. There have been almost no indigenous trees on the islands for 2,000–3,000 years, apart from relict patches protected from sheep in steep valleys or on small islets. There are, however, small areas of planted woodland, the largest stand being at Kergord on Mainland. The lack of cover means that migrants must use what they can find, and any crops, trees or bushes are worth investigating. In particular, any and all areas of cover on the east coast can hold migrants; which you choose to work is a question of accessibility and convenience.

Species

Fulmars first bred on Foula in 1878 and now number over 180,000 pairs (making it the most abundant bird in Shetland); they include occasional 'blue' Fulmars. There are two known colonies of Manx Shearwaters, Storm Petrel is widespread, and Leach's Petrel breeds on Foula and on Gruney in Yell Sound. The Gannet colonies at Noss and Hermaness are increasing, and Shag outnumbers Cormorant by 20 to 1. Common Guillemot, Razorbill and Puffin are all common (and easily seen at Sumburgh Head), with smaller numbers of Black Guillemots. Common and Arctic Terns, six species of gulls, and Arctic and Great Skuas complete the complement of breeding seabirds.

Several hundred pairs of Red-throated Divers nest on the lochs and pools, and the marshes and wetlands also support Mute Swan, Greylag Goose, Teal, Tufted Duck and Red-breasted Merganser, with Common Eider on coastal moorland. There are also a number of pairs of Wigeons and c. five pairs of Whooper Swans. Twelve species of waders breed, notably up to 500 pairs of Whimbrels (although this species has declined significantly in recent years), a few pairs of Black-tailed Godwits and up to 40 pairs of Red-necked Phalaropes, which usually arrive in late May and leave by early August. Corncrake has sadly all but disappeared, although a few singing males are heard every year. Red Grouse has been introduced, and there are small numbers of Merlins and two or three pairs of Peregrines. Skylark and Rock and Meadow Pipits are common, but Swallow breeds only in small numbers, House Martin only irregularly, and Pied and White Wagtails and Redwing occasionally. Blackbird, Wheatear, Wren (the rather dark endemic Shetland form *zetlandicus*), Starling (subspecies *zetlandicus*, which also occurs on the Outer Hebrides), Raven and Hooded Crow are common, and Rook breeds at Kergord. Twite and House Sparrow are both common around crofts, and Linnet is now a regular breeder in very small numbers.

Occasionally individuals of species that breed further north, in Iceland or Scandinavia, summer on Shetland. These include Great Northern Diver, Long-tailed Duck (believed to have bred three times in the 19th century), Common and Velvet Scoters, Purple Sandpiper, Turnstone, Glaucous Gull (has hybridised with Herring Gull), Snowy Owl (bred 1967–75 on Fetlar), Fieldfare (bred most years 1968–73 and a few times since) and Snow Bunting.

Sooty Shearwater is regularly seen on passage in late summer/autumn, especially from Sumburgh Head. Other seabirds to look for include Pomarine and Long-tailed Skuas in both spring and autumn, and there is a notable late-autumn passage of 'blue' Fulmars. Flocks of Pink-footed, Greylag and Barnacle Geese occur, mainly in autumn, as well as Gadwall, Pintail and Shoveler. Migrant raptors include Long-eared and Short-eared Owls, and less often Honey-buzzard, Hen Harrier, Sparrowhawk, Common Buzzard, Rough-legged Buzzard and Osprey. Grey Plover, Sanderling, Bar-tailed Go dwit and Knot occur exclusively on passage, joining the usual array of migrant waders. Woodcock, thrushes, chats, Blackcap, Willow

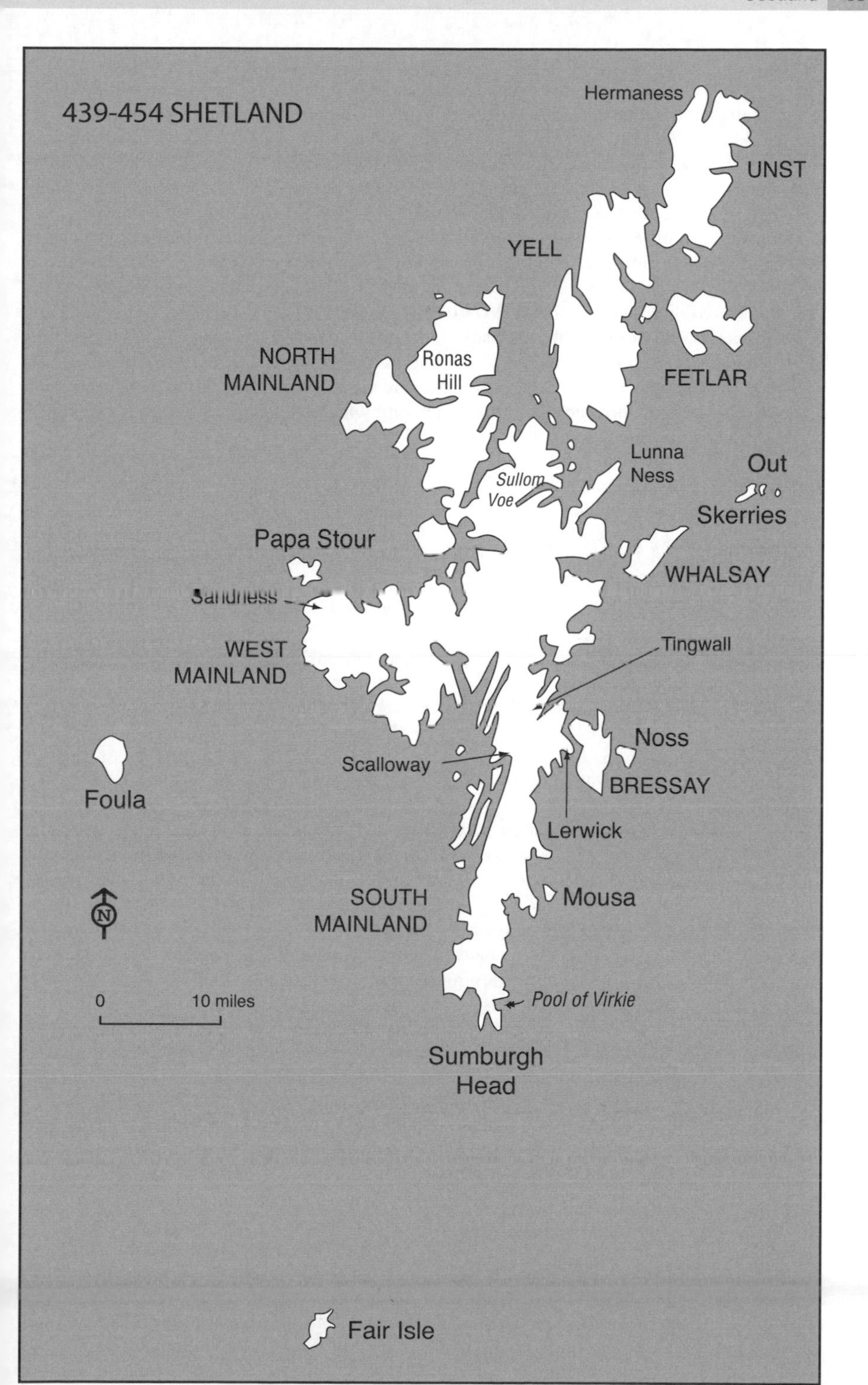
439-454 SHETLAND
Hermaness
UNST
YELL
FETLAR
NORTH
MAINLAND
Ronas
Hill
Lunna
Ness
Out
Skerries
Sullom
Voe
Papa Stour
WHALSAY
Sandness
WEST
MAINLAND
Tingwall
Noss
Scalloway
BRESSAY
Foula
Lerwick
Mousa
SOUTH
MAINLAND
0
10 miles
Pool of Virkie
Sumburgh
Head
Fair Isle

Warbler, Goldcrest, Pied Flycatcher etc. all occur in quite large numbers given suitable weather. The best conditions are south-east winds. Among scarcer migrants, those usually more frequent in spring than autumn are Bluethroat, Marsh Warbler, Golden Oriole and Red-backed Shrike. Short-toed Lark, Red-breasted Flycatcher, Waxwing, Lapland, Ortolan and Little Buntings and Common Rosefinch are much more frequent in autumn, though they do occur in spring, but Richard's Pipit, Barred and Yellow-browed Warblers and Yellow-breasted Bunting are virtually unknown in spring. Wryneck and Icterine Warbler are recorded equally at both seasons.

True rarities recorded with some regularity in spring include Thrush Nightingale, Subalpine Warbler and Rustic Bunting, while in autumn Olive-backed Pipit, Citrine Wagtail, Arctic and Greenish Warblers and Rustic Bunting are annual. More erratic specialities include Great Snipe, Pechora Pipit, Booted and Lanceolated Warblers and Yellow-breasted Bunting.

Some species common on the mainland but rare or absent as breeding birds on Shetland occur on migration. These include Dunnock, Robin and finches (apart from Twite), while Great Spotted Woodpecker, tits and northern Bullfinches are rare and erratic visitors. Foula, Unst, South Mainland and Out Skerries are probably the most productive areas for migrants, but birds can be, and are, found almost anywhere.

A total of 500–600 Great Northern Divers winter around the islands, with many fewer Red-throated. Slavonian Grebe occurs in small numbers, especially in Tresta Voe. Up to 400 Whooper Swans pass through on migration, but in general there are few in winter unless it is very mild. They, like many ducks (with the exception of Wigeon, Tufted Duck, Red-breasted Merganser, Common Eider and Long-tailed Duck) move south early in the season. King Eider is seen most winters, with records widely scattered around the islands. Large numbers of Turnstones and Purple Sandpipers winter on shores, and away from the coast there may be Golden Plovers, Curlews and Common and Jack Snipes and sometimes numbers of Woodcocks in late autumn. Small flocks of Little Auks can occur October–February, and Glaucous and Iceland Gulls are regular visitors, especially to fishing harbours. Small numbers of Snow Buntings sometimes remain and may be found almost anywhere.

Access

Full details of all transport and accommodation are available from Visit Shetland (see Further Information).

By Sea Northlink Ferries offer departures seven nights a week in both directions on the Aberdeen–Lerwick route, with three sailings a week calling in at Kirkwall, Orkney, en route; their 'cruise ferries', the *Hjaltland* and *Hrossey*, are comfortable and extremely punctual, arriving in Lerwick at 7.30am after the 12 hour voyage (14 hours via Orkney). Advance booking is recommended on all sailings, most especially for vehicles.

By Air Flybe's franchise-holder Loganair operates frequent daily flights to Shetland from Aberdeen, Edinburgh, Glasgow, Inverness and Orkney, with onward connections to London and international destinations.

Ground Transportation Public transport on Shetland is limited, so it is best to hire a car or bring your own. Most roads on Shetland are narrow and consideration should always be exercised when parking. In addition, many of the areas described below are private. Especially when searching for migrants, the nature of the habitat calls for discretion. Do not enter gardens or crofts without permission, and even when looking in from the outside remember that not everyone may appreciate you peeping over their garden wall. In general, however, you will receive a friendly reception if you ask first.

Inter-island Inter-island ferries link the larger islands with the Shetland Mainland. The services are fast and frequent, and passenger fares are nominal. Drive-on/drive-off services operate to Bressay, Whalsay, Yell, Unst, Fetlar and Papa Stour and Out Skerries (booking essential for vehicles on all but the Bressay ferry, but foot passengers do not usually need to book for any of these services apart from Papa Stour and Out Skerries, for which booking is essential). Passenger-only services operate to Foula (and Fair Isle – see page 648). For a full schedule visit www.shetland.gov.uk or tel. 01595 744225. Foula, Papa Stour and Out Skerries are also served by flights operated by Directflight from Tingwall, near Lerwick.

Accommodation Available on all islands detailed below (except Mousa), but away from Lerwick and Scalloway accommodation is limited. A variety of budget accommodation is available, and camping is not discouraged, but ask permission before setting up.

439 SOUTH MAINLAND — OS Landranger 4

RSPB Sumburgh Head Turn south-east off the A970 at Grutness (0.3 miles north-east of the turn to Sumburgh Airport), signed to Sumburgh Head and, after 1.25 miles, park in the main car park. Disabled visitors may continue down and park next to the lighthouse. The cliffs can be viewed from here as well as along the final 500 m of road between the car park and the lighthouse. The Head has significant colonies of seabirds (including Puffin) that are readily accessible, especially on the west side. The Head is a good seawatching site, with Sooty Shearwater regular in late summer and autumn, when Pomarine and Long-tailed Skuas can occur, and Little Auk is regular in the late autumn. Watch from just south of the lighthouse in south and west winds, and from the car park in east winds. The stone dykes and quarries around the Head often hold migrant passerines, and the patch of roses just south-west of the lighthouse is particularly attractive. West Voe of Sumburgh holds seaducks and Great Northern Diver in winter.

Sumburgh Hotel The gardens have some small trees and shrubs offering cover to migrants, and the tiny marsh behind the hotel is also worth checking (but beware disturbing breeding waders in summer).

Grutness Turn north-east off the A970 0.3 miles north-east of the turn to Sumburgh Head and park by the pier. The beach attracts waders and gulls, with breeding Arctic Tern north of the pier, while Grutness Voe has seaducks and Great Northern Diver in winter. There are areas of cover attractive to migrants around the gardens, but note that these are strictly private with no access. The *Good Shepherd* leaves for Fair Isle from the small harbour.

Loch of Gards, Scatness Turn off the A970 at the sign to Scatness, pass through the settlement, park carefully at the roadside after c. 250 yards and walk south to the gate overlooking the pools. Access on foot from here is unrestricted. The Loch harbours wildfowl, gulls and waders (which may roost here at high tide or on the beach at the west end of the peninsula), with breeding Arctic Terns on the Scatness Peninsula, while the adjacent stone walls and gardens attract migrants.

Pool of Virkie One of Shetland's few tidal basins, usually holding a wider variety of waders than elsewhere, as well as gulls and terns. Turn east off the A970 to Eastshore (just north of the north set of airport warning lights) and view the pool from the road, preferably in the 2 hours either side of high tide (with the 2 hours *after* high water usually the best). The best site for Shelduck in the islands, there are also gardens on its northern shore.

Toab Turn west off the A970 just north of the northern set of airport warning lights (or turn west off the A970 just north of the Exnaboe turning). The gardens and fields are productive areas for migrants.

Exnaboe Turn east off the A970 c. 0.5 miles north of the Pool of Virkie turning. The gardens are good for migrants.

Bay of Quendale View from the road on the west side. Good for wintering seaducks and Great Northern Diver, which may linger into spring. The streams that run from Quendale and Loch of Hillwell to the sea attract migrants, and are favoured by Bluethroat and Marsh Warbler in spring.

Loch of Hillwell Turn west off the A970 to Quendale, just south of the G & S Mainland shop, and continue for 1.25 miles, parking carefully to view the Loch from the road. The only machair loch in Shetland, with marshy edges, it holds numbers of wildfowl, including Whooper Swan, Greylag Goose and occasionally Smew in winter, and attracts passage wildfowl, waders and passerines.

RSPB Loch of Spiggie A fertile eutrophic loch surrounded by areas of marshland. Leave the A970 west on the B9122 at Skelberry and after c. 1.3 miles turn west onto the minor road to Spiggie. There are good views from this road of the north, west and south shores of the loch. Wildfowl include Greylag Geese and up to 100 Whooper Swans in the late autumn/early winter, with small spring gatherings of Long-tailed Ducks. Breeding birds include

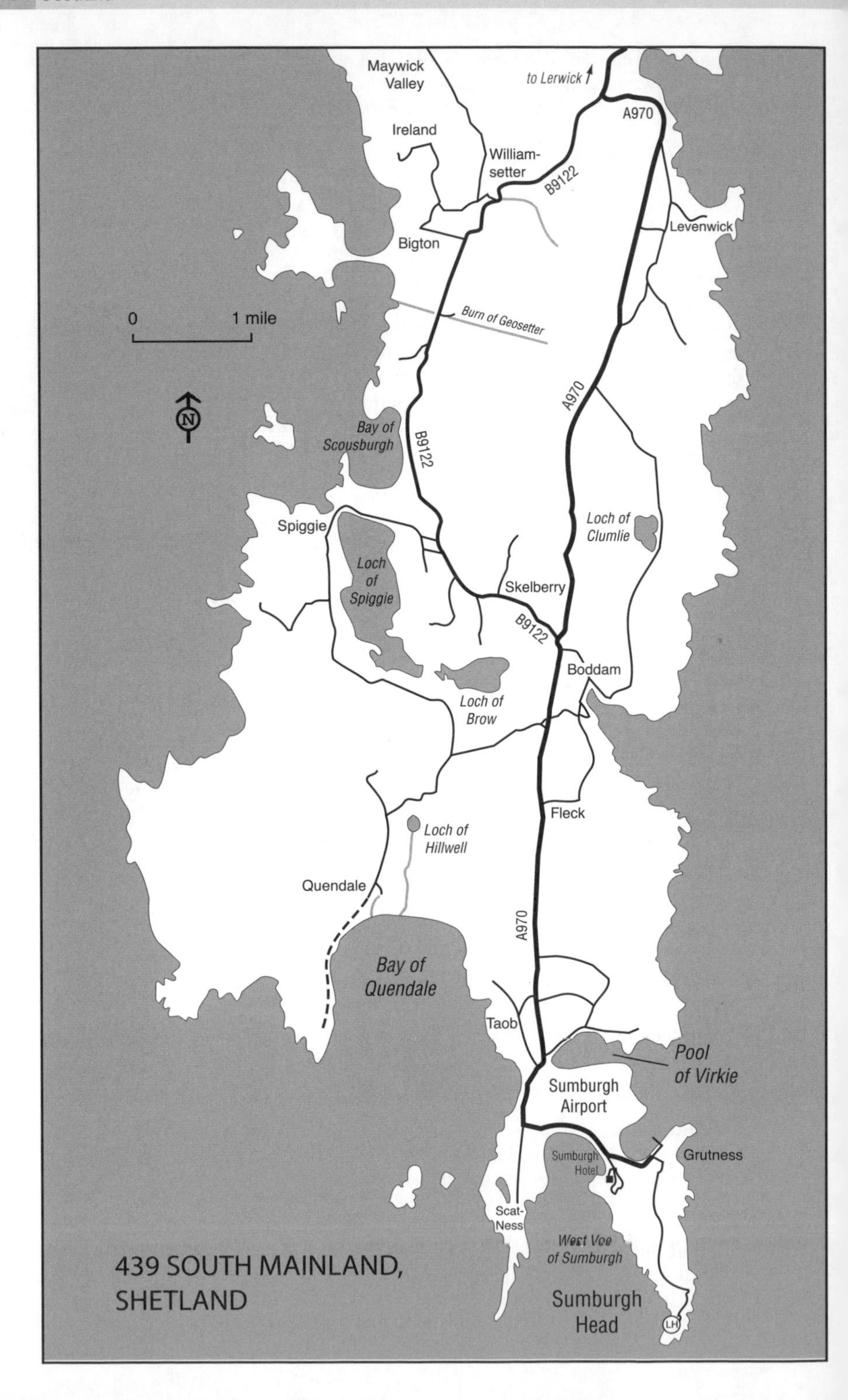

Maywick Valley
to Lerwick
A970
Ireland
William-setter
B9122
Levenwick
Bigton
0
1 mile
Burn of Geosetter
A970
Bay of Scousburgh
B9122
Spiggie
Loch of Clumlie
Loch of Spiggie
Skelberry
B9122
Boddam
Loch of Brow
Fleck
Loch of Hillwell
Quendale
A970
Bay of Quendale
Taob
Pool of Virkie
Sumburgh Airport
Grutness
Sumburgh Hotel
Scat-Ness
West Voe of Sumburgh
Sumburgh Head
LH
439 SOUTH MAINLAND, SHETLAND

several species of ducks and waders, while the loch is used by bathing gulls, terns and skuas, and waders occur on passage (favouring the north-west corner). In winter the dunes to the north hold Snow Bunting, and Bay of Scousburgh hosts seaducks and Great Northern Divers.

Boddam Turn east off the A970 just south of the B9122 near Skelberry or just south of the G & S Mainland shop, or at the turning to Fleck. There are seaducks, waders and gulls in the voe and the gardens shelter migrants, especially along the Fleck road.

Burn of Geosetter A very good area for migrants. Park carefully by the B9122 and explore the patches of cover along the stream east of the road; the gardens are strictly private.

Bigton and Ireland Access along minor roads off the B9122. The gardens and fields are attractive to migrants.

Maywick Valley Access along minor road off B9122 north-east of Bigton. An underwatched valley, the gardens and fields are attractive to migrants and Loch of Vatsetter holds small numbers of wintering wildfowl.

Levenwick Access from minor roads east off the A970. The many gardens attract migrants.

440 LERWICK AREA, MAINLAND — OS Landranger 4

Lerwick Harbour Lerwick is the point of arrival for visitors coming by sea and the harbour is well worth a look, with resident Black Guillemot and gulls, including wintering Glaucous and Iceland. The Shetland Catch fish factory, just to the north at Gremista, is one of the best areas for white-winged gulls. Otherwise, the gulls often perch on warehouse roofs.

Sandy Loch Reservoir View from the A970 just south of Lerwick, and it is possible to walk around the reservoir. Loafing gulls and passage waders are the attraction.

Loch of Clickimin Gulls favour the sewage outfall directly south of Loch of Clickimin, and the short grass on the east shore is also frequented by loafing gulls. The Loch is used by diving ducks in winter.

Helendale Immediately west of Loch of Clickimin at Helendale there are several large wooded gardens that attract migrants. Please Stay on the tarmac track and respect resident's privacy.

Gremista Just north of Lerwick there are a few gardens and a dense sycamore copse west of the road along the North Burn of Gremista (just north of the A970) which holds migrants.

441 CENTRAL MAINLAND — OS Landranger 4

Scalloway This has similar attractions to Lerwick. The best area for gulls is the fish dock pier on the east side of the harbour near the castle and the adjacent fish factories. During westerly gales seabirds may seek shelter here, occasionally including Pomarine Skua and, more rarely, Storm Petrel. Long-tailed Duck is present in winter, together with small numbers of auks, Tufted Duck and occasionally Scaup. The gardens in the town hold migrants, and formerly, wintering Long-eared Owls.

Veensgarth Approximately 6 miles north of Lerwick. Turn south at HU 428 444 and follow the road until it ends. Search the trees from the road for migrants and wintering Long-eared Owl.

Lochs of Tingwall and Asta Adjacent to the B9074, these are good for diving ducks, passage Whooper Swans and passage waders. The crofts on the west side of the B9074 may hold migrants.

Kergord On the B9075 in the Weisdale Valley, this area has plantations of sycamore, larch and spruce, which support breeding Woodpigeon, Rook and sometimes Goldcrest or even Redwing, Robin or Siskin. They also attract migrants but due to the extent of the cover are difficult to work. There is free access to the plantations but the walled garden at Kergord House is strictly private.

442 WEST MAINLAND — OS Landranger 3 and 4

Tresta Voe, Sandsound Voe and Reawick Good for Great Northern Diver, Slavonian Grebe and seaducks (often producing a King Eider, especially in Tresta Voe), the area can be viewed from the A971 and B9071, with Reawick accessible via a minor road off the B9071. Weisdale Voe has a similar range of birds.

Tresta A good area for migrants, the village lies on the A971 and has gardens and Sycamore trees by the chapel.

Sandness Mainland's westernmost village, the nearby Loch of Norby and Loch of Melby (view from the track just north-west of Norby) and Loch of Collaster (view from the minor road to the north) have potential for vagrant American waders and ducks.

Ness of Melby At the terminus of the A971, this makes a good seawatching point, with the possibility of Pomarine or Long-tailed Skuas.

Wats Ness The most westerly point on Mainland. From Walls on the A971 take the minor road west to Dale and, after c. 3.5 miles, turn at the sign to Wats Ness. From the end of the road walk south-west to find a sheltered part of the cliff. The best seawatching station in spring in the entire archipelago, flocks of both Long-tailed and Pomarine Skuas are regular (in May 1992 Pomarine peaked at the astonishing day total of 2,093 birds). Strong west or north-west winds coupled with squalls are the best conditions, and May the month.

443 NORTH MAINLAND — OS Landranger 3

Voe The gardens at Voe, just off the A970 on the B9071, are very good. It is a large area with lots of houses.

Voxter and Sullom Plantations More good areas for migrants, the former north of the B9076 c. 1 mile from its junction with the A970, the latter adjacent to the minor road to Sullom, c. 2 miles from the A970.

Lunna Ness This peninsula in the north-east is an excellent area for migrants, especially Lunna itself with its line of Sycamores (on the left of the road 300 yards before Lunna House). The gardens and stream at the head of Swining Voe, west of Lunna Ness, can also be productive.

Sullom Voe Wintering Great Northern Diver, Slavonian Grebe and seaducks, sometimes including Velvet Scoter. The sewage outfall just before the main gate of the oil terminal is good for gulls, and the Houb of Scatsta, north of the B9076 just beyond the airfield, attracts waders.

Ronas Hill In the far north of Mainland, this is the largest wilderness area in Shetland. The west flank has spectacular cliffs and, at 1,475 feet, is the archipelago's highest point, with areas of arctic/alpine habitats. The terrain is very rough and it is quite easy to get lost, especially in the very changeable weather. Snowy Owl has occasionally been seen, the best area being around Mid Field and Roga Field. Dotterel is irregular in spring and Snow Bunting has been seen in summer and may be numerous in autumn. Access is from the A970. The track west of the road just north of North Collafirth goes to Collafirth Hill, accessing Mid Field and Roga Field.

Ronas Voe This scenic fjord has the usual species, with King Eider a possibility. A minor road runs along the shore from the A970.

444 MOUSA — OS Landranger 4

This 180 ha uninhabited island lies just over 0.5 miles off the east coast of Mainland, opposite Sandwick, and passenger ferries run from Leebitten: turn south off the A970 at the north Sandwick junction (signed to Sandwick and Mousa Ferry) and then follow signs to Mousa Broch; the crossing takes 15 minutes. Mousa Broch is the best-preserved Pictish broch in Britain and this archaeological attraction is not without its birds – many of the island's 11,000 pairs of Storm Petrels nest in the broch, as well as in the surrounding stone walls and boulder beaches. It is necessary to visit at night to see and hear the petrels, and special petrel-watching boats trips operate from late May to mid-July. The petrels are best viewed by standing next to the broch; do not, however, lift rocks to look for nesting birds! The daytime ferry service operates 1 April to mid-September: ring Tom Jamieson on 01950 431367 for details of ferries and nocturnal trips or visit www.mousaboattrips.co.uk. An RSPB reserve, the path to the broch is less than 0.5 miles long and forms part of a 1.5 mile circular trail. Apart from the petrels, the several small pools are good for migrant waders and the island holds breeding Arctic Terns, Arctic and Great Skuas and good numbers of Black Guillemots.

445 BRESSAY — OS Landranger 4

Bressay lies immediately east of Lerwick and is accessible by frequent ro-ro ferry (5 minutes).

There are colonies of Great and Arctic Skuas on the south-east side of the island, but the main attraction is migrants. The best areas are the crofts on the west side. However, a lot of walking is required to cover them without a car.

446 NOSS

OS Landranger 4

Lying just a few hundred yards east of Bressay, Noss is an NNR covering 313 ha. This green and fertile island rises to 594 feet at the Noup. There are spectacular seabird colonies: over 100,000 individuals of 12 species breed, including 8,600 pairs of Gannets, 45,000 Common Guillemots, 400 pairs of Great Skuas but now just one or two pairs of Arctic Skuas. The island is accessible late April to late August, with the ferry operating 10am–5pm daily except Monday and Thursday. On Bressay follow signs to Noss and proceed to the end of the road, walking from the car park for 0.5 miles down a steep farm track to the 'wait here' sign. The resident wardens provide a ferry service using an inflatable dinghy from Bressay (3 minutes) for a small fee, but in bad weather the island is closed (tel. 0800 107 7818 for daily updates during the ferry season). There is a visitor centre and toilets on the island. Allow at least 4 hours for exploration. Private boat operators also run trips around Noss, contact the tourist office for details.

447 WHALSAY

OS Landranger 2

Whalsay has breeding Red-throated Divers, Whimbrels (view from the road between Brough and Isbister) and Great and Arctic Skuas, but its main attraction is passage migrants. A large island, over 5 miles long, a certain amount of legwork is necessary to cover all of the better areas. The Whalsay ro-ro ferry leaves Mainland from Laxo (30 minutes, though in some weathers it sails from Vidlin, the crossing taking 50 minutes; advance booking recommended).

Dury Voe The ferry regularly passes small groups of Little Auks in winter, and a White-billed Diver wintered for 13 years in the Voe.

Symbister Bay Good for gulls in winter, attracted by the fish factory; the harbour is best viewed from the raised bank by the ferry terminal. The gardens and crops are also very attractive to migrants.

Skaw, Brough and Isbister Skaw in the far north-east of the island is where many of the most exciting migrants have been found, and all of the crofts should be checked, as well as the airfield (from the perimeter) and the close-cropped grass of the golf course. The gardens and crofts on the seaward side of Isbister and around Brough are also worth a look.

448 OUT SKERRIES

OS Landranger 2

Out Skerries is a group of small, rocky islands. Their position at the easternmost point in Shetland and their ease of coverage has resulted in a list of rare birds to rival Fair Isle and Foula. The best areas for migrants are the iris beds and gardens on Bruray and Housay. Out Skerries are accessible by car ferry, which runs daily from Vidlin (1.5 hours, or Laxo if the weather is bad from the north) and twice a week from Lerwick (c. 4 hours); booking is essential. Directflight operate flights from Tingwall Airport, near Lerwick.

449 PAPA STOUR

OS Landranger 3

This small island off West Mainland has the largest Arctic Tern colony in Shetland, as well as Great and Arctic Skuas, seabird cliffs, spectacular coastal scenery and massive sea caves. Several small lochs are good for wildfowl and waders, and though under-watched it has great potential for interesting migrants. A passenger ferry runs from West Burrafirth, weather permitting, the journey taking 35 minutes (booking essential), and Directflight operate services from Tingwall, near Lerwick.

450 FOULA

OS Landranger 4

Foula, with a population of 30, is one of the most isolated inhabited islands in Britain. Fourteen miles west of Mainland, it has a spectacular, precipitous coastline, and at 1,220 feet the Kame is the second-highest sea cliff in Britain. The island is served by the *New Advance*, operated by Atlantic Ferries from Walls in West Mainland, and there are also flights from Tingwall Airport, just outside Lerwick, operated by Directflight. Both air and sea services are very dependent on the weather. All accommodation must be booked in advance and sufficient provisions for the duration of the stay must be taken if self-catering or camping.

Foula's main attraction, apart from its splendid isolation and superb atmosphere, is the large number of breeding seabirds on the west cliffs, including Manx Shearwater, Storm and Leach's Petrels (the last almost impossible to see) and c. 200 pairs of Gannets, while the moorland in the west of the island holds the world's largest colony of Great Skuas, comprising around 2,500 pairs. Over the last 15 years Foula has been well-covered during the autumn and has developed a reputation for producing outstanding rarities. The best areas are the crofts in the south and the well-vegetated Ham Burn.

451 YELL — OS Landranger 1 and 2

Yell is the second-largest island in Shetland and is predominantly covered by blanket-bogs on low, rounded hills. There are many areas of cover which are seldom visited by birdwatchers, and are thus fertile hunting grounds for more adventurous rarity hunters. Access is by ro-ro ferry across Yell Sound from Toft on Mainland (20 minutes), while to the north there are ferries to Unst and Fetlar.

Yell Sound Ramna Stacks and Gruney in the north of the Sound are managed by the RSPB and have breeding Storm and Leach's Petrels. There are, however, no visiting arrangements. In winter Great Northern Diver, Long-tailed Duck and, sometimes, Little Auk may be seen from the ferry between Yell and Mainland.

Copister Broch Off the far south of the island, this holds a large colony of Storm Petrels.

Burravoe, Otterswick and Mid Yell There are good areas for migrants on Yell, including the gardens in Burravoe and Otterswick in the south-east, and Mid Yell in the centre of the island.

RSPB Lumbister The RSPB has a reserve at Lumbister covering over 1,600 ha west of the A968 and north-east of Whale Firth, with other areas at Black Park and Kirkabister. Moorland, several lochs and sea cliffs support breeding Red-throated Diver and both species of skua, as well as Merlin, seabirds and occasionally Whimbrel. Access is from the lay-by 4 miles north of Mid Yell. The reserve is always open; contact the Fetlar warden for details.

452 FETLAR — OS Landranger 1 and 2

Accessed by ro-ro ferry from either Gutcher on Yell or Belmont on Unst (sailing times 25 minutes). The west of Fetlar, especially Lamb Hoga, is Heather moorland and peat bog, while the east has grassy moorland and dry heath, all interspersed with lochs, pools, marshes and patches of cultivation. There is a fine coastline of cliffs and beaches, and seabirds breed in small numbers, the south tip of Lamb Hoga having a good selection.

Fetlar's specialities are Red-necked Phalarope (see below) and Whimbrel, which are quite widespread, good areas being the roadside moorland west of Loch of Funzie and moorland west of the airstrip and east of the school. With such rare birds, you should take special care to avoid disturbance. Do not linger in Whimbrel territories and do not try to find phalarope breeding areas. Fetlar is also the place to see Manx Shearwater and Storm Petrel (see below), and there are large colonies of Arctic Terns, as well as Great and Arctic Skuas. Specific sites include:

RSPB Fetlar/Loch of Funzie Originally established to protect breeding Snowy Owls (now sadly long gone), the reserve extends over Vord Hill, Stackaberg and the surrounding slopes, and the coastline from East Neap to Urie Ness. From the ferry terminal at Hamars Ness take the B9088 east for 6 miles. There is a small car park to the west of Loch of Funzie; walk east from here for 100 m and then follow signs for 300 m (not wheelchair-accessible) to the hide overlooking the Mires of Funzie. Between 20 and 40 pairs of Red-necked Phalaropes breed on Fetlar, several of which are in the Mires of Funzie, though they are most easily viewed from the road along the north shore of the loch (the best tactic is to sit and wait for these confiding birds to come to you!). The Loch also holds Red-throated Diver. Entrance to the reserve is otherwise only by arrangement with the summer warden, whose house is signposted at Baelans, 2.5 miles along the road from the pier (tel: 01957 733246).

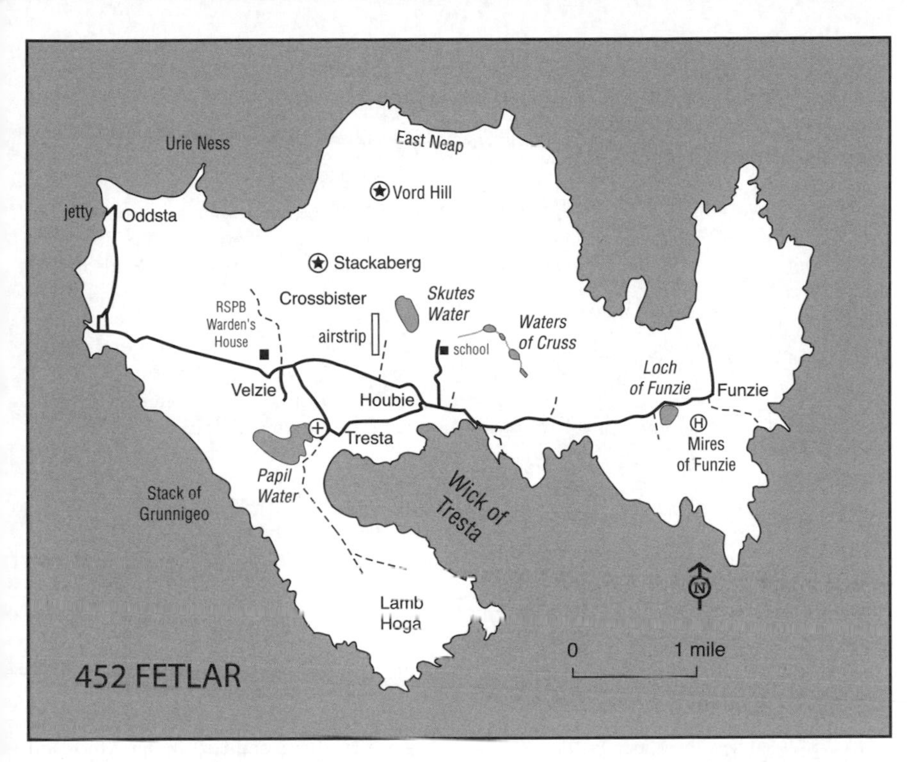

Papil Water Used by large numbers of bathing seabirds in summer, notably Great Skua, and also Red-throated Diver. It attracts wildfowl in winter.

Wick of Tresta A few Manx Shearwaters breed on Lamb Hoga and a good place to watch the evening gatherings in the Wick of Tresta is the beach adjacent to Papil Water, especially the Tresta end.

Grunnigeo Storm Petrel cannot be seen in daylight but the colonies along the west slopes of Lamb Hoga are accessible with great care, especially at Stack of Grunnigeo. From Velzie walk past the north shore of Papil Water, directly over the saddle of the hills to Grunnigeo. Descend on the north side of Grunnigeo to the grassy ridge that connects the stack to the main slope. The ridge makes a good vantage and petrels can be seen and heard around midnight. They favour overcast nights, and in clear, bright conditions none may come ashore. Note that the slopes here are steep and slippery and should only be attempted by the fit and agile; it is best to go out in daylight, and a good torch is essential for the return journey.

Tresta, Houbie and Funzie Fetlar attracts its fair share of migrants and there are productive crofts in the south and east of the island.

453 UNST

OS Landranger 1

Unst is the third-largest and northernmost main island in Shetland. Access from Gutcher on Yell (10 minutes) and Oddsta on Fetlar (30 minutes) is by ro-ro car ferry.

Hermaness NNR Unst's star attraction is the magnificent seabird colony on Hermaness, at the northern tip of Unst. Here, the cliffs rise to 400 feet and overlook Muckle Flugga and Out Stack, Britain's most northerly point. From the end of the B9086 at Burrafirth take the minor road north to the car park; nearby, a visitor centre is sited at the old Muckle Flugga Shore Station, where a summer warden is based. From the car park a marked trail leads across the moor to the cliffs, joining a cliff-top path extending from Saito north to the Greing, while the path from the Greing back to the visitor centre completes the circuit; allow 3–4 hours. While crossing the moors, beware of dive-bombing skuas. They seldom press home an attack,

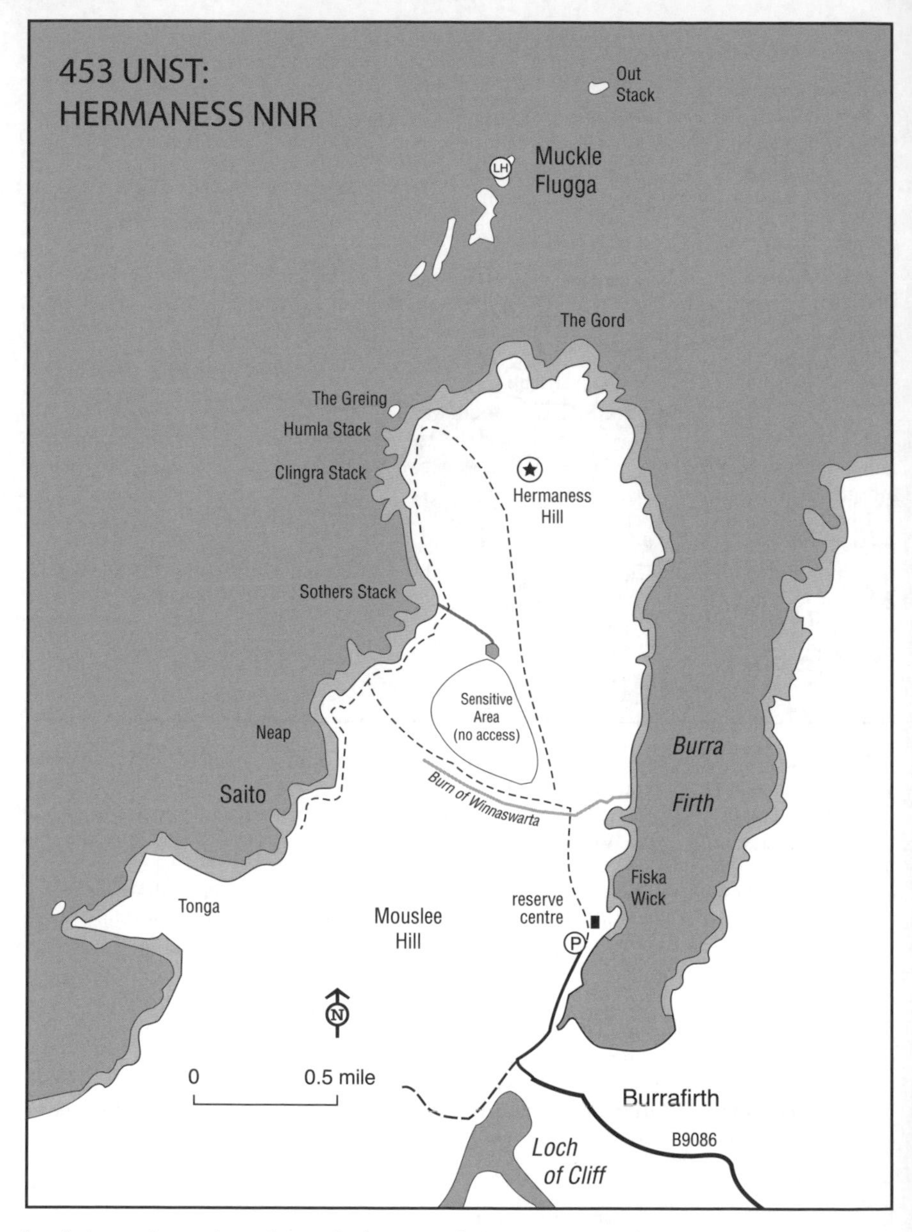

but it is worth wearing a hat and taking a stick (or using your telescope and tripod) to hold above your head. Take great care on the steep and slippery grass slopes above the cliffs, especially when it is wet – there have been accidents here (it can be dangerous to wear waterproof trousers near the cliff edge and on steep seaward facing slopes because if you slip you will slide more easily). Note also that the weather is very changeable and you should be prepared for the worst. Hermaness is an NNR with no restrictions on access except to the sensitive area.

Breeding seabirds include over 16,300 pairs of Gannet, 14,000 pairs of Fulmar, 22,000–28,000 pairs of Puffin, 10,000 pairs of Common Guillemot, and smaller numbers of Shags, Razorbills, Black Guillemots and Kittiwakes. The moorland holds Red-throated Diver, Golden

Plover, and Arctic and Great Skuas (650 pairs). In 1972–1995 a solitary Black-browed Albatross (known affectionately as 'Albert') returned almost every year from late March–August to build its nest among the Gannets.

Burra Firth The beach at Burrafirth often holds Glaucous or Iceland Gulls in the winter, while to the south, Loch of Cliff attracts bathing seabirds and in winter may hold a few roosting white-winged gulls.

Norwick and Skaw Good spots for migrants in the extreme north-east of the island, access is off the B9087. Of note are the well-vegetated streams. Seek permission to venture off the road.

Haroldswick The village gardens attract migrants, the beach waders, and the bay wintering divers and seaducks. Between Norwick and Haroldswick there are marshy areas and several shallow pools by the roadside which sometimes attract migrant waders.

Baltasound This settlement on the east coast has some of the best areas of cover for passerine migrants, notably at the school and at Halligarth.

Uyeasound and Skuda Sound Access south off the A968 via the B9084 or the minor road to Uyeasound. The sounds hold seaducks and divers, with Great Northern Diver often staying late in spring. Nearby Easter Loch holds Whooper Swan and the village gardens at Uyeasound are attractive to passerine migrants.

Bluemull Sound This separates Unst from Yell and Fetlar. Large numbers of seabirds pass through the area on their way to and from the colonies. In winter there are large numbers of Common Eiders and Long-tailed Ducks as well as divers, Black Guillemot and sometimes Little Auk, and White-billed Diver and King Eider are both regularly recorded. View from the ferry or from the ferry terminals on Unst, Yell and Fetlar.

FURTHER INFORMATION

Grid refs: RSPB Sumburgh Head HU 408 083; Sumburgh Hotel HU 401 095; Grutness HU 405 101; Loch of Gards HU 300 096; Pool of Virkie HU 396 115; Toab HU 388 115; Exnaboe HU 397 118; Bay of Quendale HU 368 127; Loch of Hillwell HU 377 142; Loch of Spiggie RSPB reserve HU 371 176; Boddam HU 397 156; Burn of Geosetter HU 379 201; Bigton HU 377 210; Maywick valley HU 378 246; Levenwick HU 409 212; Lerwick Harbour (Shetland Catch) HU 464 429; Sandy Loch Reservoir HU 450 396; Loch of Clickimin HU 464 406; Helendale HU 462 408; Gremista HU 461 431; Scalloway HU 403 393; Veensgarth HU 426 436; Lochs of Tingwall HU 416 433; Kergord HU 395 542; Tresta HU 361 511; Sandness (Loch of Norby) HU 198 576; Ness of Melby HU 186 579; Wats Ness HU 181 507; Voe HU 407 631; Voxter HU 371 698; Lunna HU 485 691; Sullom Voe (Houb of Scatsta) HU 397 728; Ronas Hill (North Collafirth) HU 352 835; Ronas Voe HU 291 809; Mousa (Leebitten) HU 435 249; Noss HU 525 408; Symbister Bay HU 536 623; Skaw HU 591 665; RSPB Fetlar/Loch of Funzie HU 655 900, Papil Water HU 655 900; Hermaness NNR (Burrafirth) HP 611 149; Skaw HP 658 163; Haroldswick HP 637 122; Baltasound HP 622 089; Uyeasound and Skuda Sound HP 596 012

RSPB Shetland: tel: 01950 460800

RSPB Fetlar: tel: 01957 733246; email: fetlar@rspb.org.uk

Scottish Natural Heritage, Shetland: tel: 01595 693345; web: snh.org.uk

Tourist Information (VisitShetland): tel: 01595 693434 email: info@visitshetland.com; web: www.visitshetland.com

Flybe/Loganair: web: www.flybe.com

Directflight Ltd: tel: 01595 840246 (Tingwall Airport); web: www.directflight.co.uk

Northlink Ferries, reservations and booking enquiries: tel: 0845 6000 449; web:www.northlinkferries.co.uk

Atlantic Ferries: web: www.atlanticferries.co.uk

Useful websites: www.nature-shetland.co.uk

454 FAIR ISLE (Fair Isle)

OS Landranger 4

Britain's most remote inhabited island, Fair Isle is justifiably famous for the number of rare and unusual birds that appear annually, and an additional attraction in spring and summer are the excellent seabird colonies. Following the pioneering studies of Eagle Clark in the early years of the 20th century, George Waterston bought the island and established a Bird Observatory in 1948; the island now belongs to the National Trust for Scotland.

Habitat

The island is 3.5 miles long by 1.5 miles wide and covers 765 ha, with the cliffs of the magnificent coastline reaching around 500 feet. The south of the island is crofted, while the north is mostly Heather moor, rising to 712 feet at Ward Hill (with nesting skuas). Migrants can occur anywhere but generally favour the crops, ditches and any sheltered area around the crofts, as well as the geos (inlets) on the coast, which can be very good.

Species

Spring migration begins in March, but despite the occurrence of several outstanding rarities in early spring, it is in late May–early June that the most exciting range of species has been recorded. Golden Oriole, Bluethroat, Icterine and Marsh Warblers, Red-backed Shrike, Common Rosefinch and Ortolan Bunting are annual, and sometimes occur in comparatively large numbers. Thrush Nightingale is something of a speciality in the last week of May, and Red-throated Pipit, Short-toed Lark, Subalpine Warbler and Rustic Bunting are almost annual at this time.

Breeding seabirds include a few pairs of Storm Petrels, and many more non-breeders come ashore at night. There are c. 28,000 pairs of Fulmars, 2,000 pairs of Gannets, 500 pairs of Shags, rather less than 5,000 pairs of Kittiwakes (a dramatic decline from the 20,000 in 1988), four other species of gull, 200–400 pairs of Arctic Terns (although the colonies have produced few young or been abandoned in some recent seasons); a few pairs of Common Terns and all four auks also breed. There are as many as 300 pairs of Great but far fewer Arctic Skuas (the former increasing in recent years, apparently at the expense of the latter), and in midsummer single Long-tailed Skuas are occasionally seen in the colonies. The commonest breeding landbirds are Starling, Skylark, Meadow and Rock Pipits and Wheatear, while the Fair Isle subspecies of Wren (*fridariensis*) is reasonably ubiquitous, with c. 35 pairs on the island.

Corncrake, Wryneck, Richard's Pipit, Bluethroat, Icterine, Barred and Yellow-browed Warblers, Red-breasted Flycatcher, Red-backed and Great Grey Shrikes, Common Rosefinch and Lapland Bunting are annual in autumn, and several rarities are nearly annual. Aquatic Warbler and Black-headed Bunting have appeared several times in August–early September, and there has also been several Booted Warblers during this period. Arctic and Greenish Warblers and Red-throated Pipit can be seen from early September, but it is in the second half of September that interest usually peaks. However, the quality of birds is very weather-dependent. Just a hint of a south-east wind can, and does, produce rarities, but west or south-west winds may spell disaster for the visiting rarity hunter, with very few birds at all. It is a risk you must take. Given favourable winds regulars include Great Snipe, Short-toed Lark, Olive-backed and Pechora Pipits, Citrine Wagtail, Siberian Stonechat, Pallas's Grasshopper and Lanceolated Warblers, and Yellow-breasted, Little and (more erratically) Rustic Buntings. If these are not sufficient, even more sought-after species, for example Siberian Rubythroat, Red-flanked Bluetail, and Pallas's Reed and Yellow-browed Buntings have also occurred, some more than once. There are common migrants too, and sometimes large numbers of Woodcocks, Redwings and Fieldfares. Others include Pink-footed and Greylag Geese, Whooper Swan, a trickle of Merlins, and some interesting subspecies, e.g. Greenland Redpoll and northern Bullfinch.

Access

By air Direct Flight operate several flights a week from the Shetland Mainland to Fair Isle, departing from Tingwall airstrip near Lerwick.

By sea The MV *Good Shepherd IV* mail boat, operated by Shetlands Island Council, sails between Fair Isle and Shetland (Grutness) on Tuesday, Thursday and Saturday, May–September, Tuesdays only during the rest of the year. The voyage takes 2.5 hours and there is a nominal charge. Alternate Thursdays in summer the Good Shepherd sails between Lerwick and Fair Isle, the trip taking 4.5 hours. Bookings are essential. Sailings are affected by bad weather and this crossing is definitely to be avoided if you are a bad sailor (although Storm Petrel is regular from the boat in summer). For details of transport to Shetland, see page 638.

Accommodation The newly re-built Fair Isle Lodge and Bird Observatory offers ensuite accommodation from mid-April to late October.

FURTHER INFORMATION

Fair Isle Lodge and Bird Observatory: tel: 01595 760258; email: fibo@btconnect.com; web; www.fairislebirdobs.co.uk
Direct Flight, bookings: tel: 01595 840246; web: www.directflight.co.uk
***Good Shepherd IV*, bookings:** tel: 01595 760222

TOP 100 SPECIES TO SEE IN BRITAIN

Our choice of the top 100 species includes many scarce and sought-after species, as well as some commoner birds that require a bit of help to find. Hopefully we have included most of the harder-to-find species in Britain, but there are inevitably a few omissions (such as Garganey, Goosander and Hobby which, although they can be elusive, actually occur at a great many sites). We have generally excluded species for which there are no regular sites, such as Grey Phalarope (which is a scarce non-breeding visitor at the best of times) or Brambling; although the latter occurs widely in Britain in winter, sometimes in large flocks, its presence is erratic and there are no guaranteed sites. Likewise, former or irregular breeders such as Wryneck and Red-backed Shrike are also excluded; even though migrants of these species are recorded frequently in spring or autumn, there are no reliable sites for them. A number of other scarce migrant species such as Aquatic, Barred, Icterine and Melodious Warblers, Bluethroat and Ortolan Bunting all turn up with some regularity each year, but again there are no reliable sites for them. However, one increasingly familiar scarce migrant (Yellow-browed Warbler) is included in the list, and the advice presented for this species could equally apply to other less common migrants. Woodcock is another species that is not included but for a different reason. It is true that this bird can be hard to find and may be keenly sought-after, but it is widespread throughout Britain; a diligent search of many woodlands is likely turn up a sighting, especially in winter when migrants from the continent augment the resident population or in the spring and early summer when breeding birds perform their 'roding' display flight.

Bewick's Swan
Regular winter visitor to southern Britain, with notable concentrations at a handful of sites such as the Ouse Washes (182) and Slimbridge (202). Regular sites include: 23, 30, 32, 39, 50, 67, 69, 70, 77, 90–92, 105, 108, 109, 122–124, 128–132, 154, 159, 182, 183, 202, 244, 250, 269, 336, 340, 358, 387.

Whooper Swan
A few pairs breed annually in Scotland, but mainly a widespread winter visitor, with a more northerly distribution than Bewick's. Regular sites include: 23, 32, 105, 108, 109, 122-124, 159, 182, 244, 250, 258, 259, 264, 268, 269, 285, 296, 298, 308, 322, 328, 339, 340, 355, 357, 358, 361, 363, 366, 367, 370, 376, 385, 387, 393, 394, 401, 404, 406, 418, 419, 422–425, 439, 453.

Bean Goose
Scarce winter visitor with only a few regular sites in Britain. Two forms occur, sometimes regarded as separate species. Taiga Bean Goose (nominate *fabalis*) occurs at two principal sites, the Slamannan Plateau near Falkirk, Stirling (369) and the Yare Valley, Norfolk (156). Small numbers of Tundra Bean Goose (*rossicus*) are most likely to be seen in SE England, e.g. at sites 100, 108, 109, 149, 152.

Pink-footed Goose
Increasing winter visitor, principally to eastern and southern Scotland, Lancashire and East Anglia, often in large concentrations. Key sites: 159, 166, 169, 170, 223, 230, 244, 250, 251, 256, 264, 273, 357, 358, 363, 370, 376, 385, 387.

White-fronted Goose
The nominate race winters mainly in southern England, from East Anglia to Hampshire and Gloucestershire, but numbers are declining. Regular sites include: 23, 30, 32, 39, 66, 67, 69, 77, 78, 90–92, 100, 106, 108, 109, 122–124, 128–132, 151, 152, 154, 156, 159, 166, 202. The race *flavirostris* (Greenland White-fronted Goose) winters mainly in Ireland and western Scotland. Key sites: 328, 352, 355, 406, 416–419, 425.

Barnacle Goose
Small flocks of naturalised birds may be seen throughout the year, especially in southern and eastern England (e.g. 108). Genuinely wild birds are winter visitors, mostly to northern Britain. Greenland birds winter in

Ireland and western Scotland. Sites: 328, 387, 417–419. The Svalbard population winters almost exclusively on the Solway Firth. Sites: 264, 357, 358.

Brent Goose
Dark-bellied Brent Geese are common on the coasts of eastern and southern England (but very rare inland). The much less common pale-bellied race *hrota* from Svalbard and Greenland winters mainly around Lindisfarne (298) and Denmark, with small flocks also now at Hilbre (245) and Loch Ryan (350); passage birds occur at Loch of Strathbeg (387), but it is worth checking dark-bellied flocks for stragglers of the pale-bellied form. The Canadian population of the latter winters almost exclusively in Ireland.

Egyptian Goose
Well-established introduction from Africa, mainly to East Anglia, but also spreading into the Midlands. Sites: 133, 166, 189, 234.

Mandarin Duck
Well-established
introduction from east Asia; its stronghold is in SE England, but it also occurs in other parts of the country. Sites: 38, 95, 134, 203, 207, 234.

Scaup
Rather localised winter visitor to the coasts of Scotland, with concentrations in SW Scotland, Islay and the Moray Firth. Only small numbers elsewhere and uncommon and irregular inland and in southern England. Sites include: 14, 18, 26, 33, 39, 42, 43, 49, 53, 54, 65, 70, 86, 107, 108, 110, 125–128, 132, 171, 222, 230, 251, 256, 258, 259, 273, 288, 302, 350, 357, 358, 364, 371, 375, 383, 384, 392, 401, 403, 419.

Long-tailed Duck
Winter visitor to coasts of Scotland and northern England; much less frequent elsewhere and scarce inland. Larger concentrations occur in the Moray Firth, Firth of Forth and on Orkney and Shetland. Regular sites include: 33, 36, 39–43, 51, 107–109, 168, 169, 230, 259, 281, 287, 296, 298, 342, 361, 363, 364, 371, 375, 383–385, 387, 391–393, 401, 404, 406, 417, 422, 425, 439, 453.

Velvet Scoter
Winter visitor to coasts of Britain but more thinly scattered than Common Scoter; rare inland. Larger concentrations occur in the Moray Firth, St Andrews Bay and the Firth of Forth; elsewhere small numbers are often found in flocks of Common Scoter. Regular sites include: 19, 42, 43, 51, 68, 87, 104, 106, 124, 166, 168, 169, 230, 258, 281, 287, 296, 298, 342, 361, 363, 364, 371, 375, 384, 385, 391–393, 403, 404.

Smew
Rather localised and uncommon winter visitor to lakes and gravel pits, especially in SE England. Usually only occurs in small numbers, but often more during severe weather. Very reliable sites include: 97, 107, 109; also occurs regularly at 26, 27, 33, 36, 49, 50, 70, 77, 86, 88, 93, 96, 98, 102, 106, 108, 110, 128-132, 139, 143, 152, 157, 201, 215, 221, 222, 235, 242, 296, 302, 356, 367, 387.

Red Grouse
Widespread but localised gamebird of heather moorland, mainly in northern Britain. Sites include: 237, 252, 253, 263, 284, 285, 292, 324, 326, 327, 346, 378, 385, 395, 396, 398, 405, 407, 409, 412, 413, 415, 416, 419, 422, 425.

Ptarmigan
Uncommon resident of the Scottish highlands, mainly in the Grampians and western Highlands. Sites include: 380, 395, 407, 409, 412–414.

Black Grouse
Declining resident of woodland edges, rushy pastures and moorland, mainly in Scotland, northern England and Wales. Sites include: 263, 284, 292, 327, 346, 368, 378, 380, 395, 398, 399, 419.

Capercaillie
Localised resident of Scottish pine forests and one of Scotland's most threatened birds. Vulnerable to disturbance and best looked for at approved localities in Speyside (395).

Quail
Secretive summer visitor, varying greatly in abundance from year to year. Best located by listening for calling males but very hard to

see. Few sites can be regarded as regular or reliable, but try: 32, 45, 196, 197, 237, 239, 269.

Golden Pheasant
Spectacular introduction from SW China, now only persisting in Breckland (Norfolk/ Suffolk) with no recent records from SW Scotland. Sites: 167, 172, 335.

Black-throated Diver
Scarce breeding bird in Scotland (perhaps only c. 200 pairs). Thinly distributed around coasts in winter with concentrations off western Scotland and Cornwall; very scarce inland. Regular sites include: (summer) 368, 405, 407, 410–412; (winter) 8, 9, 11–13, 16, 19–23, 34, 36, 44, 51, 59, 60, 104, 168, 169, 170, 228, 279, 281, 335, 337, 350, 363, 371, 387, 388, 390, 392, 393, 395, 404, 406, 413, 419, 420, 424, 425.

Great Northern Diver
Regular winter visitor to the Northern Isles and west coasts of Britain. Less common in southern areas and off the east coast but frequent in SW England; very scarce inland. Regular sites include: 1-5, 8, 9, 11–13, 16, 19–23, 34–36, 44, 51, 59, 60, 125-127, 138, 161, 168–170, 213, 222, 228, 279, 281, 337, 338, 340, 342, 345, 350, 387, 388, 390, 392, 393, 404, 406, 409, 411, 417–420, 422–424, 425, 439, 442, 443, 451, 453.

Cory's Shearwater
Scarce visitor off headlands in the south and west, mainly Jul-Oct, occasionally in large numbers. Regular sites include: 1, 2, 7, 8, 11, 67.

Great Shearwater
tatus similar to Cory's, but numbers vary from year to year, and perhaps less likely to be seen from headlands. Regular sites include: 1, 2, 7, 8, 11.

Sooty Shearwater
Regular visitor to coastal areas Apr–Nov, but mainly Aug-Sep. Largest numbers occur off SW England, northern Britain and in the North Sea from Norfolk northwards. Regular sites include: 1, 2, 7, 8, 11, 12, 17, 22, 28, 34–36, 44, 52, 63, 67, 87, 161, 163, 170, 258, 275, 279, 281, 289, 290, 294, 295, 298, 306, 317, 328, 333, 334, 338, 341, 345, 359, 360, 374, 382, 387, 388, 391, 392, 402, 408, 418, 419, 421, 422–424, 436, 437, 439.

Balearic Shearwater
Regular visitor to southern and western coasts in summer and autumn, but also recorded off the east coast. Critically endangered, the world population is fewer than 10,000 individuals. Regular sites include: 8, 9, 11, 12, 17, 34, 35, 44, 52, 68, 109, 317; also possible off northern North Sea coasts: 279, 281, 290, 294, 295, 421.

Storm Petrel
Breeds on small offshore islands off the west coast and in the Northern Isles, but rarely seen from land. The best chances to encounter this species are on pelagics or on boat trips to offshore islands. Sites: (breeding) 6, 312, 313, 444, 450, 452, 454; (passage) 7, 11, 12, 22, 31, 44, 63, 67, 306, 328, 334, 338, 416, 418, 419, 421, 422-424.

Leach's Petrel
Scarcer as a breeding bird than Storm Petrel, with the bulk of the population on St Kilda. Most frequently encountered off headlands in western and SW Britain after NW gales and storms. Regular sites include: 1, 2, 11, 12, 22, 64, 161, 163, 170, 223, 245, 246, 249, 255, 258, 281, 294, 317, 328, 333, 334, 338, 340, 341, 345, 347, 416, 418, 419, 421.

Bittern
Rare breeding bird of extensive reedbeds at just a handful of sites, but showing signs of a slow increase in numbers. Somewhat easier to see in winter when the population is augmented by visitors from the continent, but numbers are still small. Sites: (resident/ breeding) 46, 116-119, 149–152, 155, 157, 163, 168, 181, 257, 272; (winter) 21, 36, 46, 61-63, 70, 78, 79, 93, 97, 99, 107, 109, 110, 121, 133, 187, 192–194, 197, 208, 211, 212, 217, 230, 266–268, 270, 271, 296, 299, 301, 309, 316, 320, 387.

Spoonbill
Colonisation as a breeding bird probably imminent. Non-breeding visitors occur widely, especially in eastern and southern Britain. Small numbers winter in SW England. Regular sites include: 15, 19, 32,

39-43, 44, 46, 56, 61-63, 76, 100, 116-119, 148, 150, 154, 157, 163, 166, 168, 257.

Red-necked Grebe
Although recorded occasionally in summer, mainly an uncommon winter visitor to coastal areas, especially the east and south coasts of England and southern Scotland (e.g. Firth of Forth); very scarce inland. Numbers increase in severe weather. Regular sites include: 21, 88, 96, 102, 109, 125-127, 161, 166, 169, 170, 222, 228, 235, 281, 282, 298, 335, 337, 363, 371.

Slavonian Grebe
Rare breeding bird at just a handful of sites in Scotland. Regular winter visitor to coastal waters in small numbers, especially in Scotland. Sites: (breeding) 395, 396; (winter) 1-5, 12, 13, 16, 19, 20–23, 28, 32, 36, 42, 43, 51, 55, 59, 77, 80–81, 87, 88, 96, 107–109, 125-127, 161, 166, 168–170, 200, 222, 228, 235, 282, 296, 298, 307, 335, 342, 350, 363, 364, 367, 370, 371, 401, 404, 419, 420, 422–425, 442, 443.

Black-necked Grebe
Breeds at a few scattered sites across Britain, but is easier to find in winter. More southerly in its distribution than Slavonian Grebe, but generally scarcer. Only a few sites hold regular wintering birds. Sites: (passage/winter) 19, 51, 55, 59, 77, 80–81, 96, 107–109, 125-127, 200, 214, 218, 220–222, 233, 248, 266, 269, 271, 285, 342, 350, 370.

Honey-buzzard
Rare breeding summer visitor with fragmentary distribution, mainly in southern England but occurring north to the Scottish highlands. Generally secretive on its breeding grounds and most easily seen when displaying, or (if lucky) on migration. Regular sites include: 1-5, 35, 71-75, 100, 113, 115, 160, 234, 283.

Red Kite
Formerly a rare breeder confined to central Wales, the population there has made a spectacular recovery. Simultaneously, a number of reintroduction schemes in England and Scotland have been very successful, especially in the Chilterns. Key sites include: 205, 206, 217, 222, 237, 268, 303, 319, 321–325, 327, 328, 332, 355, 394, 395.

White-tailed Eagle
Successfully reintroduced to western Scotland, mainly the Inner and Outer Hebrides. Young birds roam widely. Very rare winter visitor to England and Wales; these are presumably birds from the Continent. Key sites include: 414–416, 419, 424.

Hen Harrier
Scarce breeding bird of uplands of northern Britain with pitifully low breeding success in many areas due largely to illegal persecution. Rather more widespread in winter, occurring in the lowlands and coastal areas of southern Britain. Sites: (summer) 253, 263, 327, 346, 380, 397, 398, 405, 417, 419, 422–425, 427, 429; (winter) 8-10, 17, 24, 26, 27, 30, 32, 33, 45–47, 57, 58, 66, 69, 71–77, 90-92, 94, 100, 101, 105–109, 113–119, 121–128, 132, 138, 141, 148–152, 154, 157, 164–166, 168, 169, 181–183, 187, 196, 223, 224, 226, 227, 237, 244, 250, 251, 253, 264, 269, 272, 288, 299, 307, 308, 316, 322, 328, 335, 336, 338–340, 347, 352, 355, 357, 358, 367, 370, 379, 387, 394.

Montagu's Harrier
Rare summer visitor breeding very sparsely but widely, mainly in southern Britain. There are currently no regular breeding sites, nor reliable sites for passage birds, but the species has been recorded recently at sites such as 100, 196, 223, 224 and 272.

Goshawk
Scarce but overlooked breeding resident; most if not all of the population is derived from falconers' birds. Now well established in widely scattered areas including Wales, parts of Scotland, the Midlands and East Anglia. Always hard to see, no sites can be regarded as guaranteed, but the following are worth a try: (summer) 71-75, 176, 178, 205–207, 237, 283, 324, 327, 332, 346, 370, 377, 380, 395; (winter) 24, 100, 151, 152, 269.

Rough-legged Buzzard
An uncommon to scarce winter visitor whose numbers fluctuate from year to year (in recent years the species has been very scarce in Britain). Most frequent on the east

coast. Possible sites include: 113, 115, 122-124, 148–152, 237.

Golden Eagle
Widespread resident breeder in the Scottish highlands with a population of around 450 pairs. Occasionally breeds in northern England (Cumbria). Suggested sites: 262, 368, 380, 396, 398, 407, 412–416, 419, 422-424; also possible at 405.

Osprey
Breeding summer visitor, mainly to Scotland, with a population of c. 150 pairs and increasing. Successfully introduced to Rutland Water (222). Widespread on passage in southern Britain. Sites: (breeding/summer) 221, 261, 332, 370, 377, 379, 380, 393–398, 401, 403. More or less regular sites for passage birds include: 1-5, 23, 29, 39–44, 46, 49, 61-63, 65, 66, 70, 76, 77, 100, 102, 104, 113, 115–119, 133, 134, 155, 157, 184, 222, 234, 237, 239, 240, 242, 254, 257, 328, 355, 385, 387, 404.

Merlin
Scarce breeding species of upland areas, mainly in northern Britain. More widespread in winter when birds move to lowland areas, especially coasts and open farmland. Sites: (breeding) 327, 346, 395, 396, 398, 405, 412, 413, 415–417, 419, 425, 427, 429; (winter) 1-5, 8-10, 17, 24, 26, 27, 30–34, 45–47, 66, 71–77, 90–92, 105–109, 115, 121–132, 138, 141, 151, 152, 154, 157, 164–166, 168, 169, 181–183, 223, 224, 226, 227, 233, 237, 244, 245, 247, 250, 251, 253, 256, 263, 264, 267, 269, 272–274, 285–289, 292, 296, 298, 299, 301, 307, 308, 316, 328, 335, 336, 338, 339, 347, 352, 355, 357, 358, 370, 385, 387.

Spotted Crake
Rare breeding species, occurring widely and sporadically, and probably under-recorded. Sites: (summer, calls at night but impossible to see) 155, 182, 183, 187, 257, 269, 355, 394. Most likely to be found on passage, though elusive and with few reliable sites; the following perhaps offer the best chances: 1, 2, 3, 15, 21, 26, 53, 54, 61-63, 116–119, 163, 210, 244, 247, 268, 277, 286.

Corncrake
Declining summer visitor now largely restricted to the Hebrides and Orkney, with only small numbers on mainland Scotland. Formerly bred in every county in Britain and there is a current reintroduction programme in the Nene Washes (183). Sites: (breeding) 269, 417–419, 422, 424, 430.

Crane
Formerly only a scarce visitor to Britain with occasional influxes. Since 1981, a small population has become established in the Norfolk Broads but numbers remain very small, with some dispersal noted. Sites: 157, 159, 181.

Stone-curlew
Scarce breeding summer visitor of dry grassland and open areas whose numbers are slowly increasing (currently around 350 pairs). The key breeding site, which can be visited without fear of disturbance to the birds, is Weeting Heath in Norfolk (174). Very rarely recorded on passage.

Dotterel
Breeds on the tops of the highest hills, mainly in Scotland, but probably easier to find on passage at one of their regular stop-over sites, often in arable fields. Sites: (breeding) 380, 395; (passage) 1, 10, 24, 31, 104, 113, 227, 250, 252, 338, 345, 387.

Purple Sandpiper
Not uncommon winter visitor to rocky shores, mainly in the north and west (but with a contraction of the winter range northwards in recent years). Very scarce on the coasts of East Anglia and southern England. Regular sites include: 1-5, 12, 16, 21, 34, 42, 43, 52, 63, 67, 89, 105, 112, 115, 153, 170, 245, 278, 281, 282, 287, 289, 290, 294–296, 298, 301, 305–307, 328, 335, 345, 358–361, 363, 374, 375, 382, 388, 390–392, 404, 406, 419.

Jack Snipe
Easily overlooked winter visitor to wet marshes and waterlogged grassland. Regular sites include: 15, 27, 30, 53, 54, 61, 62, 66, 70, 89, 90, 96, 98, 99, 106, 182, 184, 192, 194, 198, 208–210, 212, 222, 226, 227, 229, 230, 233, 235, 244, 249, 251, 257, 264, 266, 271, 272, 276, 277, 286, 288, 301, 308, 367, 385.

Red-necked Phalarope
Very rare breeding species largely confined to Shetland (and occasionally still the Outer Hebrides, Sutherland and Orkney); also a scarce passage migrant but no regular sites. Sites: (summer) 424, 452.

Pomarine Skua
Scarce passage migrant with regular pattern of occurrence. In spring occurs off the Hebrides and through the English Channel. In autumn, most birds are juveniles with the majority off the east coast. Regular sites include: 7, 8, 11, 12, 22, 44, 52, 67, 68, 87, 103, 104, 109, 115, 161, 163, 170, 264, 275, 279, 281, 289, 290, 294, 295, 306, 317, 334, 338, 341, 345, 347, 360, 365, 374, 382, 421–424, 439, 442.

Long-tailed Skua
Scarce but regular passage migrant, notably off the Hebrides in spring. In autumn more regular off the east coast. Sites: 7, 11, 12, 151, 161, 163, 281, 287, 289, 290, 294, 338, 341, 347, 365, 422-424, 436, 439, 442.

Great Skua
In Britain, confined mainly to Orkney and Shetland as a breeding species. On passage, found around all coasts in small numbers. Sites: (summer) 385, 406, 409, 411, 425, 427, 429, 436, 444, 446, 450, 452–454; (passage) 1, 2, 7–9, 11, 12, 17, 34, 35, 42–44, 52, 63, 68, 87, 109, 115, 124, 161, 163, 168–170, 223, 255, 258, 264, 275, 279, 281, 289, 290, 294, 295, 298, 317, 334, 338, 341, 345, 347, 359, 360, 361, 364, 374, 382, 384, 391, 392, 418, 419, 421.

Sabine's Gull
Annual visitor to Britain with about 100 records per year, mainly Aug–Nov. Regular sites include: 7, 11, 12, 22, 279, 294, 334, 347, 421.

Little Gull
Although occasional breeding has been noted, mainly an uncommon passage migrant, especially on the east coast. Can turn up anywhere, but regular sites include: 12, 13, 15, 16, 18, 21, 22, 27, 39-43, 51–54, 61-63, 77, 86, 88, 96, 104, 107, 109, 110,163, 168, 170, 184, 225, 235, 242, 245, 246, 249, 254, 255, 268, 275, 276, 279, 281, 282, 288, 290, 294, 304, 317, 359, 364, 365, 372.

Mediterranean Gull
Increasing breeding species with around 500 pairs in Britain, mainly on the south coast. More widespread at other times of year. The best sites are: 67, 107, 112, 125, 150, 164, 171, 254; other regular sites (passage/winter) include: 12–14, 16, 21, 22, 36, 39–44, 48, 51, 53, 54, 61-63, 65, 78, 80-81, 87, 89, 104, 106, 109, 110, 145, 146, 154, 163, 184, 191, 194, 195, 208, 214, 218, 220–222, 224, 232, 233, 235, 236, 239, 240, 242, 245, 246, 248, 249, 255, 259, 265, 268, 271, 282, 285, 287, 290, 294, 300, 302, 304, 311, 328, 364.

Yellow-legged Gull
Increasing visitor to Britain, mainly non-breeding birds in southern Britain (Jun–Dec, peaking Jul–Oct), but a few breeding records since 1995. Regular sites include: 44, 48, 55–63, 99, 100, 145, 150, 157, 163, 191, 199, 200, 202, 213, 214, 217, 218, 220, 221, 222, 224, 232, 233, 235, 236, 239, 240, 242, 249, 271, 320.

Caspian Gull
First recorded in 1995, this species has been found in increasing numbers, mainly in the south and east in autumn and winter. Regular sites include: 100, 109, 150, 200, 218, 232.

Iceland Gull
Regular winter visitor in variable but usually small numbers, especially in the north and west. Regular sites include: 12–14, 16, 21, 22, 32, 44, 53, 89, 100, 191, 218, 222, 232, 236, 239, 240, 242, 248, 249, 259, 265, 267, 268, 282, 290, 294, 367, 388, 390, 406, 419, 422–425, 440, 453.

Glaucous Gull
Regular winter visitor to Britain, with largest numbers in the north, and relatively scarce in the south. Usually more frequent than Iceland Gull. Regular sites include: 12–14, 16, 21, 22, 32, 36, 44, 53, 87, 89, 100, 191, 213, 218, 222, 232, 236, 239, 240, 242, 248, 249, 259, 265, 267, 268, 281, 282, 285, 290, 294, 367, 382, 383, 388, 390–392, 404, 406, 418–420, 422-424, 425, 440, 453.

Little Tern
Uncommon breeding tern of coastal shingle with colonies scattered around much of Britain. Sites may move from year to year. Regular sites include: 12, 13, 32, 67, 76, 80, 81, 88, 109, 114, 124–127, 137, 146–148, 150–152, 154, 157, 158, 162–164, 166, 169, 226, 227, 245, 258, 259, 275, 289, 347, 385, 418, 419.

Roseate Tern
Rarest breeding tern whose numbers have declined sharply in recent years. Now only breeds at a handful of sites in Britain, notably at 296 and 297. Passage birds regularly seen at: 12, 13, 23, 42, 61-63, 104, 107, 109, 154, 163, 164, 289.

Black Guillemot
Resident on rocky coasts. Widespread in small numbers around Scotland and occurs at a very few sites in northern England and Wales; absent from the south. Sites: (breeding) 260, 350, 351, 388, 393, 406, 408, 409, 411, 412, 418, 419, 422–426, 436, 440, 444, 453, 454; (winter) 19, 337.

Little Auk
Regular visitor to coasts of northern Scotland, mainly Sep–Mar, with smaller numbers further south. Occasional large movements are noted off eastern coasts, usually after strong northerly winds. Regular sites include: 11, 12, 22, 64, 77, 115, 161, 163, 170, 224, 275, 279, 281, 282, 289, 294, 295, 334, 360, 364, 365, 390, 439.

Puffin
Patchily distributed around coasts of Britain, with the majority in Scotland. The largest numbers are found in the Northern Isles and Hebrides. Outside Scotland, the main colonies are on the Farne Islands (297), Coquet Island (296) and Skomer (313). Very localised in the south. Key sites include: 6, 31, 52, 60, 68, 104, 280, 296, 297, 306, 309, 312, 313, 338, 351, 362, 373, 381, 386, 389, 406, 408, 409, 411, 425, 426, 435, 436, 439, 453, 454.

Ring-necked Parakeet
Well-established introduction and a familiar bird in some parts of SE England, with a few populations further afield. Regular sites include: 95, 97, 99, 115, 133.

Long-eared Owl
Widespread breeding species of mature woodland, especially in conifers. Usually occurs at low density and very hard to locate on the breeding grounds (listen for 'squeaky-gate' juveniles). Easier to find in winter when immigrant birds visit coastal sites, sometimes on a regular basis. Key sites include: (breeding) 130, 346, 393, 395, 422; (winter) 3, 107, 223, 229, 231, 235, 248, 253, 266, 269, 275, 288, 293, 363, 376, 441.

Nightjar
Summer visitor to heathland and open woodland, best located by distinctive churring call. Regular sites include: 38, 57, 58, 71-75, 90-92, 94, 101, 111, 120, 150, 151, 158, 162, 176, 205, 207, 234, 283.

Lesser Spotted Woodpecker
Scarce and declining species of woodland, foraging on smaller branches than Great Spotted Woodpecker, often higher in the canopy. Best located in late winter or early spring, before the leaves are out and when it is most vocal. Good sites include: 38, 71-75, 90–95, 101, 118, 120, 130, 133, 134, 144, 155, 167, 175, 177, 185, 188, 190, 194, 204, 205, 207, 220, 222, 233–235, 240, 242, 266, 271, 321, 323, 324, 328, 331.

Golden Oriole
Rare summer visitor to poplar plantations in East Anglia, mainly in the Suffolk fens at Lakenheath (181). Breeds only sporadically elsewhere. Also occurs as a scarce passage migrant, such as at 1, 31, 65, 113, 114, 226, 454.

Great Grey Shrike
Regular winter visitor to Britain in very small numbers, favouring open habitats with scattered trees or bushes. Often very mobile and can be hard to locate, even at regular sites. More or less regular sites for passage birds or wintering individuals include: 71-75, 93, 94, 101, 116-119, 150, 151, 172, 224–226, 234, 254, 287, 289, 290, 298, 355, 374, 386, 395, 454.

Chough
Scarce and localised breeding bird of cliffs on the west coast of Britain. Occurs primarily in west Wales and the southern Inner

Hebrides. Recently returned to Cornwall as a breeding species. Sites: 17, 306, 309, 310, 312, 313, 315, 317, 319, 330, 333, 334, 338, 345, 419.

Firecrest
First colonised Britain as a breeding species in 1962, and now at least 350 pairs in southern England. Also an uncommon passage migrant to the east and south coasts, especially in autumn, with some birds wintering. Regular sites include: (breeding) 71-75, 111, 173, 178, 205, 207, 234; (passage/winter) 1-5, 8-10, 22, 31, 34–36, 42, 44, 51, 52, 55-60, 63, 64, 67, 87–89, 104, 107, 109, 112–114, 133, 164, 166, 169, 224, 226, 279, 286, 301, 305, 342.

Crested Tit
Restricted to mature pine forest in northern Scotland, where resident and not uncommon. Reliable sites include: 392, 393, 395, 398, 399.

Willow Tit
Declining species of woodland and damp scrub, and now much less common than Marsh Tit (from which it needs careful separation). Regular sites include: 30, 33, 93, 151, 177, 187, 205, 207, 209–212, 216, 222, 231, 233, 235, 238, 242, 243, 248, 266, 268, 271, 285, 291, 293, 305, 321–323, 328, 343, 353, 355 (but the decline is so marked that it may soon vanish from some of these).

Bearded Tit
Localised reedbed specialist occurring at about 50 sites in Britain, mainly in England. Some birds disperse in winter when they may occur in reedbeds far from the breeding areas. Regular sites include: 36, 46, 50, 51, 53, 54, 57, 58, 67, 77, 78, 88, 100, 106, 116-119, 121, 149–152, 155, 157, 163, 168, 181, 187, 198, 230, 257, 272, 299.

Woodlark
Following serious declines, the British population has made a spectacular recovery as a result of favourable forestry practices and sympathetic habitat management. The population in central and southern England is now in excess of 3,000 pairs. Regular sites include: 71-75, 93, 94, 100, 149–152, 174, 176, 178, 234.

Shore Lark
Breeds very rarely on mountain tops in Scotland, but primarily a winter visitor in small numbers to the east coasts of Britain; very rare elsewhere. Reliable sites include: 146, 162, 164, 166. Less regular sites include: 32, 107, 124, 137, 149–152, 168, 169, 224, 226, 294, 298, 347, 361, 363.

Cetti's Warbler
Breeding first confirmed in 1972, and now well established as a resident species is southern England, with over 1,400 singing males, but susceptible to cold weather (thus numbers may have fallen following hard winter of 2009–10); now most frequent in SW England where climate is generally less severe. Regular sites include: 15, 21, 36, 39, 44, 46, 53, 54, 61-63, 65, 67, 69, 70, 77, 78, 100, 105, 116-119, 133, 150, 155–157, 163, 168, 181, 187, 192, 194, 208, 211, 212, 216, 299, 301, 305, 318, 320, 336.

Yellow-browed Warbler
Scarce but regular autumn visitor with over 300 records per year; a few birds overwinter. Most are found at coastal migration hotspots. Regular sites include: 1-5, 8-10, 34, 60, 64, 100, 104, 113, 115, 166, 169, 255, 275, 279, 281, 286, 294, 359, 386, 454.

Wood Warbler
Breeding summer visitor to deciduous woodland in western Britain, but has declined recently. Uncommon migrant at coastal migration sites. Regular sites include: (breeding) 38, 71-75, 205–207, 237, 262, 263, 292, 303, 321–328, 331, 343, 353–355, 368, 378, 380, 394, 395, 412, 413, 416, 419; (passage) 63, 104.

Dartford Warbler
A resident heathland specialist, favouring areas of gorse and heather, mainly in southern England. Very susceptible to prolonged cold weather and the population after the 1962–63 winter was down to around 12 pairs. Recent mild winters have resulted in a population of over 3,000 pairs, but cold spells in the winter of 2008–09 and severe weather in winter 2009–10 have reduced numbers significantly. Regular sites include: 35, 38, 57, 58, 61-64, 67, 71-75, 93, 94, 100, 101, 104, 106, 110, 149, 150, 306.

Grasshopper Warbler
Uncommon and declining summer visitor, mainly to southern England. Regular sites (breeding/passage) include: 8-10, 22, 33, 35, 38, 53, 54, 100, 104, 113, 116-119, 149–152, 155–158, 166, 167, 169, 184–187, 192, 197, 198, 205, 211, 212, 217–219, 222, 225, 226, 229, 232, 233, 235, 236, 238, 240, 244, 245, 247, 248, 254, 259, 264, 266, 271, 272, 277, 280, 296, 301, 305, 308, 310, 316–318, 322, 327, 328, 331, 335, 336, 339, 353, 355, 357, 358, 363, 366, 367, 387, 394, 395.

Marsh Warbler
Formerly an uncommon breeding summer visitor to a number of counties in southern England, with its stronghold in Worcestershire until the late 1980s. A recent, very limited colonisation in SE England offers some hope, but as it is only a scarce migrant at best, this is a tough bird to find. There are no regular breeding sites but recently migrants have been recorded at 65, 99, 100, 113, 275.

Dipper
A distinctive resident of fast-flowing streams and rivers, absent from lowlands of England. Although widespread, knowledge of key sites is invaluable in finding this species: 26, 206, 207, 237, 241, 252, 253, 262, 263, 282, 284, 292, 303, 318, 321–328, 331, 343, 346, 353–355, 361, 367, 368, 378, 380, 394–396, 399, 400, 405, 413, 416.

Ring Ouzel
Declining summer visitor of upland areas, mainly in the north and west. Migrants at coastal sites in Oct–Nov on east coast are probably from Scandinavia. Regular sites include: (breeding) 237, 253, 262, 263, 284, 326, 327, 343, 346, 380, 395, 396, 412, 413, 416; (passage) 1-5, 8-10, 17, 31, 34, 44, 52, 63, 64, 76, 87, 89, 100, 104, 109, 112, 113, 162, 164, 166, 169, 170, 224, 244, 246, 251, 279, 286, 290, 292, 294, 296.

Nightingale
Uncommon, declining summer visitor, largely restricted to southern and eastern England where patchily distributed. Regular sites include: 45, 47, 66, 90–94, 104, 111, 120, 130, 134, 142–145, 149, 150, 151, 152, 162, 184, 185, 190, 200, 201, 203, 204, 217, 222, 232.

Black Redstart
Scarce breeding visitor to southern England, with a population of around 100 pairs. Also an uncommon passage migrant at coastal sites and a few birds overwinter (mainly in the south and west). Regular sites include: (breeding) 109, 146, 149, 153, 154; (passage/winter) 1-5, 8-10, 12, 17, 28, 31, 34–36, 42, 44, 52, 55-60, 63, 64, 67, 76, 87–89, 104, 107, 110, 112, 114, 115, 164, 169, 170, 226, 275, 279, 286, 290, 294, 296, 345, 359, 360.

Pied Flycatcher
Summer visitor to deciduous woodland in the north and west; largely absent as a breeding bird from northern Scotland and SE England. Passage birds may be seen at coastal migration sites. Regular sites include: (breeding) 38, 205, 207, 237, 240, 254, 262, 263, 292, 303, 321–328, 331, 343, 353, 354, 368, 394, (passage) 1-5, 8-10, 17, 28, 31, 35, 44, 52, 63, 64, 87, 88, 89, 100, 104, 107, 109, 164, 166, 275.

Tree Sparrow
Formerly a common resident, this species has undergone a population crash in the last three decades. It is now decidedly uncommon (or absent) throughout most of its former range. Current sites include: 107, 108, 182, 211, 215, 218, 221–223, 239, 240, 243, 244, 250, 268, 270, 272, 280, 291, 296, 357, 358, 363, 387.

Water Pipit
Uncommon winter visitor Oct–Apr, mainly to wetlands in southern and eastern England. Regular sites include: 15, 44, 50, 54, 63, 69, 100, 105, 116-119, 150, 163, 168, 181, 182, 211, 216, 227, 244, 267, 347.

Twite
Breeds in uplands of northern England and Scotland, but has declined recently. Some populations (e.g. from the Pennines) winter on coastal saltmarshes of eastern England, but has declined there and now almost absent from the south coast in winter. Regular sites include: (breeding) 351, 397, 406, 407, 416–419, 422–425; (winter) 32, 100, 103, 107, 114, 121, 124, 151, 154, 166, 169, 223, 225–227, 244, 251, 264, 288, 289, 296, 347, 348, 350, 357, 361, 363, 364, 368, 412, 413.

Lesser Redpoll
Declining resident, partial to birches and conifers in the breeding season; population is now but a fraction of the peak in the 1970s. 'Redpolls' are winter visitors to many sites, especially where waterside alders are present; most or all are likely to be Lesser Redpolls. Specific mention of Lesser Redpoll as follows: 33, 49, 79, 95, 111, 157, 167, 178, 187, 188, 205, 207, 222, 224, 234, 237, 240, 248, 253, 254, 257, 268, 271, 283, 305, 318, 321–323, 326, 327, 331, 335, 367, 370, 377, 395, 397, 399, 413, 416, 419.

Common Crossbill
Breeding resident of conifer woodlands, subject to occasional irruptions. Most frequent in northern England and Scotland, with good numbers in East Anglia and Wales. In the Scottish highlands tends to prefer larch and spruce. Regular sites (some only in invasion years) include: 71–75, 111, 151, 173, 176, 205, 207, 237, 253, 254, 283, 324, 326, 327, 335, 343, 346, 354, 355, 370, 378, 380, 393, 395, 398, 413.

Scottish Crossbill
Endemic to older coniferous forest in the highlands of Scotland, but other crossbill species may breed in the same areas (care is needed to separate them). No particular dietary preferences. Sites include: 380, 392, 393, 395, 398, 399, 404, 413.

Parrot Crossbill
Recently discovered to breed sympatrically with Common and Scottish Crossbills in forests of Strathspey and Deeside. The population is resident and estimated to be around 100 pairs. Generally prefers to feed in Scots Pines. Sites: 395, 398.

Hawfinch
Secretive and declining resident of deciduous woodland; partial to hornbeams. Regular sites include: 71–75, 95, 101, 111, 144, 173, 180, 205, 234, 257, 346.

Snow Bunting
Very small numbers breed on the highest Scottish mountains (50-100 pairs at most), but mainly a winter visitor to saltmarshes and sand dunes, especially on North Sea coasts. Regular sites include: (summer) 395, 412, 413, 419; (winter) 1–5, 10, 12, 17, 22, 32, 76, 114, 124, 128–132, 146, 147, 150, 151, 152, 154, 162–164, 166, 168, 169, 171, 224–227, 230, 245, 264, 275, 281, 287, 289, 296, 298, 335, 347, 361, 363, 364, 384, 385, 387, 392, 395, 404, 406, 412.

Lapland Bunting
Rare and erratic breeder in the Scottish highlands, but mainly a passage migrant and winter visitor to coastal areas, especially in eastern England. Wintering birds are usually hard to find, frequenting extensive areas of rough coastal pasture where they simply vanish unless 'walked up'. Regular sites include: 1-5, 10, 17, 31, 32, 124, 128-132, 154, 161–163, 169, 223, 224, 226, 227, 244, 281, 288, 289, 296.

Cirl Bunting
Formerly widespread in southern England, serious declines resulted in the population being confined to South Devon. There has been a slight recovery in Devon and the species may also be found at a few sites in surrounding areas. Key sites: 34, 36, 37, 39, 42.

INDEX OF SITES

C

D

E

F

I

J

K

L

M

Q

R

S